# Electricity and magnetism

# Electricity and magnetism

**B. I. BLEANEY**

FORMERLY FELLOW OF ST. HUGH'S COLLEGE, OXFORD

AND

**B. BLEANEY**

DR. LEE'S PROFESSOR OF EXPERIMENTAL PHILOSOPHY
UNIVERSITY OF OXFORD

THIRD EDITION

OXFORD UNIVERSITY PRESS
1976

*Oxford University Press, Ely House, London W. 1*

GLASGOW NEW YORK TORONTO MELBOURNE WELLINGTON
CAPE TOWN IBADAN NAIROBI DAR ES SALAAM LUSAKA ADDIS ABABA
DELHI BOMBAY CALCUTTA MADRAS KARACHI LAHORE DACCA
KUALA LUMPUR SINGAPORE HONG KONG TOKYO

Casebound ISBN 0 19 851140 X
Paperback ISBN 0 19 851141 8

© OXFORD UNIVERSITY PRESS 1976

FIRST EDITION 1957
SECOND EDITION 1965
THIRD EDITION 1976

SET IN N. IRELAND AT THE UNIVERSITIES PRESS, BELFAST
AND PRINTED IN GREAT BRITAIN BY REDWOOD BURN LTD., TROWBRIDGE

# Preface to the third edition

*De manera que acordé, aunque contra mi voluntad, meter segunda vez la pluma en tan estraña lavor é tan agena de mi facultad, hurtando algunos ratos á mi principal estudio, con otras horas destinadas para recreación, puesto que no han de faltar nuevos detractores á la nueva edicion.*
*1499*

*Fernando de Rojas*

So I agreed, albeit unwillingly (since there cannot fail to be fresh critics of a new edition), again to exercise my pen in so strange a labour, and one so foreign to my ability, stealing some moments from my principal study, together with other hours destined for recreation.

FOR the third edition of this textbook the material has been completely revised and in many parts substantially rewritten. S.I. units are used throughout; references to c.g.s. units have been almost wholly eliminated, but a short conversion table is given in Appendix D. The dominance of solid-state devices in the practical world of electronics is reflected in a major change in the subject order.

Chapters 1–9 set out the macroscopic theory of electricity and magnetism, with only minor references to the atomic background, which is discussed in Chapters 10–17. A simple treatment of lattice vibrations is introduced in Chapter 10 in considering the dielectric properties of ionic solids. The discussion of conduction electrons and metals has been expanded into two chapters, and superconductivity, a topic previously excluded, is the subject of Chapter 13. Minor changes have been made in the three chapters (14–16) on magnetism. The discussion of semiconductor theory precedes new chapters on solid-state devices, but we have endeavoured to present such devices in a manner which does not presuppose a knowledge in depth of the theory. The remaining chapters, on amplifiers and oscillators, vacuum tubes, a.c. measurements, noise, and magnetic resonance, bring together the discussion of electronics and its applications.

The authors are grateful to many colleagues in Oxford and readers elsewhere for helpful comments on previous editions which have been incorporated in the present volume. In particular we are indebted to Dr. G. A. Brooker for numerous and detailed comments and suggestions; to Drs F. V. Price and J. W. Hodby, whose reading of new material on electronics in draft form resulted in substantial improvement of the presentation; to Drs F. N. H. Robinson and R. A. Stradling for several helpful suggestions; and to Messrs. C. A. Carpenter and J. Ward for the considerable trouble taken in producing Fig. 23.3. We are indebted to Professors M. Tinkham and O. V. Lounasmaa for generously sending us material in advance of publication; and to Professor L. F. Bates, F.R.S., Drs R. Dupree, and R. A. Stradling for their kindness in providing the basic diagrams for Figs 15.6, 6.15, and 17.9. We wish to thank Miss C. H. Bleaney for suggesting the quotation which appears above.

*Clarendon Laboratory,*                                        B. I. B.
*Oxford*                                                       B. B.
*February 1975*

# Acknowledgements

THE authors are indebted to the following for permission to use published diagrams as a basis for figures in the text: the late Sir K. S. Krishnan; G. Benedek; R. Berman; A. H. Cooke; B. R. Cooper; G. Duyckaerts; D. K. Finnemore; M. P. Garfunkel; R. V. Jones; A. F. Kip; C. Kittel; D. N. Langenberg; D. E. Mapother; K. A. G. Mendelssohn; R. W. Morse; D. E. Nagle; N. E. Phillips; H. M. Rosenberg; J. W. Stout; R. A. Stradling; W. Sucksmith; W. P. Wolf; American Institute of Physics; American Physical Society; Institution of Electrical Engineers; Institute of Physics and the Physical Society (London); Royal Society (London); Bell Telephone Laboratories; G.E.C. Hirst Research Centre.

# Contents

# 1. Electrostatics I

## 1.1. The electrical nature of matter

THE fundamental laws of electricity and magnetism were discovered by experimenters who had little or no knowledge of the modern theory of the atomic nature of matter. It should therefore be possible to present these laws in a textbook by dealing at first purely in macroscopic phenomena and then introducing gradually the details of atomic theory as required, developing the subject almost in the historical order of discovery. Instead, in this book a basic knowledge of atomic theory is assumed from the beginning, and the macroscopic phenomena are related to atomic properties throughout.

*Conductors and insulators*

For the purpose of electrostatic theory all substances can be divided into two fairly distinct classes: *conductors*, in which electrical charge can flow easily from one place to another; and *insulators*, in which it cannot. In the case of solids, all metals and a number of substances such as carbon are conductors, and their electrical properties can be explained by assuming that a number of electrons (roughly one per atom) are free to wander about the whole volume of the solid instead of being rigidly attached to one atom. Atoms which have lost one or more electrons in this way have a positive charge, and are called ions. They remain fixed in position in the solid lattice. In solid substances of the second class, insulators, each electron is firmly bound to the lattice of positive ions, and cannot move from point to point. Typical solid insulators are sulphur, polystyrene, and alumina.

When a substance has no net electrical charge, the total numbers of positive and negative charges within it must just be equal. Charge may be given to or removed from a substance, and a positively charged substance has an excess of positive ions, while a negatively charged substance has an excess of electrons. Since the electrons can move so much more easily in a conductor than the positive ions, a net positive charge is usually produced by the removal of electrons. In a charged conductor the electrons will move to positions of equilibrium under the influence of the forces of mutual repulsion between them, while in an insulator they are fixed in position and any initial distribution of charge will remain almost indefinitely. In a good conductor the movement of charge is almost instantaneous, while in a good insulator it is extremely slow. While there is no such thing as a perfect conductor or perfect insulator, such concepts are useful in developing electrostatic theory; metals form a good approximation to the former, and substances such as sulphur to the latter.

## 1.2. Coulomb's law and fundamental definitions

The force of attraction between charges of opposite sign, and of repulsion between charges of like sign, is found to be inversely proportional to the square of the distance between the charges (assuming them to be located at points), and proportional to the product of the magnitudes of the two charges. This law was discovered experimentally by Coulomb in 1785. In his apparatus the charges were carried on pith balls, and the force between them was measured with a torsion balance. The experiment was not very accurate, and a modern method of verifying the inverse square law with high precision will be given later (§ 1.3). From here on we shall assume it to be exact.

If the charges are $q_1$ and $q_2$, and $r$ is the distance between them, then the force $F$ on $q_2$ is along $r$. If the charges are of the same sign, the force is one of repulsion, whose magnitude is

$$F = C \frac{q_1 q_2}{r^2}.$$

The vector equation for the force is

$$\mathbf{F} = C \frac{q_1 q_2}{r^3} \mathbf{r}. \tag{1.1}$$

Here $\mathbf{F}$, $\mathbf{r}$ are counted as positive when directed from $q_1$ to $q_2$. Eqn (1.1) is the mathematical expression of Coulomb's law.

The units of $\mathbf{F}$ and $\mathbf{r}$ are those already familiar from mechanics; it remains to determine the units of $C$ and $q$. Here there are two alternatives: either $C$ is arbitrarily given some fixed numerical value, when eqn (1.1) may be used to determine the unit of charge, or the unit of charge may be taken as some arbitrary value, when the constant $C$ is to be determined by experiment. The *Système International* (S.I.), which will be used throughout this book, makes use of the second method. The force $F$ is in newtons, the distance $r$ in metres, and an arbitrary unit, the coulomb, is used to measure the charges $q_1$ and $q_2$. The coulomb is directly related to the unit of current, the ampere, which is one coulomb per second; the ampere is defined by the forces acting between current-carrying conductors (see § 4.1). Eqn (1.1) for Coulomb's law is then analogous to that for gravitational attraction, except that it deals with electrical charges instead of masses; the constant of proportionality $C$ must be determined by experiment. In the S.I., the constant $C$ is written as $1/4\pi\epsilon_0$, the factor $4\pi$ being introduced here so that it occurs in formulae involving spherical rather than plane geometry. Eqn (1.1) therefore becomes

$$\mathbf{F} = \frac{1}{4\pi\epsilon_0} \frac{q_1 q_2}{r^3} \mathbf{r},\tag{1.2}$$

where $\mathbf{F}$ is in newtons (N), $\mathbf{r}$ in metres (m), and $q$ in coulombs (C). The quantity $\epsilon_0$ is known as the 'permittivity of a vacuum' (see § 1.5); its experimental value (see § 5.8) is $8 \cdot 85 \times 10^{-12}$ coulomb$^2$ newton$^{-1}$ metre$^{-2}$ ($C^2 N^{-2} m^{-2}$)—a more convenient name for this unit (see § 1.6) is farad metre$^{-1}$ ($F m^{-1}$).

*Electric field and electric potential*

The force which a charge $q_2$ experiences when in the neighbourhood of another charge $q_1$ may be ascribed to the presence of an 'electric field' of strength $\mathbf{E}$ produced by the charge $q_1$. Since the force on a charge $q_2$ is proportional to the magnitude of $q_2$, we define the field strength $\mathbf{E}$ by the equation

$$\mathbf{F} = \mathbf{E} q_2.\tag{1.3}$$

From this definition and Coulomb's law it follows that $\mathbf{E}$ does not depend on $q_2$, and is a vector quantity, like $\mathbf{F}$. From eqn (1.2) we find that

$$\mathbf{E} = \frac{q_1}{4\pi\epsilon_0 r^3} \mathbf{r}\tag{1.4}$$

is the electric field due to the charge $q_1$.

If a unit positive charge is moved an infinitesimal distance $d\mathbf{s}$ in a field of strength $\mathbf{E}$, then the work done by the field is $\mathbf{E} \cdot d\mathbf{s}$, and the work done against the field is $-\mathbf{E} \cdot d\mathbf{s}$. This follows from the fact that the force on unit charge is equal to the electric field strength $\mathbf{E}$. The work done against the field in moving a unit positive charge from a point A to a point B will therefore be

$$V = -\int_A^B \mathbf{E} \cdot d\mathbf{s}.$$

This is a scalar quantity known as the electric potential. If the field strength $\mathbf{E}$ is due to a single charge $q$ at O, as in Fig. 1.1, then the force on unit charge at an arbitrary point P is along OP, and $d\mathbf{s}$ is the vector element $P_1 P_2$. Now $\mathbf{E} \cdot d\mathbf{s} = E \cos \theta \, ds = E \, dr$, and hence

$$V_B - V_A = -\int_A^B E \, dr = -\frac{q}{4\pi\epsilon_0} \int_{r_1}^{r_2} \frac{dr}{r^2} = \frac{q}{4\pi\epsilon_0}\left(\frac{1}{r_2} - \frac{1}{r_1}\right).$$

Thus the difference of potential between A and B depends only on the positions of A and B, and is independent of the path taken between them.

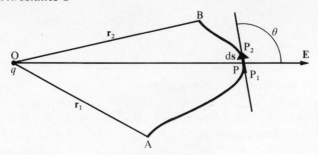

FIG. 1.1. Calculation of the potential difference between points A and B due to the field of a point charge $q$ at O.

The potential at a point distance $r$ from a charge $q$ is the work done in bringing up unit charge to the point in question from a point at zero potential. By convention, the potential is taken as zero at an infinite distance from all charges, that is, $V = 0$ for $r = \infty$. Therefore the potential at a point distance $r$ from a charge $q$ is

$$V = q/4\pi\epsilon_0 r. \tag{1.5}$$

The difference in potential $dV$ between $P_1$ and $P_2$ (Fig. 1.1) distance $ds$ apart is

$$dV = -\mathbf{E}.\mathbf{ds} = -(E_x \, dx + E_y \, dy + E_z \, dz).$$

Hence

$$\mathbf{E} = -\text{grad } V = -\nabla V, \tag{1.6}$$

where in Cartesian coordinates grad $V = \mathbf{i} \, \partial V/\partial x + \mathbf{j} \, \partial V/\partial y + \mathbf{k} \, \partial V/\partial z$ and $\mathbf{i}$, $\mathbf{j}$, $\mathbf{k}$ are unit vectors parallel to the $x$-, $y$-, and $z$-axes. The components of $\mathbf{E}$ along the three axes are

$$E_x = -\frac{\partial V}{\partial x}, \qquad E_y = -\frac{\partial V}{\partial y}, \qquad E_z = -\frac{\partial V}{\partial z}.$$

The negative sign shows that of itself a positive charge will move from a higher to a lower potential, and work must be done to move it in the opposite direction. (For vector relations, see Appendix A.)

The work done in taking a charge $q$ round a closed path in an electrostatic field is zero. This can be seen from Fig. 1.2. The work done in taking the charge $q$ round the path ABCA is

$$W = -q \oint \mathbf{E}.\mathbf{ds} = q(V_B - V_A) + q(V_C - V_B) + q(V_A - V_C) = 0,$$

and is independent of the path taken provided it begins and ends at the same point. Therefore the electric potential is a single-valued function of the space coordinates for any stationary distribution of electric charges; it has only one value at any point in the field.

FIG. 1.2. The work done in taking an electric charge round a closed path in an electrostatic field is zero.

From the vector identity curl grad $V = 0$ or $\nabla \wedge (\nabla V) = 0$ (see Appendix § A.9, eqn (A.19)) it follows that curl $\mathbf{E} = 0$. Here curl $\mathbf{E}$ is a vector whose components are

$$\left(\frac{\partial E_z}{\partial y} - \frac{\partial E_y}{\partial z}, \quad \frac{\partial E_x}{\partial z} - \frac{\partial E_z}{\partial x}, \quad \frac{\partial E_y}{\partial x} - \frac{\partial E_x}{\partial y}\right).$$

These components can be shown to be zero by the use of elementary circuits (cf. Appendix § A.6), and the fact that no work is done in taking a charge round a closed path. The relation curl $\mathbf{E} = 0$ holds because $\mathbf{E}$ can be expressed as the gradient of a scalar potential: $\mathbf{E} = \text{grad } V$. This is true in electrostatics, but does not hold when a changing magnetic flux threads the circuit (see § 5.1).

Since potential is a scalar quantity the potential at any point is simply the algebraic sum of the potentials due to each separate charge. On the other hand, $\mathbf{E}$ is a vector quantity, and the resultant field is the vector sum of the individual fields. Hence it is nearly always simpler to work in terms of potential rather than field; once the potential distribution is found, the field at any point is found by using eqn (1.6).

*Units*

From eqn (1.3) we obtain the unit of electric field strength. An electric field of 1 unit exerts a force of 1 newton on a charge of 1 coulomb. Electric field strengths can therefore be expressed in newton coulomb$^{-1}$ (N C$^{-1}$).

The unit of potential is defined as follows: When 1 joule (J) of work is done in transferring a charge of 1 C from A to B, the potential difference between A and B is 1 volt (V). From eqn (1.6) $E$ can be expressed in volt metre$^{-1}$ (V m$^{-1}$), and this is the unit which is customarily used. It is easily verified that the two alternative units for $E$ are equivalent.

*Lines of force*

A line drawn in such a way that it is parallel to the direction of the field at any point is called a line of force. Fig. 1.3 shows the lines of force for

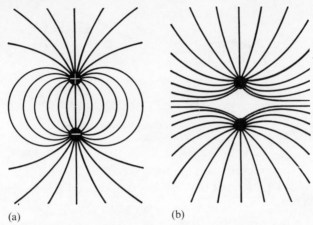

(a)                                    (b)

FIG. 1.3. (a) Lines of force between equal charges of opposite sign. (b) Lines of force
between equal charges of the same sign.

two equal charges. Lines of force do not intersect one another since the
direction of the field cannot have two values at one point; they are
continuous in a region containing no free charges, and they begin and end
on free charges. The number of lines of force drawn through unit area
normal to the direction of **E** is equal to the value of **E** at that point.

If a series of curves is drawn, each curve passing through points at a
given potential, these equipotential curves cut the lines of force orthogon-
ally. Equipotential curves are generally drawn for equal increments of
potential; then **E** is greatest where the equipotentials are closest together.

### 1.3. Gauss's theorem

Let $S$ be a closed surface surrounding a charge $q$, and let $q$ be distant $r$
from a small area d$S$ on the surface $S$ at A (Fig. 1.4(a)). The electric field
strength **E** at A has the value

$$E = q/4\pi\epsilon_0 r^2.$$

The number of lines of force passing through an element of area d$S$ is

$$\mathbf{E}.\mathbf{dS} = E\cos\theta\, dS = \frac{q\cos\theta\, dS}{4\pi\epsilon_0 r^2},$$

where the outward normal to the surface element makes an angle $\theta$ with
**E**. Now the solid angle subtended by d$S$ at O is $d\Omega = \cos\theta\, dS/r^2$, and the
value of $E\cos\theta\, dS$ is therefore $q\, d\Omega/4\pi\epsilon_0$. Hence the total number of
lines of force passing through the whole surface is

$$\int E\cos\theta\, dS = \frac{q}{4\pi\epsilon_0}\int d\Omega = \frac{q}{\epsilon_0}, \tag{1.7a}$$

since a closed surface subtends a total solid angle of $4\pi$ at any point within the volume enclosed by the surface. If there are a number of charges $q_1, q_2,..., q_n$ inside $S$, the resultant intensity of $\mathbf{E}$ at any point is the vector sum of the intensities due to each separate charge, and the integration of eqn (1.7a) may be carried out separately for each charge. In this way it is found that $\int E \cos \theta \, dS = \sum q/\epsilon_0$. On the other hand, the contribution of any charge outside $S$ is zero, as may be seen from Fig. 1.4(b), since in this case

$$\int E \cos \theta \, dS = \frac{q}{4\pi\epsilon_0}\left(\int \frac{dS_1 \cos \theta_1}{r_1^2} - \int \frac{dS_2 \cos \theta_2}{r_2^2}\right) = 0.$$

We may summarize these results in the form

$$\int E \cos \theta \, dS = \int \mathbf{E} \cdot d\mathbf{S} = \sum q/\epsilon_0, \tag{1.7b}$$

where the summation is to be taken only over the charges lying within the closed surface $S$. This is known as Gauss's theorem. We see that the integral of the normal component of $\mathbf{E}$ over the surface is equal to the total charge enclosed, divided by $\epsilon_0$, irrespective of the way in which the charge is distributed.

If there exists throughout a volume enclosed by a surface $S$ a charge distribution of varying density $\rho_e$, we have

$$\frac{1}{\epsilon_0}\int \rho_e \, d\tau = \int \mathbf{E} \cdot d\mathbf{S} = \int \text{div } \mathbf{E} \, d\tau, \tag{1.7c}$$

where $d\tau$ is an element of volume. The two volume integrals must be equal whatever the volume over which the integration takes place, and it therefore follows that the integrands themselves must be equal. Hence

$$\text{div } \mathbf{E} = \frac{\partial E_x}{\partial x} + \frac{\partial E_y}{\partial y} + \frac{\partial E_z}{\partial z} = \frac{\rho_e}{\epsilon_0}, \tag{1.8}$$

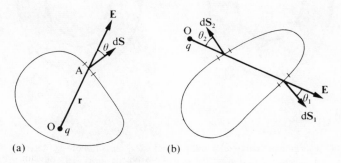

(a)                               (b)

FIG. 1.4. Illustrating Gauss's theorem.

and this is the expression in differential form of Gauss's theorem. The transformation from a surface to a volume integral used above is due to Gauss (see Appendix § A.7).

One of the consequences of Gauss's theorem is that there can be no field within a conductor, nor can there be any volume distribution of charge within it. For, if there were such a charge distribution, a field would exist within the conductor, which would act on the charges. Since they are free to move in a conductor, they cannot then be in a state of equilibrium. Thus no electrostatic field can exist within the body of a conductor, and all parts of it must be at the same potential. If the conductor has a total charge different from zero, then this charge must reside entirely in a thin layer on the outer surface (see Fig. 1.5).

The fact that there can be no electric field within the body of a conductor has an important consequence in the case of a hollow closed conductor. If we apply Gauss's theorem to a surface $S$ lying entirely within the conducting substance, as in Fig. 1.5(b), then $\int E \cos \theta \, dS = 0$, since $\mathbf{E} = 0$ everywhere over the surface. Hence the net charge inside the surface must be zero. This can be realized in two ways: (1) if there is a total charge $q$ in the hollow space within the conductor, the lines of force from the charges comprising $q$ must end on a distribution of charge on the inner surface of the conductor, and the total charge in this layer must be equal to $-q$; (2) if there is no charge in the hollow space, then there can be no field in this space. This last result is important, for many proofs

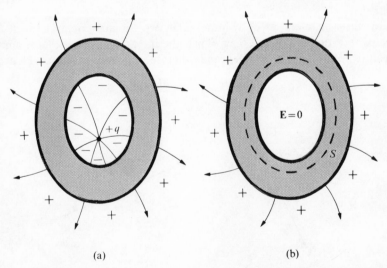

(a)                                              (b)

FIG. 1.5. Distribution of charge on a hollow conductor. (a) A hollow conductor with a point charge $+q$ inside, and induced charges $-q$ and $+q$ on the inside and outside surfaces. (b) The same conductor with no charge inside, and total charge $+q$ on the surface.

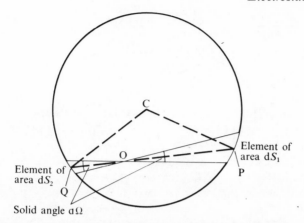

FIG. 1.6. The field inside a spherical conductor at a point O. Distances $OP = r_1$, $OQ = r_2$. From the geometry of the circle, $\angle OPC = \angle OQC = \theta$. Hence

$$d\Omega = \frac{dS_1 \cos \theta}{r_1^2} = \frac{dS_2 \cos \theta}{r_2^2}.$$

of the inverse square law (see below) depend on it. It means that if we put a closed conductor into a field, a charge distribution on the outer surface will be set up such that the field inside remains exactly zero.

### Experimental proof of the inverse square law

Coulomb's attempts to check the inverse square law using a torsion balance were not capable of great accuracy, and most subsequent attempts have relied on the fact that the field inside a closed conductor is only zero if the inverse square law holds. We shall prove this for the special case of a spherical conductor.

In Fig. 1.6 let an elementary cone of solid angle $d\Omega$ be drawn with vertex at the point O within the sphere. This cone intersects the surface of the sphere in the elementary areas $dS_1$, $dS_2$ at distances $r_1$, $r_2$ from O. If the charge on the sphere has a uniform density $\sigma_e$ per unit area, then the field at O due to the elements $dS_1$ and $dS_2$ will be

$$dE = \frac{\sigma_e}{4\pi\epsilon_0}\left(\frac{dS_1}{r_1^n} - \frac{dS_2}{r_2^n}\right)$$

assuming that the field of a point charge falls off as $r^{-n}$. But the solid angle $d\Omega = dS_1 \cos \theta/r_1^2 = dS_2 \cos \theta/r_2^2$, and we can therefore write

$$dE = \frac{\sigma_e \, d\Omega}{4\pi\epsilon_0 \cos \theta}\left(\frac{1}{r_1^{n-2}} - \frac{1}{r_2^{n-2}}\right).$$

This gives a resultant field towards the nearer element if $n < 2$, and towards the further element if $n > 2$. Clearly, the whole surface of the

'FIG. 1.7. Apparatus of Plimpton and Lawton for verifying the inverse square law.

sphere can be divided into elementary areas in this way, and the vector resultant of the fields at O will not be zero unless all the individual $dE$ are zero, since there will be a resultant towards the nearer portion of the spherical surface if $n < 2$, and vice versa. Thus, if it is shown experimentally that there is no field inside a charged sphere, it follows that the power of $n$ in the inverse power law must be exactly 2.

This result was used to test the validity of the inverse square law by Cavendish and, later, by Maxwell. Maxwell had a spherical air condenser consisting of two concentric insulated spherical shells. The outer sphere had a small hole in it so that the inner one could be tested for charge by inserting through the hole an electrode connected to an electrometer. The two spheres were initially connected by a wire and charged to a high potential, and then insulated from one another. After earthing the outer sphere, the inner one was tested and found to have no charge. In this way Maxwell found that the value of $n$ did not differ from 2 by more than one part in 21 600.

The experiment has been repeated by Plimpton and Lawton (1936) with a more sensitive detector, the electrometer being replaced by an amplifier and galvanometer. The detecting apparatus was placed inside the sphere A (see Fig. 1.7), and its conducting case, together with the hemisphere B, formed the inner conductor (it was shown that this does not necessarily have to be spherical in shape). The galvanometer deflection was observed through a small hole in the sphere A, covered by a wire grid immersed in salt solution so that A was effectively a closed conductor. It was found that this was essential for the field inside to be rigorously zero when $n = 2$. The galvanometer was undamped so that it could swing at its natural period (0·5 s), and an alternating voltage of 3 kV, whose frequency was adjusted to synchronism with the galvanometer, was applied to the outer sphere. No potential difference between the inner and outer spheres was found, though a voltage of $10^{-6}$ V could have been detected. It was found that this was only true if the hole in A was covered with the salt solution. The galvanometer deflection observed if this was removed was used to check that the frequency of the alternating voltage was equal to the natural period of the galvanometer. From this experiment, Plimpton and Lawton concluded that $n$ did not differ from 2 by more than one .part in $10^9$.

### 1.4. Electric dipoles

An electric dipole consists of two charges equal in magnitude but of opposite sign, separated by a small distance. Fig. 1.8 shows such a dipole with charges $+q$ and $-q$ separated by a distance $a$. Then the potential at a point $P$ is

$$V = \frac{q}{4\pi\epsilon_0}\left(\frac{1}{r_2} - \frac{1}{r_1}\right). \qquad (1.9)$$

If $a \ll r$, the quantities $1/r_1$ and $1/r_2$ may be expanded in a power series. Since $r_2^2 = r^2 - ra \cos\theta + \frac{1}{4}a^2$, we have

$$\frac{1}{r_2} = \frac{1}{r} + \frac{a}{2r^2}\cos\theta + \frac{a^2}{8r^3}(3\cos^2\theta - 1) + \ldots,$$

and the expansion of $1/r_1$ differs only in that the signs are reversed of all terms with odd powers of $\cos\theta$. Hence

$$V = \frac{qa\cos\theta}{4\pi\epsilon_0 r^2} = \frac{p\cos\theta}{4\pi\epsilon_0 r^2} \qquad (1.10a)$$

provided that the next term, which is smaller by a factor $(a/r)^2$, and all higher terms can be neglected. This approximation is justified if we let

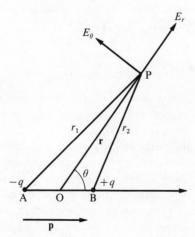

FIG. 1.8. The electric field due to an electric dipole.

$a/r \rightarrow 0$, keeping the product $qa = p$ constant; this approximation is that of a 'point dipole', and the electric dipole moment **p** is a vector whose direction is from the negative to the positive charge.

If **r** is a vector drawn from O as origin to P, then $p \cos \theta = \mathbf{p} \cdot \mathbf{r}/r$, and

$$V = \frac{\mathbf{p} \cdot \mathbf{r}}{4\pi\epsilon_0 r^3}. \tag{1.10b}$$

If O is a fixed point and P is regarded as a variable point, then

$$\mathbf{r}/r^3 = -\mathrm{grad}(1/r),$$

so that the formula for the potential may also be written as

$$V = -\frac{1}{4\pi\epsilon_0} \{\mathbf{p} \cdot \mathrm{grad}_\mathrm{P}(1/r)\}. \tag{1.11a}$$

Here the subscript P is added to the operator grad to denote that differentiation is with respect to P as the variable point. If we regard P as fixed, and move from A to B, then eqn (1.9) could have been written in the form

$$V = \frac{1}{4\pi\epsilon_0} \{\mathbf{p} \cdot \mathrm{grad}_\mathrm{O}(1/r)\}. \tag{1.11b}$$

Here the subscript O denotes that O is now the variable point, and there is a change of sign from eqn (1.11a) because **r** is now measured in the opposite direction, from P to O instead of from O to P.

The components of the electric field strength at *P* can be calculated by differentiating the potential given by eqn (1.10a). The radial and azimuthal components are

$$E_r = -\left(\frac{\partial V}{\partial r}\right)_\theta = \frac{1}{4\pi\epsilon_0}\left(\frac{2p \cos \theta}{r^3}\right),$$

$$\tag{1.12}$$

$$E_\theta = -\frac{1}{r}\left(\frac{\partial V}{\partial \theta}\right)_r = \frac{1}{4\pi\epsilon_0}\left(\frac{p \sin \theta}{r^3}\right).$$

These equations show that the electric field strength of a dipole falls off as $1/r^3$, and its potential as $1/r^2$, whereas the corresponding laws for a single pole are $1/r^2$ and $1/r$. The significance of this difference is that at large distances the fields of the two equal and opposite charges which comprise a dipole cancel one another in the first approximation (that is, terms varying as $1/r^2$ vanish), but terms in the next order ($1/r^3$ in the field) remain. Similarly, if two equal dipoles are placed end to end, giving a set of charges as in Fig. 1.9 (known as a quadrupole), their fields annul one another at large distances, and the potential of a quadrupole falls off as $1/r^3$ (see Problem 1.1), and its field as $1/r^4$.

FIG. 1.9. Two quadrupole moments represented as an assembly of charges or a pair of dipoles; note that the net charge and net dipole moment are zero in each case.

If a dipole consisting of two charges $-q$ and $+q$ a distance **a** apart is placed in a uniform field, its potential energy $U_P$ is (see Fig. 1.10)

$$U_P = q(V_B - V_A) = -qa\cos\theta\, E = -q\mathbf{a.E} = -\mathbf{p.E}, \qquad (1.13)$$

where **p** is the dipole moment. This shows that the energy depends only on the angle which **p** makes with **E**, and not on the position of the dipole. Hence there is no translational force acting on the dipole, but there is a couple

$$\mathbf{T} = -(dU_P/d\theta) = pE\sin\theta = \mathbf{p}\wedge\mathbf{E}, \qquad (1.14)$$

which tends to turn the dipole into a position parallel to the field.

If the dipole is placed in a non-uniform field, a translational force is exerted on it, and we shall derive an expression for the $x$-component of this force. If $E_x$ is the value of the $x$-component of the field at A, its value at B may be written as

$$E'_x = E_x + \left(\frac{\partial E_x}{\partial x}\right)a_x + \left(\frac{\partial E_x}{\partial y}\right)a_y + \left(\frac{\partial E_x}{\partial z}\right)a_z,$$

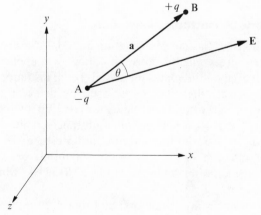

FIG. 1.10. Dipole formed by two charges $-q$, $+q$ separated by a distance **a**, in a field **E**.

where $a_x$, $a_y$, $a_z$ are the components of **a** along the three axes. The $x$-component of the force on the dipole is therefore

$$F_x = -qE_x + qE'_x = qa_x\left(\frac{\partial E_x}{\partial x}\right) + qa_y\left(\frac{\partial E_x}{\partial y}\right) + qa_z\left(\frac{\partial E_x}{\partial z}\right)$$

$$= p_x\left(\frac{\partial E_x}{\partial x}\right) + p_y\left(\frac{\partial E_x}{\partial y}\right) + p_z\left(\frac{\partial E_x}{\partial z}\right). \tag{1.15a}$$

Now

$$\frac{\partial E_x}{\partial y} = \frac{\partial}{\partial y}\left(-\frac{\partial V}{\partial x}\right) = \frac{\partial}{\partial x}\left(-\frac{\partial V}{\partial y}\right) = \frac{\partial E_y}{\partial x},$$

since the order of differentiation is immaterial, and similarly

$$\frac{\partial E_x}{\partial z} = \frac{\partial E_z}{\partial x}.$$

Hence the force component may also be written as

$$F_x = p_x\frac{\partial E_x}{\partial x} + p_y\frac{\partial E_y}{\partial x} + p_z\frac{\partial E_z}{\partial x}, \tag{1.15b}$$

with similar expressions for the other components. This result may also be derived directly from the potential energy $U_P$ of a dipole (eqn (1.13)) differentiation of which gives

$$F_x = -\left(\frac{\partial U_P}{\partial x}\right) = \frac{\partial}{\partial x}(p_xE_x + p_yE_y + p_zE_z),$$

where the components of **E** (but not those of **p**) are functions of position. The components of **F** are those of a vector which (see eqn (A.10) of Appendix A) may be written as

$$\mathbf{F} = (\mathbf{p} \cdot \mathrm{grad})\mathbf{E}. \tag{1.15c}$$

## 1.5. The theory of isotropic dielectrics

Faraday found that if a slab of insulating material was inserted between two metal plates across which a constant voltage was applied by means of a battery, the charge on the plates increased. If the insulator entirely filled the intervening space, the charge increased by a factor $\epsilon_r$ where $\epsilon_r$ is called the relative permittivity or dielectric constant of the insulator. It varies between 1 and 10 for most solid substances, being 1 for vacuum and 1·00057 for air at room temperature and pressure. To find how $\epsilon_r$ is related to the intrinsic properties of the material, or dielectric, it is necessary to consider what happens inside a dielectric when an electric field is applied to it.

Dielectric substances are insulators, and therefore do not contain free electrons. Each electron is bound to the ionic lattice by the electrostatic attraction between the negative electronic charge and the positive charges on the nuclei. In the absence of any external field, the electrons are distributed symmetrically with respect to the nuclei, but when a field is applied, the electrons are displaced in the direction opposite to that of the field, while the more massive nuclei are slightly displaced in the direction of the field. (The centre of gravity remains fixed, since there is no translational force on the system as a whole.) Each ion thus acquires an

FIG. 1.11. Effect of a dielectric in increasing the capacitance between two parallel plates.

electric dipole moment which is parallel to and in the same direction as the applied field. If a slab of dielectric is placed between parallel metal plates as in Faraday's experiment and the voltage across the plates is constant, there will be an induced negative charge on the dielectric surface near the positive plate (Fig. 1.11) and a similar positive charge on the surface near the negative plate. There will be no resultant charge density within a simple uniform dielectric, as all the individual dipoles are aligned parallel to the field and hence each negative charge of one dipole is next to the positive charge of the next dipole. The surface charges on the dielectric will induce charges of opposite sign on the plates, and the charge on the plates is increased when the dielectric is inserted, if the voltage is kept constant—as Faraday found in his experiments.

The action of the electric field in giving each atom of a dielectric an induced dipole moment is termed 'polarization'. (We shall see in Chapter 10 that electric polarization can arise in other ways in substances containing molecules.) The macroscopic polarization is defined as the electric dipole moment per unit volume. It is found by taking the vector sum of the microscopic dipole moments in unit volume, and is therefore itself a vector, denoted by the symbol $\mathbf{P}$.

The resultant moment for an element of volume $d\tau$ is $\mathbf{P}\,d\tau$, and the potential of such an element a distance $r$ away is, by eqn (1.11b),

$$dV = \frac{1}{4\pi\epsilon_0}(\mathbf{P}\,d\tau)\cdot\mathrm{grad}(1/r),$$

where the differentiation is with respect to the coordinates of the volume element containing the dipoles. The potential due to a finite volume of dielectric is then

$$V = \frac{1}{4\pi\epsilon_0} \int \{\mathbf{P} \cdot \text{grad}(1/r)\} \, d\tau.$$

But, from eqn (A.12) in Appendix A,

$$\text{div}(\mathbf{P}/r) = \frac{1}{r} \text{div } \mathbf{P} + \mathbf{P} \cdot \text{grad}(1/r).$$

Hence

$$V = \frac{1}{4\pi\epsilon_0} \int \text{div}(\mathbf{P}/r) \, d\tau - \frac{1}{4\pi\epsilon_0} \int \frac{1}{r} \text{div } \mathbf{P} \, d\tau$$

$$= \frac{1}{4\pi\epsilon_0} \int \frac{1}{r} \mathbf{P} \cdot d\mathbf{S} - \frac{1}{4\pi\epsilon_0} \int \frac{1}{r} \text{div } \mathbf{P} \, d\tau, \tag{1.16}$$

where Gauss's theorem of divergence has been used in the transformation from a volume to a surface integral. These two terms in eqn (1.16) show that the resultant potential can be attributed to an apparent surface charge of density $P \cos \theta$, where $\theta$ is the angle which $\mathbf{P}$ makes with the normal to the surface of the dielectric, and an apparent volume distribution of charge whose density is

$$-\text{div } \mathbf{P} = -\left(\frac{\partial P_x}{\partial x} + \frac{\partial P_y}{\partial y} + \frac{\partial P_z}{\partial z}\right). \tag{1.17}$$

These apparent charge densities arise from the displacement, under the action of the electric field, of bound charges from their normal equilibrium positions in zero electric field. They are often referred to as the 'polarization charge' or the 'bound charge' to distinguish them from 'free charge' which is not bound to any particular point in the dielectric.

*Gauss's theorem in dielectrics*

In § 1.3 it was shown by Gauss's theorem that the integral $\int \mathbf{E} \cdot d\mathbf{S}$ of the normal component of $\mathbf{E}$ over any closed surface is equal to the total charge within the surface, divided by $\epsilon_0$. If the surface is within a dielectric medium, the total charge must include both the free charges and the polarization charges. The volume charge density is thus $\rho_e - (\text{div } \mathbf{P})$, so that

$$\int \mathbf{E} \cdot d\mathbf{S} = \int \text{div } \mathbf{E} \, d\tau = \frac{1}{\epsilon_0} \int (\rho_e - \text{div } \mathbf{P}) \, d\tau$$

or

$$\int \text{div}(\epsilon_0 \mathbf{E} + \mathbf{P}) \, d\tau = \int \rho_e \, d\tau.$$

Comparison of this with eqn (1.7c) shows that, in effect, $\epsilon_0 \mathbf{E}$ has been replaced by $(\epsilon_0 \mathbf{E} + \mathbf{P})$. We may define a new vector $\mathbf{D}$, such that

$$\mathbf{D} = \epsilon_0 \mathbf{E} + \mathbf{P}. \tag{1.18}$$

$\mathbf{D}$ is known as the 'electric displacement', and eqn (1.18) is valid even in an anisotropic medium, where $\mathbf{P}$ is not necessarily parallel to $\mathbf{E}$. We may now write Gauss's theorem, in a dielectric medium, in the form

$$\int \mathbf{D} \cdot d\mathbf{S} = \int \mathrm{div}\, \mathbf{D}\, d\tau = \int \rho_e\, d\tau. \tag{1.19}$$

Since the volume integrals must be equal over any arbitrary volume, it follows that their integrands must be equal, that is,

$$\mathrm{div}\, \mathbf{D} = \rho_e, \tag{1.20}$$

which is the differential form of Gauss's theorem. It is easy to see that these equations reduce to those of § 1.3 *in vacuo*, where $\mathbf{P} = 0$.

In a simple isotropic dielectric, $\mathbf{P}$ is parallel to $\mathbf{E}$, and hence so also is $\mathbf{D}$. We may write $\mathbf{P} = \chi_e \epsilon_0 \mathbf{E}$, where $\chi_e$ is known as the 'electric susceptibility'. Then $\mathbf{D} = (1 + \chi_e)\epsilon_0 \mathbf{E}$, and writing

$$\epsilon = (1 + \chi_e)\epsilon_0 = \epsilon_r \epsilon_0, \tag{1.21}$$

we have

$$\mathbf{D} = \epsilon \mathbf{E}, \tag{1.22}$$

where $\epsilon$ is known as the 'permittivity'. In free space $\epsilon = \epsilon_0$, and thus $\epsilon_0$ is the 'permittivity of a vacuum'. When a medium is present $D/E$ is increased by the ratio $\epsilon_r = \epsilon/\epsilon_0$, and the factor $\epsilon_r$ is known as the 'relative permittivity' of the medium.

In crystalline solids where the crystal structure has less than cubic symmetry, the polarization vector $\mathbf{P}$ is not necessarily parallel to the electric field strength vector $\mathbf{E}$, and it follows from eqn (1.18) that the same is true for the displacement vector $\mathbf{D}$. Under these conditions the relation between $\mathbf{P}$ and $\mathbf{E}$ is given by a set of equations of the type

$$P_x/\epsilon_0 = \chi_{xx}E_x + \chi_{xy}E_y + \chi_{xz}E_z,$$
$$P_y/\epsilon_0 = \chi_{yx}E_x + \chi_{yy}E_y + \chi_{yz}E_z,$$
$$P_z/\epsilon_0 = \chi_{zx}E_x + \chi_{zy}E_y + \chi_{zz}E_z.$$

An analogous set of equations relates the components of $\mathbf{D}$ to those of $\mathbf{E}$, and both $\chi_e$ and $\epsilon_r$ are 'tensor quantities'. We shall restrict our discussion to isotropic media, where $\chi_{xx} = \chi_{yy} = \chi_{zz}$ and the other elements, such as $\chi_{xy}$, are zero, so that both $\chi_e$ and $\epsilon_r$ reduce to simple scalar quantities. Typical of such media are cubic crystals, non-crystalline solids, and fluids.

Our second restriction is to substances in which $\mathbf{P}$ is linearly proportional to $\mathbf{E}$, so that the electric susceptibility $\chi_e$ is independent of field strength. In practice the largest static or slowly varying electric fields that can be applied are small compared with the electric fields that bind the electron to the atom, and the approximation that $\chi_e$ and $\epsilon_r$ are independent of field strength is a good one (as is implied in the older name of 'dielectric constant' for $\epsilon_r$). At optical frequencies, in the intense electric fields that exist in light beams generated by high-power lasers, it is necessary to write the polarizibility as a power series

$$\chi_e = \chi_e^{(0)} + \chi_e^{(1)}E + \chi_e^{(2)}E^2 \ldots,$$

where the terms beyond $\chi_e^{(0)}$ represent non-linear effects which may be important in applications such as modulation or frequency-changing in optical beams, in the same way as non-linear effects in electric conduction are important in radio-electronics.

If we have a single point charge in a uniform dielectric of permittivity $\epsilon$, we may apply Gauss's theorem over a sphere of radius $r$ whose centre is at $q$. The surface integral reduces to $4\pi r^2 D = 4\pi r^2 \epsilon E = q$, so that

$$\epsilon \mathbf{E} = \mathbf{D} = q\mathbf{r}/4\pi r^3. \tag{1.23}$$

It follows that the force between two charges $q_1$, $q_2$ a distance $\mathbf{r}$ apart is

$$F = \frac{q_1 q_2}{4\pi \epsilon r^3}\mathbf{r}, \tag{1.24}$$

and the potential a distance $r$ from a point charge $q$ is

$$V = q/4\pi\epsilon r. \tag{1.25}$$

In a dielectric where $\chi_e$ is uniform, isotropic, and independent of field strength there is no volume distribution of charge, even when $\mathbf{E}$ is non-uniform, as in eqn (1.23). This follows from the fact that div $\mathbf{E} = 0$, which leads to the volume distribution of charge in eqn (1.17) reducing to $-\text{div}\,\mathbf{P} = -\chi_e \epsilon_0 \,\text{div}\,\mathbf{E} = 0$. At a surface where the relative permittivity changes, there will be a surface distribution of apparent charge.

### Some properties of $\mathbf{D}$ and $\mathbf{E}$

It is important to distinguish clearly between the two vector quantities electric field strength $\mathbf{E}$ and electric displacement $\mathbf{D}$. $\mathbf{E}$ is defined as the force acting on unit charge, irrespective of whether a dielectric medium is present or not. This definition is expressed in eqn (1.3). The displacement $\mathbf{D}$ is defined by eqn (1.18). The quantity $\mathbf{D} \cdot d\mathbf{S}$ is sometimes known as the electric flux through the element of area $d\mathbf{S}$. From (1.19) the total flux is $q$ through an area surrounding a charge $q$, and this flux is unaltered by the presence of a dielectric medium. The unit of flux is the coulomb, and the unit of $D$ is the coulomb metre$^{-2}$ (C m$^{-2}$).

Since $\mathbf{D}$ is a vector we may draw lines of displacement analogous to lines of force, such that the number passing through unit area is equal to

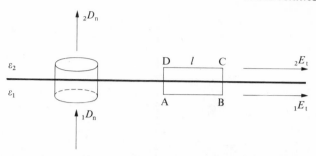

FIG. 1.12. Boundary conditions at the surface between two dielectrics.

the displacement. Also by Gauss's theorem lines of displacement are continuous in a space containing no free charges; they begin and end only on free charges. At the boundary of two dielectrics $\epsilon_1$ and $\epsilon_2$, if no free charge resides there, lines of **D** are continuous but lines of **E** are not, because lines of force end on both free and polarization charges, whereas lines of **D** end only on free charges. There is a polarization charge on the surface separating two dielectrics, since the induced moment per unit volume is different in the two media. Lines of **E** begin and end on this surface charge, but not lines of **D**. The rules governing the behaviour of **E** and **D** at the surface of a dielectric, or the boundary between two dielectrics, are embodied in two 'boundary conditions', which will now be derived.

To find the boundary condition for **D**, we apply Gauss's theorem to a small cylinder which intersects the boundary, as in Fig. 1.12, and whose axis is normal to the boundary. If the height of the cylinder is very small compared with its cross-sectional area, the only contribution to $\int \mathbf{D} \cdot d\mathbf{S}$ over its surface will come from the components of **D** normal to the boundary. Since there is no free charge on the boundary, $\int \mathbf{D} \cdot d\mathbf{S} = 0$, and hence

$$_2D_n = {_1D_n}, \tag{1.26}$$

where the symbols refer to the normal components of **D** on the two sides of the boundary.

The boundary condition for **E** is found by considering the work done in taking unit charge round a small rectangular circuit such as ABCDA in Fig. 1.12. If the sides BC, AD are very small compared with AB, CD, then the work done will be $l(_1E_t - {_2E_t}) = 0$. Hence

$$_2E_t = {_1E_t}, \tag{1.27}$$

where the symbols refer to the tangential components of **E** on either side of the boundary. Thus eqns (1.26) and (1.27) are the two fundamental boundary conditions. It follows from them that lines of **D** will in general

FIG. 1.13. Refraction of lines of displacement at the boundary between two dielectrics ($\epsilon_2 > \epsilon_1$). From eqns (1.26) and (1.27) we have

$$D_1 \cos \theta_1 = D_2 \cos \theta_2,$$

$$E_1 \sin \theta_1 = (D_1/\epsilon_1)\sin \theta_1 = E_2 \sin \theta_2 = (D_2/\epsilon_2)\sin \theta_2.$$

Hence                             $\epsilon_1 \cot \theta_1 = \epsilon_2 \cot \theta_2.$

be refracted at the boundary between two dielectrics. A typical example is shown in Fig. 1.13.

Similar boundary conditions may be applied at the surface of a conductor. Since there can be no field inside the conductor, the tangential component of **E** just outside the conductor must also be zero, and any field at the surface must be normal to the surface. If we apply Gauss's theorem to an elementary cylinder intersecting the surface, as in Fig. 1.12, we have $_2D_n\, dS = \sigma_e\, dS$ (since $_1D_n = 0$ within the conductor), where $\sigma_e$ is the charge density on the conducting surface. Since $_2\mathbf{D}$ must be normal to the surface, we have $_2\mathbf{D} = {}_2D_n$, and hence (dropping the subscripts)

$$D = \epsilon E = \sigma_e \tag{1.28}$$

at the surface of the conductor immersed in a medium of permittivity $\epsilon$.

## 1.6. Capacitors and systems of conductors

If a charge $Q$ is given to an isolated conductor its voltage is increased by an amount $V$. For a given conductor the ratio $Q/V$ is independent of $Q$ and depends only on the size and shape of the conductor. The ratio $Q/V$ is called the capacitance of the conductor, and is denoted by $C$. If a second conductor which is earthed is brought close to the first one, a charge of opposite sign is induced on it. The potential difference falls, and since $Q$ is constant on an isolated conductor, the capacitance has increased. The two conductors together form a capacitor, and the capacitance of the capacitor is defined as the ratio of the charge on either conductor to the potential difference between them. A capacitor is an instrument for storing charge, and a capacitor of large capacitance can store a correspondingly large quantity of charge for a given potential difference between the plates. The capacitance depends on the geometry of the conductors and the permittivity of the medium separating them.

In general, calculation of the capacitance of a conductor or a capacitor is difficult unless simple geometrical shapes are involved. The principle of the calculation may be illustrated by the case of an isolated sphere, of radius $a$, in an infinite dielectric. Suppose this carries a charge $Q$. Then by applying Gauss's theorem over a spherical surface of radius $r$, concentric with the sphere, we have

$$4\pi r^2 D = 4\pi r^2 \epsilon E = Q,$$

since, by symmetry, $\mathbf{D}$ and $\mathbf{E}$ are constant over the spherical surface and everywhere normal to it. Hence

$$E = \frac{Q}{4\pi\epsilon r^2},$$

which is the same as eqn (1.23) for a point charge $Q$. The potential of the sphere is

$$V = -\int_{\infty}^{a} E \, dr = \frac{Q}{4\pi\epsilon a}.$$

Hence

$$C = Q/V = 4\pi\epsilon a. \tag{1.29}$$

A more practical example is provided by a pair of circular coaxial cylinders, as used in a 'coaxial cable'. In Fig. 1.14 the radii of the cylinders are $a$, $b$; they are assumed to be infinitely long, the space between them being completely filled by a uniform dielectric. The lines of

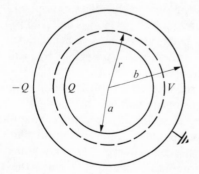

Fig. 1.14. Cross-section of two coaxial cylinders.

**D** and **E** are radial (normal to the equipotential surfaces), and we may apply Gauss's theorem to a coaxial cylindrical surface of radius $r$ and length $l$. If $+Q$ is the charge per unit length on the inner cylinder,

$$2\pi r l D = 2\pi r l \epsilon E = +Ql,$$

from which $E = Q/2\pi r \epsilon$. The potential difference between the cylinders is

$$V_b - V_a = -\int_a^b E \, dr = -(Q/2\pi \epsilon) \int_a^b r^{-1} \, dr = -(Q/2\pi \epsilon)\ln(b/a),$$

and the capacitance of a length $l$ is therefore

$$C = \frac{Ql}{V_a - V_b} = \frac{2\pi l \epsilon}{\ln(b/a)}. \tag{1.30}$$

Another simple type of capacitor is formed by two plane parallel plates of area $S$ and separation $t$. If the lateral dimensions of the plates are large compared with their separation (or if the plates are surrounded by 'guard rings' at the same potential), then the field between them is uniform and normal to the plates, being given by eqn (1.28) with $\sigma_e = Q/S$. Since the field is uniform, the potential difference between the plates is simply $V = Et$, and the capacitance is thus

$$C = Q/V = \epsilon S/t. \tag{1.31}$$

This equation (and also eqns (1.29) and (1.30)) shows that the capacitance increases by a factor $\epsilon_r$ if the space between the plates is filled with a medium of relative permittivity $\epsilon_r = \epsilon/\epsilon_0$. This agrees with the original definition of dielectric constant by Faraday, mentioned at the beginning of § 1.5.

The unit of capacitance is called the farad (F); the plates of a capacitor of 1 F carry a charge of 1 C if their potential difference is 1 V. Reference to eqns (1.29)–(1.31) above, or more fundamentally, to eqn (1.25), shows that a capacitance has the dimensions of $\epsilon_0$ multiplied by a length; hence the unit of $\epsilon_0$ is the farad metre$^{-1}$ (F m$^{-1}$). The farad is a very large unit (a sphere the size of the earth would have a capacitance of about $10^{-3}$ F), and the subdivisions microfarad ($\mu$F $= 10^{-6}$ F), and picofarad (pF $= 10^{-12}$ F) are commonly used instead.

A result often required is the net capacitance of a number of capacitors joined either in series, or in parallel, as in Fig. 1.15. If $n$ capacitors $C_1, C_2, ..., C_n$ are joined in series, and a voltage $V$ applied across them, a charge $+Q$ appears on the plate A and $-Q$ on the plate B. The plate 2 of the first capacitor will have charge $-Q$, and plate 1 of the second capacitor must therefore have charge $+Q$ since the two, though connected together, are otherwise isolated, and their total charge must

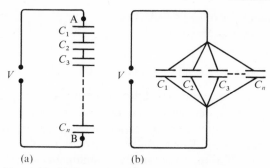

FIG. 1.15. (a) Capacitors in series. (b) Capacitors in parallel.

remain zero on connecting the battery. Thus each capacitor carries the same charge, irrespective of its size, and the potential across the $k$th capacitor is $Q/C_k$. Hence the total potential is $\sum_1^n (Q/C_k)$, and this equals $Q/C$, where $C$ is the net capacitance. Equating these two results gives

$$\frac{1}{C} = \frac{1}{C_1} + \frac{1}{C_2} + ... + \frac{1}{C_k} + ... + \frac{1}{C_n}. \tag{1.32}$$

If the capacitors are joined in parallel, the voltage across each capacitor is equal to $V$. The total charge carried by all the capacitors is $Q = Q_1 + Q_2 + ... + Q_k + ... + Q_n = V(C_1 + C_2 + ... + C_k + ... + C_n) = VC$, where $C$ is the net capacitance. Hence

$$C = C_1 + C_2 + ... + C_k + ... + C_n. \tag{1.33}$$

*The potential energy of systems of charges and charged conductors*

A system of electric charges possesses potential energy, since work must be done in bringing up any particular charge through the electrostatic field of the remaining charges. The energy depends only on the final state of the system and not on how the charges are established. We may suppose therefore that each charge is increased from zero to its final value in infinitesimal steps so that at any given instant each charge is $\alpha q_k$, where $q_k$ is its final value and $\alpha$ is a number less than unity which is the same for all charges. Then if $V_k$ is the final value of the potential at the point occupied by $q_k$, the instantaneous value of the potential will be $\alpha V_k$, and the work done in increasing the charge by $q_k \, d\alpha$ will be $(\alpha V_k)(q_k \, d\alpha)$. Thus the work done in increasing all the charges by a corresponding amount will be $\alpha \, d\alpha \sum_k q_k V_k$. The total work done is equal to the stored energy, which will therefore be

$$U = \sum_k q_k V_k \int_0^1 \alpha \, d\alpha = \tfrac{1}{2} \sum_k q_k V_k. \tag{1.34}$$

We may apply this result to a capacitor with two plates at potentials $V_1$, $V_2$ carrying charges $+Q$ and $-Q$ respectively. The energy of the capacitor will be

$$U = \tfrac{1}{2}QV_1 - \tfrac{1}{2}QV_2 = \tfrac{1}{2}QV = \tfrac{1}{2}CV^2 = \tfrac{1}{2}Q^2/C, \qquad (1.35)$$

where $V = V_1 - V_2$ is the potential difference between the plates.

The energy of the system may be expressed in a different way which implies that it is distributed over the space between the charges occupied by their electrostatic field. Consider two nearby equipotential surfaces in this space which differ in potential by a small amount $V$, and are a distance $ds$ apart. If two parallel conducting plates of area $dS$ were inserted so as to coincide with these equipotentials, they would not alter the field distribution in any way. They would form a parallel-plate capacitor of capacitance $C$ and energy $\tfrac{1}{2}CV^2$. But $C = \epsilon\, dS/ds$ and $V = -E\, ds$, where $E$ is the electric field at this point. The capacitor occupies a volume $d\tau = dS\, ds$ and its energy is $\tfrac{1}{2}\epsilon E^2 (dS\, ds) = \tfrac{1}{2}DE\, d\tau$. We may therefore regard the energy as distributed throughout the field, the energy density at any point being $\tfrac{1}{2}DE$. This equation may be derived more rigorously by vector analysis, as follows.

In Fig. 1.16 suppose there exists a volume distribution of charge of density $\rho_e$ per unit volume and a surface distribution of density $\sigma_e$ per unit area. Then from eqn (1.34), if the summation is replaced by integrations, we have for the total energy

$$U = \tfrac{1}{2}\int \rho_e V\, d\tau + \tfrac{1}{2}\int \sigma_e V\, dS,$$

where the surface integral is taken over the surfaces of all the conductors present. By Gauss's theorem, $\rho_e = \mathrm{div}\, \mathbf{D}$, and hence, using a vector transformation (see Appendix A, eqn (A.12)),

$$\rho_e V = V\, \mathrm{div}\, \mathbf{D} = \mathrm{div}(V\mathbf{D}) - \mathbf{D} \cdot \mathrm{grad}\, V.$$

FIG. 1.16. Diagram to illustrate the calculation of the energy density of a system of surface and volume charges.

Therefore

$$\tfrac{1}{2}\int \rho_e V \, d\tau = \tfrac{1}{2}\int \mathrm{div}(V\mathbf{D}) \, d\tau - \tfrac{1}{2}\int \mathbf{D}.\mathrm{grad}\, V \, d\tau$$

$$= \tfrac{1}{2}\int V\mathbf{D}.d\mathbf{S} + \tfrac{1}{2}\int \mathbf{D}.\mathbf{E} \, d\tau.$$

The first integral must be taken over a closed surface bounding the whole volume, and also over the surface of each conductor. The closed surface may be taken at an infinitely large distance from all the charges, and its contribution to the surface integral then vanishes. For, at large distances, $V$ varies as $r^{-1}$ and $D$ as $r^{-2}$, while $dS$ increases with $r^2$; thus the integral is proportional to $r^{-1}$ and tends to zero as $r$ tends to infinity. The total energy may now be written

$$U = \tfrac{1}{2}\int \mathbf{D}.\mathbf{E} \, d\tau + \tfrac{1}{2}\int V\mathbf{D}.d\mathbf{S} + \tfrac{1}{2}\int \sigma_e V \, dS,$$

where the surface integrals are taken over the surfaces of all the conductors. Since the integration is over the surface of the medium, $d\mathbf{S}$ is a vector drawn outwards from the medium and hence into the conducting surface as in Fig. 1.16. But the normal component of $\mathbf{D}$ in this direction is $-\sigma_e$, from eqn (1.28), and hence $\mathbf{D}.d\mathbf{S} = -\sigma_e \, dS$, so that the two surface integrals in our expression for $U$ cancel. Our final expression for $U$ becomes

$$U = \tfrac{1}{2}\int \mathbf{D}.\mathbf{E} \, d\tau. \tag{1.36}$$

Since $\mathbf{E}$ is zero within any conductor, we may regard the energy as distributed throughout the surrounding dielectric medium, with density $\tfrac{1}{2}\mathbf{D}.\mathbf{E}$. This expression is valid in anisotropic dielectrics, where $\mathbf{D}$ is not necessarily parallel to $\mathbf{E}$, but it assumes that $\mathbf{D}$ is always proportional to $\mathbf{E}$.

The results obtained above may easily be generalized to the case where $\mathbf{D}$ is not linearly proportional to $\mathbf{E}$ (which implies that in a capacitor containing a dielectric, $Q$ is not linearly proportional to $V$). If a charge $dQ$ is transferred from the plate at $V_1$ to that at $V_2$, the work done is

$$dW = dQ(V_1 - V_2) = V \, dQ, \tag{1.37}$$

where $V$ is the potential difference between the plates. In a parallel-plate capacitor with guard rings, with a separation $t$ between the plates, the electric field strength $E = V/t$, while for an area $S$, $Q/S = D$. If an increase $dQ$ in the charge occurs at constant $V$ because of an increase $dD$ in the dielectric, the work done per unit volume of dielectric is

$$V \, dQ/tS = E \, dD.$$

More generally, if d$\mathbf{D}$ is not parallel to $\mathbf{E}$, the increase in the stored energy per unit volume is

$$dU = \mathbf{E} . d\mathbf{D} = \epsilon_0 \mathbf{E} . d\mathbf{E} + \mathbf{E} . d\mathbf{P}, \qquad (1.38)$$

where $\epsilon_0 \mathbf{E} . d\mathbf{E}$ is the change in the field energy in the absence of the dielectric, and $\mathbf{E} . d\mathbf{P}$ is the work done in polarizing the dielectric.

A similar modification can be applied to the more general treatment given above. Suppose the volume and surface charge distribution densities increase by $\delta\rho_e$ and $\delta\sigma_e$ respectively. Then the change in stored energy is

$$\delta U = \int (V \delta\rho_e) \, d\tau + \int (V \delta\sigma_e) \, dS,$$

where the integrations are over the volume of dielectric and the surfaces of the conductors respectively. Then the same transformations as used above, and consideration of the surface integrals, leads to

$$\delta U = \int (\mathbf{E} . d\mathbf{D}) \, d\tau. \qquad (1.39)$$

In calculating such changes in the electrical energy no account is taken of any heat flow into or out of the system; in thermodynamic terms, the change is adiabatic and, since it is reversible, it is a change at constant entropy. If a reversible change of the entropy also occurs, then the total change in the internal energy $U'$ is

$$\delta U' = T \, \delta S + V \, \delta Q$$

or

$$\delta U' = T \, \delta S + \int (\mathbf{E} . d\mathbf{D}) \, d\tau, \qquad (1.40)$$

where the symbol $U'$ is used for the internal energy other than that arising from electrical causes. Since the free energy is $F = U' - TS$, we have

$$dF = dU' - T \, dS - S \, dT$$

$$= -S \, dT + \int (\mathbf{E} . d\mathbf{D}) \, d\tau, \qquad (1.41)$$

showing that at constant temperature $(\mathbf{E} . d\mathbf{D})$ is the change in the free energy per unit volume. This relation is important in considering the thermodynamics of dielectrics, and of voltaic cells, where (see § 3.10) $V \, dQ$ is the change in the free energy when a charge $dQ$ passes at constant temperature and $V$ is the potential of the cell.

## 1.7. Stress in the electrostatic field

It has already been shown that the charge on a conductor resides in a thin surface layer. This is due to the mutual repulsion between charges of

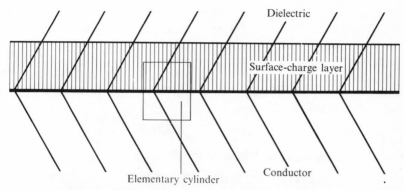

FIG. 1.17. Deduction of the tension on a charged conductor. .

like sign, so that each portion of the charge on the conductor is trying to get as far away as possible from the remainder. This results in a tension acting on the surface of the conductor, whose magnitude will now be calculated. We shall assume that the charge of surface density $\sigma_e$ is in a thin layer just outside the conducting surface, in a medium of permittivity $\epsilon$, as in Fig. 1.17. By applying Gauss's theorem to a small cylinder with its axis normal to the surface, and with one end inside the conductor and the other within the surface layer, the field strength $\mathbf{E}'$ at this latter point is found to be $\mathbf{E}' = \sigma_e\alpha/\epsilon$, where $\sigma_e\alpha$ is the portion of the surface charge density lying between the conductor and the end of the cylinder. The force on the element of charge density $\sigma_e\,\mathrm{d}\alpha$ at the end of the cylinder is therefore $\mathbf{E}'(\sigma_e\,\mathrm{d}\alpha) = \sigma_e^2\alpha\,\mathrm{d}\alpha/\epsilon$, and the total force per unit area is

$$\frac{\sigma_e^2}{\epsilon}\int_0^1 \alpha\,\mathrm{d}\alpha = \frac{\sigma_e^2}{2\epsilon}. \tag{1.42}$$

At first sight the assumption that the charge layer resides in the dielectric may appear rather artificial, but the same result may be obtained by application of the principle of virtual work to special cases. For example, consider a parallel-plate capacitor with a medium of permittivity $\epsilon$ between the plates. If the separation between the plates is $x$, and their area $S$, the capacitance $C = \epsilon S/x$, and the stored energy is $\frac{1}{2}Q^2/C = \frac{1}{2}\sigma_e^2 S^2/C$, where $\sigma_e$ is the charge density on the plates. If $\sigma_e$ is kept constant, and the separation of the plates is increased, the rate of change of the stored energy is

$$\frac{\mathrm{d}U}{\mathrm{d}x} = \frac{\mathrm{d}}{\mathrm{d}x}\left(\frac{\sigma_e^2 S x}{2\epsilon}\right) = \frac{\sigma_e^2 S}{2\epsilon}.$$

But $-(\mathrm{d}U/\mathrm{d}x)/S = -\sigma_e^2/2\epsilon$ is the tension per unit area on the plate, where the minus sign denotes that the tension acts in the opposite direction to the movement of the one plate, that is, in the direction of diminishing $x$.

It is interesting to derive the tension on the plates of this capacitor if the voltage, rather than the charge, is kept constant. In this case we must include the work done by the battery maintaining the constant potential difference. If the capacitance increases by $dC$, the charge increases by $V\,dC$, and the work done by the battery is $V(V\,dC) = V^2\,dC$. The increase in the stored energy is $dU = d(\tfrac{1}{2}CV^2) = \tfrac{1}{2}V^2\,dC$, and the external work $dW$ required is therefore

$$\tfrac{1}{2}V^2\,dC - V^2\,dC = -\tfrac{1}{2}V^2\,dC = -dU.$$

Hence the tension per unit area on the plates is

$$-\frac{1}{S}\left(\frac{\partial W}{\partial x}\right) = -\frac{1}{S}\left(\frac{\partial W}{\partial C}\right)\left(\frac{\partial C}{\partial x}\right)$$

$$= -\frac{1}{S}\left(-\frac{V^2}{2}\right)\left(-\frac{\epsilon S}{x^2}\right) = -\frac{\epsilon E^2}{2} = -\frac{\sigma_e^2}{2\epsilon},$$

which is the same as before, as we should expect. The fact that the work done by the battery in maintaining the system at constant potential is just twice the increase in the stored energy is generally true, and our example is just a particular case. It is, however, probably more instructive to remember to put in the work done by the battery in working a particular problem at constant potential rather than to make use of a general theorem.

*Stresses in dielectric media*

Both Faraday and Maxwell used the concept of tubes of force. A tube of force contains an arbitrary but very large number of lines of force, and the number of tubes crossing unit area is equal to the electric field strength; similarly, the number of tubes of displacement per unit area is equal to **D**. They imagined these tubes to be in a state of tension, so that the force of attraction between two charges of opposite sign, for example, was transmitted along the tubes of force. It was also necessary to stipulate that there was a force of repulsion between tubes of force in a direction normal to their length, otherwise the tubes would all contract until they passed straight from one charge to another.

A macroscopic theory of such forces was developed by Maxwell using the concept of a 'stress tensor' (see Panofsky and Phillips 1955; Robinson 1973). The stress is regarded as being present in the field and acting on an element of area $d\mathbf{S}$ irrespective of whether $d\mathbf{S}$ is part of a real boundary or not. If it is not, there will be an equal and opposite stress on the other side, so that equilibrium is maintained. If $d\mathbf{S}$ is part of a real boundary of a dielectric, the field components are different on the two sides and there is a net force acting at the boundary. The value of the macroscopic theory lies in providing a quantitative treatment of effects which are not easy to interpret from the viewpoint of atomic theory.

## References

PANOFSKY, W. K. H. and PHILLIPS, M. (1955). *Classical electricity and magnetism.* Addison-Wesley, New York.
PLIMPTON, S. J. and LAWTON, W. E. (1936). *Phys. Rev.* **50,** 1066.
ROBINSON, F. N. H. (1973). *Macrosopic electromagnetism.* Pergamon Press, Oxford.

## Problems

1.1. An electric quadrupole (cf. Fig. 1.9) is formed by a charge $-2q$ at the origin and charges $+q$ at the points $(\pm a, 0, 0)$. Show that the potential $V$ at a distance $r$ large compared with $a$ is approximately given by $V = +qa^2(3\cos^2\theta-1)/4\pi\epsilon r^3$, where $\theta$ is the angle between $r$ and the line through the charges.

1.2. Show that there is no translational force or couple on such an electric quadrupole in a uniform field. Prove that the couple on the quadrupole at a distance $r$ from a point charge $q'$ is $C = q(3q'a^2\sin 2\theta)/4\pi\epsilon r^3$, where $\theta$ is the angle between $r$ and the line through the charges (assume $r \gg a$).

1.3. Show that the force on an elementary dipole of moment **p**, distance **r** from a point charge $q$, has components

$$F_r = -\frac{qp\cos\theta}{2\pi\epsilon r^3}, \qquad F_\theta = -\frac{qp\sin\theta}{4\pi\epsilon r^3}$$

along and perpendicular to **r** in the plane of **p** and **r**, where $\theta$ is the angle which **p** makes with **r**.

1.4. Show that the potential energy of two coplanar dipoles $p_1$ and $p_2$ a distance $r$ apart is

$$\frac{1}{4\pi\epsilon r^3}\left\{\mathbf{p}_1\cdot\mathbf{p}_2-\frac{3(\mathbf{p}_1\cdot\mathbf{r})(\mathbf{p}_2\cdot\mathbf{r})}{r^2}\right\} = \frac{p_1p_2}{4\pi\epsilon r^3}(\sin\theta_1\sin\theta_2-2\cos\theta_1\cos\theta_2),$$

where $\theta_1$, $\theta_2$ are the angles made by $p_1$ and $p_2$ respectively with the line joining their centres.

1.5. A charge $q$ is placed at each of the four corners $(\pm a, 0, 0)$, $(0, \pm a, 0)$ of a square. Show that the potential at a point $(x, y, z)$ near the origin is

$$V = \frac{q}{4\pi\epsilon a}\left\{4+\frac{(x^2+y^2-2z^2)}{a^2}+...\right\}$$

Verify that a charge of the same sign placed at the centre of the square is in stable equilibrium against a small displacement in the plane of the square, but is unstable for a displacement normal to this plane. This is an example of Earnshaw's theorem, which shows that a charge cannot rest in stable equilibrium in an electrostatic field.

1.6. The values of the vertical potential gradient of the earth at heights of 100 m and 1000 m above its surface are 110 V m$^{-1}$ and 25 V m$^{-1}$ respectively. What is the mean electrostatic charge per cubic metre of the atmosphere between these heights?

(*Answer:* $0 \cdot 835 \times 10^{-12}$ C m$^{-3}$.)

1.7. A parallel-plate capacitor with plates of area $S$ and separation $d$ has a block of dielectric, of relative permittivity $\epsilon_r$, of cross-sectional area $S$, and thickness $t$ ($t \ll d$), inserted in between the plates. Find the values of $E$ and $D$ in the space between the plates, in both air and dielectric. Show that the capacitance of the capacitor is $C = \dfrac{S\epsilon_r\epsilon_0}{\epsilon_r d - (\epsilon_r - 1)t}$, and calculate the change in stored energy of the system when the dielectric is inserted ($a$) if the plates have a constant charge $Q$, and ($b$) if they are connected to a battery at a constant potential $V$.

$$\left( Answer:\ (a)\ \Delta U = -\frac{Q^2(\epsilon_r - 1)t}{2\epsilon_r\epsilon_0 S};\qquad (b)\ \Delta U = +\frac{\epsilon_0 S V^2(\epsilon_r - 1)t}{2d\{\epsilon_r d - (\epsilon_r - 1)t\}}. \right)$$

1.8. For the capacitor consisting of two coaxial cylinders (Fig. 1.14), show by integration of eqn (1.36) that the stored energy per unit length is

$$\left(\frac{Q^2}{4\pi\epsilon}\right)\ln(b/a),$$

and that this is equal to $\frac{1}{2}Q^2/C$, where $Q$ is the charge and $C$ the capacitance per unit length.

If $E_{max}$ is the greatest electric field strength that can exist anywhere in the dielectric without breakdown, show that the maximum voltage that can be applied between inner and outer cylinders is $aE_{max}\ln(b/a)$.

1.9. A capacitor is formed by two coaxial cylinders of radii $a$ and $b$. The axes of the cylinders are vertical and the inner cylinder is suspended from a balance so that it hangs only partly within the outer cylinder. Find an expression for the mass which must be added to the other pan of the balance to maintain equilibrium when a voltage $V$ is connected between the two cylinders.

$$\left( Answer:\ mg = \frac{\pi\epsilon V^2}{\ln(b/a)}. \right)$$

1.10. The electrometer is an instrument for measuring voltages by means of the force on a charged conductor. An attracted-disc electrometer has a moving plate of area $0\cdot01\ \mathrm{m^2}$, separated by a distance of 1 mm from the fixed plate. Calculate the force between the plates when the potential difference across them is 100 V. Calculate the sensitivity at this voltage in newtons per volt.

($Answer:\ F = 4\cdot42\times10^{-4}\ \mathrm{N},\ dF/dV = 8\cdot85\times10^{-6}\ \mathrm{N\ V^{-1}}.$)

1.11. The upper disc of such an electrometer is suspended by a spring. In equilibrium, the separation between the two discs is $x$ when a voltage $V$ is applied and $a$ when $V = 0$. Show that the equilibrium in the former case is stable provided that $x > 2a/3$.

1.12. A sphere carrying a charge density $\sigma_e$ per unit area is immersed in an infinite dielectric medium. Verify eqn (1.42) by using the principle of virtual work and allowing the radius to change infinitesimally.

1.13. If $\mathbf{n}$ is a unit vector normal to the surface, show that the boundary conditions (1.26)–(1.28) can be expressed in vectorial form as

$$\mathbf{n}\cdot(_2\mathbf{D} - _1\mathbf{D}) = \sigma_e; \qquad \mathbf{n}\wedge(_2\mathbf{E} - _1\mathbf{E}) = 0.$$

An infinitely thin surface charge density $\sigma_e$ is a mathematical abstraction; we may assume instead that near the surface there is a volume charge density $\rho_e$

which falls rapidly to zero inside a conductor. If the surface is the plane $z = 0$, and the charge density is independent of $x$, $y$, show that the value of $D_z$ just outside the surface is given by

$$D_z = \int_{-\infty}^{0} \rho_e \, dz = \sigma_e,$$

where $\sigma_e$ is the total charge per unit area of the surface.

1.14. Assuming that the total charge $Ze$ of an atomic nucleus is uniformly distributed within a sphere of radius $a$, show that the potential at a distance $r$ from the centre $(r \leqslant a)$ is

$$V = \left\{ \frac{3}{2} - \frac{1}{2} \left( \frac{r}{a} \right)^2 \right\} \frac{Ze}{4\pi\epsilon_0 a}.$$

Show that the electrostatic energy of such a nucleus is

$$U = \frac{3(Ze)^2}{20\pi\epsilon_0 a}.$$

This electrostatic energy must be provided at the expense of a small decrease $\delta m$ in the mass of the nucleus, such that $U = \delta m c^2$, where $c$ = velocity of light, by the Einstein relation. $\delta m$ is part of the 'mass defect'.

1.15. An atom with an electron in an s-state has a finite density $-\rho_e$ of electronic charge inside the nucleus. Assuming that $\rho_e$ is very small compared with the nuclear charge density, use the formula for the potential inside the nucleus given in the previous problem to show that the potential energy associated with the electron density $-\rho_e$ inside the sphere of radius $a$ is $-2Ze\rho_e a^2/5\epsilon_0$, and that this is greater by $Ze\rho_e a^2/10\epsilon_0$ than it would have been if the nucleus were a point charge.

Since isotopes of the same element have different nuclear radii, this energy forms part of the 'isotope shift', that is, the difference in frequency of spectrum lines from different isotopes.

# 2. Electrostatics II

## 2.1. The equations of Poisson and Laplace

IN a region where there exists a charge distribution of density $\rho_e$ per unit volume, the differential form of Gauss's theorem is (eqn (1.20))

$$\text{div } \mathbf{D} = \rho_e.$$

Now $\mathbf{E} = -\text{grad } V$, and in an isotropic dielectric $\mathbf{D} = \epsilon\mathbf{E}$. Hence

$$\nabla^2 V = \text{div}(\text{grad } V) = -\text{div } \mathbf{E} = -\rho_e/\epsilon. \tag{2.1}$$

This is known as Poisson's equation. If there is no free charge present, $\rho_e = 0$, and we have Laplace's equation

$$\nabla^2 V = 0. \tag{2.2}$$

The operator denoted by $\nabla^2$ is a scalar operator, which has its simplest form in Cartesian coordinates, where Poisson's equation becomes

$$\frac{\partial^2 V}{\partial x^2} + \frac{\partial^2 V}{\partial y^2} + \frac{\partial^2 V}{\partial z^2} = -\frac{\rho_e}{\epsilon}. \tag{2.3}$$

Two other coordinate systems will be considered. These are spherical polar coordinates, where Poisson's equation becomes

$$\frac{1}{r^2}\frac{\partial}{\partial r}\left(r^2\frac{\partial V}{\partial r}\right) + \frac{1}{r^2 \sin\theta}\frac{\partial}{\partial\theta}\left(\sin\theta\frac{\partial V}{\partial\theta}\right) + \frac{1}{r^2\sin^2\theta}\frac{\partial^2 V}{\partial\phi^2} = -\frac{\rho_e}{\epsilon}, \tag{2.4}$$

and cylindrical polar coordinates, where we have

$$\frac{1}{r}\frac{\partial}{\partial r}\left(r\frac{\partial V}{\partial r}\right) + \frac{1}{r^2}\frac{\partial^2 V}{\partial\theta^2} + \frac{\partial^2 V}{\partial z^2} = -\frac{\rho_e}{\epsilon}. \tag{2.5}$$

In principle, eqn (2.1) enables us to calculate the potential distribution due to any given set of charges and conductors. A formal solution of Poisson's equation can be found,

$$V = \frac{1}{4\pi\epsilon}\int\frac{\rho_e\,d\tau}{r}, \tag{2.6}$$

but this holds only in a vacuum or an infinite dielectric medium. Eqn (2.6) can be regarded as a formal extension, in a dielectric medium, of eqn (1.5) for the potential of a point charge $q$, by regarding $\rho_e\,d\tau$ as the element of charge. It can be shown (see, for example, Stratton 1941

(§ 3.7)) that the integral in eqn (2.6) is convergent although the denominator vanishes for $r = 0$, provided that $\rho_e$ remains finite, since the contribution from a sphere of radius $a$ to the potential at its centre vanishes as $a \rightarrow 0$. It can be shown also that $V$ is continuous in a region where $\rho_e$ is finite, continuous, but bounded (that is, the charge density does not stretch to infinity in all dimensions), and that it is continuous through a single surface layer of surface charge density $\sigma_e$. In the latter case this condition for $V$ clearly makes the tangential components of $\mathbf{E}$ continuous (eqn (1.27)), though from eqn (1.28) the normal components of $\mathbf{D}$ are not (that is, the gradient of $V$ changes as we cross the surface charge). In general we can take $V$ to be a continuous function, even at a boundary, an exception being the case of a double charge layer (see Problem 2.19).

If conducting surfaces are present, additional terms are needed in (2.6) to allow for the charge distribution on their surfaces. In general we do not know what this distribution is, and we have to resort to a number of special methods, but we must be sure that any solution we obtain which satisfies the boundary conditions is the correct and only answer. That this is the case is shown by an important theorem, known as the 'uniqueness theorem'. This theorem (see Stratton 1941) shows that if two different potential distributions are assumed to satisfy Laplace's equation and the boundary conditions, their difference is zero. We now discuss a number of methods for the case of no free charges, where the solutions needed are of Laplace's rather than Poisson's equation.

The required solution may be a sum of a number of functions, each of which satisfies Laplace's equation; for, if the functions $V_1, V_2, ..., V_n$ are each individual solutions of Laplace's equation, then

$$V = a_1 V_1 + a_2 V_2 + ... + a_n V_n,$$

where $a_1, a_2, ..., a_n$ are a set of numerical coefficients, is also a solution. A series of functions, each of which is a solution of Laplace's equation, may sometimes be found by making use of the fact that if $V_1$ is a solution, so also are any differentials of $V_1$ with respect to the space coordinates. Thus in Cartesian coordinates the functions $\partial V_1/\partial x$, $\partial V_1/\partial y$, $\partial V_1/\partial z$, $\partial^2 V_1/\partial x^2$ $\partial^2 V_1/\partial x\,\partial y$, etc., all satisfy eqn (2.2) if $V_1$ does. The proof of this can be seen from a single example. On partial differentiation of eqn (2.3) with respect to $x$, we have (setting $\rho_e = 0$)

$$0 = \frac{\partial}{\partial x}\left\{\frac{\partial^2 V_1}{\partial x^2} + \frac{\partial^2 V_1}{\partial y^2} + \frac{\partial^2 V_1}{\partial z^2}\right\}$$

$$= \frac{\partial^2}{\partial x^2}\left(\frac{\partial V_1}{\partial x}\right) + \frac{\partial^2}{\partial y^2}\left(\frac{\partial V_1}{\partial x}\right) + \frac{\partial^2}{\partial z^2}\left(\frac{\partial V_1}{\partial x}\right) = \nabla^2\left(\frac{\partial V_1}{\partial x}\right),$$

since the order of differentiation is immaterial when $x$, $y$, $z$ are independent coordinates. The value of this method lies in the fact that once a series of functions which satisfy eqn (2.2) is established, any linear combination of these functions may be taken, and if they can be chosen in such a way as to satisfy the boundary conditions by adjustment of the coefficients, they give the unique solution to the problem.

In theory, any problem involving electrostatic fields may be solved by finding a solution which satisfies eqn (2.2) and gives the right boundary conditions. In practice the problem is almost insoluble by ordinary mathematical methods except in cases where there is a high degree of symmetry. These may be handled by the use of a series of known functions, and some examples of this method are given below. We shall consider also another special method which can be applied to the case of one or two point charges near to a conducting surface of simple shape. Though a number of other problems may be handled by mathematical methods which are beyond the scope of this volume, most of the problems met with in practice, such as the design of electron guns to give a focused beam in a cathode-ray tube, are dealt with either by use of approximate solutions, or by numerical methods using digital computers.

## 2.2. Solutions of Laplace's equation in spherical coordinates

We shall consider first the case of spherical coordinates, and assume initially that we have symmetry about the polar axis so that $V$ is independent of $\phi$. Then Laplace's equation reduces to

$$\frac{\partial}{\partial r}\left(r^2 \frac{\partial V}{\partial r}\right) + \frac{\partial}{\partial \mu}\left\{(1-\mu^2)\frac{\partial V}{\partial \mu}\right\} = 0, \qquad (2.7)$$

where $\mu$ has been written for $\cos \theta$. This has solutions of the form $V = r^l P_l$, where $P_l$ is a function of $\mu = \cos \theta$ only, and $l$ is an integer. If we substitute such a function in eqn (2.7), and divide through by $r^l$, we obtain Legendre's differential equation for $P_l$

$$\frac{d}{d\mu}\left\{(1-\mu^2)\frac{\partial P_l}{\partial \mu}\right\} + l(l+1)P_l = 0.$$

It is readily seen that replacing $l$ by $-(l+1)$ leaves this equation unaltered, so that $P_l = P_{-(l+1)}$, that is, $V = r^l P_l$ and $V = r^{-(l+1)}P_l$ are both solutions of eqn (2.7).

Solutions of Legendre's equation may be obtained by standard methods, but a quick alternative method is as follows. We know that $V = 1/r$ is a solution, and hence so is any partial derivative of this such as $(\partial V/\partial z)$ under the conditions $x$, $y$ constant. Since $r^2 = x^2 + y^2 + z^2$, we have $2r(\partial r/\partial z) = 2z$ when $x$, $y$ are kept constant, so that

$$\left(\frac{\partial r}{\partial z}\right)_{x,y} = \frac{z}{r}.$$

Hence

$$-\left(\frac{\partial}{\partial z}\right)_{x,y}\left(\frac{1}{r}\right) = \frac{1}{r^2}\left(\frac{\partial r}{\partial z}\right)_{x,y} = \frac{z}{r^3} = r^{-2}\cos\theta$$

if we take $z$ to lie along the polar axis, so that $z = r\cos\theta$. The two functions $V = r^{-1}$ and $r^{-2}\cos\theta$ are the first two types of solution in the inverse powers $r^{-(l+1)}$, and correspond to values of $l = 0$ and 1 respectively. Thus $P_0 = 1$, and $P_1 = \cos\theta$. Further functions may be generated by successive differentiation; thus

$$\left(\frac{\partial}{\partial z}\right)\left(\frac{z}{r^3}\right) = \left(1 - \frac{3z^2}{r^2}\right)r^{-3} = (1 - 3\cos^2\theta)r^{-3}$$

gives the next function, which is proportional to $P_2$. A general formula for $P_l$ is

$$P_l = \frac{1}{2^l l!}\left(\frac{\partial}{\partial\mu}\right)^l(\mu^2 - 1)^l,\tag{2.8}$$

where the numerical coefficients are such that $P_l = 1$ at $\mu = 1$, that is, at $\theta = 0$. The first few functions are given in Table 2.1, together with the radial functions $r^{-(l+1)}$ and $r^l$ with which they combine to give solutions of Laplace's equation.

TABLE 2.1
*Some spherical harmonic functions*

| Legendre function | Function of $r$ | |
|---|---|---|
| $P_0 = 1$ | $r^{-1}$ | $1$ |
| $P_1 = \cos\theta$ | $r^{-2}$ | $r$ |
| $P_2 = \frac{1}{2}(3\cos^2\theta - 1)$ | $r^{-3}$ | $r^2$ |
| $P_3 = \frac{1}{2}(5\cos^3\theta - 3\cos\theta)$ | $r^{-4}$ | $r^3$ |
| $P_4 = \frac{1}{8}(35\cos^4\theta - 30\cos^2\theta + 3)$ | $r^{-5}$ | $r^4$ |
| $P_5 = \frac{1}{8}(63\cos^5\theta - 70\cos^3\theta + 15\cos\theta)$ | $r^{-6}$ | $r^5$ |
| $P_6 = \frac{1}{16}(231\cos^6\theta - 315\cos^4\theta + 105\cos^2\theta - 5)$ | $r^{-7}$ | $r^6$ |

*Associated Legendre functions*

Table 2.1 clearly does not contain all possible solutions of Laplace's equation, since we can find others by differentiation with respect to $x$, $y$. For example,

$$-\left(\frac{\partial}{\partial x}\right)_{y,z}\left(\frac{z}{r^3}\right) = \frac{3z}{r^4}\left(\frac{\partial r}{\partial x}\right)_{y,z} = \frac{3zx}{r^5} = 3r^{-3}\cos\theta\sin\theta\cos\phi$$

must also be a solution. The fact that it contains $\phi$ shows that it is not a solution of eqn (2.7), but of the more general equation which includes the dependence on $\phi$. This is

$$\frac{\partial}{\partial r}\left(r^2\frac{\partial V}{\partial r}\right) + \frac{\partial}{\partial\mu}\left\{(1-\mu^2)\frac{\partial V}{\partial\mu}\right\} + \frac{1}{(1-\mu^2)}\frac{\partial^2 V}{\partial\phi^2} = 0,\tag{2.9}$$

where we have again written $\mu$ for $\cos \theta$. As before, we assume that there exists a solution of the form $V = r^l \Theta \Phi$, where $\Theta$, $\Phi$ are functions only of $\theta$, $\phi$ respectively. Then the differential equation for $\Theta \Phi$ is

$$\frac{\partial}{\partial \mu}\left\{(1-\mu^2)\frac{\partial(\Theta\Phi)}{\partial\mu}\right\}+l(l+1)(\Theta\Phi)+\frac{1}{1-\mu^2}\frac{\partial^2(\Theta\Phi)}{\partial\phi^2}=0, \qquad (2.10)$$

which can be written in the form

$$\frac{(1-\mu^2)}{\Theta}\left[\frac{\partial}{\partial\mu}\left\{(1-\mu^2)\frac{\partial\Theta}{\partial\mu}\right\}+l(l+1)\Theta\right]=m^2=-\frac{1}{\Phi}\frac{\partial^2\Phi}{\partial\phi^2}, \qquad (2.11)$$

where the variables are separated. The right-hand side has the solution

$$\Phi_m = (2\pi)^{-\frac{1}{2}}e^{jm\phi}, \qquad (2.12)$$

and the equation for $\Theta$ becomes

$$\frac{\partial}{\partial\mu}\left\{(1-\mu^2)\frac{\partial\Theta}{\partial\mu}\right\}+\left\{l(l+1)-\frac{m^2}{1-\mu^2}\right\}\Theta=0. \qquad (2.13)$$

It is apparent that the functions listed in Table 2.1 are solutions of this equation for the special case $m = 0$, where there is no dependence on $\phi$. The solutions of eqn (2.13) are

$$P_{l,m} = \sin^m\theta\left(\frac{\partial}{\partial\mu}\right)^m P_l = (1-\mu^2)^{\frac{1}{2}m}\left(\frac{\partial}{\partial\mu}\right)^m P_l$$

$$= \frac{(1-\mu^2)^{\frac{1}{2}m}}{2^l l!}\left(\frac{\partial}{\partial\mu}\right)^{m+l}(\mu^2-1)^l. \qquad (2.14)$$

The functions $\Phi$ defined by eqn (2.12) are 'normalized', that is,

$$\int_0^{2\pi}\Phi_m^*\Phi_m\,d\phi=\int_0^{2\pi}(2\pi)^{-1}e^{-jm\phi}e^{jm\phi}\,d\phi=1, \qquad (2.15)$$

and they are also 'orthogonal', that is, $(m' \neq m)$

$$\int_0^{2\pi}\Phi_{m'}^*\Phi_m\,d\phi=\int_0^{2\pi}(2\pi)^{-1}e^{-jm'\phi}e^{jm\phi}\,d\phi=0. \qquad (2.16)$$

By using the Kronecker $\delta$, whose properties are that $\delta(m', m) = 1$ if $m' = m$, but $\delta(m', m) = 0$ if $m' \neq m$, we can write these two equations in the short form

$$\int_0^{2\pi}\Phi_{m'}^*\Phi_m\,d\phi=\delta(m', m). \qquad (2.17)$$

The functions $P_{l,m}$ are orthogonal but not normalized, and hence it is often convenient to work in terms of 'spherical harmonics' defined by

$$Y_{l,m} = \Theta_{l,m}\Phi_m,$$ (2.18)

where

$$\Theta_{l,m} = (-1)^m \left\{ \frac{(2l+1)(l-|m|)!}{2(l+|m|)!} \right\}^{\frac{1}{2}} P_{l,m} \quad \text{for } m \geqslant 0,$$

$$\Theta_{l,m} = (-1)^m \Theta_{l,|m|} \qquad \text{for } m < 0.$$ (2.19)

The functions $Y_{l,m}$ are both orthogonal and normalized, that is,

$$\int_0^{2\pi} \int_0^{\pi} Y_{l',m'}^* \cdot Y_{l,m} \sin\theta \, d\theta \, d\phi = \delta(l',l)\,\delta(m',m),$$ (2.20)

where the integration is over the solid angle $4\pi$. Here the significance of the $\delta$ functions is that the integral is zero unless both $l' = l$ and $m' = m$, in which case it is unity.

Expressions for the first members of the series of spherical harmonics are listed in Table 2.2 (note that the signs of the functions used by different authors sometimes differ; we have followed the definition adopted by Ramsey (1956) and Brink and Satchler (1968)). A general proof of the orthogonality and normalization relations is tedious, but the reader may verify that they are correct by evaluating the integrals for some of the functions given in Table 2.2.

The spherical harmonics have many applications, some of which are discussed in the next section. From the atomic viewpoint, their particular interest is that $Y_{l,m}$ represents the angular variation of the wavefunction

TABLE 2.2

*Some spherical harmonic functions*

| | | | |
|---|---|---|---|
| $Y_{10}$ | $+(3/4\pi)^{\frac{1}{2}}\cos\theta$ | $C_{10}$ | $+\cos\theta$ |
| $Y_{1+1}$ | $-(3/8\pi)^{\frac{1}{2}}\sin\theta e^{+j\phi}$ | $C_{1+1}$ | $-2^{-\frac{1}{2}}\sin\theta e^{+j\phi}$ |
| $Y_{1-1}$ | $+(3/8\pi)^{\frac{1}{2}}\sin\theta e^{-j\phi}$ | $C_{1-1}$ | $+2^{-\frac{1}{2}}\sin\theta e^{-j\phi}$ |
| $Y_{20}$ | $+(5/16\pi)^{\frac{1}{2}}(3\cos^2\theta-1)$ | $C_{20}$ | $+\frac{1}{2}(3\cos^2\theta-1)$ |
| $Y_{2+1}$ | $-(15/8\pi)^{\frac{1}{2}}\cos\theta\sin\theta e^{+j\phi}$ | $C_{2+1}$ | $-(3/2)^{\frac{1}{2}}\cos\theta\sin\theta e^{+j\phi}$ |
| $Y_{2-1}$ | $+(15/8\pi)^{\frac{1}{2}}\cos\theta\sin\theta e^{-j\phi}$ | $C_{2-1}$ | $+(3/2)^{\frac{1}{2}}\cos\theta\sin\theta e^{-j\phi}$ |
| $Y_{2+2}$ | $+(15/32\pi)^{\frac{1}{2}}\sin^2\theta e^{+j2\phi}$ | $C_{2+2}$ | $+(3/8)^{\frac{1}{2}}\sin^2\theta e^{+j2\phi}$ |
| $Y_{2-2}$ | $+(15/32\pi)^{\frac{1}{2}}\sin^2\theta e^{-j2\phi}$ | $C_{2-2}$ | $+(3/8)^{\frac{1}{2}}\sin^2\theta e^{-j2\phi}$ |
| $Y_{30}$ | $+(7/16\pi)^{\frac{1}{2}}(5\cos^3\theta-3\cos\theta)$ | $C_{30}$ | $+\frac{1}{2}(5\cos^3\theta-3\cos\theta)$ |
| $Y_{3+1}$ | $-(21/64\pi)^{\frac{1}{2}}(5\cos^2\theta-1)\sin\theta e^{+j\phi}$ | $C_{3+1}$ | $-(3/16)^{\frac{1}{2}}(5\cos^2\theta-1)\sin\theta e^{+j\phi}$ |
| $Y_{3-1}$ | $+(21/64\pi)^{\frac{1}{2}}(5\cos^2\theta-1)\sin\theta e^{-j\phi}$ | $C_{3-1}$ | $+(3/16)^{\frac{1}{2}}(5\cos^2\theta-1)\sin\theta e^{-j\phi}$ |
| $Y_{3+2}$ | $+(105/32\pi)^{\frac{1}{2}}\cos\theta\sin^2\theta e^{+j2\phi}$ | $C_{3+2}$ | $+(15/8)^{\frac{1}{2}}\cos\theta\sin^2\theta e^{+j2\phi}$ |
| $Y_{3-2}$ | $+(105/32\pi)^{\frac{1}{2}}\cos\theta\sin^2\theta e^{-j2\phi}$ | $C_{3-2}$ | $+(15/8)^{\frac{1}{2}}\cos\theta\sin^2\theta e^{-j2\phi}$ |
| $Y_{3+3}$ | $-(35/64\pi)^{\frac{1}{2}}\sin^3\theta e^{+j3\phi}$ | $C_{3+3}$ | $-(5/16)^{\frac{1}{2}}\sin^3\theta e^{+j3\phi}$ |
| $Y_{3-3}$ | $+(35/64\pi)^{\frac{1}{2}}\sin^3\theta e^{-j3\phi}$ | $C_{3-3}$ | $+(5/16)^{\frac{1}{2}}\sin^2\theta e^{-j3\phi}$ |

The functions $C_{lm}$ are related to $Y_{lm}$ by $C_{lm} = \{4\pi/(2l+1)\}^{\frac{1}{2}} Y_{lm}$. Note that $C_{l0} = P_l$.

for an electron in an atom which has orbital angular momentum $\sqrt{\{l(l+1)\}}(h/2\pi)$, and a component $m(h/2\pi)$ of angular momentum along the $z$-axis (the polar axis), where $h$ is Planck's constant.

## 2.3. The multipole expansion

For a charge distribution whose density is $\rho_e$ in a volume element $d\tau$ at a point where the potential is $V$, the potential energy is

$$U_P = \int \rho_e V \, d\tau, \tag{2.21}$$

where both $\rho_e$ and $V$ are functions of position. If the charge distribution extends only over a small volume, we can expand $V$ in a series

$$V = V_0 + x(\partial V/\partial x) + y(\partial V/\partial y) + z(\partial V/\partial z) +$$
$$+ \tfrac{1}{2}x^2(\partial^2 V/\partial x^2) + \tfrac{1}{2}y^2(\partial^2 V/\partial y^2) + \tfrac{1}{2}z^2(\partial^2 V/\partial z^2) +$$
$$+ \tfrac{1}{2}xy(\partial^2 V/\partial x \, \partial y) + \tfrac{1}{2}yx(\partial^2 V/\partial y \, \partial x) + .... \tag{2.22}$$

The potential energy then becomes

$$U_P = \int \rho_e V_0 \, d\tau + \int \rho_e x(\partial V/\partial x) \, d\tau + \int \rho_e y(\partial V/\partial y) \, d\tau + \int \rho_e z(\partial V/\partial z) \, d\tau + .... \tag{2.23}$$

Here the first term is simply $qV_0$, where $q = \int \rho_e \, d\tau$ is the total charge. Since $\partial V/\partial x = -E_x$, etc., we can write the next three terms as

$$p_x(\partial V/\partial x) + p_y(\partial V/\partial y) + p_z(\partial V/\partial z) = -(p_x E_x + p_y E_y + p_z E_z) = -\mathbf{p} \cdot \mathbf{E}, \tag{2.24}$$

and by comparison with eqn (1.13) we identify the quantities $p_x$, $p_y$, $p_z$ as the components of the electric dipole moment of the charge distribution. This gives a general definition of the dipole moment of a charge distribution, the components being

$$p_x = \int \rho_e x \, d\tau, \qquad p_y = \int \rho_e y \, d\tau, \qquad p_z = \int \rho_e z \, d\tau, \qquad \text{or } \mathbf{p} = \int \rho_e \mathbf{r} \, d\tau. \tag{2.25}$$

If we have a number of point charges rather than a continuous distribution, the integrals can be replaced by a summation (a dipole consisting of two equal and opposite charges separated by a small distance, as defined in Chapter 1, is thus a special case). Note that in eqn (2.24) we have implicitly assumed that the first differentials of $V$ are constant over the region occupied by charge, since only then can we take them outside the integration.

The definition of the dipole moment $\mathbf{p}$ is independent of the origin chosen for the coordinate system provided that the total charge is zero. If the origin is moved by a vector distance $\mathbf{r}_0$, so that $\mathbf{r}' = \mathbf{r} - \mathbf{r}_0$ is the position vector in the new coordinate system, then

$$\mathbf{p} = \int \rho_e \mathbf{r} \, d\tau = \int \rho_e(\mathbf{r}' + \mathbf{r}_0) \, d\tau = \int \rho_e \mathbf{r}' \, d\tau + \mathbf{r}_0 \int \rho_e \, d\tau.$$

$$= \int \rho_e \mathbf{r}' \, d\tau = \mathbf{p}',$$

provided that $\int \rho_e \, d\tau = 0$. A similar transformation applies to a set of point charges $q_i$ provided that $\sum q_i = 0$. This means that the electric dipole moment is uniquely defined for a charge distribution which has a net charge of zero, such as a neutral molecule.

If the charge distribution has reflection symmetry in the plane $z = 0$, that is, if the charge density $\rho_e$ at the point $(x, y, z)$ is the same as that at the point $(x, y, -z)$, the component $p_z$ of the dipole moment will be zero, since in the integral $\int \rho_e z \, d\tau$ the contributions from the points $(x, y, z)$ and $(x, y, -z)$ will be equal and opposite. Thus in a diatomic molecule, an electric dipole moment can exist parallel to the line joining the two nuclei if they are different (that is, if the molecule is heteronuclear, such as HCl), but not if they are identical, as in a homonuclear molecule ($H_2$, $Cl_2$).

Similar considerations apply to $p_x$, $p_y$ of course, and show that a diatomic molecule can have no electric dipole moment perpendicular to the internuclear axis. The method can be extended to more complicated molecules, such as $CH_3Cl$, $C_2H_6$, $C_6H_6$.

In an atom or nucleus the charge distribution is expected to have reflection symmetry in three mutually perpendicular planes, so that there will be no permanent electric dipole moment in any direction. Note that three such reflections change the point $(x, y, z)$ into $(-x, -y, -z)$ and are equivalent to inversion through the origin. The assumption we have made about the charge distribution is equivalent to assuming that the system is 'invariant under the parity operation', that is, that its properties are unaltered by inversion.

*Expansion in spherical harmonics*

Similar considerations can be applied to the higher terms in eqn (2.23), but it is obvious that the large number of terms makes the expansion in Cartesian coordinates clumsy to handle. We therefore turn to another method of expanding the potential of a point charge which makes use of spherical harmonics.

Assume that we have a charge at the point B and wish to know how its potential varies in the neighbourhood of the point O, for example, at the point A (Fig. 2.1(a)). This requires evaluation of the quantity $(R^2 - 2Rr \cos \theta + r^2)^{-\frac{1}{2}}$, which is the inverse of the distance AB. If $r < R$,

FIG. 2.1. Expansion of the potential at A due to a point charge $q_B$ at B (or of the potential at B due to charge $q_A$ at A) using spherical harmonics. In (b), the points A, B are not necessarily in the plane $\phi = 0$.

this function can be expanded in powers of $(r/R)$:

$$\frac{1}{|\mathbf{R}-\mathbf{r}|} = (R^2 - 2Rr\cos\theta + r^2)^{-\frac{1}{2}}$$

$$= \frac{1}{R} + \frac{r}{R^2}P_1 + \frac{r^2}{R^3}P_2 + \ldots + \frac{r^l}{R^{l+1}}P_l + \ldots . \quad (2.26)$$

Here the functions $P_l$ are the Legendre functions defined by eqn (2.8), as can readily be verified by direct expansion for the first terms.

A more general formula can be found where the points A, B in spherical coordinates are at $(r, \theta_A, \phi_A)$ and $(R, \theta_B, \phi_B)$ respectively. In Fig. 2.1(b), $Oz$ is the polar axis, and the lines OA, OB (which are not necessarily coplanar) make angles $\theta_A$, $\theta_B$ with it: the angle AOB is denoted by $\theta_{AB}$. It can be shown that the Legendre function $P_l(\cos\theta_{AB})$ can be expressed in terms of $\theta_A$, $\phi_A$ and $\theta_B$, $\phi_B$ by the formula

$$P_l(\cos\theta_{AB}) = \frac{4\pi}{2l+1} \sum_{m=-l}^{+l} (-1)^{|m|} Y_{l,m}(\theta_A, \phi_A) Y_{l,-m}(\theta_B, \phi_B)$$

$$= \sum_{m=-l}^{+l} (-1)^{|m|} C_{l,m}(\theta_A, \phi_A) C_{l,-m}(\theta_B, \phi_B), \quad (2.27)$$

where the function

$$C_{l,m} = \left(\frac{4\pi}{2l+1}\right)^{\frac{1}{2}} Y_{l,m} \quad (2.28)$$

is also listed in Table 2.2. By the use of eqn (2.27) the potential at the point A due to a charge $q_B$ at B may be written as

$$V = \frac{q_B}{4\pi\epsilon} \frac{1}{|\mathbf{R}-\mathbf{r}|}$$

$$= \frac{q_B}{4\pi\epsilon} \sum_{l=0}^{\infty} \sum_{m=-l}^{+l} (-1)^{|m|} \frac{r^l}{R^{l+1}} C_{l,m}(\theta_A, \phi_A) C_{l,-m}(\theta_B, \phi_B). \quad (2.29)$$

This shows that if we take A as a variable point, the potential at A due to the charge $q_B$ at B has a series of components, and the magnitude of the component which varies as $r^l C_{l,m}(\theta_A, \phi_A)$ is determined by the value of $R^{-(l+1)} C_{l,-m}(\theta_B, \phi_B)$ at the point B. We may equally well take B to be a variable point at which we wish to find the potential due to a charge $q_A$ at A; this is given by eqn (2.29) on replacing $q_B$ by $q_A$. The potential at B then has a series of components, where the magnitude of the component varying as $R^{-(l+1)} C_{l,-m}(\theta_B, \phi_B)$ is determined by the value of $r^l C_{l,m}(\theta_A, \phi_A)$ at the point A.

If we have a number of point charges $q_B$ at large distances, the magnitude of a given component in the potential at A near O can be found by the summation $\sum_B q_B R^{-(l+1)} C_{l,-m}(\theta_B, \phi_B)$, an example of which is given in Problem 2.18. Conversely, if we have a number of point charges $q_A$ close to O, the magnitude of a given component in the potential at a distant point such as B can be found by the summation $\sum_A q_A r^l C_{l,m}(\theta_A, \phi_A)$. In either case the summation is replaced by an integration if we have a continuous distribution of charge. Note that the series expansion for the potential does not assume that $r \ll R$, but the terms will only converge rapidly if this is so.

The electrostatic energy of two charges $q_A$, $q_B$ at the points A, B may be written by means of eqn (2.29) as

$$U_P = \frac{1}{4\pi\epsilon} \sum_{l=0}^{\infty} \sum_{m=-l}^{+l} (-1)^{|m|} \{q_A r^l C_{l,m}(\theta_A, \phi_A)\}\{q_B R^{-(l+1)} C_{l,-m}(\theta_B, \phi_B)\}, \quad (2.30)$$

which has the advantage that the terms involving the coordinates of the two charges have been separated. Thus if we have distributed charges with densities $\rho_A$, $\rho_B$ at the points A, B the electrostatic energy is

$$U_P = \frac{1}{4\pi\epsilon} \int \int \frac{\rho_A \rho_B \, d\tau_A \, d\tau_B}{|\mathbf{R} - \mathbf{r}|}$$

$$= \frac{1}{4\pi\epsilon} \sum_{l=0}^{\infty} \sum_{m=-l}^{l} (-1)^{|m|} \int \rho_A r^l C_{l,m}(\theta_A, \phi_A) \, d\tau_A \times$$

$$\times \int \rho_B R^{-(l+1)} C_{l,-m}(\theta_B, \phi_B) \, d\tau_B$$

$$= \frac{1}{4\pi\epsilon} \sum_l \sum_m A_{l,m} B_{l,-m}, \quad (2.31)$$

where

$$A_{l,m} = \int \rho_A r^l C_{l,m}(\theta_A, \phi_A) \, d\tau_A, \quad (2.32)$$

$$B_{l,-m} = \int (-1)^{|m|} \rho_B R^{-(l+1)} C_{l,-m}(\theta_B, \phi_B) \, d\tau_B. \quad (2.33)$$

The quantities $A_{l,m}$ may be regarded as defining the components of the multipole moments of degree $l$, of the charge distribution near O, and it will be seen that they interact only with the conjugate components $B_{l,-m}$ which have the same value of $l$, $m$. The monopole component $(l=0)$ contains only one term, while the dipole $(l=1)$ components contain three terms which may easily be shown (Problem 2.15) to give the same interaction as eqn (2.24). In general the interaction energy involving $r^l$ and $R^{-(l+1)}$ contains $(2l+1)$ terms, but the advantage of the quantities $A_{l,m}$, $B_{l,m}$ is that they can be expressed in terms of functions $C_{l,m}$ which are tabulated. We shall go no further than the quadrupole terms $(l=2)$, which can be written out using Table 2.2. It can then readily be verified that, for the particular case where either charge distribution is symmetrical about the axis $\theta = 0$, all the terms in eqn (2.31) vanish except that with $m = 0$. If $\rho_A$ has such symmetry (that is, it is independent of $\phi$), its quadrupole interaction can be expressed in terms of a single component

$$A_{2,0} = \int \rho_A \tfrac{1}{2} r^2 (3 \cos^2 \theta_A - 1) \, d\tau_A = \tfrac{1}{2} q Q, \qquad (2.34)$$

where the quantity (writing $r \cos \theta_A = z$)

$$Q = \frac{1}{|q|} \int \rho_A (3z^2 - r^2) \, d\tau_A \qquad (2.35)$$

is called the 'quadrupole moment' of the charge distribution, and has the dimensions of an area. It might be expected that one would take $q = \int \rho_A \, d\tau_A$, the total charge in the distribution, but for a nucleus by convention, $q$ is taken as the charge on a single proton (not the total nuclear charge), and $Q$ is expressed in terms of the unit of a barn (b) $= 10^{-28} \text{ m}^{-2}$. This unit is chosen because it is of the same order as the square of the nuclear radius (for an atom the quadrupole moment would be of order $10^{-20} \text{ m}^{-2}$).

It has already been shown that invariance under the parity operation excludes the possibility of permanent electric dipole moments in atoms and nuclei. This may be expressed more generally using spherical harmonics. Inversion through the origin is equivalent to changing the point $(r, \theta, \phi)$ into $(r, \pi - \theta, \pi + \phi)$. Since

$$P_{l,m}(\pi - \theta) = (-1)^{l-m} P_{l,m}(\theta), \qquad e^{jm(\pi+\phi)} = (-1)^m e^{jm\phi},$$

we have

$$C_{l,m}(\pi - \theta, \pi + \phi) = (-1)^l C_{l,m}(\theta, \phi), \qquad (2.36)$$

and it follows that

$$\int \rho r^l C_{l,m} \, d\tau = 0 \quad \text{if } l \text{ is odd.} \qquad (2.37)$$

Thus invariance under the parity operation excludes the possibility of electric multipole moments of any odd degree.

The form in which the interaction energy is expressed in eqn (2.31) is very suitable to a case where one charge distribution (such as that of a nucleus) is confined to a small volume, but interacts with another charge distribution which is comparatively far away (such as the atomic electrons). The series then converges very rapidly, since $(r/R) \ll 1$; experimentally, nuclear electric multipole interactions higher than quadrupole are very difficult to detect. The convergence is much less rapid in atomic cases, such as the interaction between electrons within an atom, or between atomic electrons and the surrounding ions in a solid (which strongly affects their magnetic properties—see Chapter 14). However, integrals involving all but the first few values of $l$ can be shown to vanish by means of orthogonality theorems, since the wavefunctions are themselves spherical harmonics.

## 2.4. Some electrostatic problems

*Conducting sphere in a uniform field*

Suppose an earthed conducting sphere of radius $R$ is placed in a uniform field of strength $\mathbf{E}_0$. Then the field immediately around the sphere will become distorted owing to the induced charges on the surface of the sphere (Fig. 2.2), but the field at large distances will approach the value $\mathbf{E}_0$. In general the potential distribution can be expressed as a sum of terms of the type given in Table 2.1, with the condition that $V = 0$ over the surface of the sphere. If we take a whole series of terms, it would turn out that the coefficients of most of them are zero. It is simpler to try a possible solution with a few terms whose nature is suggested by the symmetry of the problem; if it is then possible to satisfy the boundary condition $V = 0$ at $r = R$ (taking the centre of the sphere as origin of coordinates), then this is the only correct solution.

If the polar axis is taken parallel to the uniform field strength $\mathbf{E}_0$, then

FIG. 2.2. The lines of force near a conducting sphere in a uniform electric field.

the potential at large distances is that of this field alone, so that

$$V = -E_0 r \cos \theta \quad \text{for } r \to \infty.$$

In order to make $V = 0$ at $r = R$ for all values of $\theta$, it seems likely that we can only add terms which vary with the same power of $\cos \theta$. Hence we try as a solution

$$V = -E_0 r \cos \theta + A r^{-2} \cos \theta. \tag{2.38}$$

It is clear that this satisfies our boundary condition if $E_0 R = A R^{-2}$; that is, $A = E_0 R^3$. Hence we have

$$V = -E_0 r \cos \theta (1 - R^3/r^3).$$

This shows that the potential outside the sphere is that due to the uniform field together with that of a dipole of moment $\mathbf{p} = 4\pi\epsilon E_0 R^3$ situated at the centre of the sphere. Inside the sphere a solution of the type (2.38) is not acceptable, since it would give an infinite potential at the origin. Instead we must add a term $E_0 r \cos \theta$, which will just cancel the potential of the external field so that $V = 0$ everywhere inside the sphere.

The magnitude of the induced charge at any point on the sphere can be found from the normal component of the field at the surface. This is

$$E_r = -\partial V/\partial r = E_0 \cos \theta + 2E_0 (R^3/r^3)\cos \theta$$
$$= 3E_0 \cos \theta \quad \text{at } r = R.$$

Hence the charge $\sigma_e$ per unit area will be (from eqn (1.28))

$$\sigma_e = \epsilon E_r = 3\epsilon E_0 \cos \theta,$$

where $\epsilon$ is the permittivity of the medium surrounding the sphere.

*Dielectric sphere in a uniform field*

A slightly harder problem is that of a dielectric sphere of radius $R$ and relative permittivity $\epsilon_1$, surrounded by a medium of relative permittivity $\epsilon_2$, and placed in a uniform field of strength $\mathbf{E}_0$, as in Fig. 2.3. Two separate potential functions must now be taken, one for inside and the other for outside the sphere (in effect, this was also required for the conducting sphere, but then inside we had just $V = 0$). We must also satisfy the boundary conditions at the surface of the sphere, which are, from eqns (1.26) and (1.27),

$$\epsilon_1 E_{1r} = \epsilon_2 E_{2r}, \quad \text{and} \quad E_{1t} = E_{2t},$$

where subscripts 1 and 2 refer to inside and outside the surface, and subscripts $r$ and $t$ refer to normal and tangential components respectively. Guided by the potential function for the case of the conducting sphere,

(a) $\varepsilon_1 < \varepsilon_2$                                    (b) $\varepsilon_1 > \varepsilon_2$

FIG. 2.3. The lines of electric displacement **D** due to a dielectric sphere of relative permittivity $\epsilon_1$, in a uniform electric field in a medium of relative permittivity $\epsilon_2$.

we shall assume

$$V_2 = -E_0 r \cos \theta + A_2 r^{-2} \cos \theta,$$

$$V_1 = B_1 r \cos \theta + B_2 r^{-2} \cos \theta$$

for outside and inside the sphere respectively. Clearly we cannot have $V_1 \to \infty$ for $r \to 0$, so that the coefficient $B_2$ must be zero. It is also obvious that $V$ must be continuous at the boundary, since a discontinuity would give an infinite electric field there. Thus $V_1 = V_2$ at $r = R$; this automatically satisfies our second boundary condition since

$$E_t = E_\theta = -\frac{1}{r}\left(\frac{\partial V}{\partial \theta}\right),$$

and gives

$$B_1 R \cos \theta = -E_0 R \cos \theta + A_2 R^{-2} \cos \theta,$$

or

$$B_1 = A_2 R^{-3} - E_0. \tag{2.39}$$

The normal components of **E** at the surface are

$$E_{1r} = -(\partial V_1/\partial r)_{r=R} = -B_1 \cos \theta$$

and

$$E_{2r} = -(\partial V_2/\partial r)_{r=R} = E_0 \cos \theta + 2A_2 R^{-3} \cos \theta.$$

Hence the first boundary condition gives

$$-B_1 = (\epsilon_2/\epsilon_1)(E_0 + 2A_2 R^{-3}). \tag{2.40}$$

The solution of eqns (2.39) and (2.40) is

$$B_1 = -\left(\frac{3\epsilon_2}{\epsilon_1 + 2\epsilon_2}\right)E_0 \quad \text{and} \quad A_2 = \left(\frac{\epsilon_1 - \epsilon_2}{\epsilon_1 + 2\epsilon_2}\right)R^3 E_0,$$

so that the potential functions inside and outside the sphere are

$$V_1 = -\left(\frac{3\epsilon_2}{\epsilon_1 + 2\epsilon_2}\right)E_0 r \cos \theta, \tag{2.41}$$

$$V_2 = -\left(1 - \frac{R^3}{r^3}\frac{\epsilon_1 - \epsilon_2}{\epsilon_1 + 2\epsilon_2}\right)E_0 r \cos \theta. \tag{2.42}$$

These equations show that the field strength $\mathbf{E}_1$ inside the sphere is parallel to $\mathbf{E}_0$, and of magnitude

$$\mathbf{E}_1 = \left(\frac{3\epsilon_2}{\epsilon_1 + 2\epsilon_2}\right)\mathbf{E}_0. \tag{2.43}$$

We can regard the field strength $\mathbf{E}_1$ inside the sphere as the sum of $\mathbf{E}_0$ and a field strength $\mathbf{E}_d$ associated with the polarization of the sphere, whose value is

$$\mathbf{E}_d = \mathbf{E}_1 - \mathbf{E}_0 = -\left(\frac{\epsilon_1 - \epsilon_2}{\epsilon_1 + 2\epsilon_2}\right)\mathbf{E}_0. \tag{2.44}$$

When $\epsilon_1 > \epsilon_2$, $\mathbf{E}_d$ is in the reverse direction to $\mathbf{E}_0$, and it is known as the 'depolarization field'. It is proportional to $(\mathbf{P}_1 - \mathbf{P}_2)$, where $\mathbf{P}_1 = (\epsilon_1 - 1)\epsilon_0\mathbf{E}_1$ is the actual polarization inside the sphere and $\mathbf{P}_2 = (\epsilon_2 - 1)\epsilon_0\mathbf{E}_0$ is the value it would have if the permittivity of the sphere were the same as that of the surrounding medium. The value of $(\mathbf{P}_1 - \mathbf{P}_2)$ is readily found to be

$$\mathbf{P}_1 - \mathbf{P}_2 = \epsilon_0\mathbf{E}_0(\epsilon_1 - \epsilon_2)\left(\frac{1 + 2\epsilon_2}{\epsilon_1 + 2\epsilon_2}\right) \tag{2.45}$$

and

$$\mathbf{E}_d = -\frac{(\mathbf{P}_1 - \mathbf{P}_2)}{\epsilon_0(1 + 2\epsilon_2)}. \tag{2.46}$$

Obviously both $(\mathbf{P}_1 - \mathbf{P}_2)$ and $\mathbf{E}_d$ vanish when $\epsilon_1 = \epsilon_2$.

Outside the sphere, apart from the term $-\mathbf{E}_0 r \cos\theta$ arising directly from the field strength $\mathbf{E}_0$, the potential is just that of a point dipole $\mathbf{p}$ situated at the centre of the sphere and immersed in a uniform and infinite medium of relative permittivity $\epsilon_2$. The value of $\mathbf{p}$ is

$$\mathbf{p} = V\frac{3\epsilon_2}{(1 + 2\epsilon_2)}(\mathbf{P}_1 - \mathbf{P}_2), \tag{2.47}$$

where $V = 4\pi R^3/3$ is the volume of the sphere.

In the special case of a sphere in a vacuum, $\epsilon_2 = 1$ and $\mathbf{p}_2 = 0$; then $\mathbf{E}_d = -\frac{1}{3}(\mathbf{P}_1/\epsilon_0)$, and $\mathbf{p} = V\mathbf{P}_1$.

A depolarization field is present in all dielectric bodies but (1) only for a simple mathematical figure such as the ellipsoid is it a uniform field, and (2) it is only parallel to the polarization $\mathbf{P}$ when $\mathbf{P}$ is along a principal axis of the ellipsoid. Parallel to a principal axis such as the $a$-axis there exists a depolarization field strength $\mathbf{E}_d$ such that

$$\mathbf{E}_d = -\frac{\mathbf{P}_1 - \mathbf{P}_2}{\epsilon_0\{1 + (d_a^{-1} - 1)\epsilon_2\}}, \tag{2.48}$$

where $d_a$ is a pure number which depends only on the shape of the ellipsoid. The principal axes are mutually perpendicular, and any uniform polarization **P** can be resolved into components $P_a$, $P_b$, $P_c$; the factors $d_a$, $d_b$, $d_c$ are unequal except for a sphere, so that the depolarizing field will be uniform but not in general parallel to **P**.

The ellipsoid is a useful example because it includes not only the sphere, but also the thin infinite slab and the long infinite cylinder as limiting cases. Mathematically it is rather tedious and the results are given in Appendix B. An important point is that for a finite slab and a finite cylinder the internal field and the polarization cannot be uniform, because of the boundary conditions at the corners.

*Problems with cylindrical symmetry—conducting cylinder in a uniform field*

In some three-dimensional problems the potential may be independent of one coordinate, and the problem then reduces to a two-dimensional one. It is often convenient to use cylindrical coordinates in such a case, taking the $z$ direction as that in which the potential is invariant. Then putting $\rho_e = 0$ and $\partial^2 V/\partial z^2 = 0$ in eqn (2.5), we have for Laplace's equation

$$r\frac{\partial}{\partial r}\left(r\frac{\partial V}{\partial r}\right) + \frac{\partial^2 V}{\partial \theta^2} = 0. \tag{2.49}$$

This is satisfied by a function of the form $V = r^n D_n$, where $D_n$ is a function of $\theta$ alone (known as a cylindrical harmonic) which must satisfy the differential equation

$$\frac{\partial^2 D_n}{\partial \theta^2} + n^2 D_n = 0. \tag{2.50}$$

This equation is unchanged by the substitution of $-n$ for $n$, so that if $V = r^n D_n$ is a solution of Laplace's equation, so also is $V = r^{-n} D_n$. One solution is $V = \ln r$, and other solutions may be obtained either by partial differentiation with respect to $x = r\cos\theta$, or by direct solution of eqn (2.50).

A number of the simplest functions are given in Table 2.3; note that the general form of $D_n$ will be $A_n \cos n\theta + B_n \sin n\theta$, where $A_n$ and $B_n$ are constants.

TABLE 2.3

*Some cylindrical harmonic functions*

| Cylindrical harmonic | Corresponding solutions of Laplace's equation | |
|---|---|---|
| $D_0$ | $\log r$ | $1$ |
| $D_1$ | $r^{-1}(A\cos\theta + B\sin\theta)$ | $r(A\cos\theta + B\sin\theta)$ |
| $D_2$ | $r^{-2}(A\cos 2\theta + B\sin 2\theta)$ | $r^2(A\cos 2\theta + B\sin 2\theta)$ |
| $D_3$ | $r^{-3}(A\cos 3\theta + B\sin 3\theta)$ | $r^3(A\cos 3\theta + B\sin 3\theta)$ |

The type of problem to which the solutions may be applied is illustrated by the case of a conducting circular cylinder, initially uncharged, lying with its axis at right-angles to a uniform field of strength $E_0$. If the axis of the cylinder is taken as the $z$-axis, it is clear that the potential distribution will be independent of $z$. At large distances the potential will tend to $V = -E_0 r \cos \theta$, and we will assume that other terms required must also vary as $\cos \theta$. Then the potential outside the cylinder will be of the form

$$V = -E_0 r \cos \theta + A r^{-1} \cos \theta.$$

To satisfy the boundary condition $V = 0$ at $r = R$ for all values of $\theta$, we must have $E_0 R = A R^{-1}$, so that the potential is

$$V = -E_0 r \cos \theta (1 - R^2 / r^2). \tag{2.51}$$

The first term is the potential of the external field, the second that of an extended dipole consisting of two parallel lines of positive and negative charge close to the $z$-axis.

In these problems, we have assumed a potential containing just the required number of terms. This is a matter of intelligent anticipation rather than guesswork or knowing the answer beforehand. If we had taken fewer terms, we could not have satisfied the boundary conditions. If we had taken more terms, the coefficients of the additional terms would have been found to be zero. In the case just considered, terms such as $\cos n\theta$ or $\sin n\theta$ could not satisfy the condition $V = 0$ at $r = R$ for *all values of* $\theta$, because the potential of the external field varies only as $\cos \theta$. We are justified in assuming that the solution we have found is the correct and only solution because of the uniqueness theorem. This theorem also justifies the use of another special method, which we shall now consider.

## 2.5. Electrical images

If we have two equal point charges of opposite sign separated by a certain distance $2a$, the plane passing through the midpoint of the line joining them and normal to this line is an equipotential surface at zero potential. Therefore if the negative charge (say) is replaced by a plane conducting sheet AB in Fig. 2.4, the field to the right of AB will remain unaltered. Conversely, if a point charge is placed in front of an infinite conducting plane, the resultant electric field to the right of AB will be the same as that produced by the original charge plus a negative charge an equal distance from the plane on the opposite side. The negative charge is the 'electrical image' of the original charge in the plane AB.

The method of images thus consists in replacing a conductor by a point charge such that the conducting surface is still an equipotential surface. Then Laplace's equation is still satisfied at all points outside the conductor, and by the principle of uniqueness, the problem of a point charge and its image is identical with that of a point charge and an infinite conducting surface as regards the region outside the conducting surface. Electrical

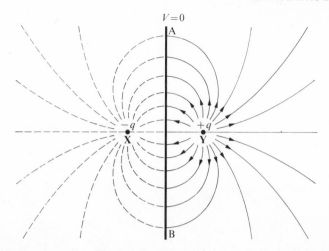

FIG. 2.4. A point charge $+q$ and its image charge $-q$ in an infinite conducting plane AB, showing the lines of force from $+q$ on the right of the plane which end on the surface charge on AB. The distance $XY = 2a$.

images are entirely virtual; a field on one side of a closed equipotential surface cannot be represented by an image on the same side of the surface, since this would give a singularity at the point occupied by the image charge. The field on the one side of the surface is identical with that which would be produced if the surface were replaced by an image charge on the other side of it.

### Point charge and infinite conducting plane

The method of images will now be applied to a number of special cases, the simplest of which is that of a point charge $q$ placed a distance $a$ from an infinite conducting plane at zero potential. In this case it is obvious that the image must be a charge $-q$ at a distance $a$ behind the plane, as in Fig. 2.5. The potential at an arbitrary point P is then

$$V = \frac{q}{4\pi\epsilon}\left\{\frac{1}{r} - \frac{1}{(r^2+4a^2+4ar\cos\theta)^{\frac{1}{2}}}\right\},$$

where $\epsilon$ is the permittivity of the medium outside the conductor. In order to calculate the charge density at any point on the plane, we must find the component of the electric field normal to the plane. This is (see Fig. 2.5)

$$E_x = E_r\cos\theta - E_\theta\sin\theta = -\frac{\partial V}{\partial r}\cos\theta + \frac{1}{r}\frac{\partial V}{\partial \theta}\sin\theta$$

$$= \frac{q}{4\pi\epsilon}\left\{\frac{\cos\theta}{r^2} - \frac{r\cos\theta+2a}{(r^2+4a^2+4ar\cos\theta)^{\frac{3}{2}}}\right\}$$

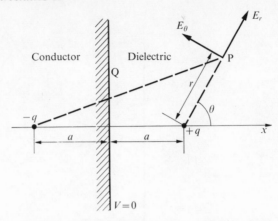

FIG. 2.5. Image of point charge in a conducting plane.

at the point P. At the point Q on the plane, $r \cos \theta = -a$, so that at Q

$$E_x = -qa/2\pi\epsilon r^3$$

(cf. Problem 2.6), where $r$ is the distance of Q from the charge $+q$. The induced charge per unit area at Q is then

$$\sigma_e = \epsilon E_x = -qa/2\pi r^3. \tag{2.52}$$

The force exerted on the point charge by the induced charge on the plane is just equal to the force exerted on the charge by its image. That is, $F = -q^2/16\pi\epsilon a^2$, where the negative sign indicates that the charge is attracted towards the plane (see Problem 2.6).

*Point charge and conducting sphere*

A more difficult problem is that of a point charge $q$ placed (*in vacuo*) a distance $a$ from the centre of a conducting sphere of radius $R$ (Fig. 2.6).

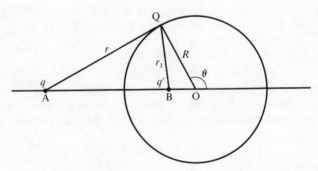

FIG. 2.6. A point charge $q$ and its image $\acute{q}$ in a conducting sphere of radius $R$.

$$OA = a, \qquad OB = b.$$

We shall consider first the case where the sphere is earthed and at zero potential. By symmetry, the image charge must be on the line through $q$ to the centre of the sphere O, and we will assume that it consists of a single charge $q'$ at a distance $b$ from O. The potential at a point Q on the surface of the sphere is then

$$V = \frac{1}{4\pi\epsilon}\left(\frac{q}{r} + \frac{q'}{r_1}\right)$$

$$= \frac{1}{4\pi\epsilon}\left\{\frac{q}{(R^2+a^2+2aR\cos\theta)^{\frac{1}{2}}} + \frac{q'}{(R^2+b^2+2bR\cos\theta)^{\frac{1}{2}}}\right\}.$$

It is only possible to make $V = 0$ over the whole surface of the sphere (that is, for all values of $\theta$) if the functions in the denominators are similar functions of $\theta$. This requires that we choose $b$ so that $b/R = R/a$; that is, B is the inverse point to A in the sphere. Then the potential at Q is

$$V = \frac{q+(a/R)q'}{4\pi\epsilon(R^2+a^2+2Ra\cos\theta)^{\frac{1}{2}}},$$

and this will be zero if we make $q' = -q(R/a)$. Hence the image charge is of magnitude $-q(R/a)$ at the inverse point in the sphere, and the reader may verify, by integrating the charge density on the sphere, that the total charge on the sphere is equal to the image charge.

If the sphere is insulated and initially uncharged, the total charge on it must remain zero. It is therefore necessary to add a second image charge $-q'$ at such a point that the surface of the sphere remains an equipotential surface. This is accomplished by placing a charge $+q(R/a)$ at the centre of the sphere in addition to the charge $-q(R/a)$ at the inverse point. If the sphere is insulated but carries an initial charge Q the total charge at the centre would of course be $Q+q(R/a)$.

## 2.6. Line charges

Just as we have considered the mathematical abstraction of a point charge, so we may postulate a 'line charge' in which charge is uniformly distributed along an infinite straight line. Its strength is denoted by $\lambda$, the charge per unit length. To find the field of such a line charge, immersed in a medium of permittivity $\epsilon$, we apply Gauss's theorem to a section of length $t$ of a cylinder of radius $r$ whose axis coincides with the line charge. This gives

$$\epsilon E(2\pi rt) = \lambda t,$$

since by symmetry the field strength **E** is everywhere normal to the axis. Hence $E = \lambda/2\pi\epsilon r$, and the potential at a distance $r$ from the axis is

$$V = -\frac{\lambda}{2\pi\epsilon}\int\frac{dr}{r} = -\frac{\lambda}{2\pi\epsilon}\ln r + V_0. \tag{2.53}$$

FIG. 2.7. Two parallel line charges normal to the plane of the paper, with charge $+\lambda$ and $-\lambda$ per unit length.

Here the constant $V_0$ cannot be defined by assuming $V = 0$ at $r = \infty$ since the line charge itself extends to infinity.

If we have two parallel line charges of equal strength but opposite sign, as in Fig. 2.7, the potential at a point whose perpendicular distances from the line charges are $r_1$, $r_2$ respectively is

$$V = \frac{\lambda}{2\pi\epsilon} \ln (r_2/r_1) + V_0. \tag{2.54}$$

The equipotential surfaces given by this equation are shown in Fig. 2.8. They have the form of cylinders whose cross-sections form a set of coaxial circles with limiting points at the line charges. The surface whose potential is $V_0$ is the median plane (for which $r_1 = r_2$) between the two line charges. From this it follows that the problem of a line charge parallel to a conducting plane can be solved by the method of images, using a line charge of opposite sign as image. We shall apply our results to a more realistic problem.

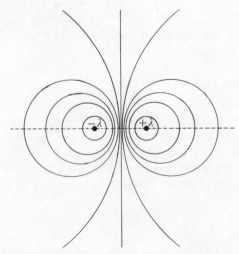

FIG. 2.8. The lines of constant potential for two parallel infinite line charges $+\lambda$ and $-\lambda$, normal to the plane of the paper. They form systems of coaxial circles with limiting points at the charges.

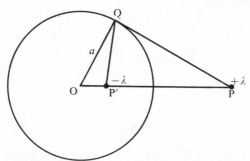

FIG. 2.9. The image of an infinite line charge in an infinite conducting cylinder.

## Capacitance between two parallel infinite circular cylinders

Consider first an infinite line charge of strength $\lambda$ which is parallel to an infinite cylinder of radius $a$. In the cross-section shown in Fig. 2.9, the line charge is at P and the axis of the cylinder at O. We imagine an image line charge of strength $-\lambda$ to be placed at P', the position of P' being chosen so that the circle formed by the cross-section of the conducting cylinder coincides with one of the family of coaxial circles which are the equipotentials of the line charge and its image. Then the potential at any point Q on the surface of the cylinder is

$$V = -\frac{\lambda}{2\pi\epsilon}\ln\left(\frac{\text{QP}}{\text{QP}'}\right) + V_0.$$

Since P and P' are the limiting points of the family of coaxial circles, it follows that they are inverse points with regard to any one of these circles. Hence

$$\text{QP/QP}' = \text{OP/OQ} = \text{OP}/a,$$

and the potential at Q is

$$V = -\frac{\lambda}{2\pi\epsilon}\ln\left(\frac{\text{OP}}{a}\right) + V_0,$$

which is independent of the position of Q on the surface of the cylinder, showing that this is an equipotential.

We turn now to the case of two infinite parallel cylinders, each of radius $a$, whose axes are a distance $2d$ apart. Then in Fig. 2.10 the distance OO' is $2d$, and P, P' are the limiting points of a family of coaxial circles. P and P' are chosen so that two of the circles coincide with the surfaces of the cylinders, making each of these an equipotential if line charges of strength $\lambda$ and $-\lambda$ were placed at P and P' respectively. Then the potential at an arbitrary point whose distance is $r_1$ from P and $r_2$ from P' is

$$V = -\frac{\lambda}{2\pi\epsilon}\ln(r_1/r_2),$$

FIG. 2.10. Two infinitely long parallel conducting cylinders.

where the constant $V_0$ is zero if we take the median plane $(r_1 = r_2)$ between the cylinders to be at zero potential. Then the potentials of the two cylinders are

$$V_Q = -\frac{\lambda}{2\pi\epsilon}\ln(OP/a) \quad \text{and} \quad V_{Q'} = +\frac{\lambda}{2\pi\epsilon}\ln(OP/a)$$

respectively. But $OP + OP' = 2d$, and $OP \cdot OP' = a^2$, since P, P' are inverse points in the circle of radius $a$ centre O. Hence

$$OP = d + \sqrt{(d^2 - a^2)},$$

and the capacitance per unit length between the two cylinders is

$$C = \lambda/(V_{Q'} - V_Q)$$

$$= \frac{\pi\epsilon}{\ln[\{d + \sqrt{(d^2 - a^2)}\}/a]}. \tag{2.55}$$

When $d \gg a$, this approaches the limiting value

$$C = \frac{\pi\epsilon}{\ln(2d/a)}. \tag{2.56}$$

A similar problem is the capacitance of a horizontal telegraph wire with respect to the earth. This may be treated as an infinite cylinder of radius $a$ a distance $d$ above an infinite conducting plane. It is clear that the potential distribution will be the same as in the case of the two parallel cylinders if we assume that the conducting plane coincides with the median plane between the cylinders, which is the equipotential surface $V = 0$. Then the charge on the wire per unit length is $\lambda$, and the potential difference between it and the plane is just half that between the two cylinders in the previous problem. Hence the capacitance per unit length will be (assuming $d \gg a$)

$$C = \frac{2\pi\epsilon}{\ln(2d/a)}. \tag{2.57}$$

Note that the approximation $d \gg a$ is tantamount to assuming that the wire behaves as if it had a line charge $\lambda$ per unit length along its axis, since as $(a/d)$ approaches zero the point $P'$ moves towards $O$ in Fig. 2.10 and $OP \rightarrow 2d$.

## 2.7. Images in dielectrics

The potential distribution due to a point charge near a dielectric surface may sometimes be found by the method of images. We shall illustrate this type of problem by considering the case of a point charge $q$ a distance $a$ from a semi-infinite dielectric bounded by a plane surface. This problem is more complex than that of a point charge and conducting plane since a second image system is required to represent the field within the dielectric. It is not obvious that the field can be represented by that of a single point charge, but we shall assume that this is possible (if our assumption is wrong we shall not be able to satisfy the boundary conditions). We take therefore a single charge $q_2$ at a point $B$, as in Fig. 2.11, and the potential at point $Q$ in the dielectric will then be

$$V_Q = \frac{q_2}{4\pi\epsilon_0 r_2}. \qquad (2.58)$$

The field of the point charge $q$ will polarize the dielectric and there will therefore be a surface charge on the dielectric which affects the field outside. We assume that this can be represented by an image charge $q_1$ at $C$ in the dielectric, and the potential at point $P$ outside the dielectric is then

$$V_P = \frac{1}{4\pi\epsilon_0}\left(\frac{q}{r} + \frac{q_1}{r_1}\right). \qquad (2.59)$$

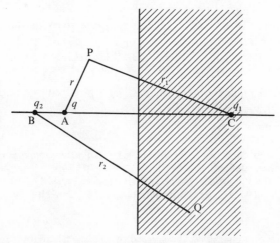

FIG. 2.11. Image systems for a point charge $q$ and a semi-infinite dielectric. The field outside the dielectric is that of $q$ and $q_1$; the field inside the dielectric is that of $q_2$.

By symmetry, $q$, $q_1$, and $q_2$ will all lie on a normal to the dielectric surface. Note that the relative permittivity $\epsilon_r$ does not appear in these equations since the effect of the dielectric is replaced by two image systems *in vacuo*.

To avoid an infinite electric field at the boundary we must assume that $V_P = V_Q$ at the boundary, and this automatically satisfies our first boundary condition, that the tangential components of **E** must be the same on either side of the boundary. It is clear that the condition $V_P = V_Q$ everywhere on the boundary can only be satisfied if $r$, $r_1$, and $r_2$ vary at the same rate as we move along the boundary, and we must therefore have $r = r_1 = r_2$, so that B coincides with A, and C is as far behind the surface as A is in front. In addition, $q + q_1 = q_2$. Our second boundary condition is that the normal components of **D** must be continuous at the boundary, that is, $(\partial V_P/\partial z) = \epsilon(\partial V_Q/\partial z)$ at the surface, which we take to be the plane $z = 0$. Now at an arbitrary point $(x, y, z)$, $r_2 = \{x^2 + y^2 + (a+z)^2\}^{\frac{1}{2}}$, and $r_1 = \{x^2 + y^2 + (a-z)^2\}^{\frac{1}{2}}$, so that

$$\frac{\partial}{\partial z}\left(\frac{1}{r}\right) = \frac{\partial}{\partial z}\left(\frac{1}{r_2}\right) = -\frac{a+z}{r^3}, \qquad \frac{\partial}{\partial z}\left(\frac{1}{r_1}\right) = \frac{a-z}{r_1^3}.$$

Using these relations for the case $z = 0$, our second boundary condition becomes

$$\frac{a(q - q_1)}{4\pi\epsilon_0 r^3} = \frac{\epsilon_r a q_2}{4\pi\epsilon_0 r^3}.$$

Hence we have the relations

$$q + q_1 = q_2 \quad \text{and} \quad q - q_1 = \epsilon_r q_2,$$

which give $q_2 = 2q/(\epsilon_r + 1)$ and $q_1 = -q(\epsilon_r - 1)/(\epsilon_r + 1)$ for the image charges. The force of attraction on the charge $q$ towards the dielectric is therefore

$$F = -\frac{qq_1}{4\pi\epsilon_0(2a)^2} = \frac{q^2(\epsilon_r - 1)}{16\pi\epsilon_0 a^2(\epsilon_r + 1)}. \tag{2.60}$$

The lines of displacement for the case of a point charge and an infinite dielectric are shown in Fig. 1.13.

### References

BRINK, D. M. and SATCHLER, G. R. (1968). *Angular momentum* (2nd edn). Clarendon Press, Oxford.

RAMSEY, N. F. (1956). *Molecular beams.* Clarendon Press, Oxford.

STRATTON, J. A. (1941). *Electromagnetic theory.* McGraw-Hill, New York.

### Problems

2.1. The polarization charge on the surface of a spherical cavity is $-\sigma_e \cos\theta$, at a point whose radius vector from the centre makes an angle $\theta$ with a given axis $Oz$. Prove that the field strength at the centre is $\sigma_e/3\epsilon_0$, parallel to $Oz$.

If the cavity is in a uniform dielectric subject to a field of strength $\mathbf{E}_0$ parallel to the direction $\theta = 0$, show that $\sigma_e = 3E_0\epsilon_0(\epsilon_r - 1)/(1 + 2\epsilon_r)$, where $\epsilon_r$ is the relative permittivity of the dielectric. Verify that this gives the correct value for the field strength at the centre of the cavity (eqn (2.43)) and note that $\sigma_e$ is not simply $(\epsilon_r - 1)\epsilon_0 E_0$ because of the distortion of the field in the dielectric caused by the presence of the cavity.

2.2. A dipole $\mathbf{p}$ is situated at the centre of a spherical cavity of radius $a$ in an infinite medium of relative permittivity $\epsilon_r$. Show that the potential in the dielectric medium is the same as would be produced by a dipole $\mathbf{p}'$ immersed in a continuous dielectric, where

$$\mathbf{p}' = \mathbf{p}\frac{3\epsilon_r}{2\epsilon_e + 1},$$

and that the field strength inside the cavity is equal to that which the dipole would produce in the absence of the dielectric, plus a uniform field $\mathbf{E}_r$

$$\mathbf{E}_r = \frac{2(\epsilon_r - 1)}{2\epsilon_r + 1}\frac{\mathbf{p}}{4\pi\epsilon_0 a^3}.$$

$\mathbf{E}_r$ is known as the 'reaction field strength'. These formulae are used in Onsager's theory of dielectrics (see Chapter 10).

2.3. An infinite circular cylinder of relative permittivity $\epsilon_r$ is subjected in vacuum to a transverse uniform field of strength $\mathbf{E}_0$. Show that the field inside the cylinder is

$$\frac{2}{\epsilon_r + 1}\mathbf{E}_0,$$

and that the depolarizing field is $-\mathbf{P}_1/2\epsilon_0$.

2.4. Find an expression for the surface density of charge on an infinitely long conducting cylinder of radius $a$, placed with its axis at right-angles to a uniform electric field of strength $\mathbf{E}_0$, as a function of the polar angle $\theta$.

(*Answer*: $\sigma_e = 2\epsilon E_o \cos\theta$.)

2.5. A uniform electric field of strength $\mathbf{E}$ is set up in an infinite dielectric. Show that ($a$) if a long needle-shaped cavity, whose lateral dimensions are very small compared with its length, is cut in the dielectric with its axis parallel to $\mathbf{E}$, then the field strength in this cavity is $\mathbf{E}$; ($b$) if a flat disc-shaped cavity, whose lateral dimensions are very large compared with its thickness, is cut with its plane normal to the direction of $\mathbf{E}$, then the field strength in the cavity is $\mathbf{D}/\epsilon_0$, where $\mathbf{D}$ is the displacement in the dielectric.

Verify that the field in an intermediate shape of cavity (such as in Problem 2.1) lies between these extreme values.

2.6. A point charge $q$ is placed near an infinite conducting plane. Verify that the total charge on the plane is $-q$ by integrating eqn (2.52), and calculate the total force on the plane by integration of the tension per unit area (eqn (1.42)) over the area of the plane. Verify the expression given in the text for the field strength at $Q$ (Fig. 2.5) by vector addition of the fields of the point charge and its image.

2.7. Show that the work done in bringing up a charge $q$ from infinity to a distance $a$ from a conducting plane at zero potential is $-q^2/16\pi\epsilon a$. Verify that the same result is obtained using eqn (1.34) (remember that the induced charge on the plane is at zero potential).

2.8. Show that the force on a charge $q$ distance $a$ *in vacuo* from the centre of an insulated and uncharged conducting sphere of radius $R$ is $(a > R)$

$$F = \frac{q^2}{4\pi\epsilon_0}\left\{\frac{R}{a^3} - \frac{Ra}{(a^2 - R^2)^2}\right\}.$$

2.9. A point charge is placed in a hollow metal sphere of radius $R$. If the charge is $q$ and is a distance $b$ from the centre of the sphere show that the force on it is

$$\frac{q^2}{4\pi\epsilon_0}\frac{Rb}{(R^2 - b^2)^2}.$$

In what direction does the force act?

(*Hint.* Find the image of $q$ outside the sphere such that the sphere is an equipotential surface.)

2.10. A point charge $q$ is placed at a distance $3R$ from the centre of an isolated conducting sphere of radius $R$ which already has a charge equal to $q$. Prove that the surface densities at points on the sphere nearest to and farthest from the point charge are in the ratio $8 : 29$.

2.11. An elementary dipole of strength $p$ is placed at a point P, outside and at a distance $a$ from the centre C of an earthed conducting sphere of radius $R$. The axis of the dipole is in the direction CP. Prove that its image system consists of a point charge $pR/a^2$ and a dipole of strength $pR^3/a^3$, both situated at the point P′ which is inverse to P in the sphere.

2.12. Show that if an infinite line charge $\lambda$ per unit length is at a distance $d$ from an infinite conducting plane in a medium of permittivity $\epsilon$, the surface density of charge in the plane is $\sigma_e = -d\lambda/\pi r^2$, where $r$ is the shortest distance from the line of charge to the point in question.

2.13. Calculate the force per unit length on the infinite line charge of the last question.

$$(\textit{Answer: } F = -\lambda^2/4\pi\epsilon d.)$$

2.14. Electric charge is distributed over a thin spherical shell with a density which varies in proportion to the value of a single function $P_l(\cos\theta)$ at any point on the shell. Show, by using the expansions (2.26) and (2.27) and the orthogonality relations for the Legendre functions, that the potential varies as $r^l P_l(\cos\theta)$ at a point $(r, \theta)$ inside the sphere and $r^{-(l+1)}P_l(\cos\theta)$ at a point $(R, \theta)$ outside.

2.15. Show that, for $l = 1$, the quantities defined by eqn (2.32) are

$$A_{1,0} = p_z, \qquad A_{1,1} = -2^{-\frac{1}{2}}(p_x + jp_y), \qquad A_{1,-1} = 2^{-\frac{1}{2}}(p_x - jp_y),$$

while for a point charge at B in Fig. 2.1 (b),

$$\frac{1}{4\pi\epsilon}B_{1,0} = -E_z, \qquad \frac{1}{4\pi\epsilon}B_{1,1} = -2^{-\frac{1}{2}}(E_x + jE_y), \qquad \frac{1}{4\pi\epsilon}B_{1,-1} = 2^{-\frac{1}{2}}(E_x - jE_y),$$

and verify that this gives the same interaction energy as eqn (2.24).

2.16. For an atom in a p-state, the wavefunction is $\psi = f(R)Y_{1,0}$, and the charge density is $-e\psi^2$. Show that the atomic quadrupole moment (as defined by eqn

(2.35)) is $\frac{4}{5}\langle R^2 \rangle$, where $\langle R^2 \rangle$ is the mean-square distance of the electron from the nucleus.

2.17. The isotope of mass 35 of chlorine has a nuclear electric quadrupole moment $Q$. Show that for this nucleus in a chlorine atom whose wavefunction $\psi = f(R)Y_{1,0}$, the energy of interaction between the nuclear electric quadrupole moment and the electron is

$$U_P = -\frac{\frac{1}{5}e^2 Q\langle R^{-3}\rangle}{4\pi\epsilon_0},$$

where $\langle R^{-3}\rangle$ is the mean inverse cube of the distance of the electron from the nucleus and $e$ the electronic charge.

If $Q = -0\cdot079\,\text{b} = -7\cdot9\times10^{-30}\,\text{m}^2$, and $\langle R^{-3}\rangle = 5\times10^{31}\,\text{m}^{-3}$, show that the energy of interaction ($U_P/h$) expressed in frequency units ($h$ is Planck's constant) is about 27 MHz.

2.18. Six equal charges $q$ are placed at the points $(\pm R, 0, 0)$, $(0, \pm R, 0)$, $(0, 0, \pm R)$. Show that the terms of lowest degree in the potential at a point $(x, y, z) = (r, \theta, \phi)$ near the origin are

$$V = \frac{6q}{4\pi\epsilon R} + \frac{q}{4\pi\epsilon} \cdot \frac{r^4}{R^5} \cdot \frac{7}{2}\{C_{4,0} + (\tfrac{5}{14})^{\frac{1}{2}}(C_{4,4} + C_{4,-4})\},$$

where $C_{4,0} = P_4$ (which is given in Table 2.1), and

$$C_{4,\pm4} = (\tfrac{35}{128})^{\frac{1}{2}}\sin^4\theta\,e^{\pm j4\phi}.$$

(*Hints.* Since the system has inversion symmetry through the origin, terms in odd powers of $r$ must vanish. Also $V$ must have fourfold symmetry about the polar ($z$-)axis, so that rotations changing $\phi$ by $\frac{1}{2}\pi$ must leave $V$ unchanged: only functions with $m = 0$ or 4 satisfy this condition.)

Note that in Cartesian coordinates

$$V = \frac{6q}{4\pi\epsilon R} + \frac{q}{4\pi\epsilon} \cdot \frac{1}{R^5} \cdot \frac{35}{4}(x^4 + y^4 + z^4 - \tfrac{3}{5}r^4),$$

showing that the $x$-, $y$-, $z$-axes are all equivalent (cubic symmetry).

2.19. An electric double layer is formed of two parallel planes a distance $t$ apart, one with a surface charge density $+\sigma_e$ and the other with a surface charge density $-\sigma_e$. Show that the normal components of $\mathbf{D}$ are the same on either side of the double layer (but not in between), but that the values of the potential $V$ differ by

$$V_2 - V_1 = P_s/\epsilon_0,$$

where $P_s = \sigma_e t$ is the dipole moment per unit area of the double layer, directed from side 1 to side 2.

# 3. Current and voltage

## 3.1. Introduction

IN the previous chapters on electrostatics we have been concerned with stationary electric charges. If a free charge is placed in an electric field it will be acted on by a force, and will move in the direction of the lines of force. Thus, if an initial difference of potential exists in a conductor, the charges will move until they reach positions of equilibrium, and the whole of the conductor becomes an equipotential surface. But by connecting a battery between two points of a conductor, a permanent difference of potential may be maintained between these two points, and there will then be a continuing flow of charge. This constitutes an electric current, and the strength of the current $I$ is defined by the rate at which charge passes any given point in the circuit. If we are dealing with a current extended in space, then we may define the current density $\mathbf{J}$ as the quantity of charge passing per second through unit area of a plane normal to the line of flow. The total current flowing through any surface is found by integrating the normal component of the current density: that is

$$I = \int \mathbf{J} . d\mathbf{S}. \tag{3.1}$$

If the current is carried by particles of charge $q$ with density $n$ per unit volume and velocity $\mathbf{v}$, then the current density is

$$\mathbf{J} = nq\mathbf{v}. \tag{3.2}$$

Thus $\mathbf{J}$ is a vector whose direction is that of the velocity $\mathbf{v}$ of the carriers if $q$ is positive.

In early experiments on electricity there was no evidence for the sign of the charges forming the current, since there was no means of distinguishing between a flow of positive charges in one direction and a flow of negative charges in the opposite direction. The positive direction of current flow was therefore taken as that in which a positive charge would move in an electric field. Thus in a circuit, the conventional direction for the flow of current is from the higher potential to the lower potential: for example, from the positive pole of a battery round the external circuit to the negative pole. It is customary to retain this convention, although the modern theory of metallic conduction shows that the positively charged ions are fixed, while a certain number of electrons are free to move about the body of the metal. Since the electrons are negatively charged, their direction of movement is opposite to that of the conventional current flow.

*Measurement of $q/m$ for carriers of electric current in a metal*

The first direct experimental evidence that the carriers of electricity in a metal are electrons was supplied by the measurements of Tolman and Stewart (1917). The principle of this experiment depends on a comparison of the electric current with the momentum carried by the particles. If the current density is $nqv$, and the particles have mass $m$, then the momentum associated with the current crossing unit area of a plane normal to the direction of flow is $nmv$. Thus the ratio of the electric current density to the momentum 'current density' is simply equal to the ratio of charge to mass ($q/m$) of the carriers. The sign of $q/m$ is obtained from comparison of the directions of flow of the electric current and momentum current. The method we shall now describe is that of a later experiment by Kettering and Scott (1944).

A circular coil is suspended by a thin fibre so that its plane is horizontal and it forms a torsional pendulum with very small damping. The coil, consisting of $N$ turns of radius $r$, carries a current $I$. If the number of electrons per unit length of the wire is $n$, and they move with a mean velocity $v$, their angular momentum about the axis of the coil is

$$\Gamma = mrvn(2\pi rN),$$

since the total number of electrons in the coil is $n(2\pi rN)$. The current $I = nqv$, and hence we have

$$\Gamma = 2\pi r^2 N(m/q)I = 2AN(m/q)I,$$

where $A$ is the area of the coil. In the experiment, a current $I$ is maintained in the coil, and then suddenly reversed. This imparts an impulse $2\Gamma$ to the coil, whose angular momentum is thus altered by $4AN(m/q)I$. In practice the amplitude of swing of the coil is observed with the current flowing in one direction, and the current is reversed at the moment when the coil passes through its equilibrium position. This changes the amplitude of swing $\theta_0$ by an amount

$$\Delta\theta_0 = 2\Gamma(T/2\pi\mathfrak{I}),$$

where $2\Gamma$ is the change of angular momentum due to the current reversal, $T$ is the time of swing, and $\mathfrak{I}$ the moment of inertia of the coil. The value of $q/m$ can thus be determined by measurement of the change in amplitude of oscillation for a given current $I$.

Although the theory of the experiment is simple there were many practical difficulties. Leads to the coil had to be brought in so that the free suspension by the torsion fibre was not affected. To eliminate vibration and disturbance due to changing magnetic fields, the apparatus was installed in an underground vault. The experiment was performed both with coils made of copper and aluminium. The values of $m/q$ obtained were 5·64, 5·67, and 5·79×10$^{-12}$ kg C$^{-1}$ for three different copper coils, and 5·66×10$^{-12}$ kg C$^{-1}$ for an aluminium coil. The mean of these results, 5·69×10$^{-12}$ kg C$^{-1}$, is in very close agreement with the reciprocal (5·686×10$^{-12}$ kg C$^{-1}$) of the most accurate determinations of $|q/m|$ for free electrons. The sign of $m/q$, obtained from the direction of the change of amplitude, corresponded to the carriers being negatively charged.

## 3.2. Flow of current in conductors

Since electric charge can neither be created nor destroyed, it follows that the rate of increase of the total charge inside any arbitrary volume must be equal to the net flow of charge into this volume. We have therefore

$$\frac{\partial}{\partial t}\left(\int \rho_e \, d\tau\right) = \int \frac{\partial \rho_e}{\partial t} \, d\tau = -\int \mathbf{J} \cdot d\mathbf{S},$$

where the integrals are taken respectively over the volume and the surface bounding it. On transforming the surface integral into a volume integral, we have

$$\int \frac{\partial \rho_e}{\partial t} \, d\tau = -\int \text{div} \, \mathbf{J} \, d\tau$$

or

$$\int \left(\text{div} \, \mathbf{J} + \frac{\partial \rho_e}{\partial t}\right) d\tau = 0.$$

This integral must be zero whatever the volume over which we integrate, and this can only be true if the integrand is itself zero. We may therefore write

$$\text{div} \, \mathbf{J} = -\frac{\partial \rho_e}{\partial t}, \tag{3.3}$$

which is known as the equation of continuity. In the steady state $\partial \rho_e/\partial t = 0$, and therefore

$$\text{div} \, \mathbf{J} = 0 \tag{3.4}$$

in any region of current flow which does not contain a source or sink of current. Such a source or sink by which current may be injected into or withdrawn from a conducting region is known as an electrode. The total current flow to or from an electrode may be found by integrating the current density over any surface which totally encloses the electrode.

*Ohm's law*

It is found experimentally that in a metallic conductor at constant temperature the current density is linearly proportional to the electric field strength. This is expressed by the equation

$$\mathbf{J} = \sigma\mathbf{E}, \tag{3.5}$$

where the constant $\sigma$ is known as the specific conductivity. Its reciprocal is known as the specific resistance or resistivity, and denoted by the symbol $\rho$. Clearly

$$\mathbf{E} = \rho\mathbf{J}, \tag{3.5a}$$

and it is obvious that the quantities denoted by $\sigma$ and $\rho$ have no connection with those denoted by $\sigma_e$ and $\rho_e$ in the previous chapters.

If a conducting wire of cross-section $A$ carries a current $I$, then $I = JA$; if the current enters at a point where the potential is $V_1$ and leaves at a point a distance $l$ away where the potential is $V_2$, then $E = -(V_2 - V_1)/l$. Hence

$$I = \sigma E A = \sigma A (V_1 - V_2)/l = (V_1 - V_2)/R, \tag{3.6}$$

where $R = l/A\sigma = \rho l/A$ is known as the resistance of the wire. Eqn (3.6), which expresses the fact that the voltage between the ends of a conductor is proportional to the current flowing, is known as Ohm's law. In S.I. units $V$ is measured in volts (V), $I$ in amperes (A) (1 A is that current which when flowing in two parallel straight conductors of infinite length and negligible cross-section, placed 1 m apart in a vacuum, would produce a force $2 \times 10^{-7}$ N per m of length between the two conductors—see § 4.1), and the unit of resistance is the ohm ($\Omega$). Its reciprocal, the unit of conductance, is the reciprocal ohm (or siemens = S). Since $\rho = R(A/l)$, the dimensions of specific resistance are those of resistance × length; its unit is the ohm-metre ($\Omega$ m), while that of $\sigma$ is (ohm-metre)$^{-1}$ ($\Omega^{-1}$ m$^{-1}$).

Values of the resistivity at room temperature are given in Table 3.1 for

TABLE 3.1

*Specific resistance of some typical materials*

| Substance | | Specific resistance at 290 K ($\Omega$ m) | Temperature coefficient (K$^{-1}$) |
|---|---|---|---|
| Pure metals | silver | $1\cdot6\times10^{-8}$ | $3\cdot8\times10^{-3}$ |
| | copper | $1\cdot72\times10^{-8}$ | $3\cdot9\times10^{-3}$ |
| | aluminium | $2\cdot83\times10^{-8}$ | $3\cdot9\times10^{-3}$ |
| | platinum | $10\times10^{-8}$ | $3\cdot9\times10^{-3}$ |
| Alloys | constantan | $44\cdot2\times10^{-8}$ | $\sim10^{-6}$ |
| | manganin | $44\times10^{-8}$ | $\sim10^{-6}$ |
| Semiconductors | pure silicon | $\sim2\times10^{3}$ | negative |
| | pure germanium | $\sim0\cdot5$ | negative |
| Insulators | glass plate | $2\times10^{11}$ | — |
| | sealing wax | $\sim10^{14}$ | — |
| | sulphur | $\sim10^{15}$ | — |
| | fused quartz | $>5\times10^{16}$ | — |

a number of substances, and it will be seen that they vary over an enormous range. For most metals the resistivity is of order $10^{-8}\,\Omega\,\mathrm{m}$, being generally higher for alloys than for pure metals. The monatomic semiconductors silicon and germanium have resistivities higher by several orders of magnitude, and it is a characteristic of all semiconductors that the resistivity is very sensitive to the presence of extremely small amounts of impurity. Finally there is again an enormous increase in resistivity to the substances which are classed as insulators.

By combining eqns (3.2) and (3.5) we find that

$$\sigma = nq(\mathbf{v}/\mathbf{E}) = nqu. \tag{3.7}$$

Here $u$ is a quantity known as the mobility; it is equal to the mean drift velocity which the carriers acquire in unit electric field. For positive charges, $\mathbf{v}$ has the same direction as $\mathbf{E}$, and their ratio is positive. For negative charges, though $\mathbf{J} = nq\mathbf{v}$ has the same direction as $\mathbf{E}$, the drift velocity $\mathbf{v}$ has the opposite direction; however, it is not usual to give the mobility a sign, and this may be expressed by writing $u = |(\mathbf{v}/\mathbf{E})|$. In S.I. the mobility is expressed in units of $\mathrm{m\,s^{-1}}$ per $\mathrm{V\,m^{-1}}$ or $\mathrm{m^2\,V^{-1}\,s^{-1}}$.

If all current is carried by particles whose charge is numerically equal to that of the electron, variations in $\sigma$ or $\rho$ must come from changes in $n$, the number of carriers per unit volume, or in the mobility $u$. In a metal $n$ is constant; at room temperature the resistivity of pure metals varies roughly as the absolute temperature, and this arises from the fact that the mobility $u$ (and hence also the conductivity) varies as $T^{-1}$. For semiconductors the conductivity rises with temperature, and it is often possible to represent it by a relation of the type

$$\sigma = \sigma_0 \exp(-b/T). \tag{3.8}$$

The mobility is not drastically different from that in a metal, and both the low value of $\sigma$ and its temperature variation can be explained by assuming that the number of free carriers varies as $\exp(-b/T)$. At low temperatures all charges are firmly bound, and the conductivity is very small. As the temperature rises, some bound charges are released to become free carriers, but this requires a certain amount of extra energy, so that the number released varies as $\exp(-b/T)$. The larger the value of $b$, the smaller the number of carriers; an 'insulator' may be regarded as an extreme case in which $b$ is so large that at room temperature the number of free carriers is negligible. This view is supported by the fact that at very high temperatures all 'insulators' acquire a certain amount of conductivity.

### Theoretical basis of Ohm's law

When an electric field of strength $\mathbf{E}$ acts on a free particle of charge $q$, the particle is accelerated under the action of the force and its velocity increases at the rate

$$dv/dt = (q/m)E. \qquad (3.9)$$

If there are $n$ particles per unit volume, the current density $J = nqv$, and (3.9) therefore leads to a current increasing at the rate

$$dJ/dt = n(q^2/m)E, \qquad (3.9a)$$

an equation which is clearly at variance with the experimental observation, expressed in Ohm's law, that $J$ is proportional to $E$. The latter is, however, a steady-state result, and it can be reconciled with eqn (3.9a) if we assume that this expresses only a transient effect, and that there are other forces which prevent the current from increasing indefinitely. That such other forces must be present follows also from the observation that a current ceases to flow when the electric field driving it is removed, since the particles carrying the current have a net forward momentum $nmv = J(m/q)$ which must be destroyed in some way. A natural assumption is that this occurs through collisions between the charged particles and other constituents of the conducting medium.

If the particle last made a collision at time $t = 0$, after which its velocity was $v_0$, then its displacement $s$ at time $t$ is

$$s = v_0 t + \tfrac{1}{2}(q/m)Et^2.$$

On averaging over a distribution of times, we have

$$\bar{s} = \tfrac{1}{2}(q/m)E\overline{t^2},$$

where we have assumed that the velocity after a collision is completely random, and averages to zero, so that $\overline{v_0} = 0$ and $\overline{v_0 t} = 0$. This gives an average drift velocity under the influence of the field strength $E$ of

$$\frac{\bar{s}}{t} = \tfrac{1}{2}(q/m)E\left(\frac{\overline{t^2}}{t}\right).$$

If we now assume that the chance of a free time $t$ between collisions is proportional to $\exp(-t/\tau)$, simple integrations show that $\overline{t^2} = 2\tau$ and $\bar{t} = \tau$. Thus $\tau$ is the mean free time between collisions, and the drift velocity is

$$\bar{s}/\bar{t} = (q/m)E\tau.$$

The mobility is therefore

$$u = |(q/m)|\,\tau, \qquad (3.10)$$

while the current density and conductivity become

$$J = n(q^2/m)E\tau \qquad (3.11)$$

and

$$\sigma = n(q^2/m)\tau. \qquad (3.12)$$

These equations show that both $\mathbf{J}$ and $\sigma$ are independent of the sign of $q$.

The effects of collisions are formally similar to the presence of a viscous medium in which the particle is subjected to a retarding force proportional to its velocity. If such a force is added to eqn (3.9) we obtain

$$d\mathbf{v}/dt + \mathbf{v}/\tau = (q/m)\mathbf{E}, \tag{3.13}$$

where the constant $\tau$ clearly has the dimensions of time. If this equation is multiplied by $nq$, we have a differential equation for the current density $\mathbf{J} = nq\mathbf{v}$,

$$d\mathbf{J}/dt + \mathbf{J}/\tau = n(q^2/m)\mathbf{E}, \tag{3.13a}$$

whose solution is

$$\mathbf{J} = \sigma\mathbf{E}\{1 - \exp(-t/\tau)\}. \tag{3.14}$$

This result shows that if the electric field of strength $\mathbf{E}$ is switched on at $t = 0$, when the current is zero, then the new equilibrium current $\mathbf{J} = \sigma\mathbf{E}$, corresponding to Ohm's law, is established in a time of order $\tau$, the mean time between collisions. For a metal such as copper, assuming one charge carrier per atom, we obtain from the conductivity at room temperature a value for $\tau$ of order $10^{-14}$ s. This shows that for most practical purposes the steady state is reached almost instantaneously after the electric field is switched on or off, or changed in any way. For this reason the constant $\tau$ is known as the relaxation time for current flow. It should not be confused with the time constant derived below for a different process, the establishment of a uniform *charge* distribution within a conductor.

*Charge distribution and current flow*

Although a large current may be flowing in a conductor the net charge density is everywhere the same, and if, as is generally the case, the conductor as a whole carries no charge, the net charge density is zero. If the charge density is not uniform, local electric fields are set up and charges move in such a way as to equalize the charge distribution. The time required for such a process is extremely short for all substances except the best insulators. From eqns (1.20), (3.3), and (3.5) we have

$$-\partial\rho_e/\partial t = \operatorname{div} \mathbf{J} = \operatorname{div}(\sigma\mathbf{E}) = (\sigma/\epsilon)\rho_e, \tag{3.15}$$

a differential equation whose solution is

$$\rho_e = \rho_{e0} \exp(-t/\tau'), \tag{3.16}$$

where $\tau' = \epsilon/\sigma$. For a metal (taking $\epsilon = \epsilon_0$) this time is of order $(10^{-11}/10^8)$ s $= 10^{-19}$ s, and only in an insulator whose conductivity is less than about $10^{-12}\ \Omega^{-1}\ m^{-1}$ can a non-uniform charge distribution exist for more than a few seconds. This is the reason why in experiments on electrostatics extreme precautions are needed to achieve high insulation.

In the steady state, for a conducting medium where $\sigma$ is uniform and isotropic, $\text{div } \mathbf{J} = \sigma \text{ div } \mathbf{E} = 0$, and since $\mathbf{E} = -\text{grad } V$ we have

$$\nabla^2 V = 0,$$

so that Laplace's equation holds, as in electrostatics. If two perfectly conducting electrodes are immersed in an infinite medium of finite conductivity, the potential distribution in the medium is the same as in a condenser whose plates are formed by the two conductors; for the potential must satisfy Laplace's equation in each case with the boundary conditions $V = \text{constant}$ on the surface of the conductors. In the medium, the lines of current flow are orthogonal to the lines of constant $V$, and coincide with the lines of $\mathbf{E}$. Since the resistance $R$ of the medium between the conductors, and the capacity $C$ of the capacitor formed when the medium is replaced by a dielectric, depend essentially on the distribution of the lines of $\mathbf{E}$, there is a simple relation between them.

The analogous equations for the two cases are

$$\left.\begin{array}{cc} \mathbf{D} = \epsilon\mathbf{E} & \mathbf{J} = \sigma\mathbf{E} \\[4pt] \epsilon \text{ div(grad } V) = 0 & \sigma \text{ div(grad } V) = 0 \\[4pt] \displaystyle\int \mathbf{D}.d\mathbf{S} = Q & \displaystyle\int \mathbf{J}.d\mathbf{S} = I. \end{array}\right\} \quad (3.17)$$

Since $C = Q/V$, and $1/R = I/V$, $C/\epsilon$ is equivalent to $(1/R)/\sigma$; that is,

$$RC = \epsilon/\sigma. \tag{3.18}$$

The product $RC$, like the quotient $\epsilon/\sigma$ above, has the dimensions of a time constant (see § 5.3).

### 3.3. The voltaic circuit

The mechanism by which a battery acts as the source of a constant potential will be discussed in § 3.10. For the present purpose the battery will merely be regarded as maintaining a potential difference between its two terminals. Fig. 3.1 illustrates the usual notation for the case of a battery B which causes a current $I$ to flow through the resistance $R$ connected across the terminals P, Q. Since the battery drives the current round any circuit attached to it, the potential difference it produces is

FIG. 3.1. A battery B sending a current $I$ through a resistance $R$.

often called the 'electromotive force' or e.m.f. The total e.m.f. is equal to the line integral $\oint \mathbf{E}.\mathrm{d}\mathbf{s}$ taken round the circuit; since no work is done by any external agency we must have

$$0 = V - \oint \mathbf{E}.\mathrm{d}\mathbf{s}.$$

Hence, using eqn (3.6),

$$V = \oint \mathbf{E}.\mathrm{d}\mathbf{s} = El = I(l/\sigma A) = IR \tag{3.19}$$

if the battery of e.m.f. $V$ is connected to a single conductor of conductivity $\sigma$, cross-section $A$, and length $l$. Eqn (3.19) is the basis of all calculations on resistance networks.

When a current flows through a conductor of finite resistance, charge is being transferred from a point at one potential to a point at a different potential. The direction of positive current flow is to a place at a lower potential so that there is a loss of electrical energy which appears as heat in the conductor. If a charge $\mathrm{d}Q$ flows between two points differing in potential by $V$ in a time $\mathrm{d}t$, the energy lost per second (power) is

$$-\mathrm{d}W/\mathrm{d}t = V(\mathrm{d}Q/\mathrm{d}t) = VI = V^2/R = I^2R, \tag{3.20}$$

where $R$ is the resistance between the two points. The unit of power is the volt-ampere, known as the watt (W). In an extended medium of conductivity $\sigma$, the power dissipated in an element of cross-section $\mathrm{d}S$ and length $\mathrm{d}s$ is $VI = (E\,\mathrm{d}s)(J\,\mathrm{d}S) = (E\,\mathrm{d}s)(\sigma E\,\mathrm{d}S) = \sigma E^2\,\mathrm{d}\tau$, where $\mathrm{d}\tau$ is the volume of the element. This result can also be obtained directly from eqn (3.2), since the power expended by a force $q\mathbf{E}$ moving $n$ particles per unit volume with velocity $\mathbf{v}$ is

$$n(q\mathbf{E}).\mathbf{v} = \mathbf{J}.\mathbf{E} = \sigma E^2 = \rho J^2. \tag{3.21}$$

The total power dissipated is found by integration over the whole volume of the conductor.

In practice it is found that the e.m.f. produced by a battery is not quite constant, but drops slightly when a current is drawn from it. The variation is the same as would be produced by an ideal source of e.m.f. $V_0$ equal to that produced by the battery on open-circuit, less the potential drop in a resistance $R_i$. The equivalent circuit is shown in Fig. 3.2, and $R_i$ is known as the 'internal resistance' of the battery. If the battery is used to supply power to a load of resistance $R$, as in Fig. 3.2, the current $I$ which flows is given by the equation $V_0 = I(R_i + R)$. The power dissipated in the load is therefore

$$-\mathrm{d}W/\mathrm{d}t = I^2R = V_0^2R/(R_i + R)^2. \tag{3.22}$$

If the load can be varied, so that $R$ is adjustable, then differentiation of this expression shows that it has a maximum value when $R = R_i$. This

FIG. 3.2. Representation of a battery by an open-circuit e.m.f. $V_0$ and an internal resistance $R_i$.

is an example of the 'maximum power theorem', which states that, if a variable load is to be matched to a source of power so that the maximum power is to be dissipated in the load, its resistance must be adjusted to be equal to the internal resistance of the source. The greatest power which can be obtained is thus $\frac{1}{4}V_0^2/R_i$, and this is known as the 'available power' of the source. It should be noted that with a given load, and a range of batteries of the same voltage but of different internal resistance, maximum power is obtained with the battery of lowest internal resistance, so that the maximum power theorem does not apply to the converse problem.

It is often convenient to work in terms of conductance $G$ rather than resistance $R$, where $G = 1/R$. Instead of (3.19), we have then

$$I = GV, \tag{3.23}$$

and the power dissipation in $G$ is

$$-\mathrm{d}W/\mathrm{d}t = IV = I^2/G = GV^2. \tag{3.24}$$

In dealing with sources which produce an almost constant current rather than an almost constant e.m.f. it may be convenient to use the concept of a 'current generator', which produces a constant current $I_0$, shunted by an internal conductance $G_i$, as in Fig. 3.3. When connected to an external conductance $G$, the potentials across $G$ and $G_i$ must be the same, so that $I_i/G_i = I/G$ and the total current $I_0 = I_i + I$. Hence the current through $G$ is $I = I_0 G/(G+G_i)$, and the power dissipated in $G$ is

$$-\mathrm{d}W/\mathrm{d}t = I_0^2 G/(G+G_i)^2, \tag{3.25}$$

whose maximum value, if $G$ is variable, is $\frac{1}{4}I_0^2/G_i$. This is another example of the maximum power theorem.

The current generator shunted by an internal conductance is completely equivalent to the voltage generator with an internal resistance. In Fig. 3.2 the current through $R = 1/G$ is $V_0/(R+R_i)$, and this is identical

FIG. 3.3. The 'current-generator' circuit. The source provides a constant current $I_0$, shunted by a conductance $G_i$. It is equivalent to the 'voltage–generator' circuit of Fig. 3.2 provided $I_0 = G_i V_0$ and $R_i = 1/G_i$.

with $I_0 G/(G+G_i)$ for all values of $R$ provided that $I_0 = G_i V_0$ and $R_i = 1/G_i$.

## 3.4. Resistance networks

In a complicated network of resistances containing many branches, the calculation of the currents in the various branches is based on two laws due to Kirchhoff. They are extensions of the principles used above:

(1) the algebraic sum of all the currents meeting at a point is zero;
(2) the algebraic sum of the potential differences across the resistances in any closed circuit is equal to the total e.m.f. in that circuit.

The first law follows from the equation of continuity, since there can be no accumulation of charge at any point. It can be written

$$\sum I_k = 0. \tag{3.26}$$

The second law is an extension of eqn (3.19), and can be written

$$\sum_j V_j = \sum_k I_k R_k = \sum_k I_k/G_k, \tag{3.27}$$

where $I_k$ is the current in the resistance $R_k$ or conductance $G_k$.

These laws can immediately be applied to find the equivalent resistance of a number of resistances $R_1, R_2,..., R_n$ in series or in parallel, as in Fig. 3.4. In the former case the voltage across all the resistances is

$$V = IR_1+IR_2+...+IR_n = I(R_1+R_2+...+R_n) = IR,$$

where $R$ is the equivalent resistance. Hence

$$R = R_1+R_2+...+R_n = \sum_k R_k. \tag{3.28a}$$

When the resistances are in parallel, the voltage across each is the same.

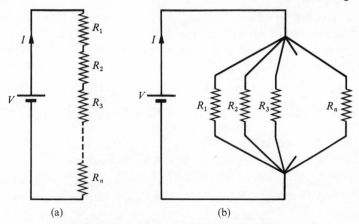

(a)                        (b)

FIG. 3.4. Resistance arranged (a) in series, (b) in parallel.

The total current $I$ is

$$I = \frac{V}{R_1} + \frac{V}{R_2} + \dots + \frac{V}{R_n} = V\left(\frac{1}{R_1} + \frac{1}{R_2} + \dots + \frac{1}{R_n}\right) = \frac{V}{R},$$

where $R$ is the equivalent resistance. Hence in this case

$$\frac{1}{R} = \frac{1}{R_1} + \frac{1}{R_2} + \dots + \frac{1}{R_n} = \sum_k \frac{1}{R_k}. \tag{3.28b}$$

In terms of conductances, it is clear that for a set of conductances in series

$$\frac{1}{G} = \sum_k \frac{1}{G_k}, \tag{3.29a}$$

while for conductances in parallel

$$G = \sum_k G_k. \tag{3.29b}$$

If a network has many branches, the problem of finding the current in each branch is best solved by the method of cyclic currents. Fig. 3.5 is

FIG. 3.5. A general network with cyclic currents.

part of a network in which there are $n$ cyclic currents all flowing in an anticlockwise sense. Such a system of currents satisfies the first of Kirchhoff's laws automatically. Then the current through $R_1$ is $I_1$, but the current through $R_{12}$ is $(I_1-I_2)$. For circuit (1), for example, we have

$$V_1 = I_1R_1+(I_1-I_3)R_{13}+(I_1-I_2)R_{12} = I_1R_{11}-I_2R_{12}-I_3R_{13},$$

where $R_{11} = R_1+R_{12}+R_{13}$. In general, if $R_{qq} = R_q + \sum_p R_{qp}$,

$$V_1 = +R_{11}I_1\pm R_{12}I_2\pm R_{13}I_3\pm...\pm R_{1q}I_q\pm...\pm R_{1n}I_n,$$
$$V_2 = \pm R_{21}I_1+R_{22}I_2\pm R_{23}I_3\pm...\pm R_{2q}I_q\pm...\pm R_{2n}I_n,$$
$$\cdot \quad \cdot \quad \cdot \quad \cdot \quad \cdot \quad \cdot \quad \cdot \quad \cdot \quad \cdot \quad \cdot \quad \cdot \quad \cdot$$
$$V_q = \pm R_{q1}I_1\pm R_{q2}I_2\pm R_{q3}I_3\pm...+R_{qq}I_q\pm...\pm R_{qn}I_n,$$
$$\cdot \quad \cdot \quad \cdot \quad \cdot \quad \cdot \quad \cdot \quad \cdot \quad \cdot \quad \cdot \quad \cdot \quad \cdot \quad \cdot$$
$$V_n = \pm R_{n1}I_1\pm R_{n2}I_2\pm R_{n3}I_3\pm...\pm R_{nq}I_q\pm...+R_{nn}I_n.$$

$V_q$ is the total e.m.f. in the $q$th circuit, and is positive if it acts in the direction of $I_q$. The subscripts to the resistances denote the currents which flow through them. Then if $\Delta_{pq}$ is the cofactor of $R_{pq}$ in the determinant

$$\Delta = \begin{vmatrix} +R_{11} & \cdot & \cdot & \pm R_{1q} & \cdot & \cdot & \pm R_{1n} \\ \cdot & \cdot & \cdot & \cdot & \cdot & \cdot & \cdot \\ \pm R_{p1} & \cdot & \cdot & \pm R_{pq} & \cdot & \cdot & \pm R_{pn} \\ \cdot & \cdot & \cdot & \cdot & \cdot & \cdot & \cdot \\ \pm R_{n1} & \cdot & \cdot & \pm R_{nq} & \cdot & \cdot & +R_{nn} \end{vmatrix},$$

$\Delta_{pq} = \Delta_{qp}$, since $R_{pq}$ is identical with $R_{qp}$.

### The reciprocity theorem

If we solve these equations for the current $I_q$ when there is only one source of e.m.f. $V_p$ in the circuit, then we find

$$I_q = V_p\Delta_{pq}/\Delta.$$

Similarly, the current $I_p$, when there is only the e.m.f. $V_q$ in circuit, is

$$I_p = V_q\Delta_{qp}/\Delta.$$

Since $\Delta_{pq} = \Delta_{qp}$, we have the 'reciprocity theorem', which states that a given e.m.f. in the $p$th branch will produce the same current in the $q$th branch of a circuit as the same e.m.f. in the $q$th branch would produce in the $p$th branch.

### Thévenin's theorem

The analysis of resistance networks may often be simplified by the use of a theorem due to Thévenin, which may be stated as follows: Suppose two terminals A, B emerge from a network for connection to an external circuit. Let $V_{AB}$ be the voltage measured across these terminals when the external circuit is an open-circuit, and let $I_{AB}$ be the current flowing through the external circuit when it is a short-circuit of zero resistance. Then the resistance of the network measured across the terminals A, B when all internal sources of e.m.f. have been short-circuited is given by $R_{AB} = V_{AB}/I_{AB}$. Thus, as far as the external circuit is concerned, the network behaves as a generator of e.m.f. equal to $V_{AB}$ with an internal resistance $R_{AB}$.

A general proof of this theorem, which depends on the components of the network obeying Ohm's law, can be found in Duffin (1965). There is an equivalent theorem, due to Norton, which states that the network can also be represented by a current generator shunted by an internal conductance.

### 3.5. The potential divider and resistance bridge

If a pair of resistances $R_1$, $R_2$ are connected in series to a source of potential $V_0$, as in Fig. 3.6(a), the potential across the resistance $R_2$ is $V_0 R_2/(R_1+R_2)$, and this potential may be varied by adjusting $R_1$ or $R_2$. Such a device is called a potential divider and is the basis of methods of comparing two potentials. The equivalent circuit is shown in Fig. 3.6(b). The open-circuit potential across the terminals A, B is clearly $V_{AB} = V_0 R_2/(R_1+R_2)$, while if $V_0$ is short-circuited the resistance measured between terminals A, B corresponds to $R_1$ and $R_2$ in parallel. By Thévenin's theorem, this gives the value of $R_{AB}$ as

$$R_{AB} = \frac{R_1 R_2}{R_1+R_2},$$
(3.30)

and it is easy to verify that the current through a resistance $R$ connected across A, B is the same as would result from the equivalent circuit in Fig. 3.6(b). The equivalent current circuit is given in Fig. 3.6(c), where $I_{AB} = V_0/R_2 = G_{AB} V_{AB}$ with $G_{AB} = 1/R_{AB}$.

(a)  (b)  (c)

FIG. 3.6. The potential divider (a) and its equivalent circuits in terms of a voltage generator (b) or a current generator (c).

FIG. 3.7. Wheatstone's bridge. The current $I$ through the detector (resistance $R_G$) is zero if $R_1/R_2 = R_3/R_4$. In the nomenclature of Callendar (1910) used in Problem 3.3, $R_4 = R$ (the unknown resistance), $R_2 = mR'$, $R_3 = nR'$, and $R_1 = nmR'$, so that the bridge is balanced when $R' = R$.

The value of a resistance $R$ can be determined by measuring the voltage across it and the current through it, but this requires instruments for the accurate measurement of both voltage and current. In Wheatstone's bridge an unknown resistance can be determined in terms of other resistances without an accurate measurement of current or voltage. The circuit is shown in Fig. 3.7; $R_4$ is the unknown resistance, while $R_1$, $R_2$, and $R_3$ are known resistances, one of which must be variable. A current detector of resistance $R_G$ is connected between A and B, but this is required only to determine when no current flows through $R_G$, that is, when the voltage at terminals A and B is the same. At this null point $R_1$, $R_2$ form a simple potential divider and the voltage across $R_2$ is $V_{AD} = VR_2/(R_1+R_2)$; similarly that across $R_4$ is $V_{BD} = VR_4/(R_3+R_4)$. These two voltages are equal provided that

$$R_1/R_2 = R_3/R_4. \tag{3.31}$$

This result is independent of any internal resistance of the battery and of the battery voltage itself, since $V$ is just the voltage applied between terminals C and D.

### 3.6. Electron optics

In the following sections of this chapter we consider the flow of current due to the motion of charged particles, first in a vacuum, then in a gas, and finally in liquids and solids (metals). In §§ 3.6 and 3.7 the vacuum is assumed to be such that ions travel directly from one electrode to another without making collisions with gas molecules, and their equation of motion is given by eqn (3.9). On multiplication by $m\mathbf{v}$ we obtain

$$m\mathbf{v} \cdot \left(\frac{d\mathbf{v}}{dt}\right) = \frac{d}{dt}\left(\tfrac{1}{2}mv^2\right) = q\mathbf{E} \cdot \mathbf{v},$$

which on integration gives

$$\int_1^2 d(\tfrac{1}{2}mv^2) = q\int_1^2 (\mathbf{E}\cdot\mathbf{v})\,dt = q\int_1^2 \mathbf{E}\cdot d\mathbf{s}. \tag{3.32}$$

If the particle starts from rest at a point where the potential $V = 0$, we have

$$\tfrac{1}{2}mv^2 + qV = 0. \tag{3.33}$$

This result, which represents the conservation of energy, shows that the particle velocity $v$ is proportional to $|V|^{\frac{1}{2}}$ irrespective of the direction of the velocity.

In an evacuated field-free space electrons travel in straight lines, and a beam of electrons leaving an electrode will eventually diverge. In an electronic vacuum device this must be counteracted in order to direct the electrons to a particular electrode or to a spot of small dimensions, as in a cathode-ray tube. The required convergence of the beam can be achieved by means of electric (or magnetic) fields, and is known as 'focusing'. The technology used is known as 'electron optics', though it is equally applicable to any charged particles; only the principles are outlined below.

As the name implies, there is a close analogy between the behaviour of electrons in an electrostatic field and that of light in a refracting medium. The basis of geometrical optics is Snell's law: when a light ray passes from a medium of refractive index $n_1$ through a plane boundary to another medium $n_2$, the angles of incidence and refraction obey the relation

$$n_1 \sin \alpha_1 = n_2 \sin \alpha_2.$$

In electron optics the corresponding case is that of an electron in a field-free space (that is, a region of constant potential $V_1$) crossing into a region at another potential $V_2$, as in Fig. 3.8. At the boundary there exists an electric field which accelerates the electron in the direction normal to the boundary, while the component of velocity parallel to the boundary remains unchanged. If the initial and final velocities are $v_1$ and $v_2$, the components parallel to the boundary are $v_1 \sin \alpha_1$ and $v_2 \sin \alpha_2$, so that we have

$$v_1 \sin \alpha_1 = v_2 \sin \alpha_2.$$

If the electron started from rest at a point where the potential is zero, this may be written

$$V_1^{\frac{1}{2}} \sin \alpha_1 = V_2^{\frac{1}{2}} \sin \alpha_2, \tag{3.34}$$

since $v$ is proportional to $V^{\frac{1}{2}}$. Note that the electron velocity plays the same role as the refractive index and does not correspond to the velocity of light in the medium.

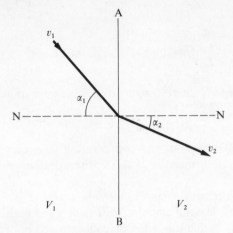

FIG. 3.8. 'Refraction' of an electron on crossing a boundary AB between two regions of
potential $V_1$ and $V_2$. NN is the normal to this boundary.

The example just given is a special case of a more general correspondence based on Fermat's principle of least time in optics and Hamilton's
principle of least action in mechanics. The former states that the path of a
light ray is such that the time taken between any two points of the path is
an extremum (generally a minimum). Thus we have

$$t = \int dt = \int \frac{ds}{v} = \frac{1}{c} \int n \, ds = \text{minimum}. \tag{3.35}$$

Hamilton's principle states that the path of a particle is such that the line
integral of its momentum is a minimum, that is,

$$\int mv \, ds = \text{mimimum}. \tag{3.36}$$

So long as the mass of the particle is constant (that is, so long as relativity
corrections are negligible) the analogy between electron velocity and
refractive index is complete.

Although the formula for the focal length of a thin lens can be
calculated quite simply in optics, the equivalent calculation for electron
optics is generally very difficult. We shall content ourselves by showing
how a focusing action can be obtained in a simple case. Fig. 3.9 shows a
pair of parallel conducting planes each with a small circular aperture. The
planes are at potentials $V_1$, $V_2$, and **the** potentials outside the planes away
from the aperture are constant and **equal** to $V_1$ and $V_2$. Near the aperture
the equipotentials are curved and bulge out as shown. If an electron
travelling parallel to the $z$-axis (that is, normal to the planes) enters the
aperture, it finds itself in a region where the lines of electric field, which

are normal to the equipotentials, have a radial component. This gives the electron an acceleration normal to the axis, and it emerges into the field-free region with a component of velocity to or away from the axis. If the electron is in the $xz$ plane as shown in the diagram, the $x$ component of velocity given to it is

$$v_x = -\int_0^t \frac{e}{m} E_x \, dt = +\frac{e}{m} \int_0^t \left(\frac{\partial V}{\partial x}\right) dt. \tag{3.37}$$

If the electron enters the aperture at a small distance $h$ from the axis, then we may make the approximation

$$\left(\frac{\partial V}{\partial x}\right)_{x=h} = \left(\frac{\partial V}{\partial x}\right)_{x=0} + h\left(\frac{\partial^2 V}{\partial x^2}\right).$$

Since the potential satisfies Laplace's equation

$$\frac{\partial^2 V}{\partial x^2} + \frac{\partial^2 V}{\partial y^2} + \frac{\partial^2 V}{\partial z^2} = 0,$$

where, by symmetry, $\partial^2 V/\partial x^2 = \partial^2 V/\partial y^2$, and $(\partial V/\partial x) = (\partial V/\partial y) = 0$ on the axis, we have

$$\left(\frac{\partial V}{\partial x}\right)_{x=h} = -\frac{h}{2}\left(\frac{\partial^2 V}{\partial z^2}\right),$$

from which

$$v_x = -\int \frac{1}{2} \frac{e}{m} h\left(\frac{\partial^2 V}{\partial z^2}\right) dt = -\frac{1}{2}\frac{e}{m} h \int \left(\frac{\partial^2 V}{\partial z^2}\right)\frac{dz}{v_z},$$

where $v_z = dz/dt$ is the $z$ component of the instantaneous velocity at any point. On emerging from the aperture the electron has a velocity $v_2 = (2eV_2/m)^{\frac{1}{2}}$ which is independent of $h$, and it moves at an angle $\theta$ with the

FIG. 3.9. A simple electron lens, consisting of two parallel conducting planes at potentials $V_1$, $V_2$ with apertures. The bulging of the equipotential surfaces near the apertures is shown.

axis where $\sin \theta = v_x/v_2$. Since $v_x$ is proportional to $h$, for small values of $\theta$ (where the difference between $\sin \theta$ and $\tan \theta$ is negligible) all electrons will move towards (or appear to diverge from) a particular point on the axis, whose distance from the aperture is

$$-h/\tan \theta = -hv_2/v_x.$$

As the electrons were assumed to enter parallel to the axis, this is one focal point of the lens. If $v_x$ is negative when $h$ is positive we have a converging lens whose focal length is given by

$$\frac{1}{f_2} = -\frac{v_x}{hv_2} = \frac{e/2m}{(2eV_2/m)^{\frac{1}{2}}} \int \left(\frac{\partial^2 V}{\partial z^2}\right)\frac{\mathrm{d}z}{v}$$

$$= \frac{e/2m}{(2eV_2/m)^{\frac{1}{2}}} \int \left(\frac{\partial^2 V}{\partial z^2}\right)\frac{\mathrm{d}z}{(2eV/m)^{\frac{1}{2}}} = \frac{1}{4V_2^{\frac{1}{2}}} \int \frac{1}{V^{\frac{1}{2}}}\left(\frac{\partial^2 V}{\partial z^2}\right) \mathrm{d}z, \qquad (3.38)$$

where we have assumed that $v_z = v$, the actual velocity at any point, and substituted $(2eV/m)^{\frac{1}{2}}$ for it inside the integral.

If the electrons had entered the lens from the right, they would have been brought to a focus at a point $f_1$, where

$$\frac{1}{f_1} = -\frac{1}{4V_1^{\frac{1}{2}}} \int \frac{1}{V^{\frac{1}{2}}}\left(\frac{\partial^2 V}{\partial z^2}\right) \mathrm{d}z. \qquad (3.39)$$

Hence $f_1/f_2 = -(V_1/V_2)^{\frac{1}{2}}$, a formula which is exactly analogous to the optical case of a thin lens with initial and final media of different refractive indices $n_1$ and $n_2$.

Inspection of eqns (3.38) and (3.39) for the focal lengths shows that they require a knowledge of the variation of the potential on the axis of the lens. It is only possible in very simple cases to derive an analytical expression for $V$ in the aperture, and in general the variation of $V$ is calculated by numerical methods, by which an approximate solution of Laplace's equation with the required boundary conditions may be found. The variation of $V$ always occupies a finite distance, and if it occupies a distance comparable with either of the focal lengths the electron lens is a 'thick lens' rather than a 'thin lens'. The behaviour of the system is again similar to that of an optical system, and is defined if the cardinal points are determined. These can be found by tracing the paths of a number of electrons through the system. One method of doing this is to divide the potential field into thin slices along the equipotential surfaces, and treat each slice as a thin lens.

The approximations made above in expanding $\partial V/\partial x$ near the axis and retaining only the first term are equivalent to the approximations made in 'Gaussian optics' in treating only rays near the axis. It is to be expected, therefore, that electron-lens systems will suffer from defects similar to those of optical systems, such as spherical aberration, etc. The equivalent of 'chromatic aberration' arises when not all electrons enter the system with the same velocity, since this corresponds to a variable refractive index. If the spread in velocity is due only to the Maxwellian distribution of velocity on emission from the cathode, chromatic aberration is small.

## 3.7. Space charge and Child's law

In the previous section the mutual repulsion of the electrons was neglected; this is justifiable only when the charge density is small. One result of the mutual repulsion is that a beam of electrons, initially moving parallel to one another, will gradually spread sideways, but this can be overcome by 'focusing' with a suitable electron-optical system. When the charge density is large, a much more important effect arises from the repulsive forces in the direction of motion, which can change the forward velocity and hence the total current. Such high charge densities occur in thermionic vacuum tubes, where a copious stream of electrons is emitted from a hot cathode (see Chapter 21). Under normal conditions the current flow in such devices is controlled by the charge density or 'space charge' near the cathode.

To examine the effect of this space charge on the flow of current, we shall consider the case of a diode where the cathode and anode form parts of parallel planes denoted respectively by the equations $x = 0$ and $x = d$. We shall further assume that the potential of the cathode is zero, while that of the anode is $V_a$. The potential between the electrodes can be determined by solving Poisson's equation

$$\frac{d^2 V}{dx^2} = -\frac{\rho_e}{\epsilon_0} = \frac{ne}{\epsilon_0}, \tag{3.40}$$

where $-e$ is the electronic charge, and $n$ the number of electrons per cubic metre. If the mass of an electron is $m$, and its velocity $u$ at the point $x$ where the potential is $V$, then the energy equation gives

$$\tfrac{1}{2}mu^2 = eV, \tag{3.41}$$

while the current density is

$$|J| = neu. \tag{3.42}$$

(Here the flow is unidirectional and it is not necessary to treat $J$ as a vector quantity; we have avoided the negative sign which denotes that the direction of positive current flow is from anode to cathode.) The velocity can be eliminated between these equations, giving

$$ne = |J| \left(\frac{m}{2eV}\right)^{\frac{1}{2}}.$$

Substitution of this in Poisson's equation gives

$$d^2 V/dx^2 = aV^{-\frac{1}{2}},$$

where $a = (|J|/\epsilon_0)(m/2e)^{\frac{1}{2}}$. This equation may be integrated if both sides are multiplied by $2(\mathrm{d}V/\mathrm{d}x)$, giving

$$\left(\frac{\mathrm{d}V}{\mathrm{d}x}\right)^2 - \left(\frac{\mathrm{d}V}{\mathrm{d}x}\right)_0^2 = 4aV^{\frac{1}{2}}.$$

$(\mathrm{d}V/\mathrm{d}x)_0$ is the electric field at the cathode, where $V$ and $x$ are zero. Since the constant $a$ is proportional to $|J|$, it is evident that the maximum current density is attained when $(\mathrm{d}V/\mathrm{d}x)_0 = 0$. Then we may write

$$\mathrm{d}V/\mathrm{d}x = 2a^{\frac{1}{2}}V^{\frac{1}{4}},$$

integration of which gives

$$V^{\frac{3}{4}} = \tfrac{3}{2}a^{\frac{1}{2}}x, \tag{3.43}$$

where the constant of integration is zero because $V = 0$ at $x = 0$. Since at the anode $x = d$ and $V = V_a$, we have

$$V_a^{\frac{3}{2}} = 9ad^2/4 = (9/4\epsilon_0)(m/2e)^{\frac{1}{2}}d^2 |J|, \tag{3.44}$$

showing that the current density $|J|$ is proportional to the three-halves power of the anode voltage, and inversely proportional to the square of the separation between cathode and anode. This relation was first derived by Child, and is sometimes known as Child's law.

Since the current density is independent of $x$, it follows from eqn (3.42) that the density of electrons is greatest where their velocity is smallest, that is, near the cathode. It is this concentration of electrons which reduces the electric field at the cathode, since their electric field is oppositely directed at this point to that due to the positive potential on the anode. The electron concentration cannot rise to a greater value than that required to make $\mathrm{d}V/\mathrm{d}x$ zero at the cathode, since no electrons could then leave the cathode, and the space charge would fall as electrons move away to the anode, without their being replenished from the cathode. Near the anode the electric field is greater than that due to the anode potential alone, because the field is here increased by the repulsive force due to the negative space charge near the cathode. The potential variation is shown in Fig. 3.10. Curve A shows the linear potential gradient which would exist in the absence of space charge, while curve B is that calculated above, on the assumption that the electrons are emitted from the cathode with zero velocity. It is easily seen from eqn (3.43) that the equation of curve B may be written in the form $V/V_a = (x/d)^{\frac{4}{3}}$. Owing to the finite velocity of emission, electrons can leave the cathode even when there is a small reverse electric field, and the space charge can then increase to the extent of setting up a potential minimum, as shown by curve C. The depth of this minimum is of the order $W/e$, where $W$ is the average energy of the emitted electrons, since only those electrons with

FIG. 3.10. Potential distribution in a diode with plane-parallel electrodes. Curve A, no space charge; curve B, space charge limited, electrons emitted with zero velocity; curve C, space charge limited, electrons emitted with finite velocity.

sufficient energy to penetrate the potential minimum will eventually reach the anode.

If numerical values are inserted the equation for the current density becomes

$$|J| = 2 \cdot 34 \times 10^{-6}\, V_a^{\frac{3}{2}}/d^2, \tag{3.45}$$

where $|J|$ is in amperes per square metre, $V_a$ in volts, and $d$ in metres. For example, if $V_a = 100$ V and $d = 10^{-2}$ m, then $|J| = 23 \cdot 4$ A m$^{-2}$. The three-halves power law has been shown to hold also for electrodes in the shape of coaxial circular cylinders and of concentric spheres. By a dimensional argument it may be shown to hold for any electrode geometry, assuming always that the electrons are emitted with zero velocity, though the numerical constant will obviously change. It does not hold indefinitely as $V_a$ is increased, and the current will depart from the three-halves power law as soon as the space charge is no longer sufficiently dense to nullify the electric field at the cathode. Ultimately the current will be determined solely by the rate of emission of electrons from the cathode and will become independent of the anode voltage.

### 3.8. Conduction of electricity through gases

Under perfect conditions a gas consists of uncharged molecules, and therefore behaves as an insulator since there are no charged particles present to carry a current. In practice, owing to cosmic rays and radio-active background (especially in the walls of the containing vessel), there are always a few ions present, which are sufficient to initiate a spark discharge at sufficiently high electric fields (of the order of $3 \times 10^6$ V m$^{-1}$ in air at atmospheric pressure), but at low fields the current passing is negligibly small unless ions are deliberately produced in the gas, or electrons are liberated at one of the electrodes (the cathode). The essential distinction between the two cases is that at low fields the current is limited by the supply of ions through external action (X-rays or ultraviolet light releasing electrons from the electrodes or from the gas molecules) while at high fields new ions are created by collisions between charged particles (accelerated by the applied electric field) and neutral molecules. A typical current–voltage characteristic is shown in Fig. 3.11. At very low voltages the current is proportional to the voltage, but at higher voltages it rises less rapidly and reaches a constant value, independent of the voltage over a wide range.

These results can be understood with the aid of a simple model. Suppose the electrodes consist of two infinite parallel planes; in Fig. 3.12 an electrode at $x = +d/2$ is maintained at potential $+V_0/2$, while that at $x = -d/2$ is at $-V_0/2$. We consider the numbers of positive and negative charges in a volume of unit cross-section, bounded by the planes $x$ and $x+dx$. Through the action of X-rays or a similar source of ionization, an equal number ($\beta\, dx$) of positive and of negative ions is formed per unit time. Some of these are lost by recombination, the rate of which (the same for positive and for negative ions) is $-\alpha np\, dx$, where $n$ and $p$ are the numbers of negative and positive ions per unit volume. The rate at which negative ions move in across the plane $x$ is $n\dot{x}_n$, and the rate at

FIG. 3.11. Current–voltage characteristic of a gas.

OA    Ohm's law is obeyed; most ions formed are lost by recombination.
BC    All ions formed are swept to electrodes before recombination can take place.
CD    Fresh ions are formed by collision when electrons can reach the ionization potential of the gas molecules between collisions (Townsend discharge).

FIG. 3.12. Flow of current in an ionized gas. (a) Coordinate system for two parallel plane electrodes with potentials $\pm V_0/2$ at $x = \pm d/2$. (b) Potential distribution between the electrodes.

which they move out across the plane $(x+\mathrm{d}x)$ is $-n\dot{x}_n - \mathrm{d}(n\dot{x}_n)$, so that the net rate of gain is $-\mathrm{d}(n\dot{x}_n)$. Similarly the net rate of gain of positive ions is $-\mathrm{d}(p\dot{x}_p)$. In the steady state we have

$$\beta - \alpha np - \frac{\mathrm{d}}{\mathrm{d}x}(n\dot{x}_n) = 0.$$

$$\beta - \alpha np - \frac{\mathrm{d}}{\mathrm{d}x}(p\dot{x}_p) = 0.$$

(3.46)

Subtraction of these two equations gives

$$\frac{\mathrm{d}}{\mathrm{d}x}(p\dot{x}_p - n\dot{x}_n) = 0.$$

If the ions carry charges $q$ and $-q$ this means that the current density $J$, given by

$$J = q(p\dot{x}_p - n\dot{x}_n),$$

is a constant independent of $x$. This result is true only in so far as the loss of ions by recombination at the side walls, which we have not included, can justifiably be neglected.

An exact solution of our equations in the general case is difficult, and we shall consider only limiting cases. When the applied field is sufficiently high (about $2 \times 10^3$ V m$^{-1}$ at atmospheric pressure), the ions move quickly to the electrodes and we may neglect the effect of recombination. Eqns (3.46), with $\alpha = 0$, may then be integrated to give

$$p\dot{x}_p = \beta(x - \tfrac{1}{2}d),$$
$$n\dot{x}_n = \beta(x + \tfrac{1}{2}d). \tag{3.47}$$

Here the constants of integration are obtained from the fact that no ions are emitted by the electrodes, and they are formed only in the body of the gas; thus the positive ion density will be zero at the positive electrode ($p = 0$ at $x = +d/2$), and similarly $n = 0$ at $x = -d/2$. Then the current density is

$$J = q(p\dot{x}_p - n\dot{x}_n) = -q\beta d. \tag{3.48}$$

This result shows that the current is independent of the voltage, corresponding to the portion BC of the current–voltage characteristic in Fig. 3.11. The current density is proportional to $\beta d$, the total number of ions formed in unit cross-section between the electrodes per second. (The negative sign in (3.48) reflects the fact that the direction of positive current is towards the negative electrode.)

The effect of space charge may be deduced qualitatively from Poisson's equation, which gives

$$d^2V/dx^2 = -\epsilon_0\rho_e = \epsilon_0 q(n - p). \tag{3.49}$$

For small densities of charge $dV/dx$ will be constant as indicated by the broken line in Fig. 3.12(b), corresponding to a uniform electric field, but this will not be true at higher degrees of ionization. Since $p = 0$ at the positive electrode, $d^2V/dx^2$ is positive near $x = +d/2$ and, since $n = 0$ at the negative electrode, $d^2V/dx^2$ will be negative near $x = -d/2$. This gives a potential distribution of the type indicated by the solid line in Fig. 3.12(b), with the electric fields strongest near the electrodes.

When the voltage applied between the electrodes is small, the ionic velocities will also be small and we can neglect the last term in each of eqns (3.46). Then $p = n = (\beta/\alpha)^{\frac{1}{2}}$ and the space-charge density is zero everywhere, because the densities of positive and negative ions are equal. The field is uniform and the strength is $E = -(V_0/d)$. If the mobilities of positive and negative ions are $u_p$, $u_n$ then, allowing for their opposite directions of motion,

$$\dot{x}_p = u_p E = -u_p(V_0/d),$$
$$\dot{x}_n = -u_n E = +u_n(V_0/d).$$

The current density is

$$J = -q(\beta/\alpha)^{\frac{1}{2}}(u_p + u_n)(V_0/d). \tag{3.50}$$

Since $J$ is proportional to $V_0$, Ohm's law is obeyed and we have the straight line corresponding to the initial part OA of the current–voltage characteristic in Fig. 3.11.

The mobility of an ion may be estimated in the following way. After collision with a gas molecule the ion is initially moving in a random direction with random velocity, and is then accelerated by the electric field strength $E$. The ions crossing the plane at which we wish to calculate the current density have come, on average, a distance $L$ since the last collision, and the average time since the last collision is $\tau = L/c$, where $c$ is the random velocity. The average velocity acquired under the action of the field of strength $E$ is therefore $(q/M)E\tau$, and the mobility is

$$u = (q/M)\tau = (q/M)(L/c), \tag{3.51}$$

where $M$ is the mass of the ion. Here $L$ is equal to the mean free path, and not half the mean free path as is sometimes assumed; for a discussion of this point see, for example, Kennard (1938) (section 82) or Morse (1964) (Chapter 14). For ions of molecular size, $L$ and $c$ have the usual values given by kinetic theory: $L = 1/\sqrt{2}n\pi\sigma^2$, where $n$ is the number of molecules per unit volume of diameter $\sigma$; and $c = (8kT/\pi M)^{\frac{1}{2}}$, where $k$ is Boltzmann's constant and $T$ is the absolute temperature.

At atmospheric pressure, both positive and negative ions have mobilities of the order of $10^{-4}\,\mathrm{m^2\,V^{-1}\,s^{-1}}$, in fair agreement with values calculated using eqn (3.51). As the pressure is lowered, the mobility increases inversely with the density for positive ions, corresponding to the expected increase in mean free path, but for negative ions it increases much more rapidly. This is due to the fact that at low pressures most of the negative ions are electrons rather than heavy charged molecules. At a given pressure, the ratio of electrons to heavy negative ions varies markedly from gas to gas; some molecules, such as $Cl_2$, readily attach electrons to form negative ions, while others such as $H_2$ do not. The mobility of electrons is much greater than that of heavy ions, mainly owing to their small mass, but also partly owing to their longer mean free paths. Since the diameter of an electron is negligible, its collision diameter with a gas molecule is only $\frac{1}{2}\sigma$, and since its velocity is much greater than that of the gas molecules the factor $\sqrt{2}$ introduced by Maxwell to allow for the relative velocities is absent, so that the electron mean free path is $4/n\pi\sigma^2$, or $4\sqrt{2}$ times that of a heavy ion. In addition, the average loss of energy by an electron in an elastic collision with a molecule is very small (see Problem 3.12), and the average energy of the electrons when a field is applied is much higher than that of the gas molecules or heavy ions. We may express this by saying that the 'mean temperature' of the electrons is higher than that of the gas. As the pressure is reduced, and the mean free path increases, the energy gained by an electron from the applied field increases and the effective electron temperature rises. The energy gained is proportional to the product of the mean free path and the applied field strength $E$, and since the mean free path is inversely proportional to the pressure $p$ it follows that the conditions are a function of $E/p$. At low values of $E/p$ the energy gained by an electron between collisions is small, and it makes only elastic collisions with the gas molecules, but at high values of $E/p$ the mean electron temperature rises and the number of electrons in the high energy tail of the energy distribution increases rapidly. Those which have a few electron volts of energy can make inelastic collisions in which most of the energy is transferred to the colliding molecule. The effect on the molecule will now be discussed.

On quantum theory the total energy, kinetic plus potential, of an electron bound in an atom can only have certain allowed values, and in the normal state the electrons in an atom are in the lowest allowed levels, this is the 'ground state' of the atom. The different energy levels can be plotted on an 'energy-level diagram' (such as Fig. 14.2). The atom cannot exist with intermediate values of the energy, and if it is in an excited state (one of the higher energy levels) it may return to the ground state by emitting its excess energy as a quantum of light whose frequency $\nu$ is defined by the equation

$$W_1 - W_2 = h\nu. \tag{3.52}$$

For a molecule the energy-level diagram is similar to but rather more complicated than that of an atom.

If an electron with sufficient energy collides with an atom or molecule in its ground state, it may transfer some of its kinetic energy to the molecule and raise it to an excited state. For this to be possible the electron must have at least as much energy as the difference between the ground state and the first excited state of the molecule, and the potential through which the electron must be accelerated to obtain this energy is called the 'resonance potential' of the molecule. In general the molecule will get rid of this extra energy by emitting a photon (light quantum) within about $10^{-8}$ s, and the gas thus becomes luminous when the electrons gain sufficient energy from the applied electric field to raise the molecules into these excited states. As the energy of the electrons increases, the molecules are raised into higher excited states, corresponding to a bound electron being in an orbit of larger radius, and finally the molecule may be ionized, that is, an electron is completely removed, leaving the molecule as a positively charged ion. The energy required to do this (expressed in electronvolts) is called the ionization potential of the molecule. This process of ionization through electron impact increases the number of charged ions and electrons, and when the value of $E/p$ is large enough for it to occur, the current through the gas is greatly increased. The steep rise in current CD with applied voltage shown after the saturation plateau BC in Fig. 3.11 is due to the formation of ions by collision; it was extensively investigated by Townsend and is known as the Townsend discharge. With specially designed electrodes the voltage in this region becomes almost independent of current, and small gas-filled tubes have been used as voltage stabilizers.

Ionization by electron collision is the primary process in producing fresh ions in the body of the gas. Experiments have shown that collisions with positive ions are much less effective in causing ionization (owing to their shorter mean free paths, heavy ions pick up less energy from the applied field than electrons), and this process can be neglected in comparison. The most important secondary processes for producing further charged particles occur at the cathode, from which electrons are emitted under the action of (1) bombardment by positive ions, (2) the photoelectric effect caused by photons emitted by excited molecules, and (3) bombardment by excited molecules. The relative importance of these three processes varies with the conditions; in general (1) is more important with cathode surfaces of high work function, and (2) with surfaces of low work function. The main types of excited molecules reaching the cathode are those in 'metastable states', that is, molecules in certain excited states which cannot return to the ground state by emitting a photon, and so have much longer lives than the $10^{-8}$ s mentioned above.

Secondary ionization processes occurring within the body of the gas, which (except at high pressure) appear to be less important than those at the cathode, are (4) photo-ionization, in which high-energy photons emitted by one molecule are absorbed by another, and may have sufficient energy to ionize it. This occurs mostly with the high-frequency ultraviolet radiation which is found in high-voltage discharge tubes; (5) as the temperature of the gas rises owing to the conversion of electrical energy to heat energy through collisions between molecules and ions, neutral molecules may have sufficient kinetic energy of random motion to ionize other molecules by collision. This process is sometimes called thermal ionization.

### The Townsend discharge

Suppose we have two plane parallel electrodes a distance $d$ apart and an electric field is applied between them. Let the negative electrode (the cathode) be illuminated with ultraviolet light which causes $n_0$ electrons to be emitted per second. These electrons are accelerated, and if the value of $E/p$ is sufficiently high, they will produce further ions by collision. If $n$ electrons cross a plane at a distance $x$ from the cathode per second, then the number formed by ionization in the next element of distance $dx$ will be proportional both to $n$ and $dx$, so that we write

$$dn = C_1 n \, dx,$$

which on integration gives $n = n_0 \exp(C_1 x)$, and the current at the anode is therefore

$$I = n_0 q \exp(C_1 d) = I_0 \exp(C_1 d), \tag{3.53}$$

where $I_0$ is the current due to the original $n_0$ electrons alone. For low currents this equation is in good agreement with experiment, but at high currents the current shoots up rapidly towards infinity. This is due to the secondary processes, which increase the supply of ions, principally by causing the emission of more electrons from the cathode. If now the total emission of electrons from the cathode is $n_0'$, the number of extra electrons produced in the gas by primary ionization must be $n_0'\{\exp(C_1 d)-1\}$, and this will also be the number of positive ions. The number of excited molecules emitting photons, and the number of molecules in excited states, will also be proportional to this number, and hence so also will be the number of secondary electrons emitted from the cathode, whatever the mechanism. Hence

$$n_0' = n_0 + C_2 n_0'\{\exp(C_1 d)-1\}, \tag{3.54}$$

from which

$$n_0'/n_0 = [1 - C_2\{\exp(C_1 d)-1\}]^{-1},$$

and the total current will be

$$I = n_0' q \exp(C_1 d) = I_0 \frac{\exp(C_1 d)}{1 - C_2\{\exp(C_1 d)-1\}}. \tag{3.55}$$

The total current will become infinite when the denominator is zero, that is, when

$$\exp(C_1 d) = (C_2+1)/C_2. \tag{3.56}$$

This implies that a finite current will pass when this condition is satisfied, even if $I_0$ is zero. The voltage at which this occurs is known as the sparking or breakdown potential $V_s$. Experimentally it was discovered by Paschen that for a given gas $V_s$ depends only on the product $pd$ of the gas pressure $p$ and the electrode separation $d$. This is known as Paschen's law, and it holds up to very high pressures; it follows from the Townsend theory (above), for the constant $C_1$ is the number of ions produced by an electron in going unit distance. This number must be proportional to the number of molecules per unit volume, and hence to the pressure, and it also depends on the average energy gained by an electron between collisions. This energy varies as $EL$, where $L$ is the mean free path, and since $E = V/d$ and $L$ is inversely proportional to $p$, we have

$$C_1 d = (pd) \, F(V/pd), \tag{3.57}$$

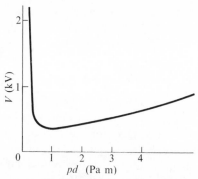

FIG. 3.13. Breakdown voltage $V$ for air plotted against $pd$. $V$ in volts, $p$ in pascals, and $d$ in metres.

where $F(V/pd)$ is some single-valued function of $(V/pd)$. Since $C_2$ is a constant it follows from eqn (3.56) that the sparking potential $V_s$ is a function only of $pd$ for a given gas.

Inspection of eqn (3.54) shows that when the condition of eqn (3.56) is fulfilled, $C_2\{\exp(C_1 d)-1\}=1$, that is, the secondary processes produce all the electrons leaving the cathode. These electrons increase at an exponential rate, and the discharge current rises very rapidly (in a time of the order $10^{-7}$ s at atmospheric pressure) and a spark passes. A typical curve for the variation of $V_s$ with the product (pressure × electrode separation) is shown in Fig. 3.13. The sharp rise at the low-pressure end is due to the low density, when the chance of an electron encountering a molecule is small and few ions are formed by collision. In the high-pressure region collisions are frequent, but the mean free path is small so that few electrons gain sufficient energy from the field between collisions to cause ionization. Thus for any given electrode separation $d$, there is always a certain pressure at which the sparking potential is a minimum.

Later work on the Townsend discharge (see Llewellyn Jones 1953) shows that the theory given above holds over a very wide range of values of $pd$. At the higher pressures ($pd$ somewhat greater than $10^2$ Pa×metres and sparking potentials of 10–100 kV) (Pa = pascal) positive ions cannot reach the cathode in the duration $\approx 10^{-7}$ s found experimentally for a spark, and most of the secondary emission from the cathode is due to photons; however, cathode emission is then probably less important than ionization in the body of the gas. At pressures of the order of $10^7$ Pa, and with gaps of the order of centimetres, Paschen's law breaks down. This is due to the high fields at the cathode ($\sim 10^8$ V m$^{-1}$), which cause appreciable field emission (see §11.5), a process which does not depend on the number of ions formed in the body of the gas, as assumed in eqn (3.54). Since both field emission and photoelectric emission depend on the work function (see §§ 11.4, 11.5), the nature of the cathode surface becomes increasingly important at high pressures.

### 3.9. Conduction of electricity through liquids

Certain liquids, such as hydrocarbons, are extremely good insulators, while others, such as water, have an appreciable conductivity. Solutions of some salts in water have a conductivity of the order of $10^{-5}$ times that of metals, and such salts are known as ionic compounds. An example is sodium chloride, which in a simple picture is formed by the transfer of one electron from the sodium atom to the chlorine atom, so that the molecule consists of a positively charged sodium ion and a negatively charged chlorine ion. Such a molecule has a permanent electric dipole moment, and is called a 'polar molecule'; in a non-polar molecule, such as hydrogen, the electrons are shared equally between the two atoms, and there is no permanent dipole moment. The water molecule is itself strongly polar, and when a substance such as sodium chloride is dissolved in it, the electric fields of the water molecules are strong enough to dissociate the solute molecules into separate $Na^+$ ions and $Cl^-$ ions. There are also composite ions formed by groups of molecules which have gained or lost an electron. The solution is called an 'electrolytic solution', the solute being known as the 'electrolyte'. The degree of dissociation of the electrolyte in solution is determined by the dynamic equilibrium between recombination and dissociation. With 'strong electrolytes', such as sodium chloride, the dissociation is practically complete at all ordinary concentrations; for 'weak electrolytes', such as acetic acid, the degree of dissociation is greatest at high dilution and falls steadily with increasing concentration according to the law

$$\alpha^2/(1-\alpha) = K/c,$$

where $\alpha$ is the fraction of solute molecules dissociated, $c$ the concentration, and $K$ is a constant which depends on the temperature.

If a potential difference is applied between two electrodes in an electrolytic solution, a current will flow through the solution. The current is carried by both positive and negative ions, which are produced by the decomposition of the electrolyte. If $n_1$, $n_2$ are the numbers of ions per unit volume with charges $q_1$, $q_2$ and mobilities $u_1$, $u_2$, the current density is

$$\mathbf{J} = (n_1 |q_1| u_1 + n_2 |q_2| u_2)\mathbf{E} = \sigma\mathbf{E}. \tag{3.58}$$

The specific conductivity $\sigma$ is found to be proportional to the concentration for very weak solutions, but increases less rapidly at higher concentrations. For weak electrolytes, the conductivity is determined by the degree of dissociation—that is, the numbers of ions present—indicating that the mobility is independent of concentration. For strong electrolytes, where dissociation is practically complete at all concentrations, the mobility falls at high concentrations, because each ion attracts round itself an 'atmosphere' of ions of opposite sign, effectively making it bigger and retarding its progress through the solution. The mobility of the ions in a liquid, as we should expect, is always very small compared with that in a gas.

### Electrolysis

Metallic ions and hydrogen ($H^+$) always travel to the cathode, or negative electrode, and negative ions to the anode. This transfer of charged ions by an electric current is called electrolysis. For example, if two copper electrodes are immersed in copper-sulphate solution, copper is dissolved from the anode, and deposited on the cathode. With carbon electrodes in a brine solution, hydrogen appears at the cathode and chlorine at the anode. There are two fundamental laws, discovered by Faraday, which are obeyed by all electrolytes.

(1) The mass of a given substance liberated at one electrode is proportional to the total charge which has passed.

(2) The mass of a given substance liberated at an electrode by unit charge is proportional to the chemical equivalent of that substance.

These two laws can be condensed into the following form: If the mass liberated at an electrode is $m$ when a current $I$ passes for a time $t$, then $m = ZIt$, where $Z$ is a constant for a given element, called the *electrochemical equivalent*. The chemical equivalent of an ion is the atomic weight divided by the valency, the valency being equal to the number of electronic charges carried by the ion. Then if $A$ is the atomic weight and $v$ the valency,

$$m = (A/v)\frac{It}{F},  \tag{3.59}$$

where $F$ is a universal constant known as the Faraday. The Faraday is the charge of electricity which liberates one gram equivalent $(A/v)$ of an ion in electrolysis. If $N_A$ (Avogadro's number) is the number of atoms in a mole, the total charge carried by a gram equivalent is $(N_A/v)(ve)$, since each atom (more correctly, each ion) has a charge $ve$, and this total charge $N_Ae$ must just be equal to $F$. Hence we have the important relation

$$F = N_Ae.  \tag{3.60}$$

Basically, the value of the Faraday is found by determining the mass of electrolyte liberated when a known current is passed through a solution for a measured length of time. The most reliable methods are those in which silver is dissolved in perchloric acid, and the 'iodine coulometer', in which the basic reaction is

$$3I^- \rightleftharpoons I_3^- + 2e.$$

Results obtained at the National Bureau of Standards (see Bower 1972) are

$$9 \cdot 64865 \times 10^7 \text{ C kg}^{-1} \text{ (silver)},$$

$$9 \cdot 64831 \times 10^7 \text{ C kg}^{-1} \text{ (iodine)},$$

which agree well with the value $9 \cdot 64846 \times 10^7$ C kg$^{-1}$ obtained indirectly from measurements of $e/m_p$ for the proton.

Although the primary reaction in electrolysis is simply the flow of ions of positive and negative sign to the cathode and anode respectively, secondary processes may occur at the electrodes, so that different products appear there which do not correspond to the primary ions. The first product may be unstable, as in the case

$$NH_4 + H_2O \rightarrow NH_4OH + H,$$

or it may react with the solvent, the solute, or the electrode in a chemical reaction such as

$$Na + H_2O \rightarrow NaOH + H.$$

In these examples the hydrogen atoms or ions will combine to form hydrogen molecules which appear as a mass of small bubbles of gas covering the electrode. In a cell where gas appears at the electrodes, the phenomenon of 'polarization' of the electrodes is generally observed. For example, in the electrolysis of acidulated water using polished platinum electrodes, no electrolysis occurs until the applied e.m.f. $V$ exceeds a certain critical value $V'$, known as the decomposition potential of the electrolyte. If $V > V'$, the current $I$ flowing obeys a modified Ohm's-law relation

$$V - V' = IR,$$

but if $V < V'$, no current flows. The cell therefore becomes irreversible owing to 'polarization' of the electrodes. This is probably due to the presence of positively charged hydrogen ions in the gas bubbles, which repel other positive ions and so give the effect of a back e.m.f. Practical cells include a 'depolarizer', a substance which reacts chemically with hydrogen ions appearing at the electrode and so prevents the formation of gas bubbles.

### 3.10. Voltaic cells

When two electrodes are placed in an electrolyte, it is generally found that there exists an e.m.f. between them. The combination is known as a 'voltaic cell' or battery and can be used to drive a current round an external circuit. There are a number of types in large-scale commercial use, from which many amperes of current can be drawn for short periods. These include the 'dry cell', a zinc electrode (generally forming the container), with $NH_4Cl$ solution in the form of a paste with $MnO_2$ as depolarizer round a carbon rod, giving 1·5 V; the lead–dilute sulphuric acid–lead oxide ($PbO_2$) battery, giving 2·05 V when fully charged; and (with a more favourable stored-energy-to-weight ratio) the heavy-duty iron–nickel cell (Fe–KOH solution–Ni), giving 1·4 V.

In contrast, special cells have been developed to maintain a very constant e.m.f. for long periods, provided virtually no current is drawn; the 'Weston standard cell' is widely used as a laboratory standard for precision measurements of voltage. Its construction is rather complicated; basically, at one electrode (an amalgam of mercury and cadmium) cadmium atoms go into solution as $Cd^{2+}$ ions in an aqueous solution of $CdSO_4$–$Hg_2SO_4$, while $Hg^+$ ions leave the electrolyte to become neutral mercury atoms at the second electrode (pure mercury). Thus the second electrode becomes positively charged, while the first becomes negatively charged because the cadmium atoms must leave two electrons behind when going into solution as $Cd^{2+}$ ions. When current flows the changes are

$$Cd - 2e^- \rightarrow Cd^{2+},$$

$$2Hg^+ + 2e^- \rightarrow 2Hg,$$

and chemical energy is available because the latter reaction releases more energy than is required to turn a cadmium atom into a $Cd^{2+}$ ion. Electrically two electrons are transferred in this process from the second to the first electrode; on open-circuit the e.m.f. between the two electrodes at 290 K is just over 1·018 V.

*Application of thermodynamics to voltaic cells*

Since in a reversible cell the chemical reactions taking place when a current is passed through it in one direction may be reversed by sending the current through it in the opposite direction, the standard equations of thermodynamics may be applied to the cell. In an ideal case we may suppose the current to be infinitely small, so that Joule heat losses, which depend on the square of the current, can be made negligibly small in comparison with the chemical-energy changes which vary with the first power of the current. Then the equation for the change in the free energy $F$ of a cell when a charge $Q$ is passed at constant temperature is

$$F = U + T(\partial F/\partial T),$$

where $U$ is the change in internal energy and $F$ is the energy available for external work provided the cell volume is constant (that is, no gases are liberated at the electrodes). If $V$ is the e.m.f. of the cell, then $F = VQ$, and we have

$$V = (U/Q) + T(\partial V/\partial T)$$
$$= h + T(\partial V/\partial T),\qquad (3.61)$$

where $h = U/Q$ is the heat of reaction when unit charge is passed. The reason why $VQ$ is not just equal to $U$ is because it may be necessary for the cell to exchange heat with its surroundings in order to remain at constant temperature, and this flow of heat may be related, by the second law of thermodynamics, to the temperature coefficient of the e.m.f.

### 3.11. Metallic conduction: classical theory

In a theory of metallic conduction put forward by Drude in 1900 it was assumed that there are electrons (known as the 'conduction electrons') which are free to move throughout the metal in the same way as the molecules of a perfect gas move within a container. In the absence of an electric field these electrons move in random directions and with random velocities so that there is no net current flow. Under the action of an electric field they are accelerated, but the additional momentum thus acquired is dissipated by a process resembling collisions, presumably with the positive ions which are fixed in the lattice. The conduction mechanism is clearly similar to that in an ionized gas at low field strengths (§ 3.8). If the mean free time between 'collisions' is $\tau$, Then the mobility is given by eqn (3.51) and the electrical conductivity is

$$\sigma = n(q^2/m)\tau.$$

These formulae are identical with eqns (3.10) and (3.12) if we equate the mean free time to the relaxation time.

The theory holds only if the drift velocity is small compared with the random velocity. If the mean free path $L$ is independent of velocity then $\tau$ is inversely proportional to the average speed of the carriers. If this is appreciably altered by the additional velocity acquired between collisions through acceleration by the electric field, $\tau$ would depend on $\mathbf{E}$ and the conductivity would be dependent on field strength. No such effects are observed in metals.

Since metals are much better conductors of heat than most electrical insulators, it is plausible to assume that thermal conduction in a metal is also mainly due to the free carriers. Application of the ordinary kinetic theory formula for the thermal conductivity $K$ of a gas to the 'electron gas' in a conductor gives

$$K = \tfrac{1}{3}ncL(d\bar{W}/dT),\qquad (3.62)$$

where $c$ is the random thermal velocity and $\bar{W} = \tfrac{1}{2}mc^2$ the average kinetic energy of a carrier. Taking the mean free path $L$ as equal to $c\tau$, we obtain an expression for the ratio of the thermal to the electrical conductivity

$$\frac{K}{\sigma} = \frac{mc^2}{3q^2}\frac{d\bar{W}}{dT} = \frac{2\bar{W}}{3q^2}\frac{d\bar{W}}{dT}.$$  (3.63)

If we assume that the carriers are electrons ($q = -e$) and obey classical statistics so that $\bar{W} = \frac{1}{2}mc^2 = \frac{3}{2}kT$, then

$$\frac{K}{\sigma T} = \frac{3}{2}\left(\frac{k}{e}\right)^2$$  (3.64)

should be the same for all metals and thus a universal constant. This is in accordance with an empirical law discovered by Wiedemann and Franz in 1853, and the numerical value predicted by eqn (3.64) is in good agreement with the experimental values for copper, silver, and gold over the limited temperature range of their experiments. As in ordinary kinetic theory, the exact numerical value depends on corrections made for the velocity spread; Drude originally obtained a factor 3 rather than $\frac{3}{2}$, while Lorentz deduced a factor 2.

In spite of this remarkable success of Drude's theory, neither $\sigma$ nor $K$ can be derived separately, since on classical theory there is no way of estimating $L$ or $\tau$. A much more important objection arises from the specific heat; from the average energy $\bar{W}$ assumed above, the free electrons should provide a contribution $\frac{3}{2}R$ per mole in addition to the value $3R$ from the law of Dulong and Petit. This should make the value for a metal some 50 per cent greater than that for an insulator, but experimentally there is no significant difference. This difficulty was not overcome until it was realized that electrons must be treated using quantum statistics rather than the classical Maxwell–Boltzmann statistics. In fact, since electrons have an intrinsic spin angular momentum of $\frac{1}{2}\hbar = \frac{1}{2}h/2\pi$, the correct type of statistics is that associated with the names of Fermi and Dirac. The use of these statistics is the starting point of the discussion in Chapter 11.

### References

BOWER, V. E. (1972). *Atomic masses and fundamental constants* 4 (Eds J. H. Sanders and A. H. Wapstra). Plenum Press, London.
CALLENDAR, H. L. (1910). *Proc. phys. Soc.* **22**, 220.
DUFFIN, W.J. (1965). *Electricity and Magnetism*. McGraw-Hill, London.
KENNARD, E.H. (1938). *The kinetic theory of gases*. McGraw-Hill, New York.
KETTERING, C. F. and SCOTT. G. C. (1944). *Phys. Rev.* **66**, 257.
LLEWELLYN JONES, F. (1953). *Ann. Rep. Progr. Phys.* **16**, 216.
MORSE, P. W. (1964). *Thermal physics*. Benjamin, New York.
SMYTHE, W. R. (1939). *Static and dynamic electricity*. McGraw-Hill, New York.
TOLMAN, R. C. and STEWART, T. D. (1917). *Phys. Rev.* **9**, 64.

**Problems**

3.1. For the Wheatstone's bridge circuit (Fig. 3.7) show that the current through the detector is

$$I = \frac{V(R_1R_4 - R_2R_3)}{D + R_G(R_1 + R_2)(R_3 + R_4)},$$

where $D = R_1R_2R_3 + R_2R_3R_4 + R_3R_4R_1 + R_4R_1R_2$.

3.2. Show from the results of the previous problem that if $R_G = \infty$ the open-circuit voltage across the terminals AB is

$$V_{AB} = V(R_1R_4 - R_2R_3)/(R_1 + R_2)(R_3 + R_4),$$

while if the battery is short-circuited the resistance measured at the terminals AB would be

$$R_{AB} = \frac{R_1R_2}{(R_1 + R_2)} + \frac{R_3R_4}{(R_3 + R_4)}.$$

Verify that with detector and battery in place, the bridge behaves as a voltage generator of e.m.f. $V_{AB}$ with internal resistance $R_{AB}$ working into a load $R_G$ (Thévenin's theorem).

3.3. Using the nomenclature of Callendar (Fig. 3.7) for Wheatstone's bridge, show that

$$\frac{I}{I_4} = \frac{R - R'}{R_G(1 + 1/n) + R'(1 + m)}.$$

The ratio $I/I_4$ is a measure of the sensitivity which is useful when the allowed current $I_4$ through the unknown resistance $R$ is limited.

3.4. Resistances $P$, $Q$, $R$ each of 10 $\Omega$ are placed in three arms of a Wheatstone's bridge, and a resistance $S$ is adjusted in the fourth arm so that the bridge is balanced. The resistance $R$ is now replaced by a resistance $X$, and the balance is restored by shunting $S$ with a resistance of $10 \cdot 123$ $k\Omega$. What is the value of $X$?

Discuss the advantages and disadvantages of this method of measuring resistances when high accuracy is required.

3.5. In Fig. 3.14, show that a balance condition for the bridge (that is, no current through the galvanometer) which is independent of the value of $r$ is

$$R_1/R_2 = R_3/R_4 = R_5/R_6.$$

This is Kelvin's double bridge for measuring small resistances of the order of $0 \cdot 01$ $\Omega$. The resistance $r$ represents the contact resistance between the two small resistances $R_1$, $R_2$, and its value does not affect the balance. If readings are taken with the currents flowing in both directions, so that errors due to thermoelectric voltages at the junctions are avoided, an accuracy of about $0 \cdot 02$ per cent can be achieved.

3.6. An accurate method, due to Hamon, of obtaining a known resistance ratio is to use $n$ equal resistances, alternately connected all in series or all in parallel. Show that the ratio of the net resistance in series to that in parallel is $n^2$ and that, if the fractional error in each resistance is of order $\delta$, then the fractional error in the ratio is of order $\delta^2$.

3.7. Six 1 $\Omega$ resistances are joined to form a regular tetrahedron, and a potential

FIG. 3.14. Kelvin's double bridge.

of 1 V is maintained across one of the resistances. Find the current flowing in each conductor.

(*Answer:* 1, $\frac{1}{2}$, $\frac{1}{2}$, $\frac{1}{2}$, $\frac{1}{2}$, $\frac{1}{2}$, and 0A.)

3.8. Two long parallel circular copper rods of diameter 2·5 mm are placed with their axes 0·2 m apart in a large tank of copper-sulphate solution. If the conductivity of the solution is $4\cdot1\ \Omega^{-1}\ m^{-1}$, find the conductance between the rods of a section 1 m long.

(*Answer:* $3\cdot0\ \Omega^{-1}$.)

3.9. A solid sphere of mass density $\rho_d$ and radius $R$ is composed of hydrogen atoms. If the proton and electron differ in the magnitude of their charges by a fraction $\delta$, show that the electric field strength at the surface of the sphere is

$$\delta(RF\rho_d/3\epsilon_0),$$

where $F$ is the Faraday constant.

At the surface of the earth ($R = 6\times10^3$ km, $\rho_d = 5\cdot5\times10^3\ kg\ m^{-3}$) the normal electric field strength is about 100 V m$^{-1}$. Show that this gives a value of $\delta \sim 10^{-22}$, if ascribed to such an effect.

3.10. Find the equation of motion of a free electron in an electric field strength $E = E_0 \sin \omega t$.

If the field strength $E_0$ is $10^4$ V m$^{-1}$, and the frequency is 100 MHz, show that the amplitude of oscillation of the electron is 0·0045 m and that its maximum energy is 22·5 eV.

3.11. A spark passes between two electrodes 10 mm apart in air at atmospheric pressure when a uniform field of $10^6$ V m$^{-1}$ is applied across the gap. If the mean free path of an oxygen molecule in air is $6\times10^{-8}$ m show that the time required for a singly ionized oxygen molecule to cross the gap is $4\cdot5\times10^{-5}$ s.

3.12. An electron of mass $m$ collides with a molecule of mass $M$. Show that if the molecule is stationary the fraction of the electron energy which is transferred to the molecule in a head-on collision is $4Mm/(m+M)^2$, and evaluate this for the case $M = 200\times$mass of the proton.

(*Answer:* $\approx 1\cdot1\times10^{-5}$.)

# 4. The magnetic effects of currents and moving charges; and magnetostatics

## 4.1. Forces between currents

THE first experimental investigation of the interaction between coils carrying electric currents was performed by Ampère during the years 1820–5, and the work was continued by Oersted, Biot, and Savart. They found that two long parallel wires carrying currents in opposite directions repel one another, whereas when carrying currents in the same direction they attract one another, so that the direction of the force is reversed when one current is reversed. Ampère used circular coils, the leads to the coils being twisted together, and as these leads each carried equal currents in opposite directions they exerted no force on other circuits, so any forces observed were due only to the coils. He found that, if the dimensions of the coils were small compared with their distance apart, one coil exerts a force and a couple on another coil exactly similar to the force and couple which one electric dipole exerts on another. In magnitude, this force and couple are proportional to the current through the coil, the number of turns, and the area. If the plane of each coil is normal to the line joining the centre of each coil, the force is along this line. It is found also that if a coil carrying a current is placed near a magnet it experiences both a force and a couple. At distances large compared with the dimensions of either coil or magnet, this force and couple are similar in nature to those due to a second coil carrying a current. Thus both a magnet and a current-carrying coil are said to produce a magnetic field described by a flux density **B**, which exerts forces on other coils or magnets. **B** is a vector quantity and lines of **B** can be drawn whose direction at any point is that of **B**, in the same way as lines of electric force are drawn in an electric field. The number of lines of **B** passing normally through any area is known as the 'magnetic flux' through that area, and the term 'magnetic flux density' refers to the number of lines of **B** passing normally through unit area.

The force exerted on an element of wire $ds_1$ carrying a current $I_1$ at a place where the magnetic flux density is **B** can be expressed in the simple form (see Fig. 4.1)

$$d\mathbf{F} = I_1(d\mathbf{s}_1 \wedge \mathbf{B}).  \tag{4.1}$$

This equation can be used to define a unit for **B** as that flux density which exerts unit force on unit length of a wire carrying one unit of current. In

FIG. 4.1. Diagram to illustrate eqn (4.1). $ds_1$ and **B** are in the plane of the paper, the angle between them being $\theta$. d**F** is normal to the paper, towards the reader, and has magnitude $ds_1 B \sin\theta$.

S.I., the unit of force is the newton, that of length is the metre, and the unit of current is the ampere, thus the unit of **B** is the newton ampere$^{-1}$ metre$^{-1}$ (N A$^{-1}$ m$^{-1}$); it is known as the tesla (T).

The experiments of Ampère and others showed that the force on an element $ds_1$ carrying a current $I_1$ due to another element $ds_2$ carrying a current $I_2$ is

$$d\mathbf{F}_1 = \frac{\mu_0 I_1 I_2}{4\pi r^3}\{d\mathbf{s}_1 \wedge (d\mathbf{s}_2 \wedge \mathbf{r})\}, \qquad (4.2)$$

where $\mu_0$ is a constant. **r** is the vector joining the two elements, being positive when drawn from $d\mathbf{s}_2$ to $d\mathbf{s}_1$, as in Fig. 4.2. The force $d\mathbf{F}_2$ on the element $d\mathbf{s}_2$ due to $d\mathbf{s}_1$ is given by a similar expression with $d\mathbf{s}_1$ and $d\mathbf{s}_2$ interchanged, and **r** must then be taken as positive when drawn from $d\mathbf{s}_1$ to $d\mathbf{s}_2$. The directions of the forces for the special case of two coplanar elements are shown in Fig. 4.2, and it will be seen that they are not equal

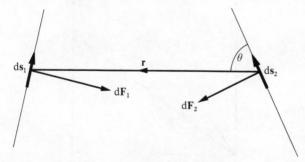

FIG. 4.2. Diagram showing the direction of the forces between two current elements. $d\mathbf{s}_1$ and $d\mathbf{s}_2$ are in the plane of the paper. The vector $(d\mathbf{s}_2 \wedge \mathbf{r})$ is normal to the plane of the paper, and the vector $\{d\mathbf{s}_1 \wedge (d\mathbf{s}_2 \wedge \mathbf{r})\}$ is in the plane of the paper, normal to $d\mathbf{s}_1$. The flux density d**B** due to $d\mathbf{s}_2$ is parallel to $(d\mathbf{s}_2 \wedge \mathbf{r})$), and the force $d\mathbf{F}_1$ on $d\mathbf{s}_1$ due to it is parallel to $\{d\mathbf{s}_1 \wedge \mathbf{r}\}$.

and opposite unless the current elements are parallel. This apparent violation of Newton's third law of motion has caused much discussion, but Page and Adams (1945) have shown that there is no real violation, since the electromagnetic field of the current elements possesses momentum which is changing at a rate just equal to the difference of the two forces. Ampère's original formulation of the law of force between two current elements was different from eqn (4.2), but gave the correct result when integrated over a closed circuit carrying a constant current.

Comparison of eqns (4.1) and (4.2) shows that we may say that the current $I_2$ in the element $ds_2$ produces a magnetic flux density $dB$ at a distance $r$ given by the formula

$$dB = \frac{\mu_0}{4\pi r^3} I_2(ds_2 \wedge r). \tag{4.3}$$

These equations may be used to calculate the value of $B$ produced by a current in an infinite straight wire, and hence the force between currents in two parallel infinite straight wires. In Fig. 4.3 we have two such wires a distance $a$ apart, carrying currents $I_1$, $I_2$; we choose a coordinate system where the first wire lies along the $z$-axis, and the second is parallel to it but passes through the point $x = a$, $y = 0$. We first calculate $B$ at the point $(0, 0, 0)$ due to the current $I_2$ in the second wire, using eqn (4.3). Then the element $ds_2$ has components $(dx, dy, dz) = (0, 0, dz)$ and $r$ has components $(-a, 0, -z)$ since it is defined by the coordinates of the point $(0, 0, 0)$ relative to the point $A(a, 0, z)$ at which $ds_2$ is placed. The components of $ds_2 \wedge r$ are $(0, -a\,dz, 0)$, showing that $dB$ at $O$ is antiparallel to the y-axis, wherever the point A lies along the second wire. Hence $B = \int dB$ will also be antiparallel to the y-axis, so that $B_x = B_z = 0$, and integration yields for

FIG. 4.3. Parallel wires carrying currents.

$B_y$ the result

$$B_y = \frac{\mu_0 I_2}{4\pi} \int\limits_{-\infty}^{+\infty} \frac{-a \, dz}{(a^2+z^2)^{\frac{3}{2}}} = -\frac{\mu_0 I_2}{4\pi a} \int\limits_{-\frac{1}{2}\pi}^{+\frac{1}{2}\pi} \cos\theta \, d\theta = -\frac{\mu_0 I_2}{2\pi a}, \quad (4.4)$$

where we have used the substitution $z = a \tan \theta$. Eqn (4.4) shows that the field of a current $I_2$ in an infinite wire is proportional to $I_2$ and inversely proportional to the distance $a$ from the wire. **B** is normal to the plane containing the wire and the radius vector **r**, so that lines of constant $B$ form closed circles centred on the wire.

We can now use eqn (4.1) to find the force d**F** on an element $ds_1$ of the first wire. Since **B** is in the y direction, and $ds_1$ in the z direction, the force is in the x direction, its only component being

$$dF_x = \frac{\mu_0 I_1 I_2 \, ds_1}{2\pi a}. \quad (4.5)$$

If the currents are in the same direction the force is one of attraction, if the currents are opposed the force is one of repulsion, as stated above.

The value of the constant $\mu_0$ depends on the system of units employed. We regard the ampere (or coulomb, since 1 A is a current of $1 \, \mathrm{C\,s^{-1}}$) as a standard of current (or charge) defined in an arbitrary way, a base unit similar to the kilogram and metre. The value of $\mu_0$ is defined to be exactly $4\pi \times 10^{-7} \, \mathrm{N\,A^{-2}}$. Eqn (4.5) then shows that for two parallel wires 1 m apart, each carrying 1 A of current, the force per metre length of wire is $2 \times 10^{-7} \, \mathrm{N}$. This is a convenient way of defining the ampere. The quantity $\mu_0$ is known as the 'permeability of a vacuum' (see § 4.4), and its unit is generally called the henry metre$^{-1}$ ($\mathrm{H\,m^{-1}}$) (see § 5.2) rather than the newton ampere$^{-2}$ ($\mathrm{N\,A^{-2}}$); the two units are equivalent.

We can use eqn (4.3) to evaluate the quantity div d**B**. In this operation $ds_2$ is a constant vector, and the variable is $\mathbf{r}/r^3 = -\mathrm{grad}\,(1/r)$; the quantity div $(ds_2 \wedge \mathbf{r}/r^3)$ is identically zero (see eqns (A.15) and (A.19)), so that div d**B** = 0. If this is true for every contribution d**B**, it must also be true for the vector **B** itself. Thus the variation of the magnetic flux density in space obeys the differential equation

$$\mathrm{div}\,\mathbf{B} = \frac{\partial \mathbf{B}_x}{\partial x} + \frac{\partial \mathbf{B}_y}{\partial y} + \frac{\partial \mathbf{B}_z}{\partial z} = 0. \quad (4.6)$$

## 4.2. Magnetic dipole moment and magnetic shell

The investigations by Ampère of the forces between two small coils showed that they were similar to those between two dipoles. Comparison with eqns (1.14) and (1.15) shows that we should expect such a dipole, if placed in a field of uniform flux density, to experience a couple but no translational force, while in a non-uniform flux density it will experience a translational force. We now derive expressions for the couple and force, starting from eqn (4.1).

A set of elementary current circuits is shown in Fig. 4.4; they are similar but lie in three mutually orthogonal planes, each normal to one axis of a Cartesian coordinate system. We begin with the case of a uniform magnetic flux density **B**. Then, for the current circuit in Fig. 4.4(a), the forces on the sides CD, EF are clearly equal and opposite to one another, and so are those on the sides DE, FC. Thus there is no net translational force. The couple on the current element can be found as follows. The force $F_z = -I\,dxB_y$ on the side EF has a moment about the $x$-axis equal to

$$dT_x = +F_z\,dy = -I\,dxdy\,B_y = -dm_zB_y,$$

where $m_z = I\,dx\,dy$. Similarly, the force $F_z = -I\,dyB_x$ on DE has a moment about the $y$-axis equal to

$$dT_y = -F_z\,dx = +I\,dxdy\,B_x = +dm_zB_x,$$

while $dT_z = 0$. If we write

$$\mathbf{dm} = I(\mathbf{dS}), \tag{4.7a}$$

the moments are the components of a couple

$$d\mathbf{T} = I(\mathbf{dS} \wedge \mathbf{B}) = \mathbf{dm} \wedge \mathbf{B}. \tag{4.8a}$$

The quantity $dx\,dy$ is the area of the current loop, which can be represented by a vector normal to it, whose direction is that in which a right-handed screw would move if turned in the sense in which current flows round the loop; in our case this vector has the single component $(dS)_z = dx\,dy$ and the vector $\mathbf{dm} = I\,\mathbf{dS}$ has the single component $dm_z$. Similar relations hold for Fig. 4.4(b) and (c), where the components are $dm_x = I'\,dy\,dz$ and $dm_y = I''\,dz\,dx$. The quantity $\mathbf{dm}$ is known as the

FIG. 4.4. Elementary current circuits in three mutually perpendicular planes. The equivalent magnetic dipole moments are $m_z = I\,dx\,dy$ in (a), $m_x = I'\,dv\,dz$ in (b), and $m_y = I''\,dz\,dx$ in (c).

'magnetic dipole moment' of the elementary current loop. For a loop of finite size we can write

$$\mathbf{m} = \int d\mathbf{m} = I \int d\mathbf{S} = I\mathbf{S}, \tag{4.7b}$$

where the vector $\mathbf{S}$ must be found by a vector integration, except in the case of a plane loop, and $\mathbf{m}$ is the equivalent magnetic dipole moment. In a field of uniform flux density $\mathbf{B}$ the couple is then

$$\mathbf{T} = \mathbf{m} \wedge \mathbf{B}. \tag{4.8b}$$

The translational force on a current loop in a non-uniform flux density may also be derived with the help of Fig. 4.4(a). The $x$-component of this force is

$$dF_x = I\left\{ dy\left(B_z + \frac{\partial B_z}{\partial x}\,dx\right) - dyB_z \right\}$$

$$= I\,dx\,dy(\partial B_z/\partial x) = dm_z(\partial B_z/\partial x),$$

where $B_z$ is the component of the flux density acting on the element FC and $(B_z + (\partial B_z/\partial x)\,dx)$ that on DE. Similarly

$$dF_y = I\left\{ -dxB_z + dx\left(B_z + \frac{\partial B_z}{\partial y}\,dy\right) \right\}$$

$$= dm_z(\partial B_z/\partial y),$$

while

$$dF_z = I\left\{ dxB_y - dy\left(B_x + \frac{\partial B_x}{\partial x}\,dx\right) - dx\left(B_y + \frac{\partial B_y}{\partial y}\,dy\right) + dyB_x \right\}$$

$$= -dm_z\left\{ \frac{\partial B_x}{\partial x} + \frac{\partial B_y}{\partial y} \right\}.$$

Since, from eqn (4.6), div $\mathbf{B} = 0$, this last equation is equal to

$$dF_z = dm_z(\partial B_z/\partial z).$$

By considering the circuits in Fig. 4.4(b) and (c) it is easily shown that altogether, for a dipole $\mathbf{m}$ in a constant gradient of magnetic flux density,

$$F_x = m_x \frac{\partial B_x}{\partial x} + m_y \frac{\partial B_y}{\partial x} + m_z \frac{\partial B_z}{\partial x}, \tag{4.9}$$

with similar relations for the other force components $F_y$, $F_z$.

Comparison of eqns (4.8b) and (4.9) with (1.14) and (1.15b) shows that the magnetic dipole moment $\mathbf{m}$ plays a similar role to the electric dipole moment $\mathbf{p}$ and the magnetic flux density $\mathbf{B}$ plays the same role as the electric field strength $\mathbf{E}$. We may therefore expect that it can be written as the gradient of a scalar potential $\phi$, such that

$$\mathbf{B} = -\mu_0 \,\text{grad}\, \phi. \tag{4.10}$$

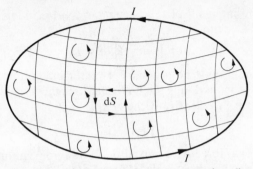

FIG.4.5. A finite current circuit divided into a number of small current loops.

Whatever its shape, a finite current circuit can be divided up as in Fig. 4.5 into a number of current loops which lie on any surface which is bounded by the current circuit. Each element of area d$S$ may be regarded as having a current $I$ flowing round its edge, and summation of the currents in all the elements comprising the entire surface leaves only the current in the circuit as the resultant. If d$S$ is infinitesimal in size, it can be regarded as a plane element and will have an associated magnetic dipole moment given by eqn (4.7a), and the couple on the whole circuit is obtained by integration of eqn (4.8a). The surface forms a magnetic 'double layer', or 'magnetic shell', with a certain dipole moment per unit area. The potential due to such a shell will now be calculated using eqn (4.3).

The field at a point P due to the current circuit is found by integration of eqn (4.3) round the circuit. In Fig. 4.6 if the point P is displaced a distance $\delta\mathbf{s}$ the change in potential will be

$$\delta\phi = -\frac{1}{\mu_0}\mathbf{B}.\delta\mathbf{s} = -I\,\delta\mathbf{s}.\int\frac{(\mathrm{d}\mathbf{a}\wedge\mathbf{r})}{4\pi r^3} = -I\int\frac{\delta\mathbf{s}.(\mathrm{d}\mathbf{a}\wedge\mathbf{r})}{4\pi r^3}, \quad (4.11)$$

where $\delta\mathbf{s}$ can be taken inside the integral sign because it is a constant during the integration. It is clear that we should obtain the same change

FIG. 4.6. Displacement of a current element, in order to calculate the potential at P due to a magnetic shell.

in potential if the point P were kept fixed and the circuit were displaced by an amount $-\delta\mathbf{s}$. In such a displacement the circuit element $\mathbf{da}$ sweeps out an area $-(\delta\mathbf{s}\wedge\mathbf{da})$, and this area subtends a solid angle at the point P of

$$-\frac{\mathbf{r}\cdot(\delta\mathbf{s}\wedge\mathbf{da})}{r^3} = -\frac{\delta\mathbf{s}\cdot(\mathbf{da}\wedge\mathbf{r})}{r^3}.$$

Hence the line integral in eqn (4.11) is the total solid angle subtended at P by the area swept out by the circuit when it is displaced by $-\delta\mathbf{s}$, and this is equal to the change $\delta\Omega$ in the solid angle due to the displacement of P by $\delta\mathbf{s}$. Hence we may write the change in potential as

$$\delta\phi = I\frac{\delta\Omega}{4\pi},$$

and the potential at P is

$$\phi = I\Omega/4\pi, \tag{4.12}$$

where $\Omega$ is the solid angle which the circuit subtends at P. But

$$\Omega = \int\frac{\mathbf{dS}\cdot\mathbf{r}}{r^3},$$

where $\mathbf{dS}$ is an element of area of any surface bounded by the circuit, and hence

$$\phi = \int\frac{I\,\mathbf{dS}\cdot\mathbf{r}}{4\pi r^3} = \int\frac{\mathbf{dm}\cdot\mathbf{r}}{4\pi r^3}. \tag{4.13}$$

Here the integration is over the surface of the magnetic shell, and the potential of an individual dipole $\mathbf{m}$ must therefore be

$$\phi = \frac{\mathbf{m}\cdot\mathbf{r}}{4\pi r^3}. \tag{4.14}$$

This is of the same form as eqn (1.10b), except that $\mu_0$, which we might expect to replace the constant $\epsilon_0$, does not occur here but in eqn (4.10). The reason for this choice will appear later (see eqn (4.32)).

The quantity $\phi$ is related to the line integral of $\mathbf{B}$ between two points, since

$$\int\mathbf{B}\cdot\mathbf{ds} = -\mu_0\int\operatorname{grad}\phi\cdot\mathbf{ds} = -\mu_0\phi. \tag{4.15}$$

By analogy with electromotive force, which is the line integral of $\mathbf{E}$ (see eqn (3.19)), the quantity $-\phi$ is sometimes known as the 'magnetomotive force', or m.m.f.

In the electrostatic case, the work done in traversing a closed circuit is zero, and this would also be the case for a true double layer of magnetic charge. If we take the integral $\int\mathbf{B}\cdot\mathbf{ds}$ from a point P very close to a magnetic shell round to a point P' just on the other side of the magnetic shell, the difference $\Delta\Omega$ in the solid angle which the shell subtends at these two points is $-4\pi$, and the m.m.f. between these two points is

$$-\Delta\phi = -I \,\Delta\Omega/4\pi = I,$$

from eqn (4.12). With a real magnetic shell, if we now move from P′ to P through the shell, there would be a contribution to the m.m.f. which would just make the total zero, but with a current circuit there is no such contribution. We have therefore an important difference, that the m.m.f. increases by $I$ every time we go round a closed path which threads the coil positively (that is, in the same direction as the lines of **B**). Thus the magnetostatic potential is not single-valued and cannot be used in a region where there are currents flowing. On the other hand, if the path does not thread a current circuit, the change in solid angle is zero, and the potential is single-valued. If the path does encircle a current $I$, we have

$$\oint \mathbf{B}.\,\mathrm{d}\mathbf{s} = \mu_0 I. \tag{4.16}$$

In a region of distributed current flow, the total current threaded by the path is $\int \mathbf{J}.\,\mathrm{d}\mathbf{S}$, where **J** is the current density in an element d**S** of a surface bounded by the path. Hence

$$\oint \mathbf{B}.\,\mathrm{d}\mathbf{s} = \int \mathrm{curl}\,\mathbf{B}.\,\mathrm{d}\mathbf{S} = \mu_0 \int \mathbf{J}.\,\mathrm{d}\mathbf{S},$$

where the transformation from a line integral of **B** to a surface integral of curl **B** is an example of Stokes's theorem (see Appendix § A.8). Since the integrals must be equal over any surface, the integrands must be equal, and we have

$$\mathrm{curl}\,\mathbf{B} = \mu_0 \mathbf{J}. \tag{4.17}$$

The relation given by eqn (4.16) is known as Ampère's law, and eqn (4.17) is its representation in differential form. Since any function of the form curl(grad $\phi$) is identically zero (see Appendix A, eqn (A.19)), we stress again that **B** can be derived only from a scalar potential $\phi$ in a region where the current density $\mathbf{J} = 0$. In such a region, we can derive a general expression for the translational force on a current loop in a non-uniform flux density. From the vector relation (A.27) in Appendix A, we can write

$$\mathbf{F} = I \int \mathrm{d}\mathbf{s} \wedge \mathbf{B} = -I \int \mathbf{B} \wedge \mathrm{d}\mathbf{s}$$

$$= -I \int (\mathrm{div}\,\mathbf{B})\,\mathrm{d}\mathbf{S} - I \int (\mathrm{curl}\,\mathbf{B}) \wedge \mathrm{d}\mathbf{S} + I \int (\mathrm{d}\mathbf{S}.\,\mathrm{grad})\mathbf{B}$$

$$= I \int (\mathrm{d}\mathbf{S}.\,\mathrm{grad})\mathbf{B} = \int (\mathrm{d}\mathbf{m}.\,\mathrm{grad})\mathbf{B}. \tag{4.18a}$$

Here we have used both the relation $\operatorname{curl}\mathbf{B} = 0$ when $\mathbf{J} = 0$, and also $\operatorname{div}\mathbf{B} = 0$ from eqn (4.6). If $\operatorname{grad}\mathbf{B}$ is constant in the region occupied by the loop we have

$$\mathbf{F} = I(\mathbf{S}.\operatorname{grad})\mathbf{B} = (\mathbf{m}.\operatorname{grad})\mathbf{B}, \tag{4.18b}$$

where $\mathbf{m} = I\mathbf{S}$ is the equivalent dipole moment of the current loop of area $\mathbf{S}$. The components of the force $\mathbf{F}$ in (4.18b) do not have the same form (see Appendix A, eqn (A.10)) as those in eqn (4.9) and its analogues, but they are in fact identical when $\operatorname{curl}\mathbf{B} = 0$, since then $\partial B_y/\partial x = \partial B_x/\partial y$, etc. (cf. eqns (1.15).

Eqns (4.8b) for the couple and (4.18b) for the translational force on a magnetic dipole are exact analogues of the relations given by eqns (1.14) and (1.15c) for an electric dipole. It follows that both the torque and the force can be derived from a potential energy

$$U_P = -\mathbf{m}.\mathbf{B} \tag{4.19}$$

which is the analogue of eqn (1.13). For a current circuit we have, from eqn (4.8a),

$$\mathbf{T} = I\int d\mathbf{S} \wedge \mathbf{B}$$

for the couple, while the translational force is given by eqn (4.18a),

$$\mathbf{F} = I\int (d\mathbf{S}.\operatorname{grad})\mathbf{B}.$$

The integrands are of the same form as the relations for electric dipoles, and the potential energy of the current circuit in the magnetic flux density $\mathbf{B}$ is

$$U_P = -I\int d\mathbf{S}.\mathbf{B}. \tag{4.20}$$

The quantity

$$\Phi = \int d\mathbf{S}.\mathbf{B} = \int \mathbf{B}.d\mathbf{S} \tag{4.21}$$

is the magnetic flux through the circuit, and the potential energy may be written as

$$U_P = -I\Phi. \tag{4.22}$$

The unit of flux is the weber (Wb); it is equal to $1 \, \mathrm{J} \, \mathrm{A}^{-1}$. The unit of $\mathbf{B}$, the tesla, is equal to 1 weber metre$^{-2}$ ($1 \, \mathrm{Wb} \, \mathrm{m}^{-2}$). The unit of magnetic dipole moment $\mathbf{m}$ can be written either as $\mathrm{A} \, \mathrm{m}^2$, from eqn (4.7), or $\mathrm{J} \, \mathrm{T}^{-1}$, from eqn (4.19). The unit of scalar magnetic potential $\phi$ is the ampere (A), from eqn (4.12).

## 4.3. Magnetostatics and magnetic media

The theory so far has been concerned with the magnetic effects of currents *in vacuo*, that is, in the absence of any magnetizable media. It is found experimentally that a material substance acquires a magnetic polarization when placed in a magnetic field, just as a dielectric medium acquires an electric polarization in an electric field. The magnetic dipole moment per unit volume is called the intensity of magnetization (or often simply the magnetization), and is represented by a vector **M**. All such magnetic effects are produced by current loops of atomic dimensions, which arise from the circulation of electric charge within the atom. The relation between the magnetic moment of such a loop and the circulating current is given by eqn (4.7b). If a magnetic medium has a magnetization **M**, which is not necessarily uniform throughout the substance, the equivalent current flow can be found by considering elementary current loops, as in Fig. 4.5. There the currents were all equal, and cancelled one another except at the perimeter, but in general this will not be the case. In Fig. 4.7 both **M** and $I$ are functions of the space coordinates. We consider an element of volume $d\tau = dx\ dy\ dz$ at the point $(x, y, z)$, for which the magnetic moment has a component $M_z\ dx\ dy\ dz$ in the $z$ direction. This is equivalent to a current flowing round the loop, the strength of the current being

$$I = (M_z\ dx\ dy\ dz)/(dx\ dy) = M_z\ dz,$$

since the area of the loop is $dx\ dy$. The adjacent loop at the point $(x+dx, y, z)$ has a current

$$I' = I + (\partial I/\partial x)\ dx = M_z\ dz + (\partial M_z/\partial x)\ dx\ dz.$$

Hence the net current flow on the interface between the two elements has a component in the y direction of magnitude

$$I - I' = -(\partial M_z/\partial x)\ dx\ dz$$

FIG. 4.7. Representation of a non-uniform magnetization by circulating currents,

$$M_z' = M_z + \frac{\partial M_z}{\partial x}\ dx.$$

and if $\mathbf{J}$ is the current density, this component must equal $J_y\,dx\,dz$. By considering similar current loops in the $yz$ plane, we find there is another current component in the $y$ direction of magnitude $(\partial M_x/\partial z)\,dx\,dz$ associated with the component of magnetization $M_x$. Hence the total $y$ component of the current density is $J_y = (\partial M_x/\partial z) - (\partial M_z/\partial x)$, with similar expressions for the other components of $\mathbf{J}$. These are the components of the vector curl $\mathbf{M}$, so that we can write for the current density $\mathbf{J_m}$ associated with a magnetization $\mathbf{M}$

$$\mathbf{J_m} = \text{curl } \mathbf{M}. \tag{4.23}$$

It follows that if $\mathbf{M}$ is uniform in space, $\mathbf{J_m} = 0$, so that an equivalent current flow exists only in regions where $\mathbf{M}$ is varying.

In a medium which is both electrically conducting and magnetizable, the total current density will be the sum of the real current density $\mathbf{J}$ and the equivalent magnetization current density $\mathbf{J_m}$, both of which must be counted in Ampère's law. Hence eqn (4.17), which was derived for a vacuum, must be replaced by

$$\text{curl } \mathbf{B} = \mu_0(\mathbf{J} + \mathbf{J_m}) = \mu_0(\mathbf{J} + \text{curl } \mathbf{M}),$$

or

$$\text{curl}(\mathbf{B} - \mu_0\mathbf{M}) = \mu_0\mathbf{J}. \tag{4.24}$$

This is the form which Ampere's law takes in the presence of a magnetizable medium, and the quantity $(\mathbf{B} - \mu_0\mathbf{M})$, whose curl is related to the flow of real current, is used to define a new vector such that

$$\mathbf{B} - \mu_0\mathbf{M} = \mu_0\mathbf{H},$$

or

$$\mathbf{B} = \mu_0(\mathbf{H} + \mathbf{M}). \tag{4.25}$$

The vector $\mathbf{H}$ is called the 'magnetic field strength'. It is clear from eqn (4.25) that the dimensions of $\mathbf{H}$ are different from those of magnetic flux density $\mathbf{B}$; it is important to remember this distinction, because magnetic flux density is often loosely referred to as 'the field $\mathbf{B}$'.

Ampère's law now takes the simple form

$$\text{curl } \mathbf{H} = \mathbf{J}, \tag{4.26}$$

which is more general than eqn (4.17), since it holds both *in vacuo* and in a medium. Obviously, *in vacuo* $\mathbf{B} = \mu_0\mathbf{H}$, so that eqn (4.17) is a special case of (4.26), which is the general differential form of Ampère's law. Similarly, eqn (4.16) must be replaced by the more general equation

$$\int \mathbf{H} \cdot d\mathbf{s} = \int \text{curl } \mathbf{H} \cdot d\mathbf{S} = \int \mathbf{J} \cdot d\mathbf{S} = I. \tag{4.27}$$

It is clear from this equation that the dimensions of $\mathbf{H}$ must be amperes metre$^{-1}$ (A m$^{-1}$), since the line integral of $\mathbf{H}$ round a circuit is equal to the total current threading the circuit. From eqn (4.25), **M** must have the same dimensions as **H**, and this can be readily verified, since **M** = magnetic moment per unit volume, and magnetic moment = current×area, from eqn (4.7).

The process by which we have introduced a new vector **H** in modifying our equations to allow for the presence of a polarizable medium is analogous to that in electrostatics, where a new vector **D** was introduced. There, this vector followed from the modification of Gauss's theorem needed in the presence of a polarizable medium; the force vector **E** is related by Gauss's theorem to the sum of the real charge density and the polarization charge density, and the advantage of **D** is that div **D** is related only to the real charge density. In the magnetic case, the force vector **B** is related by Ampère's law to the sum of the real current density and the magnetization current density, and the advantage of **H** is that curl **H** is determined solely by the real current density.

For a volume distribution of current, eqn (4.3) may be written

$$d\mathbf{B} = \frac{\mu_0}{4\pi}\left(\frac{\mathbf{J}\wedge\mathbf{r}}{r^3}\right)d\tau = -\frac{\mu_0}{4\pi}\left(\mathbf{J}\wedge\operatorname{grad}\frac{1}{r}\right)d\tau. \tag{4.28}$$

Since the only change is the replacement of $I\,d\mathbf{s}_2$ by $\mathbf{J}\,d\tau$, it is clear that the proof of eqn (4.6) given in §4.2 is again valid, and div **B** = 0. This relation was derived only for a current *in vacuo*, but we have shown that any magnetization **M** present can be replaced by an equivalent current density $\mathbf{J}_m$ for which it will also be true that div **B** = 0. Hence eqn (4.6) holds also in a magnetizable medium.

This equation is similar to that derived for div **D** in electrostatics, except that div **D** = $\rho_e$, where $\rho_e$ is the density of true electric charge, while div **B** = 0 because we have no true magnetic charges. Again, as in electrostatics, we can use Gauss's theorem applied to an elementary flat box surrounding the boundary between two magnetic media as in Fig. 4.8 to show that $\int\mathbf{B}.d\mathbf{S} = 0$ over the surface of the box. If the height of the

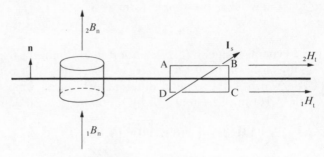

FIG. 4.8. Boundary conditions at the surface between two media. $I_s$ is the surface current per unit width normal to the plane of the circuit ABCDA.

box is very small compared with its cross-section, the only contributions to $\int \mathbf{B} \cdot d\mathbf{S}$ come from the components of $\mathbf{B}$ normal to the boundary. Hence we have

$$_1B_n = {}_2B_n. \tag{4.29}$$

The boundary conditions for $\mathbf{H}$ are found by using Ampère's law applied to a small rectangular circuit ABCDA whose sides BC, AD are very small compared with AB, CD. If there is a surface current $I_s$ per unit length of the surface normal to the circuit, then eqn (4.27) leads immediately to the result

$$_2H_t - {}_1H_t = I_s$$

for the components of $H_t$ normal to $I_s$, while the components of $H_t$ parallel to $I_s$ are continuous. If the surface current is represented by a vector $\mathbf{I}_s$, the boundary conditions can be expressed as

$$\mathbf{n} \wedge ({}_2\mathbf{H} - {}_1\mathbf{H}) = \mathbf{I}_s, \tag{4.30}$$

where $\mathbf{n}$ is a unit vector normal to the surface as in Fig. 4.8. Similarly, eqn (4.29) can be written as

$$\mathbf{n} \cdot ({}_2\mathbf{B} - {}_1\mathbf{B}) = 0. \tag{4.31}$$

Thus the normal components of $\mathbf{B}$ are always continuous, while the tangential components of $\mathbf{H}$ are continuous if there is no surface current. These boundary conditions are similar to those in electrostatics, but note that the formal equivalence here is between $\mathbf{B}$ and $\mathbf{D}$, and between $\mathbf{H}$ and $\mathbf{E}$. This equivalence can be carried a stage further, since if no currents are present, we have curl $\mathbf{H} = 0$, and we can therefore write

$$\mathbf{H} = -\text{grad } \phi, \tag{4.32}$$

which is analogous to $\mathbf{E} = -\text{grad } V$. Eqn (4.32) is true both *in vacuo* and in a magnetizable medium, our earlier eqn (4.10) being limited to the special case of a vacuum.

## 4.4. Solution of magnetostatic problems

In many materials it is found that the magnetization $\mathbf{M}$ is linearly proportional to the magnetic field strength $\mathbf{H}$, so that we can write

$$\mathbf{M} = \chi\mathbf{H}. \tag{4.33}$$

Here $\chi$ is known as the magnetic susceptibility; if we wish to distinguish it from the electric susceptibility (§ 1.5) we may write them as $\chi_m$ and $\chi_e$ respectively, but where there is no danger of confusion the subscripts may be omitted. Representative values of $\chi$ for different substances vary widely, and will be discussed in Chapter 6. At ordinary temperatures $\chi$ is small and independent of $\mathbf{H}$ for most substances, the exceptions being ferromagnetics, where $\chi$ is large and very dependent on field strength; $\mathbf{M}$ may even be non-zero when $\mathbf{H} = 0$.

From eqn (4.25) we have

$$\mathbf{B} = \mu_0(\mathbf{H} + \mathbf{M}) = \mu_0\mathbf{H}(1 + \chi)$$

$$= \mu_r\mu_0\mathbf{H} = \mu\mathbf{H}, \qquad (4.34)$$

where the quantity $\mu$ is known as the magnetic permeability at the medium, and $\mu_0$ is the magnetic permeability of vacuum. The quantity $\mu_r$ given by

$$\mu/\mu_0 = \mu_r = 1 + \chi \qquad (4.35)$$

is known as the relative magnetic permeability of the medium. It is clear that $\mu_r$ plays a similar role in magnetostatics to that played by the relative permittivity $\epsilon_r$ in electrostatics.

Since $\mathrm{div}\,\mathbf{B} = 0$, we have, when $\mu$ is independent of position,

$$\mathrm{div}\,\mathbf{B} = \mathrm{div}(\mu\mathbf{H}) = -\mu\,\mathrm{div}\,\mathrm{grad}\,\phi$$

$$= -\mu\,\nabla^2\phi = 0,$$

or

$$\nabla^2\phi = 0, \qquad (4.36)$$

showing that the magnetostatic potential obeys Laplace's equation. The theory of Chapter 2 may therefore be adapted to magnetostatic problems, and we shall illustrate this by treating a special case.

The problem of a polarizable sphere in a uniform electric field was solved by means of spherical harmonics in § 2.4. The corresponding magnetic problem may be approached in the same way, but we shall extend it slightly by assuming that $\mathbf{M}$ is not necessarily proportional to $\mathbf{H}$, though still parallel to it. Then, from eqn (4.25),

$$\mathrm{div}\,\mathbf{B} = -\mu_0\,\mathrm{div}\,\mathrm{grad}\,\phi + \mu_0\,\mathrm{div}\,\mathbf{M} = 0,$$

whence

$$\nabla^2\phi = \mathrm{div}\,\mathbf{M}. \qquad (4.37)$$

In the corresponding electrostatic case we found that the sphere was uniformly polarized, and we shall assume that this is true also in the magnetic case. Then $\mathrm{div}\,\mathbf{M} = 0$, and the potentials required are solutions of Laplace's equation.

In Fig. 4.9, the potentials inside and outside the sphere are assumed to be

$$\phi_1 = -H_1 r \cos\theta \qquad (r < a),$$

$$\phi_2 = -H_0 r \cos\theta + A r^{-2} \cos\theta \qquad (r > a).$$

As in the electrostatic case, there can be no term in $r^{-2}\cos\theta$ inside the sphere, since it would become infinite at $r = 0$; thus the field inside is

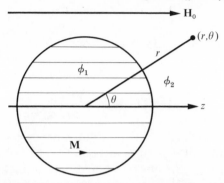

FIG. 4.9. A magnetizable sphere of radius $a$ in a uniform field $H_0$. $\mu_1$, $\mu_2$ are the relative permittivities inside and outside the sphere.

uniform and equal to $H_1$. Outside the sphere the field at large distances is uniform and equal to $H_0$; we take $H_0$ to be given, so that $H_1$ and $A$ are the unknowns to be determined from the boundary conditions.

The first boundary condition, that the tangential components of $\mathbf{H}$ be continuous at the boundary, is equivalent to making $\phi_1 = \phi_2$ at $r = a$, giving

$$H_1 = H_0 - Aa^{-3}.$$

We now assume that the magnetization of the sphere is parallel to $\mathbf{H}_0$, but that it has a permanent component $\mathbf{M}_0$ as well as an induced component $(\mu_1 - 1)\mathbf{H}_1$. Then the magnetization inside the sphere is

$$\mathbf{M}_1 = (\mu_1 - 1)\mathbf{H}_1 + \mathbf{M}_0,$$

so that $\mathbf{B}_1 = \mu_1\mu_0\mathbf{H}_1 + \mu_0\mathbf{M}_0$, while outside $\mathbf{B}_2 = \mu_2\mu_0\mathbf{H}_2$. Then the radial components of $\mathbf{B}$ are

$$-\mu_1\mu_0(\partial\phi_1/\partial r) + \mu_0\mathbf{M}_0 \cos\theta = \mu_0(\mu_1\mathbf{H}_1 + \mathbf{M}_0)\cos\theta \quad \text{(inside)}$$

and

$$-\mu_2\mu_0(\partial\phi_2/\partial r) = \mu_2\mu_0(\mathbf{H}_0 + 2Ar^{-3})\cos\theta \quad \text{(outside)}.$$

On equating the two at $r = a$ and eliminating $A$, we have

$$\mathbf{H}_1 = \mathbf{H}_0 \frac{3\mu_2}{\mu_1 + 2\mu_2} - \mathbf{M}_0 \frac{1}{\mu_1 + 2\mu_2}. \tag{4.38a}$$

Here the first term is the analogue of eqn (2.43) and the second arises from the permanent moment $\mathbf{M}_0$. The difference between $\mathbf{H}_1$ and $\mathbf{H}_0$ can be ascribed (cf. Appendix B) to a 'demagnetizing field' $\mathbf{H}_d = \mathbf{H}_1 - \mathbf{H}_0$ which is proportional to the difference between the actual magnetization $\mathbf{M}_1$ and the value $\mathbf{M}_2 = (\mu_2 - 1)\mathbf{H}_0$ which would have existed there if the sphere were absent and replaced by the medium $\mu_2$. The difference between the two is

$$\mathbf{M}_1 - \mathbf{M}_2 = \mathbf{H}_0(\mu_1 - \mu_2)\frac{1+2\mu_2}{\mu_1+2\mu_2} + \mathbf{M}_0\frac{1+2\mu_2}{\mu_1+2\mu_2}, \qquad (4.38b)$$

and on combining this with eqn (4.38a) we obtain

$$\mathbf{H}_d = \mathbf{H}_1 - \mathbf{H}_0 = -\frac{\mathbf{M}_1 - \mathbf{M}_2}{1+2\mu_2}. \qquad (4.38c)$$

This result is the analogue of eqn (2.46), where no permanent moment was assumed. If the sphere is in a vacuum, then $\mathbf{M}_2 = 0$ and $\mu_2 = 1$, so that $\mathbf{H}_d = -\mathbf{M}_1/3$. These results obviously hold both in the special cases of no induced moment ($\mu_1 = 1$) or no permanent moment ($\mathbf{M}_0 = 0$).

We conclude the discussion of magnetostatics by finding a general expression for the magnetic potential due to a magnetized substance. In an element $d\tau$ the dipole moment is $\mathbf{M}\,d\tau$, and the potential eqn (4.13) may be written in the form

$$\phi = \frac{1}{4\pi}\int \mathbf{M}\cdot\text{grad}(1/r)\,d\tau, \qquad (4.39)$$

where the differentiation is here with respect to the coordinates of the volume element $d\tau$ (cf. eqn (1.11b)). Then, by a vector transformation similar to that used in deriving eqn (1.16), we find

$$\phi = \frac{1}{4\pi}\int\frac{1}{r}(\mathbf{M}\cdot d\mathbf{S}) - \frac{1}{4\pi}\int\frac{1}{r}(\text{div}\,\mathbf{M}\,d\tau), \qquad (4.40)$$

showing that the potential can be attributed to an apparent surface distribution of magnetic charge of surface density $M\cos\theta$, where $\theta$ is the angle $\mathbf{M}$ makes with the normal $d\mathbf{S}$ to the surface, and an apparent volume distribution of volume density $-\text{div}\,\mathbf{M}$, which, since div $\mathbf{B} = 0$, is equal to $+\text{div}\,\mathbf{H}$. Thus the field lines of $\mathbf{H}$ terminate on the polarization charges, while the field lines of $\mathbf{B}$ are all closed loops since there are no real magnetic charges. If the substance is uniformly magnetized, div $\mathbf{M} = 0$ and there are no volume charges, but there is a surface distribution corresponding to the 'magnetic poles' of classical magnetic theory.

## 4.5. Steady currents in uniform magnetic media

In § 4.3 the effects of the presence of a magnetizable medium were considered, and it was shown that Ampère's law takes the simple form

$$\text{curl}\,\mathbf{H} = \mathbf{J} \qquad (4.26)$$

or in integral form

$$\int \mathbf{H}\cdot d\mathbf{s} = I. \qquad (4.27)$$

It follows from these equations that in an infinite uniform magnetizable medium of relative permeability $\mu_r$ the value of the magnetic field strength $\mathbf{H}$ is unaltered by the presence of the medium, provided the current flow is unaltered, and is independent of $\mu_r$. Returning to eqn (4.3), which holds *in vacuo* where $\mathbf{B} = \mu_0\mathbf{H}$, we see that it may be rewritten in terms of $\mathbf{H}$ as

$$d\mathbf{H} = \frac{1}{4\pi r^3} I(d\mathbf{s} \wedge \mathbf{r}) \tag{4.41}$$

$$= \frac{1}{4\pi r^3} (\mathbf{J} \wedge \mathbf{r}) \, d\tau, \tag{4.42}$$

where the first form refers to a current $I$ in an element $d\mathbf{s}$ and the second to a current density $\mathbf{J}$ in an element $d\tau$. From the preceding remarks it is obvious that these equations are unaltered in a magnetizable medium; they are known as the law of Biot and Savart.

Since $(\mathbf{r}/r^3) = -\text{grad}(1/r)$, we may also express these relations in the form

$$d\mathbf{H} = -\frac{1}{4\pi} I\{d\mathbf{s} \wedge \text{grad}(1/r)\} = -\frac{1}{4\pi} \{\mathbf{J} \wedge \text{grad}(1/r)\} \, d\tau. \tag{4.43}$$

From the vector identity (A.13) of Appendix A,

$$\text{curl}(\mathbf{J}/r) = (1/r)\text{curl}\,\mathbf{J} - \mathbf{J} \wedge \text{grad}(1/r) = -\mathbf{J} \wedge \text{grad}(1/r), \tag{4.44}$$

since the curl operator acts only on the field point and $\mathbf{J}$ is invariant in this operation. Hence we can write in integral form

$$\mathbf{H} = \frac{1}{4\pi} \int \text{curl}(\mathbf{J}/r) \, d\tau \tag{4.45}$$

$$= \frac{1}{4\pi} \text{curl} \int (\mathbf{J}/r) \, d\tau, \tag{4.46}$$

where the order of the curl operation and the integration can be interchanged because the integration is over the current distribution, while the curl operation refers to the field point.

We have seen that in a region where there is no current flow, $\mathbf{H}$ can be derived from a scalar potential (eqn (4.32)). It is useful to find another potential function which is not subject to this limitation and which is valid in regions where $\mathbf{J}$ is finite. Such a potential must satisfy the fundamental relation div $\mathbf{B} = 0$ given by eqn (4.6), which, as we have seen, is valid both in the presence of currents and of magnetizable media. From the vector identity div curl $\equiv 0$, (see eqn (A.20) of Appendix A), it is clear that (4.6) is automatically satisfied if we write

$$\mathbf{B} = \text{curl}\,\mathbf{A}, \tag{4.47}$$

where $\mathbf{A}$ is known as the 'magnetic vector potential'. Since $\mathbf{A}$ is essentially derived from $\mathbf{B}$ by an integration, this definition is not complete, for we could add another term (equivalent to a constant of integration) such as grad $\psi$, and still have, using eqn (A.19),

$$\text{curl}(\mathbf{A} + \text{grad } \psi) = \text{curl } \mathbf{A} = \mathbf{B}. \tag{4.48}$$

We therefore add a supplementary condition, and define $\mathbf{A}$ by the relations

$$\text{curl } \mathbf{A} = \mathbf{B}, \qquad \text{div } \mathbf{A} = 0. \tag{4.49}$$

The magnetic vector potential enables us, using Stokes's theorem, to derive expressions in terms of line integrals for some quantities which hitherto have appeared only as surface integrals. The magnetic flux $\Phi$ through a circuit (eqn (4.21)) becomes

$$\Phi = \int (\mathbf{B} \cdot d\mathbf{S}) = \int (\text{curl } \mathbf{A}) \cdot d\mathbf{S} = \int (\mathbf{A} \cdot d\mathbf{s}), \tag{4.50}$$

and, if the circuit carries a current $I$, its potential energy is

$$U_P = -I\Phi = -I\oint (\mathbf{A} \cdot d\mathbf{s}). \tag{4.51}$$

For a volume current density $\mathbf{J}$, using the relation $\mathbf{I} \, d\mathbf{s} = \mathbf{J} \, d\tau$, we have

$$U_P = -\int (\mathbf{A} \cdot \mathbf{J}) \, d\tau. \tag{4.52}$$

We shall use these relations in the next chapter.

A differential equation for $\mathbf{A}$ can be found by combining eqns (4.26) and (4.48). In a medium where $\mathbf{B} = \mu\mathbf{H}$ we obtain

$$\mathbf{J} = \text{curl } \mathbf{H} = \text{curl}(\mathbf{B}/\mu) = \text{curl}\left(\frac{1}{\mu} \text{curl } \mathbf{A}\right). \tag{4.53}$$

If the medium is uniform, so that $\mu$ does not vary with position, then using the vector relation (A.22) (Appendix A), we have

$$\mu\mathbf{J} = \text{curl}(\text{curl } \mathbf{A}) = \text{grad div } \mathbf{A} - \nabla^2\mathbf{A},$$

and since div $\mathbf{A} = 0$,

$$\nabla^2\mathbf{A} = -\mu\mathbf{J}. \tag{4.54}$$

This equation is similar to Poisson's equation (2.1), except that the operand is a vector instead of a scalar quantity. This should not cause any difficulty if one remembers that eqn (4.54) implies that each of the components of the vector separately must satisfy the differential equation. A formal solution similar to eqn (2.6) can then be found for each of the components

$$A_x = \frac{\mu}{4\pi} \int \frac{J_x}{r} \, d\tau, \text{ etc.,}$$

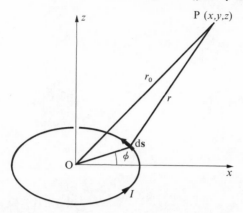

FIG. 4.10. The magnetic vector potential due to a plane circular coil carrying a current $I$.

which may be expressed in vector form as

$$A = \frac{\mu}{4\pi} \int \frac{J}{r} \, d\tau. \tag{4.55}$$

Since $H = B/\mu = (1/\mu)\mathrm{curl}\, A$ if $\mu$ is constant in space, we have a solution for $A$ which is obviously consistent with eqn (4.46).

For a current circuit carrying a current $I$, the solution for $A$ is

$$A = \frac{\mu I}{4\pi} \int \frac{ds}{r}, \tag{4.56}$$

and we can use this to find the value of the magnetic vector potential for an elementary current circuit and hence for a point dipole. We take a small circular current of radius $a$ and calculate the value of $A$ at a point P. For convenience we use Cartesian coordinates, whose $z$-axis is normal to the plane of the coil and whose origin is at the centre of the coil. Then P is at $r_0 \equiv (x, y, z)$, and $r$ is the distance of P from an element $ds$ of the circuit. In terms of the azimuthal angle $\phi$ (see Fig. 4.10), the components of $ds$ are $(-a \, d\phi \sin \phi, \; a \, d\phi \cos \phi, \; 0)$ so that $A_z = 0$. Since

$$r^2 = (x - a \cos \phi)^2 + (y - a \sin \phi)^2 + z^2$$

$$= r_0^2 - 2ax \cos \phi - 2ay \sin \phi + a^2,$$

and $a \ll r$, we have

$$\frac{1}{r} = \frac{1}{r_0} + \frac{ax \cos \phi + ay \sin \phi}{r_0^3} + \cdots$$

and hence

$$A_x = \frac{\mu I}{4\pi} \int_0^{2\pi} \frac{-a \sin \phi \, d\phi}{r}$$

$$= -\frac{a\mu I}{4\pi} \int_0^{2\pi} \left(\frac{1}{r_0} + \frac{ax \cos \phi + ay \sin \phi}{r_0^3}\right) \sin \phi \, d\phi$$

$$= -\frac{\mu}{4\pi} \left\{ \pi a^2 I\left(\frac{y}{r_0^3}\right)\right\}.$$

Similarly,

$$A_y = \frac{\mu I}{4\pi} \int_0^{2\pi} \frac{a \cos \phi \, d\phi}{r} = +\frac{\mu}{4\pi} \left\{ \pi a^2 I\left(\frac{x}{r_0^3}\right)\right\}.$$

The dipole moment equivalent to the current circuit can be represented by a vector $\mathbf{m}$ of size $\pi a^2 I$ directed along the $z$-axis, and the components of $\mathbf{A}$ are then proportional to those of the vector $\mathbf{m} \wedge \mathbf{r}_0$. Hence, dropping the subscript on $\mathbf{r}_0$, we can write ·

$$\mathbf{A} = \frac{\mu}{4\pi r^3} (\mathbf{m} \wedge \mathbf{r}) = -\frac{\mu}{4\pi}\left(\mathbf{m} \wedge \text{grad}_P \frac{1}{r}\right), \tag{4.57}$$

where the subscript P denotes that O is fixed and P is the moveable point. From eqns (A.15) and (A.20) it can be seen that eqn (4.57) satisfies the relation $\text{div} \, \mathbf{A} = 0$. It can readily be verified that the lines of constant $A$ are circles about the $z$-axis, and it is generally true that for simple current circuits the lines of $\mathbf{A}$ are similar in their geometry to those of the current flow, as in the case just discussed.

Eqn (4.57) may be compared with eqn (4.14). The vector potential is proportional to the vector product $(\mathbf{m} \wedge \mathbf{r})$, the scalar potential to the scalar product $(\mathbf{m} \cdot \mathbf{r})$; in addition the quantity $\mu$ appears in the vector potential but not in the scalar potential because the former is connected with $\mathbf{B}$ and the latter with $\mathbf{H}$.

In § 2.3 a general expression was found for the equivalent electric dipole moment of a distributed charge, and some applications on the atomic scale were given. A similar expression may be found for the equivalent magnetic dipole moment of a current distribution, and it is convenient to do this from the formula for the energy in a uniform flux density $\mathbf{B}$, using the magnetic vector potential. For a uniform $\mathbf{B}$,

$$\mathbf{A} = \tfrac{1}{2}(\mathbf{B} \wedge \mathbf{r}) \tag{4.58}$$

as can readily be verified by calculating the components of curl $\mathbf{A}$ in Cartesian coordinates. Then from eqn (4.52) the potential energy of a distributed current is

$$U_P = -\int (\mathbf{A} \cdot \mathbf{J})\, d\tau = -\tfrac{1}{2}\int \{(\mathbf{B} \wedge \mathbf{r}) \cdot \mathbf{J}\}\, d\tau$$

$$= -\tfrac{1}{2}\int \mathbf{B} \cdot (\mathbf{r} \wedge \mathbf{J})\, d\tau = -\mathbf{B} \cdot \int \tfrac{1}{2}(\mathbf{r} \wedge \mathbf{J})\, d\tau,$$

where **B** can be taken out of the integral because it is constant and independent of the space coordinates. The equivalent magnetic dipole moment may be found by equating this expression for the energy to that in eqn (4.19), $U_P = -\mathbf{m} \cdot \mathbf{B}$, giving

$$\mathbf{m} = \int \tfrac{1}{2}(\mathbf{r} \wedge \mathbf{J})\, d\tau. \tag{4.59}$$

We may check that this agrees with our earlier definition of the dipole moment equivalent to a current circuit since, for the latter, eqn (4.59) becomes

$$\mathbf{m} = I\int \tfrac{1}{2}(\mathbf{r} \wedge d\mathbf{s}) = I\int d\mathbf{S} = I\mathbf{S},$$

in agreement with eqn (4.7b).

The formulae derived in this chapter (with the exception of the relation $U = \int \tfrac{1}{2}(\mathbf{B} \cdot \mathbf{H})\, d\tau$ which is derived in § 5.5) are summarized in Table 4.1 in a form which gives a ready comparison with electrostatics. It is assumed that the permeability $\mu$ is uniform, isotropic and independent of **H** and that there is no spontaneous magnetization.

### Non-linear magnetic media

In a ferromagnetic medium where $\mu$ is not independent of **H**, formulae must be derived using the relation $\mathbf{B} = \mu_0(\mathbf{H}+\mathbf{M})$ instead of $\mathbf{B} = \mu\mathbf{H}$. Instead of eqns (4.53) and (4.54) we have

$$\text{grad div } \mathbf{A} - \nabla^2 \mathbf{A} = \mu_0 \text{ curl}(\mathbf{H}+\mathbf{M}) = \mu_0 \mathbf{J} + \mu_0 \text{ curl } \mathbf{M},$$

so that

$$\nabla^2 \mathbf{A} = -\mu_0 \mathbf{J} - \mu_0 \text{ curl } \mathbf{M}. \tag{4.60}$$

This has the formal solution

$$\mathbf{A} = \frac{\mu_0}{4\pi}\int \frac{\mathbf{J}}{r}\, d\tau + \frac{\mu_0}{4\pi}\int \frac{\text{curl } \mathbf{M}}{r}\, d\tau, \tag{4.61}$$

where, from eqn (4.23), we see that the second term is equivalent to

$$\frac{\mu_0}{4\pi}\int \frac{\mathbf{J}_m}{r}\, d\tau$$

and is thus similar in form to the first term, $\mathbf{J}_m$ being the equivalent current density of the magnetization. By using the vector relations (A.13)

TABLE 4.1
*Comparison of various formulae*

| Electrostatics | Magnetostatics | Currents |
|---|---|---|
| $\mathbf{D} = \epsilon_0\mathbf{E} + \mathbf{P}$ | $\mathbf{B} = \mu_0(\mathbf{H} + \mathbf{M})$ | $\mathbf{H} = (\mathbf{B}/\mu_0) - \mathbf{M}$ |
| $= \epsilon_r\epsilon_0\mathbf{E} = \epsilon\mathbf{E}$ | $= \mu_r\mu_0\mathbf{H} = \mu\mathbf{H}$ | $= \mathbf{B}/\mu_r\mu_0 = \mathbf{B}/\mu$ |
| $\text{div }\mathbf{D} = \rho_e$ | $\text{div }\mathbf{B} = 0$ | $\text{div }\mathbf{B} = 0$ |
| $\text{curl }\mathbf{E} = 0$ | $\text{curl }\mathbf{H} = 0$ | $\text{curl }\mathbf{H} = \mathbf{J}$ |
| $\mathbf{E} = -\text{grad }V$ | $\mathbf{H} = -\text{grad }\phi$ | $\mathbf{B} = \text{curl }\mathbf{A}$ |
| $\nabla^2 V = -\rho_e/\epsilon$ | $\nabla^2\phi = 0$ | $\nabla^2\mathbf{A} = -\mu\mathbf{J}$ |
| $V = \dfrac{1}{4\pi\epsilon}\displaystyle\int\dfrac{\rho_e\,\mathrm{d}\tau}{r}$ | | $\mathbf{A} = \dfrac{\mu}{4\pi}\displaystyle\int\dfrac{\mathbf{J}\,\mathrm{d}\tau}{r}$ |
| $\mathbf{D} = \dfrac{q\mathbf{r}}{4\pi r^3}$ | | $\mathbf{H} = \dfrac{I\,\mathrm{d}\mathbf{s}\wedge\mathbf{r}}{4\pi r^3}$ |
| $\mathbf{p} = \displaystyle\int\rho_e\mathbf{r}\,\mathrm{d}\tau$ | | $\mathbf{m} = \displaystyle\int\tfrac{1}{2}(\mathbf{r}\wedge\mathbf{J})\,\mathrm{d}\tau$ |
| $V = \dfrac{\mathbf{p}\cdot\mathbf{r}}{4\pi\epsilon r^3}$ | $\phi = \dfrac{\mathbf{m}\cdot\mathbf{r}}{4\pi r^3}$ | $\mathbf{A} = \dfrac{\mu}{4\pi r^3}(\mathbf{m}\wedge\mathbf{r})$ |
| $U_P = -\mathbf{p}\cdot\mathbf{E}$ | $U_P = -\mathbf{m}\cdot\mathbf{B}$ | $U_P = -I\Phi = -I\displaystyle\int\mathbf{A}\cdot\mathrm{d}\mathbf{s}$ |
| $U_P = \displaystyle\int\rho_e V\,\mathrm{d}\tau$ | | $U_P = -\displaystyle\int(\mathbf{A}\cdot\mathbf{J})\,\mathrm{d}\tau$ |
| $U = \displaystyle\int\tfrac{1}{2}\mathbf{D}\cdot\mathbf{E}\,\mathrm{d}\tau$ | | $U = \displaystyle\int\tfrac{1}{2}\mathbf{B}\cdot\mathbf{H}\,\mathrm{d}\tau$ |
| $\mathbf{F} = q\mathbf{E} = \displaystyle\int\rho_e\mathbf{E}\,\mathrm{d}\tau$ | | $\mathbf{F} = I\displaystyle\int(\mathrm{d}\mathbf{S}\wedge\mathbf{B}) = \displaystyle\int(\mathbf{J}\wedge\mathbf{B})\,\mathrm{d}\tau$ |
| $\mathbf{F} = (\mathbf{p}\cdot\text{grad})\mathbf{E}$ | $\mathbf{F} = (\mathbf{m}\cdot\text{grad})\mathbf{B}$ | $\mathbf{F} = I\displaystyle\int(\mathrm{d}\mathbf{S}\cdot\text{grad})\mathbf{B}$ |
| $\mathbf{T} = \mathbf{p}\wedge\mathbf{E}$ | $\mathbf{T} = \mathbf{m}\wedge\mathbf{B}$ | $\mathbf{T} = I\displaystyle\int(\mathrm{d}\mathbf{S}\wedge\mathbf{B})$ |

and (A.24) (from Appendix A) we have

$$\frac{\mu_0}{4\pi}\int\frac{\text{curl }\mathbf{M}}{r}\,\mathrm{d}\tau = \frac{\mu_0}{4\pi}\int\text{curl}\left(\frac{\mathbf{M}}{r}\right)\mathrm{d}\tau + \frac{\mu_0}{4\pi}\int\mathbf{M}\wedge\text{grad}_0\left(\frac{1}{r}\right)\mathrm{d}\tau$$

$$= \frac{\mu_0}{4\pi}\int\frac{\mathbf{M}\wedge\mathrm{d}\mathbf{S}}{r} + \frac{\mu_0}{4\pi}\int\mathbf{M}\wedge\text{grad}_0\left(\frac{1}{r}\right)\mathrm{d}\tau. \qquad (4.62)$$

In the second term, the vector $\mathbf{r}$ is drawn from a fixed field point P to a variable point O where the element of dipole moment $\mathbf{M}\,\mathrm{d}\tau$ is located, and the integration is over the magnetization. This is indicated by the subscript O in $\text{grad}_0(1/r)$. In fact the second term is equivalent to eqn (4.57), with $\mathbf{M}\,\mathrm{d}\tau$ instead of $\mathbf{m}$, since there P is the variable point and O a fixed point, and

$$\text{grad}_0(1/r) = -\text{grad}_P(1/r).$$

This change of sign is exactly the same as that in § 1.4, eqns (1.11a) and (1.11b); we note also that in the latter equations for the scalar potential $V$ we have the scalar product $(\mathbf{p}\cdot\mathrm{grad})$, while in eqn (4.62) for the vector potential $\mathbf{A}$ we have the vector product $(\mathbf{M}\wedge\mathrm{grad})$.

The first term in eqn (4.62) is the contribution to $\mathbf{A}$ from a surface layer of magnetization, which is equivalent to a surface current density $(\mathbf{I}_s)_m$ tangential to the surface and normal to $\mathbf{M}$, which can be written as

$$(\mathbf{I}_s)_m = \mathbf{M}\wedge\mathbf{n}, \tag{4.63}$$

where $\mathbf{n}$ is a unit vector normal to the surface. In such a surface layer the magnetization must be parallel to the surface (unlike the magnetic shell considered in § 4.2), a phenomenon which does not often arise in practice.

### Boundary conditions

The relations which express the boundary conditions at a surface between two dielectric or two magnetic media are summarized in Table 4.2. They include the cases where there is a surface density $\sigma_e$ of true electric charge, and a true surface current of $\mathbf{I}_s$ per unit width. The boundary conditions for $\mathbf{D}$ and $\mathbf{E}$ come from eqns (1.26), (1.27), and (1.28) expressed in vectorial form (see Problem 1.13) with $\mathbf{n}$ as a unit vector normal to the surface; those for $\mathbf{B}$ and $\mathbf{H}$ come from eqns (4.30) and (4.31). Since $\mathbf{B} = \mathrm{curl}\,\mathbf{A}$, the boundary conditions for $\mathbf{B}$ and $\mathbf{H}$ give relations only for the spatial derivatives of $\mathbf{A}$. However, it can be shown

TABLE 4.2
*Summary of boundary conditions*

| | |
|---|---|
| Surface charge, $\sigma_e$ per unit area (single layer) | $\mathbf{n}\cdot(_2\mathbf{D}-_1\mathbf{D}) = \sigma_e$ $\mathbf{n}\wedge(_2\mathbf{E}-_1\mathbf{E}) = 0$ $_2V-_1V = 0$ |
| Surface current, $\mathbf{I}_s$ per unit width (single layer) | $\mathbf{n}\cdot(_2\mathbf{B}-_1\mathbf{B}) = 0$ $\mathbf{n}\wedge(_2\mathbf{H}-_1\mathbf{H}) = \mathbf{I}_s$ $_2\mathbf{A}-_1\mathbf{A} = 0$ |
| Electric double layer, dipole moment $\mathbf{P}_s$ per unit area | $_2V-_1V = (\mathbf{P}_s\cdot\mathbf{n})/\varepsilon_0$ |
| Magnetic double layer, magnetic moment $\mathbf{M}_s$ per unit area | $_2\phi-_1\phi = \mathbf{M}_s\cdot\mathbf{n}$ $_2\mathbf{A}-_1\mathbf{A} = \mu_0(\mathbf{M}_s\wedge\mathbf{n})$ |

The unit vector $\mathbf{n}$, the electric dipole moment $\mathbf{P}_s$, and the magnetic dipole moment $\mathbf{M}_s$ per unit area of the double layer are directed from side 1 to side 2.

(see Stratton 1941) that both the normal and tangential components of **A** are continuous through a single layer of surface current; for a magnetic double layer (a 'magnetic shell') this is true for the normal components of **A**, and also for the tangential components of **A** if the magnetic moment is normal to the surface. On the other hand the scalar potential $\phi$ has different values on the two sides of a magnetic shell (see § 4.2 and cf. Problem 2.19), the difference being equal to the magnetic dipole moment per unit area.

## 4.6. Calculation of the magnetic field strengths of simple circuits

The magnetic field strength of a circuit of simple shape may be found in a number of ways, the chief of which are:

(1) use of eqn (4.27). This is possible only when the field strength has a high degree of symmetry as in the rather similar use of Gauss's theorem in electrostatics;

(2) use of the potential of the equivalent magnetic shell, eqn (4.12);

(3) use of the Biot–Savart law (eqn (4.41));

(4) use of the magnetic vector potential (eqns (4.49) and (4.56)).

Simple illustrations will be given of the use of the various methods.

The field strength due to an infinite straight wire carrying a current $I$ was calculated by method (3) in §4.1, but is quickly found by method (1). By symmetry, **H** is a function only of the radial distance from the wire, and by applying eqn (4.27) to a circle of radius $r$ about the wire we find $\int \mathbf{H} \cdot \mathrm{d}s = 2\pi r H = I$. Hence the azimuthal component of **H** is $I/2\pi r$; since this depends only on $r$, the lines of force are concentric circles about the wire, and no other components of **H** exist.

If the wire has radius $a$, and the current density is uniform, the field inside the wire can be found by a similar application of eqn (4.27). In this case the current threading a circle of radius $r$ is $I(r^2/a^2)$, so that $\int \mathbf{H} \cdot \mathrm{d}s = 2\pi r H = I(r^2/a^2)$ and $H = Ir/2\pi a^2$, showing that $H$ increases linearly from the centre to the surface of the wire.

The case of a straight wire serves also as a simple example where the vector potential can be found by solving eqn (4.54). Taking the axis of the wire to be along the $z$-axis, it is obvious that the only component of current density is $J_z$, and hence the only component of **A** is $A_z$, so that the lines of **A** are parallel to the wire. Inside the wire (permeability $\mu_r \mu_0 = \mu_1$)

$$\nabla^2 A_z = -\mu_1 J_z = -\mu_1 I/\pi a^2.$$

Since $J_z$ is independent of $z$ and $\theta$ (making use of cylindrical coordinates $r$, $\theta$, $z$), so also is $A_z$ and the differential equation becomes

$$\frac{1}{r}\frac{\partial}{\partial r}\left(r\frac{\partial A_z}{\partial r}\right) = -\frac{\mu_1 I}{\pi a^2}.$$

Integration gives

$$r\left(\frac{\partial A_z}{\partial r}\right) = -\frac{\mu_1 I r^2}{2\pi a^2},$$

where the constant of integration vanishes, because $r(\partial A_z/\partial r) = 0$ at $r = 0$ since we cannot have a discontinuity in $A_z$ on crossing the axis. A second integration gives

$$A_z = \frac{\mu_1 I}{4\pi}\left(1 - \frac{r^2}{a^2}\right) \quad \text{(inside)}, \tag{4.64}$$

where for convenience we make $A_z = 0$ at $r = a$.

Outside the wire (assuming a medium of permeability $\mu_2$)

$$\frac{1}{r}\frac{\partial}{\partial r}\left(r\frac{\partial A_z}{\partial r}\right) = 0,$$

whence $\partial A_z/\partial r = c/r$, and

$$A_z = c\,\ln(r/a) \quad \text{(outside)},$$

where the second constant of integration is chosen to make $A_z$ continuous at the boundary, that is, $A_z = 0$ at $r = a$. The constant $c$ is determined by the boundary condition for $\partial A_z/\partial r$ at $r = a$. By writing $r^2 = x^2 + y^2$ and finding the components of curl $\mathbf{A}$, it can be verified that

$$B_x = (\partial A_z/\partial r)(y/r), \qquad B_y = -(\partial A_z/\partial r)(x/r),$$

so that $B_\theta = -\partial A_z/\partial r$. Since $B_\theta$ is purely tangential (the other components are zero), the boundary condition is that $H_\theta$ must be continuous at $r = a$, and it is easily shown then that

$$A_z = -\frac{\mu_2 I}{2\pi}\ln\left(\frac{r}{a}\right) \quad \text{(outside)}. \tag{4.65}$$

Methods (2) and (3) may be compared in finding the magnetic field strength on the axis of a plane circular coil. We assume that the coil has $n$ turns each carrying a current $I$, and the radius of the coil is $a$. Then, at a point on the axis a distance $z$ away, the solid angle subtended by the coil is

$$2\pi\left\{1 - \frac{z}{(z^2 + a^2)^{\frac{1}{2}}}\right\}$$

(this formula may be verified by the integration $\Omega = \int (d\mathbf{S}\cdot\mathbf{r}/r^3)$, taken over the plane surface bounded by the coil). Hence

$$H_z = -d\phi/dz = -(nI/4\pi)(d\Omega/dz) = \tfrac{1}{2}nIa^2/(z^2 + a^2)^{\frac{3}{2}}. \tag{4.66}$$

FIG. 4.11. The magnetic field due to a circular coil at a point on its axis.

When $a \ll z$, this formula is the same as that for the field of a dipole of moment $\mathbf{m} = nI(\pi a^2)$ at a point on its axis (cf. eqn (4.7b)).

In applying the formula of Biot and Savart we consider first the field strength $d\mathbf{H}$ due to an element of wire $d\mathbf{s}$, as in Fig. 4.11. Since $d\mathbf{H}$ is normal both to $d\mathbf{s}$ and to $\mathbf{r}$, it will have the direction shown in the figure. On integrating round the coil it is clear that the sum of all the components normal to the axis will be zero. The components parallel to the axis sum to

$$H_z = nI \int (\cos \psi / 4\pi r^2) \, ds = 2\pi a nI \cos \psi / 4\pi r^2 = \tfrac{1}{2} nIa^2 / (z^2 + a^2)^{\frac{3}{2}}.$$

This formula may be extended to the case of a solenoid with $m$ turns per unit length, uniformly wound round a cylinder of radius $a$ (Fig. 4.12). If the turns are closely wound, we may regard them as being equivalent to a uniform current flowing round the cylinder, so that an element $dz$ of it forms a plane coil with a current $mI \, dz$. At the point O this gives a field along the axis equal to

$$dH_z = \tfrac{1}{2} mIa^2 \, dz / (z^2 + a^2)^{\frac{3}{2}} = -\tfrac{1}{2} mI \sin \phi \, d\phi,$$

FIG. 4.12. Magnetic field on the axis of a solenoid.

where $\phi$ is the angle AOZ. Hence

$$H_z = -\tfrac{1}{2}mI \int_{\phi_2}^{\phi_1} \sin \phi \, d\phi = \tfrac{1}{2}mI(\cos \phi_1 - \cos \phi_2). \qquad (4.67)$$

For an infinite solenoid, $\phi_1 = 0$ and $\phi_2 = \pi$, so that

$$H_z = mI \qquad (4.68)$$

and is uniform inside the solenoid.

We end this section by calculating the force between two small plane circular coils, each of one turn of radius $a$ carrying a current $I$, with a common axis, and separated by a distance $z$ ($a \ll z$). From eqn (4.66) the field strength on the axis at the centre of the second coil is

$$H_z = \tfrac{1}{2}Ia^2/z^3.$$

Since $a \ll z$, $H_z$ will not vary appreciably for a small distance off the axis, and the flux through the second coil is therefore

$$\Phi = \int \mathbf{B}.d\mathbf{S} = \mu(\tfrac{1}{2}Ia^2/z^3)(\pi a^2) = \tfrac{1}{2}\mu\pi Ia^4/z^3.$$

The potential energy of the second coil is $U_P = -\Phi I$, and hence the force on it is

$$F = -dU_P/dz = -\tfrac{3}{2}\mu\pi a^4 I^2/z^4. \qquad (4.69)$$

It is instructive to see just how this force arises. Since div $\mathbf{B} = 0$, in the region around the second coil we have

$$\frac{\partial B_x}{\partial x} + \frac{\partial B_y}{\partial y} + \frac{\partial B_z}{\partial z} = 0,$$

where $x$ and $y$ are normal to the common axis of the coils. By symmetry, $\partial B_x/\partial x = \partial B_y/\partial y$, and hence each equals $-\tfrac{1}{2}(\partial B_z/\partial z)$. At a small distance $a$ from the axis there will be a radial component of field equal to $a(\partial B_x/\partial x) = -\tfrac{1}{2}a(\partial B_z/\partial z) = 3\mu a^3 I/4z^4$. There will therefore be a force $I(ds \wedge \mathbf{B})$ on each element of the coil, of which the components due to $B_z$ are radial and sum to zero over the whole coil, while the force components due to the radial component of $B$ all act in the negative $z$ direction (assuming the currents in each coil flow in the same sense). These sum to $-I(2\pi a)\mu(3a^3 I/4z^4)$, which gives the same result as in eqn (4.69) (see also Problem 4.4).

## 4.7. Moving charges in electric and magnetic fields

The field strength of the magnetic field produced by a current density $\mathbf{J}$ is given by eqn (4.42). If the current density arises from the movement of an electrical charge density $\rho_e$ with velocity $\mathbf{v}$, then $\mathbf{J} = \rho_e \mathbf{v}$ and

$$\mathbf{H} = \int \frac{(\mathbf{J} \wedge \mathbf{r})}{4\pi r^3} \, d\tau = \int \frac{(\mathbf{v} \wedge \mathbf{r})}{4\pi r^3} \, \rho_e \, d\tau.$$

Similarly the force $\mathbf{F}$ on a moving charge density is

$$\mathbf{F} = \int (\mathbf{J} \wedge \mathbf{B}) \, d\tau = \int (\mathbf{v} \wedge \mathbf{B}) \rho_e \, d\tau.$$

For a point charge $q$ these equations reduce to

$$\mathbf{H} = \frac{q(\mathbf{v} \wedge \mathbf{r})}{4\pi r^3}, \tag{4.70}$$

and

$$\mathbf{F} = q(\mathbf{v} \wedge \mathbf{B}). \tag{4.71}$$

If an electric field of strength $\mathbf{E}$ is also present, the total force is

$$\mathbf{F} = q(\mathbf{E} + \mathbf{v} \wedge \mathbf{B}). \tag{4.72}$$

It may be remarked that though eqn (4.71) has here been introduced as an additional postulate, it follows as a consequence of eqn (1.3) when we apply the special theory of relativity. An observer in whose system a charge is at rest will ascribe the forces on it to a purely electrostatic field of strength $\mathbf{E}$. On applying the laws for the transformation of mechanical force we find that a moving observer would measure a force of the type given by eqn (4.72), that is, he would ascribe the effects to the action of both electric and magnetic fields. In a similar manner, eqn (4.70) can be deduced from the electrostatic formula for $\mathbf{D}$, (eqn (1.23)).

The motion of charged particles, usually electrons or positive ions, under the action of electric and magnetic fields is the basis of many fundamental experiments in physics, a few of which will be used as illustrations here. The motion in purely electrostatic fields has already been discussed (§ 3.6), and we shall begin by considering the motion of a charge in a uniform magnetic flux density $\mathbf{B}$. If the charge is initially moving in a plane normal to $\mathbf{B}$, then the force on it (assuming $\mathbf{E} = 0$) is also in this plane and normal to its direction of motion. Thus no work is done on the particle, since $\mathbf{F} \cdot \mathbf{v} = q\mathbf{v} \cdot (\mathbf{v} \wedge \mathbf{B}) = 0$, and its velocity remains constant in magnitude. The charge will therefore move in a circle in this plane, the force towards the centre being

$$Mv^2/r = qvB,$$

where $M$ is the mass of the charged particle, and $r$ the radius of the circle. Hence we have

$$r = Mv/qB \quad \text{and} \quad \omega_c = v/r = B(q/M), \tag{4.73}$$

where $\omega_c$ is the angular velocity. This equation shows that $\omega_c$, and hence the time taken to make one revolution, is independent of the velocity of the particle (so long as the relativistic change of mass with velocity can be neglected). This fact is made use of in many applications.

*Magnetic focusing*

If the initial velocity of the charge is not normal to **B**, but makes an angle $\theta$ with the direction of **B**, then we can resolve the velocity into a component $v \cos \theta$ parallel to **B**, and a component $v \sin \theta$ normal to **B**. The vector product $\mathbf{v} \wedge \mathbf{B}$ has no component parallel to **B**, and the component of velocity $v \cos \theta$ will therefore continue unaltered. The projection of the motion on a plane normal to **B** will be a circle of radius $r = Mv \sin \theta / qB$, and the actual path of the particle will be a helix. One revolution of the helix is completed in a time $2\pi/\omega = 2\pi M/qB$, and the particle has then moved a distance $z = 2\pi Mv \cos \theta / qB$ in the direction of **B**. For small values of $\theta$, this distance is independent of $\theta$ in the first approximation (since $\cos \theta \approx 1 - \frac{1}{2}\theta^2$), and this is the principle used in magnetic focusing.

In Fig. 4.13 electrons leave a point F, and it is desired to focus them so that they all reach a point P a distance $z$ away. If the electrons emerge from a gun with electrostatic focusing, they all have closely the same velocity $v$, but are not moving quite parallel to the line FP. By means of a solenoid, a magnetic field is applied in the direction FP, and the current in the solenoid adjusted so that the time taken to reach P is equal to one or more periods of revolution in the helical motion caused by the magnetic field. It is often impracticable to use a long solenoid, and one or more short solenoids, encased in iron with a small annular gap round the inner circumference, as shown in Fig. 4.13, are used instead. Such coils give a localized, non-uniform field, which acts like a thin lens; their design is largely empirical.

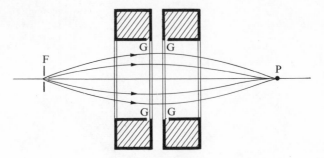

FIG. 4.13. Double solenoid encased in iron, for magnetic focusing. G, G are small annular gaps in the iron casing.

*Measurement of specific charge*

Determination of the ratio of charge to mass (or 'specific charge') of atomic particles is of prime importance in atomic physics. All such particles carry a charge (positive or negative) equal to the electronic charge $e$, or a small integral multiple of it, and the ratios of the charge to the mass for the electron and the proton are fundamental constants. From eqn (4.73) it will be seen that an accurate measurement of $\omega_c$ and $B$ for particles moving in a circle suffices to determine $q/M$, and recent methods based on this principle are described in Chapter 24. For positive ions of heavier nuclei, the main interest lies in the measurement of the mass, and instruments for measuring the specific charge $q/M$ for this purpose are known as 'mass spectrometers'. In general they make use of both electric and magnetic fields, and two measurements are required since both the velocity of the particle and the value of $q/M$ are unknown. The positive ions are usually formed in a gaseous discharge and more than one type of ion with varying velocity may be present; the instruments are therefore designed to sort these out, and bring all particles with the same specific charge to a common focus.

Many such instruments have been designed, but since our purpose here is just to illustrate the principles, we shall describe only one, due to Bainbridge. It makes use of a 'velocity selector', formed by the flat plates $P_1$, $P_2$ in Fig. 4.14, which have a very small separation. Ions enter these plates from a source through slits $S_1$, $S_2$, so that they are travelling with velocity $v$ parallel to the $z$-axis of the coordinate system shown in the figure. A voltage is maintained between the plates, so that there is an

FIG. 4.14. Diagram illustrating the principle of Bainbridge's mass spectrometer.

electric field of strength **E** in the x direction. By means of a pair of Helmholtz coils (see Problem 4.2) a uniform magnetic flux density **B** is maintained in the y direction, and the total force on an ion between the plates is therefore $q(E-vB)$ in the x direction. If the plates are long and close together, only ions for which this force is zero will emerge, and their velocity must therefore be $v = E/B$. The device thus selects ions of a particular velocity determined by this ratio. On emerging from the plates the ions travel in a semicircular path under the influence of the field **B** alone until they strike a detector at C. The distance DC is twice the radius of the orbit and is thus

$$2r = \frac{2Mv}{qB} = 2(M/q)\frac{E}{B^2}. \tag{4.74}$$

Hence the distance DC is linearly proportional to the mass of the ion. In Problem 4.9 it is shown that the distance DC is independent (to the first order) of the angle to the z-axis at which an ion emerges from the plates, provided this is small, so that we have 'first-order' focusing of the ions with a given value of $q/M$.

### References

PAGE, L. and ADAMS, N. I. (1945). *Am. J. Phys.* **13**, 141.
ROBINSON, F. N. H. (1973). *Macroscopic electromagnetism*, pp. 95–6. Pergamon Press, Oxford.
STRATTON, J. A. (1941). *Electromagnetic theory*. McGraw–Hill, New York.

### Problems

4.1. Two short magnets are attached to a cork so that they float on water with their axes horizontal. One magnet lies with its centre on the axis of the other, but with its own axis perpendicular to the line joining them. Assuming that the distance between the magnets is large compared with their lengths, so that they can be treated as point dipoles, calculate the force and the couple on each magnet, and satisfy yourself that there is no resultant force or couple on the system as a whole.

4.2. Two identical circular coils, each of n turns of radius a, are placed with their planes parallel and normal to the line joining them, a distance r apart. Calculate the magnetic field on the axis midway between the coils due to a current I through each coil. Show that, if $r = a$, the field midway between the coils is uniform over a considerable region; that is, $\partial H/\partial r$, $\partial^2 H/\partial r^2$ and $\partial^3 H/\partial r^3$ are all zero at the midpoint. Such an arrangement of coils was used by Helmholtz for a galvanometer, a small magnetic needle suspended by a torsion fibre at the centre being deflected by the current through the coils.

4.3. Two infinite cylindrical conductors are placed parallel to one another at a distance $2a$ apart. They carry equal and opposite currents. Show that in the equatorial plane the gradient of the magnetic field is greatest at a distance $a/\sqrt{3}$ from the plane through the axes of the cylinders.

4.4. Deduce eqn (4.69) by treating each coil as a point dipole of moment $I(\pi a^2)$, and using the formula for the force on a dipole in a non-uniform field.

4.5. A uniform flux density $\mathbf{B}_0$ is present in a liquid of relative permeability $\mu_2$. Show that inside a circular cylinder of relative permeability $\mu_1$, inserted in the liquid with its axis normal to $\mathbf{B}_0$, the flux density is

$$\mathbf{B}_1 = \frac{2\mu_1}{\mu_1 + \mu_2} \mathbf{B}_0.$$

*Important.*    Note that if the cylinder carries a current $I$, the force on it per unit length is $(\mathbf{I} \wedge \mathbf{B}_0)$ and not $(\mathbf{I} \wedge \mathbf{B}_1)$ (see, for example, Robinson 1973).

4.6. A spherical shell has radii $a$ and $b$ respectively $(b > a)$, and is made of a material of permeability $\mu_r$. It is placed in a uniform field of strength $H$. Show that the field inside the shell is

$$H_i = \frac{9\mu_r H}{(2\mu_r + 1)(\mu_r + 2) - 2(\mu_r - 1)^2 (a/b)^3}$$

and that for large values of $\mu_r$, this approximates to

$$H_i = 9H/\{2\mu_r (1 - a^3/b^3)\}.$$

Thus if $\mu_r$ is large, $H_i$ is much smaller than $H$, and an instrument can be shielded from stray magnetic fields by placing it in an iron case. Magnetic shielding is much less efficient than electrostatic shielding (especially if $a/b$ is close to unity), for the effective value of $\epsilon_r$ in the equivalent expression for a conductor is infinite.

4.7. A hollow sphere of internal radius $a$, external radius $b$, has a uniform spontaneous magnetization $\mathbf{M}$ per unit volume. Show that the field strength in the internal cavity $(r < a)$ is zero, and that the external field strength $(r > b)$ is the same as that of a dipole moment $\mathbf{m} = 4\pi\mathbf{M}(b^3 - a^3)/3$, the total moment of the hollow sphere.

Show also that the square of the field strength outside the sphere at a point $(r, \theta)$, measured from the centre of the sphere and with respect to the direction of magnetization, is

$$H^2 = (3\cos^2\theta + 1)\left\{\frac{M(b^3 - a^3)}{3r^3}\right\}^2.$$

If the angle of dip $\delta$ is defined as the angle which the lines of force at a point on the external surface of the sphere make with the tangent at that point, show that

$$\tan\delta = 2\cot\theta.$$

4.8. A particle of mass $M$ and charge $q$ is rotating in a circular orbit of radius $r$ with angular velocity $\omega$. Show that a magnetic dipole moment $\mathbf{m}$ is associated with the motion of the charge, such that

$$\mathbf{m} = (q/2M)\mathbf{G},$$

where $\mathbf{G} = Mr^2\omega$ is the angular momentum of the particle. (The orbit may be regarded as a small circuit carrying a current $I = q \times$ the frequency at which the charge passes any point in the orbit per unit time.)

4.9. In Bainbridge's mass spectrometer the ions emerge in a wedge-shaped beam of small semi-vertical angle $\theta$ (see Fig. 4.14). If the resolving power of the instrument as a mass spectrometer is defined as the reciprocal of the smallest fractional change of mass which will produce non-overlapping traces on the plane CD, show that the resolving power is $2/\theta^2$.

4.10. A charged particle starts from rest at the origin of coordinates in a region where there is a uniform electric field of strength $\mathbf{E}$ parallel to the $x$-axis, and a uniform magnetic flux density $\mathbf{B}$ parallel to the $z$-axis. Show that the coordinates of the particle at a time $t$ later will be

$$x = (E/\omega B)(1-\cos \omega t),$$
$$y = -(E/\omega B)(\omega t - \sin \omega t),$$
$$z = 0,$$

where $\omega = qB/m$. (The path of the particle is a cycloid.)

A charge $-q$ is liberated with zero velocity from the negative plate of a parallel-plate capacitor, to which is applied a flux density $\mathbf{B}$ parallel to the plates. Show that it will not reach the positive plate if the plate separation $d$ is greater than $2mE/qB^2$, where $\mathbf{E}$ is the field strength between the plates.

# 5. Electromagnetic induction and varying currents

## 5.1. Faraday's laws of electromagnetic induction

THE experiments of Oersted and others showed that 'electricity can produce magnetism', and established the laws governing the magnetic field set up by a current. Many experiments were devised to detect the inverse effect, the flow of electric current due to a magnetic field, without success, mainly because a steady current flow was looked for. In 1831 Faraday found that a transient flow of current occurred in a closed circuit when the magnetic flux through the circuit was changed. The change of flux could be brought about in a number of ways: in his first experiment two coils of wire were wound on a ring of soft iron as in Fig. 5.1. The presence of a current in the second coil was detected by connecting it to another coil near a small suspended magnet. When the first coil was connected to a battery, a momentary oscillation of the magnet occurred, after which it settled in its original position. A similar oscillation, though with an initial kick in the opposite direction, was observed on disconnecting the battery. In other experiments Faraday showed that similar effects were observed if a permanent magnet was moved near the second coil, or if the coil was moved in the neighbourhood of a magnet. His results were summed up in the two laws:

(1) when the magnetic flux through a circuit is changing, an electromotive force is induced in the circuit;

(2) the magnitude of this e.m.f. is proportional to the rate of change of the flux.

FIG. 5.1. Faraday's experiment on electromagnetic induction.

Fɪɢ. 5.2. Current induced in a loop by a moving magnet (broken lines represent lines of magnetic field produced by the induced current when the magnet moves towards the loop).

The sign of the e.m.f. is given by Lenz's law, which states that it is such that any current flow is in the direction which would oppose the flux change causing the e.m.f. Thus, in Fig. 5.2, if the magnet is moved towards the closed loop of wire so that the magnetic flux through the coil is increased, the induced current will flow in such a direction that its own flux opposes the increased flux of the magnet threading the loop.

These laws are expressed in the equation

$$V = -d\Phi/dt,\qquad(5.1)$$

where $V$ is the e.m.f. round the circuit, and $\Phi$ is the instantaneous value of the magnetic flux through the circuit. Now $\Phi = \int \mathbf{B} \cdot d\mathbf{S}$ and $V = \int \mathbf{E} \cdot d\mathbf{s}$, where the former integral is taken over any area bounded by the circuit and the latter integral is taken round the circuit. Hence we have

$$\int \mathbf{E} \cdot d\mathbf{s} = -\frac{d}{dt}\int \mathbf{B} \cdot d\mathbf{S}.\qquad(5.2)$$

Using the transformation $\int \mathbf{E} \cdot d\mathbf{s} = \int \operatorname{curl} \mathbf{E} \cdot d\mathbf{S}$, and the fact that the time and space coordinates are independent variables, this can be rewritten in the form

$$\int \operatorname{curl} \mathbf{E} \cdot d\mathbf{S} = -\int \frac{d\mathbf{B}}{dt} \cdot d\mathbf{S},$$

and since this must hold over any surface area, the integrands must be equal, giving the differential form

$$\operatorname{curl} \mathbf{E} = -d\mathbf{B}/dt.\qquad(5.3)$$

At first sight it would have been expected that eqn (5.1) would have contained a multiplying constant to be determined either experimentally or from theory. This constant is in fact unity, as can be seen in the following way. Let us assume we have a very thin conductor carrying no current, which is moved with a velocity **v** in a uniform field of magnetic flux density **B**. Since the wire is a conductor, it carries charges (electrons) which are free to move along the wire; let the velocity of a charge $q$ in the conductor be **u** relative to the conductor. Since the conductor is very thin **u** must be parallel to the direction of the wire at any point. The velocity of the charge $q$ relative to the observer is **v**+**u**, and the force on it will therefore be $\mathbf{F} = q(\mathbf{v}+\mathbf{u})\wedge\mathbf{B}$. If the charge moves a distance d**r** along the wire, the work done is $\mathbf{F}.\mathrm{d}\mathbf{r}$; since $(\mathbf{u}\wedge\mathbf{B}).\mathrm{d}\mathbf{r} = 0$ because **u**, d**r** are parallel, this is the same as if there existed in d**r** an e.m.f.

$$\mathrm{d}V = (\mathbf{v}\wedge\mathbf{B}).\mathrm{d}\mathbf{r}. \tag{5.4}$$

For a closed circuit in a uniform flux density **B**, since $(\mathbf{v}\wedge\mathbf{B})$ is constant, the total e.m.f. $V = \int (\mathbf{v}\wedge\mathbf{B}).\mathrm{d}\mathbf{r} = 0$. If the flux density is not uniform, the flux through the circuit will change as the circuit moves, and we can relate $V$ to the rate of change of flux. Consider a small rectangular circuit ABCD (Fig. 5.3) with sides $a$, $b$ parallel to the $x$-, $y$-axes of a Cartesian system. The e.m.f. in an anticlockwise direction (the sense in which a right-handed screw would turn to advance along the $z$-axis) is

$$V = a(v_y B_z - v_z B_y) + b\left\{ v_z\left(B_x + a\frac{\partial B_x}{\partial x}\right) - v_x\left(B_z + a\frac{\partial B_z}{\partial x}\right)\right\} -$$

$$- a\left\{ v_y\left(B_z + b\frac{\partial B_z}{\partial y}\right) - v_z\left(B_y + b\frac{\partial B_y}{\partial y}\right)\right\} - b(v_z B_x - v_x B_z)$$

$$= -ab\left\{ v_x\frac{\partial B_z}{\partial x} + v_y\frac{\partial B_z}{\partial y} - v_z\left(\frac{\partial B_x}{\partial x} + \frac{\partial B_y}{\partial y}\right)\right\}$$

$$= -ab\left\{ v_x\frac{\partial B_z}{\partial x} + v_y\frac{\partial B_z}{\partial y} + v_z\frac{\partial B_z}{\partial z}\right\} \tag{5.5}$$

since div **B** = 0. But $abB_z = \Phi$, the flux through the circuit, and since $v_x = \mathrm{d}x/\mathrm{d}t$, etc.,

$$V = -\left(\frac{\partial\Phi}{\partial x} v_x + \frac{\partial\Phi}{\partial y} v_y + \frac{\partial\Phi}{\partial z} v_z\right) = -\frac{\mathrm{d}\Phi}{\mathrm{d}t},$$

in agreement with eqn (5.1). The unit of magnetic flux $\Phi$ is the weber, and an e.m.f. of 1 V is generated in a circuit where the flux is changing at the rate of $1\ \mathrm{Wb\ s^{-1}}$.

Eqn (5.3) may be combined with the relation $\mathbf{B} = \mathrm{curl}\ \mathbf{A}$ (eqn (4.47)) to give

$$\mathrm{curl}\ \mathbf{E} = -\frac{\partial}{\partial t}\ \mathrm{curl}\ \mathbf{A} = -\mathrm{curl}(\partial\mathbf{A}/\partial t),$$

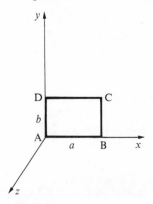

FIG. 5.3. Relation between Faraday's law and the force on a moving conductor.

whence

$$\mathbf{E} = -\frac{\partial \mathbf{A}}{\partial t} + \text{constant} = -\frac{\partial \mathbf{A}}{\partial t} - \text{grad } V. \qquad (5.6)$$

This is a more general equation than eqn (1.6) which applies only to steady fields. As an example of its use we consider a particle accelerator such as the betatron, in which a particle of charge $q$ is constrained to move in a circle of radius $R$ through which there is a changing magnetic flux. If the flux density, apart from varying with time, depends only on the radius, then the lines of $\mathbf{A}$ are concentric circles, and from eqn (4.50), $A = A_\theta = \Phi/2\pi R$. If grad $V = 0$, the rate of change of momentum of the particle is

$$d\mathbf{p}/dt = -q(d\mathbf{A}/dt)$$

and

$$\mathbf{p} - \mathbf{p}_0 = -q\mathbf{A} = -(q/2\pi R)\Phi, \qquad (5.7)$$

if the momentum was $\mathbf{p}_0$ when $\Phi$ was zero. The negative sign reminds us that the magnetic flux must link the circuit in the correct sense in order to produce an increase and not a decrease in momentum.

## 5.2. Self-inductance and mutual inductance

When a current $I$ flows in a circuit, a magnetic field is set up and there will be a magnetic flux $\Phi$ through the circuit arising from its own magnetic field. The magnetic field strength at any point is proportional to the current $I$; so also is the flux $\Phi$, and we may therefore write

$$\Phi = LI, \qquad (5.8)$$

where $L$ is a constant which depends on the geometry of the circuit and the permeability of the medium in which it is immersed. $L$ is called the self-inductance of the circuit, and is equal to the total flux through the circuit when unit current is flowing. A circuit has unit self-inductance (one henry) if it is threaded by one weber of flux when one ampere of current is flowing.

If a second coil is brought near to a coil carrying a current $I$, there will in general be a magnetic flux $\Phi_2$ through the second coil due to the current in the first coil. Since $\Phi_2$ is again linearly proportional to $I_1$, we may write

$$\Phi_2 = L_{21} I_1, \tag{5.9a}$$

where $L_{21}$ is called the mutual inductance between the two circuits. The unit of mutual inductance is again the henry. There will also be a flux $\Phi_1$ through the first circuit due to a current $I_2$ in the second circuit, given by

$$\Phi_1 = L_{12} I_2. \tag{5.9b}$$

The coefficients $L_{12}$ and $L_{21}$ are equal, as can be seen from energy considerations. The potential energy of the system can be found from the flux of either coil due to the field of the other; from eqn (4.51)

$$U_P = -\Phi_2 I_2 = -L_{21} I_1 I_2 = -\Phi_1 I_1 = -L_{12} I_2 I_1,$$

showing that

$$L_{12} = L_{21}. \tag{5.10}$$

By using eqns (4.51) and (4.55) we can derive a formula for the mutual inductance, since

$$U_P = -I_1 \int \mathbf{A}_{21} \cdot d\mathbf{s}_1 = -I_1 \int \left( \frac{\mu}{4\pi} \int \frac{I_2 \, d\mathbf{s}_2}{r} \right) \cdot d\mathbf{s}_1$$

$$= -\frac{\mu}{4\pi} I_1 I_2 \int \int \frac{d\mathbf{s}_1 \cdot d\mathbf{s}_2}{r} = -L_{12} I_1 I_2, \tag{5.11}$$

where, by symmetry,

$$M = L_{12} = L_{21} = \frac{\mu}{4\pi} \int \int \frac{d\mathbf{s}_1 \cdot d\mathbf{s}_2}{r}. \tag{5.12}$$

This result is known as Neumann's formula. Since the unit of mutual inductance is the henry, this formula shows that the dimensions of $\mu_0$ are henry metre$^{-1}$ (H m$^{-1}$).

If the two coils are closely wound, so that all the flux generated by the first coil passes through the second, and vice versa, then the ratio of the two fluxes $\Phi_1$ and $\Phi_2$ will just be equal to the ratio of the number of turns $n_1$, $n_2$ on the two coils. For a current $I_1$ in the first coil we have

$$\Phi_1 / \Phi_2 = (L_1 I_1)/(M I_1) = L_1 / M = n_1 / n_2,$$

while for the flux generated by a current $I_2$ in the second coil

$$\Phi_2/\Phi_1 = (L_2 I_2)/(MI_2) = L_2/M = n_2/n_1.$$

Hence

$$L_1/M = M/L_2 = n_1/n_2 \quad \text{and} \quad M^2 = L_1 L_2. \tag{5.13}$$

If the flux through the two coils is changing, the voltages induced in the two coils will be in the ratio

$$V_1/V_2 = (d\Phi_1/dt)/(d\Phi_2/dt) = n_1/n_2 = 1/n.$$

Hence such a device may be used as a transformer, since if a changing voltage $V_1$ is applied to the 'primary' coil, a changing voltage of different magnitude will be induced in the 'secondary' coil. The voltage transformation ratio is $n = V_2/V_1$, the 'turns ratio' of secondary to primary. In general not all the flux of one circuit passes through the other, and $M$ is less than $(L_1 L_2)^{\frac{1}{2}}$; it may be written as

$$M = k(L_1 L_2)^{\frac{1}{2}} \quad (0 \leqslant k \leqslant 1), \tag{5.14}$$

where $k$ is called the 'coefficient of coupling'. The theory of transformers is considered further in § 7.5.

The magnitude of an inductance may be calculated from first principles by computing the field produced by a given current in the coil, and then finding the total flux through the same or another coil, according to whether the self-inductance of the first coil or the mutual inductance between the two coils is required. The calculations are illustrated below for a number of simple shapes of coil.

*Long solenoid*

For an infinitely long solenoid, wound with $m$ turns per unit length and carrying a current $I$, the magnetic field strength inside is uniform and given by eqn (4.68):

$$H = mI.$$

If the core of the solenoid has permeability $\mu$, the flux through each turn is $\Phi' = \mu A m I$, where $A$ is the cross-sectional area of the solenoid. The self-inductance per unit length is therefore $L' = m\Phi'/I = \mu m^2 A$; for a solenoid of length $l$, large compared with its diameter, this formula is still very nearly correct, and we may write for the total self-inductance

$$L = \mu m^2 A l. \tag{5.15}$$

If a second short coil of $n$ turns, insulated from the first, is wound on the solenoid, as in Fig. 5.4, the mutual inductance is

$$M = \mu m n A. \tag{5.16}$$

FIG. 5.4. Solenoid with primary and secondary coils.

## Two coaxial coils

Another simple case is that of two plane coaxial coils A and B as in Fig. 5.5, of radii $a$ and $b$, and total numbers of turns $n_1$ and $n_2$ respectively, whose centres are a distance $z$ apart, where $z \gg a, b$. The field strength at the centre of B due to a current $I$ in A is, from eqn (4.66),

$$H = \tfrac{1}{2} I a^2 n_1 / z^3,$$

and the total flux through B is $\pi b^2 n_2 (\mu H)$, since the field through the coil will be uniform in the first approximation when the inequality $z \gg a, b$ holds. Hence the mutual inductance is

$$M = \frac{\mu \pi a^2 b^2 n_1 n_2}{2 z^3}. \tag{5.17}$$

## Pair of coaxial circular cylinders

An important method of carrying radio-frequency alternating currents is by means of a pair of coaxial cylinders of radii $a$, $b$ $(b > a)$, as in Fig. 5.6. At any point the current in the inner cylinder is $I$, while that in the outer cylinder is $-I$, that is, it is exactly equal in magnitude but flowing in the opposite direction. The magnetic field at a distance $r$ from the axis when $r < b$ is the same as that due to a straight wire, so that

$$H = I/2\pi r.$$

Application of the same equation shows that there will be no field outside the larger cylinder, since any circuit drawn round it is threaded by two

FIG. 5.5. Mutual inductance between two plane coaxial coils. $AB = z$.

FIG. 5.6. Self-inductance of coaxial circular cylinders.

equal and opposite currents. To compute the self-inductance of a length $l$, we find the flux through a circuit such as ABCD in Fig. 5.6. Instead of calculating this directly from the equation above for $H$, we illustrate an alternative method using the magnetic vector potential **A**. Since the current flow is entirely parallel to the $z$-axis, the vector potential has just the single component $A_z$, whose value in the space between the cylinders is, from eqn (4.65),

$$A_z = -(\mu I/2\pi)\ln(r/a).$$

The flux through the circuit ABCD of length $l$ is therefore

$$\int \mathbf{A} \cdot \mathbf{ds} = l(A_z)_{r=a} - l(A_z)_{r=b} = (l\mu I/2\pi)\ln(b/a)$$

and the inductance $L$ per unit length is

$$L = \frac{\{(A_z)_{r=a} - (A_z)_{r=b}\}}{I} = \frac{\mu}{2\pi} \ln(b/a). \tag{5.18}$$

These equations may be compared with the corresponding equation (1.30) for the capacity per unit length of such a pair of coaxial cylinders,

$$C = \frac{Q}{(V_a - V_b)} = \frac{2\pi\epsilon}{\ln(b/a)}.$$

On combining the two sets of equations we obtain

$$LC = \mu\epsilon.$$

This rather surprising result (to which we shall return in § 9.3 when discussing the propagation of electromagnetic waves along parallel cylinders) is not accidental. It arises when we are dealing with pairs of parallel conductors whose dimensions do not change in one direction (in this case, the $z$-axis), and the electric and magnetic fields are entirely transverse to this direction. If the current flow is confined to this direction, then the sole component of **A** is, from eqn (4.56),

$$A_z = \frac{\mu I}{4\pi} \int\limits_{-\infty}^{+\infty} \frac{dz}{r}. \tag{5.19a}$$

Clearly this is independent of $z$, and in the $xy$ plane it varies in the same way as the electric potential $V$, which from eqn (2.6) is

$$V = \frac{Q}{4\pi\epsilon} \int\limits_{-\infty}^{+\infty} \frac{dz}{r}, \tag{5.19b}$$

where $Q$ is the charge per unit length on the outer cylinder (also independent of $z$). As a result, similar geometrical factors appear in the calculation of $L$ and $1/C$, so that they disappear in the product $LC$ (see also Problem 5.2).

## 5.3. Transient currents in circuits containing inductance, resistance, and capacitance

If a circuit containing a battery of e.m.f. $V$ and a resistance $R$ is connected to a coil in which the magnetic flux $\Phi$ is changing, the net e.m.f. $V'$ applied to $R$ will be the sum of the battery voltage $V$ and the e.m.f. developed in the coil. Hence $V' = V - d\Phi/dt = RI$, or

$$V = RI + d\Phi/dt. \tag{5.20a}$$

We can apply this equation to a number of problems, the first being a circuit containing a self-inductance $L$ and a resistance $R$, as shown in Fig. 5.7, which is connected at time zero to a battery of constant e.m.f. $V$. Since $d\Phi/dt = L(dI/dt)$, we have

$$V = IR + L(dI/dt), \tag{5.20b}$$

which is the fundamental differential equation relating the current $I$ to the voltage $V$. Integration of this equation, with the condition $I = 0$ at $t = 0$, gives

$$I = \frac{V}{R}(1 - e^{-(R/L)t}), \tag{5.21}$$

FIG. 5.7. Battery driving current through $R$ and $L$.

Fig. 5.8. Rise of current in circuit of Fig. 5.7.

showing that the current approaches exponentially the value $V/R$ which it would have if there were no inductance present (see Fig. 5.8). The rate of approach to this steady value depends on the ratio of resistance to inductance. If $R = 0$, the steady state, corresponding to infinite current, is never reached, but the current rises linearly according to the equation $I = (V/L)t$, obtained by direct integration of eqn (5.20b) with $R = 0$. When $R$ is finite, the initial rate of rise of current, given by the tangent at the origin in Fig. 5.8, is $\mathrm{d}I/\mathrm{d}t = V/L$, but as the current through the resistance increases, the voltage across the inductance falls, with a corresponding decrease in $\mathrm{d}I/\mathrm{d}t$.

The converse problem, in which a battery has been connected to the circuit for a long time so that a steady current $I_0$ is flowing, and then at time zero the battery is replaced by a short-circuit, leads to the same differential eqn (5.20b), but with $V = 0$. Its solution is

$$I = I_0 e^{-(R/L)t}, \tag{5.22}$$

showing that the effect of the inductance is to prevent the current from falling instantaneously to zero. If the battery is suddenly open-circuited, the sudden cessation of the current produces a large impulse voltage $-L(\mathrm{d}I/\mathrm{d}t)$ in the inductance, which may be sufficient to cause a spark across the point at which the circuit is broken. With large inductances such as are found in electromagnets (see § 6.5) very high voltages may arise in this way which can damage the insulation if the circuit is broken suddenly.

In both the cases considered above the exponential is of the form $e^{-t/\tau}$, and the exponential rate of change of the current is the same; the quantity $\tau = L/R$ is called the time constant of the circuit.

*Circuit with capacitance and resistance*

An analogous problem is that of a capacitance $C$ in series with a resistance $R$, to which a battery of e.m.f. $V$ is connected at time zero (see Fig. 5.9). At any instant the charge on the capacitor is $Q$, and the voltage across it is $Q/C$. Since the current $I = \mathrm{d}Q/\mathrm{d}t$, we have

$$V = Q/C + RI = Q/C + R(\mathrm{d}Q/\mathrm{d}t). \tag{5.23}$$

FIG. 5.9. Battery charging capacitance $C$ through resistance $R$.

The solution of this equation gives

$$Q = CV(1 - e^{-t/RC}) \tag{5.24}$$

showing that the charge on the capacitor builds up in a manner similar to the current in the previous problem. The time constant of the circuit is now $\tau = RC$, and the current at any instant is

$$I = dQ/dt = \frac{V}{R} e^{-t/RC} = \frac{V}{R} e^{-t/\tau}. \tag{5.25}$$

It therefore falls exponentially from its initial value $(V/R)$ to zero, as shown in Fig. 5.10. The voltage across the capacitor increases from zero to its steady value $V$, when no more current can flow in the circuit.

If we have an isolated capacitor initially at a voltage $V_0$, and a resistance $R$ is then connected across it at time $t = 0$, the differential equation for the charge at a subsequent time is given by eqn (5.23) with $V = 0$. The solution is

$$Q = CV_0 e^{-t/RC} = CV_0 e^{-t/\tau}, \tag{5.26}$$

showing that the charge decays exponentially to zero.

*Circuit with inductance, capacitance, and resistance*

An inductance, a capacitance, and a resistance are connected in series, as in Fig. 5.11, and the circuit is closed at an instant $t = 0$ when the charge on the capacitor is $Q_0$. Since the total voltage in the circuit is always zero,

FIG. 5.10. Current in circuit of Fig. 5.9 after switch is closed.

FIG. 5.11. Discharge of a capacitance $C$ through resistance $R$ and inductance $L$.

we have

$$L\frac{dI}{dt} + IR + \frac{Q}{C} = 0.$$

Since $I = dQ/dt$, $Q$ may be eliminated by differentiation, giving

$$L\frac{d^2I}{dt^2} + R\frac{dI}{dt} + \frac{I}{C} = 0. \tag{5.27}$$

This equation has a general solution of the form

$$I = e^{-(R/2L)t}(Ae^{nt} + Be^{-nt}), \tag{5.28}$$

where

$$n^2 = (R/2L)^2 - (1/LC)$$

and $A$, $B$ are constants determined by the initial conditions. Since $I = 0$ at time $t = 0$, we must have $B = -A$, and the value of $A$ can be found by integration of (5.28) and setting $Q = Q_0$ at $t = 0$.

The nature of the solution depends on whether $n$ is real or imaginary. Three cases can be distinguished.

(1) $n$ real, that is, $(R/2L) > 1/(LC)^{\frac{1}{2}}$. The discharge of the capacitor is aperiodic, as shown in Fig. 5.12.

FIG. 5.12. Charge on capacitor and current in circuit of Fig. 5.11 (non-oscillatory discharge).

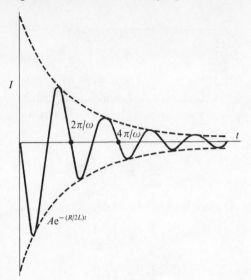

FIG. 5.13. Current in circuit of Fig. 5.11 (oscillatory discharge).

(2) $n = 0$, that is, $(R/2L) = 1/(LC)^{\frac{1}{2}}$. In this case the solution is of the form

$$Q = Q_0(1 + tR/2L)e^{-(R/2L)t} \tag{5.29}$$

and the aperiodic discharge is most rapid.

(3) $n$ imaginary, that is, $(R/2L) < 1/(LC)^{\frac{1}{2}}$. The discharge is now oscillatory, as in Fig. 5.13, and the current may be written as

$$I = Ae^{-(R/2L)t}\sin \omega t, \tag{5.30}$$

where $\omega^2 = 1/LC - R^2/4L^2$. If the resistance is small, so that

$$(R/2L) \ll 1/(LC)^{\frac{1}{2}} \quad \text{(that is, } \tfrac{1}{2}R(C/L)^{\frac{1}{2}} \ll 1),$$

the angular frequency $\omega$ will be close to $1/(LC)^{\frac{1}{2}}$. We may call

$$f_0 = \frac{1}{2\pi(LC)^{\frac{1}{2}}}$$

the natural frequency of oscillation of the circuit in the absence of damping. When the damping is small the amplitude of the oscillations decays slowly, and the maxima, which lie on the exponential curve

$$I = A\left\{1 + \left(\frac{R}{2\omega L}\right)^2\right\}^{-\frac{1}{2}}e^{-(R/2L)t},$$

occur very nearly at the points where $\sin \omega t = 1$. These points occur at intervals of time $T$, where $T = 2\pi/\omega = 2\pi(LC)^{\frac{1}{2}}$ (where we have again

FIG. 5.14. Battery of e.m.f. $V$ charging capacitance $C$ through resistance $R$ and inductance $L$.

assumed that the damping is small), and the amplitude of successive maxima (of the same sign) therefore decreases by the constant ratio $e^{-TR/2L}$. If we denote such successive maxima by $I_m$, $I_{m+1}$, the quantity

$$\ln(I_m/I_{m+1}) = TR/2L = \pi R(C/L)^{\frac{1}{2}} = \pi/Q_F \qquad (5.31)$$

is known as the 'logarithmic decrement' of the circuit. The quantity

$$Q_F = \frac{1}{R}\left(\frac{L}{C}\right)^{\frac{1}{2}}$$

is known as the 'quality factor' of the circuit. Either of these quantities serves as an important criterion for the performance of an oscillatory circuit, but in electricity it is customary to use the quality factor $Q_F$ rather than the logarithmic decrement. Further relations involving $Q_F$ will be obtained in Chapter 7.

The converse problem, where a battery of e.m.f. $V$ is connected at time zero to a circuit containing $L$, $C$, $R$ in series (Fig. 5.14), is left as an exercise for the reader. The differential equation is the same as (5.27), and the solutions are similar to those above, with an aperiodic or oscillatory approach to the equilibrium state where no current flows and the voltage across the capacitance is equal to $V$. An important point is that the approach to the equilibrium state in the $L$, $C$, $R$ circuit is most rapid at the change-over point from an aperiodic to an oscillatory condition, that is, when $(R/2L) = (LC)^{-\frac{1}{2}}$ or $Q_F = (L/C)^{\frac{1}{2}}/R = \frac{1}{2}$.

## 5.4. Magnetic energy and mechanical forces in inductive circuits

In the circuit of Fig. 5.14 the relation between the applied voltage $V$ and the current is given by the equation

$$V = L\frac{dI}{dt} + RI + \frac{Q}{C},$$

where $Q$ is the charge on the capacitance. The rate at which work is done is found by multiplying by $I$, giving

$$VI = LI\frac{dI}{dt} + RI^2 + \left(\frac{dQ}{dt}\right)\frac{Q}{C}$$

since $I = dQ/dt$. The work done in a time interval $t$, in which the current changes from $I_1$ to $I_2$, and the charge on the capacitance from $Q_1$ to $Q_2$, will be

$$W = \int_0^t VI \, dt = \tfrac{1}{2}L(I_2^2 - I_1^2) + \int_0^t RI^2 \, dt + \frac{1}{2C}(Q_2^2 - Q_1^2).$$

Now $W$ is the total work done by the battery, and the integral of $RI^2$ is the energy dissipated as heat in the resistance, which is always positive. The last term, $\tfrac{1}{2}(Q_2^2 - Q_1^2)/C$, represents the change in the stored energy of the capacitance, and we interpret the first term $\tfrac{1}{2}L(I_2^2 - I_1^2)$ as the change in the energy stored in the inductance. We note that this change is reversible, since if the current is first increased to $I_2$ and then returned to its initial value $I_1$, the change in the stored energy is zero. If $I_1 = 0$, and a current $I$ is established in the inductance, an energy $\tfrac{1}{2}LI^2$ will be associated with it.

An expression for the energy stored in a series of inductances will now be derived in a more general way. If $\Phi_k$ is the flux through the $k$th circuit, the voltage induced in it is $V_k = d\Phi_k/dt$, and the rate of doing work is $I_k V_k = I_k(d\Phi_k/dt)$. We assume that currents in all the circuits were initially zero and increased proportionally with time, so that at an intermediate instant $t'$ the values are $I'_k = \alpha I_k$, $\Phi'_k = \alpha\Phi_k$. Then the total work done is

$$\int_0^t I'_k V'_k \, dt' = I_k\Phi_k \int_0^t \alpha(d\alpha/dt') \, dt'$$

$$= I_k\Phi_k \int_0^1 \alpha \, d\alpha = \tfrac{1}{2}I_k\Phi_k. \tag{5.32}$$

On summing over all the circuits, we have

$$U = \tfrac{1}{2}\sum_k I_k\Phi_k. \tag{5.33}$$

If each coil has both self- and mutual inductance,

$$\Phi_k = L_k I_k + \sum_{j \neq k} L_{kj}I_j. \tag{5.34}$$

Hence eqn (5.33) can be written in the form

$$U = \tfrac{1}{2}\sum_k L_k I_k^2 + \tfrac{1}{2}\sum_j\sum_k^{j \neq k} L_{kj}I_k I_j. \tag{5.35}$$

FIG. 5.15. Two circuits with a mutual inductance $M$. Circuit (2) is moving relative to circuit (1) so that $M$ is varying, but $L_1$, $L_2$ and $I_1$, $I_2$ are constant.

Since $L_{kj} = L_{jk}$, each term in the second summation occurs twice. This is clearly seen if we consider just two coils, for which (writing $M$ for $L_{12}$)

$$\Phi_1 = L_1 I_1 + M I_2, \qquad \Phi_2 = L_2 I_2 + M I_1,$$

so that the energy becomes

$$U = \tfrac{1}{2} I_1 (L_1 I_1 + M I_2) + \tfrac{1}{2} I_2 (L_2 I_2 + M I_1)$$
$$= \tfrac{1}{2}(L_1 I_1^2 + L_2 I_2^2) + M I_1 I_2. \qquad (5.36)$$

It is important to realize the distinction between this stored energy $U$ and the potential energy of one circuit carrying an invariant current $I_1$ in the field of a second circuit carrying an invariant current $I_2$, given by eqn (5.11). The difference lies in the fact that the formula for $U_P$ assumes that the currents are invariant, and no account is taken of any work done by the batteries in maintaining the currents, whereas the stored energy $U$ includes the work done by the batteries in setting up the current flow, starting from zero current. Thus $U_P$ is the potential-energy function from which any mechanical force can be calculated by the usual formulae such as

$$F_x = -\mathrm{d}U_P/\mathrm{d}x \qquad (5.37)$$

for the $x$-component of the force, under the condition that the currents are kept constant.

The distinction between $U$ and $U_P$ can be seen from a simple example, that of two rigid circuits as in Fig. 5.15, one of which is moving with velocity $\mathrm{d}x/\mathrm{d}t$ with respect to the other. Then the rate at which work is done by the batteries is

$$\mathrm{d}W/\mathrm{d}t = (\mathrm{d}U/\mathrm{d}t) + F_x(\mathrm{d}x/\mathrm{d}t) + R_1 I_1^2 + R_2 I_2^2, \qquad (5.38)$$

where $\mathrm{d}U/\mathrm{d}t$ is the rate at which the stored energy changes. Now, from eqn (5.37), $F_x(\mathrm{d}x/\mathrm{d}t) = -(\mathrm{d}U_P/\mathrm{d}x)(\mathrm{d}x/\mathrm{d}t) = -\mathrm{d}U_P/\mathrm{d}t$ and hence

$$\mathrm{d}W/\mathrm{d}t = (\mathrm{d}U/\mathrm{d}t) - (\mathrm{d}U_P/\mathrm{d}t) + R_1 I_1^2 + R_2 I_2^2.$$

The flux through each circuit is

$$\Phi_1 = L_1 I_1 + M I_2, \qquad \Phi_2 = L_2 I_2 + M I_1,$$

and the rate at which work is done by the batteries

$$dW/dt = I_1(R_1 I_1 + d\Phi_1/dt) + I_2(R_2 I_2 + d\Phi_2/dt)$$
$$= 2(dM/dt)I_1 I_2 + R_1 I_1^2 + R_2 I_2^2,$$

since $I_1$, $I_2$ are constant. Comparing these two equations, we see that

$$dU/dt = dU_P/dt + 2(dM/dt)I_1 I_2. \tag{5.39}$$

From eqn (5.36), since only $M$ is changing,

$$dU/dt = (dM/dt)I_1 I_2$$

and hence

$$dU_P/dt = -(dM/dt)I_1 I_2 = -dU/dt.$$

Thus the component of the mechanical force is

$$F_x = -dU_P/dx = +dU/dx. \tag{5.40}$$

From eqn (5.39) we see that the difference in sign between the two expressions for the force in (5.40) arises from the fact that the batteries do work at the rate $2(dM/dt)I_1 I_2$ (apart from the irreversible Joule heating represented by the terms $R_1 I_1^2$, etc.) which is just twice the rate of increase $(dU/dt)$ of the stored energy. This situation with regard to the magnetic energy, and the calculation of the mechanical forces in a system where the currents are kept constant, is similar to that in electrostatics when changes take place in a system where the conductors are maintained at constant potential. It was shown in § 1.7 that in a change where the stored energy increases by $dU$, the batteries do work $2\,dU$, so that the net amount of external work $dW$ done on the system is $-dU$. Hence the component of the force in the $x$ direction is given by

$$F_x = -dU_P/dx = +(dU/dx)_V,$$

where the subscript $V$ denotes that this formula is to be used under the condition $V = $ constant. Similarly, we may emphasize that in equations such as (5.40) the current is kept constant by writing them in the form

$$F_x = +(dU/dx)_I, \text{ etc.}$$

As we should expect, the torque on a coil can be shown to be

$$T = -(dU_P/d\theta) = +(dU/d\theta)_I = (dM/d\theta)I_1 I_2, \tag{5.41}$$

where $\theta$ is the angle which the coil makes with some fixed axis.

## 5.5. Magnetic energy in magnetic media

Since for any circuit

$$\Phi = \int \mathbf{B} \cdot d\mathbf{S} = \int \text{curl} \, \mathbf{A} \cdot d\mathbf{S} = \oint \mathbf{A} \cdot d\mathbf{s} \qquad (5.42)$$

we can transform eqn (5.33) as follows:

$$U = \tfrac{1}{2} \sum_k I_k \Phi_k = \tfrac{1}{2} \sum_k I_k \oint \mathbf{A} \cdot d\mathbf{s}$$

For a system of distributed currents, we write $I \, d\mathbf{s} = \mathbf{J} \, d\tau$, and take the integral over all space, since contributions arise only from regions where $\mathbf{J}$ is finite, giving

$$U = \tfrac{1}{2} \int (\mathbf{A} \cdot \mathbf{J}) \, d\tau. \qquad (5.43)$$

However, $\mathbf{J} = \text{curl} \, \mathbf{H}$, and using the vector identity (eqn (A.15) of Appendix A)

$$\text{div}(\mathbf{A} \wedge \mathbf{H}) = \mathbf{H} \cdot \text{curl} \, \mathbf{A} - \mathbf{A} \cdot \text{curl} \, \mathbf{H},$$

we have

$$U = \tfrac{1}{2} \int (\mathbf{H} \cdot \text{curl} \, \mathbf{A}) \, d\tau - \tfrac{1}{2} \int \text{div}(\mathbf{A} \wedge \mathbf{H}) \, d\tau$$

$$= \tfrac{1}{2} \int (\mathbf{H} \cdot \mathbf{B}) \, d\tau - \tfrac{1}{2} \int (\mathbf{A} \wedge \mathbf{H}) \, d\mathbf{S}.$$

If the volume integral is extended over all space, the surface integral is taken over the sphere at infinity. For a finite system of closed circuits, the field strength $\mathbf{H}$ and the flux density $\mathbf{B}$ will fall off at large distances at least as rapidly as those of dipoles; hence $\mathbf{H} \sim r^{-3}$ and $\mathbf{A} \sim r^{-2}$, so that the integrand diminishes as $r^{-5}$ and the integral vanishes as $r \to \infty$. Thus we have

$$U = \tfrac{1}{2} \int (\mathbf{H} \cdot \mathbf{B}) \, d\tau. \qquad (5.44)$$

This formula shows that the energy may be regarded as distributed throughout the region occupied by the field with density $\tfrac{1}{2}\mathbf{H} \cdot \mathbf{B}$, which is clearly analogous to the result $\tfrac{1}{2}\mathbf{D} \cdot \mathbf{E}$ in electrostatics, eqn (1.36). In each case it has been derived under the assumption that $\mathbf{B}(\mathbf{D})$ is linearly proportional to $\mathbf{H}(\mathbf{E})$, that is, that the media have values of $\mu(\epsilon)$, which are independent of field strength. We may relax this assumption by considering infinitesimal changes of flux, for which $\delta U = \sum_k I_k \, \delta\Phi_k$. Then by applying transformations exactly similar to those above, we find

$$\delta U = \int \mathbf{J} \cdot \delta\mathbf{A} \, d\tau = \int \mathbf{H} \cdot \delta\mathbf{B} \, d\tau, \qquad (5.45)$$

a result which can then be integrated over any finite change in $\mathbf{B}$ if we know how $\mathbf{H}$ and $\mathbf{B}$ are related at every point during the change. If $\mathbf{B}$ is linearly proportional to $\mathbf{H}$, this result clearly gives again eqn (5.44).

We close this discussion by finding the work done when the magnetization changes by $\delta\mathbf{M}$ in a magnetic flux density $\mathbf{B}_0$ which is due to other sources; that is, $\mathbf{B}_0$ does not include the effect of the field due to the magnetic material itself. A simple example would be a piece of magnetizable material inside a solenoid; $\mathbf{B}_0$ is then just equal to the field $\mu_0\mathbf{H}_0$ which the solenoid would produce in the absence of the magnetizable substance, and can be calculated from the standard formulae. We start by considering two circuits, for which, from eqn (5.39), in an infinitesimal change,

$$\delta U = \delta U_P + 2\delta(MI_1I_2).$$

But $U_P = -\mathbf{m}_2 \cdot \mathbf{B}_1 = -I_2\int\mathbf{B}_1 \cdot d\mathbf{S}_2$, and the flux through the second circuit is $MI_1 = \int \mathbf{B}_1 \cdot d\mathbf{S}_2$. Hence

$$\delta U = +\delta I_2\int \mathbf{B}_1 \cdot d\mathbf{S}_2 = \delta(\mathbf{B}_1 \cdot \mathbf{m}_2),$$

where $\mathbf{m}_2$ is the equivalent magnetic moment of the second coil, and we assume that $\mathbf{B}_1$ is constant over the area of this coil. If $\mathbf{B}_1$ is fixed, and $\mathbf{m}_2$ increases by $\delta\mathbf{m}$, then $\delta U = \mathbf{B}_1 \cdot \delta\mathbf{m}$. Clearly it does not matter what is the source of $\mathbf{B}_1$, so that we can write in general

$$\delta U = \mathbf{B}_0 \cdot \delta\mathbf{m} = \int(\mathbf{B}_0 \cdot \delta\mathbf{M})\,d\tau, \tag{5.46}$$

where $\mathbf{B}_0$ excludes the field contribution from the magnetized substance itself. This formula has an important limitation; it assumes that $\mathbf{B}$ is the same inside the magnetized region as it was in the absence of the magnetized body. This neglects the demagnetizing field, and is a valid approximation only for weakly magnetic substances. In general the energy, and hence also any force or torque on the body (see Appendix B), depends on its shape.

## 5.6. Galvanometers and galvanometer damping

A galvanometer is an instrument for detecting and measuring electric current; if provided with a scale already calibrated, it is called an ammeter. The only type now in common use has a 'moving coil', usually rectangular in shape, through which the current passes while it is suspended in a uniform and constant magnetic flux density $\mathbf{B}$. The suspension is adjusted so that the plane of the coil is parallel to the lines of magnetic flux when no current is passing. Flow of current through the coil gives it a magnetic moment, on which there is a couple due to $\mathbf{B}$; this is balanced by the restoring torque of the suspension, equal to $c\theta$, where $\theta$ is the angle through which the coil turns. If the coil has $N$ turns, each of area $A$, and the current is $I$, the magnetic moment $|\mathbf{m}| = NAI$, and at the

FIG. 5.16. Pole pieces of a permanent magnet in a moving-coil galvanometer.

equilibrium deflection

$$c\theta = |\mathbf{m} \wedge \mathbf{B}| = (NAB)I \cos \theta. \tag{5.47}$$

In practice the lines of $\mathbf{B}$ are produced by a small permanent magnet, with pole faces shaped as in Fig. 5.16. The centre of the gap is filled with iron, so that the flux lines in the gap are as nearly radial as possible. The term $\cos \theta$ will then be unity, so that the equilibrium deflection $\theta$ will be linearly proportional to the current.

In another class of moving-coil instruments, known as dynamometers, the flux density $B$ is provided by a second set of fixed coils carrying a current $I_1$ surrounding the moving coil which carries a current $I_2$. If $M$ is the mutual inductance between the fixed and moving coils, by eqn (5.41) the torque is $I_1 I_2 (dM/d\theta)$, and hence the equilibrium deflection will be given by

$$c\theta = I_1 I_2 (dM/d\theta),$$

where $c\theta$ is the restoring torque due to the suspension (usually a spiral spring). The factor $(dM/d\theta)/c$ is found by calibration with a known current.

If a current $I$ to be measured is allowed to flow through the fixed and moving coils in series, so that $I = I_1 = I_2$, the instrument will act as a milliammeter with a deflection proportional to $I^2$; with a pointer-type dynamometer the full-scale reading is about 10 mA. A more important application is its use as a wattmeter; the coils are then connected as follows. In Fig. 5.17, the moving coil is connected in series with a resistance $R$, large compared with the load, so that the current $I_1$ is nearly equal to $I$, the current through the load. The current $I_2 = V/R$, where $V$ is the voltage across the load, and the deflection is proportional to $I_2 I_2 = IV/R = P/R$, where $P$ is the power being dissipated in the load. Thus the

FIG. 5.17. Circuit of a wattmeter.

instrument may be calibrated to measure power directly. Note that with the coils connected as in Fig. 5.17, the current through the fixed coil is really $I + I_2$, so that the power measured is the sum of that in the load and in $R$ (see Problem 5.17).

So far we have considered the equilibrium deflection of a galvanometer only when a steady current flows; the behaviour during the time interval between switching on the current and the attainment of the final deflection must also be discussed. When the current starts to flow through the galvanometer coil, a couple is exerted on it which gives it an angular acceleration. This couple diminishes as the coil nears its equilibrium position, owing to the reverse couple exerted by the suspension, but if the damping is small, the coil may overshoot and oscillate about its final position. If the damping is large there is no overshoot, and the coil approaches the final position very slowly. In either case some time must elapse before the equilibrium deflection can be observed, and the effect of the damping is thus of considerable importance. Though air resistance and other sources of energy loss contribute to the damping, the most important source is electromagnetic damping, due to the motion of the coil in the magnetic field of the permanent magnet. With the notation of eqn (5.47), the flux through the coil is $\Phi = NAB \sin \theta$, and there will therefore be an e.m.f. induced in the coil equal to

$$-d\Phi/dt = -NAB \cos \theta (d\theta/dt).$$

If $\theta$ is small, or if the lines of magnetic flux are always radial (as in Fig. 5.16), we can put $\cos \theta = 1$. If $V$ is the external source of e.m.f. and $R$ is the total resistance in the circuit (including the galvanometer coil resistance), then the current $I$ will be given by the equation

$$V - \frac{d\Phi}{dt} = RI$$

at any instant. The equation of motion of the coil will therefore be

$$\Im \frac{d^2\theta}{dt^2} + b \frac{d\theta}{dt} + c\theta = NABI = \frac{NAB}{R}\left(V - \frac{d\Phi}{dt}\right) =$$
$$= \frac{NAB}{R} V - \frac{(NAB)^2}{R} \frac{d\theta}{dt}, \qquad (5.48)$$

where $\Im$ is the moment of inertia of the coil, $b(d\theta/dt)$ the torque due to air damping, etc., and $c\theta$ the restoring torque due to the suspension. Rearranging the equation gives

$$\Im \frac{d^2\theta}{dt^2} + \left\{ b + \frac{(NAB)^2}{R} \right\} \frac{d\theta}{dt} + c\theta = \frac{NAB}{R} V, \qquad (5.49)$$

from which it is seen that the effect of the e.m.f. induced by the motion of the coil is to increase the damping torque by an amount

$$(NAB)^2(d\theta/dt)/R.$$

The differential equation is similar to that for the oscillatory circuit of § 5.3, and the same analysis may be applied. There are three types of solution, according to whether the damping is large or small.

(1) *Small damping.* $b' < 2(\Im c)^{\frac{1}{2}}$, where $b' = b + (NAB)^2/R$. The motion is a damped harmonic motion, of period nearly equal to $2\pi(\Im/c)^{\frac{1}{2}}$. The coil overshoots its equilibrium deflection $\theta = (NAB)V/cR$, and then oscillates with diminishing amplitude about this position.

(2) *Critical damping.* $b' = 2(\Im c)^{\frac{1}{2}}$; the coil approaches its equilibrium without overshoot.

(3) *High damping.* $b' > 2(\Im c)^{\frac{1}{2}}$; the coil approaches its final position without overshoot, but more slowly than in case (2).

In general the electromagnetic damping is much greater than other sources of damping, unless the total resistance $R$ of the galvanometer circuit is very large. As $R$ decreases the electromagnetic damping increases, and critical damping will be obtained with a certain value of resistance $R_c$ given by the relation

$$R_c = \tfrac{1}{2}(NAB)^2/(\Im c)^{\frac{1}{2}}, \tag{5.50}$$

assuming that $b$ is negligible under these conditions. It is important to operate a galvanometer at or near critical damping since it then takes up its equilibrium deflection (or, more precisely, comes to a deflection within a certain fraction of its equilibrium value) in the shortest possible time. If the galvanometer is either highly overdamped or very much underdamped it will take much longer to settle down, and this reduces the rate at which readings can be taken. If the source of voltage varies within the time required for the galvanometer to settle, the reading will always lag behind the true value; in detecting the balance point of a bridge, quite misleading indications can be obtained if the bridge arms are altered too rapidly.

Since the critical-damping resistance is of such importance, it is always specified for a galvanometer, together with the coil resistance, the sensitivity, and the period. As a rough rule, the critical-damping resistance is 10–20 times the coil resistance in a reflecting galvanometer. In a pointer instrument the coil may carry a short-circuited turn, or the coil former may be made of metal, in order to give adequate damping.

## 5.7. The ballistic galvanometer and fluxmeter

The ballistic galvanometer is a suspended moving-coil instrument with very light damping, which can be used for the measurement of charge by observing the maximum deflection in its oscillatory motion. For this purpose the charge must pass through the galvanometer in a time short compared with its period of swing, so that the coil does not deflect appreciably during the passage of the charge. If the current at any instant during this passage is $I$, the couple exerted on the coil is $NABI$, and the impulse of angular momentum given to the coil will be

$$\int NABI \, dt = NABQ,$$

where $Q$ is the total charge flowing through the coil. The effect of this impulse is to give the coil an initial angular velocity $(d\theta/dt)_{t=0} = NABQ/\Im$, where $\Im$ is the moment of inertia of the coil. We analyse the subsequent motion of the coil assuming it to remain connected to a circuit of total resistance $R$ (including the coil resistance) but with no external e.m.f. The differential equation of the motion will be the same as (5.49) but with $V = 0$, so that we have

$$\Im \frac{d^2\theta}{dt^2} + b' \frac{d\theta}{dt} + c\theta = 0, \tag{5.51}$$

where $b' = b + (NAB)^2/R$ is the total effective damping, which we shall assume to be small. Then the solution is

$$\theta = D \exp(-b't/2\Im)\sin \omega t,$$

where $\omega = (c/\Im)^{\frac{1}{2}}$ to a good approximation. The constant $D$ is found by differentiation, and setting the initial angular velocity equal to $NABQ/\Im$, giving

$$NABQ/\Im = (d\theta/dt)_{t=0} = D\omega,$$

whence

$$\theta = (NABQ/\Im\omega)\exp(-b't/2\Im)\sin \omega t. \tag{5.52}$$

If there were no damping the deflection would oscillate between maximum and minimum values $\pm\theta_0$ occurring at the instants when $\sin \omega t = 1$, and the charge $Q$ would be given by

$$Q = \frac{\Im\omega\theta_0}{NAB} = \frac{c\theta_0}{\omega NAB} = \theta_0 \frac{\tau}{2\pi} \cdot \frac{c}{NAB}, \tag{5.53}$$

where we have used in succession the relations $(\Im/c)\omega^2 = 1$ and $\omega = 2\pi/\tau$, where $\tau$ is the period of swing of the galvanometer. This can be measured experimentally, and the constant $c/NAB$ can be found from the deflection produced by passing a steady current through the galvanometer.

The effect of the damping is to make successive deflections smaller (cf. Fig. 5.13), and also to make the first throw $\theta_1$ (at $t = \frac{1}{4}\tau$) smaller than $\theta_0$,

Fig. 5.18. Apparatus for calibrating a ballistic galvanometer.
G  galvanometer,
M  standard mutual inductance,
P  potentiometer connected across standard 1 Ω resistance.

since from eqn (5.52)

$$\theta_1 = \theta_0 \exp(-b'\tau/8\Im) \approx \theta_0(1-\tfrac{1}{4}\Lambda).\qquad(5.54)$$

The size of the correction can be found by measuring the logarithmic decrement $\Lambda$ of successive swings on the same side, since

$$\Lambda = \ln(\theta_1/\theta_2) = \ln(\theta_2/\theta_3) = \ldots = b'\tau/2\Im.$$

For accurate use the damping of the ballistic galvanometer should be small and the resistance of the external circuit must therefore be high. If used to measure the charge on a capacitor by connecting the capacitor to the galvanometer and allowing it to discharge through the galvanometer, this condition is fulfilled because the capacitor is effectively an open-circuit for the slowly varying induced e.m.f. produced by the galvanometer swing. A convenient method of calibrating the ballistic galvanometer directly is by discharging through it a known capacitance charged to a known voltage. An alternative method is by means of a standard mutual inductance, using the circuit of Fig. 5.18. The secondary of the mutual inductance is connected to the galvanometer, and a known current is reversed in the primary coil. If $L$ is the inductance of the secondary windings and $R$ the total resistance in the secondary circuit, the secondary current $I$ flowing at any instant is given by the equation

$$L\frac{dI}{dt} + RI = -\frac{d\Phi}{dt},$$

where $d\Phi/dt$ is the rate at which the flux is changing through the secondary coil. Integrating over the duration of time occupied by the flux change (assuming that the time constant $R/L$ is very short compared with the period of the galvanometer) gives

$$\int -\frac{d\Phi}{dt}\,dt = \Phi_1 - \Phi_2 = L\int\frac{dI}{dt}\,dt + R\int I\,dt =$$
$$= L[I]_0^0 + RQ = RQ,\qquad(5.55)$$

showing that the charge measured by the galvanometer is just equal to the total flux change in the secondary divided by the total resistance. If $I_0$ is the current reversed in the primary, $\Phi_1 - \Phi_2 = 2MI_0$, and so $Q = 2MI_0/R$. The current $I_0$ may either be measured by an accurate ammeter, or by means of the potential drop it produces across a known resistance. To reduce the damping of the ballistic galvanometer, either $R$ must be large, which in general will mean adding considerable resistance to the circuit and thus reducing the sensitivity of the galvanometer, or the secondary circuit must be broken immediately after the primary current is reversed and before the galvanometer has deflected appreciably.

Once the ballistic galvanometer has been calibrated, it may be used to measure a flux change or an unknown mutual inductance if the resistance $R$ is known. An alternative method is to use a fluxmeter, which gives a direct reading of the flux change. This instrument consists of a moving coil, suspended in such a way as to give almost zero restoring torque, in a strong magnetic flux density of a permanent magnet. The electromagnetic damping is very strong, and if other forms of damping can be neglected, the equation of motion is

$$\mathfrak{I}\frac{d^2\theta}{dt^2} + \frac{(NAB)^2}{R}\frac{d\theta}{dt} = NABI = \frac{NAB}{R}\left(-\frac{d\Phi}{dt} - L\frac{dI}{dt}\right), \qquad (5.56)$$

where $I$ is the instantaneous current produced by the e.m.f. $-d\Phi/dt$ induced by the flux change to be measured, and $L$ is the self-inductance. The effect of this current is to impart an angular momentum to the fluxmeter coil, which is then brought to rest under the action of the electromagnetic damping. Integration of eqn (5.56) over the entire time occupied by the motion of the coil gives

$$\mathfrak{I}\left[\frac{d\theta}{dt}\right]_0^0 + \frac{(NAB)^2}{R}[\theta]_0^{\theta_0} = \frac{NAB}{R}(\Phi_1 - \Phi_2) - \frac{NABL}{R}[I]_0^0,$$

and, since the angular velocity and current are zero both initially and finally,

$$NAB\theta_0 = \Phi_1 - \Phi_2, \qquad (5.57)$$

where $\theta_0$ is the ultimate deflection of the fluxmeter. The flux change through the fluxmeter coil is just $NAB\theta_0$, and so this equation shows that the total flux threading the system is the same at the beginning and the end. The fluxmeter is generally calibrated directly in terms of flux; since there is no restoring torque on the coil it has no stable position of equilibrium but can rest anywhere in neutral equilibrium. It is therefore necessary to level the instrument carefully to prevent the coil drifting during the measurement. As a rough rule, a fluxmeter is accurate to about 1 per cent, while a ballistic galvanometer, properly calibrated, is accurate to about 0·1 per cent.

## 5.8. Absolute measurements

Relative measurements of electrical quantities can be made using bridges or potentiometers to an accuracy of some parts in $10^5$. In national Standards Laboratories the legal standard of current has usually been based on the mass of an electrolyte deposited in electrolysis in a given time (see § 3.9), while the legal standard of e.m.f. is based on a standard cell (see § 3.10). A standard resistance can then be established by passing a known current through it and comparing the potential drop across it with that of a standard cell.

For absolute measurements it is necessary to use methods in which an electrical quantity can be determined in terms of mechanical quantities using the fundamental laws of electromagnetism. One such method, for the absolute determination of current, is based on the equations of § 4.1—a current is passed through a pair of conductors, and the force between them is measured. The method of doing this, by means of a 'current balance', is illustrated in Fig. 5.19. A, B, C, D, E, F are six single-layer coils wound on marble formers and connected in series. Coils A, C, D, and F are fixed, while B and E are carried on a balance arm; the current flows through the various coils in such a direction that the force on E is upwards while that on B is downwards. The balance arm is brought back to its equilibrium position by moving a standard mass along a calibrated scale on the arm. From the distance of the mass from the fulcrum, the torque and hence the force between the coils can be evaluated. If M is the mutual inductance between B or E and either of the coils A and C or D and F, then the total torque due to the coils when

FIG. 5.19. Current balance.

FIG. 5.20. Lorenz's apparatus for determining a resistance $R$ in terms of $\mu_0$ and the metre. CC', DD' are coils through which $I$ flows (in opposite senses) producing the magnetic fields in which the plates PP' are rotated.

carrying a current $I$ is $4I^2(\mathrm{d}M/\mathrm{d}\theta)$, where $\theta$ is the angle defining the rotation of the balance arm. Since $\mathrm{d}M/\mathrm{d}\theta$ is calculated from the geometry, there must be no iron near the apparatus. Effects due to the more remote coils (that is, for example, the force between coil A and coil E) are eliminated by repeating the readings with the current through all the coils on one side reversed. An accuracy of one part in $10^5$ can be achieved, but the measurements are very laborious.

The resistance of a coil was determined by Rayleigh in terms of the constant $\mu_0$ and the standard of length, by rotating the coil in a magnetic field (see Problem 5.7). Here we shall describe in some detail the method used by Lorenz, which has the advantages first of being a null method and secondly of not needing the resistance to be in the form of a coil. If a disc-shaped conductor is rotated in a magnetic flux density **B** with a frequency of $f$ revolutions per second, and **B** is parallel to the axis of rotation, a voltage is induced in the disc between the axis and the rim of magnitude $V = kfB$, where $k$ is a constant (with the dimensions of an area: see Problem 5.6) determined by the geometry of the apparatus. In Lorenz's experiment (Fig. 5.20) the flux density **B** is produced by coils carrying a current $I$ which also flows through the resistance $R$ to be standardized. Then $B = \mu_0 I/k'$, where $k'$ is a quantity with the dimensions of a length, again fixed by the geometry of the apparatus. The voltage $V$ is balanced against the potential drop $RI$ across the resistance, so that $RI = (k/k')\mu_0 fI$, and at the balance point

$$R = (k/k')\mu_0 f. \tag{5.58}$$

The frequency $f$ can be measured accurately, $\mu_0 = 4\pi \times 10^{-7}\,\mathrm{H\,m^{-1}}$ by definition and $k/k'$ has the dimensions of a length. In fact $\mu_0(k/k')$ is the mutual inductance between the field coil and the rotating conductor, which can be calculated from the dimensions and the geometry. Fundamentally the comparison is between the resistance and the product inductance × frequency, the rotating conductor being a device by which a steady voltage is developed across the inductance for balance against the voltage drop in the resistance.

The arrangement of the apparatus is shown in Fig. 5.20. The rotating conductor is formed by the discs P and P' mounted on a shaft driven by a motor. The galvanometer G reads zero when the voltage induced in the circuit *abcd* is exactly equal to the potential across the resistance $R$, which is in series with the two coils CC' and DD'. The current flows through these coils in the opposite sense so that the voltages induced in the two discs add round the circuit *abcd*. Contact with the discs at $a$ and $d$ is made by brushes, and the coils are designed so that the magnetic flux density at these contacts is zero; this minimizes any error in determining the actual radius of the disc, that is, the distance from the centre at which contact is made. Some trouble arises from thermal e.m.f.s at the brush contacts, which would give a finite galvanometer reading at the true balance point. This difficulty is overcome by reversing the current throughout the whole system, which reverses any true unbalance current but not that due to the thermal e.m.f. Hence the correct balance point is when the galvanometer reading remains unchanged on reversing the current, and this procedure also eliminates the effect of any voltages induced by stray magnetic fields. With adequate precautions it is possible to obtain an accuracy of a few parts in $10^5$; other resistances can then be determined relative to the standard by means of a Wheatstone's bridge.

In these methods the value of a resistance is found in absolute units by comparing it with an inductance whose value is $\mu_0$ times a factor with the dimensions of a length. The capacitance $C$ of a capacitor is $\epsilon_0 l$, where $l$ is a factor with the dimensions of a length which is determined by the geometry of the capacitor. Thus if $l$ is calculated, and the capacitance $C$ is determined by comparison with a standard resistance $R$, the value of the constant $\epsilon_0$ can be found. An accurate experiment of this type was carried out by Rosa and Dorsey in 1907; the apparatus consists essentially of the bridge circuit shown in Fig. 5.21. The contact X vibrates between P and Q at a known frequency, and the capacitor of capacitance $C$ is alternately charged and discharged $f$ times a second. The voltage $V$ across it is $Q/C$, and the charging of the condenser $f$ times a second sends a current $fQ$ through the arm AB, so that the effective resistance of this arm is $V/(fQ) = 1/(fC)$. Hence at the balance point, $R_2/R_1 = fCR$. Rosa and Dorsey used both spherical and cylindrical capacitors, and took many

FIG. 5.21. Determination of a capacitance $C$ in terms of a standard resistance $R$.

readings to correct for a large variety of possible errors. Their final result was $\epsilon_0 = (8\cdot8547)10^{-12}$ F m$^{-1}$, with an accuracy of about 4 parts in $10^5$.

### Reference

HEINE, V. (1956). *Proc. Camb. Phil. Soc.* **52**, 546.

### Problems

5.1. Two infinite coaxial cylinders carry equal and opposite currents $I$. Calculate the self-inductance $L$ per unit length (eqn (5.18)) by equating the stored energy per unit length to $\frac{1}{2}LI^2$.

5.2. Prove that the inductance per unit length of two infinite parallel wires of radius $a$ separated by a distance $2d$ ($2d \gg a$) is

$$L = (\mu/\pi)\ln(2d/a).$$

Verify that $LC = \mu\epsilon$, where $C$ is the capacitance per unit length (§ 2.6).

5.3. Prove that the inductance of a long solenoid of length $l$, radius $a$, and with $m$ turns per unit length is approximately

$$L = \pi m^2 a^2 \mu\{(l^2+a^2)^{\frac{1}{2}}-a\}$$

(assume that the field is uniform over any cross-section: if $l > 10a$, this formula is accurate within 2 per cent).

5.4. Show that the mutual inductance between two coplanar coaxial coils of radii $a$ and $b$ ($b \gg a$), with turns $n_1$ and $n_2$ respectively, is approximately (*in vacuo*)

$$M = \mu_0\pi a^2 n_1 n_2/2b.$$

Use this result to show that the flux through the larger coil due to a small magnet of magnetic dipole moment $\mathfrak{m}$ placed at its centre and pointing along its axis is

$$\Phi = \mu_0 \mathfrak{m} n_2/2b.$$

5.5. If the magnet in the last problem is withdrawn along the axis at a uniform velocity $v$, show that the e.m.f. induced in the coil when the magnet is at a distance $z$ from the centre of the coil is

$$V = 3\mu_0 m n_2 z v b^2 / 2 (b^2 + z^2)^{\frac{5}{2}}.$$

If the resistance of the coil is $R$, show that the total charge which flows when the magnet is removed from the centre to infinity is

$$Q = \mu_0 m n_2 / 2bR.$$

5.6. A plane circular disc of radius $a$ rotates at a speed of $f$ revolutions per second about an axis through its centre normal to its plane. A uniform flux density **B** exists parallel to this axis. Show from first principles that there is an e.m.f. between the centre of the disc and its rim of magnitude $V = fB\pi a^2$. (Lorenz's method of determining the unit of resistance depends on this result; see § 5.8)

5.7. A circular coil of $n$ turns of radius $a$, total resistance $R$, and no self-inductance is rotated with uniform angular velocity $\omega$ about a vertical diameter in a horizontal flux density **B**. Prove that the mean power required to maintain the coil in motion is $\frac{1}{2}n^2\pi^2 a^4 B^2 \omega^2 / R$, and that this is equal to the power dissipated in the resistance of the coil.

A small magnetic needle, which is free to turn slowly in a horizontal plane, is placed at the centre of the coil. Show that it will set at an angle $\phi$ to **B**, where

$$\cot \phi = 4R / \pi n^2 \mu_0 \omega a.$$

(Rayleigh's method of determining the unit of resistance is based on an experiment of this type.)

5.8. A torsional pendulum consists of two spheres of 1 cm diameter at either end of a thin rod 10 cm long suspended at its midpoint. It swings in a horizontal plane so that its instantaneous angular deflection is $\pi \cos \frac{1}{2}\pi t$. Find the magnitude and direction of the current flowing in the rod at any instant, assuming that the capacity of each sphere is the same as if each were isolated in space, and the rod and suspension have negligible resistance. Show that there will be damping of the swing if the suspension has a finite resistance, but not otherwise. (Vertical component of earth's magnetic flux density $= 6 \times 10^{-5}$ T.)

(*Answer:* $3 \cdot 2 \times 10^{-19} \cos \frac{1}{2}\pi t$ A.)

5.9. An aeroplane is in level flight at a ground speed of 300 km h$^{-1}$. Its metal propeller measures 3 m from tip to tip and rotates at 3000 revolutions per minute. Find an expression for the potential difference between the ends of the propeller when the aeroplane is flying along the magnetic meridian. (Vertical component of earth's magnetic flux density $= 6 \times 10^{-5}$ T.)

(*Answer:* $0 \cdot 015 \cos 100\pi t$ V)

5.10. Show that for a uniformly magnetized spherical permanent magnet, $\int \frac{1}{2}\mathbf{B}.\mathbf{H}\,d\tau$, integrated over the volume inside the sphere, is just equal and opposite to the value of the integral over the volume outside the sphere. Thus $\int \frac{1}{2}\mathbf{B}.\mathbf{H}\,d\tau$ over the whole of space is zero. (This follows because there are no currents—see the derivation of $U = \int \frac{1}{2}\mathbf{B}.\mathbf{H}\,d\tau$ in § 5.5.)

5.11. The magnetostatic energy of a permanent magnet is $\int \frac{1}{2}\mu_0 H^2\,d\tau$ taken over all space. A solid spherical permanent magnet of radius $b$ is uniformly magnetized with a magnetization $M$ per unit volume. Show that the total energy stored in the field outside the sphere is $\mu_0 VM^2/9$, where $V$ is the volume of the sphere. Show also that the energy (thus defined) stored in the field inside the sphere is $\mu_0 VM^2/18$. Thus the total energy is $\mu_0 VM^2/6$, and this is the magnetostatic energy of the sphere.

(*Hint.* Use the result of Problem 4.7. The question of the magnetostatic energy of a permanent magnet is discussed by Heine (1956).)

5.12. A bridge consists of a self-inductance $L$ and resistance of $\sim 10\,\Omega$ in one arm and three non-inductive resistances in the remaining arms, so that an accurate steady balance is obtained. If, with the galvanometer in circuit, the battery key is depressed, a ballistic deflection of 10 cm of the galvanometer is recorded. If in a second experiment a resistance of $0.02\,\Omega$ is connected in series with $L$, a steady deflection of 12 cm is obtained with the key depressed. The galvanometer is a moving-coil instrument with a period of 9 s. A switch in the galvanometer circuit is opened immediately after the flow of charge through it when it is used ballistically. Show that $L$ is approximately 24 mH.

5.13. A ballistic galvanometer is calibrated by putting it in series with a 2-V battery and a resistance of $10^6\,\Omega$. A steady deflection of 17 cm is observed. The time of swing is 3.8 s. A capacitor is charged by a 4-V battery and when discharged through the galvanometer gives a throw of 24.2 cm. Find its capacitance.

(*Answer:* $C = 0.43\,\mu\text{F}$.)

5.14. A small search coil with 8 turns of mean area 1.5 cm$^2$ is placed between the poles of an electromagnet with its plane normal to the magnetic field. It is connected to a ballistic galvanometer and the total resistance of the circuit is $1000\,\Omega$. When the coil is suddenly removed to a place where the field is negligible, the throw of the galvanometer is 23 divisions. When a capacitance of $1\,\mu\text{F}$ charged to 1 V is discharged through the galvanometer, the throw is 25 divisions. Calculate $B$ between the poles of the magnet.

(*Answer:* $B = 0.77\,\text{T}$.)

5.15. A ballistic galvanometer gives a throw of 10 cm when a charge of $3.5 \times 10^{-7}$ C is passed. Its period is 2.2 s. Calculate the deflection when a steady current of $3\,\mu\text{A}$ is passed.

(*Answer:* 30 cm.)

5.16. A capacitance of $10\,\mu\text{F}$ is connected across a moving coil ballistic galvanometer. If the galvanometer coil turns through 0.05 rad for $1\,\mu\text{A}$ of current, has a moment of inertia of $10^{-6}$ kg m$^2$, a period of 10 s, and negligible resistance, show that the fractional increase in the period is $\sim 2 \times 10^{-3}$.

5.17. In the circuit of Fig. 5.17, the moving coil (in series with $R$) may be connected either as shown or between the points A and C. Show that the fractional error in the reading of the power in the load (resistance $Z$) is less with the latter method of connection if $Z > (RR')^{\frac{1}{2}}$, and vice versa, where $R'$ is the resistance of the fixed coil.

5.18. A 100-V dynamo is connected to a magnet whose resistance is $10\,\Omega$ and self-inductance 0.1 H. Show that the percentage increase in the heating of the magnet caused by the presence of a 100 Hz ripple voltage of 5 V amplitude in the output of the dynamo is 0.0031 per cent.

# 6. Magnetic materials and magnetic measurements

## 6.1. Origins of magnetism

THE fact that a substance placed in a magnetic field acquires a magnetic moment has been discussed in the theory of magnetostatics in § 4.3. The ratio of the magnetic moment per unit volume to the magnetic field strength $H$ is known as the susceptibility $\chi$, and substances are classed as diamagnetic, paramagnetic, or ferromagnetic according to the nature of their susceptibility. In the first two of these classes the induced magnetization is proportional to the applied field under ordinary conditions, so that the susceptibility is independent of the field strength. In diamagnetic substances the magnetization is in the opposite direction to the applied field, so that $\chi$ is negative, while in paramagnetic substances it is in the same direction, giving a positive value of $\chi$. Ferromagnetic substances are distinguished by very large (positive) values of $\chi$, which are not independent of the field strength; in addition they may possess a magnetic moment even in the absence of an applied field, as in a permanent magnet. In general magnetic properties are both non-linear and anisotropic; in a crystal of less than cubic symmetry the magnetic susceptibility is a tensor quantity, as in the case of the electric susceptibility (see § 1.5). Anisotropic effects are considered in Chapters 14–16; in this chapter we give a brief description of the origins of the magnetization, and describe methods of measuring magnetic properties.

After the discovery that a small coil carrying a current behaves like a magnet, Ampère suggested that the origin of all magnetism lay in small circulating currents associated with each atom. These 'amperean currents' would each possess a magnetic dipole moment, and the total magnetic moment of any substance would be just the vector sum of the magnetic dipole moments of its constituent atoms. This gave a natural explanation of the fact that no isolated magnetic pole had ever been observed, since even on the atomic scale only dipoles existed, and these were due to electric currents and did not consist of two actual magnetic poles of opposite sign separated by a small distance. Ampère's theory is essentially the same as that of modern atomic physics, the origin of his elementary current circuits being the motion of the negatively charged electrons in closed orbits round the positively charged atomic nucleus.

When a particle is moving in a closed orbit in a system where no external torque is acting, its angular momentum is constant. If the particle is charged, a magnetic moment will be associated with its motion, and there is a linear relation between the angular momentum and the magnetic moment (see Problem 4.8 for the simple case of a circular orbit). A general expression for the magnetic dipole moment associated with a distributed current has already been found in § 4.5. In eqn (4.59) we can replace $\mathbf{J}\,d\tau$ by $\mathbf{v}\,dq$ to find the equivalent expression for a moving charge, giving

$$\mathbf{m} = \int \tfrac{1}{2}(\mathbf{r} \wedge \mathbf{v})\,dq. \tag{6.1}$$

In the absence of an external force, the quantity $(\mathbf{r} \wedge \mathbf{v})$ is a constant, since $m(\mathbf{r} \wedge \mathbf{v})$ is equal to $\mathbf{G}$, the angular momentum for a particle of mass $m$. Hence

$$\mathbf{m} = \mathbf{G} \int (dq/2m) = (q/2m)\mathbf{G} = \gamma \mathbf{G}, \tag{6.2}$$

where $q = \int dq$ is the total charge circulating in the orbit. The quantity $\gamma = q/2m$ is called the magnetogyric ratio.

This close relation between magnetic moment and angular momentum is of great importance in atomic theories of magnetism, because on quantum theory the angular momentum of an electron in an atom, which is a constant of the motion, can only have discrete values. The electron possesses angular momentum not only in respect of its orbital motion round the nucleus, but also in respect of its intrinsic rotation (spin) about its own axis. The resultant angular momentum of an atom is the vector sum of the individual angular momenta of its electrons, and the resultant magnetic moment is a similar vector sum of the individual magnetic moments of the electrons. Because of the linear relation between the two, an atom, ion, or molecule will have no resultant permanent magnetic moment if the total angular momentum is zero. If the total angular momentum is not zero, the atom, ion, or molecule will have a permanent magnetic dipole moment. Most free atoms possess a permanent magnetic dipole moment because they have a resultant electronic angular momentum. Experiments such as the gyromagnetic effect (§ 15.4) in solids, and magnetic resonance (Chapter 24) for free particles and solids, give precise determinations of the ratio of the magnetic moment to the angular momentum for nuclei, and atoms and molecules, as well as magnetic ions in solids.

Such magnetic dipole moments are fundamentally different from electric dipole moments, where it has been shown (§ 2.3) that if parity is conserved in an atom or nucleus, no electric dipole moment can exist. The difference is clear if we write out the components of the magnetic

(a)                                                    (b)

FIG. 6.1. (a) Reflection of circulating currents in a plane. (b) Reflection of dipoles in a plane. Note that the circulating currents behave differently from the equivalent dipoles on reflection: the magnetic dipole is an axial vector, the electric dipole a polar vector (see Appendix A.1).

dipole moment given by eqn (6.1); for example,

$$m_z = \int \tfrac{1}{2}\left(x\,\frac{\partial y}{\partial t} - y\,\frac{\partial x}{\partial t}\right) dq.$$

If parity is conserved, we have inversion symmetry, that is, the value of $dq$ is the same at the point $(-x, -y, -z)$ as at the point $(x, y, z)$. So also is the quantity $\{x(\partial y/\partial t) - y(\partial x/\partial t)\}$, so that the integrand retains the same sign under the inversion operation, and the integral can have a finite value. This differs from the electric dipole moment in that the latter involves the first power of the coordinates (see eqn (2.25)), while the magnetic dipole moment involves the second power. The difference shows clearly under reflection in a plane, see Fig. 6.1; also, magnetic moments change sign under time reversal, $t \to -t$.

In general, however, atoms do not exist in the free state but are combined into molecules, and it so happens that the forces responsible for chemical binding strongly affect the arrangement of the individual magnetic moments of the various electrons in the molecule. As a result the stable state of the molecule is nearly always one in which the vector sum of these individual moments is just zero, so that the molecule as a whole has no resultant permanent magnetic moment. Similarly, in the majority of solids and liquids, the atomic constituents are ions with no permanent magnetic dipole moments. When such a substance is placed in a magnetic flux density **B**, each individual electron, being a moving charge, experiences a force. Its orbital motion round the nucleus is altered in such a way that it acquires an angular momentum and hence a magnetic dipole moment which is proportional to the applied field, but in the opposite direction. This gives a negative susceptibility, whose value will now be calculated. It should be noted that this diamagnetic effect is present in all substances, but in paramagnetic substances there is a much larger positive contribution to the susceptibility which generally far outweighs the diamagnetic contribution. The positive contribution arises from molecules (or ions) where the vector sum of the individual electron moments is not zero, so that the molecule or ion possesses a permanent magnetic moment. Thus all substances show a diamagnetic effect, giving a susceptibility of the order of $10^{-5}$ in S.I. units, but in paramagnetic substances there is a positive susceptibility contribution which is much greater.

## 6.2. Diamagnetism

The fact that an atom placed in a magnetic flux density $\mathbf{B}$ acquires a magnetic moment parallel to $\mathbf{B}$ can be shown by finding the effect on the atom of establishing the field $\mathbf{B}$ from zero. In § 5.1 it was shown that this changes the momentum of a particle of charge $q$ from $m\mathbf{v}_0$ to $m\mathbf{v}$, where

$$m\mathbf{v} = m\mathbf{v}_0 - q\mathbf{A}$$

and $\mathbf{A}$ is the vector potential associated with $\mathbf{B}$. The corresponding current density is given by

$$\mathbf{J} = \rho_e \mathbf{v} = \rho_e \mathbf{v}_0 - \rho_e (q/m)\mathbf{A}. \tag{6.3}$$

If $\mathbf{A}$ is increased by $\delta\mathbf{A}$, the increase in potential energy is, by eqn (4.52),

$$\delta U_P = -\int (\mathbf{J} \cdot \delta\mathbf{A})\, d\tau = -\int \rho_e (\mathbf{v}_0 \cdot \delta\mathbf{A})\, d\tau + \int \rho_e (q/m)(\mathbf{A} \cdot \delta\mathbf{A})\, d\tau. \tag{6.4}$$

For simplicity we take $\mathbf{B}$ to be uniform (which it will be over atomic dimensions), so that

$$\mathbf{A} = \tfrac{1}{2}(\mathbf{B} \wedge \mathbf{r}), \quad \text{and} \quad (\mathbf{A} \cdot \delta\mathbf{A}) = \tfrac{1}{4}(B\,\delta B)r^2 \sin^2\theta,$$

where $\theta$ is the angle between $\mathbf{B}$ and $\mathbf{r}$. Hence

$$\delta U_P = -\int \tfrac{1}{2}\rho_e \mathbf{v}_0 \cdot (\delta\mathbf{B} \wedge \mathbf{r})\, d\tau + (q/4m)\int \rho_e B\,\delta B r^2 \sin^2\theta\, d\tau$$

$$= -\delta\mathbf{B} \cdot \int \tfrac{1}{2}\rho_e(\mathbf{r} \wedge \mathbf{v}_0)\, d\tau + (q/4m)B\,\delta B \int \rho_e r^2 \sin^2\theta\, d\tau$$

$$= -\delta\mathbf{B} \cdot \mathbf{m}_0 - \delta\mathbf{B} \cdot \mathbf{m}_i = -\delta\mathbf{B} \cdot \mathbf{m}. \tag{6.5}$$

This result shows that the magnetic moment $\mathbf{m}$ consists of two parts, of which the first, $\mathbf{m}_0$, is clearly the permanent magnetic dipole moment (see eqn (6.1)), while the second $\mathbf{m}_i$ is an induced dipole moment which is proportional to the flux density $\mathbf{B}$. The value of $\mathbf{m}_i$ is

$$\mathbf{m}_i = -\mathbf{B}(q/4m)\int \rho_e r^2 \sin^2\theta\, d\tau = -\mathbf{B}(q^2/4m)\langle r^2 \sin^2\theta\rangle,$$

where $q = \int \rho_e \, d\tau$ is the total charge. If $\rho_e$ is independent of $\theta$,

$$\langle r^2 \sin^2\theta \rangle = \langle r^2 \rangle \frac{1}{4\pi} \int_0^\pi \sin^2\theta . 2\pi \sin \theta \; d\theta = \tfrac{2}{3}\langle r^2 \rangle,$$

where $\langle r^2 \rangle$ is the mean-square radius of the orbit. On summing over all electrons in the atom or ion, $\mathbf{m}_0$ must be replaced by the resultant (if any) of the permanent dipole moments, while the induced moment per atom becomes

$$\mathbf{m}_i = -(e^2/6m_e)\mathbf{B} \sum \langle r^2 \rangle, \tag{6.6}$$

where the electronic charge $-e$ and mass $m_e$ have been inserted. Hence the diamagnetic susceptibility of a sample containing $n$ atoms (or ions) per unit volume will be

$$\chi = n\mathbf{m}_i/\mathbf{H} = -n\mu_0(e^2/6m_e) \sum \langle r^2 \rangle. \tag{6.7}$$

This result can be derived in another way using Larmor's theorem (Appendix A, § A.11). The result of establishing a field of flux density $\mathbf{B}$ is to set up a precessional motion of the electronic orbits with angular velocity $\boldsymbol{\omega} = -(q/2m)\mathbf{B}$, as a result of which each electron acquires an additional angular momentum

$$\mathbf{G}_i = m\langle a^2 \rangle \boldsymbol{\omega} = -m\langle a^2 \rangle(q/2m)\mathbf{B},$$

where $\langle a^2 \rangle = \langle r^2 \sin^2\theta \rangle$ is the mean-square distance of the electron from an axis parallel to $\mathbf{B}$ through the centre of gravity of the atom (the nucleus). Associated with $\mathbf{G}_i$ is an additional magnetic moment of magnitude (from eqn (6.2))

$$\mathbf{m} = (q/2m)\mathbf{G}_i = -(q^2/4m)\langle r^2 \sin^2\theta \rangle \mathbf{B}.$$

On writing $q = -e$, $m = m_e$ and summing over all electrons in the atom, this gives eqn (6.7) above.

This equation shows that the diamagnetic susceptibility is inherently negative in sign, and does not depend, for example, on the sign of the electronic charge. Fundamentally, the negative sign follows from Lenz's law. When the external magnetic flux density $\mathbf{B}$ is switched on, there is a change in the flux through the electron orbits which induces a momentary e.m.f. The change in the orbits which this causes gives an induced magnetic moment to the atom which is in such a direction as to oppose the change in flux through the orbit, that is, the magnetic moment is due to an induced current whose own lines of magnetic flux through the atom are in the opposite direction to those of the external flux. We have assumed that $\langle r^2 \rangle$ is unaltered by the presence of the magnetic field; this is justified because the magnetic forces at ordinary field strengths are negligible compared with the internal atomic forces (see Appendix § A.11).

As defined in Chapter 4, the susceptibility refers to unit volume of substance, and $n$ is then the number of atoms in unit volume. Since $\chi$ is linearly proportional to $n$, it is permissible to take samples of different size, and refer to the 'susceptibility per unit mass', or 'susceptibility per gram atom'. These are simply equal to the susceptibility per unit volume (or 'volume susceptibility' for short) multiplied by the volume of the sample. Thus the 'mass susceptibility' $\chi_m = \chi/\rho$, where $\rho$ is the density, since unit mass (1 kg) occupies a volume of $1/\rho$ m$^3$. Similarly, the molar susceptibility will be $\chi_M = 10^{-3} M\chi_m$, where $M$ is the atomic or molecular weight in grams. Here the factor $10^{-3}$ occurs because our $\chi_m$ refers to a kilogram of substance. For a mole eqn (6.7) gives the numerical value

$$\chi_M = -3.55 \times 10^9 \sum \langle r^2 \rangle. \tag{6.8}$$

In some reference tables the susceptibility is given in electromagnetic units, and for solids and liquids the diamagnetic volume susceptibility (per cm$^3$) is of the order $-10^{-6}$ e.m.u. (for gases it is much smaller owing to the lower number of atoms per unit volume). In S.I. units, however, the volume susceptibility (per m$^3$) is greater by a factor $4\pi$ and so is of order $-10^{-5}$. As a rough rule, the molar susceptibility in S.I. is of order $-10^{-11}$ Z, where $Z$ is the atomic number, equal to the number of electrons in the atom. This indicates that the mean value of $\langle r^2 \rangle$ is about $10^{-21}$ m$^2$, as expected from atomic theory. The susceptibility of a diamagnetic substance is substantially independent of temperature, since $\sum \langle r^2 \rangle$ is practically unaltered by thermal motion.

In the first approximation $\sum \langle r^2 \rangle$ is constant for a particular type of atom or ion, and is not greatly altered by its surroundings. Thus aqueous solutions of alkali and alkaline earth halides obey quite accurately (and most substances approximately) an additivity rule known as Wiedemann's law. According to this rule the mass susceptibility $\chi_m$ of a solution containing a mass $m_1$ of a salt of mass susceptibility $\chi_1$ in a mass $m_2$ of solvent of mass susceptibility $\chi_2$ is

$$\chi_m = \frac{m_1\chi_1 + m_2\chi_2}{m_1 + m_2}. \tag{6.9}$$

Similar additivity rules are approximately valid for chemical compounds. Thus, for compounds which ionize in solution, the molar susceptibility of the compound is generally close to the sum of the ionic susceptibilities of its constituent ions in solution. By assuming a theoretical value for the susceptibility of one type of ion, approximate values for the susceptibilities of other ions can be obtained from measurements on compounds, and the validity of the additivity rules tested. (For a review of the diamagnetism of ions see Myers (1952).)

## 6.3. Paramagnetism

As already pointed out, paramagnetism occurs in those substances where the individual atoms, ions, or molecules possess a permanent magnetic dipole moment. In the absence of an external magnetic field, the atomic dipoles point in random directions and there is no resultant magnetization of the substance as a whole in any direction. This random orientation is the result of thermal agitation within the substance. When an external field is applied, the atomic dipoles tend to orient themselves parallel to the field, since this is a state of lower energy than the antiparallel position. This gives a net magnetization parallel to the field, and a positive contribution to the susceptibility. Since the thermal agitation, which tends to give a random orientation, is less at low temperatures, a bigger proportion of the dipoles are able to align themselves parallel to the field, and the magnetization is greater for a given field. It was discovered by Curie that, for ordinary fields and temperatures, the susceptibility of many substances follows the equation

$$\chi = M/H = C/T, \tag{6.10}$$

where $C$ is a constant, and $T$ is the absolute temperature; this is known as Curie's law. For large fields at low temperatures the magnetization produced is no longer proportional to the applied field, and tends to a constant value. This saturation effect is produced when all the atomic dipoles are aligned parallel to the field, so that the magnetization reaches a limiting maximum value.

The theoretical explanation of Curie's law was given by Langevin, using the classical statistics of Boltzmann. He assumed that each atom had a permanent magnetic moment $m$, and that the only force acting on it was that due to the external field **B**. Then, if a given atomic dipole is pointing in a direction making an angle $\theta$ with **B**, its magnetic potential energy is $W = -mB \cos \theta$. Now, on classical statistics, the number of atoms making an angle between $\theta$ and $\theta + d\theta$ is

$$dn = ce^{-W/kT} \sin \theta \, d\theta, \tag{6.11}$$

where $k$ is Boltzmann's constant and $T$ is the absolute temperature. $c$ is a constant defined by the fact that integration of (6.11) over the whole possible range of energies must give just $n$, the total number of atoms in the system. Hence for our case

$$dn = ce^{mB \cos \theta/kT} . \sin \theta \, d\theta$$

and $n$ is equal to this integrated over all angles from 0 to $\pi$. The component of each dipole moment parallel to **B** is $m \cos \theta$, and hence the average component per atom is $\bar{m}$, where

$$n\bar{m} = m \int \cos \theta \, dn.$$

Hence

$$\frac{\overline{m}}{m} = \frac{\displaystyle\int_0^\pi \cos\theta \, . \, c e^{mB\cos\theta/kT} \, . \, \sin\theta \, d\theta}{\displaystyle\int_0^\pi c e^{mB\cos\theta/kT} \, . \, \sin\theta \, d\theta}$$

On writing $mB/kT = y$, $\cos\theta = x$, this takes the form

$$\frac{\overline{m}}{m} = \frac{\displaystyle\int_{-1}^{+1} x e^{xy} \, dx}{\displaystyle\int_{-1}^{+1} e^{xy} \, dx} = \coth y - \frac{1}{y} = L(y), \tag{6.12}$$

where $L(y)$ is known as the Langevin function. It is plotted as a function of $y$ in Fig. 6.2. For large values of $y$ the function tends to unity, saturation being reached when all the atomic dipoles are parallel to **B**. For small values of $y$ the curve is linear, and $L(y) = y/3 = mB/3kT$. Then the susceptibility is

$$\chi = n\overline{m}/H = \mu_0 nm^2/3kT, \tag{6.13}$$

where $n$ is the number of atoms per unit volume. This is the same as Curie's law, eqn (6.10), if we identify the Curie constant $C$ with $\mu_0 nm^2/3k$. The only unknown quantity in (6.13) is the atomic dipole moment $m$ so that by measuring the susceptibility as a function of the absolute temperature in a region where $mB/kT$ is small, the magnitude of the atomic dipole moments may be found. In general these are of the

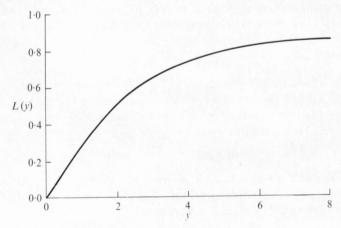

FIG. 6.2. The Langevin function $L(y) = \coth y - 1/y$.

order of $10^{-23}$ A m$^2$ or slightly greater, and the volume susceptibility (per m$^3$) at room temperature of solid paramagnetic substances which obey Curie's law $\approx +10^{-3}$. Thus the paramagnetism considerably outweighs the diamagnetism which is also present.

Langevin's theory applies strictly only to gases, where the molecules are sufficiently far apart for their mutual interactions to be negligible. In liquids and solids such interactions may be large, and many substances obey the modified Curie–Weiss law

$$\chi = C/(T-\theta). \tag{6.14}$$

$\theta$ is called the 'Weiss constant' and is characteristic of the substance; it may be either positive or negative (see Chapters 15 and 16). Eqn (6.14) holds only at temperatures where $T > |\theta|$, and for many substances no single equation represents the susceptibility variation adequately over a wide temperature range.

As already noted, the tendency in chemical combination is towards zero resultant angular momentum of the electrons and hence to zero permanent magnetic moment. Of the common gases, only oxygen ($O_2$) and nitric oxide (NO) are paramagnetic. In the solid state paramagnetism occurs in salts of the 'transition group' ions (see Chapter 14), and the magnetic moment is associated with the metallic ion itself. Thus in compounds such as $CrK(SO_4)_2.12H_2O$ ('chrome alum') or $CuSO_4.5H_2O$ (copper sulphate) only the $Cr^{3+}$ ion and $Cu^{2+}$ ions respectively have permanent magnetic moments, the other ions ($K^+$, $SO_4^{2-}$) and water molecules giving only a diamagnetic contribution to the susceptibility. In most metals the outer electrons are detached from the individual atoms, which are thus left as diamagnetic ions. The detached electrons are free to move through the metal and form the conduction electrons; these give rise to a diamagnetic and a paramagnetic effect, both of the same order of magnitude and both independent of temperature (see § 14.11). The outstanding exceptions are the ferromagnetic metals, iron, cobalt, nickel, the lanthanide metals, and many alloys.

### 6.4. Ferromagnetism

Ferromagnetic substances are all solids, and each is characterized by a certain temperature known as the Curie point at which its properties change abruptly. Above the Curie point the susceptibility is independent of field strength, and follows approximately a Curie–Weiss law (eqn (6.14)) with a Weiss constant $\theta$ whose value is close to that of the Curie point. Below the Curie temperature the behaviour is quite different; very large values of magnetization are produced by quite small fields, and the magnetization varies quite non-linearly with the field strength. This is shown by a characteristic plot of the flux density $B$ as a function of the

FIG.6.3. Magnetization curve (full line OS) and hysteresis loop (broken line) for iron.

field strength $H$ in a sample of iron (Fig. 6.3). If the iron is initially unmagnetized, and a field of slowly increasing magnitude is applied, $B$ follows the full line in Fig. 6.3, known as the 'magnetization curve'. In a field of a few hundred amperes per metre the value of $B$ becomes practically constant at about $1\cdot5$ T. If the magnetic field strength $H$ is now reduced, $B$ does not return along the magnetization curve but follows the broken line, and even at $H = 0$, corresponding to the point R in the figure, $B$ is still near the saturation value. The value of $B$ at this point is known as the 'residual flux density', and the retention of magnetization in zero field is known as 'remanence'. On applying a reverse field the value of $B$ falls and finally becomes zero (point C in Fig. 6.3); the value of the field strength at this point is called the 'coercive force'. As the magnitude of the reverse field is further increased, a reverse flux density is set up which quickly reaches the saturation value. Finally, if the reverse field is gradually removed and a positive field applied, the flux density traces out the lower broken curve in the direction S'S. The whole curve is called the 'hysteresis curve'. It shows that the change in the flux density always lags behind the change in the applied magnetic field strength.

The magnetization curve may also be represented in the form of a permeability curve, showing the variation of $\mu_r = B/(\mu_0 H)$ as a function of either $B$ or $H$. Such a curve ($\mu_r$ against $B$) is shown in Fig. 6.4. When a field is applied $\mu_r$ goes through a maximum and then falls rapidly as the material becomes saturated. The values of $\mu_r$, of the order of $10^4$, are enormously greater than in paramagnetism ($1\cdot001$ or so at ordinary temperatures).

To explain this behaviour, Weiss suggested that a ferromagnetic substance contains atoms with permanent magnetic moments, as in a paramagnetic substance, but that there are large forces acting between neighbouring atomic dipoles which cause groups of them all to point in the same direction. The substance would then be permanently magnetized

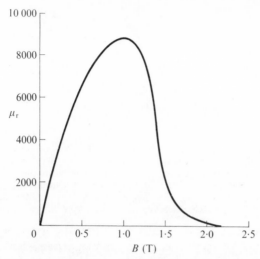

FIG.6.4. Curve of relative permeability $\mu_r$ against flux density $B$ for iron.

within each group; such groups are called 'domains' and their size is now known to vary from about $10^{-3}\,\mathrm{mm^3}$ to $1\,\mathrm{mm^3}$, or greater in single crystals. In an unmagnetized polycrystalline specimen the domains are oriented at random, so that there is no resultant magnetic moment in any direction. When a field is applied, domains where the magnetization is parallel or at a small angle with the field grow at the expense of those where the magnetization is antiparallel or nearly so, so that the boundary between domains is displaced. Initially, in the full curve of Fig. 6.3, the magnetization of the substance as a whole proceeds by small (reversible) boundary displacements, but the steeper part of the magnetization curve is due to larger (irreversible) displacements. Above the knee of the curve, magnetization proceeds by rotation of the direction of magnetization of whole domains; such a process is rather difficult and the increase in magnetization is relatively slow. These processes are shown schematically in Fig. 6.5. When the applied field is reduced, there is little change in the domain structure so that the magnetization remains quite high until reverse fields are applied, thus giving rise to the hysteresis described above.

Ferromagnetic substances may be broadly divided into two classes: (1) magnetically soft materials, which have high permeability, and are easily magnetized and demagnetized, and (2) magnetically hard materials, which have a relatively low permeability and are difficult to magnetize or demagnetize (high coercive force). The chief uses of the former are in electromagnetic machinery and transformers, and of the latter in permanent magnets. To obtain a soft magnetic material the domain walls must

FIG. 6.5. Schematic representation of domains in a ferromagnetic substance: (a) unmagnetized; (b) magnetization through movement of domain boundary wall; domains oriented parallel to **H** grow at the expense of antiparallel domains; (c) magnetization by rotation of the magnetization of whole domains. The domains remain magnetized along a preferred direction in each crystallite; very large fields are required to swing the magnetization away from such a direction towards the applied field.

be able to move easily and reversibly, so that the magnetization changes by large amounts for small changes in the magnetizing field. This requires a material as free as possible from irregularities in the crystal structure due to strains or small particles of impurities. The main treatment of such materials consists therefore of heating to a temperature where sufficient movement of the atoms is possible for them to settle into an ordered lattice, followed by a slow cooling (annealing) so as not to disturb it. On the other hand, a hard magnetic material is one in which domain wall movement is difficult owing to lattice imperfections. These are produced by heating the material and then plunging it suddenly into cold oil (quenching), which sets up internal stresses. Some alloys are then reheated to a lower temperature to cause one of the constituents partially to separate out in small particles dispersed through the alloy. In an alternative process magnets are constructed from compressed powders of very fine particles. If the particles are below a certain size, each forms a single domain and there are no domain walls within the particle. Magnetization and demagnetization can only be accomplished by rotation of the direction of magnetization of each particle. With high anisotropy, which may arise either from the shape of the particles or from their crystal structure, the only possible rotation is reversal through 180°. This requires a higher field than wall movement, and so gives a higher coercive force. To produce a good permanent magnet, the particles must consist of a material with the highest possible saturation magnetization, and be aligned so that their 'easy' directions of magnetization are all parallel. Obviously a high Curie temperature is also required.

## 6.5. Production of magnetic fields

For many purposes it is necessary to maintain a large magnetic field which is constant over a certain volume. By 'large' is meant values of *B* ranging from $0 \cdot 1$ T to 10 T, and the volume may vary from $10^{-4}$ m³ up to many cubic metres in a nuclear accelerator or a plasma-containment apparatus. According to the particular application, either permanent magnets or electromagnets (with or without iron) may be employed, and the principles of their construction are outlined below

A modern permanent magnet of typical shape is shown in Fig. 6.6. The important quantities are the values of $B$ in the air gap, and the volume of the air gap; from eqn (5.44) the energy density at any point is $\frac{1}{2}\mathbf{B}.\mathbf{H}$, and the total energy stored in the air gap is $(\frac{1}{2}BH)V_a$, where $V_a$ is the volume of the gap, and the vectorial representation of $B$ and $H$ can be dropped since they are parallel to one another. The energy stored in the gap is simply related to the quantities involved in the magnet design, as follows. If we assume that there is no leakage, so that all the lines of magnetic flux within the magnet pass through the gap, we have

$$BA_a = B_m A_m, \tag{6.15}$$

where $A_a$, $A_m$ are the cross-sections of the air gap and the magnet, respectively, and $B_m$ is the value of the flux density in the magnet. Now, if we consider a circuit through the gap and the magnet as indicated by the broken line in Fig. 6.6, the total magnetomotive force (see § 4.2) is zero, since no electric currents are involved. Hence

$$\int \mathbf{H}.d\mathbf{s} = Hd_a + H_m d_m = 0, \tag{6.16}$$

where $d_a$ and $d_m$ are the path lengths in the air gap and the magnet respectively, and $H_m$ is the magnetic field strength in the magnet. On

FIG. 6.6. Permanent magnet with coned pole pieces. The magnetomotive force is evaluated along the broken line.

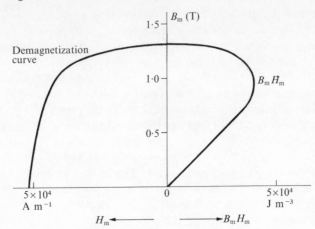

FIG. 6.7. Demagnetization curve (to the left) and plot of $(B_m H_m)$ against $B_m$ (to the right), for Alcomax III.

combining eqns (6.15) and (6.16) we have

$$(\tfrac{1}{2}BH)V_a = (\tfrac{1}{2}BH)A_a d_a$$
$$= -(\tfrac{1}{2}B_m H_m)A_m d_m = -(\tfrac{1}{2}B_m H_m)V_m. \qquad (6.17)$$

This important relation shows that for a magnet of given volume $V_m$, the greatest amount of energy stored in the gap is obtained if the product $B_m H_m$ has its maximum value.

The variation of the product $B_m H_m$ with $B_m$ for a typical material is shown in Fig. 6.7, together with the 'demagnetization curve', that is, the part of the hysteresis loop corresponding to the application of a reverse field. In a permanent magnet the field inside is the 'demagnetizing field' due to the 'free magnetic poles' near the ends of the magnet; this field is in the opposite direction to $B_m$, and the negative sign in eqn (6.17) arises from this. The size of the demagnetizing field is determined by the shape of the magnet, and this must be designed so that the material is at the point where $B_m H_m$ is a maximum. A good working rule is that at this point the ratio of $B_m$ to $H_m$ is equal to the ratio $B_r/H_c$, where $B_r$ is the residual flux density and $H_c$ the coercive force.

In practice there is always some leakage of lines of magnetic field so that the energy stored in the gap is less than the theoretical value given by eqn (6.17). The best shape of the magnet is one in which $B_m H_m$ at every point in the material is closest to its maximum value. Since the product $B_m H_m$ is a characteristic of the material the volume of material required increases linearly with the energy stored in the gap. Many special materials are available with high values of the energy product, and their

FIG. 6.8. Demagnetizing curves for some permanent magnet materials. Alnico 5 (alcomax III) is an alloy of 50% Fe, 25% Co, 13·5% Ni, 8% Al, 3% Cu, ½% Nb.

demagnetization curves are shown in Fig. 6.8. The intermetallic compound $SmCo_5$ has a value of $B_mH_m$ of order $2 \times 10^5$ J m$^{-3}$, and a coercive force of over $5 \times 10^5$ A m$^{-1}$.

### Iron-cored electromagnets

When fields of greater magnitude, or adjustable fields, are required, an electromagnet is used. For fields of flux density up to about 2 T the normal type of construction is as shown in Fig. 6.9. It consists of a yoke

FIG. 6.9. Weiss type of electromagnet. C, C coils carrying electric current, P, P pole tips.

FIG. 6.10. Coned pole tips for magnet.

of soft iron or, more commonly, mild steel, which has a reasonably high magnetization for small values of the applied field. C and C are coils of copper wire, or copper tube through which cooling water may flow, carrying current from a d.c. generator. If the coils have a total of $n$ turns each carrying a current $I$, the m.m.f. is

$$nI = \int \mathbf{H} \cdot d\mathbf{s} = Hd_a + H_m d_m \qquad (6.18)$$

round a circuit through the air gap and yoke similar to that shown in Fig. 6.6, with the same nomenclature as before. Since eqn (6.15) still holds, we have

$$nI = BA_a\left(\frac{d_a}{\mu_0 A_a} + \frac{d_m}{\mu A_m}\right), \qquad (6.19)$$

where the quantity in parenthesis, is known as the magnetic 'reluctance' of the system. In this analogy between the m.m.f. and the e.m.f. in a circuit, the flux $BA_a$ is analogous to the electric current, and the flux passes through the two components—the air gap and the yoke—in series. The total reluctance of the 'magnetic circuit' is the sum of the two parts due to the gap and the iron. Since $\mu_r$ for iron is very large, the greater part of the reluctance is generally associated with the air gap.

In most iron-cored magnets the pole faces may be removed and replaced by others of different shape for special investigations, and the width of the air gap may be altered. To obtain higher values of $B$, coned pole pieces may be used. In the design of Fig. 6.10, if it is assumed that the magnetization is everywhere parallel to the axis and has the saturation value $M_s$, the field at the centre of the gap due to the conical portions indicated by shading may be shown to be (see Problem 6.7)

$$B = \mu_0 H = \mu_0 M_s \sin^2 \phi \cos \phi \ln(b/a). \qquad (6.20)$$

This has a maximum value at $\phi = 54 \cdot 7°$; in practice the pole pieces are not everywhere completely saturated, and a value of about 60° gives the best results. The pole tips may be made of a special cobalt steel which has a higher saturation flux density than ordinary mild steel. A plot of the field in the gap against exciting current is usually fairly linear (apart from a small initial field due to the remanence) until the steel becomes saturated; the rate of increase then becomes much lower, any extra field being that due to the current in the coils themselves.

It was pointed out by Bitter (1936) that this arrangement, in which the iron is magnetized everywhere in the same direction, does not make the best use of the iron. In Fig. 6.11 the field at the origin O due to the dipole at A, oriented as shown, has a component parallel to the $x$-axis equal to

$$H_x = \frac{m}{4\pi r^3} (2 \cos \theta \cos \phi + \sin \theta \sin \phi).$$

If the orientation of the dipole is variable, then $H_x$ has a maximum value when $dH_x/d\theta = 0$, that is, $\tan \theta = \frac{1}{2} \tan \phi$, and this value is then

$$(H_x)_{max} = \frac{m}{4\pi r^3} (1 + 3 \cos^2 \phi)^{\frac{1}{2}}. \tag{6.21}$$

If, on the other hand, the dipole points parallel to the $x$-axis, so that $\theta = -\phi$, the value of $H_x$ is only

$$H_x = \frac{m}{4\pi r^3} (3 \cos^2 \phi - 1). \tag{6.22}$$

Both the angular functions in (6.21) and (6.22) have the value 2 at $\phi = 0$, but the latter falls to zero at $\phi = 54 \cdot 7°$ and then changes sign (the reason why the coned pole pieces discussed above have this as their optimum angle is that dipoles at a larger angle oriented parallel to the axis would give a reverse field and thus reduce the field in the gap). On the other hand, at this angle $(1 + 3 \cos^2 \phi)^{\frac{1}{2}}$ has fallen only to $\sqrt{2}$, and its smallest value is 1 at $\phi = 90°$.

At first sight it would seem impracticable to set up a magnetization in the iron with the angular distribution of the direction of magnetization

FIG. 6.11. Illustrating the calculation of the field at the origin O due to a dipole **m** at A.

required by the equation $\tan \theta = \frac{1}{2} \tan \phi$. The orientation of the magnetization at any point is, however, just the same as the direction of the lines of force set up by a point dipole at the origin pointing along the $x$-axis, and it follows that such a dipole would magnetize the iron in just the right direction. Bitter therefore designed a magnet of new type in which the magnetizing coil, whose field is approximately that of a point dipole, is at the centre, and surrounded by soft iron. This gave an appreciably better performance than the older type of design.

### Air-cored electromagnets

Since iron and other magnetic materials saturate at 1–2 T, the contribution they make in magnets for fields of flux density above 3 T is not sufficient to justify the expense, and solenoids with no iron are used instead. Bitter has considered the question of the current distribution, and concluded that higher efficiency is obtained if the current density is not uniform, but falls off inversely with the radius. A non-uniform current distribution of this type can be obtained by using flat conductors of the shape shown in Fig. 6.12; current is led in along the edge AA and out along BB, so that the current density varies inversely with the radius because the resistance along a path of radius $r$ increases with $r$. The solenoid is constructed of a number of such discs mounted one above the other, with holes drilled in them so that cooling water can be forced through. The low resistance of such a design ($\sim 0.01 \, \Omega$) reduces the chance of any breakdown in the insulation, and corrosion through electrolysis of the cooling water. In one such coil Bitter (1940) obtained a flux density of 10 T, uniform to 1 per cent over a volume of 25 cm$^3$ with a supply of 10 000 A at 170 V. It can be shown that the field obtained from an air-cored coil can be expressed as $(P\lambda/\rho r)^{\frac{1}{2}}$ times a factor depending only on the shape of the coil and the current distribution; here $P$ is the

Fig. 6.12. Shape of copper disc known as a 'Bitter pancake'.

FIG. 6.13. Stress and cooling limitations for disc magnets with 1-mm-diameter cooling passages, and a surface-cooling rate in these passages of $1.5 \times 10^7 \, \mathrm{W\,m^{-2}}$. $\lambda$ is the fraction of coil volume occupied by the conductor. (Bitter 1962.)

power dissipation, $\lambda$ the fraction of the coil volume occupied by conductor of resistivity $\rho$ (the remainder being insulation and coolant), and $r$ a linear dimension such as the inner radius. Thus to double the field over the same volume requires four times the power.

The ultimate performance of an air-cored coil, whatever shape of conductor is used, is limited by the rate at which heat generated in the conductors can be removed and by the mechanical strength of the material, which must withstand enormous magnetic forces. These design limitations are illustrated in Fig. 6.13.

### Superconductive magnets

Below its transition temperature a superconductive material (see Chapter 13) offers zero resistance to the flow of electric current, provided that any magnetic field present does not exceed the critical field above which normal resistivity is restored. This field is 0·1 T or less for most Type I superconductors, but in a number of Type II superconductors it lies between 5 T and 20 T. The discovery of materials such as niobium and the alloys niobium–titanium (NbTi), niobium–tin (Nb₃Sn), niobium–zirconium (NbZr), and vanadium–gallium (V₃Ga) has made it possible to construct superconductive magnets which, for fields above 2 T, are much cheaper to build than conventional magnets, and can be energized by small laboratory power supplies. The alloy NbTi can be used for fields up to 10 T at 2 K, while Nb₃Sn can reach 12 T at 4·2 K, but the critical current density falls rapidly as the field is increased. Immersion of the coil in liquid helium complicates the design of the magnet, but the cost of refrigeration is much less than that of the power supply for a conventional magnet. Once a superconductive magnet has been cooled and energized, the circuit can be closed by a superconducting link; a 'persistent current mode' is then established in which the current through the coil and link flows in a continuous superconducting circuit, and no external power supply is required to maintain the current, which decays at less than one part in $10^6$ in an hour. To alter the field, the superconducting link is driven normal by a small heating strip while the external power supply is reconnected to adjust the current.

Many superconducting alloys have poor mechanical strength, and are fabricated in the form of a composite wire containing about 60 filaments of superconductor in a matrix of a good conductor such as copper or aluminium. The filaments are less than 0·05 mm in diameter to reduce the penetration of magnetic flux into the superconductor. Such penetration must be avoided, since it causes local heating, setting up an instability which can spread through the superconductor and heat it to the point at which it reverts to a normal conductor. The geometry of the coils is designed to maximize the field at the centre, but minimize the stray field in the region of superconducting current.

*Hybrid magnets*

In a coil where a number of turns are wound one on top of the other, the inner turns are subjected to the largest field. For a given available power, a higher field in the central experimental region can be achieved by replacing the inner turns of a superconductive magnet by a conventional water-cooled magnet. If the two sections of such a 'hybrid magnet' are designed to be used separately as well as in combination, the range of fields available from one set of equipment is greatly increased. Problems arise from the enormous mechanical forces between the two sections; these cannot be rigidly connected since the outer superconductive magnet is at liquid helium temperature and the inner at room temperature. These problems have been overcome and fields of 15–20 T are feasible with hybrid magnets. The optimum design of such equipment presents a complicated choice, where the capital cost of the superconductive alloy and of the power supply must be balanced together with the running costs.

## 6.6. Measurement of magnetic fields

The classical method of measuring the value of $B$ at any point in air is by means of a 'flip coil' and an instrument for measuring flux. The flip coil consists of a number $N$ of turns of wire wound on a small former of known area $A$; this is mounted on a handle and the leads to the coil are twisted and brought out through the handle. When the coil is placed with its axis parallel to a field of flux density $\mathbf{B}$, the flux through it is $NAB$; if the coil is then quickly removed to a point where $\mathbf{B} = 0$, the flux change is just $NAB$ and this can be measured either by a ballistic galvanometer or a fluxmeter (see § 5.7). For accurate results the ballistic galvanometer should be standardized as described in § 5.7 using a mutual inductance, with the secondary winding and the flip coil in series with the galvanometer throughout all the measurements so that the total resistance of the circuit remains constant. Flip coils can be made with different values of the product $NA$ (turns×area), so that by choice of the right coil for the field to be measured a suitable deflection can be obtained on the ballistic galvanometer or fluxmeter. As noted in § 5.7, the accuracy obtainable with an average fluxmeter is about 1 per cent, and with a ballistic galvanometer about 10 times greater.

If the coil is rotated rapidly in the field, an alternating voltage is set up which is proportional to $NAB$ times the angular velocity; measurement of this voltage requires a less sensitive instrument than the fluxmeter because the energy available is much greater than with a single throw. In a typical instrument, a coil of 3 mm outer diameter is rotated at 30 Hz, and gives full scale deflection in a field of about 0·05 T; with larger and smaller coils, fields ranging from the earth's field to 10 T can be measured quickly with an accuracy of about 1 per cent.

An absolute method which is capable of giving higher accuracy is the electromagnetic balance of Cotton, in which the force due to the flux density $\mathbf{B}$ on a length of wire carrying a known current is measured directly. This method can only be used for rather strong fields which are uniform over a fair volume. A long rectangular coil is suspended from an analytical balance with the lower end of the coil in the field to be measured, this field being directed horizontally. The long sides of the coil are vertical, so that no force is exerted on them in the vertical direction; they act as leads for the current in the horizontal lower edge of the coil, and the force measured is just that on this lower edge, assuming that the value of $\mathbf{B}$ along the upper edge is negligible. If the lower edge is directed perpendicular to the lines of $\mathbf{B}$, the net vertical force is $F = I \int B \, dx$, where the integral is measured along the lower edge. Thus the integrated value of the flux density along this edge is determined, and to find the value at any point the flux density distribution must be known. The current is measured with a standard resistance and potentiometer, and the force by the change in the balance reading when the current through the rectangular coil is reversed in direction. With a balance of this type Thomas, Driscoll, and Hipple (1950) were able to measure a field of about $0 \cdot 5$ T with an accuracy of a few parts in $10^5$.

For most purposes the most accurate way of measuring a magnetic flux density which is reasonably uniform is to determine the precession frequency $\omega/2\pi$ of a nuclear moment, such as that of the proton, by means of 'nuclear magnetic resonance' (see Chapter 24). This method is based on the equation

$$\omega = |\gamma| \, B,$$

where $\gamma$ is the magnetogyric ratio (see eqn (6.2)). The frequency is linearly proportional to $B$, and may readily be determined with high accuracy (see Chapter 22); from the known value of $\gamma$ this gives an accurate measure of $B$. For protons in a spherical sample of liquid $H_2O$, the value of $\gamma$ is $2 \cdot 6751 \times 10^8 \, \mathrm{s}^{-1} \, \mathrm{T}^{-1}$, corresponding to a frequency $\omega/2\pi$ of $42 \cdot 576$ MHz in a flux density of 1 T.

A convenient but less precise method, which has the advantage that it can be used in less uniform fields, is to measure the voltage in a 'Hall probe'. This contains a small sample of a semiconductor such as n-type indium antimonide or indium arsenide, in which the Hall voltage (see § 12.11) varies almost linearly with the magnetic flux density.

## 6.7. Measurement of magnetic moment and susceptibility

Most methods of determining the susceptibility of weakly magnetic substances depend on measuring the force on the substance in an inhomogeneous magnetic field. From eqn (4.9), the force on a magnetic dipole **m** has an $x$ component

$$F_x = m_x\left(\frac{\partial B_x}{\partial x}\right) + m_y\left(\frac{\partial B_y}{\partial x}\right) + m_z\left(\frac{\partial B_z}{\partial x}\right). \tag{6.23}$$

Now, if instead of a permanent dipole we have a particle of magnetizable matter of susceptibility $\chi$ and volume $v$, its moment will be

$$\mathbf{m} = \chi v \mathbf{H}$$

and the $x$ component of the force on it is

$$F_x = \chi\mu_0 v\left(H_x \frac{\partial H_x}{\partial x} + H_y \frac{\partial H_y}{\partial x} + H_z \frac{\partial H_z}{\partial x}\right) = \tfrac{1}{2}\chi\mu_0 v\left(\frac{\partial}{\partial x} H^2\right).$$

If the particle has a susceptibility $\chi_1$ and is immersed in a medium (such as the atmosphere) with susceptibility $\chi_2$, then the force on it is

$$F_x = \tfrac{1}{2}(\chi_1 - \chi_2)\mu_0 v \frac{\partial H^2}{\partial x}. \tag{6.24}$$

This can be seen from the fact that any displacement of the particle in the $x$ direction requires an opposite displacement of an equal volume of the surrounding medium.

This equation may also be derived by considering the stored energy. The effect of the presence of the particle of volume $v$ is to increase the stored energy by

$$U = v(\tfrac{1}{2}\mathbf{B}_1 \cdot \mathbf{H} - \tfrac{1}{2}\mathbf{B}_2 \cdot \mathbf{H})$$
$$= \tfrac{1}{2}v\{(1+\chi_1) - (1+\chi_2)\}\mu_0 H^2$$
$$= \tfrac{1}{2}v(\chi_1 - \chi_2)\mu_0 H^2.$$

From eqn (5.40) the force component is given by $F_x = +(\partial U/\partial x)$, giving the same formula as eqn (6.24). In both these derivations we have assumed that the magnetic field inside the specimen is the same as the value measured before the specimen was introduced. The difference due to the demagnetizing field, of order $M = \chi H$, is negligible in diamagnetic substances, and in many other cases only relative measurements, such as the change with temperature, rather than absolute measurements, are required. If the sample is in a vacuum, as in the apparatus described below, $\chi_2$ is of course zero.

The force equation (6.23) or (6.24) is the basis of a number of methods of determining magnetic moment and susceptibility, but we describe only a modern version of a method due to Faraday. The sample is suspended by a quartz fibre from a vacuum microbalance which can support samples up to a weight of several grams, and measure force-changes from less than 1 $\mu$g up to 1 g weight. The sample hangs inside a cylindrical dewar vessel (Fig. 6.14) so that its temperature can be controlled and measured down to 1 K. Surrounding the dewar vessel is a superconductive solenoid giving a uniform vertical flux density of 5 T or more. The force on the sample is provided by a pair of coils carrying equal currents flowing in opposite senses; these give a controlled vertical gradient $dB/dx$ of up to $10\,\mathrm{T\,m^{-1}}$, so that $B(dB/dx)$ can have values up to about $50\,\mathrm{T^2\,m^{-1}}$. The geometry of the coils is designed to make the first three derivatives of the gradient zero (Problem 6.9), giving the largest possible volume in which the gradient is nearly uniform. A diamagnetic susceptibility of about $-10^{-9}\,\mathrm{kg^{-1}}$ can be measured to one part in $10^4$ with a sample of 1 g. The paramagnetic susceptibility of very dilute samples (see Fig. 6.15) and the value of the magnetization in the cooperative magnetic state can be determined.

FIG. 6.14. Measurement of magnetic moment by Faraday's method. The force on the samples is measured by a vacuum microbalance. The superconductive magnet (giving a uniform field) and the gradient coils GG, G'G' are immersed in liquid helium; a cryostat and heater (not shown) surround the sample at AA.

FIG. 6.15. Mass susceptibility (perpendicular to the hexagonal axis) of a single crystal of zinc containing 135 p.p.m. of chromium. At this low concentration the chromium ions are paramagnetic and give a positive susceptibility which follows Curie's law, but above 25 K this is outweighed by the diamagnetic contribution from the host metal. (See Appendix D for connection between e.m.u. and S.I. units) (P. L. Camwell and R. Dupree).

An alternative method of measuring a magnetic moment is to determine the flux change when a sample is moved in and out of a coil, or from one coil to another, in a known magnetic field. Movement of the sample is superior to movement of the coil, because the latter gives a flux change even in the absence of a sample due to residual inhomogeneity in the field. A null method is often used; one way of achieving this with a cylindrical sample is to wind a small coil round the sample and adjust the current through it until the magnetic moment of the coil just cancels that due to the sample, as shown by the zero flux change in a pick-up coil from which the (sample+coil) are suddenly removed. This is a good example of the equivalence of a current circuit and a magnetic shell, as discussed in § 4.2.

A sensitive magnetometer, due to Foner (1959), in which the sample is vibrated at 90 Hz, is shown in Fig. 6.16. The sample S, at the end of a long support which reaches down into a dewar vessel for work at low temperatures, is vibrated vertically by a loudspeaker transducer T fed by alternating current at 90 Hz. The sample is at the midpoint between two

FIG. 6.16. Foner's vibrating magnetometer. The pair of coils A, B (and similarly C, D) are in series but opposing, thus reducing spurious voltages due to magnetic field instability or unwanted mechanical vibration.

S    sample, producing 90 Hz voltage in coils A, B,
M    permanent magnet producing comparison voltage in coils C, D,
T    loudspeaker transducer driving sample S and magnet M in vertical vibration at
      90 Hz.

coils AA, BB whose axes are vertical; the vertical component of the flux from the horizontal magnetic moment of the sample threads the top coil and the bottom coil in opposite sense, so that the net flux is zero if the sample is exactly opposite the midpoint. When the sample is displaced vertically during vibration, the net flux through the coils becomes finite and in first approximation is linearly proportional to the displacement of the sample. Thus an e.m.f. alternating at 90 Hz is induced in the coils, which can be balanced against a similar e.m.f. induced by a small permanent magnet M in the coils CC, DD. This gives a null method where the magnetic moment of the sample is read off on the calibrated voltage divider (the phase shifter is required because the two alternating voltages may have a small phase difference—see Chapter 7). The accuracy is about 1 per cent, and changes in magnetic moment of about $10^{-7}$ A m$^2$ can be detected in a field of flux density 1 T. Rotation of the whole assembly about a vertical axis enables the magnetic moment to be measured throughout a horizontal plane; permanent moments can be measured as well as induced moments.

## 6.8. Experimental investigation of the hysteresis curve

In order to determine the hysteresis curve of a substance it is necessary to know the values of $B$ and $H$ inside the substance. For this purpose the most satisfactory shape of the substance is in the form of an anchor ring, or toroid, since then there is no demagnetizing field due to the 'free poles' at the ends of the specimen. If the radius of the ring is large compared with the dimensions of its cross-section, a coil wound uniformly round the ring will produce a uniform field strength $H$ everywhere within the ring. If $n_1$ is the number of turns per unit length in this (the 'primary' coil), and $I$ the current flowing, then $H = n_1 I$. To measure $B$, a small secondary coil is wound over the primary at some point on the ring; if this coil has $n_2$ turns, and the cross-section of the ring is $A$, the flux through the coil is $n_2 A B$. Changes in the flux through the secondary coil are measured by connecting it to a ballistic galvanometer; the galvanometer is calibrated at the same time as described in § 5.7.

For many purposes sufficient accuracy can be obtained by the very convenient method of displaying the hysteresis curve on a cathode-ray oscilloscope. A small resistance is inserted in the primary of the circuit above, and the voltage across this resistance is amplified and used to deflect the electron beam horizontally by an amount proportional to the current in the primary circuit, and hence to $H$. A primary current alternating at a low frequency is used, and the changing flux in the secondary circuit produces an alternating voltage which is amplified and used to deflect the electron beam vertically by an amount proportional to $B$.

The area enclosed by the hysteresis curve is of importance because it represents the work done in taking the material once round the hysteresis curve. Thus in a transformer for alternating current of frequency $f$, the hysteresis curve is traversed $f$ times per second and power is dissipated which appears as heat in the magnetic material of the core. From eqn (5.45) the work done per unit volume in moving from one point on the hysteresis curve to another is the integral, taken along the curve,

$$W = \int H \, dB. \qquad (6.25)$$

It is readily seen that in a complete cycle the value of the integral is just given by the area enclosed by the hysteresis curve. Since this curve is traversed once per cycle, the energy dissipated rises linearly with the frequency of the alternating current.

To minimize losses, a soft magnetic material is required, whose hysteresis loop is as thin as possible; ideally the permeability should be high and constant, with the area of the hysteresis loop zero, so that the $B$–$H$ curve is a straight line of high slope through the origin. Alloys of iron with a few per cent of silicon approach this ideal more closely than pure iron, and are used in power transformers for supply frequencies. For small transformers and other uses at higher frequencies, more expensive alloys requiring lengthy heat treatment give improved performance. For example, an alloy of about 78 % Ni and 22 % Fe (permalloy), if slowly cooled from 1150 K and then rapidly cooled from 850 K, gives an initial permeability of nearly $10^4$ and a maximum permeability of nearly $10^5$. Supermalloy, with the composition 80 % Ni, 15 % Fe, 5 % Mo (molybdenum), after heating in very pure hydrogen at over 1500 K, gives initial and maximum permeabilities some 10 times higher. These high permeabilities are accompanied by very low values of the coercive force, $\sim 4$ A m$^{-1}$ for permalloy and $0 \cdot 3$ A m$^{-1}$ for supermalloy.

Another effect of the alternating magnetic flux in the core is to set up induced voltages and produce power loss through eddy currents. The core is therefore constructed of very thin laminations, insulated from each other, and oriented so that the insulation lies across the path of the eddy current. This is therefore constrained to flow within the lamination, and the losses are reduced (see Problem 8.12). In this respect the special magnetic alloys mentioned above have the further advantage of a high electrical resistivity. At radio-frequencies ($10^5$–$10^8$ Hz) the use of magnetic cores also reduces the size of inductors and improves the coupling between the coils of transformers; such cores are made from magnetic oxides ('ferrites') whose resistivity may be as high as $10^8 \, \Omega$ m at room temperature.

## 6.9. Terrestrial magnetism

To a good approximation the magnetic flux density at the surface of and outside the earth is the same as that of a magnetic dipole at the centre of the earth. More detailed measurements show that the magnetic potential must be analysed using a series of spherical harmonics, of which the dipole term is much the largest. Leaving aside localized distortions due to iron-bearing minerals in the earth's crust, one finds a dipole term which has decreased in magnitude by about 5 per cent in the last hundred years, together with quadrupolar and higher terms which have strong and fairly rapid secular variations with lifetimes less than a hundred years. The field at any point also contains daily variations which are irregular and unpredictable; these are caused by currents in the ionosphere due to solar and lunar perturbations. Charged particles approaching the earth, including the stream of charged particles from the sun known as the 'solar wind', are deviated by the earth's magnetic field, and give rise to many fluctuating phenomena in the upper atmosphere, including such dramatic effects as the aurorae.

Archaeomagnetic data have revealed the unexpected result that the equivalent magnetic dipole moment of the earth had a maximum, about 50 per cent greater than the present value, around the beginning of the Christian era, with a minimum at about 4000 B.C. preceded by an earlier maximum at about 6000 B.C. Much more dramatic changes, including many reversals of the earth's field, are revealed by palaeomagnetic studies (see, for example, Aitken 1970; McElhinney 1973; Stacey and Banerjee 1974).

These 'sudden' changes, in which the geomagnetic field is reversed in a time of 1000–10 000 years, are difficult to understand except on a 'dynamo theory', in which electric currents flow round the core of the earth. This core, with a radius of $(3473\pm4)$ km, is assumed to be liquid, since no transverse seismic waves are transmitted through it, and to consist mainly of liquid silicates of iron, magnesium, and calcium at high temperatures and pressures. The electric currents are associated with convective heat currents caused by radioactive or chemical sources; the mathematics of the process (energy source $\rightarrow$ kinetic energy of fluid $\rightarrow$ electrical energy) has been studied by Elsasser, Bullard, and others, and it seems probable that electric currents can be maintained in this way. For detailed accounts reference should be made to Chapman and Bartels (1940) and Elsasser (1950, 1955–6).

## References

AITKEN, M. J. (1970). *Rep. Prog. Phys.* **33**, 941.
BITTER, F. (1936). *Rev. scient. Instrum.* **7**, 479; **7**, 482.
— (1940). *Rev. scient. Instrum.* **11**, 373.

BITTER, F. (1962). *Proceedings of the International Conference on high magnetic fields*, p. 85, M.I.T. Press and Wiley, New York.
CHAPMAN, S. and BARTELS, J. (1940). *Geomagnetism*. Clarendon Press, Oxford.
ELSASSER, W. M. (1950). *Rev. mod. Phys.* **22**, 1.
— (1955–6). *Am. J. Phys.* **23**, 590; **24**, 85.
FONER, S. (1959). *Rev. mod. Phys.* **30**, 548.
MCELHINNEY, M. W. (1973). *Palaeomagnetism and plate tectonics*. Cambridge University Press.
MYERS, W. R. (1952). *Rev. mod. Phys.* **24**, 15.
STACEY, F. D. and BANERJEE, S. K. (1974). *The physical principles of rock magnetism*. Elsevier, Amsterdam.
THOMAS, H. A., DRISCOLL, R. L., and HIPPLE, J. A. (1950). *Phys. Rev.* **78**, 787.

## Problems

6.1. The susceptibility of a mole of helium gas is $-2\cdot4\times10^{-11}$. Show that this corresponds to a value for the mean-square radius of each electronic orbit in the helium atom of $1\cdot22a_0^2$, where $a_0 = 0\cdot528\times10^{-10}$ m is the radius of the first Bohr orbit in the hydrogen atom.

6.2. If the mutual repulsion of the two electrons in the helium atom is neglected, the wavefunction of each electron in the ground state is

$$\psi = (Z^3/\pi a_0^3)^{\frac{1}{2}}\exp(-Zr/a_0),$$

where $-e\psi^2$ is the density of electronic charge at a distance $r$ from the nucleus, and $Ze$ is the effective charge of the helium nucleus. Show that this wavefunction leads to a value for the mean-square radius of each electronic orbit of

$$\overline{r^2} = 3a_0^2/Z^2.$$

Verify that agreement with the value of $1\cdot22a_0^2$ in the previous problem is obtained if we take $Z$ as about $1\cdot6$ (we should expect it to be less than 2 because each electron partially shields the other from the field of the nucleus).

6.3. When $23\cdot15$ g of $NiCl_2$ are dissolved in $100$ g water, the density of the solution is $1255$ kg m$^{-3}$. Show that the maximum height of a column of the solution which can be supported by magnetic force when one surface of the column is in a uniform flux density of $1$ T is $3\cdot0$ mm. (Susceptibilities of $1$ kg of $NiCl_2$ and water are $0\cdot438\times10^{-6}$ and $-0\cdot0090\times10^{-6}$ respectively; volume susceptibility of air = $+0\cdot4\times10^{-6}$.)

6.4. The susceptibility of a mole of $NiK_2(SO_4)_2.6H_2O$ is found to be $1\cdot6\times10^{-5}$ $T^{-1}$. Assuming that the diamagnetic contribution is negligible, and that the only paramagnetic contribution comes from the $Ni^{2+}$ ion, calculate the size of the permanent dipole moment on each $Ni^{2+}$ ion.
(*Answer:* $3\cdot0\times10^{-23}$ A m$^2$.)

6.5. An iron anchor ring of large mean radius $R$ and uniform cross-section has a gap of thickness $d$ cut in it. It is wound with a single-layer coil. Over the range under consideration the permeability of the iron is $(1+a/H)$, where $H$ is the field strength in the iron. Show that, if $d \ll R$, 4 times as much power is required to maintain a field strength $3a$ in the gap as is required for a field strength $2a$.

6.6. It was shown by Rayleigh that at low values of the magnetic flux density the hysteresis loop with tips at $B_0$, $H_0$ and $-B_0$, $-H_0$ is described by the equations

$$B = \mu H + \tfrac{1}{2}a(H_0^2 - H^2) \quad \text{(upper half of loop)}$$

and

$$B = \mu H - \tfrac{1}{2}a(H_0^2 - H^2) \quad \text{(lower half of loop)},$$

where $\mu = B_0/H_0$. Show that the energy loss per cycle represented by the area of the loop is

$$W = \tfrac{4}{3}aH_0^3$$

in each unit volume of the substance.

This relation is valid only for low values of $B_0$ (in iron below about 0·05 T). At high values $W$ varies approximately as $B_0^{1\cdot6}$, an empirical law due to Steinmetz.

6.7. Deduce eqn (6.20) by the use of equivalent magnetic shells and integration of eqn (4.66), or from integration of the magnetic flux density due to the 'free poles' (polarization charges) on the surfaces of the cones.

6.8. An air-cored solenoid of length $2l$ is constructed from $n$ equally spaced Bitter pancakes each of thickness $t$, inner radius $a$, outer radius $b$, and resistivity $\rho$. If $n$ is large, and $P$ is the power supplied, show that the field strength $H$ at the centre is approximately

$$H = \left(\frac{Ptn}{8\pi l^2 \rho \, \ln(b/a)}\right)^{\frac{1}{2}} \ln \frac{b\{l + (l^2 + a^2)^{\frac{1}{2}}\}}{a\{l + (l^2 + b^2)^{\frac{1}{2}}\}}.$$

Note that this is of the form $(P\lambda/\rho r)^{\frac{1}{2}}$ times a factor depending on the shape of the solenoid (as given in § 6.5), since $\lambda = nt/2l$.

6.9. Two circular coaxial coils, each carrying $N$ turns of radius $a$, carry equal currents but in opposite senses. Show that at the midpoint on the axis between the coils, the value of $H_z$ and the derivatives $\mathrm{d}^2H_z/\mathrm{d}z^2$, $\mathrm{d}^4H_z/\mathrm{d}z^4$ along the axis are zero, and that $\mathrm{d}^3H_z/\mathrm{d}z^3 = 0$ provided that the distance between the coils is $\sqrt{3}a$.

6.10. The iron core of a transformer has laminations of thickness $a$ and resistivity $\rho$; it is subject to a sinusoidally varying flux density with a maximum value $B$ and frequency $f$. Show that the power loss per unit volume (neglecting skin effects, see Problem 8.12) due to eddy currents is approximately $\pi^2B^2a^2f^2/6\rho$. If $a = 0\cdot1$ mm, $\rho = 4\times10^{-7}\ \Omega$ m, $B = 0\cdot5$ T, and $f = 50$ Hz, show that the power loss is approximately $26$ W m$^{-3}$. Compare this with the power loss through hysteresis, if the energy dissipated per cycle for permalloy at this flux density is $20$ J m$^{-3}$.

(*Answer:* Hysteresis loss $= 10^3$ W m$^{-3}$.)

# 7. Alternating current theory

## 7.1. Forced oscillations

IN § 5.3 we considered the transient currents which flow when a capacitor, initially charged, is allowed to discharge through a circuit containing both inductance and resistance, and found that an oscillatory current of decaying amplitude flowed through the circuit provided that the resistance in the circuit was not too high. The theory of such transients is due to Lord Kelvin, and its correctness was verified by early experimenters. With the invention of the dynamo and, later, the electronic vacuum tube, it became possible to produce continuous alternating currents whose frequency of oscillation may be anything up to about $10^{11}$ Hz. In the simplest case the form of the current is that of a simple sine wave, and may be written as

$$I = I_0 \cos \omega t,$$

where $I$ is the value of the current at time $t$. The maximum value of $I$ is $I_0$, known as the 'amplitude' of the current, and the frequency of alternation is $f = \omega/2\pi$ Hz. The current generated by a dynamo or other device may or may not have a simple sinusoidal waveform, but whatever the actual waveform it may be resolved by Fourier analysis into a sum of sine and cosine terms whose frequencies are integral multiples of the fundamental frequency. This frequency is given by the inverse of the period between instants at which the whole waveform is repeated. Since the behaviour of a circuit is in general different at different frequencies, it is necessary to consider each component of such a Fourier series separately, and in the theory that follows we shall assume that the waveform is sinusoidal, varying at one frequency only. Except in non-linear circuits where the behaviour of a circuit element depends on the size of the current or voltage applied to it (that is, elements in which the amplitudes of current and voltage are not linearly proportional to one another) any non-sinusoidal fluctuations may be resolved into their Fourier components, and the required solution is simply a sum of such components.

In the circuit of Fig. 7.1, a voltage $V_0 \cos \omega t$ is applied to an inductance, a resistance, and a capacitance in series. If $I$ is the current flowing at any instant, the e.m.f. set up in the inductance is $-L(dI/dt)$, and the voltage drop across the capacitance is $Q/C$, where $Q$ is the charge on the

FIG. 7.1. Forced oscillations in a circuit containing $L$, $C$, $R$.

capacitor. We have therefore for the circuit

$$V_0 \cos \omega t - L\frac{dI}{dt} - \frac{Q}{C} = RI$$

or

$$L\frac{dI}{dt} + RI + \frac{Q}{C} = V_0 \cos \omega t. \tag{7.1}$$

Now the rate of increase of the charge on the capacitor is $dQ/dt = I$, the current flowing, and hence by differentiation we have

$$L\frac{d^2I}{dt^2} + R\frac{dI}{dt} + \frac{I}{C} = \frac{dV}{dt} = -\omega V_0 \sin \omega t. \tag{7.2}$$

This is a differential equation whose solution consists of two parts. The first of these, known as the 'complementary function', is found by solving the equation obtained by setting the right-hand side equal to zero; that is, it is a solution of eqn (5.27), and hence is of the form given by eqn (5.28). This solution represents a transient flow of current produced by the act of applying the e.m.f. $V_0 \cos \omega t$, it being assumed that this starts to act at the instant $t = 0$. In all practical applications this transient current decays rapidly in amplitude, owing to the exponential term $e^{-tR/2L}$ and becomes negligible within a few seconds or less of the circuit being closed. If conditions are such that the transient current is oscillatory, its frequency is the natural frequency determined by the values of $L$, $C$, and $R$, and not that of the applied e.m.f.

The second part of the solution is known as the 'particular integral', and for eqn (7.2) it may be written as

$$I = (V_0/Z)\cos(\omega t - \phi), \tag{7.3}$$

where

$$Z = \{R^2 + (\omega L - 1/\omega C)^2\}^{\frac{1}{2}} \tag{7.4}$$

and the phase angle $\phi$ is given by

$$\tan \phi = \frac{(\omega L - 1/\omega C)}{R}. \tag{7.5}$$

This is also known as the 'steady-state' solution, since it gives the current flow at any time after the transient current has become negligible. The frequency of the current is the same as that of the applied e.m.f., so that the circuit is in 'forced' oscillation. In general the phase of the current is different from that of the applied voltage, except when $\omega L - 1/\omega C = 0$. This occurs when

$$\omega = \omega_0 = 1/(LC)^{\frac{1}{2}}, \tag{7.6}$$

and hence is the same as the angular frequency of natural oscillation of the circuit in the absence of any damping resistance. The circuit is then said to be in 'resonance', and the amplitude of the current is, by eqns (7.3) and (7.4), a maximum.

The quantity $Z$ in eqn (7.4) is called the impedance of the circuit, and at resonance the value of $Z$ is just equal to $R$, the total resistance in the circuit. At other frequencies the value of $Z$ is related to the quantities $R$, $L$, and $C$, but is not given by the simple additive relation that holds (§ 3.4) for resistances in series. The reason for this is that the voltages across the different elements are not in phase, and the total voltage amplitude is therefore not just the sum of the individual amplitudes. In the following sections we shall see how this difficulty can be overcome by the introduction of complex numbers to represent the impedances. The use of such complex impedances enables us to apply Kirchhoff's laws (§ 3.4) to alternating current (a.c.) networks, and we can find the steady-state values of the current and voltage in any branch without having to solve a differential equation.

An important consideration in a.c. circuits is the rate of doing work. At any instant the rate at which work is done by the generator in Fig. 7.1 is

$$dW/dt = VI = (V_0 \cos \omega t) \times (V_0/Z)\cos(\omega t - \phi)$$
$$= (V_0^2/Z)(\cos^2\omega t \cos \phi + \cos \omega t \sin \omega t \sin \phi).$$

To find the mean rate $d\bar{W}/dt$ of doing work, this expression must be averaged over one or more periods of oscillation. Now the mean value of $\cos^2\omega t$ averaged in this way is just $\frac{1}{2}$, while that of $\cos \omega t \sin \omega t$ is 0. Hence the mean power drawn from the generator is

$$d\bar{W}/dt = \frac{1}{2}(V_0^2/Z)\cos \phi = \frac{1}{2}V_0 I_0 \cos \phi = \frac{1}{2}I_0^2 Z \cos \phi, \tag{7.7}$$

where $I_0 = V_0/Z$ is the amplitude of the current given by eqn (7.3). From eqn (7.5) it is readily shown that $\cos \phi = R/Z$, and the expression for the mean power can therefore be written

$$d\bar{W}/dt = \frac{1}{2}I_0^2 R. \tag{7.8}$$

Now the mean rate at which power is dissipated in the resistance of the circuit is the average value of $RI_0^2 \cos^2(\omega t - \phi)$ taken over a complete period, and this is just equal to $\frac{1}{2}RI_0^2$. Hence all the power delivered by the generator, averaged over a period, is dissipated in the resistance of the circuit. The term in $\cos \omega t \sin \omega t$ in the expression for $VI$ represents work done by the generator in increasing the energy stored in the inductance and capacitance; since the product $\cos \omega t \sin \omega t$ is as often negative as positive, this work is returned to the generator in other parts of the cycle and no mean power is drawn from the generator for this purpose in the steady state. Of course, power was drawn from the generator initially to provide the stored energy, and this is represented by the transient current; when this has decayed, the mean stored energy remains constant and no further work is done by the generator on the average except to supply that dissipated in the resistance.

Since the rate at which power is dissipated is proportional to the square of the current, it is convenient to specify the root-mean-square (r.m.s.) value $\tilde{I}$ of the current, defined by the fact that

$$(\tilde{I})^2 = \langle I^2 \rangle,$$

where the average of the square of the current is taken over a whole period. The r.m.s. value of the voltage $\tilde{V}$ may be defined in a similar way, $(\tilde{V})^2 = \langle V^2 \rangle$. If the waveform is sinusoidal, the r.m.s. value is just $(1/\sqrt{2})$ times the amplitude, and we can write eqn (7.7) as

$$d\bar{W}/dt = (\tilde{V}^2/Z)\cos \phi = \tilde{V}\tilde{I} \cos \phi = \tilde{I}^2 Z \cos \phi \qquad (7.9)$$

and the ratio $(d\bar{W}/dt)/\tilde{V}\tilde{I} = \cos \phi$ is called the 'power factor' of the circuit. It represents the fraction of the product $\tilde{V}\tilde{I}$ which is dissipated as Joule heat. If the circuit behaves as a pure resistance, as occurs when the resonance condition (7.6) is fulfilled in the circuit of Fig. 7.1, the power factor is unity, while if the circuit contains no resistance the power factor is zero.

The general practice in a.c. circuits is to specify the r.m.s. values of the current and voltage, and it should be understood that any values quoted are r.m.s. values unless the contrary is specifically stated.

### 7.2. Use of vectors and complex numbers

The values of the voltage across the individual components of the circuit in Fig. 7.1 will now be considered in more detail. For this purpose we assume that a current $I = I_0 \cos \omega t$ flows through them all in series, as in Fig. 7.2. The voltages across the three circuit elements are:

across $R$:     $V = IR = RI_0 \cos \omega t$

across $L$:     $V = L(dI/dt) = -\omega L I_0 \sin \omega t = \omega L I_0 \cos(\omega t + \frac{1}{2}\pi)$

across $C$:     $V = Q/C = (1/\omega C)I_0 \sin \omega t = (1/\omega C)I_0 \cos(\omega t - \frac{1}{2}\pi)$

where in the last case we have used the relation $Q = \int I \, dt$. It will be seen

$I = I_0 \cos \omega t$

FIG. 7.2. Current $I$ flowing through $R$, $L$, $C$ in series.

that the voltage across the resistance is in phase with the current, the voltage across the inductance leads the current in phase by 90°, while that across the capacitance lags behind by 90°. We may represent these voltages by vectors such that, if the voltage across the resistance is represented by a vector drawn parallel to the $x$-axis, that across the inductance is represented by a vector parallel to the $y$-axis, and that across the capacitance by a vector parallel to the latter but in the opposite sense. The lengths of the vectors are proportional to $R$, $\omega L$, and $1/\omega C$, respectively.

If we require the voltage across two of the elements, say the resistance and inductance, it may be found by adding the two individual voltage vectors together, as in Fig. 7.3(a). For the total voltage will be

$$V = IR + L(\mathrm{d}I/\mathrm{d}t) = RI_0 \cos \omega t - \omega L I_0 \sin \omega t$$
$$= I_0(R^2 + \omega^2 L^2)^{\frac{1}{2}} \cos(\omega t + \phi) = I_0 Z \cos(\omega t + \phi), \qquad (7.10)$$

where $\tan \phi = \omega L/R$. From Fig. 7.3(a) it will be seen that $(R^2 + \omega^2 L^2)^{\frac{1}{2}}$ is just the length of the hypotenuse of the triangle, while the phase angle $\phi$ is just the angle between the vectors representing $R$ and $(R^2 + \omega^2 L^2)^{\frac{1}{2}}$. Thus the magnitude of the total voltage is represented in amplitude by the hypotenuse, and its phase relative to the current is given by the angle through which this vector is rotated with respect to vector representing $R$. Similarly it can be shown that for a resistance and capacitance in series the total voltage may be found by adding the vectors for $R$ and $-(1/\omega C)$ as in Fig. 7.3(b), while the case of a resistance $R$, inductance $L$, and capacitance $C$ all in series is represented by Fig. 7.3(c). Here the amplitude of the resultant vector is $\{R^2 + (\omega L - 1/\omega C)^2\}^{\frac{1}{2}}$, which is just the value of $Z$ (eqn (7.4)) and the voltage leads the current by the phase

(a)                         (b)                         (c)

FIG. 7.3. Vector diagram for impedance. (a) $R$, $L$ in series; (b) $R$, $C$ in series; (c) $R$, $L$, $C$ in series. The voltage vectors are the same as the impedance vectors if the current is represented by a unit vector parallel to the direction of $R$.

angle $\phi$, where

$$\tan \phi = \frac{(\omega L - 1/\omega C)}{R},$$

as in eqn (7.5). The term 'impedance' has already been introduced for $Z$, which represents the ratio of the amplitude of voltage to current for the whole circuit. The quantities $\omega L$ and $1/\omega C$ associated with inductance and capacitance respectively are known as reactances, and are usually denoted by the symbol $X$. Thus the total reactance of the circuit of Fig. 7.2 is $X = \omega L - 1/\omega C$, and the impedance is given by the vector sum of $R$ and $X$, represented by mutually perpendicular vectors. Thus $Z = (R^2 + X^2)^{\frac{1}{2}}$, and the phase angle is given by $\tan \phi = X/R$.

The vector representation of the impedance is similar to the representation of a complex number on the Argand diagram. Using the relation

$$\exp j\omega t = \cos \omega t + j \sin \omega t,$$

where $j^2 = -1$, we may replace our cosine and sine functions by complex exponentials on the understanding that we are interested only in the real or imaginary parts respectively. Then we may write the current $I_0 \cos \omega t$ as $I_0 \mathrm{Re}(\exp j\omega t)$, and quantities such as $dI/dt$ or $Q = \int I \, dt$ become

$$\frac{dI}{dt} = I_0\left(\frac{d}{dt}\right)\cos \omega t = I_0 \mathrm{Re}\left(\frac{d}{dt}\exp j\omega t\right) = I_0 \mathrm{Re}(j\omega \exp j\omega t)$$

and

$$Q = I_0 \int \cos \omega t \, dt = I_0 \mathrm{Re}\left(\int \exp j\omega t \, dt\right) = I_0 \mathrm{Re}\left(\frac{1}{j\omega}\exp j\omega t\right)$$

respectively. Hence the total voltage across the circuit of Fig. 7.2 is

$$V = RI + L(dI/dt) + (1/C)\int I \, dt$$

$$= \left(R \exp j\omega t + j\omega L \exp j\omega t + \frac{1}{j\omega C}\exp j\omega t\right)I_0$$

$$= \left(R + j\omega L + \frac{1}{j\omega C}\right)I_0 \exp j\omega t$$

$$= \left(R + j\omega L + \frac{1}{j\omega C}\right)I. \tag{7.11}$$

If the current is $I_0 \cos \omega t$, then to find the voltage we take the real part of (7.11). Thus

$$V = I_0 \mathrm{Re}\left(R \cos \omega t + jR \sin \omega t + j\omega L \cos \omega t - \omega L \sin \omega t - \frac{j}{\omega C}\cos \omega t + \right.$$

$$\left. + \frac{1}{\omega C}\sin \omega t\right)$$

$$= I_0\left(R \cos \omega t - \omega L \sin \omega t + \frac{1}{\omega C}\sin \omega t\right),$$

where the three component terms are just the voltages across the three circuit elements derived at the beginning of this section. Similarly, if the current had been $I_0 \sin \omega t = I_0 \operatorname{Im}(\exp j\omega t)$, the voltage would be found by taking the imaginary part of (7.11), giving

$$V = I_0 \left( R \sin \omega t + \omega L \cos \omega t - \frac{1}{\omega C} \cos \omega t \right).$$

The importance of eqn (7.11) lies in the fact that it shows we may represent the inductance and capacitance by impedance operators $j\omega L$ and $1/j\omega C$ respectively, and these operators may be added to one another when the elements are in series in a similar way to resistances. The phase of the voltage, which leads the current by $\frac{1}{2}\pi$ in the inductance and lags by $\frac{1}{2}\pi$ in the capacitance, is taken care of by the presence of j. If, as is usual in the Argand diagram, real quantities are represented by vectors drawn parallel to the $x$-axis, and imaginary quantities by vectors parallel to the $y$-axis, then the complex impedance operator is as shown in our vector diagram (Fig. 7.3(c)). The circuit impedance is given by the modulus of the complex impedance operator, and the phase angle by its argument. If we write the impedance operator as

$$\mathbf{Z} = R + j\omega L + \frac{1}{j\omega C} = R + jX = Z \exp j\phi$$

then $Z = (R^2 + X^2)^{\frac{1}{2}}$, and $\tan \phi = X/R$, as before. Then

$$V = \mathbf{Z} I = Z \exp j\phi \times I_0 \exp j\omega t = ZI_0 \exp j(\omega t + \phi).$$

If $I = I_0 \cos \omega t$, the real part of this gives $V = ZI_0 \cos(\omega t + \phi)$, as in eqn (7.10). If, on the other hand, the voltage is given as $V_0 \cos \omega t$, as in § 7.1, the current is found by taking the real part of

$$I = V/\mathbf{Z} = V_0 \exp j\omega t/(Z \exp j\phi) = (V_0/Z)\exp j(\omega t - \phi),$$

which gives

$$I = (V_0/Z)\cos(\omega t - \phi),$$

as in eqn (7.3).

From an extension of the treatment given above it may readily be shown that the impedance operator for a circuit consisting of a number of impedances $\mathbf{Z}_1, \mathbf{Z}_2, ..., \mathbf{Z}_n$ in series is

$$\mathbf{Z} = \mathbf{Z}_1 + \mathbf{Z}_2 + ... + \mathbf{Z}_n. \qquad (7.12)$$

The corresponding formula for a number of impedances in parallel is found by noting that the voltage across each is the same. The current through the impedance $\mathbf{Z}_k$ is $V/\mathbf{Z}_k$, and the total current is the sum of a number of similar terms. Hence the net impedance is given by

$$\frac{I}{V} = \frac{1}{\mathbf{Z}} = \frac{1}{\mathbf{Z}_1} + \frac{1}{\mathbf{Z}_2} + ... + \frac{1}{\mathbf{Z}_n}. \qquad (7.13)$$

When a number of elements are in parallel it is generally convenient to work in terms of the reciprocal of the impedance, known as the admittance **Y**. Thus $I = V/Z = YV$, and eqn (7.13) may be written as

$$Y = Y_1 + Y_2 + ... + Y_n. \tag{7.14}$$

In general **Y** is complex; thus $Y = G + jS$, where $G$ is called the conductance and $S$ the susceptance. For the three simple circuit elements we have conductance $G = 1/R$; susceptance of an inductance is $S = (j\omega L)^{-1} = -j/\omega L$; susceptance of a capacitance is $S = j\omega C$; but note that for a circuit containing both resistance and reactance these simple reciprocal relations do not hold. For

$$Y = 1/Z = 1/(R+jX) = (R-jX)/(R^2+X^2),$$

so that $Y = |Y| = |Z^{-1}| = Z^{-1}$, but

$$\left. \begin{array}{ll} G = \dfrac{R}{(R^2+X^2)} = Y \cos \phi, & S = -\dfrac{X}{(R^2+X^2)} = -Y \sin \phi \\ \text{and similarly} \\ R = \dfrac{G}{(G^2+S^2)} = Z \cos \phi, & X = -\dfrac{S}{(G^2+S^2)} = Z \sin \phi \end{array} \right\} . \tag{7.15}$$

In order to calculate the power consumed in a circuit we must use the relation $dW/dt = \text{Re}(V) \times \text{Re}(I)$; note that this is not the same as $\text{Re}(VI)$, for $\text{Re}\{V_0 \exp j\omega t\} \times \text{Re}\{I_0 \exp j(\omega t - \phi)\} = V_0 \cos \omega t \times I_0 \cos(\omega t - \phi)$, which does not equal $\text{Re}\{V_0 I_0 \exp j(2\omega t - \phi)\} = V_0 I_0 \cos(2\omega t - \phi)$. However, the mean energy $d\bar{W}/dt$ dissipated is given by any of the relations (where $\tilde{V}$ and $\tilde{I}$ are the r.m.s. values)

$$d\bar{W}/dt = \tilde{V}^2 \text{Re}(Y) = \tilde{V}^2 G = \tilde{V}^2 R/(R^2+X^2)$$
$$= \tilde{I}^2 \text{Re}(Z) = \tilde{I}^2 R = \tilde{I}^2 G/(G^2+S^2), \tag{7.16}$$

as can be verified by comparison with eqn (7.9). An alternative relation is given in Problem 7.13.

The unit of reactance and impedance is the same as that of resistance, the ohm; when $\omega$ is expressed in radians per second, $L$ in henries, and $C$ in farads, the corresponding reactances are in ohms. Similarly, the unit of admittance and susceptance is the same as that of conductance, the reciprocal ohm (or siemen).

We may illustrate the use of admittance by considering the case of a lossy capacitor, represented by a pure capacitance $C$ shunted by a resistance $R$, as in Fig. 7.4(a). The admittance operator is

$$Y = (1/R) + j\omega C = G + j\omega C.$$

This may be represented by the admittance diagram shown in Fig. 7.4(b). The phase angle of the admittance is given by $\tan \psi = \omega C/G = \omega CR$; by

FIG. 7.4. (a) Imperfect capacitor with conductance $G$, represented by capacitance $C$ shunted by resistance $R = 1/G$; (b) corresponding admittance diagram.

use of eqn (7.15) or by plotting the corresponding impedance diagram it is readily shown that $\tan \psi = -\tan \phi$, so that $\psi = -\phi$. Hence the power factor $\cos \phi = \cos \psi = (1+\omega^2 C^2 R^2)^{-\frac{1}{2}}$. For a perfect capacitance, as for a perfect inductance, the power factor is zero; in practice this is not quite true, since at radio-frequencies power is lost through eddy currents in the plates and imperfections in the dielectric. The latter are caused by hysteresis effects which cause the vector $\mathbf{D}$ to lag behind the vector $\mathbf{E}$ in the dielectric as in the corresponding case of $\mathbf{B}$ and $\mathbf{H}$ for a ferromagnetic substance. The quality of a dielectric is expressed in terms of its 'loss tangent' $\tan \delta$, where

$$\delta = \tfrac{1}{2}\pi - \psi = \tfrac{1}{2}\pi + \phi;$$

hence

$$\tan \delta = \cot \psi = -\cot \phi = G/\omega C = (\omega CR)^{-1}.$$

In an oscillating electric field the hysteresis loop is traversed once per cycle, so that the loss conductance $G$ is proportional to the frequency. The value of $\tan \delta$ should therefore be independent of frequency, and this is generally true except for strongly polar dielectrics (see Chapter 10); for a good dielectric, such as quartz, the value of $\tan \delta$ at room temperature is about $10^{-4}$.

## 7.3. Tuned circuits

The circuit of Fig. 7.1 consisting of an inductance, resistance, and capacitance in series is known as a 'series resonant' circuit. The impedance operator is

$$\mathbf{Z} = R + j(\omega L - 1/\omega C), \tag{7.17}$$

and, as pointed out in § 7.1, the modulus of this has a minimum value if the frequency is adjusted to make $\omega L = 1/\omega C$. This is the resonant frequency of the circuit, given by eqn (7.6)

$$\omega_0 = 1/(LC)^{\frac{1}{2}}.$$

At this frequency the current through the circuit is a maximum and is in phase with the applied voltage, since $\mathbf{Z}$ is real; the magnitude of the current is $V/R$. The voltage across the resistance $R$ is thus equal to the voltage across the whole circuit, and the voltages across the inductance and capacitance are therefore equal and exactly $180°$ out of phase with one another, making the voltage across the two zero. The voltage across the capacitance alone at resonance is

$$V_C = I/j\omega_0 C = V/j\omega_0 CR,$$

and hence, using (7.6),

$$\left|\frac{V_C}{V}\right| = (\omega_0 CR)^{-1} = \frac{1}{R}\left(\frac{L}{C}\right)^{\frac{1}{2}} = \frac{\omega_0 L}{R} = Q_F \qquad (7.18)$$

where $Q_F$ is the 'quality factor' of the circuit as defined in § 5.3. It is also known as the 'circuit magnification factor', since from eqn (7.18) we see that it equals the ratio of the voltage across the capacitor to the voltage across the whole circuit. Thus the tuned circuit acts as a transformer; since the current through the capacitor is a maximum at resonance, the ratio of the voltages $V_C/V$ is also a maximum at this point. The voltage across the inductance $L$ is equal to that across the inductance at resonance, but in general $R$ is associated with $L$ and only the voltage across the combination $R + j\omega L$ can be measured in practice.

The quality factor $Q_F$ is very useful both in theory and in practice. Eqn (7.17) can be written as

$$\mathbf{Z} = R\left\{1 + jQ_F\left(\frac{\omega}{\omega_0} - \frac{\omega_0}{\omega}\right)\right\}, \qquad (7.19)$$

and near resonance, on making the substitution $\omega = \omega_0 + \delta\omega$, we have

$$\mathbf{Z} = R\{1 + 2jQ_F(\delta\omega/\omega_0)\}, \qquad (7.20)$$

which can also be written as

$$\mathbf{Z} = R + 2j\,\delta\omega L. \qquad (7.21)$$

The amplitude of the impedance is

$$Z = |\mathbf{Z}| = R\{1 + 4Q_F^2(\delta\omega/\omega_0)^2\}^{\frac{1}{2}} = \{R^2 + 4(\delta\omega L)^2\}^{\frac{1}{2}}. \qquad (7.22)$$

The reactance passes through zero at $\delta\omega = 0$ or $\omega = \omega_0$, but it is varying rapidly if $Q_F$ is large. Eqn (7.22) shows that if a given current flows through the circuit, the voltage across the circuit rises by a factor $\sqrt{2}$ when the frequency deviates from the resonant frequency by a fraction $\delta\omega/\omega_0 = \pm 1/(2Q_F)$. Similarly, if a given voltage is applied to the circuit, the current through it falls to $1/\sqrt{2}$ of its maximum value when the frequency deviates by this amount; the power dissipated in the circuit falls to one-half of the maximum, and these points are therefore often referred to as the 'half-power' points. The form of eqn (7.21) shows that the variation of impedance with frequency gives a curve of universal shape but whose spread in frequency is determined by the value of $Q_F$. Since in general the ratio of $|V_C|$ to $|V|$ is $(\omega CZ)^{-1}$, the voltage step-up obtained falls sharply away on either side of the resonance point. Thus the circuit is selective in its response to signals of different frequency; its 'selectivity' is determined by the value of $Q_F$ since $Q_F$ is the ratio of the resonant frequency to the difference of frequency between the two half-power points:

$$Q_F = \omega_0/2\delta\omega = f_0/2\delta f.$$

A plot of the amplitude of the current in a series resonant circuit when a signal of given voltage but varying frequency is applied to it is shown in Fig. 7.5. Since the voltage across the capacitance $V_C = I/\omega C$, and near resonance the variation in $I$ is very much more rapid (if $Q_F$ is large) than that of $\omega$, the voltage step-up obtained is also given to a good approximation by a curve of the same shape as in Fig. 7.5.

The approximation used in the deduction of eqn (7.20) is that $\delta\omega/\omega_0 \ll 1$. At frequencies where this approximation is not valid, the variation of $\mathbf{Z}$ still follows a universal curve, given by eqn (7.19). Since the reactive part of the impedance varies as $2Q_F(\delta\omega/\omega_0)$ near resonance, the current falls to quite small values before the approximation $\delta\omega/\omega_0 \ll 1$ becomes invalid, provided that $Q_F$ is fairly high. In an ordinary tuned circuit the resistance

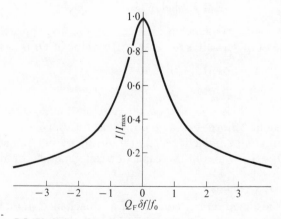

FIG. 7.5. Variation of current in series tuned circuit near resonance.

Fig. 7.6. Series resonant circuit with lossy capacitor.

is that of the wire used in winding the inductance coil; if we use a larger inductance in order to increase the value of $Q_F$ at a given frequency, the resistance goes up and so does the self-capacitance between the different parts of the coil. The latter sets up an upper limit to the size of the coil since at some point the coil will resonate at the desired frequency without any external capacitance, because of its self-capacitance. As a rough guide one may take the $Q_F$ of a coil designed to resonate at an audiofrequency as about 20; at frequencies of the order of $1\,\mathrm{MHz}$ a $Q_F$ of 100–200 may be obtained; at frequencies of $10^3$ to $10^5\,\mathrm{MHz}$, where tuned transmission lines and waveguide cavities (see Chapters 9, 21, and 22) are used instead of lumped circuits, values of 1000–10 000 are obtained for $Q_F$. Thus in most radio work with tuned circuits eqn (7.20) is a good approximation.

Hitherto we have assumed no loss in the capacitor; if we include some loss the circuit becomes that shown in Fig. 7.6. The impedance operator is now

$$\mathbf{Z} = R + \mathrm{j}\omega L + \frac{1}{\mathrm{j}\omega C + 1/r} = R + \mathrm{j}\omega L + \frac{(r - \mathrm{j}\omega C r^2)}{(1 + \omega^2 C^2 r^2)}.$$

If the power factor of the capacitor is small, $\omega C r \gg 1$ and the impedance operator is approximately (cf. Problem 7.3)

$$\mathbf{Z} = R + \mathrm{j}\omega L + 1/\omega^2 C^2 r - \mathrm{j}/\omega C. \tag{7.23}$$

Thus the resonance frequency is unaltered in the first approximation, but the resistance of the circuit is increased. The value of $Q_F$ is now given by

$$\frac{1}{Q_F} = \frac{R + 1/\omega_0^2 C^2 r}{\omega_0 L} = \frac{R}{\omega_0 L} + \frac{1}{\omega_0 C r} = \frac{1}{Q_L} + \tan \delta,$$

where $Q_L = \omega_0 L / R$ is the $Q_F$ of the circuit due to the loss in the inductance alone. Since for a good capacitor $\tan \delta < 10^{-3}$, while for lumped circuits with inductive coils (as distinct from transmission line circuits) $1/Q_L \approx 10^{-2}$, loss in the capacitor can generally be neglected.

### Parallel resonant circuits

A parallel resonant circuit consists of a capacitance shunted across an inductance+resistance, as in Fig. 7.7. The admittance operator for the

circuit is

$$Y = j\omega C + \frac{1}{R+j\omega L} = j\omega C + \frac{R-j\omega L}{R^2+\omega^2 L^2}. \tag{7.24}$$

We define the resonance point for this circuit as the point at which the admittance is real. Then, equating the imaginary terms to zero gives

$$R^2+\omega_0^2 L^2 = L/C \tag{7.25a}$$

and hence

$$\omega_0 = \left(\frac{1}{LC}\right)^{\frac{1}{2}}\left(1-\frac{R^2 C}{L}\right)^{\frac{1}{2}} = \left(\frac{1}{LC}\right)^{\frac{1}{2}}\left(1-\frac{1}{Q_F^2}\right)^{\frac{1}{2}}, \tag{7.25b}$$

while the admittance at this point is a pure conductance

$$G = R/(R^2+\omega_0^2 L^2) = RC/L = 1/(Q_F^2 R). \tag{7.26}$$

We see that the resonance frequency is not quite the same as in the series tuned circuit, but the difference is small and can usually be neglected if $Q_F$ is large (if $Q_F = 100$, the fractional difference in $\omega_0$ is $0.5\times10^{-4}$). Because of this difference, the alternative formulae $\omega_0 L/R$ and $1/\omega_0 CR$ for $Q_F$ are not exactly equivalent to $(L/C)^{\frac{1}{2}}/R$, but they are a good approximation if $Q_F$ is large. The reciprocal of the admittance at resonance is called the 'parallel resistance' of the circuit, and is equal to $Q_F^2 R$ without any approximation.

The definition of resonance as the point at which the admittance is real is to a certain extent arbitrary, but is very convenient to use; it has the further advantage of being unique. The admittance of the circuit at this point is very small, but is not necessarily a minimum. If resonance is defined as the point at which $Y$ is a minimum, then the resonance condition depends on what is being adjusted to make $Y$ a minimum. If the capacitance is altered, and other quantities are kept fixed, then, by differentiating $Y = |Y|$ with respect to $C$, it can readily be shown that $Y$ is a minimum when the value of the capacitance satisfies eqn (7.25a). On the other hand, if $L$ or the applied frequency is altered, slightly different resonance conditions are obtained; the fractional difference is, however, only of order $1/Q_F^2$, and so can be neglected if $Q_F$ is large. The calculation involved in finding these resonance conditions by differentiation is very

FIG. 7.7. Parallel resonant circuit.

tedious in comparison with the simple device of setting the imaginary part of $\mathbf{Y}$ equal to zero, and the simple definition of resonance as the point at which the circuit has unity power factor (that is, the admittance is real) has many advantages.

It is useful to have an approximate expression for $\mathbf{Y}$ near resonance similar to eqns (7.19)–(7.21) for the series resonance circuit. Writing $\omega = \omega_0 + \delta\omega$, we have from eqn (7.24)

$$\mathbf{Y} = \frac{R}{R^2 + \omega_0^2 L^2}\left(1 - \frac{2\omega_0\,\delta\omega L^2}{R^2 + \omega_0^2 L^2}\right) + j\left\{\delta\omega C - \frac{\delta\omega L}{R^2 + \omega_0^2 L^2}\left(1 - \frac{2\omega_0^2 L^2}{R^2 + \omega_0^2 L^2}\right)\right\}$$

$$= \frac{RC}{L}\{1 - 2\delta\omega(\omega_0 CL)\} + 2j\,\delta\omega C^2 L\omega_0^2,$$

where we have used eqn (7.25a). In the term of order $\delta\omega$ we may put $\omega_0^2 LC = 1$, giving

$$\mathbf{Y} = \frac{RC}{L}\left(1 - \frac{2\delta\omega}{\omega_0}\right) + 2j\,\delta\omega C.$$

In general the small change in the conductance can be neglected, so that

$$\mathbf{Y} = \frac{RC}{L} + 2j\,\delta\omega C = \frac{RC}{L}\left\{1 + 2jQ_F\frac{\delta\omega}{\omega_0}\right\}, \tag{7.27}$$

showing that the behaviour of the admittance near resonance for the parallel resonant circuit is similar to that of the impedance of the series tuned circuit. The selectivity is the same in that $Y$ rises by a factor $\sqrt{2}$ when the fractional deviation of the frequency is

$$\delta f/f_0 = \delta\omega/\omega_0 = \pm 1/2Q_F.$$

Thus the current drawn from the voltage generator is a minimum at resonance (apart from the small corrections which can be neglected when $Q_F$ is large). The current in each of the arms is larger than that drawn from the source by a factor nearly equal to $Q_F$. Thus the ratio of the current through the capacitance to that drawn from the generator is

$$\left|\frac{I_C}{I}\right| = \frac{\omega_0 CV}{RCV/L} = \frac{\omega_0 L}{R} = Q_F \tag{7.28}$$

in our approximation; this relation is complementary to that given by eqn (7.18). That equation showed that the series resonant circuit can be used as a tuned transformer where the output voltage across the condenser is larger by a factor $Q_F$ than that injected in series with the inductance. The impedance measured at the output (for example, by finding the current drawn from a generator applied across the capacitance as in Fig. 7.7) is $Q_F^2 R$ at resonance (by eqn (7.26)) whereas the impedance of the series tuned circuit is $R$. Thus the ratio of the impedances is $Q_F^2$, the square of the voltage ratio. This, which is a characteristic of all transformers, can be seen as follows: if we apply a generator of voltage $V$ to the series tuned circuit, the current drawn from it is $I$; to have the same current flowing through the capacitance in the parallel resonant circuit formed from the same elements, we must apply a generator of voltage $Q_F V$ across the capacitance, and the current drawn from it will be $I/Q_F$. The power dissipated in the circuit is $VI$, the same in each case.

The importance of the quality factor $Q_F$ in determining the properties of a resonant circuit can be readily appreciated from the following summary. For a simple resonant circuit as considered hitherto:

(1)  $Q_F = (L/C)^{\frac{1}{2}}/R \approx \omega_0 L/R \approx (\omega_0 CR)^{-1}$;

(2)  $Q_F$ is equal to the voltage step-up obtained by using the circuit as a tuned transformer;

(3)  the parallel resistance is $Q_F^2$ times the series resistance;

(4)  the fractional frequency difference $2\delta f/f_0$ between the points at which the impedance or admittance changes by a factor $\sqrt{2}$ is $1/Q_F$;

(5)  $Q_F = \pi/\Lambda$, where $\Lambda$ is the logarithmic decrement of free oscillations in the circuit (see § 5.3);

(6)  in forced resonance,

$$Q_F = \frac{\omega_0 \times \text{stored energy}}{\text{rate of energy dissipation}}.$$

This last relation can be used to define $Q_F$ in more complicated resonant circuits, and in other resonant systems such as waveguide cavities (see Chapter 9) where the values of $L$, $C$, and $R$ cannot be specified. For the series resonant circuit of Fig. 7.2 it is readily shown that this definition agrees with that in (1) above. For the stored energy at any instant is

$$\tfrac{1}{2}LI^2 + \tfrac{1}{2}Q^2/C = \tfrac{1}{2}LI_0^2\cos^2\omega_0 t + \tfrac{1}{2}(I_0^2/\omega_0^2 C)\sin^2\omega_0 t$$

$$= \tfrac{1}{2}LI_0^2 \quad \text{at resonance,}$$

while the mean rate of energy dissipation is $\tfrac{1}{2}RI_0^2$. Hence

$$Q_F = \frac{\omega_0 \times \tfrac{1}{2}LI_0^2}{\tfrac{1}{2}RI_0^2} = \frac{\omega_0 L}{R}.$$

## 7.4. Coupled resonant circuits

Two circuits are said to be coupled together if they have a common impedance. The impedance may be a resistance, inductance, or capacitance, and may be a part of each circuit, as in the example shown in Fig.

(a)

(b)

FIG. 7.8. Types of coupled circuits: (a) with common impedance $C_3$; (b) with impedance $C_3$ connected between them.

7.8(a), or connected between the two circuits as in Fig. 7.8(b). Two circuits may also be coupled together if one of them is in an electric or magnetic field set up by the other; for example, the oscillating magnetic flux due to a coil in one circuit may induce a voltage in a coil in the second circuit, so that there is a mutual inductance between the two circuits.

In Fig. 7.9 two resonant circuits 1 and 2 are coupled together by a mutual inductance $M$ between the coils $L_1$ and $L_2$. The voltage $V_1$ applied to the first circuit produces a current $I_1$, and this induces a voltage $M(dI_1/dt)$ in the secondary circuit. For a simple sinusoidal waveform this voltage may be written as $j\omega MI_1$, where $j\omega M$ is the impedance operator for the mutual inductance. Similarly, if $I_2$ is the current in the secondary, a voltage $j\omega MI_2$ will be induced in the primary. Thus for the two circuits we have

$$V_1 = I_1\mathbf{Z}_1 + j\omega MI_2,$$
$$0 = I_2\mathbf{Z}_2 + j\omega MI_1,$$

(7.29)

FIG. 7.9. Two resonant circuits coupled by mutual inductance $M$.

where $\mathbf{Z}_1$, $\mathbf{Z}_2$ are the impedance operators for the primary and secondary circuits in the absence of the mutual inductance. On eliminating $I_2$ we have

$$V_1 = I_1\left(\mathbf{Z}_1 + \frac{\omega^2 M^2}{\mathbf{Z}_2}\right). \tag{7.30}$$

The quantity $\omega^2 M^2/\mathbf{Z}_2$ is called the impedance 'reflected into the primary circuit'. For the secondary circuit, elimination of $I_1$ gives

$$-\frac{j\omega M}{\mathbf{Z}_1} V_1 = I_2\left(\mathbf{Z}_2 + \frac{\omega^2 M^2}{\mathbf{Z}_1}\right), \tag{7.31}$$

showing that the current flow is that produced by an apparent voltage $-(j\omega M/\mathbf{Z}_1)V_1$ working into $\mathbf{Z}_2$ plus the impedance 'reflected into the secondary circuit', $\omega^2 M^2/\mathbf{Z}_1$.

To investigate the behaviour of the primary circuit impedance in more detail we write $\mathbf{Z}_1 = R_1 + jX_1$, $\mathbf{Z}_2 = R_2 + jX_2$. Then

$$\begin{aligned}
\frac{V_1}{I_1} &= R_1 + jX_1 + \frac{\omega^2 M^2}{(R_2 + jX_2)} \\
&= R_1 + \frac{R_2\omega^2 M^2}{R_2^2 + X_2^2} + j\left(\frac{X_1 - X_2\omega^2 M^2}{R_2^2 + X_2^2}\right) \\
&= R_p + jX_p. \tag{7.32}
\end{aligned}$$

The current and voltage in the primary are in phase if $X_p$ is zero. This requirement is satisfied if $X_1 = X_2 = 0$, and in the following analysis we shall assume that the elements of the primary and secondary circuits are the same, so that $R_1 = R_2$, $X_1 = X_2$ at all frequencies (the latter implies $L_1 = L_2$, $C_1 = C_2$). Then $X_p$ is zero when either

$$X = 0 \quad \text{or} \quad X^2 = \omega^2 M^2 - R^2. \tag{7.33}$$

The second of these conditions can be fulfilled only if $\omega M > R$. If $\omega M = R$, the three roots are identical, and if $\omega M < R$ only one real root exists.

The significance of these three roots becomes apparent when we examine the behaviour of the secondary current. When

$$X_1 = X_2 = X = 0,$$

from eqn (7.31) we have

$$I_2 = \frac{-j\omega_0 M V_1}{R^2 + \omega_0^2 M^2}, \tag{7.34}$$

where $\omega_0 = (LC)^{-\frac{1}{2}}$ is the resonance frequency of either circuit by itself. If we can vary the mutual inductance $M$, then the value of $I_2$ at this frequency rises as $M$ is increased from zero, passes through a maximum value of $-\mathrm{j}V_1/2R$ when $\omega_0 M = R$, and then falls again. At the maximum the impedance reflected into the primary from the secondary is just equal to $R$, so that the secondary circuit is 'matched' to the primary circuit, and the power dissipated in the secondary circuit is a maximum at this point. From eqns (7.30) and (7.34) it can be seen that the currents in the primary and secondary circuits are both equal to $V_1/2R$ at this point, but they differ in phase by $\frac{1}{2}\pi$.

At the second two points given by eqn (7.33) where the effective primary impedance is real, the secondary current is

$$I_2 = \frac{-\mathrm{j}\omega M V_1}{Z^2 + \omega^2 M^2} = \frac{-\mathrm{j}\omega M V_1}{R^2 - X^2 + \omega^2 M^2 + 2\mathrm{j}RX}$$

$$= \frac{-\mathrm{j}\omega M V_1}{2R^2 + 2\mathrm{j}RX} = \frac{-\mathrm{j}\omega M V_1}{2R(R + \mathrm{j}X)},$$

where we have used the condition given by eqn (7.33). Hence

$$|I_2| = \frac{\omega M V_1}{2R(R^2 + X^2)^{\frac{1}{2}}} = \frac{V_1}{2R}.$$

These results show that when the circuits are 'over-coupled' ($\omega M > R$) the secondary current rises to $V_1/2R$ at the second two points given by eqn (7.33), while it has fallen below this value at the point $X = 0$. The primary-circuit impedance is purely resistive at all three points, but only at the second two points is it equal to $2R$, as can be seen by substituting in eqn (7.32). Thus the secondary current is a maximum at these two points because the secondary circuit is again 'matched' to the primary circuit.

The general behaviour of the secondary current is shown in Fig. 7.10, where the degree of coupling is specified in terms of the 'coefficient of coupling' defined by eqn (5.14) as

$$k = M/(L_1 L_2)^{\frac{1}{2}} = M/L \tag{7.35}$$

when the two circuits are identical.

The critical condition $\omega_0 M = R$ then occurs when the coefficient of coupling has the value

$$k_0 = R/\omega_0 L = 1/Q_F. \tag{7.36}$$

The curves in Fig. 7.10 are drawn for $Q_F = 100$, and coefficients of coupling of $k = \frac{1}{2}k_0$, $k_0$, and $2k_0$ respectively. The first curve is similar to an ordinary resonance curve, but the current is always less than the maximum possible value $V_1/2R$. The second curve has a single peak in the centre, but the peak is decidedly flattened because of the three coincident roots. As soon as $k > k_0$ we have two side maxima in the current with a central minimum. As $k/k_0$ increases these peaks move

FIG. 7.10. Secondary current in the circuit of Fig. 7.9 for two identical circuits each with $Q_F = 100$. Critical coupling occurs at $k = 0.01$.

outwards and the trough in the middle deepens. When $Q_F$ is large and the frequency is close to resonance we may write $X = 2\delta\omega L$, and eqn (7.33) gives for the separation $2\delta f$ of the peaks

$$\frac{2\delta f}{f_0} = \frac{2\delta\omega}{\omega_0} \approx \left(\frac{M^2}{L^2} - \frac{R^2}{\omega_0^2 L^2}\right)^{\frac{1}{2}} = (k^2 - 1/Q_F^2)^{\frac{1}{2}} = (k^2 - k_0^2)^{\frac{1}{2}}. \quad (7.37)$$

Although identical circuits have been assumed in this analysis, the behaviour of any pair of coupled circuits whose natural resonant frequencies are the same is similar. Thus for two circuits with the same $X$ but different $R_1$, $R_2$, optimum coupling occurs (for $X = 0$) when

$$\omega_0 M = (R_1 R_2)^{\frac{1}{2}} \qquad (k = 1/(Q_1 Q_2)^{\frac{1}{2}}),$$

and the secondary current is then $V_1/2(R_1 R_2)^{\frac{1}{2}}$. Side peaks occur when the coupling is greater than this value, and the secondary current at these peaks is $V_1/(R_1 + R_2)$, which is less than the optimum. The values of the effective primary impedance are $2R_1$ when the coupling is optimum and $X = 0$, and $R_1 + R_2$ at the side peaks when the circuits are over-coupled.

Coupled circuits have an important application in the reception of radio signals when it is desired to accept a narrow band of frequencies and reject frequencies outside this band. Clearly in radio reception the ideal receiver should have a uniform response over this band, combined with total rejection outside, so that the ideal response curve would be rectangular in shape. By the use of tuned circuits with rather more than critical coupling a response curve with steep sides is obtained, and the slight dip in the middle can be compensated by the use of somewhat under-coupled circuits (with a central peak) elsewhere in the receiver. In a tunable receiver of this kind a constant bandwidth is required independent of the central frequency. This cannot be achieved by mutual-inductance coupling alone, for eqn (7.37) shows that the bandwidth is then proportional to the central frequency. A combination of mutual inductance coupling together with capacitance coupling as in Fig. 7.8(a) may be used to give a more-or-less constant bandwidth, for the coupling through the capacitance decreases with frequency, because of the fall in the impedance common to the two circuits.

## 7.5. Low-frequency transformers

The coupled resonant circuits discussed in the last section may be regarded as a tuned transformer; they are used as such at radio-frequencies where the effects of stray capacitance can be reduced by making it part of the tuning capacitance. At low frequencies (such as those of power supplies) effects of stray capacitance are small, and transfer of power can be made very efficiently by use of the transformer whose principle was mentioned in § 5.2. It consists of two coils, the primary or input coil of $n_1$ turns, and the secondary or output coil of $n_2$ turns, which are closely wound together on an iron core so that all the flux due to one coil passes through the other. In practice there is always a small leakage of flux, so that not all the flux of one circuit passes through the other. Then the transformer may be represented by the circuit of Fig. 7.11, where $L_1$, $L_2$ are the self-inductances of the primary and secondary windings, and $M$ is the mutual inductance between them. An alternating voltage $V_1$ of angular frequency $\omega$ is applied to the primary in series with an impedance $\mathbf{Z}_1$, and the secondary is connected to a load impedance $\mathbf{Z}_2$. Then, if the primary and secondary currents are $I_1$, $I_2$, the equations for the primary and secondary circuits are

$$V_1 = (\mathbf{Z}_1 + j\omega L_1)I_1 + j\omega M I_2,$$
$$0 = (\mathbf{Z}_2 + j\omega L_2)I_2 + j\omega M I_1. \tag{7.38}$$

These equations differ from (7.29) in that the self-inductances $L_1$, $L_2$ are not included in the impedances $\mathbf{Z}_1$, $\mathbf{Z}_2$ because their reactances are normally very much larger than any other impedances in the circuits.

FIG. 7.11. Circuit diagram of transformer with load $\mathbf{Z}_2$.

Alternate elimination of $I_2$ and $I_1$ between these equations gives the following formulae for the primary and secondary circuits

$$V_1/I_1 = \mathbf{Z}_1 + j\omega L_1 + \omega^2 M^2/(\mathbf{Z}_2 + j\omega L_2) \tag{7.39}$$

and

$$-(M/L_1)V_1/I_2 = \mathbf{Z}_2 + \mathbf{Z}_1(L_2/L_1) + j\omega(L_2 - M^2/L_1), \tag{7.40}$$

where in the second equation a term $\mathbf{Z}_1\mathbf{Z}_2/j\omega L_1$ has been omitted since it is an order of magnitude smaller than $\mathbf{Z}_2$.

For a perfect transformer, there is complete coupling between the two coils, so that $M^2 = L_1 L_2$. Since the self-inductances are proportional to the squares of the number of turns, the turns ratio $n = (L_2/L_1)^{\frac{1}{2}}$, and with complete coupling we have also $n = M/L_1 = L_2/M$. Hence, assuming $\mathbf{Z}_2 \ll \omega L_2$, from eqn (7.39) the primary impedance $\mathbf{Z}_p$ may be written

$$V_1/I_1 = \mathbf{Z}_p = \mathbf{Z}_1 + j\omega L_1 + \omega^2 M^2(\mathbf{Z}_2 - j\omega L_2)/(\mathbf{Z}_2^2 + \omega^2 L_2^2)$$

$$\approx \mathbf{Z}_1 + \mathbf{Z}_2(M^2/L_2^2) + j\omega(L_1 - M^2/L_2) = \mathbf{Z}_1 + \mathbf{Z}_2/n^2. \tag{7.41}$$

The current in the secondary circuit is the same as that due to an e.m.f. $-(M/L_1)V_1 = -nV_1$ working into an impedance $\mathbf{Z}_s$, where

$$\mathbf{Z}_s = \mathbf{Z}_2 + \mathbf{Z}_1 n^2. \tag{7.42}$$

These equations show that for a perfect transformer on load ($\mathbf{Z}_2 \ll \omega L_2$) the inductive terms such as $j\omega(L_1 - M^2/L_2)$ are exactly zero when there is complete coupling between the two coils. The effect of the secondary circuit on the current in the primary is represented by the additional impedance $\mathbf{Z}_2/n^2$, known as the 'reflected impedance'. In the secondary circuit the effective e.m.f. is $-nV_1$, with an apparent internal impedance $\mathbf{Z}_1 n^2$. Hence the impedances are transformed by $n^2$ while the voltages are transformed by $n$; the current transformation ratio is $1/n$, so that the power on the two sides is the same ($V_1 I_1 = V_2 I_2$), as we should expect for a perfect transformer with no losses.

In practice $M^2$ is slightly less than $L_1 L_2$ because not all the flux from one circuit passes through the other, and we write $M = k(L_1 L_2)^{\frac{1}{2}}$, where $k$ is the coupling coefficient defined by eqn (5.14). It is useful to derive an equivalent circuit for the transformer, and in order to include the case where $\mathbf{Z}_2$ is of the same order of magnitude as $\omega L_2$, we rewrite the expression for the primary impedance (eqn (7.39)) in the following way

$$\mathbf{Z}_p = \mathbf{Z}_1 + j\omega L_1(1-k) + \frac{\omega^2 k^2 L_1 L_2 + j\omega L_1 k(\mathbf{Z}_2 + j\omega L_2)}{\mathbf{Z}_2 + j\omega L_2}. \tag{7.43}$$

The last term may be expressed in the form

$$\frac{j\omega k L_1\{\mathbf{Z}_2 + j\omega(1-k)L_2\}}{\{\mathbf{Z}_2 + j\omega(1-k)L_2\} + j\omega k L_1 n^2},$$

FIG. 7.12. Equivalent circuit of primary of imperfect transformer. $n^2 = L_2/L_1$; $k = M/(L_1L_2)^{\frac{1}{2}}$.

since in the denominator $j\omega kL_1n^2 - j\omega kL_2 = 0$. The last term is now

$$\left\{ \frac{1}{j\omega kL_1} + \frac{n^2}{\mathbf{Z}_2 + j\omega(1-k)L_2} \right\}^{-1},$$

which is equivalent to two impedances in parallel. Hence the primary circuit may be represented by Fig. 7.12. Here $(1-k)L_1$ is the 'leakage inductance' due to the imperfect coupling, and the impedance on the extreme right is the reflected impedance of the secondary, which is in parallel with the remainder of the primary inductance.

For the secondary circuit, since $M/L_1 = k(L_2/L_1)^{\frac{1}{2}} = kn$, we have, from eqn (7.40),

$$-knV_1/I_2 = \mathbf{Z}_2 + n^2\mathbf{Z}_1 + j\omega L_2(1-k) + j\omega(kL_2 - M^2/L_1)$$

$$= \mathbf{Z}_2 + n^2\{\mathbf{Z}_1 + j\omega L_1(1-k)\} + j\omega L_2 k(1-k), \tag{7.44}$$

which is represented by Fig. 7.13. The reflected impedance is the transformed value of $\mathbf{Z}_1$ plus the primary leakage inductance $L_1(1-k)$, while the secondary leakage inductance is, apart from a factor $k$, just that which appears in series with $\mathbf{Z}_2$ in the impedance reflected in the primary. If $k$ is close to unity, this approximation is a good one, for the leakage inductance will be small compared with $\mathbf{Z}_2$ except for very small values of $\mathbf{Z}_2$. The equivalent circuit of the transformer can be drawn as in Fig. 7.14, where the centre portion enclosed in the dotted rectangle is regarded as a perfect transformer. The resistances $r_1$ and $r_2$ are the resistances of the windings, which previously we have regarded as part of $\mathbf{Z}_1$ and $\mathbf{Z}_2$. The resistance $R$ in parallel with $kL_1$ allows for the dissipation of energy through hysteresis and eddy currents in the iron core.

FIG. 7.13. Equivalent circuit for secondary of imperfect transformer.

FIG. 7.14. Approximate equivalent circuit of imperfect transformer. $r_1$, $r_2$ resistance of primary and secondary windings, $R$ resistance equivalent to hysteresis and eddy-current losses in iron core. The portion within the dotted rectangle is regarded as a perfect transformer of turns ratio $n$.

The prime requirements in a transformer are therefore a high primary inductance, to keep the 'magnetizing current' through $kL_1$ in Fig. 7.14 small, and the smallest possible leakage of flux between primary and secondary. These requirements are fulfilled by winding the primary and secondary round an iron core whose magnetic circuit is completed by a yoke. If the two coils are interwound the leakage is reduced to a minimum, but if good insulation between primary and secondary is required they may be wound side by side.

## 7.6. Linear circuit analysis

It is apparent from the examples considered in the previous sections that the behaviour of a network of impedances at a given frequency can be analysed in the same way as the network of resistances considered in § 3.4. The analysis is based on the two laws of Kirchhoff:

$$\sum_k I_k = 0, \tag{7.45}$$

$$\sum_j V_j = \sum_k I_k \mathbf{Z}_k = \sum_k I_k / \mathbf{Y}_k. \tag{7.46}$$

These equations are formally similar to the equations in § 3.4 with the substitution of $\mathbf{Z}_k$ for $R_k$, and $\mathbf{Y}_k$ for $G_k$, but the fact that the impedance $\mathbf{Z}$ and the admittance $\mathbf{Y}$ are complex numbers reminds us that each equation must be satisfied independently for the real and imaginary parts. That is, both amplitude and phase must be considered in the addition of voltages and currents. Allowing for this complication, a number of important theorems from Chapter 3 can be applied at once to alternating current circuits. These include the reciprocity theorem, Thévenin's theorem, and Norton's theorem (see § 3.4). The maximum power theorem (§ 3.3) needs one modification. In Fig. 7.15 we have an a.c. voltage generator with internal impedance $\mathbf{Z}_s = R_s + jX_s$ working into a load $\mathbf{Z}_L = R_L + jX_L$. The current through the load is $I = V_s/(\mathbf{Z}_s + \mathbf{Z}_L)$, and

Fig. 7.15. The maximum power theorem for an a.c. circuit.

the power dissipated in it is

$$I^2 R_L = V_s^2 \frac{R_L}{|\mathbf{Z}_s + \mathbf{Z}_L|^2} = V_s^2 \frac{R_L}{(R_s + R_L)^2 + (X_s + X_L)^2}.$$

Since the reactances can be either positive or negative, maximum power is obtained by making first

$$X_s + X_L = 0, \tag{7.47}$$

and the problem is then the same as in § 3.3, that is, we need

$$R_s = R_L \tag{7.48}$$

provided that $R_L$ is the variable quantity.

*The four-terminal network*

If in a network, however complicated, there is only one voltage generator $v_s$, with internal impedance $\mathbf{z}_s$, and we are interested in the output at one pair of terminals into a load $\mathbf{z}_L$, we may formally represent this as in Fig. 7.16. Here the rectangular box which replaces the network has just two input terminals, through which a current $i_1$ flows and between which the voltage is $v_1$, and two output terminals, through which a current $i_2$ flows and between which the voltage is $v_2$. The box is known as a 'four-terminal network', or 'quadrupole', whose behaviour can be represented by two simple equations, relating the voltages $v_1$, $v_2$ and the currents $i_1$, $i_2$. The analysis in this section is applicable to non-linear devices provided it is

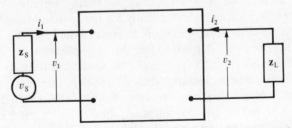

Fig. 7.16. The four-terminal network, with a voltage generator $v_s$ (internal impedance $\mathbf{z}_s$) applied to the input terminal and a load $\mathbf{z}_L$ to the output terminals.

restricted to small changes, including small alternating components, of voltage and current. Such changes are denoted here by lower-case symbols $v$, $i$; similarly, lower-case symbols are used for circuit parameters to indicate that their values must be those appropriate to such small changes.

The low-frequency transformer is a simple example of a 'four-terminal network', as can be seen by redrawing Fig. 7.14 in the form illustrated by Fig. 7.16. Here $v_s$, $z_s$ are the open-circuit voltage and internal impedance of an external generator applied to the terminals of the primary coil, while $z_L$ is a load connected across the terminals of the secondary coil. The current which flows in the primary coil depends on the impedance in the secondary circuit, and vice versa; this is shown by the presence of $z_2$ in eqn (7.43) and $z_1$ in eqn (7.44), or, for the ideal transformer, in eqns (7.41) and (7.42) respectively. These results followed from the fundamental eqns (7.38), which can be written in terms of the voltages at the input and output terminals in Fig. 7.16, and the currents to these terminals, as follows:

$$v_1 = \mathbf{z}_{11}i_1 + \mathbf{z}_{12}i_2, \tag{7.49a}$$

$$v_2 = \mathbf{z}_{21}i_1 + \mathbf{z}_{22}i_2. \tag{7.49b}$$

By a simple rearrangement these equations can be written instead as

$$v_1 = \mathbf{h}_{11}i_1 + \mathbf{h}_{12}v_2, \tag{7.50a}$$

$$i_2 = \mathbf{h}_{21}i_1 + \mathbf{h}_{22}v_2. \tag{7.50b}$$

Here the symbols $\mathbf{h}$ stand for 'hybrid parameters'; obviously $\mathbf{h}_{11}$ has the dimensions of impedance, $\mathbf{h}_{22}$ those of admittance, while $\mathbf{h}_{12}$ and $\mathbf{h}_{21}$ are dimensionless. It is clear also that we can write

$$i_1 = \mathbf{y}_{11}v_1 + \mathbf{y}_{12}v_2, \tag{7.51a}$$

$$i_2 = \mathbf{y}_{21}v_1 + \mathbf{y}_{22}v_2. \tag{7.51b}$$

Eqns (7.49) are known as the 'impedance equations' and (7.51) as the 'admittance equations'. The three forms are all equivalent, and the parameters in one set may be related to those in another by simple algebraic processes. In matrix language any of these relations may be written in the form (taking (7.49) as an example)

$$\begin{bmatrix} v_1 \\ v_2 \end{bmatrix} = \begin{bmatrix} \mathbf{z}_{11} & \mathbf{z}_{12} \\ \mathbf{z}_{21} & \mathbf{z}_{22} \end{bmatrix} \begin{bmatrix} i_1 \\ i_2 \end{bmatrix}. \tag{7.52}$$

This is called the **Z**-matrix; $\mathbf{z}_{11}$, $\mathbf{z}_{22}$ are its 'diagonal elements', and $\mathbf{z}_{12}$, $\mathbf{z}_{21}$ its 'off-diagonal elements'. The matrix equivalents of (7.50) and (7.51) are known as the **H**- and **Y**-matrices. It is often convenient to analyse combinations of networks using matrix algebra, though we shall not use such methods here.

The input impedance is $v_1/i_1$, which from eqn (7.50a) is

$$v_1/i_1 = \mathbf{h}_{11} + \mathbf{h}_{12}(v_2/i_1), \tag{7.53}$$

where the presence of $v_2$ means that the input impedance depends on the impedance $z_L$ connected to the output terminals as the load. Conversely, the output impedance $( = v_2/i_2)$ of the network depends on the impedance $z_s$ of the input source. However, in limiting cases all the parameters in the various relations between $v_1$, $v_2$, $i_1$, and $i_2$ have simple interpretations, of which we mention only the following which arise from eqns (7.50).

(1) When the output is short-circuited, $v_2 = 0$, and

$$\text{the input impedance} = \mathbf{h}_{11}; \tag{7.54}$$
$$\text{the current gain } A_i = \mathbf{h}_{21}. \tag{7.55}$$

(2) When the input is on open-circuit, $i_1 = 0$, and

$$\text{the output impedance} = \mathbf{h}_{22}^{-1}; \tag{7.56}$$
$$\text{the voltage gain } A_v = \mathbf{h}_{12}^{-1}. \tag{7.57}$$

*The passive four-terminal network*

If a network contains no voltage or current generator, it is known as a 'passive network', and for a passive four-terminal network we can derive some simple relations for the 'off-diagonal' elements of the matrices. From eqns (7.49) we have

$$i_2/v_1 = -\mathbf{z}_{21}/(\det \mathbf{z}) \quad \text{when} \quad v_2 = 0,$$

and

$$i_1/v_2 = -\mathbf{z}_{12}/(\det \mathbf{z}) \quad \text{when} \quad v_1 = 0.$$

By the reciprocity theorem (cf. § 3.4) these quantities must be equal, so that $\mathbf{z}_{12} = \mathbf{z}_{21}$. Applying similar methods to the **H**- and **Y**-matrices, we have the results

$$\mathbf{z}_{12} = \mathbf{z}_{21}, \tag{7.58}$$
$$\mathbf{y}_{12} = \mathbf{y}_{21}, \tag{7.59}$$
$$\mathbf{h}_{12} + \mathbf{h}_{21} = 0. \tag{7.60}$$

In a network containing resistance, any current must decay with time because of the dissipation of energy in the resistance, unless there is some mechanism in the circuit for creating energy. Such a mechanism is equivalent to a 'negative resistance'; if more energy is created than is dissipated, the net resistance is negative, and oscillations will increase in amplitude instead of decreasing, as is obvious from eqn (5.28). The condition under which oscillations can just be sustained is that the net resistance is zero, and under such conditions it is often useful to have a method of determining the natural frequency of oscillation which is simpler than solving a differential equation such as (5.27). This is particularly true for coupled circuits, where calculations such as those in § 7.4 are clearly rather tedious. In Fig. 7.9, if we put $R_1 = R_2 = 0$, and $V_1 = 0$, we have

$$0 = \left(j\omega L_1 - \frac{1}{j\omega C_1}\right)i_1 + j\omega M i_2,$$

$$0 = j\omega M i_1 + \left(j\omega L_2 - \frac{1}{j\omega C_2}\right)i_2.$$

$$(7.61)$$

These equations are equivalent to (7.49) with $v_1 = v_2 = 0$, and the condition for a non-trivial solution is $\det \mathbf{z} = 0$. This gives, writing $\omega_1^2 L_1 C_1 = 1$, $\omega_2^2 L_2 C_2 = 1$ for the natural resonant frequencies of the uncoupled circuits,

$$\left(1 - \frac{\omega_1^2}{\omega^2}\right)\left(1 - \frac{\omega_2^2}{\omega^2}\right) = \frac{M^2}{L_1 L_2} = k^2,$$

$$(7.62)$$

where $k$ is the coefficient of coupling defined by eqns (5.14) or (7.35). The two natural resonant frequencies are found to be given by

$$\frac{1}{\omega^2} = \frac{1}{2}\left(\frac{1}{\omega_1^2} + \frac{1}{\omega_2^2}\right) \pm \frac{1}{2}\left\{\left(\frac{1}{\omega_1^2} - \frac{1}{\omega_2^2}\right)^2 + \frac{4k^2}{\omega_1^2 \omega_2^2}\right\}^{\frac{1}{2}}.$$

$$(7.63)$$

This method can be used in dealing with the natural resonant frequencies of other coupled systems.

### The active four-terminal network

A network containing a voltage or a current generator is known as an 'active network'. In general such a generator is an electronic device, and we choose as an example the network shown in Fig. 7.17, which is applicable to thermionic triodes and to various solid-state devices.

In Fig. 7.17 the input circuit has a source of voltage $v_S$ and internal impedance $\mathbf{z}_S$, giving a voltage $v_1$ at the input to the network. The magnitude of $v_1$ depends on $i_1$ and is determined by the values of $\mathbf{h}_{11}$ and $\mathbf{h}_{12}$ as in eqn (7.50a). The output circuit contains a current generator $g_m v_1$ shunted by a resistance $r_0$, producing a voltage $v_2$ at the output terminals

FIG. 7.17. An active four-terminal network.

across a load $z_L$. This representation is chosen because the constants $g_m$ and $r_0$ are the parameters of an electronic device which are most readily determined in practice, whereas the **h** parameters cannot be directly measured. The combination of $\mathbf{h}_{11}$ and the voltage generator $\mathbf{h}_{12}v_2$ can be represented by an impedance $z_i$ which is the input impedance of the network, but it must be remembered that the magnitude of $z_i$ is thus dependent on $v_2$ and hence also on the load $z_L$. We have then the equations

$$v_S = v_1 + z_S i_1 = (z_S + z_i)i_1, \tag{7.64a}$$

$$i_2 = g_m v_1 + v_2/r_0. \tag{7.64b}$$

In the first equation $z_i = \mathbf{h}_{11}$ if $\mathbf{h}_{12} = 0$; in the second, $g_m = \mathbf{y}_{21}$ and $r_0^{-1} = \mathbf{y}_{22}$ in eqn (7.51b). If $\mathbf{h}_{12} = 0$, $\mathbf{y}_{22} = \mathbf{h}_{22}$, and $r_0$ is the output impedance of the network in the limit when $i_1 = 0$. This is a good approximation for thermionic triodes and field-effect transistors where $i_1$ is zero or very small, but is less true in junction transistors.

When a load $z_L$ is connected to the output terminals, the current generator $g_m v_1$ works into an admittance $(r_0 + z_L)/r_0 z_L$, since $r_0$ and $z_L$ are in parallel. The voltage across $z_L$ is

$$-g_m v_1 \frac{r_0 z_L}{(r_0 + z_L)}, \tag{7.65}$$

and the voltage amplification is

$$A_v = \frac{v_2}{v_1} = -g_m \frac{r_0 z_L}{(r_0 + z_L)}. \tag{7.66a}$$

When $|z_L| \ll r_0$ this reduces to

$$A_v = -g_m z_L, \tag{7.66b}$$

and under the same condition, since $v_1 = i_1 z_i$,

$$A_i = \frac{i_2}{i_1} = \frac{g_m v_1}{i_1} = g_m z_i, \tag{7.67}$$

so that

$$-A_v = A_i(z_L/z_i). \tag{7.68}$$

This relation is often used to calculate the fourth parameter when the other three are known.

In the other extreme, where $|z_L| \gg r_0$, we have

$$-A_v = g_m r_0 = \mu, \tag{7.69}$$

where $\mu$ is the voltage amplification factor of the device. The parameter $g_m$ is known as the 'mutual conductance' or 'transconductance'.

Devices such as the thermionic triode and transistors are usually three-terminal devices, but they can obviously be represented by a four-terminal network in which two terminals are joined together. A more serious difficulty is that they are inherently non-linear devices in which the currents are not simply proportional to the voltages, and the parameters in the equations above are not constants independent of the working conditions. Nevertheless, for small changes in current and voltage it is still possible to use linear circuit analysis, provided that its limitation to small signals is recognized. In Fig. 7.17, if the output current $I_2 = f(V_1, V_2)$, then for small changes $v_1$, $v_2$ in $V_1$, $V_2$ we can use a Taylor series expansion, giving

$$I_2 = (I_2)_0 + v_1\left(\frac{\partial I_2}{\partial V_1}\right)_{V_2} + v_2\left(\frac{\partial I_2}{\partial V_2}\right)_{V_1} +$$

$$+ \tfrac{1}{2}v_1^2\left(\frac{\partial^2 I_2}{\partial V_1^2}\right)_{V_2} + v_1 v_2\left(\frac{\partial^2 I_2}{\partial V_1 \partial V_2}\right)_{V_2,V_1} + \tfrac{1}{2}v_2^2\left(\frac{\partial^2 I_2}{\partial V_2^2}\right)_{V_1} + \ldots \quad (7.70)$$

The presence of second- and higher-order terms means that the change $i_2 = I_2 - (I_2)_0$ in output current is not linear with the voltage changes; this is generally undesirable because it represents distortion in the output signal. For sufficiently small changes $v_1$, $v_2$ this effect becomes negligible, and (7.70) can be written as

$$i_2 = g_m v_1 + (1/r_0)v_2, \quad (7.71)$$

where

$$g_m = (\partial I_2/\partial V_1)_{V_2}, \qquad 1/r_0 = (\partial I_2/\partial V_2)_{V_1}. \quad (7.72)$$

From the relation

$$\left(\frac{\partial V_2}{\partial I_2}\right)_{V_1}\left(\frac{\partial I_2}{\partial V_1}\right)_{V_2}\left(\frac{\partial V_1}{\partial V_2}\right)_{I_2} = -1,$$

we have

$$-\left(\frac{\partial V_2}{\partial V_1}\right)_{I_2} = g_m r_0 = \mu, \quad (7.73)$$

a result which can also be obtained directly from eqn (7.71) by putting $i_2 = 0$. The differentials in which $V_2$, $V_1$, and $I_2$ are kept constant are equivalent to equations in which $v_2$, $v_1$, and $i_2$ are respectively zero. Comparison with eqn (7.51b) shows that

$$\left(\frac{\partial I_2}{\partial V_1}\right)_{V_2} = y_{21}, \qquad \left(\frac{\partial I_2}{\partial V_2}\right)_{V_1} = y_{22} \quad \text{and} \quad \left(\frac{\partial V_2}{\partial V_1}\right)_{I_2} = \frac{y_{21}}{y_{22}}.$$

## Problems

7.1. A resistance $R$, inductance $L$, and capacitance $C$ are all connected in parallel. Show that the admittance of the circuit at frequencies near resonance is

$$\mathbf{Y} = 1/R + 2j\ \delta\omega C.$$

If $R = 3 \times 10^5\ \Omega$, $L = 10^{-3}$ H, $C = 100$ pF, calculate the current in each arm when a voltage of 10 V r.m.s. at a frequency of 0·5 MHz is applied, and the phase of the total current drawn from the generator.

(*Answers:* 0·033 mA, 3·18 mA, and 3·14 mA; 51°.)

7.2. A circuit is required to accept a signal of frequency 1·1 MHz and to reject a signal of frequency 1·2 MHz. A coil of self-inductance 200 $\mu$H and resistance 10 $\Omega$ is tuned to parallel resonance at 1·2 MHz by a capacitance $C_1$. A capacitance $C_2$ is then placed in series with the combination, so that the whole is in series resonance at 1·1 MHz. Find the values of $C_1$ and $C_2$.

(*Answers:* 88 pF and 16 pF.)

7.3. Show that a capacitance $C$ shunted by a resistance $r$ is equivalent to a capacitance $C'$ in series with a resistance $R$ at any given frequency. If $\omega Cr \gg 1$, show that approximately $R = (\omega^2 C^2 r)^{-1}$, and $C' = C$.

7.4. Four impedances $\mathbf{Z}_1$, $\mathbf{Z}_2$, $\mathbf{Z}_3$, and $\mathbf{Z}_4$, in that order, are placed in the arms of a 'generalized' Wheatstone's bridge using alternating current. Show that the balance condition is

$$\mathbf{Z}_1/\mathbf{Z}_2 = \mathbf{Z}_4/\mathbf{Z}_3.$$

Note that this is really a double balance condition, since the real and imaginary parts of this equation must be separately satisfied. This is because a null reading is obtained on the detector only if the voltages at each of its terminals are equal both in amplitude and phase.

7.5. The four arms of a Wheatstone's bridge, taken in cyclic order round the bridge, are $a$, $b$, $c$, $d$. $a$ and $b$ are equal resistances $R$; $c$ is a resistance $R$ in series with a capacitance $C$; $d$ is a resistance $R$ shunted by a capacitance $C$. Show that such a bridge will not be balanced at any frequency, but that if the resistance in arm $b$ is doubled, a balance will be obtained at a frequency

$$f = (2\pi RC)^{-1}.$$

7.6. A capacitor of capacitance $C$, a resistance $R_C$, and a coil whose inductance is $L$ and resistance is $R_L$ are connected all three in parallel. An e.m.f. of variable frequency is applied across the capacitor. Show that the frequency of parallel resonance is independent of the value of $R_C$, and prove that at resonance the conductance of the combination is

$$\frac{1}{R_C} + \left(\frac{R_L C}{L}\right).$$

If the $Q_F$ of the circuit is high, show that it is given approximately by

$$\frac{1}{Q_F} = \frac{1}{R_C}\left(\frac{L}{C}\right)^{\frac{1}{2}} + R_L\left(\frac{C}{L}\right)^{\frac{1}{2}}.$$

Fig. 7.18. Phase-shift network (see Problem 7.8).

7.7. A series resonant circuit is connected across a constant voltage generator operating at 6·50 MHz. As the capacitance is varied, the current is observed to fall to $1/\sqrt{2}$ of its maximum value when the capacitance is 12·47 pF, and again when the capacitance is 12·64 pF. Find the values of $Q_F$, $L$, $R$ for the circuit. (*Answer:* $\approx 150$; 48 $\mu$H; 13 $\Omega$.)

7.8. An alternating voltage is applied to the terminals A, B of the network shown in Fig. 7.18. Show that, as $R$ is varied, the amplitude of the potential difference between the terminals X, Y remains constant, but its phase is shifted by $\pi$ rad. Explain your results by means of a vector diagram. (This network is commonly used for producing a variable phase shift without change of the output amplitude.)

7.9. A high-frequency transformer has primary inductance 100 $\mu$H and primary resistance 5 $\Omega$, secondary inductance $2\cdot 5 \times 10^3$ $\mu$H and secondary resistance 100 $\Omega$. If the coefficient of coupling $M/(L_1 L_2)^{\frac{1}{2}}$ is 0·9, show that at high frequencies the voltage across the secondary is approximately 4·5 times that applied to the primary.

If primary and secondary are each separately tuned by capacitors to resonance at 50 kHz, show that the high-frequency power required to produce a r.m.s. voltage of 2000 V across the capacitance in the secondary circuit is approximately 650 W.

7.10. A wire-wound resistance has a small inductance and self-capacitance which may be represented by placing an inductance $L$ in series with the resistance $R$, and shunting the combination by a capacitance $C$. Show that the reactance at low frequencies is zero in the first approximation if the wire is wound so that $L/C = R^2$. Show also that under these conditions the apparent resistance is (to the second approximation)

$$R(1+\omega^2 LC)$$

7.11. An inductance $L$ with small resistance $r$ has self-capacitance which can be represented approximately by a capacitance $C$ shunted across the series combination of $L$ and $r$. Show that as the frequency increases the apparent self-inductance of the coil is increased by the factor $(1+\omega^2 LC)$ while the apparent series resistance increases by the factor $(1+2\omega^2 LC)$, in the region where $\omega LC \ll 1$.

7.12. A parallel-plate capacitor is filled with a medium which has permittivity $\epsilon$ and conductivity $\sigma$. Show that at a frequency $f = \omega/2\pi$ the power factor of the capacitor is $\cos\phi = \sin\delta$, where the loss tangent of the dielectric medium is given by the relation $\tan\delta = \sigma/\omega\epsilon$.

Note that this is independent of the shape of the capacitor, as would be expected from eqn (3.18).

(a)                                (b)

FIG. 7.19. (a) T-network with impedances **a, b, c.** (b) $\pi$-network with impedances **A, B, C.**

7.13. The voltage and current in a circuit are given by $V = A e^{j\omega t}$ and $I = A' e^{j\omega t}$, where $A = a + jb$ and $A' = a' + jb'$ are each complex quantities. Show that the mean power averaged over a cycle is

$$\tfrac{1}{2}(aa' + bb') = \overline{\mathrm{Re}(V) \times \mathrm{Re}(I)} = \tfrac{1}{2}\mathrm{Re}(VI^*),$$

where $I^*$ is the complex conjugate of $I$.

7.14. A 100-V dynamo is connected to a magnet whose resistance is $10\,\Omega$ and self-inductance is $0\cdot 1$ H. Show that the percentage increase in the heating of the magnet caused by the presence of a 100 Hz ripple voltage of amplitude 5 V in the output of the dynamo is $0\cdot 0031$ per cent.

7.15. Two networks such as those shown in Fig. 7.19 are equivalent provided that they obey any of the sets of eqns (7.49)–(7.51) with identical sets of parameters. Show that this is true provided that $\mathbf{aA} = \mathbf{bB} = \mathbf{cC} = \mathbf{ab} + \mathbf{bc} + \mathbf{ca}$.

   This formula for converting a T-network into a $\pi$-network (see § 9.1) often simplifies circuit calculations, as in the transformer ratio bridge (§ 22.2).

# 8. Electromagnetic waves

## 8.1. Maxwell's equations of the electromagnetic field

So far we have considered the propagation of electrical currents in material conductors. The possibility of the propagation of an electromagnetic wave through space was first suggested by Faraday, and this suggestion was confirmed by the work of Maxwell. Maxwell was able to show that the laws of electromagnetism could be expressed in the form of some fundamental equations which, with an important modification, lead to a differential equation whose solutions represent transverse waves travelling through free space with the velocity of light. Further work showed that the properties of these waves—reflection, refraction, diffraction—are the same as those established experimentally for light waves, and we are therefore justified in assuming that they are identical, and that light waves are a form of electromagnetic radiation.

The theory of Maxwell deals entirely with macroscopic phenomena, making the assumption that matter is continuous and has no atomistic structure. This assumption places certain limitations on the theory: thus it offers no explanation of the phenomenon of dispersion—the change of refractive index with frequency. This phenomenon will be discussed in Chapter 10; it arises from the change in the electric permittivity and magnetic permeability of a medium with frequency. These changes can be related to the effect of electromagnetic waves on the individual electrons in an atom, but for the present we shall regard the electric permittivity and magnetic permeability as macroscopic quantities whose values are obtained by experiment.

The fundamental laws of electromagnetism which have already been derived may be summarized as follows:

(1) the theorem of Gauss applied to electrostatics (eqn (1.20)):

$$\operatorname{div} \mathbf{D} = \rho_e; \tag{8.1}$$

(2) the corresponding result for magnetic fields (eqn (4.6)):

$$\operatorname{div} \mathbf{B} = 0; \tag{8.2}$$

(3) Faraday's and Lenz's law of electromagnetic induction (eqn (5.3)):

$$\operatorname{curl} \mathbf{E} = -\partial \mathbf{B}/\partial t; \tag{8.3}$$

(4) Ampère's law for magnetomotive force (eqn (4.26)):

$$\operatorname{curl} \mathbf{H} = \mathbf{J}'. \tag{8.4}$$

The reason for writing $\mathbf{J}'$ rather than the ordinary current density $\mathbf{J}$ in this last equation is as follows. Since (div curl) of any vector is identically zero, it follows that eqn (8.4) implies that div $\mathbf{J}'$ is zero. If we had written $\mathbf{J}$ instead of $\mathbf{J}'$, we should have had div $\mathbf{J} = 0$, and this conflicts with the equation of continuity (3.3) which gives

$$\text{div } \mathbf{J} = -\partial \rho_e / \partial t. \tag{8.5}$$

This equation represents the law of conservation of charge, which is confirmed by all experiments. Maxwell realized that the difficulty arose from an incomplete definition of the total current density in eqn (8.4), which is not entirely given by the current flow due to the motion of electric charges. By using eqn (8.1) we may write (8.5) in the form

$$\text{div } \mathbf{J} = -\frac{\partial}{\partial t}(\text{div } \mathbf{D}) = \text{div}\left(-\frac{\partial \mathbf{D}}{\partial t}\right),$$

or

$$\text{div}\left(\mathbf{J} + \frac{\partial \mathbf{D}}{\partial t}\right) = 0$$

Hence if we define $\mathbf{J}'$ as

$$\mathbf{J}' = \mathbf{J} + \frac{\partial \mathbf{D}}{\partial t}, \tag{8.6}$$

then div $\mathbf{J}' = 0$, and Ampère's law takes the form

$$\text{curl } \mathbf{H} = \mathbf{J} + \frac{\partial \mathbf{D}}{\partial t} = \sigma \mathbf{E} + \frac{\partial \mathbf{D}}{\partial t}. \tag{8.7}$$

The term $\mathbf{J}$ is generally called the 'conduction current' and the second term $(\partial \mathbf{D}/\partial t)$ the 'displacement current', since it arises when the electric displacement $\mathbf{D}$ is changing with time. We may obtain some physical picture of what is implied by the displacement current by considering a simple circuit such as in Fig. 8.1, where a current $I$ is flowing from a battery to charge a capacitance $C$. If we apply Ampère's law to a closed circuit such as LMNL which encircles the wire we find that $\int \mathbf{H}.\mathbf{ds}$ round this circuit is just equal to $I$. To define the current which threads the circuit we must take some surface bounded by the circuit, and integrate the normal component of the current density crossing this surface. If we take a surface intersecting the wire, this clearly gives just $I$, the current in the wire. But if we take a surface which passes between the two plates of the capacitor, then $\int \mathbf{J}.\mathbf{dS} = 0$, since no conduction current flows through the surface. The value of this integral should be independent of what surface we choose, since $\int \mathbf{H}.\mathbf{ds}$ depends only on the circuit bounding this surface; thus it is clear that we have omitted some contribution. If for simplicity we take a parallel-plate capacitor with plates of area $A$,

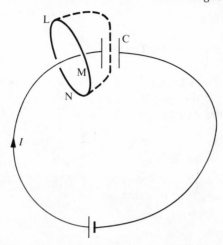

FIG. 8.1. Application of Ampère's law to calculate the magnetomotive force $\int \mathbf{H}.\mathbf{ds}$ round the circuit LMNL for the case of a current $I$ charging a capacitor C.

surrounded by a guard ring, the field in between the plates is uniform, and the displacement $\mathbf{D}$ has the value $D = Q/A$, where $Q$ is the total charge on the positive plate. Then the total displacement current between the plates is

$$A(\partial D/\partial t) = \partial Q/\partial t = I,$$

and our difficulties with the surface integral of the current density disappear if we include the displacement current.

For the case of an infinite homogeneous medium of electric permittivity $\epsilon$ and magnetic permeability $\mu$, containing no free charges ($\rho_e = 0$) and having zero conductivity ($\sigma = 0$), our equations become

$$\text{div } \mathbf{D} = \text{div}(\epsilon \mathbf{E}) = \epsilon \text{ div } \mathbf{E} = 0, \tag{8.8}$$

$$\text{div } \mathbf{B} = \text{div}(\mu \mathbf{H}) = \mu \text{ div } \mathbf{H} = 0, \tag{8.9}$$

$$\text{curl } \mathbf{E} = -\partial \mathbf{B}/\partial t = -\mu(\partial \mathbf{H}/\partial t), \tag{8.10}$$

$$\text{curl } \mathbf{H} = \partial \mathbf{D}/\partial t = \epsilon(\partial \mathbf{E}/\partial t). \tag{8.11}$$

These form a set of simultaneous partial differential equations whose solutions can be found by eliminating one of the dependent variables, $\mathbf{E}$ or $\mathbf{H}$. This can be done by means of the vector identity

$$\text{curl}(\text{curl } \mathbf{E}) = \text{grad}(\text{div } \mathbf{E}) - \nabla^2 \mathbf{E} = -\nabla^2 \mathbf{E} \quad \text{(using eqn (8.8))}.$$

Then

$$\nabla^2 \mathbf{E} = -\text{curl}(\text{curl } \mathbf{E}) = \text{curl}(\mu \partial \mathbf{H}/\partial t)$$

$$= \mu \frac{\partial}{\partial t}(\text{curl } \mathbf{H}) = \mu \epsilon \left(\frac{\partial^2 \mathbf{E}}{\partial t^2}\right). \tag{8.12}$$

Similarly it may be shown that

$$\nabla^2 \mathbf{H} = \mu\epsilon\left(\frac{\partial^2 \mathbf{H}}{\partial t^2}\right). \tag{8.13}$$

These two equations are each of the general form for a wave motion in three dimensions; if the velocity with which the waves are propagated is $v$, the general wave equation is

$$\nabla^2 \mathbf{X} = \frac{1}{v^2}\frac{\partial^2 \mathbf{X}}{\partial t^2},$$

where $\mathbf{X}$ is some scalar or vector quantity. Comparison with our case shows that the wave velocity must be

$$v = (\mu\epsilon)^{-\frac{1}{2}}, \tag{8.14}$$

with the particular value in a vacuum ($\mu_r = 1$, $\epsilon_r = 1$) of

$$c = (\mu_0\epsilon_0)^{-\frac{1}{2}}. \tag{8.15}$$

Now the value of $\mu_0$ has been defined as $4\pi \times 10^{-7}$, while the best direct measurement of $\epsilon_0$ is that of Rosa and Dorsey (see § 5.8): this leads to a wave velocity of our electromagnetic waves in a vacuum of

$$(299\,784 \pm 10)\ \text{km s}^{-1}.$$

Within the experimental error this is the same as the velocity of light determined by direct measurement, and it is accepted that light is a form of electromagnetic radiation, of the same form as radio waves, heat waves, X-rays, and $\gamma$-rays, which differ from light waves and from each other only in frequency and wavelength. The frequency range is from $10^4$–$10^{11}$ Hz for radio waves to $10^{20}$ Hz and over for $\gamma$-rays. Maxwell's theory does not give an adequate account of the interaction of electromagnetic radiation with atoms, and it has been found necessary (for example, in the photoelectric effect) to consider the electromagnetic energy as travelling about in 'packets' or 'quanta', which are indivisible. The size of a quantum for radiation of frequency $v$ is $hv$, where $h$ is a universal constant known as Planck's constant. For radio waves such quantum effects are too small to affect the interaction with matter seriously, and we shall not consider them further at present.

If we accept the identification of our electromagnetic waves with light and radio waves, then we may use one of the accurate methods of determining their velocity (see § 22.7) to infer the value of the constant $\epsilon_0$, since this gives a more accurate value than the direct comparison of a capacitance and resistance. The value obtained is given in Appendix C.

Eqn (8.14) shows that the velocity of the waves in a material medium is less than in a vacuum, since

$$v = c/(\mu_r \epsilon_r)^{\frac{1}{2}}.$$

For a light wave the velocity in a non-dispersive medium is $v = c/n$, where $n$ is the refractive index of the medium. Hence we have

$$n = (\mu_r \epsilon_r)^{\frac{1}{2}}. \tag{8.16}$$

In using this equation we must remember that it applies only if we determine the values of $n$, $\mu_r$, and $\epsilon_r$ at the same frequency, and nonsensical results may be obtained if we use values determined at different frequencies. Thus for water the optical refractive index is $1\cdot33$, but the static values of $\epsilon_r$ and $\mu_r$ are 81 and 1, leading to a refractive index (for very low frequencies only) of 9! Similarly, $\mu_r$ may be very high for a soft magnetic material at radiofrequencies, but close to unity at optical frequencies. The changes in $\epsilon_r$, $\mu_r$ with frequency give rise to 'dispersion' in the refractive index (see Chapter 10).

## 8.2. Plane waves in isotropic dielectrics

In order to examine the behaviour of the electric and magnetic fields in more detail we shall consider the case of a plane wave. For simplicity we assume that this is moving in the direction of the $z$-axis of a set of right-handed Cartesian axes $x$, $y$, $z$. Then the definition of a plane wave is one in which the quantities have the same value over any plane normal to the direction of propagation; mathematically this is expressed by setting all partial differentials with respect to the $x$ and $y$ coordinates equal to zero (that is, $\partial/\partial x = \partial/\partial y = 0$). If these conditions are put into eqns (8.8) and (8.9) we find that

$$\partial E_z/\partial z = \partial H_z/\partial z = 0,$$

and from the $z$ components of the curl equations ((8.10) and (8.11)) we have

$$\partial E_z/\partial t = \partial H_z/\partial t = 0.$$

This shows that apart from a uniform steady field in the $z$ direction, which is not part of any wave motion, both $E_z$ and $H_z$ must be zero. The wave is therefore purely transverse in that no components of the electric or magnetic fields exist in the direction of propagation. The remaining components of the curl equations are, taking the $z$ components of $\mathbf{E}$ and $\mathbf{H}$ to be zero, and $\partial/\partial y = \partial/\partial x = 0$,

$$\partial E_x/\partial z = -\mu\partial H_y/\partial t, \qquad -\partial H_y/\partial z = \epsilon\partial E_x/\partial t;$$

$$-\partial E_y/\partial z = -\mu\partial H_x/\partial t, \qquad \partial H_x/\partial z = \epsilon\partial E_y/\partial t.$$

These equations show that the $x$ component of **E** is associated with the $y$ component of **H**, while the $y$ component of **E** is associated with the $x$ component of **H**. The two components of **E** or **H** correspond to waves which are plane-polarized in directions normal to one another and normal to the direction of propagation. Thus we have two linearly independent solutions, in each of which the magnetic field is normal to the electric field. They differ only in the plane of polarization, which we shall take to be that of the electric vector; this differs from the convention accepted in light before the electromagnetic nature of the radiation was understood, which adopted the plane now known to be that of the magnetic vector. The electric vector is more important in the theory of dispersion (Chapter 10) as it determines the force on an electron in an atom.

Since we need consider only one state of polarization, we shall take the solution where the electric vector is parallel to the $x$-axis; it then follows that the magnetic vector has only a component parallel to the $y$-axis. The wave equation for $E_x$ (eqn (8.12)) becomes

$$\frac{\partial^2 E_x}{\partial z^2} - \mu\epsilon\frac{\partial^2 E_x}{\partial t^2} = 0, \tag{8.17}$$

which has a general solution of the form

$$E_x = F_1(z - vt) + F_2(z + vt), \tag{8.18}$$

where $F_1$, $F_2$ may be functions of any form. The two functions represent waves travelling with velocity $v$, whose value is given by eqn (8.14). $F_1$ represents a wave travelling in the direction of $z$ increasing, since any given point in the wave moves according to the equation

$$z = vt + \text{constant},$$

while $F_2$ is a wave travelling in the opposite direction, a given point moving as $z = -vt + \text{constant}$.

We may find the value of $H_y$ by using one of the remaining components of our curl equations. For

$$\partial H_y/\partial z = -\epsilon\, \partial E_x/\partial t = \epsilon v\{F_1'(z - vt) - F_2'(z + vt)\},$$

where $F'$ is the differential of $F$. Hence, since $v = (\epsilon\mu)^{-\frac{1}{2}}$, we find on integration

$$H_y = (\epsilon/\mu)^{\frac{1}{2}}\{F_1(z - vt) - F_2(z + vt)\}. \tag{8.19}$$

This shows that the value of $E_x$ bears a constant ratio to $H_y$ in each of the travelling waves, but has the opposite sign for a wave travelling to the left. The ratio of $E_x$ to $H_y$ in a plane wave is often useful, and it has been called the 'intrinsic impedance' $Z_0$ of the medium. In our case of a non-conducting medium, where the wave velocity is $v$,

$$Z_0 = E_x/H_y = (\mu/\epsilon)^{\frac{1}{2}} = \mu v. \qquad (8.20)$$

For a plane wave in free space $Z_0 = (\mu_0/\epsilon_0)^{\frac{1}{2}} = \mu_0 c = (\epsilon_0 c)^{-1}$, and its value is approximately $376 \cdot 7 \ \Omega$. The fact that the dimensions of $Z_0$ are the same as those of the impedance discussed in earlier chapters is readily seen from the relations

$$\int \mathbf{H}.\mathbf{ds} = I, \qquad \int \mathbf{E}.\mathbf{ds} = V,$$

where $V$ is the potential drop in a conductor; it appears also from the fact that $E$ is measured in volts per metre and $H$ in amperes per metre. The fact that the intrinsic impedance is a real quantity independent of $z$ or $t$ shows that the waveform of $H_y$ is everywhere the same as that of $E_x$ without any phase difference, in a travelling wave. This is true only of a non-conducting medium. Another important restriction is that the wave velocity must be independent of frequency if the waveform is to remain the same as the wave progresses. If it is not (that is, if we have a dispersive medium) we must perform a Fourier analysis of the initial waveform and treat each component of a given frequency separately. In the remainder of this chapter we shall assume that we are dealing with waves of one frequency only, and our field components are the real or imaginary parts of the exponential functions of the form

$$\exp\{j\omega(t \pm z/v)\}.$$

Then the general solution for waves travelling both to the right and to the left will be

$$E_x = A \exp\{j\omega(t-z/v)\} + A' \exp\{j\omega(t+z/v)\},$$
$$Z_0 H_z = A \exp\{j\omega(t-z/v)\} - A' \exp\{j\omega(t+z/v)\}. \qquad (8.21)$$

The exponentials are sometimes written in the forms $\exp\{j\omega(t \pm nz/c)\}$ or $\exp j(\omega t \pm \beta z)$; $\beta$ is called the 'phase constant', and is equal to $2\pi/\lambda$, where $\lambda$ is the wavelength in the medium. The phase velocity $v$ is given by the relation

$$v = \omega/\beta. \qquad (8.22)$$

It might be thought that a phase constant $\delta$ should be included in one of the waves in eqn (8.21), since otherwise some particular choice of zero is implied for either $z$ or $t$ which makes $\delta = 0$. We shall, however, allow $A'$ (or $A$) to be complex, and of the form $A'' \exp(j\delta)$, and thus include the phase constant in $A'$ rather than in the complex exponential.

## 8.3. The Poynting vector of energy flow

Since an electromagnetic wave consists of electric and magnetic fields, we may expect that stored energy is associated with these fields. If the energy density, for which we now use the symbol $U$, is given by the equations derived for static fields, then for a plane wave in an isotropic dielectric it will be

$$U = \tfrac{1}{2}(\mathbf{D}.\mathbf{E}+\mathbf{B}.\mathbf{H}) = \tfrac{1}{2}(\epsilon \mathbf{E}^2+\mu\mathbf{H}^2).$$

For a wave propagated in the positive $z$-direction, the energy crossing unit area per second will just be the velocity times the energy density. This is

$$vU = \frac{1}{2}\left\{\left(\frac{\epsilon}{\mu}\right)^{\tfrac{1}{2}}E_x^2+\left(\frac{\mu}{\epsilon}\right)^{\tfrac{1}{2}}H_y^2\right\}$$
$$= \tfrac{1}{2}(E_x^2/Z_0 + Z_0H_y^2) = E_xH_y = E_x^2/Z_0 = Z_0H_y^2.$$

These various relations show that the energy stored in the magnetic field is just equal to that in the electric field. Since the direction of energy flow is normal to both $\mathbf{E}$ and $\mathbf{H}$, and its magnitude is $E_xH_y$, we can represent the rate of energy flow per unit area by the vector (see Fig. 8.2)

$$\mathbf{E}\wedge\mathbf{H}. \tag{8.23a}$$

This vector is known as the Poynting vector, and it can be seen from eqns (8.21) that the direction of energy flow is reversed for a wave travelling in the opposite direction because the phase of $\mathbf{H}$ relative to $\mathbf{E}$ is reversed.

The value of $\mathbf{E}\wedge\mathbf{H}$ gives the instantaneous rate of energy flow. In a periodic wave the values of $\mathbf{E}$ and $\mathbf{H}$ at any point are oscillating functions of the time, and the mean rate of energy flow is found by averaging $\mathbf{E}\wedge\mathbf{H}$ over a complete period. In general the mean rate of energy flow is (cf. Problem 7.13) per unit area

$$\overline{\mathrm{Re}(\mathbf{E})\wedge\mathrm{Re}(\mathbf{H})} = \tfrac{1}{2}\,\mathrm{Re}(\mathbf{E}\wedge\mathbf{H}^*) \tag{8.23b}$$

$$(\mathbf{E}\wedge\mathbf{H})_z = E_x H_y$$

FIG. 8.2. Directions of the electric and magnetic field strengths $\mathbf{E}$ and $\mathbf{H}$ and the Poynting vector $\mathbf{E}\wedge\mathbf{H}$ for a plane wave.

where $\mathbf{H}^*$ is the complex conjugate of $\mathbf{H}$; this expression must be used when the phase difference between $\mathbf{E}$ and $\mathbf{H}$ is not $0$ or $\pi$.

The Poynting vector is of greater significance than might be thought from the way in which it has been introduced above, and we shall now derive it for the general case. Suppose there is a region of space where an electric field of strength $\mathbf{E}$ causes a current flow of density $\mathbf{J}$. Then the power dissipated per unit volume is $\mathbf{E}.\mathbf{J}$, and the total power dissipated will be

$$\int \mathbf{E}.\mathbf{J}\, d\tau = \int (\mathbf{E}.\text{curl }\mathbf{H})\, d\tau - \int \mathbf{E}.(\partial \mathbf{D}/\partial t)\, d\tau,$$

where we have substituted for $\mathbf{J}$ the expression given by eqn (8.7). Now, using the vector identity

$$\text{div}(\mathbf{E}\wedge\mathbf{H}) = \mathbf{H}.\text{curl }\mathbf{E} - \mathbf{E}.\text{curl }\mathbf{H}$$

and the expression for curl $\mathbf{E}$ given by eqn (8.3), we find that the power dissipated is

$$-\int \mathbf{E}.(\partial \mathbf{D}/\partial t)\, d\tau - \int \mathbf{H}.(\partial \mathbf{B}/\partial t)\, d\tau - \int \text{div}(\mathbf{E}\wedge\mathbf{H})\, d\tau.$$

The first two terms may be written as

$$-\int \left( \mathbf{E}\frac{\partial \mathbf{D}}{\partial t} + \mathbf{H}\frac{\partial \mathbf{B}}{\partial t} \right) d\tau,$$

and this is the rate at which the stored energy increases, since the work done in infinitesimal changes $d\mathbf{D}$, $d\mathbf{B}$ at constant $\mathbf{E}$, $\mathbf{H}$ is

$$\int (\mathbf{E}\, d\mathbf{D} + \mathbf{H}\, d\mathbf{B})\, d\tau.$$

If $\epsilon$, $\mu$ are independent of $\mathbf{E}$, $\mathbf{H}$ and of time, the first two terms may be written as

$$-\frac{1}{2}\frac{\partial}{\partial t}\int (\mathbf{D}.\mathbf{E} + \mathbf{B}.\mathbf{H})\, d\tau,$$

which we may interpret as the rate at which the energy stored in the electromagnetic field diminishes. The last term may be transformed using eqn (A.8) of Appendix A into the surface integral

$$-\int (\mathbf{E}\wedge\mathbf{H})\,.\,d\mathbf{S}$$

taken over the surface bounding the volume under consideration. It represents the rate at which energy flows into the volume under consideration, and we may interpret the vector $\mathbf{E}\wedge\mathbf{H}$ as the rate at which energy flows across unit area of the boundary. Strictly speaking, only the integral of this quantity over a closed surface has been shown to represent the energy flow, but in most cases the vector $\mathbf{E} \wedge \mathbf{H}$ does represent the flow of energy per unit area at each point. An obvious exception to this would be a region with an electrostatic and a steady magnetic field arising from different sources, such as a charged capacitor between the poles of a permanent magnet.

## 8.4. Plane waves in conducting media

In a medium with a finite conductivity $\sigma$ the field equations are the same as (8.8)–(8.11) except for an additional term $+\sigma\mathbf{E}$ on the right-hand side of eqn (8.11). These equations may be solved quite generally in the same way as in § 8.1, and elimination of $\mathbf{H}$ gives the differential equation

$$\nabla^2 \mathbf{E} = \mu\epsilon \frac{\partial^2 \mathbf{E}}{\partial t^2} + \mu\sigma \frac{\partial \mathbf{E}}{\partial t}. \tag{8.24}$$

An exactly similar equation holds for $\mathbf{H}$, and it may be shown that each of these equations represents a damped wave motion where the amplitude decays as the wave progresses owing to the extra term in $\partial\mathbf{E}/\partial t$ or $\partial\mathbf{H}/\partial t$. We shall not pursue this general solution, but specialize to the case of a plane wave travelling parallel to the $z$-axis. Then all partial derivatives $(\partial/\partial x)$ and $(\partial/\partial y)$ vanish, and the same methods as used in § 8.2 show that both $E_z$ and $H_z$ are zero, so that we again have a purely transverse wave. For a plane-polarized wave with a single component of $\mathbf{E}$ parallel to the $x$-axis, the field equations reduce to

$$\left.\begin{array}{l} \partial E_x/\partial z = -\mu(\partial H_y/\partial t), \\ -\partial H_y/\partial z = \epsilon(\partial E_x/\partial t) + \sigma E_x. \end{array}\right\} \tag{8.25a}$$

These equations show that, as before, $E_x$ is associated with $H_y$. (There is an independent solution, polarized in the perpendicular direction, where $E_y$ is associated with $H_x$.) Elimination of $H_y$ would give a wave equation for $E_x$ of the same type as the general equation above, but a simpler method of solution is available on assuming that we are dealing with waves of a single frequency. It will appear that the velocity of propagation is now dependent on the frequency, so that it is necessary to consider each frequency separately in any case; we therefore introduce this restriction at the beginning, and take $E_x$ and $H_y$ to vary in time as $\exp j\omega t$. We will also try a solution where the variation with $z$ takes the form of a complex exponential, so that both $E_x$ and $H_y$ vary as

$$\exp\{j\omega(t - \mathbf{n}z/c)\},$$

where **n** is a complex refractive index. Then our equations become

$$j\omega(\mathbf{n}/c)E_x = j\omega\mu H_y,$$
$$j\omega(\mathbf{n}/c)H_y = (j\omega\epsilon+\sigma)E_x.$$

(8.25b)

Elimination of $E_x$ or $H_y$ shows that these equations are satisfied provided that

$$\mathbf{n}^2 = \mu_r\epsilon_r - j(\sigma\mu_r/\omega\epsilon_0),$$

(8.26)

where we have substituted $(\mu_0\epsilon_0)$ for $1/c^2$.

To separate the real and imaginary parts of **n** we write it as

$$\mathbf{n} = n - jk,$$

(8.27)

giving

$$n^2 - k^2 = \mu_r\epsilon_r$$
$$nk = (\sigma\mu_r/2\omega\epsilon_0).$$

(8.28)

To understand the significance of $n$ and $k$ we note that the wave is propagated as

$$\exp\{j\omega(t - \mathbf{n}z/c) = \exp(-\omega kz/c)\exp j\omega(t - nz/c),$$

showing that the value of $k$ determines the rate at which the amplitude of the wave decays ($k$ appears in the argument of the real exponential) while $n$ determines the phase velocity in the medium. Thus a complex refractive index **n** means that the wave is being absorbed as it proceeds, because the finite conductivity of the medium gives rise to a power loss through the Joule heating.

The intrinsic impedance of our medium is defined as $E_x/H_y$, so that

$$\mathbf{Z}_0 = \frac{c\mu}{\mathbf{n}} = \left(\frac{\mu}{\epsilon - j\sigma/\omega}\right)^{\frac{1}{2}}.$$

(8.29)

Thus the impedance is a complex quantity, but is independent of $z$ or $t$. This means that the ratio of the amplitudes of $E_x$ and $H_y$ is everywhere the same, but there is a phase difference between them. Since also

$$\mathbf{Z}_0 = \frac{c\mu(n+jk)}{(n^2+k^2)}$$

we see that the phase difference $\phi$ is given by the relation $\tan\phi = k/n$. $(\mathbf{E}\wedge\mathbf{H})$ is also complex, and the mean power flow must be calculated using eqn (8.23b).

The full solution of eqns (8.28) to find $n$ and $k$ gives rather complicated expressions, but these simplify greatly for the case of a good conductor. For a metal the conduction current is enormously greater than the displacement current at all frequencies up to those of ultraviolet light, as will be seen by comparing the values of $\sigma$ and $\omega\epsilon$, which occur in the second of eqns (8.25b). For most metals $\sigma$ is $10^7\,\Omega^{-1}\,\text{m}^{-1}$ or greater, while for light of wavelength 500 nm, $\omega\epsilon\times10^{15}\times8\cdot85\times\epsilon_r\approx3\cdot5\times10^4\,\epsilon_r$. We do not know what $\epsilon_r$ is for a metal, but provided it is not very much greater than the ordinary values found in dielectric substances, the displacement current even for visible light will be much smaller than the conduction current. For lower frequencies the inequality increases, and a good approximation is obtained by omitting the displacement current. This is equivalent to setting $\epsilon_r = 0$, and eqns (8.28) then give

$$n = k = (\sigma\mu_r/2\omega\epsilon_0)^{\frac{1}{2}}.$$

If we introduce a quantity $\delta$ such that $n = k = c/\omega\delta$, the field components in the metal are propagated as

$$\exp(-z/\delta)\,\exp\,\text{j}(\omega t - z/\delta), \tag{8.30}$$

from which it can be seen that $\delta$ has the dimensions of a length. $\delta$ is known as the 'skin depth', and amplitude of the wave falls to $1/e$ of its initial value in a distance $\delta$, while the apparent wavelength in the metal is $2\pi\delta$. The equation for $\delta$ is (writing $\omega = 2\pi f$)

$$\delta = (\tfrac{1}{2}\sigma\omega\mu)^{-\frac{1}{2}} = (\pi\sigma f\mu)^{-\frac{1}{2}} \tag{8.31}$$

and the intrinsic impedance of our metal can be written as

$$\mathbf{Z}_0 = (1+\text{j})/(\sigma\delta) = (1+\text{j})(\rho/\delta), \tag{8.32}$$

where $\rho$ is the resistivity. The magnitude of the skin depth decreases with the inverse half-power of the frequency, and of the permeability and conductivity of the metal. An idea of its order of magnitude is obtained from the fact that for copper it has the approximate values: $6\cdot6\times10^{-5}$ m at a frequency of 1 MHz, $6\cdot6\times10^{-7}$ m at $10^4$ MHz (a wavelength of $3\times10^{-2}$ m), and $2\cdot7\times10^{-9}$ m at $6\times10^8$ MHz (green light). Thus the wavelength of the radiation in the metal is very small compared with the wavelength in free space; and $Z_0$ is also much smaller than the value for free space; for copper it is about $0\cdot026(1+\text{j})\Omega$ at $10^4$ MHz.

## 8.5. The skin effect

Since an electromagnetic wave varies in phase very rapidly inside a metal, one must ask whether this affects the distribution of alternating current inside a conductor. An electromagnetic field is associated with such a current, and Problem 8.1 shows that there is a flow of energy into the surface of a conductor which is just equal to the energy dissipated as Joule heat. This implies that an electromagnetic wave is passing through the surface, and from what has been found about the behaviour of such a wave we should expect it to be very rapidly attenuated inside. The current associated with it would therefore flow only in a thin skin of thickness of the order of δ, a phenomenon known as the 'skin effect'. For steady currents $\omega = 0$ and $\delta = \infty$, so that the current distribution would be uniform, but as the frequency rises δ decreases and the current is increasingly confined to the surface. For this reason thin tubes are just as good conductors of high-frequency currents as solid rods, but the resistance is of course higher than for a steady current in a solid rod owing to the smaller effective cross-section. We will now compute the high-frequency resistance of a wire. Neglecting displacement currents, and writing $\mathbf{J}$ for $\sigma\mathbf{E}$ in eqn (8.24), we have

$$\nabla^2\mathbf{J} = \mu\sigma(\partial\mathbf{J}/\partial t) = \mathrm{j}\omega\mu\sigma\mathbf{J} \qquad (8.33)$$

for a current oscillating with frequency $\omega/2\pi$. For a cylindrical wire we should express the operator $\nabla^2$ in cylindrical coordinates and the general solution involves Bessel functions with a complex argument. The values we have found for δ in copper show that at high frequencies it will ordinarily be very small compared with the radius $a$ of the wire. In this case we may obtain an approximate solution by neglecting the curvature of the surface of the wire and regarding it as a thin tube which can be split and unrolled to form an infinite flat strip of width $b = 2\pi a$, in which a current flows parallel to the surface of the strip in the long direction. If (as in Fig. 8.3) we take this direction as the $x$-axis and the normal to the surface of the strip as the $z$-axis, then the current density has a component $\sigma E_x$ in the $x$ direction which varies with $z$ and $t$ as given by eqn (8.30), so that

$$J_x = J_0 \exp(-z/\delta)\exp \mathrm{j}(\omega t - z/\delta)\},$$

where $J_0$ is the current density at the surface of the wire ($z = 0$). This

Fig. 8.3. Current flow at surface of cylindrical strip opened out to flat strip of width $b = 2\pi a$.

equation shows that the current changes phase as we move into the wire as well as decreasing rapidly in amplitude.

To find the effective resistance $R$ per unit length of the wire we must calculate the total current $I$ flowing in the strip at any instant and the power dissipation per unit length $P$, and set $P = \frac{1}{2}|I|^2 R$. Now

$$I = b\int_0^\infty J_x \, dz = bJ_0 \exp j\omega t \int_0^\infty \exp\{-(1+j)z/\delta\} \, dz$$

$$= \left(\frac{b\delta J_0}{1+j}\right)\exp j\omega t = \frac{1}{\sqrt{2}}b\delta J_0 \exp j(\omega t - \tfrac{1}{4}\pi).$$

This expression shows that the total current has a different phase from $J_0$; we require only the amplitude of the total current whose square is

$$|I|^2 = \tfrac{1}{2}(b\delta J_0)^2.$$

The instantaneous power dissipation per unit volume of the metal is $\rho J_x^2$, and hence the total average dissipation per unit length of the wire is

$$P = b\int_0^\infty \tfrac{1}{2}\rho\, |J_x|^2 \, dz = \tfrac{1}{2}b\rho J_0^2 \int_0^\infty \exp(-2z/\delta) \, dz = \tfrac{1}{4}b\rho\delta J_0^2 = \tfrac{1}{2}\rho\, |I|^2/b\delta,$$

where $\rho$ is the resistivity of the wire. Hence the effective resistance per unit length of the wire is

$$R = 2P/|I|^2 = \rho/b\delta = \rho/2\pi a\delta. \qquad (8.34a)$$

This equation shows that the resistance is the same as if there were a current $I$ of uniform density flowing in a thin tube of radius $a$ and thickness equal to $\delta$, and the reason for the name 'skin depth' is thus apparent. The high-frequency resistance of such a wire is greater than the d.c. resistance by a factor $a/2\delta$, when $\delta \ll a$.

This type of calculation may be applied to the more general problem of the power dissipated per unit area of a plane metallic surface when there is a tangential magnetic field whose strength has an amplitude $H_0$ at the surface. At any point in the metal the current density is

$$J_x = \sigma E_x = -\partial H_y/\partial z \quad \text{(from eqn (8.25a)),}$$

assuming that the displacement current can be neglected. It follows that the total current in the metal (per unit width) is given by $\int_0^\infty (-\partial H_y/\partial z) \, dz = H_0$, the value of the magnetic field strength at the surface. The mean power dissipation per unit area is

$$P = \tfrac{1}{4}\rho\delta J_0^2 = \tfrac{1}{2}\rho H_0^2/\delta. \qquad (8.34b)$$

This is just equal to the average value of Poynting's vector at the surface (see Problem 8.1).

## 8.6. Reflection and refraction of plane waves at the boundary of two dielectrics

The reflection and refraction of light waves at the surface separating two media of different refractive indices is a familiar phenomenon and we now inquire whether electromagnetic theory offers a simple explanation of it. We assume two non-conducting dielectric media, separated by a plane boundary which we take to be the $xy$ plane (the plane $z = 0$). We also choose the direction of the $x$-axis to be in the plane of incidence, that is, the incident ray lies in the plane $y = 0$, as in Fig. 8.4, making an angle $\theta$ with the normal to the boundary. $\theta$ is known as the 'angle of incidence'. Then all field components of the incident wave vary with the space and time coordinates as

$$\exp[j\omega\{t - n_1(x \sin \theta + z \cos \theta)/c\},$$

where $c/n_1$ is the velocity of the wave in the first medium, and $\omega/2\pi$ is its frequency of oscillation.

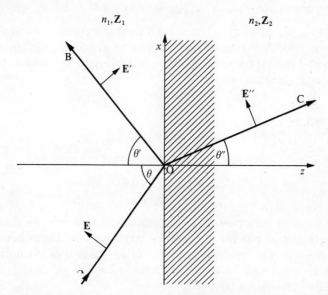

FIG. 8.4. Reflected wave OB and refracted wave OC at the boundary ($z = 0$) between two media. AO is the incident wave, and the electric vectors are in the plane of incidence as shown. The magnetic vector is parallel to the $y$-axis, and hence normal to the plane of the paper. If $n_2 > n_1$, and the permeability of both media is unity, then the direction of **E'** is as shown when the angle of incidence is greater than the Brewster angle $\tan^{-1}(n_2/n_1)$, but in the opposite direction when it is less than the Brewster angle.

When the incident wave falls on the boundary there will in general be both a reflected wave and a transmitted wave. We know nothing about these waves, either as to their frequency or their direction. We have, however, boundary conditions for the components of **E** and **H** at the surface, and we assume that the behaviour of **E** and **H** must be the same as derived earlier in Chapters 1 and 4. Hence the tangential components of **E** and **H** must be continuous on the two sides of the boundary at all times and for all values of $x$ and $y$. The first of these conditions shows that the reflected and transmitted waves must have the same frequency as the incident wave. Secondly, at all points in the boundary plane along a line for which $x$ is constant the field components of the incident wave are constant in amplitude and phase. Our boundary conditions can be satisfied everywhere only if this is true of the reflected and transmitted waves as well, and they must therefore also travel in the plane of incidence. The corresponding rays are shown in Fig. 8.4, making angles $\theta'$ and $\theta''$ with the normal, respectively; they are known as the 'angle of reflection' and the 'angle of refraction'. The field components of the two waves must be propagated as

(reflected wave)    $\exp j\omega\{t - n_1(x \sin \theta' - z \cos \theta')/c\}$,

(transmitted wave)   $\exp j\omega\{t - n_2(x \sin \theta'' + z \cos \theta'')/c\}$,

where $c/n_2$ is the velocity in the second medium. At the boundary plane $(z = 0)$, the boundary conditions can only be satisfied everywhere if the arguments of the exponentials for the incident, reflected, and transmitted waves are all identical. Hence

$$n_1 x \sin \theta = n_1 x \sin \theta' = n_2 x \sin \theta'',$$

which gives

$$\theta = \theta',$$

so that the angles of incidence and reflection are equal, and

$$n_1 \sin \theta = n_2 \sin \theta'', \tag{8.35}$$

which is Snell's law for refraction.

These laws of reflection and refraction are true not only for electromagnetic waves but for all kinds of wave motion, since they depend only on the assumption that the characteristic quantities involved in the wave motion (in this case the electric and magnetic fields) shall be continuous at the boundary.

### Intensity relations

In order to obtain the amplitudes of the reflected and refracted waves we must examine the problem in more detail, and match the amplitudes of the tangential components of the electric and magnetic fields on the two sides of the boundary. To do this we must consider the two principal directions of polarization of the incident wave separately. These directions are with the electric field in the plane of incidence and normal to the plane of incidence respectively. For the former of these there are components of $\mathbf{E}$ parallel to the $x$- and $z$-axes, but $E_y = 0$; the only component of $\mathbf{H}$ is parallel to the $y$-axis. These statements hold for all the three waves, so that we have (see Fig. 8.4) the following:

*incident wave*

$$E_x = A \cos \theta \exp j\omega\{t - n_1(x \sin \theta + z \cos \theta)/c\}$$
$$E_z = -A \sin \theta \exp j\omega\{t - n_1(x \sin \theta + z \cos \theta)/c$$
$$H_y = (A/Z_1)\exp j\omega\{t - n_1(x \sin \theta + z \cos \theta)/c\}, \quad (8.36)$$

*reflected wave*

$$E'_x = A' \cos \theta \exp j\omega\{t - n_1(x \sin \theta - z \cos \theta)/c\}$$
$$E'_z = A' \sin \theta \exp j\omega\{t - n_1(x \sin \theta - z \cos \theta)/c\}$$
$$H'_y = -(A'/Z_1)\exp j\omega\{t - n_1(x \sin \theta - z \cos \theta)/c\} \quad (8.37)$$

*refracted wave*

$$E''_x = A'' \cos \theta'' \exp j\omega\{t - n_2(x \sin \theta'' + z \cos \theta'')/c\}$$
$$E''_z = -A'' \sin \theta'' \exp j\omega\{t - n_2(x \sin \theta'' + z \cos \theta'')/c\}$$
$$H''_y = (A''/Z_2)\exp j\omega\{t - n_2(x \sin \theta'' + z \cos \theta'')/c\}, \quad (8.38)$$

where $Z_1$, $Z_2$ are the intrinsic impedances of the two media respectively.

On making the tangential components of $\mathbf{E}$ and $\mathbf{H}$ (that is, $E_x$ and $H_y$) continuous at the plane $z = 0$, we obtain the relations

$$A \cos \theta + A' \cos \theta = A'' \cos \theta'' \quad (8.39)$$

and

$$(A - A')/Z_1 = A''/Z_2, \quad (8.40)$$

which give

$$\frac{A'}{A} = \frac{Z_2 \cos \theta'' - Z_1 \cos \theta}{Z_2 \cos \theta'' + Z_1 \cos \theta} \quad (8.41)$$

and

$$\frac{A''}{A} = \frac{2Z_2 \cos \theta}{Z_2 \cos \theta'' + Z_1 \cos \theta}. \quad (8.42)$$

If the media are such that $\mu_1 = \mu_2 = 1$, as is the case for light of visible wavelengths, $Z_1/Z_2 = n_2/n_1 = \sin \theta/\sin \theta''$, and these formulae can be written in the form

$$\frac{A'}{A} = \frac{\sin 2\theta'' - \sin 2\theta}{\sin 2\theta'' + \sin 2\theta}, \quad (8.43)$$

$$\frac{A''}{A} = \frac{4 \sin \theta'' \cos \theta}{\sin 2\theta'' + \sin 2\theta}. \quad (8.44)$$

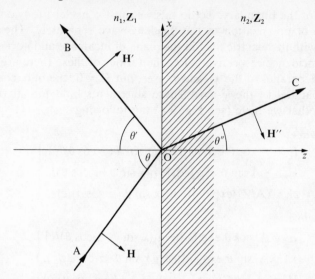

FIG. 8.5. Incident wave AO, reflected wave OB, and refracted wave OC at the boundary between two media. The electric vector is normal to the plane of incidence and hence normal to the plane of the paper; the magnetic vector is in the plane of incidence. The theory shows that if $n_2 > n_1$, and the permeability of both media is unity, the actual direction of $\mathbf{H}'$ is opposite to that given in the figure for all values of $\theta$.

The condition that there shall be no reflection $(A' = 0)$ is that

$$\sin 2\theta = \sin 2\theta'',$$

which is satisfied if $2\theta$ and $2\theta''$ are supplementary angles; that is

$$\theta + \theta'' = \tfrac{1}{2}\pi,$$

so that the reflected and refracted rays are normal to one another. Using Snell's law, we find that this occurs when

$$\tan \theta = n_2/n_1. \tag{8.45}$$

The angle which satisfies this relation is known as the Brewster angle.

The field components of the wave which is plane polarized with its electric vector normal to the plane of incidence are as follows (see Fig. 8.5):

*incident wave*

$$\left.\begin{array}{l} E_y = BF_1 \\ H_x = -(B/Z_1)\cos \theta \, F_1 \\ H_z = (B/Z_1)\sin \theta \, F_1, \end{array}\right\} \tag{8.46}$$

reflected wave

$$E'_y = B'F_2$$
$$H'_x = (B'/Z_1)\cos\theta\, F_2$$
$$H'_z = (B'/Z_1)\sin\theta\, F_2, \tag{8.47}$$

refracted wave

$$E''_y = B''F_3$$
$$H''_x = -(B''/Z_2)\cos\theta''F_3$$
$$H''_z = (B''/Z_2)\sin\theta''F_3, \tag{8.48}$$

where for simplicity we have written $F_1$, $F_2$, $F_3$ for the three complex exponentials representing the propagation of the waves, which are the same as in eqns (8.36), (8.37), and (8.38) respectively.

The boundary conditions for $E_y$ and $H_x$ now give

$$B + B' = B'',$$
$$(B - B')\cos\theta/Z_1 = B''\cos\theta''/Z_2,$$

the solutions of which are

$$\frac{B'}{B} = \frac{Z_2\cos\theta - Z_1\cos\theta''}{Z_2\cos\theta + Z_1\cos\theta''} \tag{8.49}$$

and

$$\frac{B''}{B} = \frac{2Z_2\cos\theta}{Z_2\cos\theta + Z_1\cos\theta''}. \tag{8.50}$$

These equations are similar to those obtained for the other direction of polarization but not identical, so that the magnitudes of the reflected and refracted waves are different. This is still true when we specialize to the case of $\mu_1 = \mu_2 = \mu_0$, when the formulae become

$$\frac{B'}{B} = \frac{\sin(\theta'' - \theta)}{\sin(\theta'' + \theta)}, \tag{8.51}$$

$$\frac{B''}{B} = \frac{2\sin\theta''\cos\theta}{\sin(\theta'' + \theta)}. \tag{8.52}$$

Since in general $\theta''$ is not equal to $\theta$, eqn (8.51) shows that there is no angle at which the reflected wave for this direction of polarization has zero amplitude. Thus if we start with unpolarized light, which consists of light containing a superposition of many components with their electric vectors in random orientations, and reflect it from a dielectric such as glass at the Brewster angle, the reflected wave will only contain one direction of polarization, that with the electric vector normal to the plane of incidence. This phenomenon can therefore be used to produce plane polarized light, and originally it was used to define the plane of polarization of the reflected light at the Brewster angle as being the plane of the reflected ray. Our treatment shows that this is in fact the plane of the magnetic vector in the reflected light, not of the electric vector.

The fact that the expressions for $A'/A$ and $A''/A$ are real shows that the phase changes at the boundary between two perfect dielectrics are always 0 or $\pi$. Inspection of the equations shows that for a wave with the electric vector in the plane of incidence, passing from a medium of lower refractive index to one of higher refractive index, there is a change of sign in the tangential component $E_x$ (but not in $E_z$, $H_y$) of the reflected wave when the angle of incidence is less than the Brewster angle, whereas the reverse is the case when it is greater than this angle. For the wave with the electric vector normal to the plane of incidence, there is a change of phase in $E_y$ and $H_z$ in the reflected wave (assuming $n_2 > n_1$), but not in $H_x$, whatever the angle of incidence.

Eqns (8.43), (8.44), (8.51), and (8.52) are known as Fresnel's formulae, after their discoverer. For normal incidence they give indeterminate results, but it is readily seen from eqns (8.41) and (8.42) (or (8.49) and (8.50)) that we have then

$$\frac{A'}{A} = \frac{Z_2 - Z_1}{Z_2 + Z_1} = \frac{n_1 - n_2}{n_1 + n_2} \tag{8.53}$$

and

$$\frac{A''}{A} = \frac{2Z_2}{Z_2 + Z_1} = \frac{2n_1}{n_1 + n_2}, \tag{8.54}$$

where the expressions on the extreme right apply only when

$$\mu_2 = \mu_1 = \mu_0.$$

The 'reflecting power' of the surface is equal to $(A'/A)^2$, since the power in the incident and reflected waves is proportional to the square of the amplitude. If $Z_2 > Z_1$, $A'' > A$ so that the amplitude of the electric vector is greater in the transmitted wave than the incident wave. Since the power varies as (amplitude of electric vector)$^2$/(impedance), the transmitted power is less than the incident power, and it is readily verified that the difference is equal to the reflected power.

*Total internal reflection*

When the refractive index $n_2$ of the second medium is less than $n_1$, application of Snell's law (eqn (8.35))

$$\sin \theta'' = (n_1/n_2)\sin \theta$$

to find the angle of refraction leads to values of $\sin \theta''$ greater than unity when $\sin \theta$ is greater than $n_2/n_1$. Since there is no real angle for which the sine is greater than unity, we conclude that there is no refracted wave and that all the energy is reflected. This is confirmed by inspection of eqns (8.41) and (8.49), for when $\sin \theta'' > 1$, $\cos \theta'' = (1-\sin^2\theta'')^{\frac{1}{2}}$ is a purely imaginary quantity, and the expressions for $A'/A$ and $B'/B$ are each of the form $(a-jb)/(a+jb)$ whose modulus is unity. Thus the reflection is total, but there is a change of phase on reflection which is different for the two directions of polarization. If the incident wave is plane-polarized in a direction which is not in or normal to the plane of incidence, the two components in these planes of the reflected wave will not be in phase and the wave will therefore be elliptically polarized.

The fact that there is no 'refracted wave' does not mean that there is no disturbance in the second medium, for eqns (8.42) and (8.50) show that $A''/A$ and $B''/B$ are finite. In order to find what kind of wave is propagated in the second medium, we write $\sin \theta'' = \cosh \gamma$ (since $\cosh \gamma$ is always greater than one), and then

$$\cos \theta'' = (1-\sin^2\theta'')^{\frac{1}{2}} = j(\cosh^2\gamma-1)^{\frac{1}{2}} = \pm j \sinh \gamma.$$

Hence the field components in the second medium are propagated as

$$\exp j\omega\{t-n_2(x \cosh \gamma - jz \sinh \gamma)/c\}$$
$$= \exp\{(-\omega n_2 z \sinh \gamma)/c\}\exp j\omega\{t-(n_2 x \cos \cosh \gamma)/c\}$$
$$= \exp\{(-2\pi z \sinh \gamma)/\lambda_2\}\exp j\omega\{t-(n_1 x \sin \theta)/c\} \quad (8.55)$$

where $\lambda_2$ is the wavelength of the radiation in the second medium. Eqn (8.55) shows that the wave is rapidly attenuated on the far side of the boundary, since at a distance $z = \lambda_2$ its amplitude falls by

$$\exp(-2\pi \sinh \gamma).$$

For a given value of $x$, there is no phase change as we proceed in the $z$ direction, but there is a phase change as we move along the boundary in the $x$ direction. This is because a wave front obliquely incident on the boundary arrives at points of greater $x$ at a later time.

If all the incident energy is reflected, no energy can be transmitted in the second medium. If we compute the mean value of Poynting's vector for a direction normal to the boundary (that is, the value of $\frac{1}{2}\text{Re}(E_x H_y^*)$ or $\frac{1}{2}\text{Re}(-E_y H_x^*)$ according to the direction of polarization) in the second medium, we find that it is zero, because $H_y$ is $\frac{1}{2}\pi$ out of phase with $E_x$. This means that no energy is transported away from the boundary on the average. Energy does flow into the second medium, since the stored energy must be finite if the field components are finite, but the flow is in the opposite direction during a later part of the cycle and the stored energy is returned to the first medium.

## 8.7. Reflection from the surface of a metal

When an electromagnetic wave is incident on the plane surface of a conducting medium, the amplitudes of the reflected and transmitted waves can be calculated in a manner essentially similar to that used for dielectrics. The refractive index and intrinsic impedance of a conducting medium are now complex numbers, and the formulae deduced for dielectrics may be taken over by use of complex values for $n_2$ and $Z_2$. From eqns (8.41) and (8.49) it is then apparent that the phase change on reflection is not in general 0 or $\pi$, so that a plane-polarized wave will become elliptically polarized after reflection. Here we shall only calculate the reflecting power of a metal for a wave falling on it from free space at normal incidence, using eqn (8.53). For a good conductor

$$\mathbf{Z}_2 = (1+j)/\sigma\delta \quad \text{from eqn (8.32)},$$

while for free space

$$\mathbf{Z}_1 = \mu_0 c,$$

where $\delta$ is the skin depth in the metal and $c$ is the velocity of light in free space. Thus the reflection coefficient is

$$\frac{A'}{A} = \frac{\mathbf{Z}_2 - \mathbf{Z}_1}{\mathbf{Z}_2 + \mathbf{Z}_1} = \frac{(1+j) - \mu_0 \sigma c \delta}{(1+j) + \mu_0 \sigma c \delta},$$

where the complex value shows that there is a phase change even at normal incidence, and the reflecting power is

$$\left|\frac{A'}{A}\right|^2 = \left|\frac{(1-t)+j}{(1+t)+j}\right|^2 = \frac{2+t^2-2t}{2+t^2+2t},$$

where we have written $t$ for the quantity $\mu_0 \sigma c \delta$. For a metal, where $\sigma \approx 10^7 \, \Omega^{-1} \, \text{m}^{-1}$, $t \approx 4 \times 10^9 \delta$, and is thus much greater than unity at all frequencies up to that of visible light ($f \approx 10^{15}$ Hz). Hence approximately

$$|(A'/A)|^2 = 1 - 4/t = 1 - 2 \cdot 1 \times 10^{-5} (\mu_r f/\sigma)^{\frac{1}{2}}. \tag{8.56}$$

This formula shows that metals should be almost perfect reflectors of electromagnetic radiation for all frequencies up to those of visible light. For copper at a frequency of $10^{10}$ Hz (a wavelength of $3 \times 10^{-2}$ m), the reflecting power differs from unity by about $2 \cdot 7 \times 10^{-4}$, and this formula would lead us to expect that it should only fall below unity by about $0 \cdot 09$ at a frequency of $10^{15}$ Hz (a wavelength of 300 nm). The fact that copper appears strongly coloured shows that this formula fails for optical frequencies, when electrons in the atom other than those responsible for the conductivity begin to play a role. In addition, the effective conductivity of a metal at high frequencies is less than the low frequency value (see § 12.10).

The high reflection coefficient of a metal arises from the fact that its intrinsic impedance is very much smaller than that of free space. For copper at $10^{10}$ Hz, $Z_2$ is $0 \cdot 026(1+\mathrm{j})\Omega$, while for free space $Z_1 = 376 \cdot 7 \ \Omega$. Thus to satisfy the boundary conditions the wave must be almost totally reflected with a phase change in the electric vector but not in the magnetic vector, making **E** almost zero at the metal surface, and **H** almost twice the amplitude due to the incident wave alone.

## 8.8. The pressure due to radiation

When a plane electromagnetic wave travels through a conducting medium in the $z$ direction, a conduction current flows of density $J_x = \sigma E_x$ (assuming the wave to be linearly polarized parallel to the $x$-axis). Associated with the wave is a magnetic flux density $B_y$, which will exert a force on the current whose magnitude on a volume element $\mathrm{d}\tau$ is

$$\mathrm{d}F_z = (\mathbf{J} \wedge \mathbf{B})_z \ \mathrm{d}\tau = J_x B_y \ \mathrm{d}\tau$$

in the positive $z$ direction. This force is in the direction in which the energy is travelling, and gives rise to a 'radiation pressure'. From eqn (8.7), $J_x = -\partial H_y/\partial z - \partial D_x/\partial t$, since all other components vanish in a plane wave, and hence

$$\mathrm{d}F_z/\mathrm{d}\tau = (-\partial H_y/\partial z - \partial D_x/\partial t)B_y$$
$$= -(\partial H_y/\partial z)B_y - \partial(D_x B_y)/\partial t + (\partial B_y/\partial t)D_x.$$

But from eqn (8.3), $\partial B_y/\partial t = -\partial E_x/\partial z$, so that

$$\mathrm{d}F_z/\mathrm{d}\tau = -\left(D_x\frac{\partial E_x}{\partial z} + B_y\frac{\partial H_y}{\partial z}\right) - \frac{\partial}{\partial t}(D_x B_y)$$

$$= -\frac{\partial U}{\partial z} - \epsilon\mu\frac{\partial(E_x H_y)}{\partial t}$$

in a medium where $D$, $B$ are linearly proportional to $E$, $H$. In the steady state the second term $\partial(E_x H_y)/\partial t$ vanishes when averaged over a cycle, and the significance of the minus sign before $(\partial U/\partial z)$ is that the force is in the forward direction provided that $U$ diminishes as the wave travels onwards, as it necessarily will in a conducting medium. We can interpret the volume force as due to the gradient of a pressure $P$, and, from the geometry shown in Fig. 8.6,

$$\mathrm{d}F_z = -(\partial P/\partial z) \ \mathrm{d}x \ \mathrm{d}y \ \mathrm{d}z = -(\partial P/\partial z) \ \mathrm{d}\tau.$$

Hence

$$P = U \quad \text{(normal incidence)}. \tag{8.57}$$

The pressure is exerted in the direction of Poynting's vector, and can be attributed to a momentum $G$ per unit volume which flows across unit

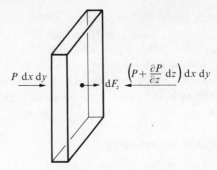

FIG. 8.6. Interpretation of volume force as gradient of a pressure.

area at the rate (see § 8.3)

$$U = vG = v^{-1} |\mathbf{E} \wedge \mathbf{H}| \qquad (8.58)$$

This result is consistent with quantum and relativity theory by which radiation consists of photons of energy $h\nu$, whose momentum *in vacuo* is $h\nu/c$. For diffuse radiation, consisting of waves travelling in all directions, only one-third of the total energy density will on average be associated with waves travelling normal to the surface, so that we have

$$P = \tfrac{1}{3}U \quad \text{(diffuse radiation)}. \qquad (8.59)$$

When a wave is totally reflected, its momentum is reversed so that the pressure is doubled; however the energy density is also doubled so that the equation $P = U$ for normal incidence and eqn (8.59) for diffuse radiation still hold. These results are easily obtained from the concept of momentum flow in the electromagnetic wave; other methods are much more complex, as it is necessary to include both the stress at the boundary and the volume force on the conduction current. In the example above there is no change of medium, and we obtain the correct answer from the volume force alone. On the other hand, when an electromagnetic wave is partially reflected at the boundary between two non-conducting dielectrics there is no force on any conduction current and the pressure at the boundary arises entirely from the difference in the Maxwell stress tensor on either side of the boundary. It can again be calculated much more easily from the momentum balance by considering the energy densities in the incident, reflected, and transmitted waves; that the two approaches give the same answer in the case of normal incidence is verified in Problem 8.13.

### 8.9. Radiation from an oscillating dipole

In Chapter 4 the concept of a vector potential $\mathbf{A}$, such that

$$\mathbf{B} = \text{curl } \mathbf{A}, \qquad (8.60)$$

was introduced. The vector potential is not of great use in elementary problems, but is of considerable assistance in calculating the radiation from an aerial. The fundamental equations of the electromagnetic field are

$$\text{div } \mathbf{D} = \rho_e, \tag{8.1}$$

$$\text{div } \mathbf{B} = 0, \tag{8.2}$$

$$\text{curl } \mathbf{E} = -\frac{\partial \mathbf{B}}{\partial t}, \tag{8.3}$$

$$\text{curl } \mathbf{H} = \mathbf{J} + \frac{\partial \mathbf{D}}{\partial t}. \tag{8.7}$$

On substituting from eqn (8.60) into (8.3) we have, as in § 5.1,

$$\text{curl}\left(\mathbf{E} + \frac{\partial \mathbf{A}}{\partial t}\right) = 0,$$

the solution of which is

$$\mathbf{E} = -\frac{\partial \mathbf{A}}{\partial t} - \text{grad } V, \tag{8.61}$$

where grad $V$ is the 'constant of integration', $V$ being some scalar function. Since curl grad of a scalar function is zero, any such function satisfies eqn (8.3). In a static problem, where $\mathbf{A}$ is constant, eqn (8.61) reduces to $\mathbf{E} = -\text{grad } V$, where $V$ is the ordinary scalar electrostatic potential, as defined in eqn (1.6).

Eqn (8.60) does not define the vector potential $\mathbf{A}$ completely, since $\mathbf{A}$ is the solution of a differential equation and we can add to this solution any vector whose curl is zero. In eqn (4.49) we added the condition div $\mathbf{A} = 0$, but for our present problem it is more convenient to generalize this condition in the form

$$\text{div } \mathbf{A} = -\mu\epsilon\frac{\partial V}{\partial t} = -\frac{1}{v^2}\frac{\partial V}{\partial t}, \tag{8.62}$$

where $v = (\mu\epsilon)^{-\frac{1}{2}}$ is the velocity of electromagnetic waves in the medium. This definition reduces to that used previously when $V$ is independent of time, and has the advantage that it enables us to separate the variables $V$ and $\mathbf{A}$. From eqn (8.61) we have (using eqn (8.1))

$$-\nabla^2 V = -\text{div grad } V = \text{div}\left(\mathbf{E} + \frac{\partial \mathbf{A}}{\partial t}\right)$$

$$= \text{div } \mathbf{E} + \frac{\partial}{\partial t}\text{ div } \mathbf{A} = \frac{\rho_e}{\epsilon} - \frac{1}{v^2}\frac{\partial^2 V}{\partial t^2},$$

so that

$$-\nabla^2 V + \frac{1}{v^2}\frac{\partial^2 V}{\partial t^2} = \frac{\rho_e}{\epsilon} \tag{8.63}$$

Again,

$$\mathrm{curl}\,\mathbf{B} = \mathrm{curl}(\mathrm{curl}\,\mathbf{A}) = \mathrm{grad}\,\mathrm{div}\,\mathbf{A} - \nabla^2\mathbf{A} = -\frac{1}{v^2}\,\mathrm{grad}\,\frac{\partial V}{\partial t} - \nabla^2\mathbf{A},$$

while from eqns (8.7) and (8.61)

$$\mathrm{curl}\,\mathbf{B} = \mu\,\mathrm{curl}\,\mathbf{H} = \mu\left(\mathbf{J} + \frac{\partial\mathbf{D}}{\partial t}\right) = \mu\mathbf{J} + \mu\epsilon\frac{\partial\mathbf{E}}{\partial t}$$

$$= \mu\mathbf{J} - \frac{1}{v^2}\left(\frac{\partial^2\mathbf{A}}{\partial t^2} + \mathrm{grad}\,\frac{\partial V}{\partial t}\right)$$

Identifying these two equations for curl **B** gives

$$-\nabla^2\mathbf{A} + \frac{1}{v^2}\frac{\partial^2\mathbf{A}}{\partial t^2} = \mu\mathbf{J}. \tag{8.64}$$

For static systems, or systems which vary with time only slowly, eqns (8.63) and (8.64) reduce to the eqns (2.1) and (4.54) obtained earlier. The general solutions of our new equations are also similar to those of the earlier equations, and may be written as

$$V = \frac{1}{4\pi\epsilon}\int\frac{[\rho_e]\,\mathrm{d}\tau}{r}, \tag{8.65}$$

$$\mathbf{A} = \frac{\mu}{4\pi}\int\frac{[\mathbf{J}]\,\mathrm{d}\tau}{r}, \tag{8.66}$$

where the square brackets round $\rho_e$ and **J** have the following significance. The values of $V$ and **A** at a time $t$ and at a point distance $r$ from the element of volume containing $\rho_e$ and **J** are related not to the values of $\rho_e$ and **J** at the origin at the same time but to those values which obtained at a time $(t-r/v)$. In other words, the disturbance set up by the values of $\rho_e$ and **J** at the origin is propagated with the velocity $v$ and reaches a point distance $r$ away at a time later by $r/v$. Thus the disturbance at this point is related to what happened at the origin a time $r/v$ earlier, just as the light reaching the earth from a star tells us what was happening on the star, not at the same instant, but at the time when the light left the star. The values of $V$ and **A** given by eqns (8.65) and (8.66) are known as 'retarded potentials'.

These equations will now be applied to the case of a short length of wire **s** at the origin of coordinates, carrying a current

$$I = I_0 \cos \omega t.$$

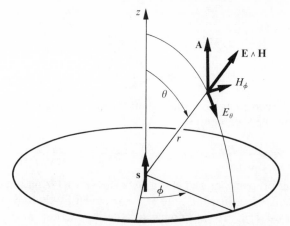

FIG. 8.7. Radiation from a short dipole **s**; at the point $(r, \theta, \phi)$ the magnetic vector potential **A** is parallel to **s**; Poynting's vector $\mathbf{E} \wedge \mathbf{H}$ is along the radius vector **r**, and the field components are $E_\theta$ (along a line of longitude) and $H_\phi$ (along a line of latitude).

Then, since $\mathbf{J}\,d\tau = I\,d\mathbf{s}$, the retarded value $[\mathbf{J}]\,d\tau = [I]\,d\mathbf{s}$, and if **s** is very short compared with $r$, so that $r$ does not change significantly during the integration, we may write

$$\mathbf{A} = \frac{\mu}{4\pi r}\int [I]\,d\mathbf{s} = \frac{\mu}{4\pi r}[I]\mathbf{s}. \tag{8.67}$$

Eqn (8.67) shows that the vector potential **A** is everywhere parallel to **s**, as shown in Fig. 8.7.

The magnetic field strength **H** can now be found at any point, for

$$\mathbf{H} = \mathbf{B}/\mu = (1/\mu)\mathrm{curl}\,\mathbf{A}$$

$$= \frac{1}{4\pi}\mathrm{curl}\left([I]\frac{\mathbf{s}}{r}\right)$$

$$= \frac{1}{4\pi}\left(\frac{[I]}{r}\mathrm{curl}\,\mathbf{s} - \mathbf{s}\wedge\mathrm{grad}\frac{[I]}{r}\right)$$

$$= -\frac{1}{4\pi}\left(\mathbf{s}\wedge\mathrm{grad}\frac{[I]}{r}\right),$$

since curl **s** = 0. Now grad$([I]/r) = \mathbf{r}_1\,\partial([I]/r)/\partial r$, where $\mathbf{r}_1$ is a unit vector in the direction of $r$, since $[I]$ varies in space only with $r$. Hence

$$\mathbf{H} = -\frac{1}{4\pi}\{\mathbf{s}\wedge\mathbf{r}_1\}\frac{\partial}{\partial r}\left\{\frac{[I]}{r}\right\},$$

showing that **H** is normal to **s** and to $\mathbf{r}_1$. It follows that it has only one component, $H_\phi$, in spherical polar coordinates.

The retarded value $[I] = I_0 \cos \omega(t - r/v)$, and hence

$$\frac{\partial}{\partial r}\left(\frac{[I]}{r}\right) = -\frac{I_0}{r^2}\cos \omega\left(t - \frac{r}{v}\right) + \frac{\omega I_0}{vr}\sin \omega\left(t - \frac{r}{v}\right)$$

$$= -\frac{I_0}{r^2}\cos \omega\left(t - \frac{r}{v}\right) + \frac{2\pi I_0}{r\lambda}\sin \omega\left(t - \frac{r}{v}\right),$$

where $\lambda$ is the wavelength of the radiation. The magnetic field strength is then given by the sole component

$$H_\phi = \frac{sI_0 \sin \theta}{4\pi r^2}\cos \omega\left(t - \frac{r}{v}\right) - \frac{sI_0 \sin \theta}{2r\lambda}\sin \omega\left(t - \frac{r}{v}\right). \tag{8.68}$$

The first term predominates at short distances ($r \ll \lambda$) from the origin, and will be recognized as just the field strength given by the law of Biot and Savart (eqn (4.41)). It is known as the 'induction field'. At large distances of many wavelengths from the origin only the second term is significant; it falls off as $(\lambda r)^{-1}$ instead of $r^{-2}$ and is known as the 'radiation field'.

The electric field may be found by using eqn (8.3), or it may be found from eqn (8.61) if the scalar potential $V$ is first computed by means of eqn (8.62). We are interested only in its value at a large distance from the origin, where our spherical wave front has very small curvature and over a small region may be taken as a plane wave. As we should expect, **E** is then normal both to **H** and to the direction of propagation, and its only component is

$$E_\theta = Z_0 H_\phi = (\mu/\epsilon)^{\frac{1}{2}}H_\phi.$$

The energy crossing unit area per second is given by Poynting's vector,

$$E_\theta H_\phi = Z_0 H_\phi^2,$$

showing that the energy flow is radially outwards. The mean value averaged over a cycle of oscillation is

$$\overline{E_\theta H_\phi} = \tfrac{1}{8}Z_0(I_0 s/r\lambda)^2 \sin^2\theta, \tag{8.69}$$

and the total energy radiated in all directions per second is

$$P = -\frac{dW}{dt} = \int_0^\pi 2\pi r^2(\overline{E_\theta H_\phi})\sin \theta \; d\theta = \frac{\pi Z_0 I_0^2 s^2}{3\lambda^2}. \tag{8.70}$$

The power radiated may be expressed in terms of a resistance $R_r$, called the 'radiation resistance', obtained by writing $P = \tfrac{1}{2}R_r I_0^2$, since this is just the mean power which would be dissipated in a real resistance $R_r$. For our current element eqn (8.70) gives

$$R_r = \frac{2\pi Z_0}{3}\left(\frac{s}{\lambda}\right)^2, \tag{8.71}$$

showing that the radiation resistance depends only on the square of the ratio of the length of the wire to the wavelength. In free space the numerical value of the radiation resistance $R_r$ is $789(s/\lambda)^2\,\Omega$.

The oscillating current element may consist of two oscillating charges $\pm Q_0 \sin \omega t$, separated by a short distance $\mathbf{s}$, which constitute an oscillating electric dipole of moment $\mathbf{s}Q_0 \sin \omega t = \mathbf{p}_0 \sin \omega t$. At any instant the current is $I = dQ/dt = \omega Q_0 \cos \omega t$, showing that $\mathbf{s}I = \omega \mathbf{p}_0 \cos \omega t$. A more realistic example is provided by a charge $Q_0$ performing a simple harmonic motion with displacement $\mathbf{s} \sin \omega t$. We may write

$$\mathbf{Is} = \int I \, d\mathbf{s} = \int (dQ/dt)(\mathbf{v}\, dt) = \int \mathbf{v} \, dQ = \mathbf{v}\, Q_0 = \omega \mathbf{s} Q_0 \cos \omega t = \omega \mathbf{p}_0 \cos \omega t.$$

In each case $\mathbf{s}I_0 = \omega \mathbf{p}_0$, and from (8.70) the total power radiated into free space is

$$P = \frac{\pi Z_0 \omega^2 p_0^2}{3\lambda^2} = \frac{4\pi^3 Z_0 p_0^2}{3\lambda^4} = \frac{Z_0 \omega^4 p_0^2}{12\pi c^2} = \frac{\omega^4}{3c^3}\left(\frac{p_0^2}{4\pi\epsilon_0}\right). \tag{8.72}$$

This shows that the power radiated is proportional to the square of the oscillating dipole moment, and the fourth power of the frequency (or the inverse fourth power of the wavelength).

Since the oscillating current element can be regarded as an oscillating electric dipole, it is interesting to calculate the scalar potential and compare it with our earlier results for a static dipole. Since $\mathbf{A}$ has a component only parallel to the polar axis (which we will call the $z$-axis), $\mathrm{div}\,\mathbf{A} = \partial A_z/\partial z$. In the differentiation $x$, $y$ are constant and

$$(\partial/\partial z)_{x,y} = (z/r)(\partial/\partial r) = \cos \theta (\partial/\partial r) \quad (\text{cf. } \S\, 2.2)$$

Hence, using eqns (8.62) and (8.67),

$$\frac{\partial V}{\partial t} = -v^2 \, \mathrm{div}\,\mathbf{A} = -\frac{v^2 \mu I_0 s \cos \theta}{4\pi} \frac{\partial}{\partial r}\left\{ r^{-1} \cos \omega\left(t - \frac{r}{v}\right)\right\}.$$

Since $r$ and $t$ are independent variables we may integrate with respect to $t$ inside the differential, giving

$$V = -\frac{v^2 \mu I_0 s \cos \theta}{4\pi\omega} \frac{\partial}{\partial r}\left\{ r^{-1} \sin \omega\left(t - \frac{r}{v}\right)\right\}$$

$$= -\frac{p_0 \cos \theta}{4\pi\epsilon} \frac{\partial}{\partial r}\left\{ r^{-1} \sin \omega\left(t - \frac{r}{v}\right)\right\}$$

$$= -\frac{1}{4\pi\epsilon}\left[\mathbf{p}_0 \cdot \mathrm{grad}\left\{ r^{-1} \sin \omega\left(t - \frac{r}{v}\right)\right\}\right]. \tag{8.73}$$

Comparison with eqn (1.11a) shows that this is of the same form except that allowance must be made for the retardation in taking the gradient.

*Magnetic dipole radiation*

An oscillating magnetic dipole also radiates electromagnetic waves, the radiation field being similar to that of an electric dipole except that the roles of **E** and **H** are interchanged. The lines of **E** (and of **A**) are concentric circles about the axis of the dipole; **H** lies in the meridian plane, and when $r \gg \lambda$ it has just a single component $H_\theta$ along a 'line of longitude' (cf. Fig. 8.7). The power radiated into free space from a magnetic dipole of moment $m = m_0 \cos \omega t$ is

$$P = \frac{\omega^4}{3c^2}\left\{ \left(\frac{\mu_0}{4\pi}\right) m_0^2 \right\}. \tag{8.74}$$

This may be compared with the power from an electric dipole. If this is formed by a current $I_0 \cos \omega t$ in a length $s$, while the magnetic dipole carries the same current in a loop of area $A$, the ratio is

$$\frac{\text{power radiated by electric dipole}}{\text{power radiated by magnetic dipole}} = \frac{(I^2 s^2/4\pi\epsilon_0 \omega^2)}{(\mu_0 I^2 A^2/4\pi)}$$

$$= \left(\frac{cs}{\omega A}\right)^2 = \left(\frac{\lambda s}{2\pi A}\right)^2. \tag{8.75}$$

This ratio is large if the linear dimensions of $s$ and $A$ are small compared to $\lambda$. The same ratio applies in the reception of electromagnetic waves (see Problem 8.8).

*Radio antennae*

The equations derived above apply only to systems where the linear dimensions are small compared with the wavelength. This does not in general hold for antennae used in radio or television transmission, but the radiative properties may be calculated by methods similar to those used above, the value of **A** being obtained by the use of eqn (8.66). For a single straight wire this gives

$$\mathbf{A} = \frac{\mu}{4\pi} \int \frac{[I]}{r} d\mathbf{s}, \tag{8.76}$$

where the integration is along the wire and $I$ is the current in the element d**s**. Two points must be noted in performing the integration: (1) the current is not in general constant along the wire, since it must fall to zero at the ends; (2) in evaluating the retarded potential, allowance must be made for the phase difference in signals coming from different parts of the wire owing to the change in the distance. The problem is similar to that of the diffraction pattern of a single slit where the amplitude varies over the aperture. The interference concepts used for light waves may be applied in finding the radiation pattern produced by an aerial array consisting of many elements; the problem is similar to that of a diffraction grating.

## Problems

8.1. A plane electromagnetic wave propagates parallel to the $z$-axis and falls at normal incidence onto a metal bounded by the plane $z = 0$. At this surface there exists a tangential oscillating magnetic field of strength $H_y = H_0 \cos \omega t$. Show that the mean value of Poynting's vector just inside the surface is equal to the power per unit area dissipated in heating the metal, as given by eqn (8.34b).

8.2. Show that the superposition of two waves of equal amplitude, one with angular frequency $\omega + d\omega$ and phase constant $\beta + d\beta$, the other with $\omega - d\omega$ and $\beta - d\beta$, gives a wave of frequency $\omega$ and phase constant $\beta$, whose amplitude varies as

$$\binom{\sin}{\cos}(t\,d\omega - x\,d\beta).$$

Hence a point of maximum amplitude moves as

$$(t\,d\omega - x\,d\beta) = \text{constant},$$

or with velocity $u = (d\omega/d\beta)$. Since the energy is proportional to the square of the amplitude, $u$ gives the rate at which energy is propagated, known as the 'group velocity'.

Show that in a good conductor, such as a metal, where the displacement current can be neglected, the phase velocity $v$ ($= \omega/\beta$) is $\omega\delta$, and the group velocity $u$ ($= d\omega/d\beta$) is $2\omega\delta$, where $\delta$ is the skin depth.

8.3. Show from eqn (8.41) that the condition for there to be no reflected wave at the boundary between two insulators ($\epsilon_1, \mu_1; \epsilon_2, \mu_2$) when the electric vector is in the plane of incidence requires that

$$\epsilon_1 \tan \theta = \epsilon_2 \tan \theta''.$$

This equation shows that at this particular angle the lines of electric field are refracted as in the electrostatic case (cf. Fig. 1.13) so that the boundary conditions are satisfied without any reflected wave. Similarly, when the magnetic vector is in the plane of incidence, the condition for no reflection is

$$\mu_1 \tan \theta = \mu_2 \tan \theta''$$

so that the lines of magnetic field are refracted as in the magnetostatic case.

Verify that in general the boundary conditions for the normal components of $\mathbf{D}$ and $\mathbf{B}$ are automatically satisfied in the theory of § 8.6 when the conditions for the tangential components of $\mathbf{E}$ and $\mathbf{H}$ are satisfied.

8.4. The rate at which solar energy falls on the earth's surface is approximately $1.4 \times 10^3 \,\mathrm{J\,m^{-2}\,s^{-1}}$. Calculate the r.m.s. values of the electric and magnetic field strengths at the earth's surface, and the pressure exerted on it, assuming it to behave as a perfect absorber.

(*Answer*: $E = 730 \,\mathrm{V\,m^{-1}}$; $H = 1.9 \,\mathrm{A\,m^{-1}}$; pressure, $4.7 \times 10^{-6} \,\mathrm{N\,m^{-2}}$.

8.5. Show that by the introduction of a complex relative permittivity $\epsilon_r = \epsilon' - \epsilon''$, where $\epsilon'' = \sigma/\omega\epsilon_0$, the eqns (8.25a) may be written in the same form as for a non-conducting dielectric. Show that this leads to the relation $\mathbf{n}^2 = \mu_r \epsilon_r$, and that $\epsilon''/\epsilon' = \tan \delta$, where $\tan \delta$ is the loss tangent of the dielectric (see Problem 7.12).

8.6. A slightly imperfect dielectric has a small loss tangent. Show that in the first approximation the velocity of electromagnetic waves is the same as if tan δ were zero, but the power falls as $\exp(-\sigma Z_0 z)$ in travelling a distance $z$, where $Z_0$ is the intrinsic impedance of the medium neglecting the conductivity $\sigma$. This result has the simple interpretation: in a thickness $dz$ the power dissipated per unit cross-section is $\sigma E^2\, dz$, while the incident power $P = E^2/Z_0$. Hence

$$-dP/dz = \sigma Z_0 P.$$

Show also that

$$\sigma Z_0 = 2\pi(\text{loss tangent of the dielectric})/\lambda,$$

where $\lambda$ is the wavelength of the radiation in the dielectric.

8.7. When a plane wave is incident on a conducting wire the electric field set up in the wire is equal to the tangential component of the electric field strength in the wave. Show that in a short straight wire the power picked up is proportional to $\sin^2\theta$, where $\theta$ is the angle between the wire and the direction of travel of the wave. This shows that the directional properties of the wire are the same for receiving as for transmitting (cf. eqn (8.69)).

8.8. When a plane wave falls on a small plane loop of wire, the e.m.f. induced in the loop is determined by the rate of change of magnetic flux through it. Show that the directional properties of the loop are the same as those of the short wire, if $\theta$ is measured from the normal to the plane of the loop, but the planes of the electric and magnetic vectors in the wave must be interchanged. This is consistent with the fact that a loop carrying an alternating current behaves as an oscillating magnetic dipole.

Show that the ratio of the e.m.f. set up in a short wire of length $s$ to that in a small loop of area $A$ is $\lambda s/2\pi A$, where $\lambda$ is the wavelength of the radiation. If the linear dimensions of the loop are roughly the same as those of the wire $(A \approx s^2)$, this shows that the loop is a much poorer aerial when $s \ll \lambda$.

8.9. A transmitter radiates a power $P$ from a short dipole aerial. Show that the r.m.s. electric field strength at a distance $D$ in the equatorial plane of the dipole aerial is

$$E = (3Z_0 P/8\pi D^2)^{\frac{1}{2}},$$

where $Z_0$ is the intrinsic impedance of free space.

If $P = 1\,\text{kW}$, show that the field strength at a distance of $10\,\text{km}$ is about $0\cdot021\,\text{V m}^{-1}$.

8.10. A transmitter radiates a power $P$ from a short horizontal dipole aerial located at a height $H$ above the sea. Show that the signal received at a target whose distance is $D$ and height above the sea is $h$ $(D \gg H, h)$ is a maximum when $h = D\lambda/4H$, where $\lambda$ is the wavelength of the transmitter. The sea may be assumed to act as a flat, perfectly reflecting (conducting) surface.

Show that if the height of the target is very much smaller than this value, the power incident on unit area of it (assuming it to lie in the equatorial plane of the dipole) is $6\pi PH^2 h^2/D^4\lambda^2$. This equation shows that the effect of the sea is to make the power fall off with the inverse fourth power of the distance instead of the inverse square; it shows also the improvement gained by using short wavelengths.

If $P = 1\,\text{kW}$, $D = 10\,\text{km}$, $H = h = \lambda = 10\,\text{m}$, show that the r.m.s. electric field strength at the target is about $2\cdot7\times10^{-4}\,\text{V m}^{-1}$.

8.11. Obtain eqn (8.56) by the use of eqn (8.34a) and the law of conservation of energy.

8.12. At high frequencies the magnetic field falls inside each lamination of a transformer owing to the skin effect; the effective permeability at high frequencies is found by calculating the total flux in a lamination in phase with that at each surface of the lamination. Show that the effective radio-frequency permeability is less than the static value by a factor $\delta/a$ when the skin depth $\delta$ is very much smaller than the thickness $a$ of the lamination.

What thickness of lamination is required to make $\delta = a$ for a material of relative permeability $10^5$ and resistivity $5\times10^{-7}\,\Omega\,m$ at a frequency of 50 Hz?

(*Answer*: 0·16 mm)

8.13. A plane wave falls at normal incidence on the boundary ($z = 0$) between two media with permittivity and permeability ($\epsilon_1, \mu_1$) and ($\epsilon_1, \mu_2$) respectively. If the field strength components at the boundary are $E_x$, $H_y$, the stress is (by Maxwell's theory)

$$\tfrac{1}{2}\{(\epsilon_1-\epsilon_2)E_x^2+(\mu_1-\mu_2)H_y^2\}.$$

Show that this is equal to the sum of the momentum flows in the incident and reflected waves less that in the transmitted wave, that is, to

$$\frac{E^2+E'^2}{Z_1v_1}-\frac{E''^2}{Z_2v_2},$$

where $E$, $E'$, and $E''$ are the electric intensities in the incident, reflected, and transmitted waves respectively, and $Z_1$, $v_1$; $Z_2$, $v_2$ are the impedance of and phase velocity in the two media. *Hint.* Note that $E_x = E''$, $H_y = E''/Z_2$ and show that both expressions can be reduced to

$$\tfrac{1}{2}E''^2\left(\frac{1}{Z_1v_1}+\frac{Z_1}{Z_2^2v_1}-\frac{2}{Z_2v_2}\right).$$

# 9. Filters, transmission lines, and waveguides

## 9.1. Elements of filter theory

A FREQUENT requirement in radio and telephony is the separation of two signals of different frequencies. Any circuit whose impedance varies with frequency can be used for this purpose, a simple example already considered being the tuned circuit, which can be employed to accept or reject a narrow band of frequencies centred on its natural resonant frequency. This acts as a 'band-pass' or 'band-stop' filter. Another type of filter may be required to pass all frequencies up to a certain value, and stop all higher frequencies; this is a 'low-pass' filter. The reverse case is a 'high-pass' filter, which rejects all low frequencies up to a certain value, and passes all higher frequencies.

The action of a filter can be understood by considering a simple example, the low-pass filter shown in Fig. 9.1. A common use of such a filter is to remove the ripple voltage from the output of a rectifier unit which is converting an a.c. voltage to a steady voltage. The output is applied to the terminals AB of the filter, which is required to pass the steady voltage component on to the terminals CD, but not the alternating component. If the latter has a frequency of, say, 100 Hz, and the capacitance $C$ is chosen to have a low impedance at 100 Hz while the inductance $L$ has a high impedance, only a small fraction of the input voltage at this frequency will appear at the terminals CD, because the inductance and second capacitance act as a voltage divider. The fraction is approximately $X_C/X_L = (1/\omega C)/\omega L = 1/\omega^2 LC$, and if $C = 10\ \mu\text{F}$, $L = 25\ \text{H}$, the fraction is about 1/100. On the other hand, if there is no leakage in the capacitor, the full steady voltage component will be passed on to CD. Thus the filter accepts the signal of zero frequency, and partially rejects the signal at 100 Hz frequency. Better rejection is obtained by

FIG. 9.1. Simple low-pass filter section.

FIG. 9.2. Chain of low-pass filter sections.

adding more sections of this kind, as in Fig. 9.2. This arrangement is called a step- or ladder-type filter. It is clear that evaluation of the currents and voltages in the different elements by application of Kirchhoff's laws would be very laborious, and it is preferable to proceed in a different way, making use of the recurrent nature of the elements.

We shall begin by considering a uniform filter, consisting of a chain of similar sections, as in Fig. 9.3, which are repeated indefinitely, forming an infinite chain. If a generator is applied at some point earlier in the chain, currents will flow in the various sections; let the currents in successive sections be $I_{n-1}$, $I_n$, $I_{n+1}$. Application of Kirchhoff's law to the central section gives

$$\mathbf{Z}_2(I_n - I_{n-1}) + \mathbf{Z}_1 I_n + \mathbf{Z}_2(I_n - I_{n+1}) = 0$$

or                                                                                     (9.1)

$$-\mathbf{Z}_2 I_{n-1} + (\mathbf{Z}_1 + 2\mathbf{Z}_2) I_n - \mathbf{Z}_2 I_{n+1} = 0.$$

We may write $I_n = aI_{n-1}$, where $a$ is a real or complex number; then in an infinite chain where we cannot distinguish between sections we must also have $I_{n+1} = aI_n$. Eqn (9.1) then gives

$$\tfrac{1}{2}(a + 1/a) = 1 + \mathbf{Z}_1/2\mathbf{Z}_2. \tag{9.2}$$

This equation determines the attentuation constant $a$. We shall confine our attention to the case where $\mathbf{Z}_1/\mathbf{Z}_2$ is real, which corresponds to $\mathbf{Z}_1$ and $\mathbf{Z}_2$ being both pure resistances or pure reactances. Then $a$ can be either real or complex, but not a purely imaginary quantity. The real roots arise when the right-hand side of eqn (9.2) lies outside the range $+1$ to $-1$, that

FIG. 9.3. Generalized type of uniform ladder filter.

is, $\mathbf{Z}_1/4\mathbf{Z}_2$ lies outside the range 0 to $-1$. We consider separately the three cases where it is greater than 0, between 0 and $-1$, and less than $-1$.

### (i) $(\mathbf{Z}_1/4\mathbf{Z}_2) > 0$

$a$ is real and positive. Since the network is 'passive' (that is, it contains no power-generating elements), the currents must decrease as we move away from the generator attached to one end of the filter. We take $a < 1$ for a wave travelling towards the right (that is a generator attached to the left-hand end of the filter) and $a > 1$ for a wave travelling towards the left. This interpretation is consistent with the fact that the roots of eqn (9.2) are reciprocal; thus the wave is attenuated by the same amount per section in the direction in which it progresses, whichever way it is going. The significance of the positive sign of $a$ is that the wave is attenuated without change of phase. If we write $a = e^{-\alpha}$, where $\alpha$ is the attenuation constant per section, eqn (9.2) becomes

$$\cosh \alpha = 1 + \mathbf{Z}_1/2\mathbf{Z}_2. \tag{9.3}$$

### (ii) $0 > (\mathbf{Z}_1/4\mathbf{Z}_2) > -1$

$a$ is complex, with modulus unity, so that we·may write $a = \exp(-j\beta)$. The wave is not attenuated at all, but suffers a phase change by an angle $\beta$ in each section, where

$$\cos \beta = 1 + \mathbf{Z}_1/2\mathbf{Z}_2. \tag{9.4}$$

### (iii) $(\mathbf{Z}_1/4\mathbf{Z}_2) < -1$

$a$ is then real and negative, so that the wave is attenuated with a phase change of $\pi$ in successive sections. If we write $a = -e^{-\alpha}$, eqn (9.2) becomes

$$-\cosh \alpha = 1 + \mathbf{Z}_1/2\mathbf{Z}_2. \tag{9.5}$$

Here we have already spoken of the current as part of a 'wave' and this terminology needs some justification. If the current in one section alternates at a given frequency, so that we can write $I_n = I_n^0 \exp j\omega t$, then the current in a later section $(n+m)$ will be

$$I_{n+m} = a^m I_n^0 \exp j\omega t.$$

If we write $a = \exp(-\gamma)$, the current in the section $n+m$ will be

$$I_{n+m} = I_n^0 \exp(-\gamma m)\exp(j\omega t) = I_n^0 \exp(j\omega t - \gamma m),$$

which is similar to the expression for a wave motion in a continuous medium except that $m$, the number of sections, which can only be an integer, replaces $x$, the distance travelled in the medium. If $\gamma$ is real, write it as $\alpha$, and we have an attenuated wave, but if $\gamma$ is a pure imaginary quantity $j\beta$ we have an unattenuated wave with a phase change $\beta$ per section. The values of $\alpha$ and $\beta$ are given by eqns (9.3)–(9.5). The

FIG. 9.4. T-section ABCD terminated by its iterative impedance.

reciprocal roots are obtained by changing the sign of $\gamma$, corresponding to a wave travelling in the opposite direction. The third case, (*iii*) above, where the attenuated wave changes sign in successive sections, could not arise in a continuous medium. The behaviour of the filter is determined immediately from the value of $(\mathbf{Z}_1/4\mathbf{Z}_2)$. If it lies within the range 0 to $-1$, we have a 'pass band' with no attenuation; outside this range we have a 'stop band'.

The word 'section' has been used so far without any precise definition of its meaning. The uniform chain can be regarded as made up of similar sections joined together, but two types of section can be obtained by cutting the chain in different ways. These two types are shown in Figs 9.4 and 9.5 and for obvious reasons are known as T-sections and $\pi$-sections respectively. It is clear that a succession of either type of section joined together gives the same chain filter (see Fig. 9.6), so that the transmission characteristics derived earlier apply to either type of section.

If a generator of voltage $V$ is connected to a semi-infinite chain and the current $I$ drawn from it is measured, a definite value of the ratio $V/I$ is obtained, known as the 'characteristic' or 'iterative' impedance of the chain. It is obvious that the same value would be obtained if a number of sections were removed and the impedance of the semi-infinite chain measured at a later point. Again, the chain may be severed at some point and the infinite tail replaced by an impedance equal to the iterative impedance without altering the impedance measured at the input terminals. This gives a method of calculating the iterative impedance. Fig. 9.4

FIG. 9.5. $\pi$-section A'B'C'D' terminated by its iterative impedance.

FIG. 9.6. Filter of T-sections such as ABCD terminated by the iterative impedance $\mathbf{Z}_T$. It may also be divided into $\pi$-sections terminated by the network to the right of C'D'.

shows a T-section terminated by an impedance $\mathbf{Z}_T$; the impedance measured at the input terminals AB can be calculated by standard methods, and on equating this to $\mathbf{Z}_T$ we have

$$\mathbf{Z}_T = \tfrac{1}{2}\mathbf{Z}_1 + \left(\frac{1}{\mathbf{Z}_2} + \frac{1}{\tfrac{1}{2}\mathbf{Z}_1 + \mathbf{Z}_T}\right)^{-1}$$

Solution of this equation gives

$$\mathbf{Z}_T = (\mathbf{Z}_1\mathbf{Z}_2 + \tfrac{1}{4}\mathbf{Z}_1^2)^{\frac{1}{2}} = (\mathbf{Z}_1\mathbf{Z}_2)^{\frac{1}{2}}(1 + \mathbf{Z}_1/4\mathbf{Z}_2)^{\frac{1}{2}}. \qquad (9.6)$$

This is the value of the iterative impedance $\mathbf{Z}_T$ for a T-section. By applying the same method to a $\pi$-section, as shown in Fig. 9.5, we find the iterative impedance of a $\pi$-section to be

$$\mathbf{Z}_\pi = (\mathbf{Z}_1\mathbf{Z}_2)^{\frac{1}{2}}(1 + \mathbf{Z}_1/4\mathbf{Z}_2)^{-\frac{1}{2}}. \qquad (9.7)$$

Thus

$$\mathbf{Z}_T\mathbf{Z}_\pi = \mathbf{Z}_1\mathbf{Z}_2. \qquad (9.8)$$

The importance of the iterative impedance lies in the fact that sections of different kinds may be joined together to form a non-uniform filter, without disturbing a wave travelling down the filter, provided that they have the same iterative impedance. If sections with different iterative impedances are used a 'reflection' of the wave occurs at the junction and the simple filter theory is no longer applicable. Similarly, the filter must be terminated by a load equal to its iterative impedance in order to avoid a reflection at the output terminals. Any such reflection will diminish the power dissipated in the load, since part of the incident energy will be reflected.

A filter correctly terminated by the iterative impedance $\mathbf{Z}_T$ is shown in Fig. 9.6, where it is regarded as made up of T-sections (ABCD). Alternatively, we may regard it as made up of $\pi$-sections (A'B'C'D'), which must still be correctly terminated. Thus the impedance of the portion to the

FIG. 9.7. Half-sections used as transformer (a) converting $\mathbf{Z}_T$ to $\mathbf{Z}_\pi$; (b) $\mathbf{Z}_\pi$ to $\mathbf{Z}_T$.

right of C'D', shown separately in Fig. 9.7(a), must be $\mathbf{Z}_\pi$. Hence the half-section acts as a transformer from the impedance $\mathbf{Z}_T$ to $\mathbf{Z}_\pi$, as may also be verified by direct calculation. Similarly, the other half-section shown in Fig. 9.7(b) transforms the impedance $\mathbf{Z}_\pi$ to $\mathbf{Z}_T$. To a limited extent such half-sections can be used to match a load to a filter, but their especial importance comes in the design of composite filters (see Problem 9.4).

## 9.2. Some simple types of filter

*Low-pass filter*

The simplest type of low-pass filter is that shown in Fig. 9.2 with capacitances in the shunt arms and inductances in the series arms. Then $\mathbf{Z}_1 = j\omega L$ and $\mathbf{Z}_2 = 1/(j\omega C)$, giving

$$\mathbf{Z}_1/4\mathbf{Z}_2 = -\omega^2 LC/4.$$

The cut-off frequency $\omega_0$ is obtained by setting this equal to $-1$, giving $\omega_0 = 2(LC)^{-\frac{1}{2}}$. Frequencies below this are passed without attenuation. while higher frequencies are attenuated. In the pass band the phase change is given by

$$\cos \beta = 1 - \omega^2 LC/2 = 1 - 2(\omega/\omega_0)^2,$$

showing that $\beta$ changes from 0 to $\pi$ as the frequency increases from zero to the cut-off value. The characteristic impedance of a T-section is

$$\mathbf{Z}_T = \left(\frac{L}{C} - \frac{\omega^2 L^2}{4}\right)^{\frac{1}{2}} = \left(\frac{L}{C}\right)^{\frac{1}{2}}\left(1 - \frac{\omega^2}{\omega_0^2}\right)^{\frac{1}{2}},$$

which is resistive in the pass band, but varies from $(L/C)^{\frac{1}{2}}$ to zero at the cut-off frequency. In the stop band $\mathbf{Z}_T$ is a pure reactance and the attenuation is given by

$$-\cosh \alpha = 1 - \omega^2 LC/2,$$

FIG. 9.8. Variation of $\alpha$, $\beta$, and the real part of $\mathbf{Z}_T$ for a simple low-pass filter.

which rises with frequency from zero at $\omega = \omega_0$. The negative sign shows that currents in successive sections are reversed in direction.

The general behaviour of $\alpha$, $\beta$, and $\mathbf{Z}_T$ is illustrated in Fig. 9.8. The variation of $\mathbf{Z}_\pi$ is easily found from the relation

$$\mathbf{Z}_\pi = \mathbf{Z}_1\mathbf{Z}_2/\mathbf{Z}_T = (L/C)\mathbf{Z}_T^{-1}$$

As an example, we may take the problem of smoothing the output of a rectifier considered at the beginning of this chapter. Then $C = 10\ \mu\text{F}$, $L = 25\ \text{H}$, and the ripple frequency is 100 Hz. The attenuation per section at this frequency is given by

$$-\cosh \alpha = -\tfrac{1}{2}(e^\alpha + e^{-\alpha}) = 1 - \tfrac{1}{2}\omega^2 LC = (1 - 49 \cdot 3).$$

Hence $e^\alpha \approx 100$, or the attenuation in power is approximately $10^4$ per section. If we have $n$ such inductances and capacitances, it is very much better to join them as a ladder filter of $n$ sections than to lump them all into one section. The latter would reduce the ripple voltage by a factor $e^\alpha \approx \omega^2 n^2 LC = 100n^2$ approximately, while the ladder type filter will reduce it by $e^{n\alpha} \approx (100)^n$. Thus three sections would reduce the ripple voltage by $10^6$, while the single section with three inductances in series and three capacitances in parallel gives only a factor 900.

### The high-pass filter

As would be expected, the simplest high-pass filter is formed by putting capacitances in the series arms and inductances in the shunt arms, and a typical T-section is shown in Fig. 9.9.

Then $\mathbf{Z}_1 = 1/j\omega C$, $\mathbf{Z}_2 = j\omega L$, and

$$\mathbf{Z}_1/4\mathbf{Z}_2 = -1/4\omega^2 LC.$$

This lies between zero and $-1$ for $\omega \geqslant \tfrac{1}{2}(LC)^{-\frac{1}{2}}$, the latter being the critical frequency $\omega_0$. The phase change per section in the pass band is given by

$$\cos \beta = 1 - 1/2\omega^2 LC = 1 - 2(\omega_0/\omega)^2,$$

and the modulus of this expression gives also the value of $\cosh \alpha$ in the stop band.

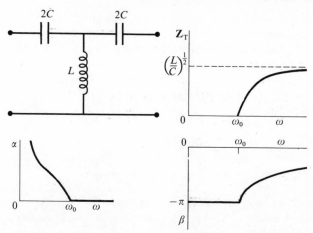

FIG. 9.9. T-section of simple high-pass filter, with variation of real part of $\mathbf{Z}_T$, $\alpha$, and $\beta$.

The iterative impedance is

$$\mathbf{Z}_T = \left(\frac{L}{C} - \frac{1}{4\omega^2 C^2}\right)^{\frac{1}{2}} = \left(\frac{L}{C}\right)^{\frac{1}{2}}\left(1 - \frac{\omega_0^2}{\omega^2}\right)^{\frac{1}{2}},$$

which is imaginary in the stop band, and rises from zero at the critical frequency to a limiting value of $(L/C)^{\frac{1}{2}}$ in the pass band. The behaviour of $\alpha$, $\beta$, and $\mathbf{Z}_T$ is illustrated in Fig. 9.9.

### Band-pass filters

A simple type of band-pass filter is shown in Fig. 9.10. Qualitatively its behaviour can be seen as follows. At very low frequencies the impedance

FIG. 9.10. Simple type of band-pass filter, showing T-section, and real part of $\mathbf{Z}_T$, $\alpha$, and $\beta$ as functions of frequency.

in the series arm will be dominated by the capacitance, so that the section will act as a simple capacitance-type attenuator (see Problem 9.1). At frequencies higher than the resonant frequency of the combination $L$ and $C_1$, $\omega L > (\omega C_1)^{-1}$ and the impedance of the series arm is inductive; then the section behaves as a low-pass filter, the highest frequencies again being stopped. Quantitatively, the analysis is

$$\mathbf{Z}_1 = j\omega L + 1/j\omega C_1,$$

$$\mathbf{Z}_2 = 1/j\omega C_2,$$

$$\mathbf{Z}_1/4\mathbf{Z}_2 = \frac{1}{4}\left(\frac{C_2}{C_1} - \omega^2 L C_2\right),$$

which is positive at zero frequency, but tends to $-\infty$ as $\omega \to \infty$. When it lies between 0 and $-1$, we have a pass band whose lowest frequency is

$$\omega_1 = (LC_1)^{-\frac{1}{2}}, \quad \text{where } \mathbf{Z}_1/4\mathbf{Z}_2 = 0,$$

and highest frequency is

$$\omega_2 = \left(\frac{C_2 + 4C_1}{LC_1C_2}\right)^{\frac{1}{2}}, \quad \text{where } \mathbf{Z}_1/4\mathbf{Z}_2 = -1.$$

The behaviour of $\alpha$, $\beta$, and $\mathbf{Z}_T$ is shown in Fig. 9.10.

*Disadvantages of the simple filter*

A simple filter, consisting of a chain of similar sections, suffers from two principal disadvantages:

(1) the iterative impedance varies with frequency in the pass band, making it impossible to terminate the filter correctly throughout the pass band;

(2) the attenuation in the stop band varies with frequency, being low near the cut-off frequencies.

FIG. 9.11. T-section and terminal half-section of an $m$-derived filter. The values must obey the equations $L_1 = mL$, $C_2 = mC$, $L_2 = L(1-m^2)/4m$.

FIG. 9.12. T-section of constant-$k$ band-pass filter ($L_2/C_1 = L_1/C_2 = k^2$).

These drawbacks can be reduced by using a composite filter, containing sections of different types, instead of a uniform filter. For example, the attenuation just above the cut-off frequency of a low-pass filter can be made high by using a section with a parallel tuned circuit in the series arm, or one with a series tuned circuit in the shunt arm (as in Fig. 9.11), adjusted to resonate at a frequency just above the cut-off frequency. If there were no resistive loss in the components, this would give infinite attenuation at the resonant frequency. Such a section will have low attenuation at the high frequencies, but it can be combined with a simple low-pass section so that the composite filter has a sufficiently high attenuation throughout the stop band. In such a composite filter each section must have the same iterative impedance at all frequencies, or reflections will occur at the junctions between sections. This can be achieved by the use of '$m$-derived' filters, an example of which is given in Problem 9.3. Half-sections may also be used to make the impedance more constant in the pass band (see Problem 9.4). The simple low-pass and high-pass filters considered earlier belong to the class of '$k$-derived' or 'constant-$k$' filters, since their impedances obey the relation

$$\mathbf{Z}_1\mathbf{Z}_2 = k^2,$$

where $k$ is a constant independent of frequency. The band-pass filter considered earlier is not of this type, but the T-section shown in Fig. 9.12 obeys this relation provided that $L_2/C_1 = L_1/C_2 = k^2$ (see Problem 9.2).

## 9.3. Travelling waves on loss-free transmission lines

In the electrical circuits considered hitherto we have been able to identify the circuit elements as inductances, resistances, capacitances, or combinations thereof, known as 'lumped impedances'. These elements are connected together and to generators and detectors by lengths of wire whose effect is assumed to be negligible. This is true only when the lengths of wire involved are very small compared with the wavelength of the radiation flowing along them. When this condition is not fullfilled, the

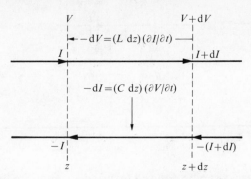

FIG. 9.13. Infinitesimal section of loss-less transmission line, showing current flow and voltage between conductors.

signal changes phase as it flows along the wires, in much the same way as in the pass band of a filter. In fact we may regard the wires as a limiting case of a low-pass filter where the elements are made infinitesimally small, but there are an infinite number of sections per unit length so that the 'distributed impedance' per unit length remains finite. Normally two wires are required to complete the circuit between any two pieces of apparatus, and we shall assume that these take the form either of two parallel wires, or two coaxial cylinders. In an element dz of two such conductors, as shown in Fig. 9.13, the voltage and current will be linearly related to one another, so that the voltage drop in a length dx will be

$$V-(V+dV)=-dV=(L\,dz)(\partial I/\partial t),$$

where $L$ is the series inductance associated with unit length of the two conductors. Since charge is conserved, $+dI$ is the change in the current flowing along the conductors in a distance dz, and the current flow between the conductors in this element is

$$-dI=(C\,dz)(\partial V/\partial t),$$

where $C$ is the shunt capacitance per unit length between the conductors. If there are no losses in the dielectric medium or in the conductors, the basic equations are simply

$$-(\partial V/\partial z)=L(\partial I/\partial t),\qquad -(\partial I/\partial z)=C(\partial V/\partial t). \qquad (9.9)$$

These form a pair of simultaneous differential equations which may be solved for either $I$ or $V$; for the latter we have

$$\frac{\partial^2 V}{\partial z^2}=-\frac{\partial}{\partial z}\left(L\frac{\partial I}{\partial t}\right)=-\frac{\partial}{\partial t}\left(L\frac{\partial I}{\partial z}\right)=LC\frac{\partial^2 V}{\partial t^2}. \qquad (9.10)$$

This is similar to eqn (8.17), and the solution is similar to (8.18):

$$V = F_1(z-vt)+F_2(z+vt). \tag{9.11a}$$

The corresponding solution for the current is

$$Z_0 I = F_1(z-vt)-F_2(z+vt) \tag{9.11b}$$

in which

$$v = (LC)^{-\frac{1}{2}} \tag{9.12}$$

and

$$Z_0 = (L/C)^{\frac{1}{2}}. \tag{9.13}$$

Here $Z_0$ is the 'characteristic impedance' of the line. It plays a role similar to the intrinsic impedance of a medium for unbounded electromagnetic waves, or the iterative impedance of a filter; it is equal to the limiting values of $Z_T$ and $Z_\pi$ for a low-pass filter in which the inductance and capacitance per section are allowed to go to zero, keeping their ratio constant.

For a pair of coaxial cylinders, radii $a$ and $b$ ($b>a$), we have

$$C = \frac{2\pi\epsilon}{\ln(b/a)} \qquad \text{(eqn (1.30)),}$$

$$L = \frac{\mu\,\ln(b/a)}{2\pi} \qquad \text{(eqn (5.18)),}$$

while for a pair of parallel cylinders radius $a$, separation $2d$ ($2d \gg a$)

$$C = \frac{\pi\epsilon}{\ln(2d/a)} \qquad \text{(eqn 2.56)),}$$

$$L = \frac{\mu\,\ln(2d/a)}{\pi} \qquad \text{(Problem 5.2).}$$

In each case the wave velocity is

$$v = (LC)^{-\frac{1}{2}} = (\mu\epsilon)^{-\frac{1}{2}} = c/(\mu_r\epsilon_r)^{\frac{1}{2}},$$

which is the same as for an electromagnetic wave in the medium when no conductors are present. This is a general result for a pair of parallel lossless conductors, whatever their shape (see below).

From the values of $L$ and $C$ for our two special types of line, we have:

$$\text{coaxial line:} \quad Z_0 = \frac{1}{2\pi}\left(\frac{\mu}{\epsilon}\right)^{\frac{1}{2}}\ln(b/a),$$

$$\text{parallel-wire line:} \quad Z_0 = \frac{1}{\pi}\left(\frac{\mu}{\epsilon}\right)^{\frac{1}{2}}\ln(2d/a).$$

Typical values of $Z_0$ are: for an air-spaced coaxial line with $b \approx 3a$, $Z_0 \approx 70\ \Omega$; for a parallel-wire line, $2d/a \approx 20$, and $Z_0 \approx 400\ \Omega$. In a coaxial cable a continuous dielectric is used to support the inner conductor, and the ratio $b/a$ is adjusted to make $Z_0$ some standard value, usually either $70\,\Omega$ or $50\,\Omega$.

An important quantity is the power flowing along the line. If we confine ourselves to a wave travelling along towards positive $z$, the energy flowing past any plane normal to the $z$ axis is, at any instant,

$$P = v(\tfrac{1}{2}LI^2 + \tfrac{1}{2}CV^2) = (\tfrac{1}{2}LI^2 + \tfrac{1}{2}CV^2)/(LC)^{\frac{1}{2}}$$
$$= \tfrac{1}{2}I^2 Z_0 + \tfrac{1}{2}V^2/Z_0 = I^2 Z_0 = V^2/Z_0 = IV. \tag{9.14}$$

Here we have given all the equivalent expressions for $P$, and we see that $Z_0$ behaves like a pure resistance, except that $P$ represents the energy flowing along the line per second rather than the energy dissipated as Joule heating of a real resistance. In a sense the energy stored in a section of the line is being dissipated, since it flows away from that section, and the stored energy would therefore diminish unless it were continually replaced by the energy flowing into it from the previous section. $P$ represents the energy crossing a given point in the line at any instant; the average flow of energy is found by using the r.m.s. values $\tilde{I}$ and $\tilde{V}$ in eqn (9.14). The fact that $Z_0$ is real shows that the phase difference between $V$ and $I$ is 0 or $\pi$, according to whether the wave is travelling towards positive or negative $x$. From eqn (9.14), $P$ has the same sign as $Z_0$, showing that the energy flows in opposite directions in the two waves, as we should expect. This result follows also from the direction of Poynting's vector (see Problem 9.5), since $V$ and $I$ are linearly related to the electric and magnetic field strengths **E** and **H**.

This treatment of parallel-wire transmission lines in terms of current and voltage provides a useful, if somewhat artificial, simplification of the mathematics. An alternative and more rigorous approach starts from Maxwell's equations or the wave equation. For two parallel conductors in which current flow is confined to one direction (the $z$-axis), the magnetic vector potential **A** has just the single component $A_z$. In the medium between the conductors this must satisfy eqn (8.64) with $\mathbf{J} = 0$; we write this in the form

$$\left\{ \left( \frac{\partial^2}{\partial x^2} + \frac{\partial^2}{\partial y^2} \right) + \left( \frac{\partial^2}{\partial z^2} - \frac{1}{v^2} \frac{\partial^2}{\partial t^2} \right) \right\} A_z = 0.$$

In the same region the electric potential $V$ must, from eqn (8.63), obey an identical equation. Each equation is satisfied if, separately,

$$\left( \frac{\partial^2}{\partial z^2} - \frac{1}{v^2} \frac{\partial^2}{\partial t^2} \right)(V, A_z) = 0, \tag{9.15a}$$

$$\left( \frac{\partial^2}{\partial x^2} + \frac{\partial^2}{\partial y^2} \right)(V, A_z) = 0. \tag{9.15b}$$

These have solutions of the form

$$V = \phi_1(x, y)F_1(z - vt) + \phi_2(x, y)F_2(z + vt),$$
$$vA_z = \phi_1(x, y)F_1(z - vt) - \phi_2(x, y)F_2(z + vt),$$

(9.16)

where the relation between $V$ and $A_z$ satisfies eqn (8.62) with $v = (\mu\epsilon)^{-\frac{1}{2}}$.

These relations show that $V$ and $A_z$ have the same variation in the $x$, $y$ plane. For a wave travelling, say, in the positive $z$-direction, each varies as $\phi_1(x, y)$, which is a solution of the time-independent eqn (9.15b) and the same as in the static case. Whatever the cross-sectional shape of the conductors, it can be shown that this leads to the relation $LC = \mu\epsilon$ for the inductance and capacitance per unit length (cf. § 5.2 and Problem 5.2). Such solutions are valid provided that (1) the current flow is unidirectional, and in the direction of propagation; (2) the conductivity of the cylinders is infinite. This second condition follows from the fact that the solutions for $V$ and $A_z$ lead, through eqn (8.61), to

$$E_z = -\partial A_z/\partial t - \partial V/\partial z = 0,$$

a condition that cannot be fullfilled if the conductors have a finite resistivity and current flows in the $z$ direction.

## 9.4. Terminated loss-free lines

Hitherto we have regarded the two waves travelling in opposite directions along a line as two quite independent solutions of the wave equation. Usually, however, there is only one generator attached to the line, producing a wave travelling, say, towards positive $x$. If the line is infinite in this direction, there will be no return wave. If the line is terminated in some way, a reflection may occur at the termination and this will generate a return wave, which will not be independent of the incident wave. If the latter has a given frequency, the return wave must have the same frequency in order that the ratio of current to voltage at the termination (assumed to consist of some constant impedance) shall be independent of the time. If $Z_L$ is the load impedance (as in Fig. 9.14), the fact that $V/I$ must equal $Z_L$ at this point for all values of the

FIG. 9.14. Line terminated by impedance $Z$ at $z = 0$.

time constitutes the boundary condition, from which one can calculate the magnitude and phase of the reflected wave relative to the incident wave.

If the terminating impedance is not a pure resistance, its value depends on the frequency, and so also will the reflection coefficient. We must therefore assume a wave of a given angular frequency $\omega$. (If the wave is not purely sinusoidal, we must perform a Fourier analysis and treat each harmonic separately.) The equations for current and voltage may then be written as complex exponentials, where, as always, the real or imaginary part must be extracted at the end, according to whether the input voltage is a cosine or sine function. Eqns (9.11) become

$$
\begin{aligned}
V &= A \exp j\omega(t-z/v) + A' \exp j\omega(t+x/v) \\
Z_0 I &= A \exp j\omega(t-z/v) - A' \exp j\omega(t+z/v)
\end{aligned}
\tag{9.17}
$$

For simplicity we shall take the termination to be at the origin of coordinates $z = 0$, noting that all points on the line will then have negative values of $z$. Inserting the boundary condition, we have

$$
\left(\frac{V}{Z_0 I}\right)_{z=0} = \frac{\mathbf{Z}_L}{Z_0} = \frac{A \exp j\omega t + A' \exp j\omega t}{A \exp j\omega t - A' \exp j\omega t} = \frac{(A+A')}{(A-A')}.
$$

Hence

$$
A'/A = (\mathbf{Z}_L - Z_0)/(\mathbf{Z}_L + Z_0),
\tag{9.18}
$$

which defines the reflection coefficient $A'/A$. If $\mathbf{Z}_L$ is complex, $A'$ will be complex (assuming $A$ real, which can always be made true by choosing the zero of the time-scale correctly), showing that there is a phase change in the reflected wave. If $\mathbf{Z}_L$ is a pure resistance, $A'$ is real and there is no phase change (we may exclude a phase change of $\pi$ by allowing negative values of $A'$).

If the line is open-circuited, the voltage at the end is $2A$, since $A' = A$. If $\mathbf{Z}_L$ is finite, the voltage across it is $A + A' = 2A\mathbf{Z}_L/(\mathbf{Z}_L + Z_0)$, showing that the line behaves as a generator of voltage $2A$, with an internal resistance $Z_0$. The power transferred to the load, if $\mathbf{Z}_L$ is a resistance $R$, is $P = \tilde{V}^2/R = 2A^2R/(R+Z_0)^2$. If $\mathbf{Z}_L = R = Z_0$, the power transferred to the load is a maximum; that is, the load is matched to the generator. Reference to eqn (9.18) above shows that under this condition $A' = 0$. Thus maximum power transfer to the load corresponds to no reflected wave; hence the matching condition has a simple physical meaning, since any reflected wave would carry energy away from the load and reduce the power dissipated in it. The power in the incident wave is $\frac{1}{2}A^2/Z_0$, and that in the reflected wave is $\frac{1}{2}A'^2/Z_0$. The difference will be found to equal the power dissipated in the load, as calculated above.

Examination of the formula for $A'/A$, when $\mathbf{Z}_L$ is a pure resistance $R$, shows that as $R$ goes from zero to infinity, $A'/A$ changes continuously from $-1$ to $+1$. When $R < Z_0$, the reflected voltage wave has opposite sign from the incident wave, while the reflected current wave has the same sign. The voltage across $R$ is then less than $A$, and the current through it greater than $A/Z_0$. The reverse is true if $R > Z_0$.

When $\mathbf{Z}_L$ is a pure reactance $jX$, $A'/A$ is complex and its modulus is unity. This is to be expected, since no energy is dissipated in a pure reactance, and the amplitude of the reflected wave must therefore equal that of the incident wave. We may write the reflection coefficient $A'/A$ in this case as $\exp j\delta$, and an algebraic reduction shows that

$$\exp j\delta = (X^2 - Z_0^2 + 2jXZ_0)/(X^2 + Z_0^2).$$

Hence $\tan \delta = 2XZ_0/(X^2 - Z_0^2)$, which may be written in the more convenient form $\tan \frac{1}{2}\delta = Z_0/X$.

In the general case, when $\mathbf{Z}_L$ is complex, the formulae are rather cumbersome, but may be reduced somewhat by writing $\mathbf{Z}_L = Z_1 \exp j\phi$, $A' = A_1 \exp j\delta$. Then, on clearing imaginary terms from the denominator, we find

$$\frac{A_1}{A} \exp j\delta = \frac{Z_1^2 - Z_0^2 + 2Z_1 Z_0 \sin \phi}{Z_1^2 + Z_0^2 + 2Z_1 Z_0 \cos \phi},$$

giving

$$\left.\begin{aligned}\frac{A_1}{A} &= \left(\frac{Z_1^2 + Z_0^2 - 2Z_1 Z_0 \cos \phi}{Z_1^2 + Z_0^2 + 2Z_1 Z_0 \cos \phi}\right)^{\frac{1}{2}} \\[4pt] \tan \delta &= 2Z_1 Z_0 \sin \phi/(Z_1^2 - Z_0^2).\end{aligned}\right\} \tag{9.19}$$

and

The maximum voltage on the line is $A + A_1$, and occurs at a voltage anti-node where the incident and reflected voltages are in phase; the minimum is $A - A_1$ at a node where they are in anti-phase. The ratio $(A + A_1)/(A - A_1)$ is called the voltage standing wave ratio (v.s.w.r.), and measurement of it together with the position of the node (which is related to $\delta$) form the basis of a method of measuring an unknown impedance $\mathbf{Z}_L$ at short wavelengths (see § 22.4).

*Transmission line terminated by another line of different impedance*

A special case of a terminated line is one with impedance $Z_1$ joined to another line of different characteristic impedance $Z_2$. In this case there will be an incident and a reflected wave on the first line, and a transmitted wave on the second line. Our equations are then:

first line:

$$V_1 = A \exp j\omega(t - z/v_1) + A' \exp j\omega(t + z/v_1),$$

$$Z_1 I_1 = A \exp j\omega(t - z/v_1) - A' \exp j\omega(t + z/v_1);$$

second line:

$$V_2 = A'' \exp j\omega(t - z/v_2),$$

$$Z_2 I_2 = A'' \exp j\omega(t - z/v_2),$$

where $v_1$, $v_2$ are the wave velocities on the first and second lines respectively. If the junction is at $z = 0$, the voltage and current at this point must be the same on the two lines, so that

$$A + A' = A'',$$

$$(A - A')/Z_1 = A''/Z_2,$$

the solutions of which are

$$A'/A = (Z_2 - Z_1)/(Z_2 + Z_1), \qquad A''/A = 2Z_2/(Z_2 + Z_1). \qquad (9.20)$$

The former of these is the same as for a line terminated by a resistance $Z_2$, as we would expect. The reflected power is the same as in the case of a real resistance $Z_2$, and the transmitted power the same as would be dissipated in a real resistance. Note that if $Z_2 > Z_1$, the voltage on the second line is greater than that in the incident wave. The power is less, however, since the characteristic impedance is higher.

Eqns (9.20) are identical with the formulae for reflection and transmission of a plane electromagnetic wave at normal incidence at the boundary of two media (eqns (8.41), (8.42), (8.49), and (8.50)). This shows that there is a close analogy between the characteristic impedance of a transmission line and the intrinsic impedance of a medium transmitting an electromagnetic wave. The similarity appears also in the expressions for the power transmitted: in a plane wave we have (power transmitted across unit area) $= E_x^2/Z_0 = Z_0 H_y^2$ (see § 8.3), while for a transmission line $P = V^2/Z_0 = Z_0 I^2$ (eqn (9.14)). The analogous behaviour makes it possible to adapt many of the formulae derived below to the case of plane waves.

*Input impedance of terminated lines*

When the reflection coefficient due to the load $\mathbf{Z}_L$ used to terminate a line is known, it is a simple matter to calculate the current and voltage, and hence the effective impedance, at any point in the line. This can be done for an arbitrary load $\mathbf{Z}_L$, but we shall limit ourselves to a few of the simpler and more interesting cases.

For a short-circuited line, $\mathbf{Z}_L = 0$ and $A' = -A$ in eqns (9.18), so that at a point $z = -l$ on the line (that is, at the terminals A, B in Fig. 9.14)

$$V = A\{\exp j\omega(t + l/v) - \exp j\omega(t - l/v)\}$$

$$= j2A \exp j\omega t \sin \omega l/v$$

$$= j2A \exp j\omega t \sin 2\pi l/\lambda$$

and

$$Z_0 I = 2A \exp j\omega t \cos 2\pi l/\lambda.$$

The impedance at this point is then

$$V/I = j Z_0 \tan 2\pi l/\lambda. \tag{9.21a}$$

This formula shows that a section of short-circuited line behaves as a pure reactance. If $l$ is less than a quarter of a wavelength, then $\tan 2\pi l/\lambda$ is positive and the line behaves like an inductance. If $l$ lies between a quarter- and a half-wavelength, the tangent is negative and the line behaves like a capacitance. These statements hold also if we increase $l$ by an integral number of half-wavelengths.

If the line is open-circuited, $A' = +A$, and the equations for $V$ and $Z_0 I$ are just interchanged. The impedance at $z = -l$ is therefore

$$V/I = -j Z_0 \cot 2\pi l/\lambda. \tag{9.21b}$$

An open-circuited line, less than a quarter-wavelength long, therefore behaves like a capacitance; if its length lies between a quarter- and a half-wavelength, it behaves like an inductance. If its length is exactly a quarter-wavelength its impedance is zero. Thus if we have an open-circuited line we can cut off a quarter-wavelength and replace it by a short circuit without affecting the conditions earlier on the line. For we then have a short-circuited line of length $(l-\frac{1}{4}\lambda)$, so the impedance at the input terminals becomes $+j Z_0 \tan(2\pi l/\lambda -\frac{1}{2}\pi) = -j Z_0 \cot 2\pi l/\lambda$, in agreement with the value found directly from eqn (9.21b).

These results show that a lumped reactance $X$ at the end of a line can be replaced by a suitable additional length of line, either open- or short-circuited. We found earlier that the wave reflected by a reactance has the same amplitude as the incident wave, but a phase change $\delta$, where $\tan \frac{1}{2}\delta = Z_0/X$. If the reactance is replaced by an open- or short-circuited length of line, then the wave reflected from the far end will have the same amplitude as the incident wave, but the phase change arises from the time taken by the wave to travel the extra distance to the end of the line and back. At metre wavelengths, suitable lengths of either coaxial or parallel-wire lines are commonly used as inductances because they have a lower resistance than a coil of wire of the same 'nominal' radio-frequency resistance. By 'nominal' radio-frequency resistance is meant the value which would be calculated from the skin depth for a straight wire. In a closely wound coil there is an additional energy loss because of eddy currents induced by the oscillatory currents in neighbouring turns; this is known as the 'proximity effect', and increases the effective radio-frequency resistance. Its effect is minimized by using straight wires, as in a section of a transmission line.

### The transmission line as a transformer

Since the voltage and current are in different ratio at the input terminals of a line from the ratio they bear at the output terminals, it follows that a line can be used as an impedance transformer. The case of greatest interest is that of a line a quarter-wavelength long. If the terminating impedance $\mathbf{Z}_L$ is at $z = 0$, then at the point $z = -\frac{1}{4}\lambda$, the voltage and current are

$$V = A \exp j(\omega t + \tfrac{1}{2}\pi) + A' \exp j(\omega t - \tfrac{1}{2}\pi) = j(A - A')\exp j\omega t,$$
$$Z_0 I = j(A + A')\exp j\omega t,$$

and

$$V/I = Z_0(A - A')/(A + A') = Z_0^2/\mathbf{Z}_L, \qquad (9.22)$$

showing that the terminating impedance has been transformed to $Z_0^2/\mathbf{Z}_L$. In this respect a quarter-wave behaves like a tuned circuit, which transforms a series resistance $R$ into a parallel resistance $L/CR$. Since $Z_0^2 = L/C$, the formulae are similar in the two cases.

The quarter-wave transformer may be used to match a resistive load $Z_L$ to a line of impedance $Z_0$ by inserting immediately before the load a $\frac{1}{4}\lambda$ section of line whose impedance $Z_1$ is such that $Z_1^2/Z_L = Z_0$. If $Z_L$ and $Z_1$ are lines of the same dimensions but in different dielectric media, the situation is exactly analogous to the 'blooming' of optical lenses. The fraction of the incident intensity reflected from an air–glass surface is about 4 per cent, and the loss of light in an optical system with 10 or 20 surfaces is serious. The reflection may be reduced by depositing on the surface a quarter-wave thick layer of material of low refractive index, ideally equal to the square-root of the refractive index of the glass. The thickness is adjusted to be correct for the middle of the optical region, and is thus not quite correct for the ends of this region. 'Bloomed' surfaces appear slightly purple, therefore, owing to reflection of the red and blue rays. Quarter-wave films are also used to produce highly reflecting layers (see Problem 9.8). Further details of optical applications are given by Lissberger (1970).

The half-wave transformer is also of interest. It may be regarded as two consecutive quarter-wave transformers, giving an impedance

$$Z_0^2/(Z_0^2/\mathbf{Z}_L) = \mathbf{Z}_L.$$

Alternatively, this result may be obtained directly, since if we move one half-wave along a line, all voltages and currents are the same except for their reversed sign. The half-wave line is therefore a $1:1$ transformer. A typical use is that of a connecting link between two pieces of apparatus, which makes the impedance of either appear unchanged. This is often useful at very short wavelengths where connecting wires sufficiently short to give no impedance transformation are not practicable.

## 9.5. Lossy lines and resonant lines

When loss is present in the conductors or in the medium the approach used above in terms of current and voltage is not rigorous, but it provides a method of analysis which is simple and sufficiently accurate for most purposes when the losses are small. In Fig. 9.15, the distributed impedance $\mathbf{Z}$ and admittance $\mathbf{Y}$ per unit length are assumed to be complex, giving the relations

$$-\partial V/\partial z = I\mathbf{Z}, \qquad -\partial I/\partial z = V\mathbf{Y}. \tag{9.23a}$$

Suppose that waves are propagated as

$$\exp j(\omega t - \gamma z),$$

where $\gamma$ is complex; as in Chapter 8, we must here consider just one frequency because we shall find that the propagation constants are not independent of frequency. Eqns (9.23a) become

$$\left.\begin{aligned} V &= I\mathbf{Z} = I(R+j\omega L), \\ I &= V\mathbf{Y} = V(G+j\omega C). \end{aligned}\right\} \tag{9.23b}$$

Elimination of $I$ or $V$ gives

$$-\gamma^2 = -(R+j\omega L)(G+j\omega C) = (LC\omega^2 - RG) - j\omega(RC+GL),$$

and on writing $\gamma = \alpha + j\beta$ we obtain

$$\beta^2 - \alpha^2 = LC\omega^2 - RG, \qquad 2\beta\alpha = \omega(RC+GL).$$

These equations show that the phase constant $\beta$ is altered by the presence of loss, and that the wave is attenuated because $\alpha \neq 0$. Both the phase velocity and the attenuation are frequency-dependent.

FIG. 9.15. Infinitesimal section of transmission line, showing current and voltage between the two conductors. With loss, $\mathbf{Z} = R + j\omega L$ and $\mathbf{Y} = G + j\omega C$.

These equations may be solved exactly for $\alpha$ and $\beta$, but it is more instructive to solve them approximately, assuming $R$, $G$ to be small. Then

$$\beta = \omega(LC)^{\frac{1}{2}}\left\{1 + \frac{1}{8\omega^2}\left(\frac{G}{C} - \frac{R}{L}\right)^2\right\}, \qquad \alpha = \frac{1}{2}\left\{R\left(\frac{C}{L}\right)^{\frac{1}{2}} + G\left(\frac{L}{C}\right)^{\frac{1}{2}}\right\}. \quad (9.24)$$

The velocity is approximately

$$\frac{\omega}{\beta} = \frac{1}{(LC)^{\frac{1}{2}}}\left\{1 - \frac{1}{8\omega^2}\left(\frac{G}{C} - \frac{R}{L}\right)^2\right\}, \qquad\qquad (9.25)$$

showing that it is altered only in the second order. The power flowing along the line decays as it travels along as (using $Z_0 = (L/C)^{\frac{1}{2}}$)

$$\exp(-2\alpha z) = \exp\{-(R/Z_0 + GZ_0)z\}.$$

The two terms in the exponential represent just the fraction of the stored energy which is dissipated per unit length in the resistance and conductance respectively. We see that if $R/Z_0$ and $GZ_0$ are both small, the line is distortionless in the first approximation, since neither the velocity of the wave nor its attenuation depend on the frequency to this order. In the next approximation, distortion arises from the change in velocity with frequency and this is most serious at low frequencies. At high frequencies (owing to skin effect) the resistance rises, increasing $\alpha$. If $G$ is negligible, distortion may be reduced by increasing $L$. On telephone land lines this is accomplished by introducing inductances in series with the line at regular intervals. This also reduces the attenuation, since it increases $Z_0$, but gives the line a periodic structure so that it behaves like a low-pass filter. The cut-off frequency must be kept above the audio-frequency range, and to do this the spacing of the inductances must be small compared with the shortest wavelength which it is required to transmit. The distortion can also be greatly reduced by transmitting a voice-modulated signal of, say, 100 kHz frequency, instead of the actual voice frequency range of 100 Hz to 10 kHz. Although the bandwidth required is the same, the fractional change in frequency involved is very much smaller.

In the laboratory, the lengths of line used are so short that attenuation is negligible except at the highest frequencies, where $R$ rises owing to the skin effect. To obtain a numerical value, let us take an air-spaced coaxial line at a frequency of $3 \times 10^9$ Hz ($\lambda = 0 \cdot 1$ m). Then for copper the skin depth $\delta$ is about $1 \cdot 2 \times 10^{-6}$ m. For a conductor whose dimensions are $a = 2 \cdot 5$ mm, $b = 8$ mm (giving a 70 $\Omega$ line),

$$R = \rho\left(\frac{1}{2\pi a\delta} + \frac{1}{2\pi b\delta}\right) = 1 \cdot 2 \ \Omega \ \text{m}^{-1}.$$

Then $\alpha = 8 \times 10^{-3}$ m$^{-1}$, and the power transmitted along the line will fall by a factor of $1/e$ in a distance of 60 m.

If the space between the conductors is filled with a dielectric, the attenuation due to dielectric loss will be important unless the dielectric is of the highest quality. If its loss tangent is $\tan \delta$, then $G = \omega C \tan \delta$, and

$$2\alpha = GZ_0 = \omega(LC)^{\frac{1}{2}} \tan \delta = (\omega/v)\tan \delta = (2\pi/\lambda_1)\tan \delta,$$

where $\lambda_1$ is the wavelength in the dielectric. If its relative permittivity is 2·2, and $\tan \delta = 2\times10^{-4}$, $\alpha = 9\times10^{-3}$ m$^{-1}$ at $\lambda = 0\cdot1$ m, which is as large as that due to the resistance. For this reason high-frequency cables are often made with some device such as an open spiral of polythene string supporting the centre conductor in order to reduce the amount of dielectric in the position of maximum electric field.

In calculating the loss on the line we have taken no account of energy lost by radiation, though we might expect each element of the conductors to radiate since it carries an alternating current. A transmission line has two conductors carrying equal and opposite currents, however, and in computing the radiation we must allow for the destructive interference between their two radiation patterns. For a parallel-wire line this does not give an exact null, but the maximum phase difference between the signals from the two wires in any direction will differ from $\pi$ at most by $2\pi(2d)/\lambda$, where $2d$ is the separation between the two conductors. Hence the radiated energy is less than that from a single wire by a factor of the order $(d/\lambda)^2$, and is small if $d \ll \lambda$. The coaxial line gives an exact null because the one current entirely encloses the other, and the magnetic field at any external point is zero. For this reason coaxial lines are to be preferred at wavelengths less than about a metre.

*Transmission lines as tuned circuits*

If we have a quarter-wave section of loss-free line, short-circuited at one end, then the impedance measured at the other end is infinite. Similarly, if it is open-circuited at one end, then the input impedance at the other end is zero. We have already seen that if a resistance $R$ is connected across one end, the impedance at the other end is

$$Z_0^2/R = L/CR.$$

Thus in all respects the section behaves like a tuned circuit. If $R$ is zero or infinity, the input impedance will only be infinity or zero so long as the line is completely loss-free. This, of course, will never occur in practice, and to assess the properties of the section as a tuned circuit we must include the effect of distributed losses. We can do this by bringing in the attenuation constant $\alpha$.

We assume that the section is open-circuited at the far end, so that $A' = A$. Then the voltage and current at a point $z = -l$ are (since $A$ is the incident voltage amplitude at $z = 0$)

$$V = A\left[\exp\left\{j\left(\omega t + \frac{2\pi l}{\lambda}\right) + \alpha l\right\} + \exp\left\{j\left(\omega t - \frac{2\pi l}{\lambda}\right) - \alpha l\right\}\right],$$

$$Z_0 I = A\left[\exp\left\{j\left(\omega t + \frac{2\pi l}{\lambda}\right) + \alpha l\right\} - \exp\left\{j\left(\omega t - \frac{2\pi l}{\lambda}\right) - \alpha l\right\}\right].$$

The impedance $\mathbf{Z}$ at this point is then given by

$$\frac{\mathbf{Z}}{Z_0} = \frac{1 + \exp\{-j4\pi l/\lambda - 2\alpha l\}}{1 - \exp\{-j4\pi l/\lambda - 2\alpha l\}}. \tag{9.26}$$

We now assume that the length of the line $l$ is close to an odd multiple of a quarter-wavelength, and examine how the impedance changes in the neighbourhood of this point. In the complex exponentials of eqn (9.26) the imaginary part of the argument gives a rapid variation and the real part (which is assumed to be small) a slow variation. We therefore treat them separately, and write $l = \{(2n+1)\lambda/4\} + \Delta l$ in the imaginary part only. Then the exponential becomes

$$\exp\{-j(2n+1)\pi - j4\pi\Delta l/\lambda - 2\alpha l\} = -\exp(-j4\pi\Delta l/\lambda - 2\alpha l)$$
$$\approx -(1 - j4\pi\Delta l/\lambda - 2\alpha l),$$

where we have assumed that both $\Delta l/\lambda$ and $\alpha l$ are small. Thus the impedance becomes

$$\mathbf{Z} = Z_0\frac{1 - (1 - j4\pi\Delta l/\lambda - 2\alpha l)}{1 + (1 - j4\pi\Delta l/\lambda - 2\alpha l)} \approx Z_0(\alpha l + j2\pi\Delta l/\lambda) = Z_0\alpha l\left(1 + j\frac{2\pi}{\alpha\lambda}\frac{\Delta l}{l}\right), \tag{9.27}$$

where only small quantities of the first order have been retained. This equation is of the same form as eqn (7.20) for a series tuned circuit near resonance, which is

$$\mathbf{Z} = R'(1 + j2Q_F\Delta\omega/\omega_0)$$

since $\Delta l/l = -\Delta\lambda/\lambda = \Delta\omega/\omega_0$ (here the minus sign is introduced in relating $\Delta l/l$ to $\Delta\lambda/\lambda$ because increasing the length of the line has the same effect as shortening the wavelength of the applied radiation). Hence at resonance the quarter-wavelength line (or a line an odd multiple of this length) behaves as a series tuned circuit with a resistance $Z_0\alpha l$ and a quality factor $Q_F = \pi/\alpha\lambda$.

In making measurements at short wavelengths it is always advisable to keep the generator frequency constant, if possible, and vary the element under test, since this avoids errors due to detectors, connecting lines, etc., being frequency sensitive. Eqn (9.27) shows that we may conveniently measure $Q_F$ by finding the fractional change in length of the line ($\Delta l/l$) required to move between the points at which the impedance rises to $\sqrt{2}$ of its minimum value.

To find a numerical value for $Q_F$ we take the value $\alpha = 8 \times 10^{-3} \, \text{m}^{-1}$ found earlier for an air-spaced coaxial line at $0 \cdot 1$ m wavelength. This gives $Q_F = 4000$, which is very much higher than can be obtained normally with a lumped circuit at medium radio-frequencies. $Q_F$ is independent of the number of quarter-wavelengths in the section, but the series resistance $R' = Z_0 \alpha l = 1 \cdot 4 \times 10^{-2}(2n+1) \, \Omega$. Thus $R'$ increases when we make $n$ larger, but $Q_F$ does not change. This is because $Q_F$ depends on the ratio of the stored energy to the energy dissipated, and both of these increase as the length of the line increases.

The impedance of a short-circuited quarter-wave line can readily be calculated from the above, since in this case $A' = -A$, and the formulae are the same, if we interchange $V$ and $Z_0 I$. This gives

$$\frac{I}{V} = \mathbf{Y} = \frac{\alpha l}{Z_0}\left(1+\mathrm{j}\frac{2\pi}{\alpha\lambda}\frac{\Delta l}{l}\right), \qquad (9.28)$$

showing that at resonance the impedance is $Z_0/\alpha l = Z_0^2/R' = L/CR'$, the same expression as for a parallel tuned circuit. For a single quarter-wavelength of the coaxial line considered previously, the parallel impedance is $350 \, \text{k}\Omega$, showing that the line makes a good load for an oscillator or amplifier. In practice the line may be rather shorter than a quarter-wavelength, so that it behaves as an inductance which can be adjusted to resonate with the stray capacitance of the circuit.

### 9.6. Guided waves—propagation between two parallel conducting planes

When an electromagnetic wave is launched from an aerial into free space (or a non-conducting medium) its amplitude falls off inversely with the distance owing to the spreading out of the wave in a spherical wave front. There is no dissipation of energy, but the power flowing through unit area normal to the wave front falls off according to the inverse square law. On the other hand, a wave sent along a coaxial line suffers no diminution in amplitude, apart from that due to resistive losses, because it is confined to the space between the conductors and does not spread out. Such a wave is a guided wave, and in the absence of loss it was shown that with two parallel conductors a solution of Maxwell's equations can be found giving a purely transverse wave (no components of $\mathbf{E}$ or $\mathbf{H}$ in the

FIG. 9.16. Coordinate system for propagation between two parallel planes.

direction of propagation) which is freely propagated at all frequencies with the same velocity as a wave in the unbounded medium. With a single hollow conductor this is no longer the case, though the fact that it is possible to see down a metal tube shows that some form of electromagnetic wave can be propagated through it. Such a wave is again a guided wave, since it must move in the direction of the tube. As a preliminary to studying propagation through such a tube (known as a 'waveguide'), we shall investigate the problem of propagation between two parallel infinite perfectly conducting planes, separated by a distance $c$, as shown in Fig. 9.16.

A Cartesian coordinate system may be defined by taking the conductors to be the planes $z = 0$ and $z = c$, and assuming that the wave is propagated parallel to the $x$-axis, which is normal to the plane of the paper. The boundary conditions now demand that any components of the electric field strength ($E_x$ or $E_y$) tangential to the planes must vanish at the planes $z = 0$ and $z = c$. We try to find the simplest possible solution of Maxwell's equations consistent with these demands, and begin by assuming a purely transverse plane wave in which both $E_x$ and $H_x$ are zero, such as we would have in the absence of the conductor. If this wave is polarized with its electric vector normal to the planes (that is, $E_y = 0$) it is readily seen that the boundary conditions are satisfied automatically, and a wave of this polarization is possible. The solutions are of the same form as for a wave in the unbounded medium, and the velocity is also the same. On the other hand, a plane wave in which the field components do not vary with $z$ is obviously impossible if the electric vector is parallel to the planes, since $E_y$ must be zero at $z = 0$ and $z = c$ and will be zero everywhere unless we allow it to vary with $z$. We therefore examine whether it is possible to have a wave in which $E_y$ is finite, but $E_z$ and $E_x$ are both zero, so that the electric field is purely transverse. The field components of **H** can then be computed from Maxwell's equations. We shall assume that the wave is propagated as $\exp(j\omega t - \gamma x)$, so that we can replace differentiation with respect to $t$ and $x$ by multiplication by $j\omega$ and $-\gamma$ respectively. Then, if the medium between the planes has permittivity $\epsilon$, magnetic permeability $\mu$, and zero conductivity, the curl eqns (8.3) and

(8.7) give the following components:

$$\left.\begin{array}{l} -j\omega\mu H_x = -\partial E_y/\partial z, \\ -j\omega\mu H_y = 0, \\ -j\omega\mu H_z = -\gamma E_y, \end{array}\right\} \qquad (9.29)$$

$$\left.\begin{array}{l} 0 = \partial H_z/\partial y - \partial H_y/\partial z, \\ j\omega\epsilon E_y = \partial H_x/\partial z + \gamma H_z, \\ 0 = -\gamma H_y - \partial H_x/\partial y, \end{array}\right\} \qquad (9.30)$$

where we have already assumed $E_x = E_z = 0$. Eqns (9.29) show immediately that $H_y = 0$, but $H_x$ cannot be zero unless $\partial E_y/\partial z$ is zero, and this is not allowed by the boundary conditions. Hence the wave will not be purely transverse, but will have a component of **H** in the direction of propagation. On putting $H_y = 0$ in eqns (9.30), we see that $\partial H_z/\partial y$ and $\partial H_x/\partial y$ are both zero, so that there is no variation in the y direction, and examination of the components of div $\mathbf{E} = 0$ shows this to be true also of $E_y$. The remaining components of eqns (9.29) and (9.30) therefore reduce to

$$\left.\begin{array}{l} j\omega\mu H_x = \partial E_y/\partial z, \\ j\omega\mu H_z = \gamma E_y, \\ j\omega\epsilon E_y = \partial H_x/\partial z + \gamma H_z. \end{array}\right\} \qquad (9.31)$$

Elimination of $H_x$ and $H_z$ between these three equations gives

$$-\omega^2\mu\epsilon E_y = \partial^2 E_y/\partial z^2 + \gamma^2 E_y$$

or

$$\partial^2 E_y/\partial z^2 = -(\gamma^2 + \omega^2/v^2)E_y, \qquad (9.32)$$

where $v = (\mu\epsilon)^{-\frac{1}{2}}$ is the velocity of an electromagnetic wave in the unbounded medium (in the absence of the conducting planes we could put $\partial/\partial z = 0$ and obtain this result directly from eqn (9.32), since $\gamma$ must then be an imaginary quantity $j\beta$, and $v = \omega/\beta$). The solution of eqn (9.32) can be written in the form

$$E_y = A \sin(\pi n z/c) + B \cos(\pi n z/c),$$

and the boundary conditions $E_y = 0$ at $z = 0$ and $z = c$ require that $B = 0$ and $n$ must be an integer. To satisfy eqn (9.32) we must have

$$(n\pi/c)^2 = \gamma^2 + \omega^2/v^2$$

and hence

$$\gamma = \left\{ \left(\frac{n\pi}{c}\right)^2 - \left(\frac{\omega}{v}\right)^2 \right\}^{\frac{1}{2}}. \qquad (9.33)$$

Here $\gamma$ is either real or purely imaginary according to whether the quantity inside the square-root is positive or negative. It is clear that at low frequencies $\gamma$ will be real, and the wave will then be attenuated. At sufficiently high frequencies $\gamma$ will be imaginary, and waves will be freely transmitted without attenuation; thus the system acts as a high-pass filter. The condition for free transmission of waves is that

$$\omega/v > n\pi/c.$$

Since $\omega/v = 2\pi/\lambda_0$, where $\lambda_0$ is the wavelength of the radiation in the unbounded medium, this condition may be written in the form

$$\lambda_0 < 2c/n.$$

Hence $2c/n$ is the cut-off wavelength $\lambda_c$, and only radiation of shorter wavelength is freely transmitted. In the pass band we may write $\gamma = j\beta = j(2\pi/\lambda_g)$, where $\lambda_g$ is the apparent wavelength of the radiation in the guide, that is, the distance between successive points along the $x$-axis where the phase differs by $2\pi$. Eqn (9.33) then reduces to

$$\frac{f^2}{v^2} = \frac{1}{\lambda_0^2} = \frac{1}{\lambda_g^2} + \frac{1}{\lambda_c^2}. \tag{9.34}$$

This is known as the 'waveguide equation', and it is found to hold for any shape of waveguide, although it has here been deduced only for a simple special case. The cut-off wavelength $\lambda_c$ depends on the shape and dimensions of the waveguide, and on the mode of propagation (that is, in the present case, on the values of $c$ and $n$ respectively).

Eqn (9.34) shows that the wavelength in the guide is always greater than that in the unbounded medium $\lambda_0$. When $\lambda_0 \ll \lambda_c$, $\lambda_g$ approaches $\lambda_0$, while when $\lambda_0 \to \lambda_c$, $\lambda_g$ tends to infinity. The phase velocity in the guide behaves in the same way as $\lambda_g$, since

$$v_g = \omega/\beta = f\lambda_g, \tag{9.35}$$

showing that $v_g$ is always greater than the velocity in the unbounded medium. If the guide contains no material medium, the phase velocity will be greater than the velocity of light ($f\lambda_0$) in a vacuum, since $\lambda_g > \lambda_0$. This does not mean that energy is transmitted with a velocity greater than that of light, since we have dispersion: the phase velocity depends on the frequency, and does not equal the group velocity $u_g = d\omega/d\beta$. From eqn (9.33)

$$-\gamma^2 = \beta^2 = \left(\frac{\omega}{v}\right)^2 - \left(\frac{n\pi}{c}\right)^2,$$

and hence

$$2\beta(d\beta/d\omega) = 2\omega/v^2,$$

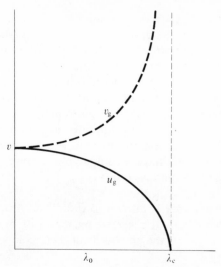

FIG. 9.17. Variation of the phase velocity $v_g$ (broken line) and group velocity $u_g$ (full line) in a waveguide.

giving

$$\left(\frac{\omega}{\beta}\right)\left(\frac{d\omega}{d\beta}\right) = v_g u_g = v^2. \tag{9.36}$$

Since $v_g$ is always greater than $v$, it follows that $u_g$ is always less than $v$, and is thus always less than the velocity of light. The behaviour of $u_g$ and $v_g$ is illustrated in Fig. 9.17; these relations hold for all waveguides, since they depend only on the waveguide equation, (9.34).

The propagation of waves between two parallel conducting planes may be considered in another way which is illuminating, particularly in respect of the group and phase velocity. The wave motion may be regarded as consisting of an ordinary plane wave, with the same properties as a wave in the unbounded medium, which is multiply reflected from the two planes. The normal to the wave front is assumed to make an angle $\theta$ with the normal to the conducting planes, as in Fig. 9.18. From Figs 8.4, 8.5, if

FIG. 9.18. Reflection of ordinary plane waves between two parallel planes. FG, GH normals to wave fronts incident and reflected at plane $z = 0$. LL′, MM′ incident wave fronts; LL″, MM″ reflected wave fronts, differing in phase by $2\pi$. LM gives the guide wavelength $\lambda_g$.

we replace $\theta$ by $\pi-\theta$, the components of such incident and reflected waves are given by eqns (8.36) and (8.37), if the electric vector is in the plane of incidence, or eqns (8.46) and (8.47) if it is normal to the plane of incidence. In either case we must satisfy the boundary conditions, that the tangential components of **E** must be zero all over the planes $z = 0$ and $z = c$. For the former case (electric vector in plane of incidence) the first of these conditions gives $A' = -A$ in order to make the $x$ component of **E** zero at $z = 0$; the same condition is obtained at $z = c$ if we take $\cos \theta = 0$ ($\theta = \frac{1}{2}\pi$). This makes the $x$ component of **E** zero everywhere, but $E_z$ and $H_y$ are finite, so that we have a simple plane wave moving in the $x$ direction; this wave is purely transverse, and moves with the same velocity as in the unbounded medium.

When the electric vector is normal to the plane of incidence, to make the $y$ component of **E** zero at $z = 0$, we must take $B' = -B$ in eqns (8.46) and (8.47). At other points the amplitude of the $y$ component of **E** is then (replacing $\theta$ by $\pi-\theta$)

$$E_y = B(F_1 - F_2)$$

$$= B \exp\{j\omega(t-x\sin\theta/v)\}\{\exp(j\omega z\cos\theta/v) - \exp(-j\omega z\cos\theta/v)\}$$

$$= 2jB\sin(\omega z\cos\theta/v)\exp j\omega(t-x\sin\theta/v), \tag{9.37}$$

where we have written $v$ for the velocity in the unbounded medium. At the plane $z = c$ the field component given by eqn (9.37) is zero provided that

$$\omega c\cos\theta/v = 2\pi c\cos\theta/\lambda_0 = n\pi$$

or

$$\cos\theta = \lambda_0(n/2c) = \lambda_0/\lambda_c, \tag{9.38}$$

where $\lambda_c$ is the cut-off wavelength as previously defined, and $\lambda_0$ is the wavelength in the unbounded medium. This equation shows clearly that no wave is possible for $\lambda_0 > \lambda_c$, for then there is no real value of $\theta$ which satisfies it.

The wavelength $\lambda_0$ is defined as the normal distance between two wave fronts such as LL′ and MM′ in the plane waves where the phase differs by $2\pi$; these wave fronts have an intercept LM on the plane $z = 0$, and the length of this intercept, which gives the apparent wavelength $\lambda_g$ of a wave propagated in the $x$ direction, is $\lambda_0/\sin\theta$. Hence

$$\sin\theta = \lambda_0/\lambda_g, \tag{9.39}$$

and on combining this with eqn (9.38) we have

$$1/\lambda_0^2 = (\sin^2\theta + \cos^2\theta)/\lambda_0^2 = 1/\lambda_g^2 + 1/\lambda_c^2,$$

which is the 'waveguide equation' already derived (eqn (9.34)). We see that it follows from the fact that only one angle $\theta$ is possible for the direction of our multiply reflected plane wave in order to satisfy both the boundary conditions. The energy flow travels with velocity $v$ in the plane wave in a direction normal to the plane wave front (that is, along FG or GH); the component of this velocity in the $x$ direction is $v \sin \theta$, and this is the speed $u_g$ at which the energy flows in the guided wave. On the other hand, the phase velocity $v_g$ of the guided wave, from eqn (9.37), is $v/\sin \theta$; hence $u_g v_g = v^2$, as shown earlier (eqn (9.36)). (It should be noted that the possibility of $v_g$ being greater than $v$ is not peculiar to electromagnetic waves; the effect can be observed by watching the movement parallel to a reflecting boundary of a crest in any wave motion, as, for example, in water waves being reflected at an angle from a breakwater.) The behaviour of our guided wave when $\lambda_0$ is equal to the cut-off wavelength $\lambda_c$ can be understood if we remember that this requires $\theta = 0$; that is, the wave motion is an ordinary plane wave being reflected at normal incidence between the two planes. Then no energy is propagated in the $x$ direction, so that $u_g = 0$; but the phase at a given value of $z$ is independent of $x$, so that the apparent phase velocity $v_g$ in the $x$ direction is infinite.

## 9.7. Waveguides

The type of wave we have been considering, propagated between two parallel planes, has the following components:

$$\left. \begin{array}{l} E_y = A \sin(2\pi z/\lambda_c)\sin(\omega t - 2\pi x/\lambda_g), \\ H_x = -A(\lambda_0/Z_1\lambda_c)\cos(2\pi z/\lambda_c)\cos(\omega t - 2\pi x/\lambda_g), \\ H_z = A(\lambda_0/Z_1\lambda_g)\sin(2\pi z/\lambda_c)\sin(\omega t - 2\pi x/\lambda_g). \end{array} \right\} \quad (9.40)$$

These components may either be obtained from eqns (8.46) and (8.47) (for example, $E_y$ is found by taking the real part of eqn (9.37) and writing $2B = -A$), or by taking $E_y$ as the appropriate solution of eqn (9.32) and using eqns (9.31); $Z_1$ is the intrinsic impedance of the medium between the conducting planes. These equations show that for a given value of $\lambda_c$, the component $H_x$ in the direction of propagation diminishes in amplitude as $\lambda_0$ is decreased, so that the wave approaches a purely transverse wave travelling along the $x$-axis. As $\lambda_0 \to \lambda_c$, $H_z \to 0$ since $\lambda_0/\lambda_g \to 0$, giving in the limit a transverse wave travelling along the $z$-axis.

Since the only electric field component is in the $y$ direction, it is possible to insert conducting planes normal to the $y$-axis without introducing any new boundary conditions. We have then a closed rectangular waveguide, as shown in Fig. 9.19(a), bounded by the perfectly conducting planes $z = 0$, $z = c$: $y = 0$, $y = b$. The field components within the guide are given by eqns (9.40), and are zero outside. This type of wave is designated $TE_{0n}$ (or $H_{0n}$); TE means 'transverse electric', indicating that

Fig. 9.19. Rectangular waveguide with $TE_{01}$ mode. (a) Lines of electric field ———. (b) Lines of magnetic field - - - - - -.

there is no electric field component in the direction of propagation; the subscripts 0, $n$ mean that there is no variation in the y direction of any field component, while in the z direction they vary as sin or $\cos(\pi nz/c)$. The simplest wave ($n = 1$) is shown in Fig. 9.19, and this is also the mode with the largest cut-off wavelength $\lambda_c = 2c$. It is therefore used as the standard mode for waveguide propagation, and the guide dimensions are chosen so that $2c > \lambda_0 > c$ for the wavelength it is desired to propagate. No higher mode (with $n = 2$ or more) can then be propagated; this has the advantage that waves of higher modes, set up by a local disturbance of the field (due to discontinuities or changes in the guide dimensions), decay exponentially along the guide. By making the dimension $b$ less than $\lambda_0/2$, no TE mode can be propagated with the electric vector polarized in the z direction; and it can be shown that all other modes have still smaller cut-off wavelengths, and so cannot be propagated.

The field components in the $TE_{01}$ mode are shown in Fig. 9.19. $E_y$ is a maximum in the centre of the guide, and zero at the planes $z = 0$ and $z = c$; it varies sinusoidally with $z$, with just one half-period of variation (modes with higher values of $n$ make $n$ half-periods of variation, and so require a correspondingly larger value of $c$ for a given $\lambda_0$). The components of the magnetic field are everywhere tangential to the boundaries, and the lines of magnetic field are shown in Fig. 9.19(b), where the guide is viewed looking down on the broad face. The lines of magnetic field encircle the points at which $\partial E_y/\partial t$ is greatest, that is, the points where the displacement current is greatest. This corresponds to Maxwell's equation curl $\mathbf{H} = \partial\mathbf{D}/\partial t$, which implies that a displacement current (a changing electric displacement) is encircled by lines of magnetic field. The flow of conduction current is always normal to the magnetic field at the metal surface. Thus on the narrow face it is everywhere parallel to the y-axis, and on the broad face it is parallel to the x-axis at the centre. Narrow slits can be cut in the metal without causing energy to leak out, providing that they do not interrupt any current flow. This is used in the construction of a standing-wave detector (cf. § 22.4), where the electric field strength inside the guide is detected by a small probe moving along a narrow slot cut along the centre of the broad face.

Study of the propagation characteristics of waves in guides of other than rectangular shape involves the use of more complex mathematics and we mention only the cylindrical waveguide. This involves the solution of the wave equation or Maxwell's equations in cylindrical coordinates, and requires the use of Bessel functions. Waves may be designated as $TM_{mn}$ or $TE_{mn}$, corresponding to no magnetic or no electric field in the direction of propagation; the first subscript indicates that the field components vary as cos or sin $m\phi$, where $\phi$ is the azimuthal angle, while the second gives the number of values of the radius at which the electric field components other than the radial component $E_r$ are zero. The simplest modes are $TE_{11}$, which is rather similar to the $TE_{01}$ mode in rectangular guide; the electric field is purely transverse, and distributed as shown in Fig. 9.20(a); the cut-off wavelength is $1.707d$, where $d$ is the diameter of the guide. The $TM_{01}$ mode has a transverse magnetic field whose lines of force are circular; the cut-off wavelength is $1.308d$, and the mode is similar to that in a coaxial line except that the conduction current in the centre conductor is replaced by displacement current, with lines of $\mathbf{E}$ running down the centre and turning outwards to terminate on the wall as in Fig. 9.20(b). The $TE_{01}$ mode is rather similar, but with the lines of electric field and magnetic field interchanged; the electric field has closed circular lines of force and is purely transverse, while the magnetic field is greatest down the axis (see Fig. 9.20(c)); the cut-off wavelength is $0.820d$.

*Cavity resonators*

If a length of waveguide is closed by conducting walls at each end, it will resonate at wavelengths such that the distance between the end walls is a multiple of half a guide wavelength. In the case of a rectangular guide closed by conducting walls normal to the x-axis a distance $a$ apart, the distance $a$ must be $\frac{1}{2}l\lambda_g$ where $l$ is an integer, in order that the electric field (which is tangential to the end walls) may be zero at the two ends. From eqn (9.34) the wavelength $\lambda_0$ in the unbounded medium at which

Sections through C-D

Lines of electric field    ------ Lines of magnetic field
• towards observer    ○ away from observer

FIG. 9.20. Approximate configurations of electric and magnetic fields in a cylindrical waveguide. Propagation is directed away from the observer or to the right. (a) $TE_{11}$ or $H_{11}$ mode; (b) $TM_{01}$ or $E_{01}$ mode; (c) $TE_{01}$ or $H_{01}$ mode. (After Southworth 1936, by courtesy of Bell Telephone Laboratories.)

the rectangular cavity will resonate in a $TE_{0n}$ mode is thus given by

$$\frac{1}{\lambda_0^2} = \left(\frac{l}{2a}\right)^2 + \left(\frac{n}{2c}\right)^2. \tag{9.41}$$

This is a special case of the more general formula for a $TE_{mn}$ mode, for which

$$\frac{1}{\lambda_0^2} = \left(\frac{l}{2a}\right)^2 + \left(\frac{m}{2b}\right)^2 + \left(\frac{n}{2c}\right)^2 \tag{9.42}$$

for a rectangular cavity of dimensions $a$, $b$, $c$; this formula may be recognized as that used in the theory of heat radiation in computing the resonant modes of a hollow rectangular cavity (see Problem 9.12).

The most important quantity for any resonant system is its quality factor $Q_F$; this may be found for a waveguide cavity by using the relation (see § 7.3)

$$Q_F = \omega(\text{energy stored})/(\text{energy dissipated per second}).$$

The total stored energy may be computed by integrating the energy density in the cavity, while the total energy dissipated in the metallic walls (owing to their finite resistivity) can be found by using eqn (8.34b). The order of magnitude of $Q_F$ can readily be found without carrying through the details of the integration in the following way. If the amplitude of the strength of the oscillating magnetic field in the cavity is $H_0$, the stored energy $\approx \frac{1}{2}\mu_0 H_0^2 V$, where $V$ is the volume of the cavity (assumed to be evacuated), while the energy lost at the walls $\approx \frac{1}{2}\rho H_0^2 A/\delta$, where $A$ is the total wall area, and $\rho$ and $\delta$ the resistivity and skin depth in the wall. Hence

$$Q_F \approx \omega(\tfrac{1}{2}\mu_0 H_0^2 V)/(\tfrac{1}{2}\rho H_0^2 A/\delta) = (V/A)(\omega\delta\mu_0/\rho) \approx V/A\delta, \qquad (9.43)$$

assuming the magnetic permeability of the wall to be unity, and using eqn (8.31). This result shows that at a given wavelength the value of $Q_F$ increases with the linear dimensions of the resonator; while at different wavelengths, if the linear dimensions are scaled in proportion to the wavelength, $Q_F$ varies as $\lambda_0^{\frac{1}{2}}$, since $\delta$ varies as $\lambda_0^{\frac{1}{2}}$, and $V/A$ as $\lambda_0$. For a given wavelength and size of cavity $Q_F$ does not vary greatly with the mode of resonance, with one exception. The $TE_{01}$ mode in a cylindrical cavity has rather a high $Q_F$, and there is no radial flow of current on the end walls; for this reason it is used in wavemeters (see § 22.5) where one end is a movable plunger. A good contact between this and the cylindrical wall is not essential to a high $Q_F$, since there is no current flow across the contact. A value of $10^4$ may be obtained for $Q_F$ at wavelengths of $10^{-2}$ m, and the sharpness of resonance is thus rather higher than for a resonant coaxial line, mainly because there is no centre conductor with its rather high current density to contribute to the dissipation of energy.

## References

LISSBERGER, P. H. (1970). *A. Rep. Prog. Phys.* **33**, 197.
SOUTHWORTH, G. C. (1936). *Bell Syst. tech. J.* **15**, 287 or (1937) *Proc. Inst. Radio Engrs*, **25**, 237.

## Problems

9.1. Show that if $\mathbf{Z}_1$ and $\mathbf{Z}_2$ in a simple filter are both pure resistances or pure capacitances the filter acts as an attenuator at all frequencies. Calculate the attenuation per section when both $Z_1$ and $Z_2$ are pure resistances of $100\,\Omega$ and find the iterative impedance of a T-section.

(*Answers*: Power falls by factor 6·8 per section; $Z_T = 112\,\Omega$.)

9.2. A filter where $Z_1Z_2 = k^2$, a constant independent of frequency, is called a constant-$k$ filter. Show that the simple low-pass filters of § 9.2 are of this type, but the band-pass filter of Fig. 9.10 is not.

Show that the filter section of Fig. 9.12 is a band-pass section of the constant-$k$ type provided that $L_1C_1 = L_2C_2$, when $k^2 = L_2/C_1 = L_1/C_2$. Sketch the variation with frequency of the quantity $Z_1/4Z_2$.

9.3. The values of the components in the $m$-derived T-section shown in Fig. 9.11(a) obey the relations $(m<1)$ $L_1 = mL$, $C_2 = mC$, $L_2 = L(1-m^2)/4m$.

Show that the section behaves as a low-pass filter with the following properties: (a) the cut-off frequency $f_0$ is independent of $m$; (b) the iterative impedance is $Z_T = (L/C - \omega^2L^2/4)^{\frac{1}{2}}$, and is thus the same as that of a simple low-pass filter section in § 9.2; (c) the attenuation in the stop band is infinite at a frequency $f = f_0(1-m^2)^{\frac{1}{2}}$.

9.4. A chain of the T-sections of Fig. 9.11(a) is terminated by the half-section shown in Fig. 9.11(b), where the values of the components obey the same relations as in the preceding problem. Prove that the impedance at the terminals CD is

$$Z = \left(\frac{L}{C}\right)^{\frac{1}{2}} \frac{\{1-(1-m^2)f^2/f_0^2\}}{(1-f^2/f_0^2)^{\frac{1}{2}}}.$$

If $m = 0\cdot6$, show that this does not depart by more than 4 per cent from the value $(L/C)^{\frac{1}{2}}$ for frequencies up to 85 per cent of the cut-off frequency $f_0$. Thus the use of a half-section as a transformer gives a more uniform impedance in the pass band.

9.5. Find expressions for the electric and magnetic field strengths in a coaxial transmission line carrying a current $I$ and a voltage $V$, and show by integrating Poynting's vector over the cross-section between the two conductors that the power flowing along the line is $IV$.

If the conductors have a finite resistivity, compute the power flowing into them per unit length by means of eqn (8.34b), and show that this gives the same attenuation as calculated in § 9.5.

9.6. In an infinite transmission line a leak develops at one point whose resistance is just equal to the characteristic impedance of the line. Show that of the power in the incident wave one-ninth is reflected, four-ninths is transmitted, and four-ninths is dissipated in the leak.

9.7. A length of lossless transmission line is first short-circuited at one end and then open-circuited; the impedance measured at the other end is $Z_1$ in the first case and $Z_2$ in the second. Show that $Z_1Z_2 = Z_0^2$, where $Z_0$ is the characteristic impedance of the line. This is a convenient way of measuring $Z_0$ for a cable of unknown electrical length.

9.8. A film of magnesium fluoride ($MgF_2$) (refractive index = 1·38) one-quarter wavelength thick is deposited on crown glass (refractive index = 1·52). Show that the reflected intensity is reduced to about 1 per cent.

Alternate layers, each $\lambda/4$ thick, of transparent media with refractive indices $n_1$, $n_2$ are deposited onto a substrate of index $n_0$. With $x$ layers of each material show that for incident light the surface appears to have a refractive index $n_0(n_2/n_1)^{2x}$. If $n_0 = 1 \cdot 52$, $n_1 = 1 \cdot 38$, and $n_2 = 2 \cdot 35$ (ZnS), how many layers are needed to produce reflection of at least 99·99 per cent of the incident energy?

(*Answer:* $x = 10$.)

9.9. A quarter-wavelength, air-spaced, parallel-wire transmission line is found to be in resonance with an oscillator when its length is 0·25 m. When a capacitance of 1 pF is connected across the open end, it is found that the length of the line must be reduced to 0·125 m to obtain resonance. Show that the characteristic impedance of the line is approximately 530 $\Omega$.

9.10. Show, either by the use of equations similar to (9.29) and (9.30) but with the assumptions $H_x = H_z = 0$, or by the use of eqns (8.36) and (8.37), that a wave can be propagated between two parallel conducting planes with the following field components:

$$H_y = (A/Z_1)\cos(2\pi z/\lambda_c)\cos(\omega t - 2\pi x/\lambda_g),$$
$$E_x = (A\lambda_0/\lambda_c)\sin(2\pi z/\lambda_c)\sin(\omega t - 2\pi x/\lambda_g),$$
$$E_z = -(A\lambda_0/\lambda_g)\cos(2\pi z/\lambda_c)\cos(\omega t - 2\pi x/\lambda_g),$$

where $\lambda_c$ and $\lambda_g$ have the same values as for the transverse electric wave derived in § 9.6. This wave is a transverse magnetic wave, and may be designated as $TM_{0n}$; note that it cannot exist in a closed rectangular guide because the tangential components of **E** must then vanish at the walls $y = 0$ and $y = b$. The lowest transverse magnetic wave then possible would be $TM_{11}$, with components $H_y$, $H_z$, $E_x$, $E_z$ each varying sinusoidally in both the $y$ and $z$ directions.

9.11. A hollow cubical box of side $a$ resonates in the $TE_{101}$ mode (that is $l = n = 1$ in eqn (9.41)). Calculate the energy stored and energy dissipated per second, and show that the value of $Q_F = a/2\delta$, where $\delta$ is the skin depth in the metal walls at the resonant frequency.

9.12. A hollow rectangular box is bounded by perfectly conducting planes at $x = 0$, $x = a$; $y = 0$, $y = b$; $z = 0$, $z = c$. Show that the standing wave system

$$E_x = A_x \cos \alpha x \sin \beta y \sin \gamma z \exp j\omega t,$$
$$E_y = A_y \sin \alpha x \cos \beta y \sin \gamma z \exp j\omega t,$$
$$E_z = A_z \sin \alpha x \sin \beta y \cos \gamma z \exp j\omega t$$

satisfies the boundary conditions provided that $\alpha a = l\pi$, $\beta b = m\pi$, $\gamma c = n\pi$, and that the wave equation is satisfied if

$$\frac{1}{4\pi^2}(\alpha^2 + \beta^2 + \gamma^2) = \left(\frac{l}{2a}\right)^2 + \left(\frac{m}{2b}\right)^2 + \left(\frac{n}{2c}\right)^2 = \left(\frac{\omega}{2\pi v}\right)^2 = \frac{1}{\lambda_0^2}.$$

how also that, to satisfy div **D** = 0,

$$\alpha A_x + \beta A_y + \gamma A_z = 0.$$

(In general there can be only two independent amplitudes, corresponding to the two possible polarizations of an electromagnetic wave.)

# 10. Dielectrics

THIS chapter is the first of several which approach electric and magnetic phenomena from an atomic viewpoint. For dielectrics the discussion is confined to linear phenomena, where the polarization is linearly proportional to the applied field. High-field effects, and effects involving spontaneous polarization (ferro- and antiferroelectricity) are excluded.

## 10.1. Macroscopic quantities in an atomic medium

A dielectric medium consists of an assembly of negatively charged particles (electrons) and positively charged particles (nuclei). We assume it to be electrically neutral, the total negative charge being equal to the total positive charge, and that all the charges are 'bound'. By this is meant that in zero electric field each charge occupies its equilibrium position, from which it may be slightly displaced when an electric field is applied, but that it is unable to move continuously from place to place: there is no 'free charge' and no conduction current.

In electromagnetic theory, a dielectric is a continuous medium which becomes electrically polarized under the action of an electric field. The concepts involved are the macroscopic quantities, charge density, and electric polarization, defined as continuous functions. The charge density $\rho_e$ is the ratio of an infinitesimal charge to an infinitesimal volume, taken to the limit as the infinitesimals are reduced to zero. The electric polarization is similarly defined, with net dipole moment instead of net charge. These concepts are clearly not immediately applicable to a set of discontinuous charges, and a method of relating the electrical properties of such an assembly to the corresponding properties of a continuous medium is required.

In the electron theory of Lorentz macroscopic quantities such as charge density and polarization are regarded as averages over loosely defined volumes which are small on a macroscopic scale but large enough to contain very many elementary charges. In considering the propagation of an electromagnetic wave through the medium, such averages must be taken over regions whose linear dimensions are small compared with the wavelength. For optical waves ($\lambda \sim 5 \times 10^{-7}$ m) such dimensions can be over 100 times the atomic spacing in a solid ($\sim 2 \times 10^{-10}$ m), so that the number of charges involved in the average may be $10^6$–$10^9$. Nevertheless there is a conceptual difficulty. The total charge in any such volume must be an integral multiple of the fundamental unit of charge, but if the boundaries are moved to include or exclude just one electronic charge, an apparent change in the charge density is obtained which is clearly not meaningful. (In fact one electronic charge of $1 \cdot 6 \times 10^{-19}$ C in a cube of dimensions $10^{-7}$ m gives a charge density of $160$ C m$^{-3}$, which is enormous compared with any charge density realizable in practice.) This difficulty is overcome by taking averages over a large number of such volumes of random size, so that fluctuations in volumes of size small compared with $\lambda^3$ are smoothed out.

The averages which we require may be defined more precisely by the use of a three-dimensional Fourier analysis to represent the distribution of charge in space (Robinson 1973). The method may be illustrated by considering a simple one-dimensional example: an infinite line of charges $q$ placed at regular intervals $a_0$, for example at the points $0$, $\pm a_0$, and all integral multiples of $\pm a_0$. The charge density, expressed as a Fourier series, is easily shown to be

$$\rho_e = \frac{q}{a_0} + \frac{q}{2a_0} \sum_{n=1}^{\infty} \cos\left(2\pi nx/a_0\right). \tag{10.1}$$

Clearly the average charge density is $q/a_0$, and the spatial repetition rates of the oscillating terms are $a_0^{-1}$ and multiples thereof. In any length $l$ reaching from $x_0 - \frac{1}{2}l$ to $x_0 + \frac{1}{2}l$ the total charge is found by integration over this interval, and the corresponding charge density is

$$\frac{1}{l}\int \rho_e \, \mathrm{d}x = \frac{q}{a_0} + \frac{q}{2a_0 l} \sum_{n=1}^{\infty} \int_{x_0 - \frac{1}{2}l}^{x_0 + \frac{1}{2}l} \cos\left(2\pi nx/a_0\right) \mathrm{d}x$$

$$= \frac{q}{a_0} + \frac{q}{2\pi nl} \sum_{n=1}^{\infty} \cos\left(2\pi nx_0/a_0\right)\sin\left(\pi nl/a_0\right). \tag{10.2}$$

For arbitrary values of $x_0$ and $l$ the oscillatory terms are as often positive as negative, and the macroscopic charge density is $q/a_0$ (which in general is not an exact multiple of $q$). The Lorentz approach is in fact equivalent to taking an average over ranges of values of $x_0$ and of $l$ which are large compared with $a_0$.

This one-dimensional example illustrates the result that with a regular array of charges, as in the lattice of a crystal, Fourier analysis produces a constant component (which is the macroscopic average we require), together with components which repeat in space at the same rate as the lattice or higher. Such oscillating components cause strong scattering, the lattice acting as a three-dimensional diffraction grating for wavelengths simply related to the lattice spacing, that is, X-rays of wavelengths $\sim 2 \times 10^{-10}$ m. In contrast, much longer wavelengths are propagated freely through the lattice and can be treated using the macroscopic theory of

Chapter 8. Imperfections on an atomic scale, such as displacements of lattice ions under the action of thermal vibrations or the irregularities of molecular position in a liquid, do not much affect this result. This is true even in a gas, since at N.T.P. the average intermolecular distance ($\sim 5 \times 10^{-9}$ m) is still small compared with optical wavelengths. The randomness in the molecular spacing will cause some scattering, but this is small for optical waves and quite negligible for radio waves.

## 10.2. Macroscopic polarization and the local field

Experimentally no net charge on an atom or molecule has ever been detected, and Dylla and King (1973) have shown that the fractional difference in the numerical values of the charge on electron and proton is less than $1 \cdot 3 \times 10^{-21}$. However, application of an electric field displaces the positive charges in the direction of the field and the negative charges in the opposite direction, so that every substance acquires an 'induced' electric dipole moment. It is usual to distinguish between two effects: (1) an 'electronic contribution', arising from the displacement of electrons relative to positively charged nuclei; (2) an 'ionic contribution', arising from the displacement of heavy positive ions relative to heavy negative ions, which occurs in molecules and in ionic solids containing ions of different charge.

In addition to induced moments there may be permanent electric dipole moments. In an atom or monatomic ion the distribution of charge is symmetric about the centre and, as discussed in § 2.3, no permanent electric dipole moment can be present. Some molecules also fall into this symmetry classification; simple examples are the homonuclear diatomic molecules such as $H_2$, $N_2$, $O_2$ and symmetric linear polyatomic molecules such as $CO_2$, which has the structure O—C—O. Such molecules are known as 'non-polar molecules'. Molecules which do not satisfy this symmetry requirement normally possess permanent electric dipole moments, and are called 'polar molecules'.

For a finite body, the macroscopic electric dipole moment is the vector sum of the individual dipole moments. We may distinguish between three types of contributions.

(1) Each atom, ion, or non-polar molecule has zero dipole moment in zero field, but acquires a dipole moment when a field is applied; such moments are known as 'induced dipoles'.

(2) When polar molecules are present, their permanent dipole moments are completely randomly oriented in zero field, so that the vector sum is zero, at any rate on a time average; in the presence of a field, orientations in which the dipole moment is parallel to the

field have a lower energy and so are statistically more favourable, giving a resultant dipole moment in the direction of the field.

(3) In an ionic lattice, the positive ions are displaced in the direction of an applied field while the negative ions are displaced in the opposite direction, giving a resultant dipole moment to the whole body.

The techniques of Fourier analysis that were illustrated above for the charge density can also be applied to the components of the dipole moments, and under similar conditions yield a macroscopic dipole moment per unit volume. This is the polarization $\mathbf{P}$, which appears in the macroscopic equations (cf. § 1.5)

$$\mathbf{D} = \epsilon_0 \mathbf{E} + \mathbf{P} = \epsilon_r \epsilon_0 \mathbf{E} \tag{10.3}$$

and

$$\mathbf{P} = (\epsilon_r - 1)\epsilon_0 \mathbf{E}. \tag{10.4}$$

In performing the Fourier analysis, we use the dipole components given by eqns (2.25). For neutral atoms and neutral molecules the result is independent of the origin of $\mathbf{r}$ in the equation $\mathbf{p} = \int \rho_e \mathbf{r} \, d\tau$, as pointed out in § 2.3. For charged ions this is not true; for the induced dipoles arising from the displacement of the electrons relative to the nucleus we must measure $\mathbf{r}$ from the nucleus, while if the ion is bodily displaced in the lattice $\mathbf{r}$ must be measured from the point occupied by the ion when no electric field is applied.

To evaluate the dipole moment, we must calculate the effect of an electric field on an electron bound in a molecule. This is a quantum-mechanical problem, whose solution is only approximate except in a simple case such as the hydrogen atom, and which we shall not pursue further. Instead we regard the medium as composed of atoms, ions, or molecules which acquire an induced dipole moment $\mathbf{p}_i$ under the action of the electric field, and which sometimes possess also a permanent dipole moment. The size of the induced dipole $\mathbf{p}_i$ is determined by the magnitude of the field $\mathbf{E}_{local}$ acting on the particle; we can write

$$\mathbf{p}_i = \alpha \mathbf{E}_{local}, \tag{10.5}$$

where $\alpha$ is a quantity known as the 'polarizability' of the atom, ion, or molecule. At ordinary field strengths $\mathbf{p}_i$ is linearly proportional to $\mathbf{E}_{local}$, and $\alpha$ is therefore a constant, independent of field strength. The polarization arising from the induced dipoles, if all are identical and the number per unit volume is $n_0$, is

$$\mathbf{P} = n_0 \alpha \mathbf{E}_{local}, \tag{10.6}$$

where we have assumed that each induced dipole $\mathbf{p}_i$ is parallel to $\mathbf{E}_{local}$ as would be the case in the absence of any anisotropy.

The field strength $\mathbf{E}_{local}$ is the actual field acting on a particle (atom, ion, or molecule) in the dielectric medium. It is not necessarily the same as the macroscopic field strength $\mathbf{E}$, which is the average field in the medium. The reason for this is that in calculating $\mathbf{E}$ we must take the average of the vector sum of the electric fields set up by all the dipoles in the medium, including that of the particle for which we require the value of $\mathbf{E}_{local}$. Obviously, the field of this dipole does not act upon itself, and it must therefore be excluded from the vector summation which gives the value of $\mathbf{E}_{local}$. The difference between the two can be written as $\mathbf{E}_{corr}$, where

$$\mathbf{E}_{local} = \mathbf{E} + \mathbf{E}_{corr}. \qquad (10.7)$$

In forming these vector sums it is found that, although the electric field of a dipole falls off as the inverse cube of the distance, the sums do not converge rapidly. There are larger numbers of more distant dipoles, so that their total contribution is as important as those of nearby dipoles, and the 'dipole field' is said to be 'long-range'. We shall consider first the case of a dielectric of infinite extent in which there exists a uniform macroscopic field of strength $\mathbf{E}$. Such a field exists inside a parallel-plate capacitor completely filled with dielectric, and then $\mathbf{E} = V/t$, where $V$ is the voltage between the plates and $t$ their separation. In capacitors of different geometry $\mathbf{E}$ is not uniform, but this does not affect the results provided that $\mathbf{D}$, $\mathbf{P}$ in eqns (10.3)–(10.4) are linearly proportional to $\mathbf{E}$.

## 10.3. The Lorentz correction and the Clausius–Mossotti relation

An approximate solution to the problem of computing $\mathbf{E}_{corr}$ was obtained by Lorentz; it applies only to induced dipoles which are all parallel to the applied field. In an infinite dielectric, the more distant dipoles can be treated as a continuous medium with a polarization $\mathbf{P}$, producing a field strength $\mathbf{E}'$, while the net field strength $\mathbf{E}''$ of the nearby dipoles must be calculated exactly (excluding the field of the dipole at which we need the value of $\mathbf{E}_{local}$). The sum of the two contributions is

$$\mathbf{E}' + \mathbf{E}'' = \mathbf{E}_{corr}. \qquad (10.8)$$

An imaginary spherical surface is drawn centred on the point at which we require $\mathbf{E}_{local}$. The radius of this sphere is sufficiently large compared with the intermolecular distance that the effects of dipoles outside it may be evaluated by macroscopic theory, while it also contains so many molecules that its centre is a representative point and summation of the fields of the dipoles inside it gives a correct average.

To evaluate $\mathbf{E}'$ we use the results embodied in eqn (1.16). The potential, and hence also the field strength $\mathbf{E}'$, of the dipoles outside the spherical surface are the same as those of a volume distribution $-\text{div}\,\mathbf{P}$

outside the sphere, and an apparent surface charge distribution on the sphere. Since $\mathbf{P}$ is uniform, div $\mathbf{P} = 0$; the surface charge density is $-P \cos \theta$, where the minus sign appears because we are considering the effect of the dielectric *outside* the sphere. From the result in the first part of Problem 2.1, we have

$$\mathbf{E}' = +\mathbf{P}/3\epsilon_0, \tag{10.9}$$

where $\mathbf{P}$ is the actual polarization of the medium. We must not include any allowance for the distortion of the field by a spherical cavity, (as in the second part of Problem 2.1) because we have not excavated a real cavity in the medium.

The value of $\mathbf{E}''$ is clearly more difficult to calculate, and it depends on how the dipoles are arranged. Lorentz showed that for a cubical array of parallel dipoles, $\mathbf{E}'' = 0$. This result is valid for induced dipoles, which are all parallel to the field, but not for permanent dipoles which are nearly randomly oriented. With $\mathbf{E}'' = 0$, the Lorentz correction is simply

$$\mathbf{E}_{\text{corr}} = \mathbf{E}' = \mathbf{P}/3\epsilon_0, \tag{10.10}$$

and the local field is given by

$$\mathbf{E}_{\text{local}} = \mathbf{E} + \mathbf{E}_{\text{corr}} = \mathbf{E} + \mathbf{P}/3\epsilon_0. \tag{10.11}$$

From eqn (10.6) the polarization is

$$\mathbf{P} = n_0 \alpha (\mathbf{E} + \mathbf{P}/3\epsilon_0), \tag{10.12}$$

and on combining this with eqn (10.4) we have the equation

$$\frac{\epsilon_r - 1}{\epsilon_r + 2} = \frac{n_0 \alpha}{3\epsilon_0}. \tag{10.13}$$

This result was first derived by Clausius and Mossotti, and is generally known by their names.

If the dielectric is not infinite in extent, the vector sum of the dipole fields depends on the shape of the body. This result has already been encountered in macroscopic theory (§ 2.4), which shows that the field inside the body differs from the external applied field. In Fig. 10.1 a dielectric body B is placed in a uniform field of strength $\mathbf{E}_0$. The field acting on a dipole at the centre of an imaginary Lorentz sphere L is $\mathbf{E}_{\text{local}}$, which is the sum of $\mathbf{E}_0$ and the field strengths due to all the surrounding dipoles. Those inside the sphere contribute a field strength $\mathbf{E}''$ which is taken (as above) to be zero. Those outside L are equivalent to a volume distribution of charge $-\text{div } \mathbf{P}$, together with distributions of charge on the surfaces L and B, by eqn (1.16). If the susceptibility $\chi_e$ is uniform, isotropic, and independent of field strength, then div $\mathbf{P} = \chi_e \epsilon_0$ div $\mathbf{E} = 0$. The inner surface charge on the sphere L contributes the field strength $\mathbf{E}'$,

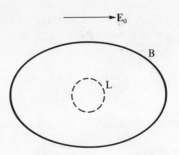

FIG. 10.1. A dielectric body B in a uniform field of strength $\mathbf{E}_0$. Inside B the field $\mathbf{E}$ differs from $\mathbf{E}_0$ because of the depolarizing field. L is the imaginary surface of a Lorentz sphere.

as above. The outer surface charge on B produces the 'depolarization field' which causes the field strength $\mathbf{E}$ in the dielectric to be different from $\mathbf{E}_0$. For a dielectric sphere in a vacuum, from § 2.4 we have $\mathbf{E} = \mathbf{E}_0 - \mathbf{P}/3\epsilon_0$, so that $\mathbf{E}_{\text{local}} = \mathbf{E} + \mathbf{E}' = \mathbf{E}_0$. This is obviously a special case; it is just the result we should obtain by increasing the size of the Lorentz sphere until it contains the whole body. There is then no dielectric outside this sphere so that the Lorentz field must vanish, that is, $\mathbf{E}_{\text{local}} = \mathbf{E}_0$ since $\mathbf{E}'' = 0$. For a dielectric of infinite extent the question of a depolarizing field does not arise; eqns (10.3) and (10.4) relate to the field $\mathbf{E}$ in the medium, which is assumed to be known.

## 10.4. Static permittivity of non-polar gases

In a gas the molecules are randomly placed and moving with molecular velocities, but it can be shown that the average value of $\mathbf{E}''$ is again zero for induced dipole moments which are all parallel to the applied field. Hence for non-polar gases, in which the molecules have no permanent dipole moments, we have $\mathbf{E}' = \mathbf{P}/3\epsilon_0$; the value of $\mathbf{E}_{\text{local}}$ is given by eqn (10.11), and we should expect the Clausius–Mossotti relation (10.13) to hold. Since the number of molecules $n_0$ per unit volume varies with the density, it is convenient to multiply both sides of (10.13) by $M/\rho$, where $M$ is the molecular weight and $\rho$ the density. This gives

$$\frac{\epsilon_r - 1}{\epsilon_r + 2} \cdot \frac{M}{\rho} = \frac{N_A \alpha}{3\epsilon_0}, \tag{10.14}$$

where $N_A$ is Avogadro's number, a universal constant. The quantity $N_A \alpha / 3\epsilon_0$ is sometimes called the 'molar polarizability'.

Values of the relative permittivity of a number of common gases at N.T.P. are given in Table 10.1. The difference between $\epsilon_r$ and unity is of order $10^{-3}$, and for non-polar gases it increases with the complexity and hence with the size of the molecule. The permittivity has been used as a method of estimating molecular size. This makes the approximation that

TABLE 10.1

*Relative permittivity of some common gases at atmospheric pressure* $(10^5 \text{ Pa})$ *and* 273 K

| Gas | $(\epsilon_r-1)10^3$ | Dipole moment (debyes) |
|-----|------|------|
| He | 0·071 | 0 |
| $H_2$ | 0·270 | 0 |
| $O_2$ | 0·531 | 0 |
| $N_2$ | 0·588 | 0 |
| $CO_2$ | 0·988 | 0 |
| $CH_4$ | 0·948 | 0 |
| $C_2H_4$ | 1·38 | 0 |
| CO | 0·692 | 0·10 |
| $N_2O$ | 1·08 | 0·17 |
| $NH_3$ | 8·34 | 1·45 |
| $SO_2$ | 9·93 | 1·59 |

TABLE 10.2

*Relative permittivity of* $CO_2$ *and the Clausius–Mossotti relation*

| Experimenters | Pressure (Pa) | Relative permittivity (at 373 K) | $\dfrac{\epsilon_r-1}{\epsilon_r+2}\cdot\dfrac{M}{\rho}$ (m³) |
|-----|------|------|------|
| Keyes and Kirkwood (1930)[†] | 1·0 ×10⁶ | 1·00753 | 7·49×10⁻⁶ |
| | 3·0 ×10⁶ | 1·0240 | 7·53×10⁻⁶ |
| | 5·0 ×10⁶ | 1·0431 | 7·57×10⁻⁶ |
| | 7·0 ×10⁶ | 1·0645 | 7·60×10⁻⁶ |
| | 10·0 ×10⁶ | 1·1041 | 7·69×10⁻⁶ |
| | 15·1 ×10⁶ | 1·1912 | 7·73×10⁻⁶ |
| Michels and Michels (1932)[‡] | 10·32×10⁶ | 1·1086 | 7·71×10⁻⁶ |
| | 19·45×10⁶ | 1·2695 | 7·75×10⁻⁶ |
| | 29·54×10⁶ | 1·3895 | 7·70×10⁻⁶ |
| | 36·50×10⁶ | 1·4375 | 7·68×10⁻⁶ |
| | 47·66×10⁶ | 1·4900 | 7·67×10⁻⁶ |
| | 58·83×10⁶ | 1·5274 | 7·66×10⁻⁶ |
| | 70·02×10⁶ | 1·5570 | 7·66×10⁻⁶ |
| | 81·23×10⁶ | 1·5812 | 7·65×10⁻⁶ |
| | 97·06×10⁶ | 1·6097 | 7·62×10⁻⁶ |

The value of $\dfrac{\epsilon_r-1}{\epsilon_r+2}\cdot\dfrac{M}{\rho}$ at N.T.P. is $7\cdot33\times 10^{-6}$ m³ for 1 mol.

The data indicate a slight rise in the molar polarizability with pressure, followed by a small decrease at the highest pressures (~1000 atmospheres).

[†] Keyes, F. G. and Kirkwood, J. G. (1930). *Phys. Rev.* **36**, 754.
[‡] Michels, A. and Michels, C. (1932). *Phil. Trans. R. Soc.* A**231**, 409.

each molecule can be regarded as a perfectly conducting sphere of radius $a$; in a field of strength $E$ such a sphere acquires a dipole moment $4\pi\epsilon_0 a^3 E$, giving a value for $\alpha = 4\pi\epsilon_0 a^3$. Values of the molecular radius $a$ obtained in this way are generally of the same order but somewhat smaller than those obtained from kinetic theory (cf. Problem 10.2).

The validity of the Clausius–Mossotti relation has been tested by various experimenters up to pressures of $10^8$ Pa. For the non-polar gas $CO_2$ the constancy of the quantity $N_A\alpha/3\epsilon_0$, calculated using eqn (10.14), is shown in Table 10.2. At low pressures the value of $\epsilon_r-1$ is so small that the approximation $\epsilon_r+2=3$ is quite good; this is equivalent to neglecting the Lorentz correction and ignoring the difference between $\mathbf{E}_{local} = \mathbf{E}(\epsilon_r+2)/3$ and $\mathbf{E}$. At higher pressures this approximation is obviously not valid, and the constancy of the values in the last column of Table 10.2 justifies the assumptions made in the Lorentz method. The Clausius–Mossotti relation also depends on the assumption that the polarizability $\alpha$ is independent of $n_0$; this implies that it does not depend on the distance to other molecules, and that short-range forces between the molecules can be neglected. This is a reasonable assumption in a gas, but we must not be surprised if it fails in a liquid or a solid.

## 10.5. Static permittivity of polar gases

The last four molecules listed in Table 10.1 have permanent electric dipole moments, and the values of $\epsilon_r-1$ are noticeably greater for $NH_3$ and $SO_2$, whose molecules have dipole moments of order 1 debye. This unit is defined as

$$1 \text{ debye} = 3\cdot336\times10^{-30} \text{ C m.}$$

It is equivalent to an electronic charge of $1\cdot6\times10^{-19}$ C multiplied by a distance of about $2\times10^{-11}$ m, which is about one-tenth of the internuclear distance in a typical molecule.

The simplest type of polar molecule is a diatomic molecule formed from two dissimilar atoms. For a molecule such as KCl, an oversimplified picture is that of two ions, $K^+$ and $Cl^-$, for which the electric dipole moment should equal the electronic charge multiplied by the internuclear distance. The observed dipole moments listed in Table 10.3 show that this crude model is not far wrong for the alkali halides; the model can be improved by allowing for the fact that the field of each charged ion polarizes the other ion (see Fig. 10.2), producing induced moments $\mathbf{p}_i$ which have the opposite direction to the main dipole moment $\mathbf{p}$. The ionic picture is clearly much less good for the hydrogen halide molecules HCl, HBr, HI, where in progression the dipole moments actually decrease while the internuclear distance increases. This shows that the picture of such a molecule as a pair of charged ions is a gross oversimplification; in

TABLE 10.3
*Internuclear distances and electric dipole moments*
*of some diatomic molecules*

| Molecule | Internuclear distance $r$ (nm) | Electronic charge$\times r$ (debye) | Observed dipole moment (debye) |
|----------|----------|----------|----------|
| CsF | 0·2345 | 11·2 | 7·88 |
| CsCl | 0·2906 | 14·0 | 10·46 |
| CsI | 0·3315 | 15·9 | 12·1 |
| KF | 0·255 | 12·2 | 7·33 |
| KCl | 0·2667 | 12·8 | 10·48 |
| KBr | 0·2821 | 13·5 | 10·41 |
| KI | 0·3048 | 14·6 | 11·05 |
| HCl | 0·127 | 6·1 | 1·03 |
| HBr | 0·142 | 6·8 | 0·78 |
| HI | 0·162 | 7·8 | 0·38 |

fact the electronic charge tends to be shared more equally between the two nuclei. This tendency increases as we go from HCl to HI, corresponding to a progressive change towards covalent binding in which the valence electrons are wholly shared between the two atoms.

For polyatomic molecules the problem is more complicated, and we illustrate it by reference to some triatomic molecules. Both $SO_2$ (1·59 debye) and $H_2O$ (1·84 debye) have large dipole moments, and this shows they cannot be symmetrical linear molecules like $CO_2$, which is O—C—O and has no dipole moment. It does not exclude the possibility that they are asymmetrical linear molecules like $N_2O$, whose structure is N—N—O, but the latter has only a small moment of 0·17 debye. In fact both $H_2O$ and $SO_2$ have the shape of isosceles triangles, as in Fig. 10.3, the angles at the apex (the O or S nucleus) being 105° and 120° respectively. The direction of the dipole moment is clearly normal to the

FIG. 10.2.  The induced dipoles $p_i$ on each ion are in the opposite direction to the main dipole $p$ formed by the charges on the two ions, so that the total dipole moment is less than $p$.

FIG. 10.3. (a) The $H_2O$ molecule has the shape of an isosceles triangle, with a large electric dipole moment. (b) The $CO_2$ molecule is linear and symmetrical, with no electric dipole moment. The $CO_2$ molecule acquires an instantaneous moment in a bending (d) or asymmetrical vibration (c). In a symmetrical vibration (not shown) where the oxygen atoms move in and out together and the carbon atom is stationary, the shape remains as in (b) and there is no oscillating dipole moment.

base of the triangle, and in the plane of the triangle, since reflection symmetry excludes other possibilities.

When an electric field is applied to a polar gas, in addition to the induced moments considered in the last section there will be a contribution to the polarization from the permanent dipole moments. The potential energy of a dipole $\mathbf{p}$ in a field of strength $\mathbf{E}$ is $-\mathbf{p}.\mathbf{E} = -pE \cos \theta$, where $\theta$ is the angle between $\mathbf{p}$ and $\mathbf{E}$. In a classical treatment using the Boltzmann distribution, since the parallel direction has the lower energy, there will be more dipoles pointing with the field than pointing against it, giving a net polarization in the direction of the field. The problem has already been treated in detail for the corresponding magnetic case, and the result can be adapted to the electrical case by writing $\mathbf{p}$ and $\mathbf{E}_{local}$ instead of $\mathbf{m}$ and $\mathbf{B}$. For small field strengths, the contribution $\mathbf{P}_d$ from the permanent dipoles is (cf. eqn (6.13))

$$\mathbf{P}_d = \left(\frac{n_0 p^2}{3kT}\right)\mathbf{E}_{local}, \tag{10.15}$$

where $k$ is Boltzmann's constant and $T$ the absolute temperature.

The relation between the local field $\mathbf{E}_{local}$ and the macroscopic field $\mathbf{E}$ in the medium is more complicated than for induced dipoles because the permanent dipoles are not all oriented parallel to the field (see § 10.7). For gases at such low densities that the difference between $\mathbf{E}_{local}$ and $\mathbf{E}$ can be neglected, the static relative permittivity $\epsilon_s$ is given by the relation

$$\epsilon_s - \epsilon_i = \frac{\mathbf{P}_d}{\epsilon_0 \mathbf{E}} = \frac{n_0 p^2}{3\epsilon_0 kT}. \tag{10.16}$$

Here $\epsilon_i$ is that part of the relative permittivity due to the induced dipoles alone; and, in the low-density limit, $\epsilon_i - 1 = n_0\alpha/\epsilon_0$.

In this derivation we have used the formula appropriate to the limit of $pE/kT \ll 1$, a condition which is generally readily fulfilled. At room temperature $kT = 4 \times 10^{-21}$ J, so that even for a dipole with a moment of 4 debye a field of $3 \times 10^7$ V m$^{-1}$ is required to make $pE/kT = 0 \cdot 1$. We cannot increase this ratio for a gas by going to low temperatures, since polar gases tend to have high liquefaction- and freezing-points owing to the large intermolecular forces between the permanent dipoles. These same intermolecular forces make eqn (10.16) quite inapplicable to a liquid or a solid. This is very different from the corresponding magnetic case, because in many paramagnetic salts the magnetic dipoles remain relatively free to orient themselves in a magnetic field, and Curie's law may hold accurately over a wide range of temperature.

The derivation of the contribution to the polarization from the permanent dipoles which we have given is a purely classical one, and the reader may wonder to what extent it is confirmed by wave mechanics. The answer to this is that exactly the same result is obtained, but in a surprisingly different way. This may be illustrated by reference to a diatomic molecule. The rotational states of such a molecule are distinguished by having quantized values of the angular momentum equal to $J(h/2\pi)$, where $J$ is zero or a positive integer and $h$ is Planck's constant. The calculation shows that in small fields the states for which $J \neq 0$ contribute nothing to the polarization in respect of the permanent dipole moment of the molecule. This is reasonable because when the molecule is turning end over end, the average projection of the dipole moment on any direction in space is zero. In the state $J = 0$, however, the molecule is not rotating, and the whole of the contribution comes from this state. At high temperatures a large number of rotational states are occupied, and the fraction of molecules which are in the state $J = 0$ is proportional to $1/T$. This gives the same temperature variation as the classical theory, and detailed calculation shows that the numerical constant is also the same (see Pauling and Wilson 1935).

Eqn (10.16) may be used to determine the electric dipole moment from measurements of the permittivity of a gas (for experimental methods see § 22.6). To separate the contributions from the induced polarization and the permanent dipoles, measurements are made over as wide a temperature range as possible. A plot of $\epsilon_r$ against $T^{-1}$ should give a straight line, from the slope of which the dipole moment can be found using eqn (10.16). The intercept at $T^{-1} = 0$ gives the value of $\epsilon_i$.

## 10.6. Dispersion in gases

The theory of electromagnetic waves (Chapter 8) shows that the refractive index $n$ of a substance should be equal to the square-root of $\epsilon_r$, if the magnetic permeability can be taken as unity, as is usually the case. A comparison of the values of $\epsilon_r$ measured at low frequencies with the

TABLE 10.4

| Gas | $(\epsilon_r - 1)\,10^6$ at N.T.P. | | | | |
|---|---|---|---|---|---|
| | 0·1 MHz† | 1 MHz‡ | 9000 MHz§ | 24 000 MHz‖ | Optical†† |
| Air | 570±0·7 | 567·0±1·0 | 575·4±1·4 | 576·0±0·2 | 575·7±0·2 |
| Nitrogen | 578±0·7 | 579·6±1·0 | 586·9±2·9 | 588·3±0·2 | 581·3 |
| Oxygen | 528±1 | 523·3±1 | 530·0±1·9 | 531·0±0·4 | 532·7 |
| Argon | 545±1 | 545·1±0·5 | — | 555·7±0·4 | 554·7 |
| Carbon dioxide | 987±1 | 987·5±2 | 985·5±3 | 988  ±2 | — |
| Hydrogen | 270±1 | 272 | — | — | 272 |

† Lovering W. F. and Wiltshire L. (1951), *Proc Inst. Elect. Engrs.* 98, Part II, 557.
‡ Hector L. G. and Woernley D. L. (1946), *Phys. Rev.* **69**, 101.
§ Birnbaum G., Kryder S. J., and Lyons L. (1951), *J. Appl. Phys.* **22**, 95.
‖ Essen L., and Froome K. D. (1951), *Proc. phys. Soc.* B**64**, 862.
†† $(n^2-1)10^6$ (various authors), extrapolated to infinite wavelength.

square of the refractive indices measured in the optical region (that is, at frequencies of the order of $10^{14}$ Hz) gives very poor agreement with this relation except in the case of simple non-polar gases. Values of $\epsilon_r$ of a few such gases measured over a wide range of frequencies are given in Table 10.4 together with the square of the optical refractive index. The latter is extrapolated to 'infinite wavelengths' to correct for dispersion in the optical region. The agreement is seen to be excellent in the cases quoted.

In the optical region, variation of the refractive index with wavelength has been known for a very long time, and is called dispersion. In general the refractive index increases as the wavelength decreases, and this is known as 'normal dispersion'. The reverse case, where the refractive index decreases with decreasing wavelength, occurs only in the vicinity of an absorption line and is difficult to observe because of the absorption. This is known as 'anomalous dispersion', but both types have a simple explanation in terms of classical theory, based on the assumption that an atom contains electrons vibrating at certain natural frequencies characteristic of the type of atom, and that the application of an alternating electric field sets such electrons into forced vibration.

Let us take the simplest possible case of a gas subjected to an oscillating electric field $\mathbf{E}_{local} = \mathbf{E}^{(0)} \exp j\omega t$. We shall assume that the wavelength of the incident radiation is very large compared with atomic dimensions (which is true up to the region of hard X-rays), so that the field acting on an electron in a given atom is independent of its position with respect to the nucleus, which is assumed to be stationary. Each electron of charge $-e$ is displaced a distance $\mathbf{s}$ by the field, and the restoring force is written as $-m\omega_p^2\mathbf{s}$, where $\omega_p/2\pi$ is the natural frequency of oscillation of the electron and $m$ its mass. In addition there will be damping due to collisions, radiation of energy, etc., which may be represented by a term $-m\gamma(ds/dt)$. Hence we have

$$m\left(\frac{d^2\mathbf{s}}{dt^2} + \gamma\frac{d\mathbf{s}}{dt} + \omega_p^2\mathbf{s}\right) = -e\,\mathbf{E}^{(0)}\exp j\omega t. \tag{10.17}$$

The solution of this is

$$\mathbf{s} = -\frac{e\,\mathbf{E}_{\text{local}}}{m\{(\omega_p^2 - \omega^2) + j\gamma\omega\}} + \exp(-\tfrac{1}{2}\gamma t)[\mathbf{A}\cos\{(\omega_p^2 - \tfrac{1}{4}\gamma^2)^{\frac{1}{2}}t\} + \mathbf{B}\sin\{(\omega_p^2 - \tfrac{1}{4}\gamma^2)^{\frac{1}{2}}t\}].$$

The terms in $\mathbf{A}$ and $\mathbf{B}$ average to zero over many atoms since $\mathbf{A}$ and $\mathbf{B}$ depend on the initial conditions and are as often positive as negative. The instantaneous electric dipole moment due to the displacement of the electron is $\mathbf{p} = -e\mathbf{s}$, and, if there are $n_0$ molecules per unit volume, the polarization $\mathbf{P}$ is

$$\mathbf{P} = n_0\mathbf{p} = \frac{n_0 e^2\mathbf{E}_{\text{local}}}{m}\frac{1}{(\omega_p^2 - \omega^2) + j\gamma\omega}. \tag{10.18}$$

For gases at higher density a correction for the difference between the local field and the external field may be applied in the same way as in § 10.3, leading to the formula

$$\frac{\epsilon_r - 1}{\epsilon_r + 2} = \frac{\mathbf{n}^2 - 1}{\mathbf{n}^2 + 2} = \frac{n_0 e^2}{3m\epsilon_0}\frac{1}{(\omega_p^2 - \omega^2) + j\gamma\omega}. \tag{10.19}$$

This formula shows that both $\epsilon_r$ and $\mathbf{n}$ must be regarded as complex. Writing $\epsilon_r = \epsilon' - j\epsilon'' = (n - jk)^2$, where $n$ is the real part of the refractive index and $k$ is the absorption coefficient, we may separate the real and imaginary parts of eqn (10.19). The formula is clumsy to handle, however, and we shall assume that we are dealing only with gases at such low pressures that we can neglect the Lorentz correction. When the value of $k$ is small, and negligible except near an absorption line, we may also make the approximation, if the line is narrow, of writing

$$(\omega_p^2 - \omega^2) = (\omega_p + \omega)(\omega_p - \omega) \approx 2\omega(\omega_p - \omega).$$

Then we obtain the formulae

$$\left.\begin{array}{l} \epsilon' = n^2 - k^2 \approx n^2 = 1 + \dfrac{n_0 e^2}{2m\omega\epsilon_0}\left\{\dfrac{\omega_p - \omega}{(\omega_p - \omega)^2 + \Delta\omega^2}\right\}, \\[3mm] \epsilon'' = 2nk \approx 2k = \dfrac{n_0 e^2}{2m\omega\epsilon_0}\left\{\dfrac{\Delta\omega}{(\omega_p - \omega)^2 + \Delta\omega^2}\right\}, \end{array}\right\} \tag{10.20}$$

where the symbol $\Delta\omega$ has been used for $\gamma/2$, and we have assumed $n \approx 1$, $k \ll n$.

The variation of $n$ and $k$ in the neighbourhood of a weak absorption line is shown in Fig. 10.4. The absorption coefficient reaches a maximum at the resonant frequency where $\omega = \omega_p$, and falls to half its maximum

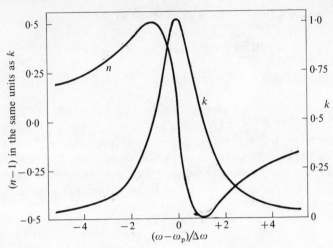

FIG. 10.4. Variation of $n$ and $k$ near a narrow absorption line (from eqn (10.20)). $(n-1)$ and $k$ are in units of $n_0 e^2/4m\omega\epsilon_0 \Delta\omega$.

value at $\omega_p - \omega = \pm\Delta\omega$. In optical usage, the quantity $2\,\Delta\nu = \Delta\omega/\pi$ is called the 'half-width' of the line, meaning the frequency difference between the points at which the absorption has dropped to half the maximum value. Microwave spectroscopists, however, prefer to call $\Delta\nu$ the half-width.

In general, each atom or molecule possesses a number of characteristic resonant frequencies, and the expressions given above for the refractive index and absorption coefficient should be replaced by others with summations over the various values of $\omega_p$. If the number of electrons per molecule which have a resonant frequency $\omega_p$ is denoted by $f_p$, we may write eqn (10.19) in the form

$$\frac{\epsilon_r - 1}{\epsilon_r + 2} = \frac{\mathbf{n}^2 - 1}{\mathbf{n}^2 + 2} = \frac{n_0 e^2}{3m\epsilon_0} \sum_p \frac{f_p}{(\omega_p^2 - \omega^2) + \mathrm{j}\gamma\omega}. \tag{10.21}$$

The value of $f_p$ is known as the 'oscillator strength' of an absorption line, and on classical theory we should expect it to be unity. In practice it generally has values less than unity, and the quantum-mechanical explanation shows that this corresponds to the fact that each electron possesses a number of possible frequencies of oscillation, and its total oscillator strength is divided between them. Then $\sum_p f_p = 1$ for each electron.

At frequencies far from resonance the absorption coefficient is negligible. At very low frequencies, where $\omega \ll$ any value of $\omega_p$, we have

$$\frac{\epsilon_r - 1}{\epsilon_r + 2} = \frac{n_0 e^2}{3m\epsilon_0} \sum_p \frac{f_p}{\omega_p^2} = \frac{n_0 \alpha}{3\epsilon_0} \tag{10.22}$$

by comparison with (10.13). This shows that the molecular polarizability $\alpha$ is intimately connected with the oscillator strengths and absorption lines. This important result follows from a simple picture in which the displacement $\mathbf{s}$ of a particle is assumed to set up a restoring force which is proportional to $\mathbf{s}$. The dynamical equation of motion is $m(\mathrm{d}^2\mathbf{s}/\mathrm{d}t^2)+m\omega_q^2\mathbf{s}=0$, corresponding to harmonic vibrations of frequency $\omega_p/2\pi$. In a static field of strength $\mathbf{E}$, if the charge on the particle is $q$, we have $q\mathbf{E}=m\omega_p^2\mathbf{E}$, and the induced dipole moment is

$$\mathbf{p}_i=q^2\mathbf{E}/m\omega_p^2=\alpha_p\mathbf{E},$$

where $\alpha_p$ is the polarizability associated with the vibration at frequency $\omega_p/2\pi$. On writing $q^2=e^2f_p$, and summing over all resonant frequencies, we obtain

$$\alpha=\frac{e^2}{m}\sum_p\frac{f_p}{\omega_p^2}. \tag{10.23}$$

As the frequency is raised to pass through an absorption line at $\omega_p/2\pi$, the refractive index goes through the anomalous variation shown in Fig. 10.4 and approaches a smaller limiting value on the high-frequency side than it had on the low-frequency side. When there are a number of absorption lines, the behaviour is as shown in Fig. 10.5, and finally, when $\omega$ is greater than all values of $\omega_p$, $n$ approaches unity, but the value of $(n-1)$ is slightly negative. This is the well-known anomaly in the refractive index in the X-ray region, and the value of $n$ is then generally calculated by assuming that the electrons are free, so that eqn (10.17)

FIG. 10.5. Schematic diagram showing variation of refractive index with frequency.

reduces to

$$m\frac{d^2s}{dt^2} = -e\mathbf{E}' \exp j\omega t.$$

This is equivalent to the assumption that $\omega \gg \omega_p, \gamma$.

Electronic resonant frequencies lie in the visible or ultraviolet regions of the spectrum; for colourless compounds they lie entirely in the ultraviolet, and the refractive index increases with frequency in the visible region, corresponding to 'normal dispersion'. The electronic contribution to the polarizability may therefore be deduced from the optical refractive index, with a correction to 'infinite wavelength' to allow for dispersion. The electronic molar polarizability is defined, by analogy with eqn (10.14), as the quantity

$$\frac{n^2-1}{n^2+2} \cdot \frac{M}{\rho} = \frac{N_A \alpha_{el}}{3\epsilon_0}. \tag{10.24}$$

For a compound this quantity is approximately equal to the sum of the contributions from the individual atoms or ions, and, from measurements on a large range of compounds, it is possible to deduce a set of atomic or ionic polarizabilities. The additive relation is not exact because the polarizability is sensitive to the nature of the chemical binding. However, a general rule is that the polarizability is larger for negative ions than for positive ions with the same number of electrons (for example, $Cl^-$ has a polarizability several times greater than $K^+$), and in any column of the periodic table it increases with atomic number (for example, the polarizability of xenon is some 20 times greater than that of helium).

In addition to the electronic contributions to the polarizability, there are 'ionic' contributions arising from the motion of heavy ions, with resonant frequencies in the infrared. We consider first internal vibrations in molecules which involve changes in the internuclear distances, with resonances in the region $\lambda \sim 3\text{--}100 \ \mu\text{m}$. Only vibrations in which an oscillating dipole moment is set up are important, as can be illustrated for the molecule $CO_2$. In equilibrium this has a linear symmetric structure with no dipole moment, but if it is bent or asymmetrically distorted as in Fig. 10.3 it acquires an electric dipole moment. Such distortions may be induced by a static electric field, and contribute to the static polarizability; when vibrating, the oscillating dipole moment causes absorption bands in the near infrared. On the other hand, the symmetric vibration in $CO_2$ gives no dipole moment and no absorption band; it is said to be 'inactive'.

If a molecule has a permanent electric dipole moment, there is an additional term (eqn (10.16)) in the static permittivity. Classically this arises from a net change in the orientation of the dipoles; in quantum mechanics it is connected with the rotational energy levels. For such a molecule, an electromagnetic field exerts a couple which, in classical mechanics, changes the rate of rotation. In quantum mechanics, transitions occur between the rotational levels which give rise to a 'pure rotational' spectrum in the far infrared (typically, $\lambda \sim 100\text{–}10^4\ \mu\mathrm{m}$). Contributions to the permittivity occur only if the dipole moment is electric, but a few molecules such as $O_2$, NO, and $NO_2$ have permanent magnetic dipole moments of the order of a Bohr magneton (1 $\mu_B$). The contribution to the static permeability from such magnetic dipoles is very small compared with the permittivity due to electric dipoles, involving $\mu_0 m^2$ (eqn (6.13)) instead of $p^2/\epsilon_0$ (eqn (10.16)).

## 10.7. Static permittivity of liquids

The static values of $\epsilon_r$ for liquids are related to absorption lines in a similar way to that outlined for gases. In simple atomic substances, such as the condensed phases of the rare gases of the atmosphere, there are no absorption bands in the infrared and $\epsilon_r$ does not differ greatly from that deduced from the Clausius–Mossotti formula, using the value of $\alpha$ measured in the gas phase. This is particularly true for liquid helium, which has an exceptionally low density and permittivity; at the boiling-point, the density is $125\ \mathrm{kg\ m^{-3}}$ (one-eighth that of water), and $\epsilon_r = 1\cdot048$. This corresponds to a molar polarizability ($N_A\alpha/3\epsilon_0$) of $0\cdot12\times10^{-6}$, very close to the value of $0\cdot123\times10^{-6}$ for the gas.

Organic liquids such as benzene ($C_6H_6$) or carbon tetrachloride ($CCl_4$) are assemblies of non-polar molecules, and similar relations hold for $\epsilon_r$, though there may be small contributions from vibrations internal to the molecules.

The situation is quite different for a 'polar liquid' in which the molecules have permanent electrical dipole moments. Because of the high density, a small departure from random orientation of the dipoles produces a large contribution $P_d$ to the polarization, and if this is included in the calculation of the local field by the Lorentz method, it yields nonsensical results. If we write

$$\mathbf{E}_{\text{local}} = \mathbf{E} + \mathbf{P}_d/3\epsilon_0$$

in eqn (10.11), on combining this with (10.15) we obtain

$$\frac{\mathbf{P}_d}{\mathbf{E}} = \frac{\mathbf{P}_d}{\mathbf{E}_{\text{local}} - \mathbf{P}_d/3\epsilon_0} = \frac{n_0 p^2}{3k(T - T_c)}, \tag{10.25}$$

with $kT_c = n_0 p^2/3\epsilon_0$. For water this gives $T_c \sim 1000\ \mathrm{K}$, which would mean that below this temperature water should be spontaneously polarized, as in a ferromagnet (Chapter 15). This unrealistic result arises as follows. The interaction of a dipole with its surroundings produces a 'reaction field' which is parallel to the dipole (cf. Problem 2.2); this field may induce an added moment parallel to the permanent moment but it cannot change the orientation of the moment. In calculating the net polarization arising from reorientation of the permanent moments in a field, it is therefore wrong to use a value of the local field which includes the contribution from the permanent dipoles, as we did above.

This problem has been treated by Onsager (1936) using an approach basically different from that of Lorentz. Each dipole is regarded as being at the centre of a *real* spherical cavity, whose size is assumed equal to the average volume occupied by each molecule. Standard electrostatic theory is then used to evaluate separately the effect of the surrounding dielectric on the randomly oriented permanent dipole, and on the induced dipole (which is parallel to the applied field). Some rather complicated algebra yields the formula

$$\frac{(\epsilon_s - \epsilon_i)(2\epsilon_s + \epsilon_i)}{\epsilon_s(\epsilon_i + 2)^2} = \frac{n_0 p^2}{9\epsilon_0 kT}, \tag{10.26}$$

where $\epsilon_s$ is the total static relative permittivity, while $\epsilon_i$ is that part arising solely from the induced dipoles. For water, using $\epsilon_i = 4\cdot9$ (see § 10.8) and $p = 1\cdot94$ debye, this formula gives $\epsilon_s = 105$ at $T = 273$ K, falling to $\epsilon_s = 78$ at 373 K. The measured values are 88 and 55 respectively; although the formula gives values on the high side, they correspond much better to reality than does the Lorentz prediction. Better agreement is hardly to be expected; $H_2O$ is a triangular molecule, while a spherical cavity is assumed, and short-range forces are neglected. Only long-range forces (forces between dipoles) are included in methods using formulae based on electrostatic theory, and, in spite of its apparent difference of approach, Onsager's theory gives the same result as that of Lorentz for induced dipoles.

In dilute solutions of polar molecules in non-polar solvents, the permanent dipoles are far apart on average and their mutual interactions can be neglected. Onsager's theory can be extended to allow for interaction with the solvent molecules, giving an effective value of the dipole moment which must be allowed for when using eqn (10.16). The permittivity for a range of concentrations is measured by a standard method (see § 22.6), and extrapolated to infinite dilution. The solvents normally employed are carefully purified benzene and carbon tetrachloride. Electric dipole moments measured in this way agree fairly well with those found using the gaseous method (§ 10.5); the latter is the more satisfactory method when it can be used, the solvent method being employed for substances of very low vapour pressure. For a number of simple molecules (such as those in Table 10.3) accurate values of the dipole moment have been obtained from microwave spectroscopy (see Townes and Schawlow 1955) or electric resonance in molecular beams (see Ramsey 1956), by measurement of the splitting of the rotational lines in an electric field.

## 10.8. Dispersion in polar liquids

For polar gases $\epsilon_r$ is greater than the square of the optical refractive index; the difference is due to dispersion in the infrared, mainly associated with the rotational spectrum of the molecules. In a polar liquid such as water, the high value of $\epsilon_r$ is due to reorientation of the electric dipoles in the applied field, and in an alternating field this reorientation must occur sufficiently quickly for the dipoles to follow the field reversals if they are to make their full contribution to the polarization. If reorientation takes a finite time $\sim \tau$, the dipoles cannot follow a field whose angular frequency is such that $\omega \tau \gg 1$. In the region where $\omega \tau \sim 1$, the permittivity falls, and energy is absorbed from the alternating field into the dielectric, finally appearing as heat.

The mechanism which inhibits reorientation is collisions between the molecules. This results from the molecular motion, which is visible in the form of Brownian motion, and results in the phenomenon of viscosity. For a spherical particle of radius $a$ suspended in a liquid of viscosity $\eta$, the mean-square value $\overline{\theta^2}$ of the rotational displacement angle $\theta$ in a time $t$ is

$$\overline{\theta^2} = \frac{kT}{4\pi\eta a^3}\, t = \frac{t}{\tau}, \tag{10.27}$$

where $\tau = 4\pi\eta a^3/kT$ is a characteristic time for the Brownian motion. If we apply this to the molecules of a liquid such as water, taking $a = 2\cdot3\times10^{-10}$ m, the value found from the viscosity of the vapour, and $\eta = 0\cdot001$ S.I. units at 290 K, we find $\tau = 3\cdot7\times10^{-11}$ s. Since this time is longer than any of the characteristic periodic times of rotation of the free water molecule, it follows that the molecule cannot rotate at any of its natural frequencies in the liquid state. Instead, the dispersion associated with the permanent dipoles will take place at frequencies such that $\omega \approx 1/\tau$, that is at a wavelength of the order of 10 mm.

In order to introduce $\tau$ into our treatment of the permittivity, we consider the effect of maintaining a steady field on a polar liquid, and then suddenly removing it. Under the influence of the Brownian motion, the preferred orientations of the dipoles will gradually disappear. It is reasonable to suppose that the rate of decay of the polarization is proportional to the instantaneous value of the polarization, and we write

$$d\mathbf{P}/dt = -\mathbf{P}/\tau,$$

giving

$$\mathbf{P} = \mathbf{P}_0 e^{-t/\tau},$$

where $\tau$ is a characteristic 'relaxation time' which we would expect to be of the same order as that found above for the Brownian motion. Here $\mathbf{P}$ is, of course, only that part of the polarization associated with the permanent dipoles. If the field is not switched off but changed suddenly to a value for which the equilibrium polarization is $\mathbf{P}_0$ then the rate of change of $\mathbf{P}$ is given by the equation

$$\frac{d\mathbf{P}}{dt} = \frac{(\mathbf{P}_0 - \mathbf{P})}{\tau}, \qquad \text{or} \quad \mathbf{P} + \tau \frac{d\mathbf{P}}{dt} = \mathbf{P}_0.$$

When an alternating field $\mathbf{E} \exp j\omega t$ is applied, we may write our equation for the polarization in the form (cf. eqn (10.15))

$$\mathbf{P} + \tau \frac{d\mathbf{P}}{dt} = \mathbf{P}_0 = \frac{n_0 \mathbf{p}^2}{3kT} \mathbf{E} \exp j\omega t$$

giving

$$\mathbf{P} = \frac{n_0 \mathbf{p}^2}{3kT} \frac{\mathbf{E} \exp j\omega t}{1 + j\omega\tau}. \tag{10.28}$$

Here $\mathbf{E}$ is the amplitude of the local alternating field, and $\mathbf{P}$ is that part of the polarization due only to the permanent dipoles. Allowing for these effects, we find for the relative permittivity $\epsilon_r$ the expression

$$\frac{\epsilon_r - \epsilon_i}{\epsilon_s - \epsilon_i} = \frac{1}{1 + j\omega\tau}, \tag{10.29}$$

where $\epsilon_i$ is that part of the relative permittivity due to induced polarization, and $\epsilon_s$ is the static relative permittivity. This result holds for the Onsager treatment for the local field, but it can be shown that the formula is similar if the Lorentz correction is used, except that we must use a modified relaxation time $\tau' = \tau(\epsilon_s + 2)/(\epsilon_i + 2)$. This difference is significant only in a comparison of the relaxation time determined from the dispersion of the relative permittivity with that from the Brownian motion.

Eqn (10.28) above shows that $\mathbf{P}$, and hence $\epsilon_r$ is complex, and we may either write $\epsilon_r = \epsilon_r' - j\epsilon_r''$, or $\epsilon_r = (n - jk)^2$, where $n$ is the refractive index and $k$ the absorption coefficient. Then

$$n^2 - k^2 = \epsilon_r' = \frac{\epsilon_s - \epsilon_i}{1 + \omega^2 \tau^2} + \epsilon_i, \tag{10.30}$$

$$2nk = \epsilon_r'' = \frac{(\epsilon_s - \epsilon_i)\omega\tau}{1 + \omega^2 \tau^2}. \tag{10.31}$$

The variation of these quantities with frequency is easily seen from Fig. 10.6. $\epsilon_r'$ falls from $\epsilon_s$ at low frequencies to $\epsilon_i$ at high frequencies, the transition taking place near $\omega = 1/\tau$. $\epsilon_r''$ has a maximum in this region at $\omega = 1/\tau$, and falls to zero at both low and high frequencies.

$$\epsilon'_r = \epsilon_i + \frac{\epsilon_s - \epsilon_i}{1 + \omega^2 \tau^2}$$

$$\frac{\epsilon''_r}{\epsilon_s - \epsilon_i} = \frac{\omega\tau}{1 + \omega^2 \tau^2}$$

$\omega\tau$ (logarithmic scale)

FIG. 10.6. Variation of $\epsilon'_r$ and $\epsilon''_r$ for a polar liquid.

The theory of the dispersion in polar liquids was given by Debye (1929), and that above is a simplified version of his treatment. It was first verified for glycerine and a number of alcohols, for which $\tau$ is much greater than for water owing to their higher viscosity and larger molecular radius. For example, at 295 K the values of $\epsilon'_r$ and $\epsilon''_r$ for glycerine at a wavelength of 9·5 m are 42 and 8·6 respectively, showing that we are already well into the region of anomalous dispersion at this wavelength. Vacuum-tube oscillators and detectors were available for such wavelengths, but the dispersion in water could not be measured accurately until centimetre-wave technique was established, owing to the small value of $\tau$. We shall here describe the measurements of Collie, Hasted, and Ritson (1948).

An accurate method of determining the permittivity of low-loss liquids, using resonant cavities, is described in § 22.6. This method cannot be used with water, since the absorption is so great if the cavity is filled with water that no resonance can be observed. This difficulty can be surmounted by using a cavity partly filled with water, the degree of filling being adjusted to give a measurable change in the resonant frequency and $Q_F$ of the cavity.

An alternative method, used by the same workers, is to determine the propagation constant in a waveguide filled with water. This constant depends on both the real and imaginary parts of the relative permittivity and on the linear dimensions of the guide. By using two guides of different sizes, the values of $\epsilon'_r$ and $\epsilon''_r$ can be found separately, since their relative contributions to the propagation constant depend on the size of the guide. From § 9.6 the field in the guide varies as

$$\exp(-\gamma x) = \exp\{-(\alpha + j\beta)x\},$$

where (writing $\lambda_0$ for the wavelength in free space)

$$\gamma = 2\pi\left(\frac{1}{\lambda_c^2} - \frac{f^2}{v^2}\right)^{\frac{1}{2}} = 2\pi\left(\frac{1}{\lambda_c^2} - \frac{\epsilon_r}{\lambda_0^2}\right)^{\frac{1}{2}} = 2\pi\left(\frac{1}{\lambda_c^2} - \frac{\epsilon_r' - j\epsilon_r''}{\lambda_0^2}\right)^{\frac{1}{2}}.$$

Hence

$$\alpha^2 - \beta^2 = 4\pi^2\left(\frac{1}{\lambda_c^2} - \frac{\epsilon_r'}{\lambda_0^2}\right),$$

$$\alpha\beta = 2\pi^2\frac{\epsilon_r''}{\lambda_0^2}. \tag{10.32}$$

Thus two separate measurements of $\alpha$, with different values of $\lambda_c$, suffice to determine $\epsilon_r'$ and $\epsilon_r''$. The value of $\alpha$ can be found by moving a detector through the liquid, and the measurement of phase (that is, $\beta$), which is very difficult when large attenuation is present, is avoided. Essentially the method adopted was to use two waveguides beyond cut-off, which act as attenuators in series, one being filled with water and the other not. The two attenuators are adjusted in opposite directions so as to keep the power reaching a receiver constant; thus no calibration of the receiver is required. The attenuation in the water is calculated from the known law of the air-filled attenuator.

The results obtained may be fitted accurately to the theory using a value of $\tau = 1\cdot01 \times 10^{-11}$ s, as shown in Fig. 10.7, where both the calculated curves and the experimental points are given. The great intensity of

FIG. 10.7. Complex permittivity of water at 293 K.

the absorption is illustrated by the fact that at a wavelength of 12·4 mm, the power in an incident wave would be diminished by a factor of $e^{-2(2\pi k)} = e^{-36}$ or about $10^{-15\cdot5}$ in passing through a thickness of 12·4 mm. Thus water is quite 'black' at such wavelengths. It should be noted that in fitting these results, the best value of $\epsilon_i$, that part of the permittivity due to induced polarization, is found to be 4·9. This is appreciably higher than the square of the optical refractive index ($n = 1\cdot33$), showing that there must be other strong absorption bands in liquid water in the infrared; these are associated with internal vibrations of the $H_2O$ molecule. Rather similar results have been obtained by Lane and Saxton (1952) for methyl and ethyl alcohols.

The values of $\epsilon_r'$ and $\epsilon_r''$ in the region of dispersion vary quite rapidly with temperature, corresponding to a variation in the relaxation time $\tau$. Saxton (1952) has shown that $\tau$ varies from about $27 \times 10^{-12}$ s at 263 K (in supercooled water) down to $4\cdot7 \times 10^{-12}$ s at 323 K. This variation is very closely parallel to that of the viscosity, and indeed the value of $\tau$ is surprisingly close to that which would be obtained using the simple formula $\tau = 4\pi\eta a^3/kT$.

## 10.9. Ionic solids

For a simple atomic solid (such as diamond) or a molecular solid (most organic compounds) the permittivity is similar to that outlined at the beginning of § 10.7. When polar molecules are present, rotation of the permanent dipoles is completely inhibited except near the melting-point. For example, the value of $\epsilon_r$ for ice, measured at $10^{10}$ Hz and low temperatures, is about 3. Near the melting-point there is some dispersion in the region of $10^6$ Hz, $\epsilon_r$ rising from the value just quoted to about 80 at zero frequency. This results from some residual dipole rotation similar to that in liquid water, but with much larger values of $\tau$, associated with a high effective viscosity which increases rapidly at lower temperatures.

The position is different for another important class of solids: the ionic solids. These consist of regular lattices of positive and negative ions; the strong binding forces between such ions result in high melting-points. In an electric field the forces on the ions cause a small displacement of the lattice of positive ions relative to the negative ions. This gives a rather large electric polarization and high relative permittivity (see Table 10.5). The values of $\epsilon_r$ are the same as the static relative permittivity at radio-frequencies, but are appreciably greater than $n^2$ at a wavelength of 1 $\mu$m. Resonant frequencies must therefore exist in the infrared; these are connected with vibrations of the ions about their equilibrium positions. For a three-dimensional lattice this is a rather complex problem, but the main feature can be understood from a one-dimensional example.

TABLE 10.5

*Values of the static relative permittivity $\epsilon_s$,
the square of the optical refractive index
$n^2$, and the frequency of the transverse
optical mode $\nu_T$ for various simple ionic
solids (all cubic)*

|      | $\epsilon_s$ | $n^2$ | $\nu_T$ (Hz) |
|------|------|------|------|
| LiF  | 9·0  | 1·9  | $9{\cdot}1\times10^{12}$ |
| NaF  | 5·1  | 1·7  | $7{\cdot}4\times10^{12}$ |
| NaCl | 6·0  | 2·3  | $4{\cdot}9\times10^{12}$ |
| KCl  | 5·0  | 2·2  | $4{\cdot}2\times10^{12}$ |
| KBr  | 4·8  | 2·4  | $3{\cdot}5\times10^{12}$ |
| KI   | 4·9  | 2·7  | $3{\cdot}0\times10^{12}$ |
| TlCl | 32   | 5·1  | $2{\cdot}0\times10^{12}$ |

### The linear diatomic chain

A linear diatomic chain (Fig. 10.8) consists of alternate masses $m_1$, $m_2$; in equilibrium the separation between successive masses is $a$. When a mass is displaced longitudinally by a small amount, a restoring force is set up which is assumed to be proportional to its instantaneous displacement relative to its immediate neighbours. For typical ions of mass $m_1$ at position $2n$ and $m_2$ at $2n+1$, the force equations are

$$m_1\ddot{x}_{2n} = F(x_{2n+1}+x_{2n-1}-2x_{2n}),$$
$$m_2\ddot{x}_{2n+1} = F(x_{2n+2}+x_{2n}-2x_{2n+1}). \tag{10.33}$$

These equations are similar to those for the electric filter (§ 9.1), and we assume that there exist solutions of a wave-like nature with the form

$$x_{2n} = A \exp \mathrm{j}\{\omega t+2nqa\},$$
$$x_{2n+1} = B \exp \mathrm{j}\{\omega t+(2n+1)qa\}. \tag{10.34}$$

Substitution into (10.33) gives

$$-\omega^2 m_1 A - FB(\mathrm{e}^{\mathrm{j}qa}+\mathrm{e}^{-\mathrm{j}qa})+2FA = 0,$$
$$-\omega^2 m_2 B - FA(\mathrm{e}^{\mathrm{j}qa}+\mathrm{e}^{-\mathrm{j}qa})+2FB = 0. \tag{10.35}$$

Apart from the trivial case of $A = B = 0$, these two simultaneous equations have a solution only if the determinant formed from the coefficients of $A$, $B$ is zero:

$$\begin{vmatrix} 2F-m_1\omega^2 & -2F\cos qa \\ -2F\cos qa & 2F-m_2\omega^2 \end{vmatrix} = 0. \tag{10.36}$$

FIG. 10.8. A linear diatomic chain (a) of masses $m_1$ (at points $2n$, $2n\pm2$, etc.) and $m_2$ (at points $2n\pm1$, etc.), and the displacements of the masses in (b) the optical mode and (c) the acoustic mode.

The solution of this determinantal equation is

$$\frac{\omega^2}{F} = \left(\frac{1}{m_1}+\frac{1}{m_2}\right) \pm \left\{\left(\frac{1}{m_1}+\frac{1}{m_2}\right)^2 - \frac{4\sin^2 qa}{m_1 m_2}\right\}^{\frac{1}{2}}, \qquad (10.37)$$

and the ratio $A/B$ can be written as

$$\frac{A}{B} = \frac{\cos qa}{1-(\omega^2 m_1/2F)} = \frac{1-(\omega^2 m_2/2F)}{\cos qa}. \qquad (10.38)$$

Limiting values occur at $qa = 0$ and $qa = \pm\pi/2$, and are summarized in Table 10.6, where it is assumed that $m_1 > m_2$.

TABLE 10.6

|  | $qa = 0$ | $qa = \pm\pi/2$ |
|---|---|---|
| Lower or 'acoustic' branch | $\omega = 0$<br>$A = B$ | $\omega = (2F/m_1)^{\frac{1}{2}}$<br>$B = 0$ |
| Upper or 'optical' branch | $\omega = \left\{2F\left(\dfrac{1}{m_1}+\dfrac{1}{m_2}\right)\right\}^{\frac{1}{2}}$<br>$A/B = -m_1/m_2$ | $\omega = (2F/m_2)^{\frac{1}{2}}$<br>$A = 0$ |

FIG. 10.9. The relation between $\omega$ and $q$ for a diatomic linear chain, given by eqn (10.37). In three dimensions the slope of the acoustic branch near $q = 0$ corresponds to the velocity of sound; the slope of the broken line $\omega/q = c$ is that of electromagnetic waves, and is larger by a factor of about $10^5$.

The two solutions corresponding to eqn (10.37) are plotted in Fig. 10.9 for the case $m_1 = 2m_2$. The lower branch starts at the origin. For values of $qa$ so small that we can replace $\sin qa$ by $qa$, the value of $\omega$ is proportional to $q$ and the phase velocity $\omega/q$ is constant; since $A = B$, adjacent masses move together (Fig. 10.8c) with the same amplitude of vibration. The wave motion is similar to that in a low-frequency elastic wave (a sound wave), and this branch of the spectrum is known as the 'acoustic' branch.

As $qa$ increases, $\omega$ no longer rises as fast as $q$, and the phase velocity $\omega/q$ becomes frequency-dependent. This corresponds to 'dispersion', and the way in which $\omega$ varies with $q$ is known as the 'dispersion relation'. At $\sin qa = \pm 1$, $B = 0$ while $A$ is finite. The masses $m_2$ are all stationary, and the frequency depends only on the masses $m_1$; successive masses $m_1$ vibrate with equal amplitude but opposite phase, since $2qa = \pi$. This is a standing wave in which no energy is propagated; the group velocity $d\omega/dq \to 0$ as $\sin qa \to \pm 1$.

For the upper branch at these points we have a similar result, except that the roles of the masses $m_1$ and $m_2$ are interchanged. The behaviour of the upper branch at $q = 0$ is of especial interest, as we shall see later. Masses $m_1$, $m_2$ vibrate in opposite directions with relative amplitudes

$A/B = -m_1/m_2$, keeping the centre of mass constant (Fig. 10.8(b)). The relative displacement of adjacent particles is large, so that the restoring forces are large, and so is $\omega$. In contrast, at $q = 0$ for the lower branch adjacent particles move together; the relative displacement is small and $\omega \rightarrow 0$ as $q \rightarrow 0$. The wavemotion resembles that on a continuous string.

In this analysis longitudinal wave motion has been assumed, but it is equally valid for transverse waves, where the vibrations can take place in two mutually perpendicular planes. The displacements of the particles shown in Fig. 10.8, are for a transverse wave. The restoring forces differ for longitudinal and transverse waves, so that different dispersion relations must be expected. Each has an upper and a lower branch; the branches are known as 'longitudinal acoustic' (LA), 'longitudinal optic' (LO), 'transverse acoustic' (TA), and 'transverse optic' (TO). In general there are two transverse dispersion relations corresponding to vibrations in orthogonal planes, each normal to the direction of propagation, but these are clearly identical for the linear chain.

So far no restriction has appeared on the number of values of $q$ which can be present. Such a restriction appears only when we consider a chain of finite length, and impose boundary conditions at the ends. A simple approach to this problem is to divide the infinite chain into sections each containing $N$ masses $m_1$ and $N$ masses $m_2$, each section being therefore of length $2Na$. The method of 'periodic boundary conditions' requires that the solution be identical at each equivalent point in each section, that is,

$$x_{2n} = x_{2n+2N}.$$

Then, for any value of $q$, we must have

$$\exp j(2nqa) = \exp j(2nqa + 2Nqa)$$

which is satisfied provided that $2Nqa$ is an integral multiple of $2\pi$. The allowed values of $q$ are therefore

$$q = l(\pi/Na), \tag{10.39}$$

where $l$ is an integer with values lying between $\pm 1$ and $\pm\frac{1}{2}N$, the latter values corresponding to the zone boundaries at $q = \pm\pi/2a$. This gives altogether $N$ acoustic modes and $N$ optical modes, that is, a total of $2N$ modes (for displacements in one single direction), as we should expect for a total of $2N$ masses.

The range of values of $q$ between $-\pi/2a$ and $+\pi/2a$ is known as the 'first Brillouin zone'. The 'zone boundaries' at $\pm\pi/2a$ occur because we have a chain of discrete masses repeating at distances $2a$, and not a continuous string. It is possible to represent the vibrational motion by means of waves of shorter wavelength, but it is unnecessary; only the displacement of the masses can be observed, and this can always be represented by wavelengths in the first zone. Note also that the points $q = \pm\pi/2a$ represent identical states of motion of the masses.

### Interaction with an electromagnetic wave

In the theory of forced oscillations the amplitude is largest at resonance; the frequency of the driving force must coincide with a natural resonant frequency of the driven system. For a system which extends over distances large compared with the wavelength, this is part of a more general requirement of a constant-phase relation between the driving force and the whole of the driven system. This implies that both $\omega$ and $q$ must be the same for an electromagnetic wave and for the vibration in the chain to produce an appreciable interaction. For an electromagnetic wave $\omega/q = c$, represented by a straight line in Fig. 10.9; its slope is the velocity of light, which is about $10^5$ times that of sound, given by $\omega/q$ for the acoustic branch at $q = 0$. A resonant interaction between a light wave and the vibrational motion can occur only at the points of intersection of $\omega/q = c$ with a dispersion curve. One such point is at $\omega = q = 0$, and corresponds to a static electric field; the other occurs at the intersection with the upper branch. This may also be taken as $q = 0$ without serious error, so that its angular frequency is

$$\omega_\mathrm{T} = \left\{ 2F\left(\frac{1}{m_1} + \frac{1}{m_2}\right) \right\}^{\frac{1}{2}}. \tag{10.40}$$

Interaction with an electromagnetic wave requires that the masses carry electric charges, and for an electromagnetic wave travelling along the chain, the forces exerted are perpendicular to the chain. Thus interaction occurs only with 'transverse modes'; this is denoted by the subscript T in eqn (10.40).

If $m_1 = m_2$, the displacements of the two sets of masses are equal and opposite in the transverse optical mode at $q = 0$. If the masses have the same charge, no net dipole moment is created by the displacements, and the vibration is not excited by an electromagnetic wave. Correspondingly, no absorption occurs in a monatomic solid through interaction with the ionic vibrations, provided that the lattice is perfectly regular.

For a diatomic chain with unequal masses, we can deduce the interaction with an electromagnetic wave by assigning charges of $+Q$ and $-Q$ to alternate masses, and adding terms $+Q\mathbf{E}_\mathrm{local}$ and $-Q\mathbf{E}_\mathrm{local}$ to eqn (10.33). Only waves for which $q$ is virtually zero are excited, so that in eqns (10.34) we can put $q = 0$ and drop the subscript $n$. For an angular frequency $\omega$ we can write $\ddot{x} = -\omega^2 x$, and eqns (10.35) become

$$\begin{aligned}
-\omega^2 A &= (2F/m_1)(B - A) + (Q/m_1)\mathbf{E}_\mathrm{local}, \\
-\omega^2 B &= (2F/m_2)(A - B) - (Q/m_2)\mathbf{E}_\mathrm{local}.
\end{aligned} \tag{10.41}$$

The quantity $(A-B)$ can be found by subtracting these two equations. For unit length of chain, containing $n_0$ masses of each type, the polarization $P = n_0 Q(A-B)$. With the help of eqn (10.40) we obtain

$$\mathbf{P} = \frac{n_0 Q^2 \mathbf{E}_{local}}{(\omega_T^2 - \omega^2)} \left( \frac{1}{m_1} + \frac{1}{m_2} \right). \tag{10.42}$$

This has a resonance at $\omega = \omega_T$, and is similar to eqn (10.18) without the damping term.

### Three dimensions

A real solid consists of an assembly of atoms or ions, vibrating in random directions about their equilibrium positions which, in a crystal, are points on a regular three-dimensional lattice. This is clearly an extremely complex problem for which no exact solution can be obtained. A simple analysis assumes that the vibrations can be represented by means of plane waves of the form

$$\exp \mathrm{j}(\omega t - q_x x - q_y y - q_z z) = \exp \mathrm{j}(\omega t - \mathbf{q} \cdot \mathbf{r}). \tag{10.43}$$

Here $\mathbf{q}$ is known as the 'wave vector'; its magnitude is $2\pi/\lambda$, and its direction is the direction of propagation. When boundary conditions are introduced, $q$ is restricted to certain allowed values, as in eqn (10.39). These values may be plotted as a set of points equally spaced along a line in 'reciprocal space'; the spacing is $\pi/Na$, which has the dimensions of $(\text{length})^{-1}$. In three dimensions this gives a reciprocal space with an inverse relationship to the unit cell of the crystal lattice; just as the atoms lie on a lattice of discrete points in real space, the allowed values of $\mathbf{q}$ end on a discrete set of points in reciprocal space. In each case the spacing reflects the symmetry of the crystal lattice. For example, in a cubic lattice, directions parallel to the cube edges (axes of fourfold symmetry) must all be equivalent; a similar condition exists for the body diagonals (axes of threefold symmetry), and face diagonals (axes of twofold symmetry).

For the linear chain the zone boundaries correspond to the points at which travelling waves cannot be propagated, and the vibrations are represented by standing waves. The same is true in three dimensions; the boundaries now consist of sets of intersecting planes in reciprocal space, which form a surface enclosing a 'Brillouin zone'. A wave vector drawn from the origin which terminates on this surface corresponds to a wavelength which cannot propagate in the direction of the wave vector, but suffers an exponential attenuation similar to that of a wave whose frequency lies in the stop band of a filter or in the 'cut-off region' of a waveguide. All possible lattice vibrations can be represented by points in the 'first zone'; at the boundary the values of the wave vector are close to

$10^{10}$ m$^{-1}$, since the distance between identical atoms in a simple crystal is a few tenths of a nanometre ($10^{-10}$ m). For electromagnetic waves, such wave vectors lie in the X-ray region; in fact the condition that a wave vector terminates at the zone boundary is identical with the Bragg 'reflection condition' for diffraction of X-rays by the crystal lattice.

The simple equations for the diatomic linear chain are similar to those for a diatomic cubic crystal in which planes can be drawn containing only the one type of ion, alternating with planes containing the other type. The force constants are those between successive planes, but the dispersion relations are similar to Fig. 10.9, with acoustic and optical branches for both longitudinal and transverse modes. We may make a rough estimate of the frequency at the zone boundary by ignoring dispersion, and writing $\nu = \omega/2\pi = qv/2\pi$, where $v$ is the velocity of low-frequency sound waves. For most ionic solids this lies in the range $2\text{--}5\times10^3$ m s$^{-1}$, and for $q = 10^{10}$ m$^{-1}$ we obtain frequencies of $10^{12}\text{--}10^{13}$ Hz, in good agreement with the values of $\nu_T = \omega_T/2\pi$ listed in Table 10.5.

### Permittivity and dispersion

From dispersion theory we expect the static permittivity of an ionic solid to be related to the value of $\omega_T$. The actual relationship is complicated because of (1) local field corrections of the Lorentz type, arising from dipolar forces; and (2) short-range forces set up by the overlap of charge clouds on adjacent ions. Szigeti (1949) has derived the formula

$$\epsilon_s - \epsilon_{opt} = \left(\frac{\epsilon_{opt}+2}{3}\right)^2 \frac{n_0 Q^2}{m_r \epsilon_0 \omega_T^2}, \tag{10.44}$$

where $\epsilon_s$ is the static relative permittivity. $\epsilon_{opt}$ is the contribution from electrons within the ions, given by $n^2$ where $n$ is the extrapolated value of the optical refractive index. $m_r$ is a reduced mass, where for a simple diatomic solid with ionic masses $m_1$, $m_2$

$$\frac{1}{m_r} = \frac{1}{m_1} + \frac{1}{m_2}, \tag{10.45}$$

and $\pm Q$ is the effective charge on each ion, an empirical quantity but close to $ze$, where $z$ is the valency of the ion and $e$ the electronic charge.

Eqn (10.44) is clearly of the same form as eqn (10.42) when allowance is made for the difference between $\mathbf{E}$ and $\mathbf{E}_{local}$. If $\mathbf{P}_s$ is the static polarization at $\omega = 0$, (10.42) can be written in the form

$$\mathbf{P}/\mathbf{P}_s = \omega_T^2/(\omega_T^2 - \omega^2), \tag{10.46}$$

where $\mathbf{P}$ is the polarization at frequency $\omega/2\pi$ due to the ionic displacements. If $\epsilon_r$ is the relative permittivity at this frequency we have

$$\frac{\epsilon_r - \epsilon_{opt}}{\epsilon_s - \epsilon_{opt}} = \frac{\omega_T^2}{\omega_T^2 - \omega^2}. \tag{10.47}$$

This result clearly gives $\epsilon_r = \epsilon_s$ at zero frequency, while $\epsilon_r$ becomes infinite at $\omega = \omega_T$ because no damping is included. When $\omega$ just exceeds $\omega_T$, $\epsilon_r$ is negative; it diminishes in magnitude as $\omega$ increases, and passes through zero at an angular frequency $\omega_L$ such that

$$\epsilon_s/\epsilon_{opt} = (\omega_L/\omega_T)^2. \tag{10.48}$$

This result is known as the Lyddane–Sachs–Teller relationship, and it can be shown that $\omega_L$ corresponds to the limiting value of a longitudinal optical mode at $q = 0$. In a number of cases values of $\omega_L$ and $\omega_T$ have been determined by neutron diffraction, and the ratio is in good agreement with that obtained from dielectric measurements.

Since $\epsilon_r = (n - jk)^2$, a value of $\epsilon_r$ which is real but negative means $n = 0$ but $k$ is finite. The wave decays as $\exp(-\omega kz/c)$, and is not propagated in the frequency range between $\omega_T$ and $\omega_L$. This corresponds to the stop band in a filter, and the intrinsic impedance of the medium $Z_0$ is imaginary. It follows from eqn (8.53) that a wave incident on the dielectric within this frequency range is totally reflected. In early work this property was used to isolate various infrared wavelengths by successive reflections from the surfaces of ionic crystals; such wavelengths were known as '*reststrahlen*' or '*residual rays*'.

## 10.10. Ionized media and plasma oscillations

In a medium containing free particles with electric charges the theory of § 10.6 can be used, putting $\omega_p = 0$ since no restoring forces act on the particles. At low densities local field corrections can be neglected, and eqn (10.19) leads to a complex permittivity

$$\epsilon_r = 1 - \frac{n_0 q^2}{m\epsilon_0 \omega}\left(\frac{\omega + j\gamma}{\omega^2 + \gamma^2}\right) \tag{10.49}$$

for a medium containing $n_0$ particles (per unit volume) of charge $q$ and mass $m$.

For free particles an alternative approach is to work in terms of current. For an oscillating field $\mathbf{E}^{(0)} \exp j\omega t$, eqn (3.13a) becomes

$$\frac{d\mathbf{J}}{dt} + \frac{\mathbf{J}}{\tau} = \mathbf{J}\left(j\omega + \frac{1}{\tau}\right) = \left(\frac{n_0 q^2}{m}\right)\mathbf{E}^{(0)} \exp j\omega t$$

for that part of the current oscillating with angular frequency $\omega$. Since $(n_0 q^2/m)\tau = \sigma_0$, the conductivity at $\omega = 0$, we can write the equation for $\mathbf{J}$ as

$$\mathbf{J}(1 + j\omega\tau) = \sigma_0 \mathbf{E}^{(0)} \exp j\omega t. \tag{10.50}$$

This gives a complex conductivity $\mathbf{J}/\mathbf{E} = \boldsymbol{\sigma} = \sigma' - j\sigma''$, at frequency $\omega$, of

$$\boldsymbol{\sigma} = \sigma' - j\sigma'' = \sigma_0 \frac{1 - j\omega\tau}{1 + \omega^2\tau^2}. \tag{10.51}$$

These two different approaches emphasize the equivalence between a complex permittivity and a complex conductivity. For a given frequency eqn (8.7) becomes

$$\text{curl } \mathbf{H} = (\sigma' + j\omega\epsilon_r'\epsilon_0)\mathbf{E},$$

where $\sigma'$ and $\epsilon_r'$ are the real parts of the conductivity and relative permittivity. The coefficient of $\mathbf{E}$ in this equation can be expressed either as $\sigma' - j\sigma''$ with $\sigma'' = -\omega\epsilon_r''\epsilon_0$, or as $j\omega\epsilon_0(\epsilon_r' - j\epsilon_r'')$ with $\epsilon_r'' = \sigma'/\omega\epsilon_0$.

In addition to oscillations driven by an applied field, natural oscillations may be present. In an assembly of charged and neutral particles, known as a plasma, the local charge density fluctuates because of the random motion of the charged particles. However, a momentary excess of charge in one region is counteracted by the mutual repulsion between the particles, which pushes them apart so that the excess quickly disappears. The momentum gained by the particles in this process causes them to overshoot; the excess is replaced by a defect in the charge density, and the particles are attracted back. Repetition of this process sets up a periodic disturbance known as a 'plasma oscillation'; the striations visible in gas discharges and electron or ion beams are due to standing waves set up by such oscillations.

A simple derivation of the frequency of such oscillations starts from the conservation of charge and Gauss's theorem. The charge motion creates a current density $\mathbf{J} = \rho_e \mathbf{v}$, and from eqn (3.3) we have

$$d\rho_e/dt = -\text{div } \mathbf{J} = -\text{div } \rho_e\mathbf{v}.$$

From Gauss's theorem $\epsilon_0 \text{ div } \mathbf{E} = \rho_e$, and hence

$$\epsilon_0 \text{ div}(d\mathbf{E}/dt) = d\rho_e/dt = -\text{div } \rho_e\mathbf{v} = -\text{div } \mathbf{J}.$$

Hence

$$\epsilon_0(d\mathbf{E}/dt) = -\rho_e\mathbf{v} = -\mathbf{J}, \tag{10.52}$$

where the constant of integration is zero since $\dot{\mathbf{E}} = 0$ where there is no motion of the particles.

In a plasma where both positive and negative ions are present, the negative ions are mostly electrons which move very much more rapidly than the more massive ionized atoms, and we can assume that all the current is carried by electrons, that is, by one type of particle, of mass $m$ and charge $q$. The equation of motion of these particles is $m(d\mathbf{v}/dt) = q\mathbf{E}$, so that (using eqn (10.52))

$$d\mathbf{J}/dt = \rho_e(d\mathbf{v}/dt) = (\rho_e q/m)\mathbf{E},$$

$$d^2\mathbf{J}/dt^2 = (\rho_e q/m)(d\mathbf{E}/dt) = -(\rho_e q/m\epsilon_0)\mathbf{J}.$$

If we assume that oscillations are of vanishingly small amplitude, the departure of $\rho_e$ from its mean value $\bar{n}q$, where $\bar{n}$ is the average number per unit volume, is negligible, and we can write

$$d^2\mathbf{J}/dt^2 = -(\bar{n}q^2/m\epsilon_0)\mathbf{J}. \tag{10.53}$$

This is the equation of simple harmonic motion, showing that the current will oscillate at a frequency

$$\omega_p = 2\pi f_p = (\bar{n}q^2/m\epsilon_0)^{\frac{1}{2}}, \tag{10.54}$$

known as the plasma frequency. We can estimate its magnitude by taking as an example a plasma of fully ionized hydrogen at a pressure of 1 Pa, for which $\bar{n} = 2 \cdot 7 \times 10^{20} \text{ m}^{-3}$. Then, with $q = -e$,

$$f_p = (\bar{n}e^2/4\pi^2\epsilon_0 m)^{\frac{1}{2}} = 9 \cdot 0\bar{n}^{\frac{1}{2}} = 1 \cdot 5 \times 10^{11}. \tag{10.55}$$

This frequency corresponds to a wavelength of 2 mm for electromagnetic waves, and its measurement is an important tool in plasma physics since it gives the density of electrons. In ordinary gas discharges the degree of ionization is relatively low, and the plasma frequency may lie in the region $10^3$–$10^8$ Hz. In metals, on the other hand, the electron density is very much higher, and the electron plasma frequency is about $10^{15}$ Hz.

### References

COLLIE, C. H., HASTED, J. B., and RITSON, D. M. (1948). *Proc Phys. Soc.* **60**, 145.

DEBYE, P. (1929). *Polar molecules*. Dover Publications, New York.

DYLLA. H. R. and KING. J. G. (1973). *Phys. Rev.* A**7**, 1224.

LANE, J. A. and SAXTON, J. A. (1952). *Proc. R. Soc.* A**213**, 400.

ONSAGER, L. (1936). *J. Am. Chem. Soc.* **58**, 1486.

PAULING, L. and WILSON, E. B. (1935). *Introduction to quantum mechanics*. McGraw-Hill, New York.

RAMSEY, N. F. (1956). *Molecular beams*. Clarendon Press, Oxford.

ROBINSON. F. N. H. (1973). *Macroscopic electromagnetism*. Pergamon Press, Oxford.

SAXTON, J. A. (1952). *Proc. R. Soc.* A**213**, 473.

SZIGETI, C. (1949). *Trans. Faraday Soc.* **45**, 155.

TOWNES, C. H. and SCHAWLOW, A. L. (1955). *Microwave spectroscopy*. McGraw-Hill, New York.

### Problems

**10.1.** The static relative permittivity of $CO_2$ and $NH_3$ is measured at 273 K and 373 K at a pressure of $10^5$ Pa (1 atm) and the values of $10^3(\epsilon_r - 1)$ are found to be:

|        | $CO_2$ | $NH_3$ |
| ------ | ------ | ------ |
| 273 K  | 0·988  | 8·34   |
| 373 K  | 0·723  | 4·87   |

Calculate the permanent electric dipole moment for each gas, and also the radius of the molecule, assuming the polarizability to be the same as that of a conducting sphere.

(*Answer:* $p = 0$ debye and 1·45 debye; radius = 0·14 nm and 0·18 nm; from viscosity data the radii are 0·23 nm and 0·22 nm respectively.)

10.2. The relative permittivity of liquid helium at its boiling-point is 1·048, and its density is 125 kg m$^{-3}$. Calculate the refractive index of the gas at N.T.P., and estimate the radius of the helium atom, assuming it to behave like a conducting sphere. Compare the radius with (*a*) that given by the Bohr theory for an atom with a nuclear charge of two units in its ground state, (*b*) with that calculated from the diamagnetic susceptibility (see Problem 6.1).

(*Answer:* $n = 1\cdot000034$; radius = 0·059 nm; Bohr-theory radius = 0·026 nm.)

10.3. In the upper regions of the atmosphere (the ionosphere) the gas molecules are ionized, mostly through the effects of ultraviolet radiation from the sun. Show that in a region where the number of free electrons per cubic metre is $n_0$, the refractive index for waves of frequency $f$ (Hz) is

$$\left(1 - \frac{n_0 e^2}{m\omega^2 \epsilon_0}\right)^{\frac{1}{2}} = \left(1 - \frac{\omega_p^2}{\omega^2}\right)^{\frac{1}{2}} \approx \left(1 - \frac{81 n_0}{f^2}\right)^{\frac{1}{2}},$$

where $\omega_p/2\pi$ is the plasma frequency.

At this frequency, the refractive index falls to zero, and the ionosphere is totally reflecting even at normal incidence. The frequency at which this occurs is called the critical frequency and at midday is about 4 MHz for the E-layer at latitude 40° N. Estimate the maximum value of $n_0$ in the E-layer from this figure.

(*Answer:* $n_0 = 2 \times 10^{11}$ m$^{-3}$. Neglect the Lorentz field.)

10.4. Show that in an ionized region such as that in the previous question the product of the group velocity and the phase velocity is equal to the square of the velocity in a vacuum.

10.5. A particle of charge $-e$ and mass $m$ performs a simple harmonic motion $s = s_0 \cos \omega t$ under the action of a restoring force. Show that through the radiation of energy (given by eqn (8.72)) the total energy of the particle falls as $W = W_0 e^{-\gamma t}$, where

$$\gamma = Z_0 e^2 \omega^2 / 6\pi mc^2 = 2\pi Z_0 e^2 f^2 / 3mc^2.$$

Here $Z_0 =$ intrinsic impedance of free space, and $f = \omega/2\pi$. On the quantum theory, the chance that an atom spends a time $t$ in an excited state has the probability of the order of $\exp(-\gamma t)$; hence show that the mean lifetime $\tau = 1/\gamma$ of a sodium atom in an excited state before emitting one of the sodium D lines ($\lambda = 590$ nm) is roughly $1\cdot6 \times 10^{-8}$ s.

10.6. By the uncertainty principle, the width $\Delta E$ of the upper energy level of the previous question is given by the relation $\tau \Delta E = h/2\pi$, where $h$ is Planck's constant. Use the relation $\Delta E = h(\Delta f)$ to show that this gives a linewidth of $\Delta f = \gamma/2\pi$. This is called the natural or radiation breadth of the line, and the same result is obtained on classical theory by a Fourier analysis of the spectrum of an oscillator whose energy is decaying exponentially as $W = W_0 \exp(-\gamma t)$.

Estimate the linewidths due to the Doppler effect and to collisions in a gas at 1000 K and 100 Pa pressure, and show that (*a*) at $f = 10^{10}$ Hz ($\lambda = 0\cdot03$ m) collision-broadening is dominant, (*b*) at $f = 10^{15}$ Hz ($\lambda = 300$ nm) Doppler effect is dominant, (*c*) at $f = 10^{18}$ Hz ($\lambda = 0\cdot3$ nm) natural line-breadth is dominant.

10.7. For a vibrating molecule, the polarizability $\alpha$ varies as $\alpha = \alpha_0(1+bx)$, where $x = a \cos pt$ is the change in the normal dimensions of the molecule due to the vibration. Show that if incident light of frequency $\omega/2\pi$ falls on the molecule, the scattered radiation will contain light of frequencies $(\omega \pm p)/2\pi$. (This is the classical explanation of the Raman effect.)

10.8. Show that for a narrow absorption line the maximum and minimum of the refractive index in the region of anomalous dispersion occur at the frequencies where the absorption coefficient has fallen to half its maximum value.

10.9. The conductivity of sea-water at 293 K is about $2\,\Omega^{-1}\,m^{-1}$. Show that the absorption at $0{\cdot}01\,m$ wavelength due to this conductivity is small compared with the Debye absorption, but the two are roughly equal at a wavelength of about $0{\cdot}1\,m$. (Use the data given for pure water in Fig. 10.7; in fact the Debye relaxation time is somewhat altered by the salts dissolved.)

10.10. Discuss the propagation in and reflection from the surface of the sea of radio waves in the light of the data given in the previous question.

10.11. In a region of space containing $\bar{n}$ particles per unit volume of charge $q$ and mass $m$, where the pressure is so low that collisions may be neglected, eqns (8.3) and (8.7) may be written

$$\text{curl } \mathbf{E} = -\mu_0(\partial \mathbf{H}/\partial t); \qquad \text{curl } \mathbf{H} = \bar{n}q\mathbf{v} + \epsilon_0(\partial \mathbf{E}/\partial t).$$

Show that by using the equation of motion $q\mathbf{E} = m(\partial \mathbf{v}/\partial t)$, and eliminating $\mathbf{v}$ between these equations (taking grad div $\epsilon_0\mathbf{E} = \text{grad } \bar{n}q = 0$), we can obtain the equation

$$\frac{\partial^2 \mathbf{E}}{\partial t^2} + \frac{\bar{n}q^2}{\epsilon_0 m}\mathbf{E} = c^2\,\nabla^2\mathbf{E}.$$

This is the equation of Tonks and Langmuir for propagation of waves in an ionized medium. Verify that this wave equation is satisfied by

$$\mathbf{E} = \mathbf{E}_0 \exp \text{j}(\omega t - kz)$$

provided that $\omega^2 = \omega_p^2 + c^2 k^2$, where $\omega_p/2\pi$ is the plasma frequency given by eqn (10.54).

Verify that this leads to the expression for the refractive index given in Problem 10.3.

# **11.** Free electrons in metals

## 11.1. Quantum theory of free electrons in metals

ON the Drude theory of free electrons in a metal, outlined in § 3.11, the conduction electrons are confined to the volume of the metal but move freely inside like gas molecules in a container. It leads to formulae for the electrical conductivity (see eqn (3.12))

$$\sigma = n(q^2/m)\tau, \tag{11.1}$$

and for the thermal conductivity (see eqn (3.62))

$$K = \tfrac{1}{3}ncL(\mathrm{d}\bar{W}/\mathrm{d}T), \tag{11.2}$$

but makes no prediction as to the magnitude of the relaxation time $\tau$ or the mean free path $L = c\tau$, where $c$ is the random thermal velocity. This problem is avoided in considering the ratio $(K/\sigma)$, which is predicted to be of order $T(k/e)^2$ in accordance with the measurements of Wiedemann and Franz. There is one serious difficulty: the specific heat. On classical theory the kinetic energy of a free particle can have any value, and there is a continuous distribution of values, though some are more probable than others, corresponding to a Boltzmann distribution. As the temperature falls, the average energy decreases to zero at $T = 0$ K; at an absolute temperature $T$ the total translational energy of $N$ particles is $\tfrac{3}{2}NkT$. The differential of this gives a contribution of $\tfrac{3}{2}R$ to the molar heat, so that the molar heat of a metal should exceed that of an insulator by this amount. No such contribution is observed.

In quantum mechanics not every value of the energy is allowed, and the continuous distribution of energies is replaced by a discrete set of allowed energy levels. The spacing is extremely small, however, and the difference between this and the assumed classical continuous distribution produces no observable effect for real gases. It would become appreciable only at temperatures so low that ordinary substances have negligible vapour pressure, and at higher temperatures it is always masked by deviations from the perfect gas laws owing to the van der Waals forces. In the case of electrons in a metal, the spacing of the translational energy levels is larger because of the smaller mass of the electron, and the number per unit volume is much larger than in any real gas. The result of this is to emphasize the role played by the Pauli exclusion principle, which states that no two electrons in a given system can have the same set of quantum numbers. When allowance is made for the intrinsic spin angular momentum of the electron, this means that only two electrons can occupy any given translational energy level. Hence the kinetic energy of the electrons cannot be zero at 0 K, since this would mean that all the electrons were in one particular energy level. In fact the electrons occupy the lowest possible set of energy levels consistent with the Pauli exclusion principle, and their mean energy is very far from zero at 0 K. At a finite temperature the energy distribution can only be found using the Fermi–Dirac statistics, that is, the quantum statistics which take account of the Pauli exclusion principle. At ordinary temperatures it turns out that the energy distribution differs very little from that at 0 K; the latter can be found when the values of the allowed energy levels are known, and a simple method of computing the levels will now be given.

The fact that electrons and other particles have a wave-like aspect is well established from experiments on electron and neutron diffraction; the wavelength associated with a particle of linear momentum $\mathbf{p}$ is given by the de Broglie relation

$$2\pi/k = \lambda = h/p \quad \text{or} \quad \mathbf{k} = \mathbf{p}(2\pi/h) = \mathbf{p}/\hbar, \tag{11.3}$$

where $\mathbf{k}$ is the 'wave vector' which is parallel to the direction in which the wave is travelling, and $h$ is Planck's constant and $\hbar = h/2\pi$. For example, the wavelength is $1.23 \times 10^{-9}$ m for an electron of kinetic energy equal to 1 electronvolt (eV). (1 eV is the energy acquired by an electron in falling through a potential difference of 1 V). Free electrons confined within a metal rebound from the surface of the metal without losing any energy; on the wave aspect this means that the waves are totally reflected at the boundaries, and standing waves are set up. These standing waves are the allowed solutions of the wave equation for a particle in a box, in the same way that standing waves of certain wavelengths are allowed in a waveguide resonator (see § 9.7). By analogy with the theory of heat radiation, the number of allowed wavelengths in the range $\lambda$ to $\lambda + d\lambda$ is $V4\pi \, d\lambda/\lambda^4$, where $V$ is the volume of the box. For particles the wavelength is determined by the momentum, and from eqn (11.3) the number of possible values of the momentum in the range $p$ to $p + dp$ is found to be

$$(V/h^3)4\pi p^2 \, dp. \tag{11.4}$$

This relation may be derived in another way by means of the uncertainty relation, which states that the momentum component $p_x$ cannot be determined more precisely than to an amount $\Delta p_x$ such that $\Delta p_x \, \Delta x = h$, where $\Delta x$ is the uncertainty in the position coordinate. Similar relations hold for the other two axes, and hence

$$\Delta p_x \, \Delta p_y \, \Delta p_z = h^3/\Delta x \, \Delta y \, \Delta z = h^3/V$$

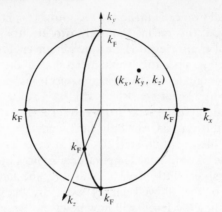

FIG. 11.1. The wave vector **k** of a particle is specified by its components $(k_x, k_y, k_z)$, corresponding to a point in '**k**-space'. At 0 K all electrons are specified by points within the sphere of radius $k_F$, since this makes the energy a minimum.

if the particle is constrained to be within the volume $V$. If we write $p_x = \hbar k_x$, etc., by using eqn (11.3), we have

$$\Delta k_x \, \Delta k_y \, \Delta k_z = 8\pi^3/V.$$

The wave vector of a particle is specified in magnitude and direction by its components $k_x$, $k_y$, $k_z$ along three Cartesian axes, and the square of the wave vector is $k^2 = k_x^2 + k_y^2 + k_z^2$. A '**k**-space' may be constructed, as in Fig. 11.1, where the coordinates are the components $k_x$, $k_y$, $k_z$ and the magnitude and direction of **k** are given by a vector drawn from the origin to the point $(k_x, k_y, k_z)$. All values of the wave vector which lie between **k** and **k**+d**k** are represented by points which lie within the spherical shell bounded by the radii $k$ and $k+dk$, and the volume of this shell is $4\pi k^2 \, dk$. Since $\Delta k_x \, \Delta k_y \, \Delta k_z$ is an element of volume in **k**-space, the Pauli exclusion principle may be stated in the form that only one point in **k**-space is allowed in an element of volume of size $8\pi^3/V$, so that in the volume $4\pi k^2 \, dk$ the allowed number of points is

$$di = (V/8\pi^3)4\pi k^2 \, dk. \tag{11.5}$$

In addition, we have to consider the angular momentum of the electron $\frac{1}{2}\hbar$ due to its intrinsic spin, which can have one of two components $\pm\frac{1}{2}\hbar$ along any axis (see § 14.2). The Pauli principle then allows two electrons, with opposite values of these angular-momentum components, to have the same translational energy, that is, we can assign a maximum of two electrons to each point in the (linear) momentum space.

The translational energy $W$ of an electron of momentum $\hbar k$ and mass $m$ is $\hbar^2 k^2/2m$, and the total kinetic energy will be smallest when $\sum (\hbar^2 k^2/2m)$ has its minimum value. It is readily seen that this occurs when the points in momentum space just fill a sphere of the least volume which will accommodate all the electrons, since if we replace any point within the sphere by one outside, its value of $\mathbf{k}$ and hence the value of the energy will be larger. If the radius of this sphere is $k_F$, its volume is $\frac{4}{3}\pi k_F^3$ and the number of possible points (that is, allowed values of the momentum) is $(V/8\pi^3)(\frac{4}{3}\pi k_F^3)$. Since only two electrons are allowed per point, if the total number of electrons in a volume $V$ of the metal is $N$, we must have

$$\tfrac{1}{2}N = (V/8\pi^3)\tfrac{4}{3}\pi k_F^3 \tag{11.6}$$

and hence

$$(W_F)_0 = \frac{\hbar^2 k_F^2}{2m} = \frac{\hbar^2}{2m}(3\pi^2 n)^{\frac{2}{3}}. \tag{11.7}$$

Here $n = N/V$ is the number of electrons per unit volume of the metal. $W_F$ is known as the 'Fermi energy' and at 0 K the energy of the highest occupied level is $(W_F)_0$.

The distribution may be expressed in terms of energy rather than momentum by using the relation $W = \hbar^2 k^2/2m$. The number $(2/V)\,di$ of states per unit volume which have energy in the range between $W$ and $W+dW$ is

$$g(W)\,dW = (2/V)\,di = \pi^{-2}k^2\,dk = CW^{\frac{1}{2}}\,dW, \tag{11.8}$$

where

$$C = \frac{1}{2\pi^2}\left(\frac{2m}{\hbar^2}\right)^{\frac{3}{2}} \tag{11.9}$$

has the value $1{\cdot}06\times10^{55}$ if $W$ is in joules, and $6{\cdot}79\times10^{27}$ if $W$ is expressed in electronvolts. Eqn (11.7) can be written as

$$n = \tfrac{2}{3}C\{(W_F)_0\}^{\frac{3}{2}} \tag{11.10}$$

and (11.8) as

$$g(W)\,dW = \frac{3n}{2}\frac{W^{\frac{1}{2}}}{(W_F)_0^{\frac{3}{2}}}\,dW. \tag{11.11}$$

A plot of $g(W)$ against $W$ gives a parabola, as in Fig. 11.2; at $T = 0$ each state is occupied by an electron up to the sharp cut-off at the Fermi energy $(W_F)_0$, and the momentum $\mathbf{p}_F = \hbar\mathbf{k}_F$ corresponding to this point is known as the 'Fermi momentum'. The mean energy $\bar{W}$ per electron is

$$\bar{W} = \frac{1}{n}\int_0^{(W_F)_0} Wg(W)\,dW = (W_F)_0^{-\frac{3}{2}}\int_0^{(W_F)_0} W^{\frac{3}{2}}\,dW = \tfrac{3}{5}(W_F)_0, \tag{11.12}$$

and the total internal energy $U_0$ of $n$ electrons at 0 K is

$$U_0 = \tfrac{3}{5}n(W_F)_0. \tag{11.13}$$

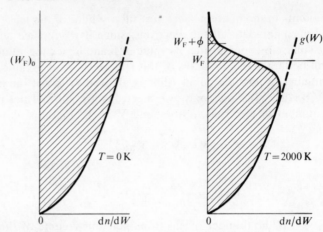

FIG. 11.2. The number of free electrons d$n$ with kinetic energy between $W$ and $W+dW$ in a metal, as given by the Fermi–Dirac distribution. At 0 K the electrons have energies only up to the Fermi level $(W_F)_0$; at a finite temperature $W_F$ is smaller than $(W)_{F0}$ (see eqn (11.19)). At room temperature the distribution is much closer to that on the left, since d$n$/d$W$ is only altered for electrons whose energy lies within $\approx kT$ of $W_F$.

The great difference between this type of energy distribution and a Maxwell–Boltzmann distribution is best demonstrated by considering numerical values. Eqn (11.7) shows that $(W_F)_0$ depends on the metal only through the quantity $n$, the number of electrons per unit volume. If we take sodium as an example, and assume that one valence electron per atom is released in the metal as a free electron, we have

$$n = 2 \cdot 5 \times 10^{28}\, \text{m}^{-3},$$

and $(W_F)_0$ is found to be $3 \cdot 1$ eV, the mean energy $\bar{W}$ being $1 \cdot 9$ eV. This is very large indeed. On the classical Maxwell-Boltzmann statistics, where the mean energy is $\frac{3}{2}kT$ at temperature $T$ in degree Kelvin, it is equivalent to $T = 20\,000$ K. The values for other metals calculated in the same way are given in Table 11.1.

For comparison, we calculate the value of $W_F$ for a monatomic gas of the rare isotope of mass 3 of helium, atoms of which should also obey the Fermi–Dirac statistics. The boiling-point of this gas is $3 \cdot 2$ K, and the number of atoms per cubic metre at atmospheric pressure in the gas at this temperature is $2 \cdot 3 \times 10^{27}$. This gives $W_F = 1 \cdot 15 \times 10^{-4}$ eV, which is equivalent only to a temperature of $1 \cdot 3$ K. This is small compared with the actual temperature, indicating that deviations from the perfect gas laws due to quantum effects will not be large. It is obvious that the low value of $W_F$ for this, or any other real gas, as compared with the electron gas in a metal, is due primarily to the difference in the mass of the

TABLE 11.1

*Values of* $W_F$ (the 'Fermi' energy), the work function $\phi$ (as deduced from measurements of the photoelectric effect and thermionic emission), and the thermionic emission constant A

| Substance | $W_F$ (eV) | Work function $\phi$ | | $A$ $(A\,m^{-2}\,K^{-2})$ |
|---|---|---|---|---|
| | | Photoelectric effect (eV) | Thermionic emission (eV) | |
| Lithium | 4·7 | 2·2 | | |
| Sodium | 3·1 | 1·9 | | |
| Potassium | 2·1 | 1·8 | | |
| Copper | 7·0 | 4·1 | 4·5 | $1·10\times10^6$ |
| Silver | 5·5 | 4·7 | 4·3 | $1·07\times10^6$ |
| Gold | 5·5 | 4·8 | 4·25 | $1·00\times10^6$ |
| Molybdenum | 5·9 | | 4·2 | $(0·50–1·15)\times10^6$ |
| Tungsten | 5·8 | 4·49 | 4·5 | $(0·20–0·60)\times10^6$ |
| Platinum | 6·0 | >6·2 | 5·3 | $0·30\times10^6$ |
| Nickel | 7·4 | 4·9 | 4·5 | $1·20\times10^6$ |
| Thoriated tungsten | | | 2·6 | $0·60\times10^6$ |
| (BaO, SrO) mixture | | | 1·8 | $0·03\times10^6$ |

particles, which comes in the denominator of eqn (11.7), and also partly to the smaller number per unit volume. When $W_F \ll kT$, the particles have all energies up to those of order $kT$, and the density of occupied points in momentum space is low, that is, the chance of finding an occupied point in the fundamental volume $h^3/V$ is small. Under these circumstances the classical statistics are a valid approximation. For an electron gas, on the other hand, $W_F \gg kT$ at all ordinary temperatures, and every fundamental volume $8\pi^3/V$ of **k**-space contains an occupied point, only one such point (or two electrons, allowing for the spin) in each such volume being allowed by the exclusion principle. If we tried to use the classical picture, with the average energy of the order $kT$, this would correspond to putting a large number of electrons in each volume $8\pi^3/V$ of **k**-space, and the exclusion principle would be violated.

## 11.2. The Fermi–Dirac distribution function

The energy distribution at a finite temperature is given by Fermi–Dirac statistical mechanics, and we quote the results. Each 'point' in **k**-space corresponds to a quantized state of translational motion, and inclusion of the electron spin gives two quantum states to each point. The number of such states $g(W)\,dW$ per unit volume in the energy range $W$ to $W+dW$ is known as the 'density of states', and is given by eqns (11.8) and (11.9). The chance that a state of energy $W$ is occupied by an electron is given by the function

$$f = \frac{1}{\exp\{(W - W_F)/kT\} + 1}. \tag{11.14}$$

At $T = 0$ the denominator is infinite for $W > W_F$ and unity for $W < W_F$, so that $f = 0$ in the former case and unity in the latter. This corresponds to the sharp cut-off in the occupation of states already discussed. At a finite temperature $f$ is still very close to zero for $W > W_F$ and to unity when $W < W_F$, except for the comparatively small range of energies which lie within a few $kT$ of $W_F$. The quantity $W_F$ is defined as the value of the energy at which the probability of a state being occupied is $f = \frac{1}{2}$, since then the denominator in (11.14) is $e^0 + 1 = 2$.

The number of electrons with energy between $W$ and $W + dW$ is

$$dn = fg(W)\, dW = \frac{CW^{\frac{1}{2}}}{\exp\{(W - W_F)/kT\} + 1}\, dW, \tag{11.15}$$

and the total number of electrons per unit volume is

$$n = C \int_0^\infty \frac{W^{\frac{1}{2}}}{\exp\{(W - W_F)/kT\} + 1}\, dW. \tag{11.16}$$

The distribution function $dn/dW$ appropriate to a temperature of about 2000 K is also shown in Fig. 11.2.

The value of $W_F$ is found from the integration in eqn (11.16), which can be done by numerical methods. In general we are interested only in the region where $kT \ll W_F$, and approximate solutions in the form of a power series in $kT/W_F$ can be obtained for this and a number of other integrals, using a method due to Sommerfeld. This gives

$$\int_0^\infty W^r f\, dW = \frac{1}{r+1} W_F^{r+1} \left\{ 1 + \frac{\pi^2}{6} r(r+1)\left(\frac{kT}{W_F}\right)^2 + \ldots \right\}, \tag{11.17}$$

where successive terms are each smaller by factors of order $(kT/W_F)^2$. Use of this result with $r = \frac{1}{2}$ leads to

$$n = \frac{2}{3} CW_F^{\frac{3}{2}} \left\{ 1 + \frac{\pi^2}{8}\left(\frac{kT}{W_F}\right)^2 + \ldots \right\}. \tag{11.18}$$

Since $n$ is independent of temperature, it follows that $W_F$ must vary slightly with temperature, and by equating (11.10) and (11.18) we obtain

$$W_F = (W_F)_0 - \frac{\pi^2}{12}\frac{(kT)^2}{W_F} + \ldots \tag{11.19}$$

in the same approximation. At room temperature $kT$ is equivalent to

0·025 eV while $W_F$ is several volts, so that the approximation is a good one. In the following sections the difference between $W_F$ and $(W_F)_0$ can be neglected for most purposes. Obviously this does not apply to quantities involving differentials with respect to the temperature, such as the specific heat (§ 11.6).

## 11.3. Work function and contact potential

Hitherto we have considered only the kinetic energy of the electrons, but they may also possess potential energy. The Fermi–Dirac distribution function $f$ must then be written in a more general form, where $W$ is the total energy of a particle (kinetic plus potential energy). The probability that an electron occupies a quantum state of total energy $W$ is

$$f = \frac{1}{\exp\{(W-G)/kT\}+1}, \tag{11.20}$$

where the quantity $G$ is the thermodynamic or chemical potential (sometimes known as the Gibb's free energy). It must be chosen in such a way that the total number of particles per unit volume is $n$, and the chance that a state with total energy $W = G$ is occupied by an electron is exactly $\frac{1}{2}$. If the potential energy is zero, then $W$ is just equal to the kinetic energy (which we now distinguish by writing it as $W_k$), and comparison with eqn (11.14) shows that $G$ is then equal to the Fermi energy $W_F$ and is determined by eqn (11.16). If a particle of charge $q$ has potential energy $qV$, its total energy is $W = W_k + qV$, and to reconcile eqn (11.20) with (11.14) we must have

$$W-G = W_k+qV-G = W_k-W_F,$$

which shows that

$$G = W_F+qV. \tag{11.21}$$

The value of $G$, like that of $W_F$, must be a function of temperature, but for a degenerate electron gas it is only a slowly varying function (cf. eqn (11.19)). The important property of $G$ is that for all systems in thermal equilibrium it must have the same value. We shall not present a detailed proof of this result, but it follows from the statistical result that systems are in dynamic equilibrium when all states of equal energy have the same probability of being occupied, combined with the thermodynamic result that systems are in thermal equilibrium when they are at the same temperature. Thus for two systems A and B for which the distribution functions are

$$f_A = \frac{1}{\exp\{(W-W_{FA})/kT\}+1},$$

$$f_B = \frac{1}{\exp\{(W-W_{FB})/kT\}+1},$$

FIG. 11.3. The jump in potential energy at the surface of a metal corresponds to the difference in energy between the level for zero kinetic energy outside the metal ($W_k' = 0$) and the level for zero kinetic energy inside ($W_k = 0$). The jump is equal to $W_F + \phi$, the sum of the Fermi energy $W_F$ and the work function $\phi$.

it is clear that the condition which makes $f_A = f_B$ for all values of the energy $W$ and the temperature $T$ is simply $W_{FA} = W_{FB}$.

When there is no current flowing in a metal, $\mathbf{E} = 0$ from eqn (3.5), and the potential $V$ is constant inside the metal; so also are the values of $W_F$ and $G$. At the surface of the metal there must exist forces which restrain the charge carriers from leaving the metal. With no detailed knowledge of the nature of these forces, the simplest model we can adopt is to suppose that the electric potential rises sharply at the edge of the metal to a value well above the Fermi energy, so that electrons inside the metal have insufficient kinetic energy to surmount this barrier. In Fig. 11.3, the potential energy is assumed to rise vertically at the surface of the metal to a value which, outside the metal, is greater than the Fermi energy by an amount $\phi$. The quantity $\phi$ is known as the 'work function' and, at $T = 0$, it represents the least amount of energy which will suffice to raise an electron from the top of the Fermi distribution and remove it to a point outside the metal with zero kinetic energy.

When two metals are placed in contact, a potential difference is found to exist between them. This was discovered by Volta in 1797, and is sometimes known as the 'Volta effect'. It was not at first generally accepted as a fundamental property of metals, but it has a natural explanation in terms of the free-electron theory. In Fig. 11.4 it is seen that if metal A has a smaller work function $\phi_A$ than that $\phi_B$ of metal B, it is energetically possible for electrons from the top of the energy band in metal A to flow into metal B when contact is made, since the top of the band in B has a lower energy. The flow of charge sets up a potential difference between B and A which increases until the tops of the two energy distributions reach the same level. Equilibrium is then attained, and no more charge is transferred. The actual number of electrons transferred is an insignificant fraction of the total, so that the shaded areas in Figs 11.4(a) and (b) are equal. Thus the contact potential is just equal to the difference of the work functions.

At ordinary temperatures the chance that an electron has sufficient energy to surmount the 'potential barrier' between the two metals in Fig. 11.4(b) is extremely small. At $T = 0$ K it is zero, but no temperature-dependence is found experimentally in the flow of electrons between two metals in contact. This is attributed to the 'tunnel effect'. In quantum mechanics (but not in classical mechanics) there is a finite chance of an electron penetrating a barrier whose height is a few electronvolts provided its width is less than about 1 nm (a few atomic spacings). At any given energy the chance of crossing the barrier is the same in either direction. The net flow is therefore zero when the number of electrons approaching the barrier per unit time is the same on either side. This can be calculated from the Fermi–Dirac distribution function, and it leads again to the condition that the Fermi levels must be the same on either side.

The work function of a metal, and hence also the contact potential between two metals, is very sensitive to the state of the surface. For this reason, measurements of $\phi$ show rather a wide scatter; the best determinations are made with metallic films newly deposited by evaporation in a very high vacuum (for metals with low boiling-points), or (for a high-melting-point metal such as tungsten) with a surface cleaned by 'flashing' the metal at a temperature close to the melting-point in a high vacuum. The contact potential between two surfaces is measured indirectly as follows. A narrow beam of electrons is directed onto the metallic surface in a high vacuum, and the current reaching the surface is plotted as a function of the retarding potential applied to the surface. When a second

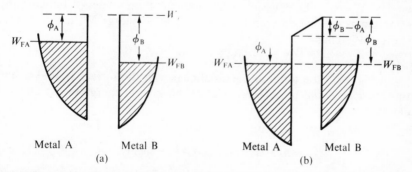

FIG. 11.4. Energy distribution of electrons in two metals A, B. (a) Before contact: $\phi_A$, $\phi_B$ are the energies required to remove an electron to rest at infinity. (b) After contact: electrons flow to metal B, changing its potential relative to A until the tops of the two energy distributions are level. The contact potential difference thus created $= \phi_B - \phi_A$ (in energy), and an electron released from B (for example by the photoelectric effect) would gain this amount of energy in moving from a point near B to a point near A.

metallic surface is substituted, the current to it gives a curve of similar shape but displaced by an amount equal to the contact potential difference between the two surfaces. The work of Mitchell and Mitchell (1951) gave the following mean values for the contact potential relative to a clean tungsten surface: copper $(-0.05\pm0.02)$V; silver $(+0.23\pm0.03)$V; aluminium $(+0.31\pm0.03)$ V.

## 11.4. The photoelectric effect and secondary emission

If an electron acquires an excess energy at least equal to the work function, it can escape from the metal and travel to an electrode held at a positive potential with respect to the emitting surface. A continuous flow of such electrons constitutes a current, and this is the basis of a number of applications. In this section we discuss first the photoelectric effect, in which the electrons acquire the excess energy from photons, and then 'secondary emission', in which the surface is bombarded with 'primary' electrons, some of whose energy is transferred by collision to electrons within the metal. Electrons are also emitted if an intense electric field is applied at the metal surface; this is closely related to thermionic emission, and is discussed in § 11.5.

*Photoelectric emission*

The energy associated with a quantum of radiation of frequency $\nu$ (a 'photon') is $h\nu$, where $h$ is Planck's constant, and in many respects the photons behave as particles with this energy. In an inelastic collision the whole of this energy may be transferred to an electron. If the electron in the metal originally had total energy $W = W_k + W_p$, where $W_k$ is its kinetic energy and $W_p$ its potential energy, then after collision it acquires an energy $W + h\nu$. On emission from the metal this energy remains the same but is divided differently between kinetic energy $W_k'$ and potential energy $W_p'$. Thus

$$W + h\nu = W_k + W_p + h\nu = W_k' + W_p'. \tag{11.22}$$

From Fig. 11.3 the jump in potential energy at the surface is $W_F + \phi$, and this must be equal to $W_p' - W_p$. Hence the kinetic energy of the emitted electron is

$$W_k' = h\nu + W_k - (W_p' - W_p)$$
$$= h\nu + W_k - (W_F + \phi) = h\nu - \phi - (W_F - W_k). \tag{11.23}$$

At any given frequency $\nu$, the maximum kinetic energy of the emitted electrons is $(h\nu - \phi)$, corresponding to electrons which come from the Fermi surface where $W_k = W_F$. No electrons are emitted unless $W_k' \geq 0$, so that the lowest frequency which can just cause photoelectric emission occurs at $h\nu = \phi$. Measurement of this frequency provides a method of

FIG. 11.5. Relative sensitivity of photoemissive tubes with cathodes of (a) caesium on oxidized silver; (b) caesium–antimony (glass window); (c) caesium–antimony (quartz window).

determining the work function $\phi$, knowing the value of $h$ (or $h/e$ if $\phi$ is expressed in volts); some values obtained by this method are given in Table 11.1. The cut-off frequency lies in the visible region only for barium, strontium, and the alkali metals, which have low work functions; for other metals it lies in the ultraviolet.

An important application of photoemission is in devices for the detection and quantitative measurement of the intensity of visible radiation. A photoemissive tube usually has a photosensitive area of a few square centimetres; a composite caesium–antimony surface reaches its maximum response at about $\lambda = 400$ nm, and with a quartz window it can be used into the ultraviolet (see Fig. 11.5). A surface of caesium on oxidized silver is 'red-sensitive', with peak sensitivity around $\lambda = 800$ nm. The photocurrent is attracted to an anode and reaches its saturation value of a few microamperes when the anode is at a positive potential of 50–100 V. The 'dark current' (the background current when the tube is in total darkness) is less than $0.1\,\mu A$. For a vacuum tube the sensitivity is 20–40 $\mu A$ per lumen; the response of the photoemissive surface is virtually instantaneous, the time lag for electron emission after light is switched on being less than $10^{-9}$ s. Gas-filled tubes have a sensitivity of over 100 $\mu A$ per lumen, the additional current arising from ionization of the molecules of an inert gas by collision with the photoelectrons. Such tubes have the disadvantages that the current sensitivity depends on the anode voltage, and the response falls at frequencies above 10 kHz.

### Secondary emission

When a solid surface is struck by electrons or ions of appreciable energy, secondary electrons are emitted from the surface. If the bombarding particles (primaries) are electrons, they must have an energy of a few electronvolts to eject an appreciable number of secondary electrons, but at higher energies the number of secondaries may be greater than the number of incident primaries. The secondary-emission ratio

$$\delta = \text{(number of secondary electrons/number of incident primaries)}$$

has maximum values of 0·5–1·6 for pure metals, and is very sensitive to impurities and contamination of the surface. For composite surfaces, such as a layer of $Cs_2O$ on a base of silver, covered with a surface film of absorbed caesium, $\delta$ may be as high as 10.

At low energies there is an initial rise in $\delta$ because the number of secondaries released within the metal increases with the energy of the incident primary. The production of secondaries is most copious near the end of the primary path, so that the chance of escape declines rapidly as the primaries penetrate deeper into the metal at higher energies, causing $\delta$ to fall again after passing through a maximum at about 200–400 eV in the primary energy.

The chief use of secondary emission is for current amplification. In a 'photomultiplier', electrons from a photosensitive cathode are accelerated to strike a secondary emitting electrode, so arranged that these secondary electrons are accelerated to another secondary-emitting electrode; the process is repeated with a succession of electrodes to give an avalanche effect. An over-all current amplification of $10^5$ is readily attained with 10 electrodes, each with $\delta$ of 3–3·5. The potential of each successive electrode is maintained at 100–200 V higher than the previous one by means of a potentiometer system.

Another type of electron multiplier consists of a curved glass tube (Fig. 11.6), about 2 mm in outside diameter and 1 mm inside, some 50 mm long. The inside wall of the tube has a continuous, highly resistive coating; the output end is maintained at 3–4 kV positive with respect to the input, and the resistance between the ends is over $10^3$ M$\Omega$. When the tube is evacuated, an electron entering at the end where the potential is negative generates secondary electrons on collision with the wall; these are accelerated to produce further electrons on successive collisions, the over-all current gain being about $10^8$. A small horn, 5–10 mm in diameter, may be attached to the input to guide primary electrons into the tube.

FIG. 11.6. The channel electron multiplier. Primary electrons, guided in by the horn, generate secondary electrons on collision with the inside wall of the tube; this process is repeated many times to give an over-all current gain of about $10^8$. The positive electrode is maintained some 3–4 kV above the negative electrode.

If the output current is passed through a resistance, so that each primary electron generates an output voltage pulse, this may be amplified and fed to a counting circuit.

## 11.5. Thermionic emission and field emission

When a metal is heated to a fairly high temperature, a small fraction of the conduction electrons will have kinetic energy (see Fig. 11.2) greater than $W_F + \phi$, and can therefore escape from the metal. This is known as 'thermionic emission'. If the metal (the 'cathode') is in a vacuum, and an electrode at a positive potential (the 'anode') is placed nearby to collect the emitted electrons, a continuous current ranging up to milliamperes or even amperes may be obtained. The magnitude of the current depends on the work function of the emitter, and varies very rapidly with temperature. For pure metals, copious emission can be obtained only at temperatures of order 2000 K, and metals with very high melting-points, such as tungsten or tantalum, must be used. Pure tungsten, with a work function of 4·5 eV (see Table 11.1), must be run at 2500 K, but thoriated tungsten ($\phi = 2.6$ eV) can operate at 1900 K. Most vacuum tubes, however, use an 'oxide-coated cathode' consisting of a barium–strontium oxide mixture, whose lower work function of 1·8 eV makes it possible to obtain current densities of order $10^4$ A m$^{-2}$ at 1100 K. Such oxide coatings, deposited on a nickel tube with an internal heater of tungsten wire (coated with a refractory insulator such as alumina), have the advantage of being equipotential surfaces. In contrast, along a tungsten filament directly heated by a current there will be a finite potential drop, so that the potential between cathode and anode varies along the cathode. This has the important disadvantage that the filament cannot be heated by alternating current, since this sets up an alternating voltage between cathode and anode which modulates the emitted current.

The close parallel between thermionic emission and evaporation was recognized by Richardson in 1901, but his formula for the emitted current density was based on classical statistics. We shall here use Fermi–Dirac statistics to consider the equilibrium between the electron gas in the metal and the 'electron gas' in the vacuum. Outside the metal, we assume there is no electrical field (since otherwise we do not have equilibrium) and the electrons are moving in random directions so that there is no net current flow. The electron distribution outside is described by the Fermi–Dirac distribution function of eqn (11.20), where in equilibrium the value of $G$ outside is just equal to the value of $W_F$ inside. From Fig. 11.3 it is clear that $W - G = W_k' + \phi$, where $W_k'$ is the kinetic energy outside the metal, and

$$f = \frac{1}{\exp\{(W_k' + \phi)/kT\} + 1} \tag{11.24}$$

Since the exponential term here is much larger than unity, we can write to a good approximation

$$f = \exp(-\phi/kT)\exp(-W_k'/kT). \tag{11.25}$$

The kinetic energy of an electron outside the metal is $W_k' = p^2/2m$; the density of states is thus given, as in eqn (11.8), by

$$g(W_k')\,dW_k' = C(W_k')^{\frac{1}{2}}\,dW_k' \tag{11.26}$$

and the number $dn$ of electrons with energies between $W_k'$ and $W_k' + dW_k'$ is found by multiplying this by eqn (11.25). We can now write simply $W$ for $W_k'$ (this is equivalent to taking $W_k' = 0$ as the zero of total energy), and

$$dn' = C'\exp(-W/kT)W^{\frac{1}{2}}\,dW, \tag{11.27}$$

where

$$C' = C\exp(-\phi/kT). \tag{11.28}$$

The total number of electrons is

$$n' = C'\int_0^\infty \exp(-W/kT)W^{\frac{1}{2}}\,dW$$

$$= C'(kT)^{\frac{3}{2}}\int_0^\infty \exp(-z)z^{\frac{1}{2}}\,dz = \tfrac{1}{2}\pi^{\frac{1}{2}}C'(kT)^{\frac{3}{2}}, \tag{11.29}$$

and their total kinetic energy is

$$\int_0^\infty W\,dn' = C'\int_0^\infty \exp(-W/kT)W^{\frac{3}{2}}\,dW = C'(kT)^{\frac{5}{2}}\int_0^\infty \exp(-z)z^{\frac{3}{2}}\,dz. \tag{11.30}$$

Since

$$\int_0^\infty \exp(-z)z^{\frac{3}{2}}\,dz = -[z^{\frac{3}{2}}\exp(-z)]_0^\infty + \frac{3}{2}\int_0^\infty \exp(-z)z^{\frac{1}{2}}\,dz,$$

where the term in square brackets is zero, the mean kinetic energy of the electrons which emerge from the metal is

$$\bar{W} = \frac{1}{n'}\int_0^\infty W\,dn' = \tfrac{3}{2}kT. \tag{11.31}$$

This result is the same as in classical theory, where the chance of an electron having kinetic energy $W$ has the Boltzmann value $\exp(-W/kT)$; this corresponds to the Fermi–Dirac distribution when the chance $f$ of a quantum state with energy $W$ being occupied is extremely small. The electron gas outside the metal is said to be 'non-degenerate', in contrast with the 'degenerate' state inside the metal where $f$ is virtually equal to unity for all values of $W$ up to $W_F$.

The number of electrons with velocity $v = (2W/m)^{\frac{1}{2}}$ leaving the surface at an angle $\theta$ to the normal is $v\,dn'(\frac{1}{2}\sin\theta\,d\theta)$. This corresponds to a current density normal to the surface equal to $q(v\cos\theta)\,dn'(\frac{1}{2}\sin\theta\,d\theta)$, and on integrating over all angles we obtain

$$q\,dn'v\int_0^{\pi/2} \tfrac{1}{2}\sin\theta\cos\theta\,d\theta = \tfrac{1}{4}qv\,dn'. \tag{11.32}$$

On substituting for $dn'$ and $v$ we obtain a total current density of

$$J = \tfrac{1}{4}qC'\int_0^\infty (2W/m)^{\frac{1}{2}}W^{\frac{1}{2}}\exp(-W/kT)\,dW$$

$$= \tfrac{1}{4}qC'(2/m)^{\frac{1}{2}}(kT)^2\int_0^\infty z\exp(-z)\,dz. \tag{11.33}$$

This last integral is unity, so that we can write

$$J = AT^2\exp(-\phi/kT), \tag{11.34}$$

where the numerical value of the constant $A$ is $1{\cdot}20\times10^6\,\text{A m}^{-2}\,\text{K}^{-2}$.

This calculation assumes equilibrium conditions between the 'electron gas' outside the metal and that inside, the flow of electrons out of the metal being balanced by a flow back into the metal. If an electric field is applied to accelerate the electrons away from the surface, the return flow is removed, and $J$ represents the current density due to thermionic

FIG. 11.7. Plot of $\ln(J/T^2)$ against $10^4/T$ for thermionic emission from copper, silver, and gold. (After Jain and Krishnan 1953.)

emission. Some experimental results are shown in Fig. 11.7, where $\ln(J/T^2)$ is plotted against $1/T$. The slope of the straight lines gives the value of the work function $\phi$; a number of values of $\phi$ and of $A$ are given in Table 11.1. In practice a considerable variation is found in the values of $A$, and this is attributed to various surface effects, though there is also a slight variation in $\phi$ with temperature, reflecting the variation in $W_F$ (eqn (11.19)); this affects the apparent value of $A$ obtained in fitting eqn (11.34). There is also evidence that the rate of emission is different from different surfaces of a single crystal, and that the values of $A$ obtained from a polycrystalline surface are too low.

*Emission of positive ions*

The emission of positive ions from a hot metal is much less probable than that of electrons since the ions are bound in the metal with no translational kinetic energy. An exception occurs in the case of some substances which have high vapour pressures and which would therefore be expected to 'boil off' from a hot surface. It was discovered by Langmuir and Kingdom (1925) that every atom of caesium (boiling-point of metal $\approx 950$ K) which strikes a hot tungsten wire comes off as a positive ion. The reason for this is that a caesium atom has an ionization potential

($W_I = 3 \cdot 87$ eV) lower than the work function of tungsten ($\phi = 4.5$ eV); thus it requires more energy to remove an electron from the tungsten metal to make a neutral caesium atom than to emit a positive caesium ion. The ratio of positive ions to neutral atoms is

$$s^+/s^0 = \exp\{(\phi - W_I)/kT\} \qquad (11.35)$$

and for $(\phi - W_I) \approx 0 \cdot 6$ eV, virtually all the atoms emerge as ions from a tungsten wire at $T = 1200$ K ($\equiv 0 \cdot 1$ eV). This method of 'surface ionization' can be extended to atoms with rather higher ionization potentials by the use of oxidized tungsten, for which $\phi \approx 6$ eV. Neutral atoms and molecules are rather difficult to detect directly, and surface ionization provides a valuable method of detection for atoms (and for molecules which dissociate on the hot wire), allowing a positive monatomic ion to emerge. After acceleration, this may be passed through a mass spectrometer to identify the ion (for example, when several isotopes are present) and eliminate contaminants. The positive ion current may be measured directly, or accelerated onto a surface to produce secondary electrons which are then passed through an electron multiplier. Further details can be found in Ramsey (1956).

*Field emission*

The emission of electrons from a metal under the influence of an applied electric field is closely related to thermionic emission. It occurs only when the direction of the field is such that negative charges are attracted away from the metal, and we must consider the effect of such a field on the potential barrier at the surface. Hitherto we have regarded the potential jump at the surface as infinitely sharp, and an applied field of the correct sign would cause the potential energy to drop linearly outside the surface, as indicated by the line OC in Fig. 11.8. However, the 'image force' (see § 2.5) outside a conducting surface gives a potential energy $W_p = -e^2/16\pi\epsilon_0 x$ for an electron at a distance $x$ from the surface, provided that $x$ is large compared with atomic separations. This potential energy corresponds to the broken line AA′, and on including a field strength $E$ the potential energy outside the metal becomes

$$W_p = (-e^2/16\pi\epsilon_0 x) - Eex. \qquad (11.36)$$

This function (the full line AB in Fig. 11.8) has a maximum, and electrons inside the metal whose energy exceeds the value corresponding to this maximum can easily escape. Effectively the work function is reduced, and the thermionic emission at a given temperature is increased. Emission will also occur by means of 'tunnelling' through the potential barrier, but for metals at ordinary temperatures this requires fields of order $10^8$ V m$^{-1}$ to produce appreciable emission. Field emission and thermionic emission

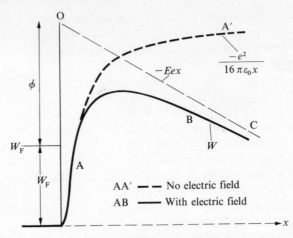

FIG. 11.8. Variation of potential energy $W$ outside a metal, allowing for the image force and an applied electric field.

may occur from a metal electrode into a dielectric, and electrons may tunnel through a thin dielectric film from one electrode to another (see, for example, O'Dwyer 1973).

## 11.6. Specific heat of the conduction electrons

The specific heat of the free electrons can be found from the Fermi–Dirac distribution function. The total energy per unit volume is

$$U = \int_0^\infty Wfg(W)\,dW = C \int_0^\infty W^{\frac{3}{2}}f\,dW$$
$$= \tfrac{2}{5}CW_F^{\frac{5}{2}}\left(1 + \frac{5\pi^2}{8}\left(\frac{kT}{W_F}\right)^2 + \ldots\right) \tag{11.37}$$

from eqn (11.17). Hence, using eqns (11.18) and (11.19), we find for the mean energy per electron $\bar{W} = U/n$,

$$\bar{W} = \tfrac{3}{5}(W_F)_0 + \frac{\pi^2}{4}\frac{(kT)^2}{W_F} + \ldots. \tag{11.38}$$

The specific heat (at constant volume) is obtained by differentiation with respect to $T$, and for a mole of electrons we have

$$C_V/R = \tfrac{1}{2}\pi^2 kT/W_F, \tag{11.39}$$

where $R$ is the gas constant and higher terms in the expansion are omitted.

Comparison with the classical value of $C_V/R = \frac{3}{2}$ shows that the quantum value is smaller by a factor of order $kT/W_F$. The specific heat is proportional to the absolute temperature, and approaches the classical value only when $kT$ approaches $W_F$, a region where higher terms in the expansion leading to (11.38) can no longer be neglected. The reason for the small specific heat given by the quantum theory is that only a small fraction $\approx kT/W_F$ of the electrons at the top of the energy distribution curve are able to increase their kinetic energy, as illustrated by the distribution curve of Fig. 11.2. The increase in energy of each electron is $\approx kT$, and so the total internal energy increases by an amount $\approx n(kT)^2/W_F$. Electrons in the middle of the band cannot be raised to higher energies unless they can reach energy levels above $W_F$, since all the available energy levels in the middle of the band are already occupied by electrons.

Since $kT \ll W_F$ at ordinary temperatures, the electronic specific heat will be only a small fraction of that predicted by classical theory, and the difficulty of the large excess specific heat predicted by that theory for metals is removed. An experimental test of eqn (11.39) is possible only at low temperatures, where the specific heat associated with the lattice vibrations of a solid falls very rapidly. According to the theory of Debye, the lattice specific heat is proportional to $T^3$ at sufficiently low temperatures, and eventually this becomes small compared with the electronic specific heat, which falls only with $T$. The specific heats of a number of metals have been measured, and below about 20 K they are found to follow a law of the form

$$C_V = \alpha T^3 + \gamma T. \tag{11.40}$$

Fig. 11.9 shows the relative magnitudes of the two contributions to the specific heat of cobalt at temperatures below 20 K. This metal is ferromagnetic, and like a number of other transition-group metals, shows an abnormally high electronic specific heat. In many metals the electronic specific heat is predominant only below about 5 K, and rather precise measurements are required to determine it accurately. Eqn (11.40) may be written in the form

$$C_V/T = \alpha T^2 + \gamma, \tag{11.41}$$

and by plotting the quantity $C_V/T$ against $T^2$ a straight-line graph should be obtained whose intercept gives the value of $\gamma$. Fig. 11.10 shows a typical graph for copper.

The measured values of the electronic specific heats of a number of metals are listed in Table 11.2. To compare them with theory we need to know the number $n$ of free electrons per unit volume, and $W_F$; for this purpose it is sufficiently accurate to use $(W_F)_0$, which, by eqn (11.10), also

FIG. 11.9. Specific heat of cobalt at low temperatures.
A   experimental curve.
B   electronic contribution; $C_V = 50 \cdot 2 \times 10^{-4} T$.
C   lattice contribution; $C_V = 1944(T/443)^3$, where 443 is the
Debye $\theta$ for cobalt.
(After Duyckaerts 1939.)

depends on $n$. For the alkali metals, and copper, silver, and gold, it is reasonable to assume there is one conduction electron per atom, and Table 11.2 shows that theory and experiment are in reasonable but by no means exact agreement. The electronic specific heats are appreciably greater for the transition metals, of which iron, cobalt, and nickel belong to the 3d group, palladium to the 4d, and platinum to the 5d group; here it is difficult to predict the value of $n$, a point to which we shall return in § 12.6.

FIG. 11.10. Plot of $C/T$ against $T^2$ for copper. The intercept at $T^2 = 0$ gives the coefficient $\gamma$ of the electronic specific heat. (Corak, Garfunkel, Satterthwaite, and Wexler 1955.)

TABLE 11.2

*Values of the coefficient $\gamma$ of the electronic specific heat $\gamma T$, in* mJ $K^{-2}$ mol$^{-1}$, *as given by Kittel* (1971).

| Metal | $\gamma$ (experiment) | $\gamma$ (theory) | Metal | $\gamma$ (experiment) | $\gamma$ (theory) |
|-------|------------|--------|-------|------------|--------|
| Li | 1·63 | 0·75 | Fe | 4·98 | |
| Na | 1·38 | 1·09 | Co | 4·73 | |
| K | 2·08 | 1·67 | Ni | 7·02 | |
| Rb | 2·41 | 1·91 | Pd | 9·42 | |
| Cs | 3·20 | 2·24 | Pt | 6·8 | |
| Cu | 0·70 | 0·51 | Pb | 2·98 | 1·51 |
| Ag | 0·65 | 0·65 | | | |
| Au | 0·73 | 0·64 | | | |

Theoretical values are omitted for the transition metals, where the number of conduction electrons per atom is uncertain. One electron per atom is assumed for the alkali and noble metals, and two electrons per atom for lead.

## 11.7. Electrical and thermal conductivity

If all electrons are contained within the 'Fermi sphere' of Fig. 11.1, as should be the case in equilibrium at 0 K, the resultant current in any direction is zero; for every electron in a quantum state corresponding to momentum components $(p_x, p_y, p_z)$ there is another with components $(-p_x, -p_y, -p_z)$, so that the net momentum is zero. In the presence of an electric field of strength **E** a charge $q$ experiences an acceleration

$$\mathrm{d}\mathbf{p}/\mathrm{d}t = \hbar(\mathrm{d}\mathbf{k}/\mathrm{d}t) = q\mathbf{E} \tag{11.42}$$

whose effect is to displace the Fermi sphere at a constant rate $q\mathbf{E}/\hbar$ in the direction of the field. This acceleration is counteracted by the effects of collisions. The mean displacement in momentum is equal to the drift momentum $\bar{\mathbf{p}} = q\mathbf{E}\tau$ in the field **E** (cf. § 3.2), and in **k**-space this means that the Fermi sphere is displaced by an amount $q\mathbf{E}\tau/\hbar$, as shown in Fig. 11.11. It is readily seen from the figure that the essential change due to **E** is a transfer of a small number of electrons from one side of the Fermi sphere to the other, and it follows that the conductivity depends only on the electrons at the Fermi level. Thus we have the important result that in a metal the transport properties involve only electrons whose velocities correspond to the Fermi energy $W_F$. Provided that the Fermi surface is a sphere, this means that the electron velocity $c$ in a formula such as eqn (11.2) is fixed in magnitude, and random only in direction.

At a finite temperature the Fermi distribution does not give a sharp boundary at the surface of the Fermi sphere. At ordinary temperatures, however, the spread in energy (of order $kT$) is so small compared with

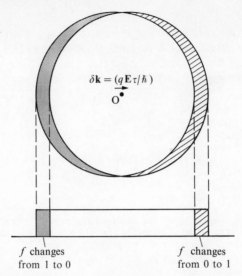

$$\delta\mathbf{k} = (q\mathbf{E}\tau/\hbar)$$

$f$ changes
from 1 to 0

$f$ changes
from 0 to 1

FIG. 11.11. Displacement of the Fermi sphere by an amount corresponding to the drift momentum $\delta\mathbf{k} = \mathbf{p}/\hbar = q\mathbf{E}\tau/\hbar$ in a field strength $\mathbf{E}$. Below is shown the change in the distribution function $f$ at $T = 0$. The current density $\mathbf{J} = n(q/m)\hbar\,\delta\mathbf{k}$.

$W_F$ that the velocity spread can be neglected, and the velocity $v_F$ at the Fermi surface assumed to be the same as at $T = 0$.

At first sight we might expect this to produce some rather dramatic changes in the formulae for the transport properties. On replacing $c$ by $v_F$ and writing $L = v_F\tau$, eqn (11.2) for the thermal conductivity becomes

$$K = \tfrac{1}{3}nv_F^2\tau(\mathrm{d}\bar{W}/\mathrm{d}T) = \tfrac{2}{3}n(W_F/m)\tau(\mathrm{d}\bar{W}/\mathrm{d}T). \tag{11.43}$$

From eqn (11.38) the value of $\mathrm{d}\bar{W}/\mathrm{d}T$ is $\tfrac{1}{2}\pi^2k^2T/W_F$, whence

$$K = \frac{\pi^2}{3} \cdot \frac{nk^2T}{m}\,\tau. \tag{11.44}$$

This equation shows that any temperature-dependence in the ratio $K/T$ must come from variation in the relaxation time $\tau$ with temperature, and it follows from eqn (11.1) that this is true also for the electrical conductivity $\sigma$. The ratio of these two conductivities is

$$\frac{K}{\sigma} = \frac{\pi^2}{3}\left(\frac{k}{e}\right)^2 T = L_0T, \tag{11.45}$$

whose numerical value is in fact very close to the value given by classical theory (§ 3.11). However, in eqn (11.45) the $T$ arises from the fact that $\mathrm{d}\bar{W}/\mathrm{d}T$ is proportional to $T$, while $v_F^2 = 2W_F/m$ is independent of $T$; in the classical formulae it is just the other way round.

The numerical value of the Lorenz number $L_0$ given by eqn (11.45) is $2 \cdot 45 \times 10^{-8} \, \mathrm{W} \, \Omega \, \mathrm{K}^{-2}$, and at room temperature the experimental results for most metals lie close to this value. The is not the case at low temperatures, as can be deduced from the variation of the electrical resistivity $\sigma^{-1}$ and thermal resistivity $K^{-1}$ shown for copper in Fig. 11.12. Near room temperature $K^{-1}$ is nearly constant, while $\sigma^{-1}$ increases almost linearly with $T$, but below 100 K each drops much faster than $T$. The quantity $K/\sigma T$, which should be equal to $L_0$, falls steadily in value as the temperature decreases. This behaviour is typical of many pure metals. It shows that

(1) $\tau$ must be a function of temperature;

(2) it is not correct to use the same value of $\tau$ in the equations for the electrical and the thermal conductivity.

In most metals the electron velocity at the Fermi surface is of the order of $10^6 \, \mathrm{m \, s}^{-1}$, and since $\tau \approx 10^{-14} \, \mathrm{s}$ at room temperature, the mean-free-path length must be of the order of $10^{-8} \, \mathrm{m}$, or about 100 times the atomic spacing in a solid. At lower temperatures, where the conductivity is much higher, the mean-free-path length must be much longer. Classical physics offered no method of estimating the path length, and on the wave theory it was first pointed out by Houston that the mean-free-path length of an electron in a perfectly regular lattice should be infinite. If an electron is in an allowed energy state, then that is a stationary state; in the absence of a perturbation the electron should continue in its state of fixed energy and fixed wave vector, (that is, fixed momentum), indefinitely. Resistance to the flow of electrical or heat current must therefore arise from the fact

FIG. 11.12. Thermal resistivity $K^{-1}$ and electrical resistivity $\sigma^{-1}$ for pure copper as a function of temperature. The units of $K^{-1}$ are $10^{-3} \, \mathrm{K \, m \, W}^{-1}$ and of $\sigma^{-1}$ are $10^{-8} \, \Omega \, \mathrm{m}$.

that real metals do not have perfect lattices, two reasons for which are:

(1) the lattice contains foreign atoms (impurities) or atoms displaced from their normal positions (point defects and dislocations);

(2) the atoms deviate from their mean positions because of thermal vibrations in the lattice.

Each of these imperfections causes scattering of the electron waves in the same way that a defect in an insulating crystal scatters a light wave, whereas a perfect crystal does not.

An alloy is an outstanding example of a disordered lattice on which the different atoms are randomly located, and we would therefore expect its resistance to be higher than that of a pure metal. Furthermore, the scattering in this case should be independent of temperature, giving rise to the constant resistance which is characteristic of alloys. On the other hand, the quantized lattice vibrations cease to be excited at low temperatures, so that both the electrical and the thermal resistance should fall towards zero, being ultimately limited by defect scattering. This provides a qualitative explanation of the behaviour illustrated in Fig. 11.12; the subject will be discussed in more detail in §§ 12.7–12.8.

## 11.8. Thermoelectricity

In 1821 Seebeck discovered that in an electrical circuit made up of two different metals, a small current flows when the junctions between the two metals are not at the same temperature. On open-circuit, an e.m.f. of the order of microvolts per degree difference in temperature of the junctions is observed. The converse effect was discovered by Peltier in 1834. If the Seebeck e.m.f. is from metal A to metal B at the hot junction (Fig. 11.13), a current driven by an external e.m.f. of the same sign produces a cooling at this junction and a heating at the other junction. Both the Seebeck and the Peltier effect are completely reversible. The Seebeck e.m.f. for many metals can be represented by an equation of the form

$$V = at + \tfrac{1}{2}bt^2, \tag{11.46}$$

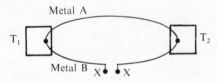

FIG. 11.13. The thermoelectric effect; the thermoelectric voltage is measured by insertion of a voltmeter at XX.

where $t$ is the temperature difference between the junctions. The Peltier coefficient $\Pi$ is defined as follows: the heat absorbed when a charge of $Q$ coulombs passes from metal A to metal B is $\Pi Q$ joules.

In 1854 Thomson (Lord Kelvin) realized that there should be a thermodynamic relation connecting the Seebeck and Peltier effects, but the relation he derived did not agree with experiment. He concluded that a further thermoelectric effect must exist, now known as the 'Thomson heat'. If an electric current of density $J_z$ flows along a conductor in which a temperature gradient $dT/dz$ is maintained, the heat produced in unit volume is

$$\rho J_z^2 - \mu_T J_z (dT/dz) + K(d^2T/dz^2). \tag{11.47}$$

The first term is the Joule heating, determined by the resistivity $\rho$, and the last term is due to heat conduction. Both are irreversible effects. The second term is a reversible effect, which can be either a cooling or a heating, depending on the relative directions of the temperature gradient and the current flow. The sign of the effect also depends on the Thomson coefficient $\mu_T$, whose dimensions are watt ampere$^{-1}$ ($=$ volt) per unit temperature gradient ($\mathrm{K\,m^{-1}}$). $\mu_T$ is positive if heat is absorbed from the lattice when current flows up the temperature gradient.

The origins of the thermoelectric effects can be understood in the following way. The energy of the free carriers depends on the temperature, and the carriers at the hot end have more energy that those at the cold end of a metal. Because of their higher velocities, carriers from the hot end diffuse down the metal more rapidly than those from the cold end; they give up their excess energy on making collisions with the lattice, and this is the Thomson heat. When two metals are joined together, the rates of heat flow in the two metals are unequal; heat therefore appears at one junction and is absorbed at the other. This is the Peltier effect.

When the metal is on 'open-circuit', the flow of carriers continues only until a potential difference is set up which just counterbalances the flow. This gives an e.m.f. which is part of the Seebeck e.m.f. A second contribution arises in the following way. The contact potential between two metals with junctions at the same temperature is equal and acts in the opposite sense at the two junctions. However, the contact potential depends slightly on the temperature, leaving a residual e.m.f. when the two junctions are at different temperatures. The contact potential is the difference between the work functions of the two metals, and by considering the temperature variation of the latter we can again reduce the problem to a question of the bulk properties of each metal. In fact theory shows that the various effects are all so interconnected that we need consider only one, such as the Thomson heat, which may be measured as follows.

### Measurement of the Thomson heat

· A thin metal wire is stretched in a good vacuum between two heavy leads (Fig. 11.14) maintained at constant temperature. When a direct current is passed through the wire, the Joule heating raises the temperature at the centre and heat flows towards each end, irrespective of the direction of the current. The Thomson effect causes heat to be released in one half and absorbed in the other; this process is reversed when the direction of current flow is reversed. The temperature of the wire is measured at two symmetrical points $P_1$, $P_2$ as in Fig. 11.14 by a pair of thermocouples, electrically insulated from the wire by means of thin paper. Two measurements are made. First the thermocouples are connected in series, so that the mean increase of temperature at the two points is found; this gives the temperature increase due to the Joule heating alone. Second, the couples are connected in opposition to give the difference in temperature between the two points due to the Thomson effect. Temperature differences arising from asymmetry in the positions of $P_1$, $P_2$ are eliminated by repeating the measurement with the direction of current flow reversed.

On the assumption that heat is lost only by thermal conduction along the wire, the differential equation including the Joule heating, the Thomson effect, and thermal conduction is

$$I^2 R - I\mu_{\mathrm{T}}\frac{\mathrm{d}T}{\mathrm{d}z} + KA\frac{\mathrm{d}^2 T}{\mathrm{d}z^2} = 0, \qquad (11.48)$$

where $I$ is the current flowing, $R$ the resistance per unit length, $A$ the cross-section of the wire, and $K$ its thermal conductivity. The solution of

FIG. 11.14. Diagram of the apparatus used by Borelius, Keesom, and Johansson (1928) to measure the absolute value of the Thomson coefficient of a metal. $P_1$, $P_2$ thermocouple junctions of platinum (P) and constantan (C).

this equation is

$$cT = \exp fz - \cosh fa - (z/a)\sinh fa, \tag{11.49}$$

where $f = I\mu_T/KA$ and $c = (\mu_T/IaR)\sinh fa$; the temperature at the ends of the wire $z = \pm a$ is taken to be zero. If the thermocouple readings are $T_1$ when connected in series and $T_2$ when in opposition then these are proportional to the sum and difference of the temperatures at $+b$ and $-b$ respectively. Hence

$$(fb)^{-1}\frac{T_2}{T_1} = \frac{(fb)^{-1}\sinh fb - (fa)^{-1}\sinh fa}{\cosh fb - \cosh fa}$$

which equals $\frac{1}{3}$ when $f$ is small. This gives

$$\mu_T = \frac{3T_2 KA}{T_1 Ib}. \tag{11.50}$$

Here we have assumed the resistivity of the wire to be independent of temperature; in practice a small correction must be made for the slight change in temperature distribution caused by the varying resistance.

*Thermoelectric theory*

When a charge passes round a circuit such as that in Fig. 11.13, there is an irreversible heating because of the resistance in the circuit, proportional to the square of the current. Thermal changes due to the Peltier and Thomson effects are proportional to the current and are reversible. By adopting the somewhat artificial device of assuming that we can neglect the irreversible heating when the current is sufficiently small, we can apply thermodynamics to the reversible changes. When a charge $Q$ is passed, the external work done is $QV$, where $V$ is the total e.m.f. in the circuit; the net heat absorbed at the junctions is $(\Pi_{T_1} - \Pi_{T_2})Q$; and the increase in internal energy is $-Q\int_{T_2}^{T_1}(\mu_{TA} - \mu_{TB})\,dT$, since this represents the energy required to heat the charge carriers from temperature $T_2$ to $T_1$ in metal A less that emitted in the reverse process in metal B. Hence from the first law of thermodynamics (dividing by $Q$),

$$V = \Pi_{T_1} - \Pi_{T_2} + \int_{T_2}^{T_1}(\mu_{TA} - \mu_{TB})\,dT. \tag{11.51}$$

To detect this e.m.f., the circuit must be broken and a voltmeter introduced at, say XX, in Fig. 11.13. Although the voltmeter may contain metals different from A and B, the total e.m.f. in the circuit will be unaffected provided that both junctions to the instrument are at the same temperature. For, if the meter M has junctions at a temperature $T$,

$$V = (\Pi_{T_1} - \Pi_{T_2})_{A \to B} + (\Pi_T)_{B \to M} + (\Pi_T)_{M \to B} + \int_{T_2}^{T_1} (\mu_{TA} - \mu_{TB}) \, dT$$

$$= (\Pi_{T_1} - \Pi_{T_2}) + \int_{T_2}^{T_1} (\mu_{TA} - \mu_{TB}) \, dT,$$

which is the same as before. The same result holds if any number of metals at different temperatures are connected in the circuit, provided that each pair of junctions to a given metal are at the same temperature.

Since no temperature changes occur in the circuit, the external work $QV$ is also the free energy of the system, so that from the second law of thermodynamics

$$QV = U + T \frac{d(QV)}{dT},$$

where

$$U = Q \int_0^T (\mu_{TA} - \mu_{TB}) \, dT.$$

Hence

$$V = \int_0^T (\mu_{TA} - \mu_{TB}) \, dT + T \frac{dV}{dT},$$

and differentiation with respect to $T$ gives

$$\mu_{TA} - \mu_{TB} = -T \frac{d^2 V}{dT^2}. \tag{11.52}$$

Then elimination of $(\mu_{TA} - \mu_{TB})$ with the help of eqn (11.51) yields

$$\Pi_{T_1} - \Pi_{T_2} = [T(dV/dT)]_{T_2}^{T_1},$$

or

$$\frac{dV}{dT} = \frac{\Pi}{T}. \tag{11.53}$$

Although the Peltier and Thomson effects are reversible, the application of reversible thermodynamics to a thermoelectric circuit is open to a number of objections, since irreversible effects such as resistive heating and thermal conduction are always present. A better method is to use irreversible thermodynamics (see Adkins 1975) or to consider the change of entropy at a junction (see Cusack 1958) and to relate this to the net heat absorbed at that temperature. A more rigorous treatment on these lines confirms that eqns (11.52) and (11.53) are valid, and they have been tested experimentally over a large temperature range for many substances.

The quantity $dV/dT$ is called the 'thermoelectric power', and from eqn (11.52) we have

$$dV/dT = -\int \frac{(\mu_{TA} - \mu_{TB})}{T} dT + \text{constant}.$$

By the third law of thermodynamics (see, for example, Wilks 1961) $dV/dT \to 0$ as $T \to 0$, and we can therefore eliminate the constant by taking the integral from 0 to $T$ (to avoid infinities in the integrand this requires $\mu_T$ to vary at least as the first power of $T$). Then

$$\frac{dV}{dT} = \int_0^T \frac{\mu_{TB}}{T} dT - \int_0^T \frac{\mu_{TA}}{T} dT = S_B - S_A, \qquad (11.54)$$

where the quantity

$$S = \int_0^T \frac{\mu_T}{T} dT \qquad (11.55)$$

is known as the 'absolute thermoelectric power' and is a property of each substance separately. Our identification of the quantity $Q \int \mu_T \, dT$ as the change in the internal energy suggests that

$$\mu_T = (1/Q)(dU/dT) = C_e/Q = \gamma T/Q \qquad (11.56)$$

where $C_e$ is the specific heat of the electrons. Then

$$S = \int_0^T \left(\frac{\mu_T}{T}\right) dT = -\left(\frac{\gamma T}{F}\right), \qquad (11.57)$$

where for 1 mol of electrons we have written $Q = -Ne = -F$, where $F$ is the Faraday constant.

The quantity $S$ is analogous to entropy per unit charge, since $\mu_T$ is the specific heat per unit charge. The theory predicts that both $S$ and $\mu_T$ should be proportional to $T$, and at temperatures above 100 K this is largely true. Values of the Thomson heat $\mu_T$ at 273 K are listed in Table 11.3 together with the values calculated from the experimental electronic specific heats given in Table 11.2. The theoretical values are generally of the right order of magnitude, but the agreement is not exact. A serious difficulty is that for a number of cases the experimental values of $\mu_T$ are positive, and not negative as would be expected when the current is carried by negatively charged electrons. (See § 12.9.)

TABLE 11.3
*Values of the Thomson heat $\mu_T$ in $\mu V\ K^{-1}$*

| Metal | $\mu_T$ (experiment) | $\mu_T$ (theory) | Metal | $\mu_T$ (experiment) | $\mu_T$ (theory) |
|---|---|---|---|---|---|
| Li | +23·3 | −4·6 | Fe | −5·5 | −14·1 |
| Na | −5·1 | −3·9 | Co | −26·7 | −13·4 |
| K | −11·3 | −5·9 | Ni | −16·5 | −19·9 |
| Rb | −9·4 | −6·8 | Pd | −16·6 | −26·6 |
| Cs | −0·8 | −9·1 | Pt | −12·0 | −19·2 |
| Cu | +1·3 | −2·0 | Pb | −0·55 | −8·4 |
| Ag | +1·3 | −1·8 | | | |
| Au | +1·6 | −2·1 | | | |

The experimental values are measured near 273 K; the theoretical values are given by eqn (11.56), using the experimental values of $\gamma$ given in Table 11.2 and assuming $T = 273$ K.

## 11.9. Electron–electron interaction

An important effect which has been neglected in our treatment is the electrostatic repulsion of the conduction electrons. This tends to keep the electrons apart, and the chance of finding two electrons close together is less than it would be on our assumption that their motion is completely independent of each other; in other words, there is a *correlation* between their motions. There is a further effect due to the fact that the over-all wavefunction for the assembly of electrons must be antisymmetric, which is similar in nature to the 'exchange interaction' discussed in Chapter 15. These two effects contribute to the 'correlation energy' and are important in calculating the cohesive energy of a metal. The electrostatic repulsion is a long-range interaction which gives rise to 'plasma oscillations' (see § 10.10), for which the characteristic frequency in a metal is of the order of $10^{15}$ Hz. At low frequencies the dynamical properties of the electrons are not greatly altered, a fortunate circumstance which makes a simple treatment neglecting the correlation energy more accurate than might have been expected.

## References

BORELIUS, G., KEESOM, W. H., and JOHANSSON, C. H. (1928). *Proc. Acad. Sci. Amster.* **31,** 1046.

CORAK, W. S., GARFUNKEL, M. P., SATTERTHWAITE, C. B., and WEXLER, A. (1955). *Phys. Rev.* **98,** 1699.

CUSACK, N. (1958). *The electrical and magnetic properties of solids.* Longmans-Green, London.

DUYCKAERTS, G. (1939). *Physica* **6,** 817.

JAIN, S. C., and KRISHNAN, K. S. (1953). *Proc. R. Soc.* **A217,** 451

KITTEL, C. (1971). *Introduction to solid state physics.* Wiley, New York.
LANGMUIR, E. and KINGDOM, K. H. (1925). *Proc. R. Soc.* **A107,** 61.
MITCHELL, E. W. J. and MITCHELL, J. W. (1951). *Proc. R. Soc.* **A210,** 70.
O'DWYER, J. J. (1973).*Electrical conduction and breakdown in solid dielectrics.* Clarendon Press, Oxford.
RAMSEY, N. F. (1956). *Molecular beams.* Clarendon Press, Oxford.
WILKS, J. (1961). *The third law of thermodynamics.* Clarendon Press, Oxford.
ADKINS, C. J. (1975). *Equilibrium thermodynamics.* McGraw-Hill, London.

## Problems

11.1. Show that eqn (11.19) leads to a value for the temperature coefficient $(dW_F/dT)$ for tungsten $(W_F = 5.8 \text{ eV})$ at 3000 K of about $-0.6 \times 10^{-5} \text{ eV K}^{-1}$. $W_F$ will also vary because of thermal expansion, since the number of electrons per unit volume changes. If $\alpha$ is the linear coefficient of expansion, show that

$$dW_F/dT = -2\alpha W_F.$$

For tungsten at 3000 K, $\alpha$ is about $6 \times 10^{-6}$, from which $dW_F/dT = -7 \times 10^{-5} \text{ eV K}^{-1}$, showing that the effect of thermal expansion is considerably more important than the second term in eqn (11.19).

11.2. Show that eqn (11.57) leads to a formula of the type (11.46) for the thermoelectric e.m.f. in a circuit where the temperature difference between the two junctions is $t$. What predictions about the coefficients $a$, $b$ follow from eqn (11.57)?

11.3. It is found that the thermoelectric power $dV/dt$ of a copper–nickel thermocouple in the range 273–373 K can be expressed as $(20.4+0.0450\,t)\mu \text{ V K}^{-1}$, where $t$ is the temperature in °C measured on a particular hydrogen gas thermometer. The Peltier coefficient is given by the expression $5560+32.6t+0.0448t^2 \,\mu\text{J C}^{-1}$. Verify that in this thermometer the hydrogen behaves as a perfect gas, and deduce the absolute temperature of the ice-point.

11.4. Show that for electrons obeying Maxwell–Boltzmann statistics the Thomson heat should be given by the formula

$$\mu_T = -\frac{3}{2}\frac{k}{|e|}$$

whose value is independent of temperature and equal to about $-130 \,\mu\text{V K}^{-1}$.

11.5. For an 'isotropic gas' of charged carriers the thermoelectric power $S$ is given by the formula

$$qS = \frac{2c_q}{3} - \frac{dW_F'}{dT},$$

where $q$ is the charge and $c_q$ the 'specific heat' of each carrier, irrespective of the statistics obeyed by the carriers. Show that for free electrons obeying Fermi–Dirac statistics this gives

$$-|e|\,S = \tfrac{1}{2}\pi^2 k^2 T/W_F.$$

# 12. The band theory of metals

## 12.1. The wave equation for free electrons

In the previous chapter the properties of free electrons in a metal were discussed using Fermi–Dirac statistics, taking account of the wave-like properties of electrons by using the de Broglie relation and the analogy with waves in a box. In a real solid the wave-like nature is important for yet another reason: the wavelength is comparable with the interatomic distance, giving rise to diffraction effects. Before considering these, we shall discuss the simpler problem of the wave equation for free electrons.

The wave equation for a free electron of total energy $W$ in an unbounded region where the potential is $V$ is

$$\frac{\hbar^2}{2m}\nabla^2\psi+(W-V)\psi = 0. \tag{12.1}$$

For simplicity we consider first the one-dimensional case, for which the wave equation is

$$\frac{\hbar^2}{2m}\frac{\partial^2\psi}{\partial z^2}+(W-V)\psi = 0, \tag{12.2}$$

and for convenience we further assume $V=0$ everywhere. Then the solutions of this equation are of the form

$$\psi = A\,\exp \mathrm{j}k_z z. \tag{12.3}$$

The momentum of the particle $p_z$ is given by

$$p_z = -\mathrm{j}\hbar\int_{-\infty}^{+\infty} \psi^*\frac{\partial\psi}{\partial z}\,\mathrm{d}z = -\mathrm{j}\hbar(\mathrm{j}k_z)\int_{-\infty}^{+\infty}\psi^*\psi\,\mathrm{d}z = \hbar k_z \tag{12.4}$$

since the normalization of the wavefunction requires $\int_{-\infty}^{+\infty}\psi^*\psi\,\mathrm{d}z = 1$. In order to satisfy eqn (12.2) we must have $(\hbar^2/2m)k_z^2 = W$, or

$$W = (\hbar^2/2m)k_z^2 = p_z^2/2m \tag{12.5}$$

so that $W$ corresponds to that part of the kinetic energy of the electron associated with its momentum $p_z$ in the $z$-direction.

The three-dimensional equation (12.1) also has a simple solution in Cartesian coordinates, corresponding to the product of three functions of

the type (12.3). This solution is

$$\psi = A \exp(jk_x x)\exp(jk_y y)\exp(jk_z z)$$
$$= A \exp j(k_x x + k_y y + k_z z) = A \exp j(\mathbf{k} \cdot \mathbf{r}) \tag{12.6}$$

since $(x, y, z)$ are the components of the vector $\mathbf{r}$, and we can similarly regard $(k_x, k_y, k_z)$ as the components of a vector $\mathbf{k}$, known as the wave vector. In order to satisfy eqn (12.1) we have

$$W = (\hbar^2/2m)(k_x^2 + k_y^2 + k_z^2) = (p_x^2 + p_y^2 + p_z^2)/2m = \mathbf{p}^2/2m, \tag{12.7}$$

where $\mathbf{p}$ is the momentum vector, with components $(p_x, p_y, p_z)$. These components of $\mathbf{p}$ are just $\hbar$ times those of $\mathbf{k}$, that is, we have the de Broglie relation

$$\mathbf{k} = \mathbf{p}/\hbar \tag{12.8}$$

which was used in § 11.1. The wavelength associated with the electron is $2\pi/k = h/p$.

In applying the free-electron model to a metal, we assume that the electrons move in a region of constant potential, with a sharp rise in the potential at the boundaries of the metal. Since the electrons do not have enough energy to surmount this barrier, they are confined within the metal (we neglect phenomena such as thermionic emission, which involve only an insignificant proportion of the electrons). Our model thus assumes a rectangular potential well, such as is shown in Fig. 12.1(a) for one dimension. Taking the floor of the well to be at $V = 0$, the solutions of the wave equation are of the form (12.6) inside the well. Outside the well, where the potential $V > W$, so that $(W - V)$ is negative, the solutions are real exponentials; the chance of finding an electron outside, which is proportional to

$$\psi^* \psi = \exp\left[\left\{\frac{8m(V - W)}{\hbar^2}\right\}^{\frac{1}{2}} z\right],$$

falls off very rapidly with distance. A proper solution of the problem requires that the wavefunctions and their derivatives be continuous at the boundary, but if we make the approximation of taking $V$ to be infinite this reduces to making $\psi$ vanish at the boundary. Then the problem is

(a)     (b)

Fig. 12.1. (a) Rectangular potential well assumed in free-electron model of a metal. (b) Actual potential variation, showing sharp fall near each positively charged ion.

similar to that of electromagnetic waves in a perfectly conducting box; if the latter is rectangular with dimensions $(a, b, c)$, the allowed solutions (cf. Problem 9.12) are a set of standing waves where $\psi$ is a product of terms such as

$$\frac{\sin}{\cos}(\pi l x/a)\frac{\sin}{\cos}(\pi m y/b)\frac{\sin}{\cos}(\pi n z/c),$$

and the wavelength is given by eqn (9.42)

$$\frac{1}{\lambda^2} = \left(\frac{l}{2a}\right)^2 + \left(\frac{m}{2b}\right)^2 + \left(\frac{n}{2c}\right)^2.$$

For electrons at the top of the Fermi distribution the wavelengths involved are of the order of a nanometre, which are very small compared with the dimensions of a metal of ordinary size. The spacing of the allowed wavelengths is therefore very close, and the number in a given wavelength range can be computed using the approximations adopted in the theory of heat radiation, as in § 11.1.

## 12.2. The energy-band approximation

At this stage we are still making the arbitrary assumption that in a metal some of the electrons are detached from their parent atoms and are merely bound to the metal as a whole by a potential well inside which they move quite freely. However, many solids are very good electrical insulators in which we must assume there are no such free electrons. There is also the intermediate class of solids, the semiconductors, which are much poorer conductors than metals and which generally possess a negative rather than a positive coefficient of resistivity. To understand why these different types of solids exist we must consider the interaction between the electrons and nuclei when they are closely packed in a solid, where the interatomic distance is of the same order as the atomic radius. The potential energy of an electron then varies rather as shown in Fig. 12.1(b), falling steeply when the electron approaches a positively charged nucleus. Obviously the motion of the electrons in such a potential is a very complicated problem and cannot be solved exactly. Approximate methods must be used, whose nature is illustrated by approaching the problem from two different standpoints.

In an isolated atom the electrons are tightly bound and have discrete, sharp energy levels. When two identical atoms are brought together, the energy levels of each atom, which are initially the same, are split into two, one higher and one lower than the corresponding levels of the separated atoms. The reason for this behaviour follows from a simple quantum-mechanical result. Suppose we have two electrons which have energies $W_1$, $W_2$ when they are on isolated atoms, with no mutual interaction; when the atoms approach one another, an energy of interaction $W_{12}$

arises between the two electrons. The new values $W$ of the energy are given by the solution of the determinant

$$\begin{vmatrix} W_1-W & W_{12} \\ W_{12} & W_2-W \end{vmatrix} = 0 \qquad (12.9)$$

and are

$$W = \tfrac{1}{2}(W_1+W_2) \pm \{\tfrac{1}{4}(W_1-W_2)^2 + W_{12}^2\}^{\frac{1}{2}}. \qquad (12.10)$$

If $W_{12} \ll W_1 - W_2$, the values of the energy are approximately

$$\begin{aligned} W_1' &= W_1 + W_{12}^2/(W_1-W_2), \\ W_2' &= W_2 - W_{12}^2/(W_1-W_2), \end{aligned} \qquad (12.11)$$

showing that the energy levels are now slightly further apart, the upper level (assumed to be $W_1$) being pushed up and the lower level $W_2$ being pushed down. The change in energy is small if $W_{12}$ is small compared with $W_1 - W_2$, but increases sharply when this inequality no longer holds. In the opposite extreme where $W_{12} \gg (W_1 - W_2)$ the energy levels approach the values

$$W' = \tfrac{1}{2}(W_1+W_2) \pm W_{12} \pm (W_1-W_2)^2/8W_{12}, \qquad (12.12)$$

and the splitting has the value $2W_{12}$ for electrons which have the same energy ($W_1 = W_2$) in the unperturbed state. In general the perturbed wavefunctions are linear combinations of the wavefunctions $\psi_1$, $\psi_2$ for the isolated atoms, and in the limit $W_1 = W_2$ they are

$$2^{-\frac{1}{2}}(\psi_1 \pm \psi_2). \qquad (12.13)$$

This means that we can no longer regard one electron as being on one atom and the other on the second atom; the electrons are shared between the two atoms.

If more atoms are brought together, further levels of slightly different energy are formed; for a solid of $N$ atoms, where $N$ is a very large number, the levels are so close together that they form an almost continuous band. The splittings depend on the interactions, which are small for atoms far apart and larger for atoms close together. The interaction arises from the electrostatic repulsion between each pair of electrons; the major contribution comes from the regions where the electron charge clouds are closest together, and in particular from the regions where the electron wavefunctions overlap. At a given distance the interactions are therefore greatest for electrons whose atomic wavefunctions spread out furthest from the parent nucleus, that is, for electrons with the highest energy, which are least closely bound to the parent atom. The over-all bandwidth depends on the degree of the overlap for electrons on adjacent atoms, since these have the largest interaction. Fig. 12.2

FIG. 12.2. Sharp energy levels in a free atom and the corresponding bands in a solid (the top band is shown only partly full). The arrows indicate allowed transitions giving X-ray emission bands when an electron has been ionized out of the lowest energy band.

is a rough diagram showing how the atomic levels develop into bands as the atoms are brought close together. The wavefunctions are linear combinations of atomic orbitals on every atom, so that the electrons must be regarded as being shared between all the atoms.

## 12.3. Electrons in a periodic potential

The problem may be approached from the opposite viewpoint, starting with electrons moving freely in the flat-bottomed potential well of Fig. 12.1(a), and investigating how the motion is modified when we allow for the variation in the potential energy near each atomic nucleus illustrated in Fig. 12.1(b). In a crystal the atoms form a regular array, and the potential has a corresponding periodic variation in three dimensions. The effect of the periodicity can be understood by considering a one-dimensional case, where a set of $N$ identical atoms form a linear chain, each atom a distance $a$ from its neighbours. We first obtain the boundary conditions. Normalization of the free-electron wavefunctions (12.3) requires

$$1 = A^2 \int_0^{Na} \psi\psi^* \, \mathrm{d}z = A^2 \int_0^{Na} \exp(jk_z z)\exp(-jk_z z) \, \mathrm{d}z = A^2(Na),$$

which gives $A = (Na)^{-\frac{1}{2}}$. We also need to make the wavefunctions orthogonal for different values of $k_z$, and this requires

$$0 = \int_0^{Na} \exp(jk_1 z)\exp(-jk_2 z) \, \mathrm{d}z = \frac{1}{j(k_2-k_1)} [\exp j(k_1-k_2)z]_0^{Na},$$

which is satisfied provided that $(k_1-k_2)Na$ is an integral multiple of $2\pi$. For this to hold for all values of $k_z$ we must have

$$k_z = 2\pi r/Na = r(G/N), \qquad (12.14)$$

where $r$ has integral values from 1 to $N$, $G = 2\pi/a$, and the number of allowed values of $k_z$ is just equal to the number of particles.

We now introduce the periodic potential, which must repeat in such a way that $V(z+a) \equiv V(z)$. Whatever the actual form of the variation of $V$ with $z$, it can be expanded in a Fourier series of the form

$$V(z) = V_0 + V_1 \cos Gz + V_2 \cos 2Gz + ..., \qquad (12.15)$$

where again $G = 2\pi/a$; the sine terms can be omitted (as here) by taking a suitable origin for $z$, since $V$ must be an even function of $z$ provided that $z = 0$ lies at the centre of an atom.

For two electrons of wave vectors $\mathbf{k}_1$, $\mathbf{k}_2$ the potential energy arising from the first term of (12.15) is

$$(Na)^{-1} \int_0^{Na} V_0 \exp(jk_1 z)\exp(-jk_2 z)\, dz, \qquad (12.16)$$

which vanishes unless $k_1 = k_2$, in which case it is equal to $V_0$. This corresponds to the steady potential at the bottom of the rectangular well in Fig. 12.1(a), and if we take this to be the zero of energy then $V_0 = 0$.

We can now find the effect of a term such as $V_s \cos sGz$ (where $s$ is a positive integer) in the periodic potential. It gives rise to an interaction energy between two electrons with wave vectors $\mathbf{k}_1$, $\mathbf{k}_2$ equal to

$$\tfrac{1}{2}(Na)^{-1} \int_0^{Na} \exp(jk_1 z)\, V_s\{\exp(jsGz)+\exp(-jsGz)\}\exp(-jk_2 z)\, dz,$$

which vanishes except when

$$k_1 - k_2 = \pm sG. \qquad (12.17)$$

Now the states $k_1$, $k_2$ have kinetic energies equal to $W_1 = (\hbar k_1)^2/2m$ and $W_2 = (\hbar k_2)^2/2m$, and the interaction energy $W_{12} = \tfrac{1}{2}V_s$ will have little effect if it is small compared with $W_1 - W_2$. On the other hand, it will have maximum effect for a pair of electrons for which $W_1 = W_2$, and this occurs for a pair for which $k_1 = -k_2$, provided that, from (12.17), $k_1 = -k_2 = \pm\tfrac{1}{2}sG$. For such a value of $k_z$, the interaction gives an additional energy which splits the levels by $\pm W_{12}$, as in eqn (12.12). The effect of this is shown in Fig. 12.3; the broken line represents the kinetic energy $W = (\hbar_z k_z)^2/2m$ in the unperturbed state. At the point $k_0 = \pm\tfrac{1}{2}G$ (that is, $k_0 = \pm\pi/a$) the periodic potential term $V_1$ causes a discontinuity in the

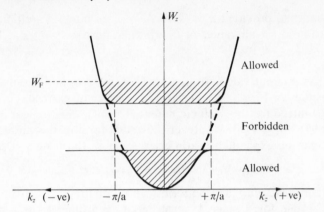

FIG. 12.3. Plot of $W_z$ against $k_z$ showing band structure due to periodic potential of lattice. - - -$W_z$ against $k_z$ in the absence of the periodic structure.

relation between $W$ and $k_z$, the curve for smaller values of $k_z$ being depressed and terminating at an energy $W = (\hbar k_0)^2/2m - \frac{1}{2}V_1$, and reappearing at a value $W = (\hbar k_0)^2/2m + \frac{1}{2}V_1$. Values of the energy between these two values are not allowed; the process is repeated at values of $k_z = \pm \frac{1}{2}sG$, whenever $s$ is an integer. We thus have a series of allowed energy bands alternating with forbidden bands, as in Fig. 12.2.

The reason why the potential energy has such a strong perturbing action at the points where the values of the wave-vector magnitude are $\pm k_0$ becomes clearer from a consideration of the wavefunctions. Since $W_1 = W_2$, the correct wavefunctions under the action of the perturbation are given by the linear combinations (12.13); these are here proportional to

$$\{\exp(jk_0z) \pm \exp(-jk_0z)\}, \quad \text{or } \cos k_0z \text{ and } \sin k_0z.$$

It is easy to see that the former corresponds to the lower energy. The charge density is proportional to $\psi\psi^*$, which for $\psi = \cos k_0z$ varies as $\cos^2 k_0z$, whose maxima lie at the points where the potential energy is smallest. For the other wavefunction, the charge density varies as $\sin^2 k_0z$, which has its maxima at points half-way between atoms, where the potential energy is greatest (see Fig. 12.4). These wavefunctions correspond to standing waves; the electrons just oscillate forwards and backwards instead of moving on.

Our simple model also leads to a suitable modification to the simple wavefunctions $\exp(jk_zz)$. The effect of the perturbing potential $\frac{1}{2}V_s\{\exp(jsGz) + \exp(-jsGz)\}$ is, by simple perturbation theory, to mix in wavefunctions of the type $\exp\{j(k_z \pm sG)z\}$. The result is to produce a

wavefunction of the form

$$a \exp(\mathrm{j}k_zz)+\sum_s b_s^{\pm} \exp \mathrm{j}(k_z \pm sG)z = \exp(\mathrm{j}k_zz)\left\{a+\sum b_s^{\pm} \exp(\pm \mathrm{j}sGz)\right\}.$$

We may write this in the simpler form

$$u(z)\exp(\mathrm{j}k_zz), \tag{12.18}$$

where $u(z)$ is a periodic function of $z$ such that $u(z+a)=u(z)$. Such a function clearly satisfies the property of translational symmetry in the lattice, and our perturbed wavefunction is of this form because $G=2\pi/a$ and $s$ is an integer.

Derivation of a general expression for the energy is clearly a complex problem, whose solution depends on the form of $V(z)$ in eqn (12.15). However, we can find readily how $W$ varies near a discontinuity such as that at $k_0=\frac{1}{2}G$, using the series expansion of eqn (12.12) with the condition that $W_1-W_2 \ll W_{12}$. The interaction energy $W_{12}=\frac{1}{2}V_1$ acts between states for which the values of $k$ can be written as $k_0+k'$ and $k_0+k'-G=k'-k_0$, since $G=2k_0$. Provided that $k' \ll k_0$ for such a pair of states, we have

$$\tfrac{1}{2}(W_1+W_2)=\hbar^2k_0^2/2m = W_0; \qquad (W_1-W_2)=\hbar^2(4k_0k')/2m.$$

Hence the values of the perturbed energy are

$$W = W_0 \pm V_1 \pm \frac{\hbar^4k_0^2k'^2}{m^2V_1}$$

$$= W_0 \pm V_1 \pm \left(\frac{\hbar^2k'^2}{2m}\right)\left(\frac{4W_0}{V_1}\right). \tag{12.19}$$

This result shows that the energy varies with the square of $k'$ near a

FIG. 12.4. Distribution of charge density for the perturbed wavefunctions at $k=k_0=\frac{1}{2}G$. In (a) the charge density is largest where the potential energy is low, reducing the over-all energy; in (b) the charge density is largest at points of high potential energy, increasing the over-all energy.

discontinuity; this corresponds to fitting the energy curve with a parabola, which is always the approximation valid for small values of $k'$.

In § 12.2 we tried to show how the energy bands in a solid are related to those in a free ion, while in this section we have derived them as a result of the periodic structure of the lattice. The use of a one-dimensional picture is not as unrealistic as might at first be supposed, since in a simple solid we can draw parallel planes through arrays of successive ions and consider the periodic variation of the potential as we move in the direction normal to the planes. Such planes are just the 'Bragg planes' which enter in the treatment of the diffraction of X-rays. Strong diffraction occurs when the X-ray wavelength satisfies the condition for interference between successive planes; for normal incidence this is $a = \frac{1}{2}\lambda$, which corresponds exactly to the relation $k_0 = 2\pi/\lambda = \frac{1}{2}G = \pi/a$, giving the wave vector at which the energy gap occurs.

## 12.4. Particle aspects

The foregoing treatment of electrons in a solid considers them as waves occupying the whole volume of the solid; these waves are the stationary states, solutions of the time-independent wave equation. We need to know how the electrons behave under the influence of a force (electric or magnetic); this is essentially a particle description, which must be related to the wave aspect of the electron. A free electron must be considered as a wave packet, where the group velocity corresponds to the particle velocity. The $z$ component of the group velocity is given by the relation

$$v_z = \frac{\partial}{\partial k_z}\left(\frac{W}{\hbar}\right) = \frac{1}{\hbar}\frac{\partial W}{\partial k_z}, \tag{12.20}$$

which is analogous to the formula $d\omega/d\beta$ used for the group velocity in § 9.6; the energy $W$ corresponds to $\hbar\omega$ and $k_z$ to the phase constant $\beta$. A rigorous analysis shows that eqn (12.20) is valid for an electron moving in the periodic potential of a crystal lattice, and also that $\hbar\,dk_z/dt = F_z$, where $F_z$ is the component of an external force. Differentiation of eqn (12.20) then gives

$$\frac{dv_z}{dt} = \frac{d}{dt}\left(\frac{1}{\hbar}\frac{\partial W}{\partial k_z}\right) = \frac{1}{\hbar}\frac{\partial^2 W}{\partial k_z^2}\frac{dk_z}{dt} = \frac{1}{m^*}F_z, \tag{12.21}$$

where

$$\frac{1}{m^*} = \frac{1}{\hbar^2}\frac{\partial^2 W}{\partial k_z^2}. \tag{12.22}$$

This last equation defines the quantity $m^*$, which from eqn (12.21) clearly has the dimensions of mass, and is known as the 'effective mass'.

For a free particle the value of $W$ is given by eqn (12.5), and it is readily verified that eqns (12.20)–(12.22) satisfy (12.5) with $m^* = m$, so that the effective mass is equal to the true mass for a free particle. For an electron in a periodic potential the effective mass may depart markedly from the true mass. The advantage of the concept of effective mass is that the dynamic behaviour of an electron in a periodic potential can be treated as if it were a particle of mass $m^*$. The difference between $m^*$ and the true mass $m$ represents the effect on the motion of the electron which results from the electric potential of the ions forming the crystal lattice; when a force is applied to the electron, its change in momentum is different from that of a free electron, and $\mathbf{p} = \hbar\mathbf{k}$ is often referred to as the 'crystal momentum'. The fact that $F_z = m^*(\mathrm{d}v_z/\mathrm{d}t)$ and not $m(\mathrm{d}v_z/\mathrm{d}t)$ does not represent a breakdown of Newton's laws of motion, since the residual momentum is taken up by the lattice. Experiments to determine the ratio of current to momentum, similar to those of Kettering and Scott described in § 3.1, have been carried out by Scott (1951) and Brown and Barnett (1951). They find that even in the case of substances where the current is carried by 'positive holes' (see below), the ratio of current to net momentum has the same sign and is numerically the same as for free electrons.

The relation between the energy $W$ and $k_z$ does not differ greatly from that for a free electron except near the edges of an allowed band. At point such as $k_0 = \pm\pi/a$ there is a discontinuity in the relation between $W$ and $k_z$, and differentiation of eqn (12.19) shows that at such points $\partial W/\partial k_z = \pm\partial W/\partial k'$ is zero since they correspond to $k' = 0$. This means that the electron velocity $v_z$ is zero at such points, a result which is exactly analogous to that for the group velocity of waves on a linear chain (§ 10.9), or in a filter at the edge of a stop band, or a waveguide at cut-off (§§ 9.1, 9.7).

From the shape of the curve of $W$ against $k_z$ (Fig. 12.3) it is clear that the effective mass $m^*$ depends on the value of $k_z$, and that it becomes negative near the top of an allowed band. From eqn (12.19) we can write

$$W = W_0 \pm \tfrac{1}{2}V_1 \pm \hbar^2(k_0 - k_z)^2/2m^*, \tag{12.23a}$$

showing that the effective mass $m^*$ is constant near $k_z = k_0$, with a negative value just below $k_0$ and a positive value just above it. This is an important result in dealing with nearly full and nearly empty bands, which occur in semiconductors.

A full band is an allowed band in which all the states are occupied. It carries no electric current, since for every electron with a positive value of $k_z$ there is another with the value $-k_z$. Suppose we have a band which is full except for one state at the top. Since $\hbar(\mathrm{d}\mathbf{k}/\mathrm{d}t) = \mathbf{F}$, under the influence of an electric field the whole state distribution moves, giving a current $I$. But this must be the negative of the current corresponding to the state which is empty, so that $I = -(-ev) = +ev$, as for a positively charged particle. Again, for the empty state, $\hbar(dk_z/dt) = F_z = -eE_z$, so that the net effect of $E_z$ is $dI_z/dt = e(dv_z/dt) = (e/m^*)F_z = e^2E_z/(-m^*)$, where we have used eqn (12.21). Now for the full band lacking one electron the net value of $\mathbf{k}$ is equal and opposite in sign to the value of $\mathbf{k}$ for the electron needed to complete the band. Since the $W$–$\mathbf{k}$ curve is symmetrical (Fig. 12.3), $m^*$ has the same value for $\mathbf{k}$ as for $-\mathbf{k}$, and near the top of the band it is negative. Hence we can write $dI_z/dt = e^2E_z/|m^*|$, showing that the net effect of all the filled states is equivalent to that of a particle whose dynamic properties correspond to a positive charge and positive mass. Such a particle is called a 'positive hole'; its properties were first derived by Peierls in 1929, and it is analogous to the hole in the filled bands of electron states in Dirac's theory of the positron.

The great advantage of the concept of positive holes is that the current and momentum in a nearly filled band can be attributed to the presence of entities which behave like ordinary particles with positive charge and positive effective mass $|m^*|$. Near the top of such a band we can write (12.19) as

$$W = W_0' - \hbar^2 k'^2/2\,|m^*|. \qquad (12.23b)$$

If we measure energy *downwards* from the top of the band $W_0'$, we can calculate the density of states for positive holes exactly as was done for free electrons in § 11.1. Furthermore, the chance of finding a state *not* occupied by an electron is

$$1 - f = \frac{1}{\exp\{(W_F - W)/kT\} + 1}.$$

This shows that for $W$ close to $W_0'$, and larger than $W_F$, $(1-f)$ is close to unity, while for $W < W_F$ the chance of finding a hole decreases exponentially, and is zero when $T = 0$. These results for holes at the top of a band are just the mirror image of those for electrons at the bottom of a band, as we should expect from the fact that moving an electron *upwards* in energy into an unoccupied state corresponds exactly to moving a hole *downwards* into a previously occupied state.

### Three dimensions

Treatment of a three-dimensional lattice corresponding to a real solid, even with simplified models of the potential variation, is very complex and will not be discussed here. The wavefunction of an electron associated with a wave vector $\mathbf{k}$ is of the form

$$\psi = u(\mathbf{r})\exp j\mathbf{k}\cdot\mathbf{r}, \qquad (12.24)$$

where

$$u(\mathbf{r} + \mathbf{a}_n) = u(\mathbf{r}).$$

Here $\mathbf{a}_n$ is a translation vector representing the repetitive property of the lattice; in particular, that the potential energy is periodic, obeying the rule that $V$ at the point $(\mathbf{r}+\mathbf{a}_n)$ is the same as at $\mathbf{r}$. From the analogy with X-rays it is clear that the values of $\mathbf{k}$ at which strong diffraction occurs will depend on the direction of $\mathbf{k}$; that is, the value of $\mathbf{k}$ at which there is a discontinuity in the energy (the boundary of a zone) is a function of direction. If we draw a vector $\mathbf{k}$ in '$\mathbf{k}$-space' whose length corresponds to this value, and repeat this process for all possible directions, the ends of the vectors will map out a three-dimensional figure, known as a 'Brillouin zone'. Its construction involves only geometry, and its symmetry is related to the symmetry of the crystal lattice.

Values of $\mathbf{k}$ whose vectors end on points inside the zone correspond to allowed energies; those which terminate at the zone boundary correspond to discontinuities in the energy. Higher zones corresponding to higher allowed energy bands exist, but it is possible to bring all wave vectors into the first zone (in the one-dimensional case this procedure corresponds to taking values of $k_z a$ only between $-\pi$ and $+\pi$). Values of $\mathbf{k}$ in the first zone then correspond to more than one allowed energy, but in general we are concerned only with the one band which is partly filled with electrons. At the absolute zero of temperature electrons fill this up to a certain energy, the Fermi energy, and it is therefore of interest to draw plots of constant $W$, or 'energy surfaces' in $\mathbf{k}$-space. Calculation of the energy surfaces is very complex: the results obtained by one method are shown in Fig. 12.5 for a simple cubic lattice. Energy surfaces well within the zone are spherical in shape, but this is by no means true near the zone boundary; in general at the boundary $\partial W/\partial \mathbf{k} = 0$ and the energy surfaces

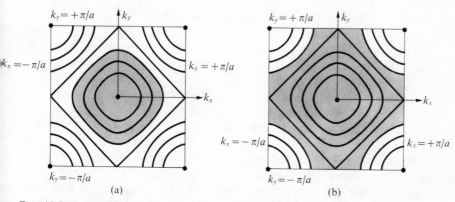

FIG. 12.5. Section through the constant-energy surfaces for a simple cubic lattice obtained by one method of calculation (the 'tight-binding' approximation). The energy surfaces end normally to the zone boundary except where two end at the same point. Shading indicates sections occupied by electrons, sufficient in number to reach the zone boundary in (b) but not in (a).

must end normal to the boundary. The boundary in **k**-space between the filled and empty states at 0 K follows the energy surface corresponding to the Fermi energy and is known as the 'Fermi surface'. Only electrons near the Fermi surface (see § 11.7) can take part in conduction processes, and many details of their behaviour are determined by the exact shape of the Fermi surface. An important property is the effective mass, which in general is a function of direction. Near a band edge the energy may be expanded in a power series in **k**, the lowest terms being quadratic; they can be reduced by a suitable choice of axes to the form

$$W = W_0 \pm \tfrac{1}{2}\hbar^2\left(\frac{k_x^2}{m_x^*}+\frac{k_y^2}{m_y^*}+\frac{k_z^2}{m_z^*}\right), \tag{12.25}$$

where $k$ is measured from the value of $k_0$ at the band edge, as in eqn (12.23b). Here the upper sign must be taken for electrons near the bottom of a band, and $W_0$ is then the energy at the bottom; while the lower sign should be used for holes near the top of a band, and $W_0$ is then the energy at the top. Thus for holes the energy appears to be measured downwards from the top of the band (see above), a point to which we shall return in considering semiconductors (Chapter 17).

## 12.5. Conductors and insulators on the band theory

In older theories it was assumed that in an insulator all the electrons belonging to each atom are firmly bound to that atom, while in a conductor some of the outer electrons are detached from their parent atoms and able to move freely throughout the whole volume of the solid. On the band theory there is no such distinction between 'bound' and 'free' electrons; the electronic wavefunctions spread out through the whole volume of the solid, through for the states of lower energy (corresponding to the inner electrons of a single atom) the electronic density ($\psi\psi^*$ of the wavefunction) is concentrated near each nucleus. How then does the band theory explain the occurrence of both conductors and insulators?

At the absolute zero of temperature the electrons in a solid will have the lowest possible energy consistent with the Pauli exclusion principle, and they will fill the energy bands from the bottom upwards. The lowest energy bands will be fully occupied, but the highest occupied level may occur in the middle of an allowed band. The state of lowest energy is one in which as many electrons have positive values of **k** as have negative values, so that the net current is zero. To establish a current flow some electrons must be transferred from negative values of **k** to positive values, but because of the exclusion principle this is possible only if they make transitions to unoccupied states of higher energy, the energy being gained by acceleration through the application of an electric field. In weak fields this can occur only if adjacent levels are unoccupied, that is, if the top of the Fermi distribution comes in the middle of a band. If, on the other hand, the highest occupied band is completely full, an electron must gain sufficient energy from a movement in the applied electric field to raise it into the next higher band. This requires enormous electric fields, and for ordinary field strengths the substance is an insulator.

In the treatment of the one-dimensional array of $N$ atoms it was shown that $k_z$ was restricted to values which are integral multiples of $G/N$, and that there are just $N$ such values. In three dimensions we have a similar result, the number of translational levels (the number of allowed values of $\mathbf{k}$ which correspond to states which are physically not identical) again being just equal to the total number of atoms. In considering the Fermi–Dirac distribution just two electrons were allowed for each value of $\mathbf{k}$, provided their spins were paired off. This clearly corresponds to s states on isolated atoms, and the treatment illustrated by Fig. 12.2 leads us to expect that there is a further multiplicity allowed for each value of $\mathbf{k}$, corresponding to p, d states, etc. on isolated atoms. Each such state must have a different wavefunction (apart from the two-fold spin degeneracy), that is, it has a different function $u(r)$, such that near the nucleus of an atom $u(r)$ resembles the wavefunction for s, p, d electrons on an isolated atom. For each value of $\mathbf{k}$ we may then have two electrons in s states, six in p states, and ten in d states.

On this picture it is readily seen that the alkali metals, such as lithium, sodium, and potassium will be good conductors, for their atoms possess only one valence electron in an s state, while the energy band corresponding to this atomic state requires two electrons per atom to fill it. The Group II (or alkaline-earth) elements, magnesium, calcium etc., have two such electrons, which we should expect to fill the s band, making these substances insulators. In fact they are rather good electrical conductors; the reason for this is that in the solid the energy bands corresponding to the outer s and p electrons on the atom are sufficiently broad that they overlap appreciably. The state of lowest energy is then one where the electrons partly fill both the s band and the p band, up to the same level, and conduction is possible. This phenomenon of overlapping bands occurs for most of the elements, particularly those of high atomic number where the outermost electrons have rather extended wavefunctions (giving wide bands in the solid) and where the atomic energy levels are rather closely spaced. The progression can be seen in Group IV, where carbon in the form of diamond is a very good insulator while tin is a metal. The intermediate elements silicon and germanium are semiconductors, where the band gap is sufficiently small that some electrons are excited into the 'conduction band' already at room temperature. The number of conduction electrons increases exponentially as the temperature rises, so that the

FIG. 12.6. Energy bands for nickel. To the right is shown the bandwidth of 4s and 3d states as a function of interatomic distance $a$ ($a_0$ is the value for solid nickel). To the left is shown g($W$), the shaded area indicating the filled parts of the bands.

conductivity rises with $T$ instead of falling, as in a pure metal. The influence of crystal structure on the energy bands is seen from the fact that carbon (as graphite) and the allotrope grey tin are both semiconductors. In Groups V and VI the light elements are insulators, but the heavy elements tend to be metallic, while in Groups VII and VIII all the elements are insulators. In transition elements the situation is more complex because there are energy bands corresponding to d-electron states. In copper the 3d band is completely filled and there is one electron per atom in the 4s band, making it a good conductor. In iron, cobalt, and nickel the 3d band is not completely filled; it is rather a narrow band, since d-electron wavefunctions do not spread so far out as s-electron functions, and interactions between d electrons on adjacent atoms are smaller than those between s electrons. The narrow 3d band is overlapped by a broad 4s band, and the 3d band (which can contain ten electrons per atom) gives an abnormally high value of the density of states g($W$), as shown in Fig. 12.6.

Information on energy bands in the solid state can be obtained directly from photoelectron spectroscopy, in which the kinetic energy is measured of the electrons emitted when electromagnetic waves (ultraviolet or X-ray) of known frequency are incident on the solid. Since the electrons can only escape if they come from within an extremely thin layer (~2 nm thick) at the surface, the latter must be specially cleaned to obtain reliable results. Results for gold are shown in Fig. 12.7; in the conduction band, the electron density stops sharply at the Fermi surface, and the broad 6s band is overlapped by the rather narrower 3d band, split by spin–orbit coupling. The inner levels $5f_{5/2}$, $5f_{7/2}$ give much narrower bands, less than 0·5 eV broad when examined with high resolution.

FIG. 12.7. Photoelectron spectrum of gold metal. (a) The conduction band, showing the broad 6s band overlapped by the narrower 5d bands, split by spin–orbit coupling. The sharp drop in electron density at the Fermi level can be seen at the left. (b) The much narrower 4f bands (width $\sim 0.5$ eV), whose ionization energies lie between 80 eV and 90 eV, again split by spin–orbit coupling.

## 12.6. Electronic specific heat

The Pauli exclusion principle restricts the number of states allowed for a given value of $\mathbf{k}$ to two for s electrons, six for p electrons, and ten for d electrons, and this determines the number of states in a given range of wave vector from $\mathbf{k}$ to $\mathbf{k}+d\mathbf{k}$. One effect of the band structure is to alter the relation between $W$ and $\mathbf{k}$, so that $g(W)$, the corresponding number of allowed states between $W$ and $W+dW$, is not the same as in the simple free-electron theory. The presence of overlapping bands provides a further complication. Since only electrons near the Fermi energy $W_F$ can be thermally excited, we expect the electronic specific heat to be determined by $\{g(W)\}_F$, the density of states at the Fermi energy. Theory shows that instead of eqn (11.38) the value of the mean energy $\bar{W}$ per electron becomes

$$\bar{W} = \bar{W}_0 + \frac{\pi^2 (kT)^2}{6n}\{g(W)\}_F + ..., \tag{12.26}$$

and the electronic specific heat is given by the fomula

$$C_V/R = \frac{\pi^2 (kT)}{3n}\{g(W)\}_F... \tag{12.27}$$

or

$$C_V = \gamma T. \tag{12.28}$$

This formula shows that the electronic specific heat gives the density of states at the Fermi energy, and the abnormally high values of $\gamma$ for the transition metals arise from the large density of states in the relatively narrow d bands. For the 'simpler' metals such as the alkali metals the reason why $\gamma$ is not equal to the free electron value (Table 11.2) can be ascribed to the effect of band structure on the relation between $W$ and $\mathbf{k}$. If $W$ varies as $k^2$ at the Fermi surface, we can write $W_F = \hbar^2 k_F^2 / 2m^*$, and the Fermi energy is then given by

$$W_F = (\hbar^2/2m^*)(3\pi^2 n)^{\frac{2}{3}}. \tag{12.29}$$

From eqn (11.11), $\{g(W)\}_F = 3n/2W_F$, so that a value of $m^*$ can be deduced from the electronic specific heat. However, this assumes that the Fermi surface is spherical; this is seldom the case in practice, and in many cases the Fermi surface touches the zone boundary, so that, at best, $m^*$ can be regarded as some kind of average over that part of the Fermi surface which does not touch the zone boundary.

For an energy band which is completely filled the electronic specific heat is obviously zero, since no electrons can be thermally excited except at temperature such that $kT$ approaches the value of the energy gap to the next allowed band. For the elements arsenic, antimony, and bismuth in Group V the highest bands are almost full, but with a small overlap to an almost empty band; the density of states is therefore rather low, giving rise to an exceptionally small electronic specific heat, and to various peculiarities which have led to the name of 'semimetals' for these elements (see, for example, Saunders 1973).

## 12.7. Variation of electrical conductivity with temperature

The detailed calculation of a quantity such as the electric current using the quantum theory of transport phenomena is quite complex. In general the conductivity must be represented by a tensor quantity, but for a cubic metal the conductivity is independent of the direction of the current flow, so that it is a simple scalar quantity. For our purposes it is sufficient to use the formula

$$\sigma = n(e^2/m^*)\tau, \tag{12.30}$$

where $m^*$ represents an average value of the mass obtained by an integration over the Fermi surface. In a metal $m^*$ is independent of the temperature, and variation in the electrical conductivity must be attributed to a variation in the relaxation time $\tau$, as previously pointed out in § 11.7. The existence of a finite relaxation time is due to scattering of the conduction electrons through 'collisions', and in this section we consider the principal causes of scattering, together with the way in which the rates of scattering vary with temperature.

The reciprocal $\tau^{-1}$ of the relaxation time is given by the rate of scattering of the electrons and, provided that the rates of scattering

FIG. 12.8. Low-temperature resistance of three specimens of sodium. (MacDonald and Mendelssohn 1950.)

arising from different mechanisms are independent of one another, the total scattering rate is just the sum of the individual scattering rates. We can then write for the electrical resistivity

$$\rho = \frac{m^*}{ne^2}\left(\frac{1}{\tau_1}+\frac{1}{\tau_2}+\dots\right), \tag{12.31}$$

a result which shows that the various contributions to the resistivity are additive. This is known as 'Matthiessen's rule', and it is used to find the resistivity of a 'perfect' crystal. Experimentally it is found that the electrical resistivity drops rapidly as the temperature is lowered (cf. Fig. 11.12); however, as $T$ tends to zero it reaches a value independent of temperature, but which varies with the amount of impurity and with the previous history of the specimen. For example, it can be reduced by annealing the specimen to remove lattice defects. The behaviour of three specimens of sodium metal is shown in Fig. 12.8, and is typical of many metals. On subtracting the residual resistivity $\rho_0$ from the resistivity at higher temperatures, results are obtained which are practically independent of the specimen used. This is assumed to be the resistivity of a perfect crystal with no impurities and no defects; it is of course zero at $T = 0$, as expected for a perfect lattice (§ 11.7).

We now consider the interaction between the electrons and the lattice vibrations, or phonons. The distortion of the lattice caused by such a vibration introduces a periodic variation in the lattice potential of the form $V_q \exp j(\omega t - \mathbf{q}\cdot\mathbf{r})$, where $\omega$ is the angular frequency of the vibrations and **q** its wave vector. This potential is oscillatory in time, and its effect on a conduction electron is somewhat different from that of a static potential, considered in § 12.3. The interaction of an electron with wavefunction exp j**k.r** with such a potential results in the generation of a new electron wave exp j**k′.r** such that **k′** = **k+q**. We say that the electron is 'scattered' from a state **k** to a state **k′** such that

$$\mathbf{k}' - \mathbf{k} = \mathbf{q}. \tag{12.32}$$

This result is analogous to eqn (12.17); in three dimensions the use of vectors makes it unnecessary to write ±**q**.

The state **k′** must not already be occupied by electrons, so that it cannot be inside the Fermi surface; in an 'elastic collision', where **k** and **k′** are electron states with the same energy, both must lie close to the Fermi surface (as in Fig. 12.9(a)). For a spherical Fermi surface, if the scattering rate from **k** to **k′** depends only on the angle $\alpha$ between **k** and **k′**, the effect of scattering on the flow of charge carriers is to give a relaxation rate

$$\tau^{-1} = \int_0^\pi (1 - \cos\alpha) f(\alpha) \sin\alpha \; d\alpha. \tag{12.33}$$

Here the factor $(1 - \cos\alpha)$ represents the diminution in the component of velocity of the electron parallel to **k** when it is scattered from **k** to **k′**, with a corresponding reduction in the contribution to the electric current. The presence of this factor means that the major part of the integral in (12.33) comes from scattering through large angles, and phonons with large values of **q** are therefore the most effective in scattering. At high temperatures the most abundant phonons are those with values of **q** close to the zone boundary, which most metals is of the same order as $\mathbf{k}_F$, the

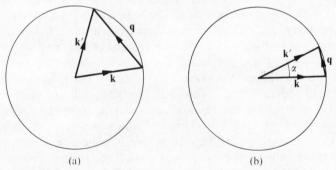

(a)                                    (b)

FIG. 12.9. Scattering of an electron with wave vector **k** into a state with wave vector **k′** by a phonon of wave vector **q**. (a) At high temperatures the predominant phonons have high values of **q**, and scattering is mostly through large angles. (b) At low temperatures only phonons with small values of **q** are excited, and scattering is limited to small values of the angle $\alpha$.

value of **k** at the Fermi surface. Scattering of electrons through large angles is therefore a rather probable process. By assuming that the predominant effect is due to phonons at the zone boundary, we can obtain a simple but rather general formula for the variation of $\tau^{-1}$ at high temperatures, using the fact that such phonons have frequencies corresponding to the Debye cut-off frequency given by $\hbar\omega_D = k\theta_D$, where $\theta_D$ is the Debye temperature.

The cross-section for scattering of electrons by the lattice vibrations is proportional to the mean-square displacement $\delta^2$ of an ion from its equilibrium position in the lattice. The mean-square displacement can be related to the vibrational energy, since, for an ion of mass $M$,

$$\text{mean kinetic energy} = \text{mean potential energy} = \tfrac{1}{2}M\omega^2\,\delta^2.$$

At high temperatures each of these has the value $\tfrac{1}{2}kT$ per vibrational mode, and for phonons of the Debye frequency

$$\delta^2 = kT/M\omega_D^2 = (\hbar^2/k)(T/M\theta_D^2). \tag{12.34}$$

Hence

$$\tau^{-1} \text{ should vary as } T/M\theta_D^2. \tag{12.35}$$

a result which is in general agreement with experimental measurements.

At lower temperatures the mean energy associated with a lattice vibration **q** is smaller than $kT$ by a factor

$$\frac{\hbar\omega_q/kT}{\exp(\hbar\omega_q/kT)-1} = \frac{x}{\exp x - 1}, \tag{12.36}$$

and the mean-square ionic displacement $\delta^2$ is smaller by the same factor. Because of the exponential factor we may assume that only lattice vibrations with values up to $\hbar\omega_q \sim kT$ are excited. By writing

$$\omega_q/\omega_D \sim q/q_D \sim T/\theta_D \tag{12.37}$$

we see that this restricts the phonons to those with wave vectors up to $\sim q_D(T/\theta_D)$. As a result, at low temperatures the electrons can only be scattered through rather small angles, as illustrated in Fig. 12.9(b); since we are concerned only with electrons at the Fermi surface, $f(\alpha)$ in eqn (12.33) is large only for values of $\alpha$ of the order of $q/k_F$. Using the relation

$$(1-\cos\alpha)\sin\alpha\,\mathrm{d}\alpha = 8\sin^3\tfrac{1}{2}\alpha\,\mathrm{d}(\sin\tfrac{1}{2}\alpha) \tag{12.38}$$

we can replace the integral in eqn (12.33) by an integral in which the variable is $\tfrac{1}{2}\alpha = q/2k_F$. On assuming that $f(\alpha)$ is proportional to $kT$ and independent of $\alpha$, we find that $\tau^{-1}$ varies as

$$kT\int_0^{(q_D/k_F)(T/\theta_D)} \frac{1}{2}\left(\frac{q}{k_F}\right)^3 \mathrm{d}\left(\frac{q}{k_F}\right) = \frac{1}{8}\left(\frac{q_D}{k_F}\right)^4 kT\left(\frac{T}{\theta_D}\right)^4. \tag{12.39}$$

This predicts that at low temperatures the resistivity of a perfect crystal should be proportional to $T^5$, a result first obtained by Bloch.

Both the high and the low temperature results are contained in a semi-empirical formula, due to Gruneisen,

$$\rho = 4A\left(\frac{T}{\theta}\right)^5 \int_0^{\theta_D/T} \frac{e^x x^5}{(e^x-1)^2} \, dx = 4A\left(\frac{T}{\theta_D}\right)^5 J_5. \qquad (12.40)$$

Here the integral $J_5$ belongs to the same family as the integral $J_4$ which appears in the Debye theory of the lattice specific heat. At high temperatures where the approximation $e^x - 1 = x$ can be used, the integral $J_5$ reduces to

$$\int_0^{\theta_D/T} x^3 \, dx = \tfrac{1}{4}(\theta_D/T)^4$$

and the resistivity becomes $\rho = A(T/\theta_D)$. At low temperatures the integration in $J_5$ can be taken to infinity, and it becomes a pure number, giving $\rho = 498A(T/\theta_D)^5$.

For the majority of real metals, apart from the alkali and the noble metals (copper, gold, silver), overlapping bands are present. Electrons in different bands, even with the same values of $W$, can have different effective masses. On a simple two-band approximation we may treat the two bands as carrying currents in parallel, and the conductivity is then

$$\sigma = \frac{n_a e^2 \tau_a}{m_a^*} + \frac{n_b e^2 \tau_b}{m_b^*}. \qquad (12.41)$$

If the total number of electrons is just sufficient to fill one band (if there were no overlap), then $n_a = n_b$, since $n_a$ is the number of holes in the nearly filled band and $n_b$ the number of electrons in the nearly empty band. If also $\tau_a \sim \tau_b$, eqn (12.41) shows that most of the current is carried by the particles with the smaller value of $m^*$; for overlapping s and d bands, as occurs in transition metals, this means that the current is predominantly carried by the s electrons.

In the previous discussion we considered only processes in which an s electron is scattered to another state in the s band, but it turns out that there is a much bigger probability of its being scattered into the d band. This reduces the mobility, and accounts for the rather high resistivity of the transition metals. Similarly, in divalent metals scattering from the s band to the overlapping p band gives a resistivity rather higher than in the monovalent metals.

## 12.8. Variation of thermal conductivity with temperature

In a solid heat is transported both by the lattice vibrations and by the conduction electrons. In a metal the phonons are scattered by collisions with electrons, and the heat current carried by the phonons should be smaller than in an insulator where this scattering process is absent. In general the thermal conductivity of metals is greater than that of insulators, apart from a few exceptional substances such as diamond (see Berman 1973). In practice the thermal transport in a metal is nearly all due to the electrons, the lattice conductivity being so small that it is difficult to measure experimentally except in superconductors, where thermal transport by the electrons is greatly reduced.

In the same approximation as eqn (12.30), the thermal conductivity can be written as

$$\frac{K}{T} = \frac{\pi^2}{3} \cdot \frac{nk^2}{m^*} \tau.$$

(12.42)

It follows that $K/\sigma T$ should be equal to the ideal Lorenz number $L_0$; in fact it can be shown that this should be true quite generally provided that the scattering of the electrons is elastic. In practice it always holds at temperatures approaching the Debye $\theta_D$, where the scattering of electrons by phonons occurs mainly through elastic collisions (as in Fig. 12.9(a)). Under these conditions $\tau^{-1}$ varies as $T$, and $K$ should be independent of temperature, as shown for copper in Fig. 11.12.

At lower temperatures 'inelastic' collisions become important for the thermal conductivity. The value of **k** for the electron changes little in a collision, but its energy is changed by amounts equal to the quantized energy of the phonons, that is, by amounts of the order of $kT$. Such collisions are effective in reducing the transport of energy, but have little effect on the transport of electric current, where a small change in energy matters little provided that **k'** has the same direction as **k**. Thus in calculating the relaxation rate $\tau^{-1}$ for the thermal conductivity, the factor $(1-\cos\alpha)$ in eqn (12.33), which allows for the change in the component of electron velocity, is absent. Instead of eqn (12.39) we obtain an integral of the form

$$kT \int_0^{(q_D/k_F)(T/\theta_D)} \left(\frac{q}{k_F}\right) d\left(\frac{q}{k_F}\right) = \frac{1}{2}\left(\frac{q_D}{k_F}\right)^2 kT\left(\frac{T}{\theta_D}\right)^2,$$

(12.43)

so that at low temperatures $T/K$ should vary as $T^3$. In practice the thermal conductivity of most pure metals can be fitted to an expression of the form (see Rosenberg 1955)

$$T/K = aT^3 + b,$$

(12.44)

where the constant $b$ represents the effect of defect scattering. When the latter is subtracted to give the thermal resistivity $K^{-1}$ of a 'perfect' lattice, it is found that it can be well represented over a wide range of temperature by the function $(T/\theta_D)^3 J_3$, where

$$J_3 = \int_0^{\theta_D/T} \frac{e^x x^3}{(e^x - 1)^2} \, dx. \tag{12.45}$$

More complex expressions have been proposed; they all give the right asymptotic behaviour at high and at low temperatures, so that experimentally it is difficult to decide between them (see White and Woods 1959). These authors also conclude that the lattice conductivity may make an appreciable contribution to the thermal conductivity at low temperatures in metals such as titanium and zirconium, which have rather a high resistance.

The general behaviour of the electrical and thermal conductivity with temperature is illustrated in Fig. 12.10. At high temperatures $\tau^{-1}$ is the same for each, so that the value of the Lorenz number $K/\sigma T$ is close to the ideal value $L_0$. At low temperatures inelastic collisions become important in reducing $K$, so that $L$ falls below the ideal value, and for a 'perfect' crystal it should approach zero as $T^2$ when $T \to 0$. In an imperfect crystal, scattering from defects is an elastic process, and the contribution to $\tau^{-1}$ should be the same for the electrical and the thermal

FIG. 12.10. Variation of electrical conductivity $\sigma$ and thermal conductivity $K$ with temperature in a metal. $\theta$ is the Debye temperature.

FIG. 12.11. Lorenz number for copper. (Berman and MacDonald 1952.)
1 experimental curve,
2 experimental curve for ideally pure copper, obtained by subtracting contributions to the electrical and thermal resistivity from impurities,
3 theoretical curve.

conductivity. Hence $L$ should again approach $L_0$ at very low temperatures, where defect scattering becomes the dominant process. The results of Berman and MacDonald for copper are shown in Fig. 12.11. $L$ rises towards $L_0$ (curve 1) at liquid-helium temperatures, but for the ideally pure metal (curve 2) it falls towards zero. The broken line represents calculations by Sondheimer; similar discrepancies between theory and experiments have been found for other metals.

## 12.9. Thermoelectric power

The presence of a temperature gradient in a metal produces a flow of charge carriers from the hot end to the cold end. This creates an electric current as well as a heat current, and formulae for the thermoelectric effects can be derived from general transport theory when both effects are included. The contributions from a given set of charge carriers to the heat current and the electric current depend respectively on their energy and their mobility, which do not necessarily vary together in the same way. Allowance for this is made in the following formula for the thermoelectric power $S$,

$$S = -\frac{\pi^2}{3} \cdot \frac{k^2 T}{|e|} \left(\frac{d \ln \sigma_W}{dW}\right)_{W=W_F}, \qquad (12.46)$$

where $\sigma_W$ is that part of the electrical conductivity arising from electrons of energy $W$. If we assume that $\sigma_W$ is proportional to $k^3$ (cf. Problem

12.4), we find that

$$S = -\frac{\pi^2}{3} \cdot \frac{k^2 T}{|e|} \left( \frac{3}{2W_F} + \frac{d \ln \tau}{dW} \right). \qquad (12.47)$$

As can be seen from eqn (11.39), the first term in this equation is just the electronic specific heat per electron, in agreement with the assumption made in § 11.8. If $\tau$ increases with $W$, there is an extra term, reflecting the fact that the more mobile electrons carry the greater part of the electric current; and since they have larger values of $W$, for a given current they transport more heat.

In these equations the negative sign for $S$ appears because the electric current is in the opposite direction to the net flow of carriers when these are negatively charged. This immediately suggests that the thermoelectric power should have a positive sign when the current is carried by positive holes. The basis for this intuitive expectation can be seen by considering the shape of the $W$–$k$ curve near the top of a band. Electrons from the hot end of the metal have higher values of $W$, and hence also of $k$, but because $(d^2 W/dk^2)$ is negative their velocity $v = \hbar^{-1}(dW/dk)$ is lower than that of electrons from the cold end instead of being greater. The net effect is that electrons diffuse more rapidly from the cold end to the hot end, setting up an e.m.f. of the opposite sign to that we should expect from free electrons.

Quantitatively, if we write as in eqn (12.23b)

$$W_0' - W = \hbar^2 k'^2 / 2 |m^*| \qquad (12.48)$$

and assume that $\sigma_W$ varies as $k'^3$ and as the power $\tau^s$ of the relaxation time, we have

$$\ln \sigma_W = \frac{3}{2} \ln(W_0' - W) + \frac{s}{2} \ln(W_0' - W),$$

and we obtain from eqn (12.46)

$$S = +\frac{\pi^2}{3} \cdot \frac{k^2 T}{|e|} \cdot \frac{(3+s)}{2(W_0' - W_F)}, \qquad (12.49)$$

showing that the thermoelectric power is reversed in sign.

At low temperatures an effect known as 'phonon drag' becomes important. As a result of the 'collisions' between the electrons and the phonons the latter are pushed along in the direction of the electron flow, creating a drift of phonons which constitutes additional heat current. If such collisions are frequent compare with any other scattering processes for the phonons, the drift velocity of the phonons will equal that of the electrons, and the heat transported will be given by the sum of the lattice specific heat $C_L$ and the electronic specific heat. The former is much larger than

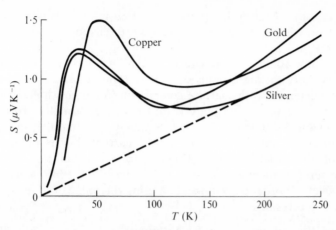

FIG. 12.12. Variation of the thermoelectric power $S$ for the metals copper, silver and gold. The large maxima at $T \sim 30$–$60$ K show the effect of 'phonon drag'. (After Pearson 1971.)

the latter, so that we should expect

$$|S| \sim C_{\mathrm{L}}/n \, |e|.$$

This suggests that at low temperatures $S$ should vary as $T^3$, and be much larger than expected from eqn (12.47). Results for the metals copper, silver, and gold are shown in Fig. 12.12; at the higher temperatures the effect disappears because the main scattering processes for the phonons are collisions with other phonons rather than with electrons.

## 12.10. The anomalous skin effect

In Chapter 8 it was shown that at radio-frequencies the current flow in a good conductor is confined to a thin layer near the surface whose thickness is determined by the skin depth $\delta$ (see § 8.5). At 300 K in copper the skin depth at $10^{10}$ Hz is about 1 $\mu$m, while the electron mean free path is only 0·01 $\mu$m. At liquid-helium temperatures the situation in a pure specimen of copper, free from defects, is reversed. The conductivity may be $10^4$ times greater, showing that the mean free path has increased to 100 $\mu$m, while the skin depth should, from eqn (8.31), have fallen to about 0·01 $\mu$m. In the skin depth both the magnitude and the phase of the oscillating electric field are changing exponentially, and the velocity acquired by an electron through the action of the electric field can only be found by an integration, starting from the point at which the electron last made a collision. This means that the current density at a point is no longer directly related to the electric field at that point, and Ohm's law in its simple form is not valid. The current density depends on the integrated effect of the electric field over a volume whose linear dimensions are of the order of the mean free path. This is a 'non-local' situation which we shall meet again in discussing superconductivity (Chapter 13).

Under these conditions the effective value of the skin depth is altered; this is known as the 'anomalous skin effect'. Only those electrons moving almost tangentially to the surface complete a free path wholly within the skin depth, and the conductivity is thus effectively smaller. To estimate the 'anomalous' skin depth $\delta'$ which results, we assume that the fraction of the electrons which contribute to the current flow is reduced by a factor of the order of $\delta'/L$, where $L$ is the mean free path. The corresponding conductivity $\sigma'$ is thus given by $\sigma'/\sigma = \beta(\delta'/L)$, where $\beta$ is a number of the order of unity, and on using this value of $\sigma'$ in eqn (8.31) we obtain the result

$$\delta' = (2L/\beta\sigma\mu\omega)^{\frac{1}{3}}. \tag{12.50}$$

If the conductivity $\sigma$ can be represented by eqn (12.30) with $\tau = L/v_F$, then eqn (12.50) shows that $\delta'$ is independent of $\tau$ or $L$. In a single crystal it can be related indirectly to the curvature of the Fermi surface, since $\delta'$ depends on the number of carriers moving parallel to the metal surface in the direction of the electric field. The velocity of a carrier is parallel to a normal to the Fermi surface in **k**-space (see Problem 12.1), and the flatter the Fermi surface the larger the number of carriers moving close to a given direction.

## 12.11. The Hall effect

When a conductor carrying a current of density **J** parallel to the $y$-axis is placed in a magnetic flux density **B** parallel to the $z$-axis, a potential difference appears across the conductor in the direction of the $x$-axis. This effect was discovered by Hall in 1879. The force exerted by **B** on the moving current carriers tends to displace them in the $x$-direction, but this sets up a non-uniform charge distribution, giving rise to an electric field in the $x$-direction. In equilibrium this produces a force which just balances the magnetic force, so that

$$F_x = qE_x + qv_yB_z = 0.$$

If we can identify $v_y$ with the drift velocity of the carriers, then $J_y = nqv_y$, where $n$ is the number of carriers per unit volume. Then

$$E_x = -(J_y/nq)B_z = -R_H J_y B_z,$$

where $R_H$ is known as the Hall coefficient. Its magnitude should be

$$R_H = 1/nq, \tag{12.51}$$

a simple result which is important because it means that measurements of the Hall effect give both the value of $n$ and the sign of $q$.

TABLE 12.1

*Observed and calculated values of the Hall effect*

| Metal | $R_H$ (in units of $10^{-11}$ m$^3$ C$^{-1}$) | |
| | Observed | Calculated, assuming $x$ electrons per atom |
| --- | --- | --- |
| Lithium | $-17.0$ | $-13.1$ $(x = 1)$ |
| Sodium | $-25.0$ | $-24.4$ $(x = 1)$ |
| Copper | $-5.5$ | $-7.4$ $(x = 1)$ |
| Silver | $-8.4$ | $-10.4$ $(x = 1)$ |
| Zinc | $+4.1$ | $-4.6$ $(x = 2)$ |
| Cadmium | $+6.0$ | $-6.5$ $(x = 2)$ |

By combining the Hall effect with the conductivity $\sigma$, the mobility $u$ can be found, since $\sigma = nqu$. The mobility is directly related to an angle $\phi$, known as the 'Hall angle'. To maintain the current density $J_y$ an electric field strength $E_y$ must be present such that $J_y = \sigma E_y$. The resultant electric field thus makes an angle $\phi$ with the $y$-axis, given by

$$\tan \phi = E_x/E_y = -R_H J_y B_z/E_y = -(\sigma R_H)B_z = -uB_z. \qquad (12.52)$$

At first sight the appearance of the density of carriers in eqn (12.51) is surprising; it reflects the fact that the force exerted by **B** depends on the carrier velocity, and the larger the mobility the smaller the number of carriers for a given current density.

A comparison of the observed values of $R_H$ for various substances with those calculated from eqn (12.51) is given in Table 12.1. In the free-electron theory we expect $q = -e$, giving a negative value for $R_H$; this is observed for the monovalent metals, and the numerical value agrees fairly well with the assumption of one conduction electron per atom. For the divalent alkaline-earth metals, $R_H$ is found to be positive, suggesting the carriers are effectively positively charged. In such metals there are over-lapping s and p bands, the s band being nearly full so that it contains positive holes (see § 12.5). These carriers have the higher mobility (see end of § 12.7), and are predominant in deciding the sign of the Hall effect.

This result draws attention to the fact that the simple derivation of eqn (12.51) given above is not correct when carriers of different mobility are present. In a thin wire, it is only the *resultant* current which must flow parallel to the wire, and the relation $E_x + v_y B_z = 0$ cannot be used when there is more than one drift velocity present. A geometrical solution using the Hall angles can be constructed if just two sets of carriers are assumed,

as in a model of a metal with two overlapping bands. Detailed evalua-
tions, assuming that $\tau$ is constant over the Fermi surface, have confirmed
that eqn (12.51) is correct if the Fermi surface is spherical or ellipsoidal.

If a velocity distribution of the Maxwellian type is assumed, an expression for $R_H$ larger by a factor $3\pi/8$ is obtained; this result is often used in
semiconductors. Because of the relatively small carrier density, the values
of $R_H$ in such materials are large (and very temperature-dependent).
Rather large values are also observed for the semimetals; for bismuth $R_H$
corresponds to about $10^{-4}$ electrons per atom.

### *Magnetoresistance*

If only one type of carrier is present, the electrical resistance is not
changed by the application of magnetic flux, because any sideways motion
induced by $B_z$ is exactly counterbalanced by the field strength $E_x$. Thus
for free electrons there should be no transverse magnetoresistance effect,
and obviously we do not expect a longitudinal effect since the drift
velocity is parallel to the magnetic flux. In general most metals do show a
resistivity which increases appreciably in high flux densities, since carriers
of more than one type are deviated by the magnetic field. The effects are
greater at low temperatures, where mean free paths are longer, and
carriers follow curved paths for longer distances before being interrupted
by collisions.

## 12.12. Conduction electrons in a strong magnetic field

The equation of motion of a free particle in a flux density $\mathbf{B}$ is

$$m\dot{\mathbf{v}} = \mathbf{F} = q(\mathbf{v} \wedge \mathbf{B}). \tag{12.53}$$

If $\mathbf{B}$ is constant and parallel to the $z$-axis, then $v_z$ is constant. In the
$x,y$-plane the particle traces out a circle with angular velocity (eqn (4.73))

$$\omega_c = (q/m)B,$$

this circle being the projection of the helical path followed by the particle
in three dimensions. No work is done on the particle, since the power
expended is $\mathbf{v} \cdot \mathbf{F} = q\mathbf{v} \cdot (\mathbf{v} \wedge \mathbf{B}) = 0$, and its kinetic energy is constant.

In a solid the corresponding problem is much more complex, and we
confine ourselves to a simplified illustration of the approach, using a
free-electron model. The equation of motion is

$$\hbar\dot{\mathbf{k}} = q(\mathbf{v} \wedge \mathbf{B}); \tag{12.54}$$

if $\mathbf{B}$ is parallel to the $z$-axis, it follows that $k_z$ is a constant, and the
movement of a representative point in $\mathbf{k}$-space is confined to a plane
normal to $k_z$. For a particle at the Fermi surface the vector $\mathbf{v}$ is always
normal to the surface (see Problem 12.1 for a special case), and $\mathbf{k}$

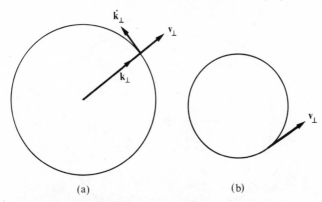

(a)                                    (b)

FIG. 12.13. Motion of a free electron in a magnetic flux density $B_z$. (a) The point in **k**-space describes a circular orbit in the plane normal to $B_z$ with angular velocity $\omega_c$. (b) In real space the particle trajectory is a helix whose projection on the $xy$-plane is a circle.

therefore lies in the surface, normal to **B**. The **k**-vector traces out a surface curve of constant energy which is just the intersection of the Fermi surface with the plane $k_z =$ constant. For the case of a free particle, we have $\hbar\mathbf{k} = \mathbf{p} = m\mathbf{v}$, so that

$$\hbar\dot{\mathbf{k}} = \hbar(q/m)(\mathbf{k}\wedge\mathbf{B}). \tag{12.55}$$

If we write $k_\perp$ for the component of **k** normal to **B**, then $\dot{k}_\perp = (q/m)(k_\perp B_z)$, showing that $\dot{k}_\perp$ is normal to and proportional in magnitude to $k_\perp$ (Fig. 12.13). Hence $k_\perp$ performs a simple rotation about the $k_z$-axis with angular velocity $(q/m)B_z$, which is just equal to the cyclotron angular velocity of eqn (4.73). $k_\perp$ traces out a circle (a section of the Fermi surface, which is spherical for free electrons) the time taken being $2\pi/\omega_c$.

The kinetic energy can be written as

$$W = \frac{\hbar^2}{2m}(k_x^2 + k_y^2 + k_z^2) = \frac{\hbar^2}{2m}(k_\perp^2 + k_z^2). \tag{12.56}$$

The component $k_z$ is unaffected by the presence of $B_z$ and the corresponding wavefunctions are of the form $\exp(jk_z z)$ as before. On the other hand, in the $x,y$-plane the energy levels turn out to be those of a two-dimensional harmonic oscillator, given by

$$W_\perp = \frac{\hbar^2}{2m}k_\perp^2 = (n+\gamma)\hbar\omega_c, \tag{12.57}$$

where $\gamma$ is a phase constant and the quantum number $n$ is an integer. The allowed electric dipole transitions are $\Delta n = \pm 1$, so that they occur just at the classical cyclotron frequency.

FIG. 12.14. In a flux density $B_z$ the representative points in **k**-space lie on the surfaces of coaxial cylinders, only one of which is shown here. The cross-sectional area $A_\perp = \pi k_\perp^2$ is proportional to $\omega_c$ and hence to $B_z$. Oscillations occur in the density of states at the Fermi surface as the radius $k_\perp$ of successive cylinders passes through the Fermi radius $k_F$.

The area of the circular orbit traced out in **k**-space is

$$A_\perp = \pi k_\perp^2 = \frac{2\pi m}{\hbar^2} W_\perp = \frac{2\pi m}{\hbar}(n+\gamma)\omega_c, \qquad (12.58)$$

showing that $A_\perp$ is proportional to $W_\perp$. Thus to each integral value of $n$ there corresponds a circle of given area in the $k_x k_y$-plane, and the allowed points in **k**-space lie on the circumference of this circle. Each time that $n$ increases by unity, the area $A_\perp$ increases by

$$\delta A_\perp = \frac{2\pi m}{\hbar}\omega_c.$$

We can use this result to find the number of allowed points in **k**-space for each value of $n$. For a cube of metal whose sides are of unit length, the number of points allowed in a two-dimensional system is $1/(2\pi)^2$ per unit area, and for each increment of area $\delta A_\perp$ the number of points is therefore

$$(2\pi m/\hbar)\omega_c \div (2\pi)^2 = (m/h)\omega_c = (qB_z)/h. \qquad (12.59)$$

These results are illustrated in Fig. 12.14. The allowed points lie on the surfaces of a set of concentric cylinders whose cross-sectional area is given by eqn (12.58); this replaces the simple cubic array of points which occurs when $B_z = 0$. Parallel to the $k_z$-axis the allowed points are equally spaced at intervals of $1/2\pi$. The surfaces of equal energy are spherical, corresponding to eqn (12.56), so that allowed points with the same energy occur at the circles where spheres and cylinders intersect.

We now consider the density of states. Each of the energy levels defined by eqn (12.57) is associated with a finite set of points defined by the value of $k_\perp^2$. Motion parallel to the $z$-axis gives points equally spaced in $k_z$, but since the associated energy is $W_z = (\hbar^2/2m)k_z^2$, the density of points $g(W_z)$ decreases as $W_z^{-\frac{1}{2}}$ (see Problem 12.2). The resultant density of states has the form shown in Fig. 12.15, being high at each value of $W$ which satisfies eqn (12.57) and then falling as $W$ increases until the next value of $W$ is reached, at which $n$ has increased by unity in this equation. The Fermi energy $W_F$ at $T = 0$ is defined by the fact that the number of states lying below it is just equal to the number of electrons which must be accommodated. As $B_z$ is increased, the vertical spacing between the energy levels in Fig. 12.15 increases in proportion to $B_z$, so that the density of states at the Fermi level oscillates as each crosses the Fermi level (which must also oscillate to fulfil the criterion for the total number of states below it). The result is a sharp change in the density whenever $W_F = (n+\gamma)\hbar\omega_c = (n+\gamma)\hbar(qB_z/m)$, and many physical properties such as the magnetic susceptibility and electrical resistance show a corresponding oscillation. This occurs at equally spaced intervals when plotted as a function of $B_z^{-1}$. To observe such effects, the sharp changes in the density of states must not be smeared over by the spread in the Fermi–Dirac distribution function around the Fermi energy; this requires that experiments must be made at low temperatures such that $kT \ll \hbar\omega_c$. For

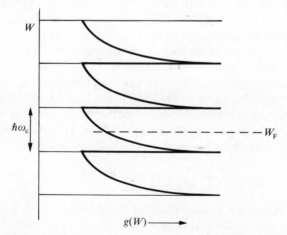

FIG. 12.15. Density of states in a magnetic flux density $B_z$. The variation is repeated at intervals of $\hbar\omega_c$, which depends linearly on $B_z$. As $B_z$ is altered, a sharp change in the density of states at the Fermi energy occurs whenever a cyclotron resonance level passes through $W_F$.

free electrons this means that a field of several tesla, and temperatures down to 1·5 K are needed.

It is important to realize that the separation of successive cyclotron resonance levels is very small compared to the Fermi energy. Even at a flux density $B_z = 10$ T the spacing for free electrons is only about $10^{-3}$ eV, and the oscillations are closely spaced as a function of $B_z^{-1}$; even at this rather high field they occur at intervals of less than $10^{-2}$ T. The changes in the density of states are correspondingly small, but are observable primarily because the density varies as $W_z^{-\frac{1}{2}}$, and so goes to infinity at each cyclotron resonance level. A further important condition is that the width of the levels $(\tau^{-1})$ must be less than the spacing, and for good resolution the criterion $\omega_c \tau \gg 1$ means that low temperatures and very pure samples are required.

The various phenomena observable in the presence of a magnetic field give important information about the Fermi surface. In the presence of band structure the cyclotron resonance frequency is given by

$$\omega_c = \frac{2\pi q B_z}{\hbar^2} \cdot \frac{\mathrm{d}W}{\mathrm{d}A_\perp}.\tag{12.60}$$

This result can also be expressed in terms of a 'cyclotron mass' $m_c$ defined by the relation $\omega_c = q B_z / m_c$, where

$$m_c = \frac{\hbar^2}{2\pi} \frac{\mathrm{d}A_\perp}{\mathrm{d}W}.\tag{12.61}$$

Direct determination of the cyclotron resonance frequency is discussed in Chapter 24. The full theory of the oscillatory effects in other properties is very complicated, and we mention only the effects in the resistivity, known as the de Haas–Shubnikov effect. The resistivity depends on the rate of scattering of the carriers, and at low temperatures the low-frequency phonons produce mainly elastic scattering to other parts of the Fermi surface, as in Fig. 12.9(b). Scattering can take place only to empty carrier states of nearly the same energy, which are most abundant when a cyclotron resonance level coincides with the Fermi level, at which point an increase in the resistivity occurs. For $k_z = 0$, $W = W_\perp$ and integration of eqn (12.60), combined with the quantization condition (12.57), gives

$$A_\perp = \frac{2\pi q B_z}{\hbar^2 \omega_c} W = \frac{2\pi q B_z}{\hbar} (n + \gamma).\tag{12.62}$$

This is the maximum cross-section of the Fermi surface, and it can be determined from the frequency of the oscillations plotted against $1/B_z$. In general oscillations are observed corresponding to either maximum or minimum values of the cross-section.

Oscillations in the magnetic susceptibility are known as the de Haas–van

Alphen effect. At fields too small for oscillations to be observable, there is a small temperature-independent susceptibilty, which will be discussed in § 14.11.

In this treatment we have omitted the electron spin, whose magnetic moment interacts with $B_z$ to give an additional energy $\pm\frac{1}{2}g_s\mu_B B_z$ in eqn (12.57). In the approximation $g_s = 2$ this is just $\pm(e\hbar/2m)B_z = \pm\frac{1}{2}\hbar\omega_c$ for a free electron, and the result is just a shift in the energy levels in Fig. 12.15. Cyclotron resonance is induced by an oscillatory electric field (see § 24.9); the spin vector remains fixed and the cyclotron frequency is not altered.

## References

BEATTIE, J. R. (1955). *Phil. Mag.* **46**, 235.
BERMAN, R. (1973). *Contemp. Phys.* **14**, 101.
—— and MACDONALD, D. K. C. (1952). *Proc. R. Soc. A*, **211**, 122.
BROWN, S. and BARNETT, S. J. (1951). *Phys. Rev.* **81**, 657.
MACDONALD, D. K. C. and MENDELSSOHN, K. (1950). *Proc. R. Soc. A*, **202**, 103.
PEARSON, W. B. (1971). *Solid St. Phys., USSR*, **3**, 1411.
ROSENBERG, H. M. (1955). *Phil. Trans. R. Soc. A* **247**, 441.
SAUNDERS, G. A. (1973). *Contemp. Phys.* **14**, 149.
SCOTT, G. G. (1951). *Phys. Rev.* **83**, 656.
WHITE, G. K. and WOODS, S. B. (1959). *Phil. Trans. R. Soc. A*, **251**, 273.

## General references

KITTEL, C. (1971). *Introduction to solid state physics.* Wiley, New York.
MACDONALD, D. K. C. (1962). *Thermoelectricity.* Wiley, New York.
ROSENBERG, H. M. (1963). *Low temperature solid state physics.* Clarendon Press, Oxford.
ZIMAN, J. M. (1970). *Electrons in metals.* Taylor and Francis, London.

## Problems

12.1. If $k_z$ = constant in eqn (12.25), show that at the point $k_x, k_y$ on the Fermi surface the slope of the tangent in the $k_x k_y$-plane is

$$dk_y/dk_x = -(k_x/k_y)(m_y/m_x)$$

and that the normal to the surface at this point is parallel to the electron velocity whose components are given by equations similar to (12.20).

12.2. Adapt the treatment of § 11.1 to the case of a one-dimensional system of unit length to show that the density of states (allowing for electron spin) is

$$g(W_z) = (1/2\pi)(m/2\hbar^2 W_z)^{\frac{1}{2}}.$$

Show similarly that for a square of unit side

$$g(W) = (1/2\pi)(2m/\hbar^2).$$

**12.3.** In a simple cubic lattice the energy surfaces given by the 'tight-binding' approximation are of the form

$$W = W_1 - W_2(\cos k_x a + \cos k_y a + \cos k_z a),$$

where $a$ is the atomic separation. Show that the width of the energy band is $6W_2$, and that near the bottom ($k_x a \to 0$, etc.) the energy is approximately.

$$W = (W_1 - 3W_2) + \tfrac{1}{2}W_2 k^2 a^2 + ...,$$

while near the top ($k_x a \to \pm\pi$, etc.) it is

$$W = (W_1 + 3W_2) - \tfrac{1}{2}W_2 k^2 a^2 + ...,$$

where $k^2 = k_x^2 + k_y^2 + k_z^2$. This shows that the energy surfaces are spheres about the centre of the zone, or the corners of the zone respectively (compare Fig. 12.5). Note that the effective mass $m^* = h^2/a^2 W_2$, and hence is inversely proportional to the bandwidth.

**12.4.** For a cubic crystal the electrical conductivity is given by

$$\sigma = \frac{e^2}{12\pi^3 \hbar} \int \tau v_F \, dS_F,$$

where $v_F$ is the velocity at the Fermi surface, and the integral is taken over the Fermi surface. Show that if $\tau$ is constant over this surface, and $m^* v_F = \hbar k_F$, this reduces to

$$\sigma = (e^2/3\pi^2)(\tau k_F^3/m^*),$$

and that this reduces to eqn (12.30) on writing $n = (2/8\pi^3)\Omega_F$ where $\Omega_F$ is the volume of the Fermi sphere.

**12.5.** Show that if $2(2l+1)$ electrons of mass $m^*$ are allowed for each point in **k**-space, the density of states $g(W) \, dW$ is still given by eqn (11.11), but that the value of $(W_F)_0$ becomes

$$(W_F)_0 = \frac{\hbar^2}{2m^*} \left( \frac{3\pi^2 n}{2l+1} \right)^{\frac{2}{3}},$$

where $n$ is the number of electrons per unit volume of the metal.

**12.6.** Using the treatment of § 8.4, find an expression for the complex refractive index $(n - jk)$ of a metal in the region where relaxation effects in the conductivity are important, that is, where the conductivity is complex as in eqn (10.51).

If the ordinary permittivity of the metal is neglected, show that

$$(n^2 - k^2)/2nk = -\omega\tau.$$

The measurements of Beattie (1955) show that for aluminium at room temperature at wavelengths between $6\,\mu\text{m}$ and $12\,\mu\text{m}$, the quantity $(n^2 - k^2)/2nk$ is roughly equal to $-11/\lambda$, where $\lambda$ is the wavelength in micrometres. Show that this gives a value of $\tau$ of about $0\cdot6 \times 10^{-14}$ s.

**12.7.** The resistivity of a sample of copper at 4 K is approximately $10^{-10}\,\Omega\,\text{m}$. Assuming that $m^*/m = 1\cdot5$, and $W_F = 4\cdot7$ eV, show that the mean free path of electrons in copper at this temperature is about $7\,\mu\text{m}$, while the classical value of the skin depth (eqn (8.31)) at a wavelength of $0\cdot03$ m is about $0\cdot05\,\mu\text{m}$. Show also that at this wavelength and temperature the value of $\omega\tau$ is about $\tfrac{1}{4}$. (Assume one electron per atom for $n$.)

# 13. Superconductivity

## 13.1. Introduction

In 1911, three years after he had first liquefied helium, Kamerlingh Onnes discovered the phenomenon of superconductivity. This quite unexpected and fascinating property of a number of metals and alloys has given rise to a vast range of experimental and theoretical work. This chapter does not set out to cover all aspects of the subject, but concentrates on the fundamental properties and their explanation in terms of the quantum-mechanical treatment of electrons in metals discussed in the two preceding chapters.

On measuring the electrical resistance of mercury at temperatures of a few degrees Kelvin, Kamerlingh Onnes found that at 4·2 K it dropped abruptly to zero, as shown in Fig. 13.1. Subsequent investigations have shown that this sudden transition to perfect electrical conductivity is a characteristic of a number of metals and alloys. The transition is very sharp, being spread over less than a thousandth of a degree in good single crystals. It occurs at a fixed temperature $T_c$ which is a characteristic of the material; values of $T_c$ range from less than 1 K to over 20 K. Below the transition temperature the electrical resistivity is so close to zero that a persistent current in a superconducting loop has run for over 2 years. The absence of any detectable decay in its size has set an upper limit of $10^{-25}\ \Omega\,\mathrm{m}$ for the resistivity (Quinn and Ittner 1962), many orders of

FIG. 13.1. The superconductive transition curve of mercury as measured by Kamerlingh Onnes in 1911; at 3 K the resistance was $10^{-7}$ of that at 300 K. Pure unstrained crystals give transitions spreading over less than $10^{-3}$ K.

magnitude lower than the residual resistivity at low temperatures of the purest copper ($\sim 10^{-12}\,\Omega\,\mathrm{m}$). Experimentally only an upper limit to the resistivity can be determined, but we shall assume that in the superconducting state it is zero.

A number of the elements which become superconducting are listed in Table 13.1, together with their transition temperatures $T_c$. In addition, the following elements of the transition groups are superconductors;

   3d;  Ti, V;

   4d;  Zr, Nb, Mo, Tc, Ru;      4f;  La (f.c.c.);

   5d;  Hf, Ta, W, Re, Os, Ir;   5f;  Th, Pa, U($\alpha$).

A few other elements become superconducting under pressure. Though the superconducting elements tend to fall into groups in the periodic table, it is difficult to deduce any obvious pattern. It must depend on the crystal structure, since it is limited to the $\alpha$-forms of mercury and uranium and the face-centred cubic structure of lanthanum; the best-known indication of this is that white tin is a superconductor but grey tin is not. However, the known superconductors are not confined to any one type of crystal lattice, though their atomic volumes fall within a fairly small range. There is no correlation with the electrical conductivity immediately above the transition temperature; in fact superconductors tend to have a rather high resistivity at room temperature.

Kamerlingh Onnes realized that the property of zero electrical resistance would have an important application in the use of superconducting coils to produce very high magnetic fields, but in 1914 he discovered another important result, that superconductivity is destroyed by a magnetic field when the flux density is greater than a certain critical value $B_c$. The value of $B_c$ depends on the temperature and on the substance. For a large number of substances the variation can be expressed, with an

TABLE 13.1

*Values of the transition temperature $T_c$ and of the critical magnetic flux density $B_0$ at $T = 0$ K for a number of superconducting elements, grouped as they occur in the periodic table*

|  | Group II | | | Group III | | | Group IV | |
|  | $T_c$ (K) | $B_0$ (T) | | $T_c$ (K) | $B_0$ (T) | | $T_c$ (K) | $B_0$ (T) |
|---|---|---|---|---|---|---|---|---|
|  |  |  | Al | 1·18 | 0·0105 | | | |
| Zn | 0·875 | 0·0053 | Ga | 1·09 | 0·0051 | | | |
| Cd | 0·56 | 0·0030 | In | 3·40 | 0·0293 | Sn (white) | 3·72 | 0·0309 |
| Hg($\alpha$) | 4·15 | 0·0412 | Tl | 2·39 | 0·0171 | Pb | 7·19 | 0·0803 |

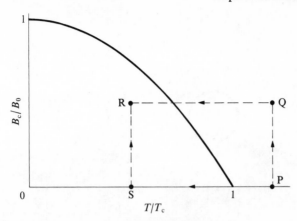

FIG. 13.2. Plot of eqn (13.1) in reduced coordinates. This equation is not exact; at $(T/T_c)^2 = \frac{1}{2}$, lead shows an upward deviation of about 2·2 per cent, and tin a downward deviation of about 3 per cent.

accuracy of a few per cent, by the relation

$$B_c = B_0\{1 - (T/T_c)^2\} \qquad (13.1)$$

This relation, which shows that $B_c$ falls to zero at the critical temperature, is illustrated in Fig. 13.2. The phase transition between the normal and superconducting state, which is very sharp at $B = 0$ for a pure sample, depends on the shape of the sample and its orientation when an external field is present. This is because of the presence of demagnetizing effects; the transition remains sharp provided that the magnetic flux density inside the specimen is uniform. The critical magnetic field may be produced by the superconducting current itself, so that at any temperature below $T_c$ there is an upper limit to the current which can be carried without causing a superconductor to revert to the normal state. The values of $B_0$, the critical flux density at $T = 0$, are also listed in Table 13.1.

## 13.2. The Meissner effect

Twenty-two years after the discovery of superconductivity, Meissner and Ochsenfeld (1933) found that a superconductor exhibits perfect diamagnetism as well as perfect electrical conductivity. This is illustrated in Fig. 13.3. In (a) the temperature is above $T_c$, and when a flux density **B** is applied, corresponding to moving from P to Q in Fig. 13.2, the magnetic flux penetrates the sample in the expected manner. If then (b) the sample is cooled through the transition temperature at constant field (that is, moving from Q to R in Fig. 13.2), the magnetic flux is completely expelled from the sample at the transition temperature. If, to avoid

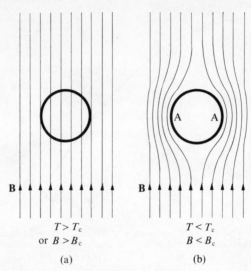

$T > T_c$ or $B > B_c$

$T < T_c$ $B < B_c$

(a)                                (b)

FIG. 13.3. The Meissner effect. In the normal state (a) the lines of magnetic flux penetrate the sample, but in the superconducting state (b) the flux density inside is zero.

problems with demagnetizing effects, we consider a sample in the form of a thin rod with **B** parallel to its axis, then we can write for the flux density inside the superconductor

$$\mathbf{B}_i = \mu_0\mathbf{H} + \mu_0\mathbf{M} = \mathbf{B} + \mu_0\mathbf{M},$$

and on putting this equal to zero we obtain

$$\mathbf{M}/\mathbf{B} = -1/\mu_0. \tag{13.2}$$

It must be emphasized that this is a new and fundamental property of superconductors and *not* a consequence of perfect electrical conductivity. For a conductor in which $\mathbf{E} = 0$, Maxwell's equation (8.3) curl $\mathbf{E} = -\partial\mathbf{B}/\partial t$ leads only to the result that **B** must remain fixed in time in the superconducting state. Thus on following the path PQR in Fig. 13.2, the flux in the metal in the normal state would remain trapped at the value which existed on entering the superconducting state. This means that the flux through a coil surrounding the metal would remain unaltered during the transition. In fact there is an abrupt change, consistent with the flux density in the superconducting state becoming zero. Fig. 13.3 shows that this results in a change in the flux distribution around the superconductor, and detailed measurements confirm that this is exactly that which would be expected if $\mathbf{B} = 0$ inside the superconductor.

Another way of proceeding from the point P to R in Fig. 13.2 is via the path PSR; the sample is first cooled in zero field to a point below the transition temperature, so that the flux density inside is zero. When an external field less than $B_c$ is applied, the flux density inside remains zero, and the same result is obtained whatever path is followed from P to R. Thus the Meissner effect suggests that the transition between the normal and the superconducting state is a reversible process, which can be treated by means of thermodynamics (see § 13.3). This would not be the case for a perfect conductor, since the flux density inside the superconductor would then be fixed at the value which existed at the moment of entering the superconducting state, and this depends on the way in which we proceed from P to R.

A second example of the Meissner effect is given by the hollow cylinder (or a hollow ring) of Fig. 13.4, in which the axis of the cylinder is parallel to the applied field. When the superconducting state is attained, either by reducing the temperature or the external field, the flux density **B** becomes zero within the body of the sample, but, in the space inside, **B** remains constant at the value it had when the transition took place. Diagrams (a) and (b) in Fig. 13.4 show that this is different, depending on whether the final state is reached by a path corresponding to PQR in Fig. 13.2 or by path PSR. These results were verified by Meissner and Ochsenfeld using a hollow lead cylinder.

Although a superconductor may be regarded as a perfect diamagnet, it would be wrong to assume that there is any volume distribution of magnetization inside the superconductor. The fact that **B** = 0 inside the body of the sample must be attributed to the presence of currents flowing

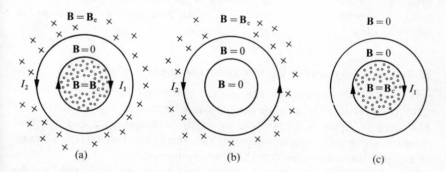

FIG. 13.4. The Meissner effect in a hollow cylinder whose axis is parallel to the magnetic flux density. (a) The cylinder is cooled into the superconducting state in a flux density **B** = **B**$_c$, and the flux density in the hole remains fixed at this value whatever the value of the external flux density $B_e$ becomes later. (b) The cylinder is cooled into the superconducting state in zero field; the flux density in the hole remains at **B** = 0, whatever the value of the external flux density $B_e$ becomes later. (c) The cylinder is cooled into the superconducting state in a flux density **B** = **B**$_c$, and the external field is subsequently reduced to zero.

The flux density in the body of the cylinder is always zero in the superconducting state.

at the surface of the sample. From the discussion of the infinite solenoid in § 4.6, we see that the situation represented in Fig. 13.4(b) could be ascribed to a current $I_2$ flowing round the outside surface of the cylinder, whose value per unit length is such that $I_2 = -B_c/\mu_0$, since this produces a field inside which just annuls the external field. Similarly, the situation in (a) can be ascribed to another current $I_1 = +B_c/\mu_0$ per unit length flowing round the inside surface.

If the hollow cylinder is cooled into the superconducting state in the presence of a flux density $\mathbf{B}_c$, and this external field is then removed, we reach the situation shown in Fig. 13.4(c). In the hollow space $\mathbf{B}$ remains equal to $\mathbf{B}_c$, while everywhere else $\mathbf{B} = 0$. The flux density $\mathbf{B}_c$ inside is just that which would be produced by the current $I_1$ per unit length, and the latter is known as the 'persistent current'. The fact that the flux density in the cylinder remains constant in the superconducting state, irrespective of changes in the external field, shows that a superconductor behaves as a perfect magnetic shield, as well as a perfect electric shield.

These surface currents are analogous to the 'magnetization currents' discussed in Chapter 4, except that they are real currents. It follows from that discussion that the discontinuity in the magnetization at the surface of an infinite cylindrical magnet may be attributed to a surface 'magnetization current', which conceptually can be an infinitely thin current sheet. With a real current this would imply an infinite current density, which intuitively seems to be unacceptable. In fact the current density decays exponentially as we go inwards and away from the surface (see § 13.4). The magnetic moment of the sample is exactly the same whether it arises from surface currents or an equivalent bulk magnetization; this enables us to use eqn (13.2) in the next section to calculate the magnetic energy.

## 13.3. Thermodynamics of the superconducting state

If the transition between the normal state and the superconducting state is reversible, it can be discussed in terms of equilibrium thermodynamics. Since it takes place at constant temperature and constant pressure, the equilibrium condition is that the Gibbs free energy $G$ must be equal in the two phases at the transition point.

In order to find the effect on $G$ of applying a magnetic field to the superconductor, we use the relation

$$dG = -S \, dT - M \, dB, \tag{13.3}$$

where $S$ is the entropy and $B$ is the flux density outside the sample. If $G_s$ is the thermodynamic potential of the superconductor at $B = 0$, and $G_s(B_c)$ the value when $B$ is increased to the transition field at constant temperature, then

$$G_s(B_c) = G_s - V \int_0^{B_c} M \, dB \qquad (13.4)$$

for a sample volume $V$. We consider again a long thin sample with its axis parallel to **B**, so that demagnetizing effects can be neglected. From eqn (13.2) we have

$$G_s(B_c) = G_s + VB_c^2/2\mu_0 \qquad (13.5)$$

and this must be equal to the thermodynamic potential of the normal state at the transition field. Hence

$$G_n = G_s + VB_c^2/2\mu_0. \qquad (13.6)$$

From this relation we can calculate the entropy difference between the normal and superconducting states. At constant $B$,

$$S_n - S_s = -\frac{d}{dT}(G_n - G_s) = -\frac{V}{\mu_0} B_c \left(\frac{dB_c}{dT}\right). \qquad (13.7)$$

From eqn (13.1) it is clear that $dB_c/dT$ is always negative, showing (see Fig. 13.5) that the entropy is lower in the superconducting state, that is, it is the state of greater order. Because of the entropy difference the transition occurs with a latent heat

$$L = T(S_n - S_s), \qquad (13.8)$$

and this is the heat that must be supplied to pass from the superconducting to the normal state at temperature $T$ and flux density $B_c$. The presence of a latent heat shows that we have a 'first-order' transition

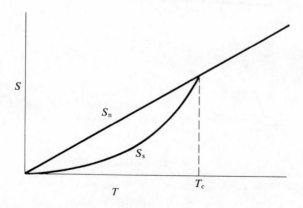

FIG. 13.5. The electronic entropy in the normal and superconducting states. At $T = T_c$ the entropy is $\gamma T_c$.

except at $T = T_c$, where $B_c = 0$. In this case there is only a jump in the specific heat at the transition between the two phases, given by

$$C_n - C_s = T\frac{d}{dT}(S_n - S_s) = -\frac{VT}{\mu_0}\left\{B_c\frac{d^2B_c}{dT^2} + \left(\frac{dB_c}{dT}\right)^2\right\}. \tag{13.9}$$

Such a jump in the specific heat is characteristic of a transition of the second order. At $T = T_c$, where $B_c = 0$,

$$C_n - C_s = -\left(\frac{VT_c}{\mu_0}\right)\left(\frac{dB_c}{dT}\right)^2, \tag{13.10}$$

showing that the specific heat is greater in the superconducting state. However, $d^2B_c/dT^2$ is negative, and eqn (13.9) changes sign at lower temperatures (see Problem 13.1); the specific heat becomes lower in the superconducting phase, the difference of course vanishing again at $T = 0$.

The specific heat of superconducting gallium (in zero magnetic field)

FIG. 13.6. Plot of $C_s/T$ and $C_n/T$ against $T^2$ for gallium.
Apart from a small nuclear hyperfine contribution near $T = 0$, the normal specific heat is well represented by the relation (in the units on the diagram)

$$C_n/T = 0\cdot596 + 0\cdot0568T^2.$$

At the lowest temperatures $C_s$ falls below $C_n$; the reverse is true as $T \rightarrow T_c$. The measurement of $C_n$ is made in an applied field of $0\cdot02$ T. (After Phillips 1964.)

and of gallium in a magnetic field large enough to restore the normal state are shown in Fig. 13.6. The quantity plotted is $C/T$ against $T^2$ (cf. Fig. 11.10), and a straight line is obtained for $C_n/T$ except at the lowest temperatures, where an additional contribution from an electrical quadrupole splitting of the nuclear levels becomes significant. On the other hand, $C_s/T$ varies rapidly with temperature and lies well above $C_n/T$ at temperatures near $T_c$. Measurements on a range of substances agree well with the assumption that the transition to the superconducting state is reversible.

When $T$ is much smaller than $T_c$ the value of $C_s-C_n$ becomes negative, and the specific heat in the superconducting state can be represented by a formula of the type

$$\frac{C_s}{\gamma T_c} \propto \exp\left(-\frac{\Delta}{kT}\right), \tag{13.11}$$

a result that is illustrated for Ga in Fig. 13.7. For this metal $\Delta = 1\cdot39kT_c$;

FIG. 13.7. Plot of $C_s/\gamma T_c$ (on logarithmic scale) against $T_c/T$ for gallium. The straight line corresponds to the exponential relation

$$C_s/\gamma T_c = 7\cdot46 \exp(-1\cdot39T_c/T).$$

The two-fluid theory predicts that $C_s$ should vary as $T^3$ when $(T/T_c) \ll 1$, but the exponential dependence suggests the presence of an energy gap. (After Phillips 1964).

similar results have been obtained for a number of superconducting elements, with a ratio of $\Delta$ to $kT_c$ of this order.

### The 'two-fluid' model

A useful phenomenological model (or working model) of a superconductor was put forward by Gorter and Casimir (1934). It assumes that in the superconducting state there exist two types of charge carrier:

(1) 'superconducting' carriers, condensed into an ordered state whose entropy is zero—they are not scattered by collisions, and give rise to perfect electrical conductivity;

(2) 'normal' carriers which are scattered by collisions—they are 'shorted out' by the 'superconducting' carriers and do not affect the conductivity.

The proportion of superconducting and normal carriers is a function of temperature. At $T = 0$ all carriers are in the superconducting state, but as the temperature rises, the proportion falls, reaching zero at the transition temperature. The superconducting state is regarded as the 'ground state' of the system, and a finite amount of energy is required to excite a carrier into the normal state; the specific heat represents the energy required to excite an increasing number as the temperature rises, and the exponential dependence in eqn (13.11) suggests that the energy 'gap' is somewhat greater than $kT_c$. The energy gap is not constant, but decreases as the temperature rises, becoming zero at $T_c$, above which the properties are those of a normal metal.

The important property of the condensed phase is that the carriers remain essentially at 0 K, with zero entropy; they carry the electrical current but their energy does not change with temperature, so that the Thomson heat is zero. They make no contribution to the transport of heat, so that in a superconductor heat is carried only by the normal carriers, whose number decreases to zero as $T \rightarrow 0$, and by the lattice. The phonons are not scattered by the condensed phase carriers, so that the lattice contribution to the thermal conductivity is increased. At low temperatures where no normal carriers are present to scatter the phonons, the scattering rate is determined solely by imperfections and crystal boundaries and is independent of temperature. The lattice conductivity is then simply proportional to the lattice specific heat that is, $K_s$ for the superconductor varies as $T^3$, while in the normal phase $K_n$ varies as $T$, following the electronic specific heat—see eqn (11.44). Hence $K_s/K_n$ varies as $T^2$, and may be as small as $10^{-5}$ if the rate of phonon scattering by imperfections is high. A superconducting wire may therefore be used as a 'heat switch' (Mendelssohn 1955), providing thermal contact only when the normal phase is restored by a local magnetic field.

## 13.4. The London equations

In the discussion of the Meissner effect in § 13.2, the fact that $\mathbf{B} = 0$ inside the sample was attributed to the presence of currents flowing at the surface. We now investigate how far such currents can be discussed in terms of the classical electromagnetic equations. In the absence of any dissipative forces, the current density $\mathbf{J}$ should obey eqn (3.9a), that is,

$$\dot{\mathbf{J}} = (nq^2/m)\mathbf{E}. \tag{13.12}$$

If $\mathbf{E}$ arises from a changing magnetic flux and grad $V = 0$, then from eqn (5.6) we can write

$$\dot{\mathbf{J}} = -(nq^2/m)\dot{\mathbf{A}}. \tag{13.13}$$

Provided that div $\mathbf{A} = 0$, we also have from eqn (4.54)

$$\nabla^2\mathbf{A} + \mu_0\mathbf{J} = 0, \tag{13.14}$$

so that

$$\nabla^2\dot{\mathbf{A}} - (\mu_0 nq^2/m)\dot{\mathbf{A}} = 0. \tag{13.15}$$

The solution of this for a semi-infinite solid bounded by the plane $z = 0$ is

$$\dot{\mathbf{A}} = \dot{\mathbf{A}}_0 \exp(-z/\lambda_L), \tag{13.16}$$

where

$$\lambda_L^{-2} = \mu_0 nq^2/m. \tag{13.17}$$

This result shows that any time variation of the field must fall off exponentially as we move into the superconductor, in a length of order $\lambda_L$, which for normal metals is of order $0\cdot01$–$0\cdot1$ $\mu$m, or about a hundred lattice spacings. This result is similar to the skin effect (§ 8.5), but at zero frequency the skin depth given by eqn (8.31) is indeterminate in a perfect conductor. All that the equations above tell us is that $\mathbf{B} = \text{curl } \mathbf{A}$ must be fixed in time after the material has entered the superconducting state at the value it had just before entering this state. This expectation is inconsistent with the Meissner effect.

The brothers F. and H. London (1935) took the bold step of proposing that instead of eqn (13.13), involving the time differentials, one should write

$$(m/nq^2)\mathbf{J} + \mathbf{A} = \mu_0\lambda_L^2\mathbf{J} + \mathbf{A} = 0. \tag{13.18}$$

The condition div $\mathbf{A} = 0$, known as the London gauge, is here equivalent to div $\mathbf{J} = 0$, and ensures continuity in the current flow. On combining eqn (13.18) with eqn (13.14) we have

$$\nabla^2\mathbf{A} - \lambda_L^{-2}\mathbf{A} = 0, \tag{13.19}$$

whose solution is

$$\mathbf{A} = \mathbf{A}_0 \exp(-z/\lambda_L). \tag{13.20}$$

TABLE 13.2

*Some important length parameters in superconductivity (a few typical values are shown in Table 13.3)*

1. $\lambda$ is the penetration depth; see eqn (13.28).
2. $\lambda_L$ is the penetration depth on the London theory; see eqn (13.23).
3. $\lambda_L(0)$ is the value of $\lambda_L$ as $T \rightarrow 0$, and $(n_s/n) \rightarrow 1$; see eqn (13.17).
4. $\xi_0$ is the Pippard coherence length, and represents the minimum size of the wave packets of superconducting charge carriers in bulk material. It is the range within which electrons can interact to form Cooper pairs (see § 13.6). In a pure superconductor such as aluminium or tin, $\xi_0 \gg \lambda_L(0)$.
5. $L$ is the mean free path for scattering of electrons in the normal state.
6. $\xi$ is the Pippard coherence length in the presence of scattering; see eqn (13.25).

This shows that **A**, and hence also **B** = curl **A** and **J**, fall off exponentially inside a superconductor. Thus **B** = 0 only at depths large compared with $\lambda_L$ from the surface. The quantity $\lambda_L$ is known as the 'penetration depth'. It is the first of several important length parameters which we shall encounter; these which are listed in Table 13.2, and some values are given in Table 13.3.

The result obtained above shows that the London equation (13.18) corresponds to the property of perfect diamagnetism in the bulk of the sample. A second London equation corresponds to perfect electrical conductivity, and is given by eqn (13.12). Through eqn (13.13) it can be seen that this follows from differentiation of the first London equation (13.18) provided that grad $V = 0$.

For a free particle in a region of changing magnetic flux we have from eqn (5.7)

$$\mathbf{p}_0 = m\mathbf{v} + q\mathbf{A}, \tag{13.21}$$

where $m\mathbf{v}$ is the momentum when the magnetic vector potential has the value **A** and $\mathbf{p}_0$ is the momentum when **A** = 0. If **v** is the mean velocity of

TABLE 13.3

*Values of the London penetration depth $\lambda_L(0)$, the measured penetration depth $\lambda$, and the Pippard coherence length $\xi_0$*

|           | $\lambda_L(0)$ (nm) | $\lambda$ (nm) | $\xi_0$ (nm) |
|-----------|---------------------|----------------|--------------|
| Aluminium | 16                  | 50             | 1600         |
| Tin       | 34                  | 51             | 230          |
| Lead      | 37                  | 39             | 83           |

the carriers, then the current density $J = nq\mathbf{v}$, and from eqn (13.18) we have

$$q(\mathbf{A} + \mu_0\lambda_L^2\mathbf{J}) = q\mathbf{A} + (m/nq)\mathbf{J} = q\mathbf{A} + m\mathbf{v} = 0. \qquad (13.22)$$

This result suggests that $\mathbf{p}_0$ is zero for the superconducting carriers, so that the condensation is into a state of zero momentum. It follows from the uncertainty principle that the wavefunctions of the superconducting carriers must be essentially unlimited in their spatial extension.

## 13.5. The Pippard coherence length

The value of $\lambda_L$ defined by eqn (13.17) involves the total number of electrons per unit volume (the total 'number density'), but in fact the current is all carried by the superconducting particles, whose density we denote by $n_s$. At $T = 0$, $n_s = n$ and eqn (13.17) thus gives only the limiting value $\lambda_L(0)$. At higher temperatures $n_s/n$ is less than unity on the two-fluid model, and we expect that the penetration depth, given by

$$\lambda_L = (m/\mu_0 n_s q^2)^{\frac{1}{2}}, \qquad (13.23)$$

should vary with temperature. Two types of experiments have been made to determine the penetration depth and its temperature dependence: (1) measurement of the reduction in the magnetic moments of thin films and colloidal particles; (2) measurement of the resonant frequency of a microwave cavity. The latter depends slightly on the depth to which radio-frequency currents penetrate and, although the frequency change is small, it can be measured accurately because of the very high quality factor $Q_F$ of the cavity; it has the advantage of giving results for the bulk metal. It was found that the effective depth was considerably greater than that calculated from eqn (13.23).

This discrepancy was resolved by Pippard (1953), who pointed out that the London equations are 'local equations' in which the current density $J$ at a point is assumed to be determined entirely by the value of $\mathbf{A}$ at that point. We have already seen (§ 12.10) that this is not true in a normal metal when the mean free path $L$ of the electrons becomes greater than the skin depth; the value of $J$ then depends on how $\mathbf{E}$ varies over a region with linear dimensions of the order of $L$. Pippard introduced the idea of a 'coherence length' $\xi_0$ for the wavefunctions of the superconducting carriers, such that a perturbation at one point influences every carrier within a distance $\xi_0$ of that point. This means we have a 'non-local' situation in which $J$ depends on the values of $\mathbf{A}$ over a region of linear dimensions $\xi_0$.

To obtain a value of $\xi_0$, Pippard argued that only electrons with a range of energy of order $kT_c$ can be responsible for a phenomenon occurring below $T_c$ in temperature. Only electrons with the Fermi velocity $v_F$ are involved, so that the corresponding range of values of momentum $\delta p$ is of order $kT_c/v_F$. The uncertainty principle then gives a corresponding length

$$\delta x \geq \hbar/\delta p = \hbar v_F/kT_c,$$

which for aluminium, tin, and lead is of order 0·1 to 1 $\mu$m. If we write

$$\xi_0 = a(\hbar v_F/kT_c), \tag{13.24}$$

where $a$ is a numerical constant to be determined, it is found that for a number of pure metals, $a$ is about 0·15. Experiments on alloys show, however, that the coherence length $\xi$ depends on the rate of scattering of the carriers, being given by the relation

$$\frac{1}{\xi} = \frac{1}{\xi_0} + \frac{1}{L}, \tag{13.25}$$

where $L$ is the mean free path in the normal phase. In the absence of scattering, $L$ tends to infinity and $\xi$ becomes equal to $\xi_0$, the value for the pure metal, but for small values of $L$ the coherence length $\xi$ is simply determined by $L$.

To allow for the effect of the mean free path, Pippard suggested that instead of eqn (13.18) one should write

$$\mu_0\lambda_L^2\mathbf{J} + (\xi/\xi_0)\mathbf{A} = 0, \tag{13.26}$$

or

$$\mu_0\lambda^2\mathbf{J} + \mathbf{A} = 0, \tag{13.27}$$

where

$$\lambda/\lambda_L = (\xi_0/\xi)^{\frac{1}{2}}. \tag{13.28}$$

Combination of (13.27) with (13.14) gives, in place of (13.19),

$$\nabla^2\mathbf{A} - \lambda^{-2}\mathbf{A} = 0, \tag{13.29}$$

whose solution is

$$\mathbf{A} = \mathbf{A}_0 \exp(-z/\lambda). \tag{13.30}$$

The penetration depth is $\lambda$, given by eqn (13.28), but eqn (13.26) is still a 'local equation' which is valid when $\mathbf{A}$ changes only slowly over the distance $\xi$, that is, when $\xi \ll \lambda$.

In the opposite limit where $\lambda \ll \xi_0$ we assume instead that the appropriate value of $\mathbf{A}$ is that averaged over a volume limited in one dimension by $\lambda$, that is, a volume $\xi_0^2\lambda$ instead of $\xi_0^3$. $\mathbf{A}$ is then effectively reduced by a factor $(\lambda/\xi_0)$, and instead of eqn (13.26) we have

$$\mu_0\lambda_L^2\mathbf{J} + (\lambda/\xi_0)\mathbf{A} = 0. \tag{13.31}$$

It is easy to show that this leads to the result

$$\lambda/\lambda_L = (\xi_0/\lambda_L)^{\frac{1}{3}}. \tag{13.32}$$

These equations can be used to deduce a value of $\xi_0$ from the measured penetration depth $\lambda$, and give a value of about $0 \cdot 15$ for the constant $a$ in eqn (13.24).

## 13.6. The Bardeen–Cooper–Schrieffer (BCS) theory

Any theory of superconductivity must account for the two fundamental properties of an ideal superconductor: perfect diamagnetism and perfect electrical conductivity. In addition there is the striking similarity between the properties of different metals; the relation between critical field and temperature (see Fig. 13.2) is represented by eqn (13.1) with departures of not more than a few per cent. The difficulties are illustrated by the fact that the energy $kT_c$ is less than a millielectronvolt, and the magnetic energy $\frac{1}{2}B_c^2/\mu_0$ is $\sim 10^{-8}$ eV per atom, while the Fermi energy is several electronvolts.

The direct interactions between electrons arising from their electric charge are repulsive, even after allowing for the screening effects of other electrons. The interactions between the electrons and the fixed positive charges of the lattice give rise to the band structure discussed in Chapter 12. An important clue was the discovery in 1950 of the 'isotope effect'; using different isotopes of mercury it was found by Maxwell (1950) and by Reynolds, Serin, Wright, and Nesbitt (1950) that the transition temperature $T_c$ depended on the isotopic mass $M$, approximately as $M^{-\frac{1}{2}}$. Similar effects have been shown to exist for many other superconducting elements, with $T_c \propto M^a$. Though the exponent $a$ is not always $-\frac{1}{2}$, it is found that for the isotopes of a given element $B_0$ is accurately proportional to $T_c$, and that $B_c/B_0$ can be represented by exactly the same function of $(T/T_c)^2$.

The lattice structure of an element is independent of its isotopic constitution; so are its chemical properties, and static electronic properties such as the Fermi energy and the electronic specific heat. The dynamical properties are different; the vibrational energy is $M\omega^2 \delta^2$, and for a given quantum of energy the deformation $\delta$ varies as $M^{-\frac{1}{2}}$. Thus the isotope effect suggests that the interaction responsible for superconductivity involves deformation of the lattice. Such a deformation could result from the attraction between an electron and the nearby positive ions of the lattice, a possibility that was investigated by Fröhlich in 1950 independently of the discovery of the isotope effect, but it remained for the theory of Bardeen, Cooper, and Schrieffer (BCS) (1957) to show how the local electrical polarization of the lattice, associated with the deformation, would interact with another electron to produce an indirect force between the two electrons which is attractive in its nature.

It was first shown by Cooper (1956) that any force of attraction between a pair of electrons, no matter how weak, would produce an instability in the Fermi sea, in which the formation of some 'bound pairs' would lead to the formation of further bound pairs. This is a co-operative effect which results in the presence of a ground state that is separated by an energy gap from the allowed excited states of the system. The pairs of electrons are called 'Cooper pairs' and they are the key to the theory of superconductivity. It turns out that the range of the interaction binding a pair together is of the order of the Pippard coherence length $\xi_0$, about 1 $\mu$m. In an ordinary metal the number of conduction electrons per cubic metre is of order $10^{28}$, and the fraction of these which lie in the energy range $kT_c$ is about $kT_c/W_F$, or about $10^{-4}$. It follows that a volume of order $\xi_0^3$ contains about $10^6$ Cooper pairs. They must therefore have some special property to prevent collisions between pairs destroying the binding process.

It is found that the lowest energy for such a Cooper pair is achieved by binding together an electron with wave vector $+\mathbf{k}$ and an electron with $-\mathbf{k}$, and with their spins antiparallel, so that the net spin is zero, as in an atomic state $S = 0$. The net linear momentum of such a pair is zero, as suggested in the original London theory, but the pair has, of course, mass $2m$ (or $2m^*$) and charge $2e$. The same is true for all pairs; each has a total wave vector $+\mathbf{k}-\mathbf{k} = 0$, corresponding to a 'condensation' in $\mathbf{k}$-space (or momentum space). The wavefunction is coherent over macroscopic distances.

When a current of density $\mathbf{J}$ is flowing, this corresponds to a displacement of the Fermi sphere (see Fig. 11.11) by an amount $\delta\mathbf{k}$, such that $\mathbf{J} = -n(e/m)\hbar\,\delta\mathbf{k}$. A Cooper pair is then formed from electrons with wave vector $+\mathbf{k}+\delta\mathbf{k}$ and $-\mathbf{k}+\delta\mathbf{k}$, so that each pair has a net wave vector $\delta\mathbf{k}$. All pairs move together, with velocity corresponding to eqn (13.22). To account for a given 'supercurrent' density of this type, it is clear that $\hbar\,\delta\mathbf{k}/m$ must be comparable with the *drift* velocity of the electrons in a normal current, so that $\delta\mathbf{k}$ will be very small compared with $\mathbf{k}_F$. The very long wavelength $2\pi/\delta\mathbf{k}$ associated with each pair means that the scattering processes which are effective in scattering normal electrons with wave vector $\sim k_F$ are ineffective in scattering Cooper pairs, thus explaining the presence of persistent currents.

Since the electron energy is proportional to the square of the wave vector, a pair with $(+\mathbf{k}+\delta\mathbf{k}, -\mathbf{k}+\delta\mathbf{k})$ has an energy which is greater by $\delta W = \hbar^2(\delta k)^2/m$ than that of a pair $(+\mathbf{k}, -\mathbf{k})$. If this energy exceeds the binding energy of the pair, the pair will 'dissociate' and the property of superfluidity will disappear. Since $\delta\mathbf{k}$ is proportional to $\mathbf{J}$, this gives an upper limit to the current density.

The BCS theory leads to an expression in which the energy gap $\Delta_0$ is proportional to the Debye frequency $\omega_D$, which for a given force constant (that is, a given lattice) varies as $M^{-\frac{1}{2}}$, where $M$ is the ionic mass. The gap energy at $T = 0$ is related to the condensation energy at $T = 0$ by the formula

$$\frac{V}{2\mu_0} B_0^2 = \frac{g(W) \Delta_0^2}{4},$$  (13.33)

and to the transition temperature $T_c$ by

$$2\Delta_0 = 3 \cdot 52 k T_c.$$  (13.34)

The density of states $g(W)$ is related to the electronic specific heat constant $\gamma$ for the normal state through eqns (12.27)–(12.28) which with $R = nk$ gives, per mole,

$$\gamma = \tfrac{1}{3}\pi^2 k^2 g(W).$$  (13.35)

Then elimination of $\Delta_0$ between (13.33) and (13.34) leads to the result

$$\frac{V}{2\mu_0} B_0^2 = 0 \cdot 235 \gamma T_c^2,$$  (13.36)

where the volume $V$ also refers (like $\gamma$) to 1 mol.

As the temperature rises, particles are excited across the gap, and the number of Cooper pairs decreases. The size of the gap $\Delta$ is correlated with the number of pairs, so that $\Delta$ also decreases, becoming zero at the transition temperature $T_c$, where the number of Cooper pairs vanishes. The BCS theory gives an explicit prediction for the way in which $\Delta$ varies with temperature, and since the condensation energy is simply proportional to $\Delta^2$ it follows from eqn (13.33) that $B_c/B_0 = \Delta/\Delta_0$. Fig. 13.8 shows how measurements of $B_c$ give a good fit to the theoretical prediction for pure samples of tin, indium, and mercury. In general the numerical value of the ratio $2\Delta_0/kT_c$, measured by various methods, lies between $2 \cdot 5$ and $4 \cdot 5$ (see Table 13.4), and is fairly close to $3 \cdot 5$ for a number of elements.

The BCS theory also leads to a relation between the Pippard coherence length and the energy gap

$$\xi_0 = \hbar v_F / \pi \, \Delta_0,$$  (13.37)

TABLE 13.4

*Numerical parameters for aluminium, tin, and lead compared with the predictions of the BCS theory*

| | $2\Delta_0/kT_c$ From tunnelling measurements | From electromagnetic absorption | $C_s/C_n$ at $T = T_c$ | $VB_0^2/2\mu_0\gamma T_c^2$ |
|---|---|---|---|---|
| Aluminium | $2 \cdot 5$–$4 \cdot 2$ | $3 \cdot 37$ | $2 \cdot 3$–$2 \cdot 6$ | $0 \cdot 233$ |
| Tin | $2 \cdot 8$–$4 \cdot 1$ | $3 \cdot 5$ | $2 \cdot 6$ | $0 \cdot 245$ |
| Lead | $4 \cdot 3$–$4 \cdot 4$ | $4 \cdot 0$–$4 \cdot 4$ | $3 \cdot 7$ | $0 \cdot 30$ |
| BCS theory | { $3 \cdot 52$ eqn (13.34) | | $2 \cdot 43$ | $0 \cdot 235$ } eqn (13.36) |

FIG. 13.8. Experimental values of $\Delta/\Delta_0$ as a function of $T/T_c$ for samples of tin, indium, and mercury, showing the good agreement with the BCS theory. The results are derived from measurements of $B_c$ as a function of temperature. (Finnemore and Mapother 1965.)

which may be combined with eqn (13.34) to give

$$\xi_0 = 2\hbar v_F / 3 \cdot 52 \pi k T_c = 0 \cdot 18 \hbar v_F / k T_c, \tag{13.38}$$

which is in good agreement with eqn (13.24). A further prediction of the theory is that at $T = T_c$, $C_s/C_n = 2 \cdot 43$ (see Table 13.4).

## 13.7. The energy gap

The energy gap in a superconductor is quite different in its origin from that in a normal metal. The band theory discussed in Chapter 12 shows that energy bands are a consequence of the static lattice structure. Discontinuities in the allowed energy occur at values of **k** determined by the spacing between the ions, that is, at the edges of the Brillouin zones. In a superconductor the energy gap is far smaller, and results from an attractive force between the electrons in which the lattice plays only an indirect role. It is present only at temperatures below $T_c$, and varies with the temperature when $T \leqslant T_c$. The gap occurs at the Fermi surface, as shown in Fig. 13.9, where the density of states is contrasted with that in a normal metal. On either side of the Fermi level there is an energy gap,

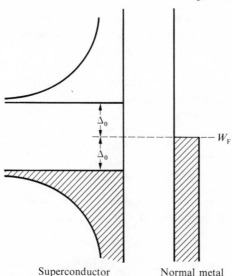

FIG. 13.9. The density of states near the Fermi energy $W_F$ in a superconductor, showing the energy gap $2\Delta_0$ at $T = 0$, and in a normal metal.

extending a distance $\Delta$ below and an equal distance $\Delta$ above $W_F$. At $T = 0$ all electrons are accommodated in states below the energy gap, and a minimum energy $2\Delta_0$ must be supplied to produce an excitation across the gap. At the lowest temperatures the specific heat in the superconducting state should have the exponential form

$$\frac{C_s}{\gamma T_c} \propto \exp\left(-\frac{\Delta_0}{kT}\right);$$

here $\Delta_0$ occurs, and not $2\Delta_0$, because this type of measurement is statistical in nature and involves the energy per particle, which is of course $\Delta_0$. At higher temperatures the behaviour of $C_s$ departs from a simple exponential because the gap energy itself varies with temperature, $\Delta$ being less than $\Delta_0$ and falling to zero at $T = T_c$.

The presence of an energy gap has been verified in a number of ways, of which we discuss only the use of electromagnetic radiation. It has been known for a long time that at optical frequencies superconductors exhibit no abnormal properties; the optical constants correspond to those for a normal metal with virtually no change at the transition temperature. This suggests that the quanta of optical radiation are large enough to excite carriers from the condensed phase into unoccupied electron states, while radio-frequency quanta are too small to reach such states. This is consistent with the idea of an energy gap in the electron states, which must obviously be expected to be of order $kT_c$. The corresponding frequency range is $10^5$–$10^6$ MHz, and much experimental work has been carried out in the microwave and far infrared regions.

The experimental results are expressed in terms of the a.c. conductivity, and normally involve comparisons between the superconducting state and the normal state, produced by imposing a magnetic field greater than the critical value, over a range of temperature. In the normal state the conductivity $\sigma_n$ is a real quantity and substantially independent of frequency. In the superconducting state it is necessary to use a complex conductivity

$$\sigma_s = \sigma_1 - j\sigma_2,$$

where both $\sigma_1$ and $\sigma_2$ vary with frequency. For an oscillatory current at frequency $\omega/2\pi$ differentiation of eqn (13.27) gives

$$\mathbf{E} = -\dot{\mathbf{A}} = \mu_0\lambda^2\mathbf{J} = j\omega\mu_0\lambda^2\mathbf{J}, \tag{13.39}$$

from which $\sigma_1 = 0$ and $\sigma_2 = (\omega\mu_0\lambda^2)^{-1}$. However, at frequencies such that $\omega \geqslant \omega_g$, where $\hbar\omega_g$ is the energy gap, carriers can be excited across the energy gap into empty states, and energy is absorbed; $\sigma_1$ then becomes finite, rising from zero at $\omega/\omega_g = 1$ to $\sigma_n$ when $\omega/\omega_g$ is $\gg 1$. The behaviour of the ratios $\sigma_1/\sigma_n$ and $\sigma_2/\sigma_n$ is shown in Fig. 13.10.

In a superconducting film whose thickness is smaller than the penetration depth sufficient energy is transmitted to measure the transmission coefficient, which can be related to the conductivity by an extension of the theory of Chapter 8. This method has been extensively explored by

FIG. 13.10. Plot of $\sigma_1/\sigma_n$ and $\sigma_2/\sigma_n$ against the reduced frequency $\omega/\omega_g$, where $\hbar\omega_g =$ superconducting energy gap. From eqn (13.40), $\sigma_2/\sigma_n$ should vary as $\omega^{-1}$ when $(\omega/\omega_g) \ll 1$, while $\sigma_1$ becomes finite only when $\omega/\omega_g \geqslant 1$. Conductivity in superconducting state $= \sigma_1 - j\sigma_2$; conductivity in normal state $= \sigma_n$.

Tinkham; for example, the transmission through films of lead and tin, some 2 nm thick, was studied by Glover and Tinkham (1957) over the frequency range $\omega/2\pi = 0\cdot3{-}40$ $(kT_c/h)$. This and later work has shown that the results for $\sigma_2/\sigma_n$ can be fitted to a universal relation

$$\sigma_2/\sigma_n = a^{-1}(kT_c/\hbar\omega), \tag{13.40}$$

with values of $a$ of the order of $0\cdot2{-}0\cdot3$. This result provides a nice confirmation of the Pippard coherence theory. In such a thin film the carrier mean free path is limited by scattering at the surfaces, so that $L$ and $\xi$ are equal (see eqn (13.25)) and $\ll \lambda$. Then, using eqn (13.28) with $\xi = L$, we have

$$\sigma_2 = (\omega\mu_0\lambda^2)^{-1} = (nq^2/m)(L/\omega\xi_0),$$

while

$$\sigma_n = (nq^2/m)\tau = (nq^2/m)(L/v_F).$$

Hence

$$\sigma_2/\sigma_n = v_F/\omega\xi_0 = a^{-1}(kT_c/\hbar\omega),$$

where the last result is derived through the use of eqn (13.24). From eqn (13.38) we see that the BCS theory predicts that $a = 0\cdot18$, in reasonable agreement with the range of experimental values given above.

When $T \ll T_c$, $\sigma_1$ in the superconducting state should be close to zero; it starts to rise only when the frequency is large enough to cause excitation across the gap. The transmission through a thin film is a function of both $\sigma_1$ and $\sigma_2$; it has a maximum in the region just above $\omega/\omega_g = 1$, where both $\sigma_1/\sigma_n$ and $\sigma_2/\sigma_n$ are less than unity, and the experimental results lead to values of the energy gap in good agreement with the BCS theory.

### Tunnelling

If only a very thin insulating film (such as oxide) separates the metal and the superconductor in Fig. 13.9, electrons can tunnel from one to the other, provided that on the far side there exists an empty allowed state of the same energy. By applying a voltage which raises the potential energy in the metal, the top of the Fermi distribution can be raised to the top of the energy gap in the superconductor, and electrons can then tunnel from the metal into the superconductor. A voltage of opposite sign depresses the Fermi distribution, and when the top falls below the bottom of the energy gap, electrons can tunnel from the superconductor into the metal. The current–voltage characteristics obtained in this way give an accurate measure of the energy gap (see Table 13.4).

### Acoustic attenuation

High-frequency sound waves are low-frequency lattice vibrations which are scattered by normal electrons but not by Cooper pairs, so that their

FIG. 13.11. Attenuation of 28–54 MHz sound waves in the superconducting state, relative to that in the normal state, as a function of $T/T_c$ for samples of tin and indium. The continuous line shows the prediction of the BCS theory. (Morse and Bohm 1957.)

attenuation is a measure of the fraction of normal electrons. The ratio of the attenuation in the superconducting state to that in the normal state is shown as a function of temperature in Fig. 13.11.

## 13.8. Type I and Type II superconductors

So far we have considered only the behaviour of superconducting materials in which the magnetic flux density in the bulk of the sample is zero up to the transition flux density $B_c$. The relation between the 'magnetization' and the applied field, given by eqn (13.2), follows the pattern shown in Fig. 13.12. Such materials are known as Type I superconductors, to distinguish them from a group of substances, mostly alloys, whose behaviour is illustrated in Fig. 13.13. As the applied field increases, all flux is excluded up to a critical flux density $B_{c1}$; above this field there is some flux penetration, but the substance remains superconducting until a second critical flux density $B_{c2}$ is reached. Above this field there is complete penetration and the substance behaves as a normal metal. The values of $B_{c1}$ and $B_{c2}$ lie respectively below and above the

FIG. 13.12. Plot of the 'magnetization' against external flux density $B$ for a Type I superconductor.

'thermodynamic' $B_c$ which appears in eqn (13.6). The transition temperatures of a number of 'Type II superconductors', as these materials are called, are given in Table 13.5. A number of the alloys are of considerable technical importance because they have high transition temperatures and high critical fields; the value of $B_{c2}$ is over 20 T for niobium-tin (Nb$_3$Sn). They are used in superconducting magnets, in which large magnetic fields can be maintained at relatively low cost (see § 6.5).

The basic difference between Type I and Type II superconductors arises from the relative sizes of the London penetration depth $\lambda$ and the coherence length $\xi$. In a Type I material $\lambda$ is small compared with $\xi$, while in a Type II material the reverse is true. The significance of this appears

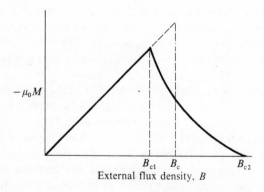

FIG. 13.13. Plot of the 'magnetization' $M$ against external flux density $B$ for a Type II superconductor, showing how $M$ starts to decrease above $B_{c1}$, the critical field for flux penetration. $M$ falls to zero only at a second critical field $B_{c2}$. The area under the magnetization curve $\int M \, \mathrm{d}B$, gives the magnetic energy $B_c^2/2\mu_0$.

TABLE 13.5

*Transition temperatures $T_c$
and upper critical flux density $B_{c2}$ for some Type II
superconductors*

|      | $T_c$ (K) | $B_{c2}$ (T) |
|------|-----------|--------------|
| V    | 5·4       | 0·14         |
| Nb   | 9·5       | 0·19         |
| NbTi | 16·0      |              |
| Nb$_3$Sn | 18·1  | 22·5         |
| V$_3$Ga | 14·0   | 19·5         |
| V$_3$Si | 16·8   | 22·5         |

when we consider the energy involved in forming a boundary between
normal and superconducting regions in a metal. There are two contributions to the energy:

(1) the magnetic energy, which rises from zero in the normal metal to a
density of $B_c^2/2\mu_0$ in the superconducting region; the change
occurs over a region determined by the flux penetration, whose
thickness is of order $\lambda$;

(2) the electron energy, which is lower in the superconducting phase
because of the negative condensation energy associated with the
formation of Cooper pairs—in this case the change occurs over a
region whose thickness is of order $\xi$.

The behaviour of the total energy density for a Type I material is
illustrated in Fig. 13.14. The sum of the magnetic energy and the electron
energy densities is greater in the boundary region, so that it is energetically unfavourable for a 'wall' to be present. The bulk of the material is
therefore superconducting, the transition region being confined to the
surface of the sample.

The consequences of the assumption that the coherence length $\xi$ is
*small* compared with the penetration depth $\lambda$ were first investigated by
Abrikosov (1957), using a theory proposed in 1950 by Ginzburg and
Landau. The behaviour of the total energy density is shown in Fig. 13.15;
the important difference is that a *lower* energy is associated with the
boundary region, and it should therefore be energetically favourable for
such boundaries to form. In fact this does not occur until the critical flux
density $B_{c1}$ is reached; below this field the behaviour is the same as for a
Type I material, with a complete Meissner effect where the flux density in
the bulk of the sample·is zero. In the region where the applied field has

FIG. 13.14. Energy associated with a boundary in a Type I material between normal and superconducting regions. For the latter the magnetic energy is greater (eqn (13.6)), but falls in the region thickness $\lambda$, where magnetic flux penetrates. The electron energy is lower in the superconducting state through the formation of Cooper pairs, whose density falls over a distance whose thickness is of order $\xi$, the coherence length. The presence of a boundary involves extra total energy.

flux density greater than $B_{c1}$ but smaller than $B_{c2}$ there is a partial flux penetration, with a mixture of normal and superconducting regions; this persists up to the second critical flux density $B_{c2}$, beyond which there is complete penetration and no superconducting regions remain. In the presence of partial flux penetration, the volume from which it is excluded becomes smaller so that the energy involved in keeping the flux out is

FIG. 13.15. Energy associated with a boundary in a Type II material, where the coherence length $\xi$ is smaller than the penetration depth $\lambda$. The formation of boundaries is favoured through the lower total energy in the boundary region.

reduced, and the critical flux density $B_{c2}$ can be greater than $B_c$. The theory leads to the result $B_{c2} = \sqrt{2}(\lambda/\xi)B_c$. In the region where $B$ lies between $B_{c1}$ and $B_{c2}$ the superconductor is said to be in a mixed or 'vortex' state (see below).

*Quantization of magnetic flux*

In a superconducting ring such as the hollow cylinder shown in Fig. 13.4 there is a finite flux parallel to the axis in the centre. A supercurrent flows round the inner surface of the cylinder, its density decreasing to zero at distances inside the metal. This current consists of a flow of Cooper pairs, each with net wave vector $\delta \mathbf{k}$ and momentum $\hbar \delta \mathbf{k} = \mathbf{p}$. According to the Bohr–Sommerfeld quantization rule, the value of $\mathbf{p}$ must satisfy the relation

$$\int \mathbf{p} \cdot d\mathbf{s} = Nh, \tag{13.41}$$

where $h$ is Planck's constant and $N$ is an integer. From eqn (13.21) we must have

$$\mathbf{p} = m\mathbf{v} + q\mathbf{A},$$

and on relating $\mathbf{v}$ to $\mathbf{J}$ as in eqn (13.22) the quantization rule leads to the result

$$Nh = \int (m\mathbf{v} + q\mathbf{A}) \cdot d\mathbf{s} = \int (m/nq)(\mathbf{J} \cdot d\mathbf{s}) + \int q(\mathbf{A} \cdot d\mathbf{s}). \tag{13.42}$$

At points well into the metal $\mathbf{J} = 0$ so that the first integral vanishes. From eqn (4.50), $\int (\mathbf{A} \cdot d\mathbf{s}) = \Phi$, the magnetic flux threading the path. Since $q = -2e$ for a Cooper pair, it follows that the flux must be quantized in multiples of $|h/2e| = 2 \cdot 07 \times 10^{-15}$ Wb. The result $\mathbf{p}_0 = 0$ in the London theory corresponds to the absence of any units of quantized flux in the interior of a Type I superconductor.

The suggestion that the flux must be quantized has an important consequence for hollow superconducting bodies such as the hollow cylinder considered in § 13.2. When such a cylinder is cooled through the superconducting transition temperature in a finite magnetic field, the experiments of Meissner showed that the flux inside the cavity remained fixed at the value existing at the moment when the cylinder entered the superconducting state. We now see that this cannot be exactly true; the current flow round the inside surface of the cylinder must adjust itself to be such that the flux inside is an exact multiple of $h/2e$. Experiments to detect this have been carried out by Doll and Näbauer (1961) and by Deaver and Fairbanks (1961). Within the errors necessarily present in the determination of such small quantities it was shown that the flux is quantized in units of $h/2e$.

FIG. 13.16. The 'vortex state' in a superconductor between $B_{c1}$ and $B_{c2}$. At the centre of a vortex there is 'normal' metal carrying lines of magnetic flux, around which flow circulating currents out to a distance equal to the coherence length $\xi$. The total flux through each vortex is just equal to one quantum $\Phi_0$, and in a perfect crystal the vortices are equally separated in a triangular array.

An idealized model of a Type II superconductor in the mixed or 'vortex' state is shown in Fig. 13.16. It consists of tiny vortices of supercurrent in the form of Cooper pairs, circulating round regions of magnetic flux. Each region contains just one unit of quantized flux, spread over an area whose radius is about equal to the penetration depth. The current density and the number of Cooper pairs decreases towards the centre of each vortex, where the metal is essentially 'normal'. The correctness of this model of a Type II superconductor has been demonstrated in a beautiful experiment of Essmann and Träuble (1971), using an electron microscope. The pattern of vortices is triangular, in agreement with the theory which predicts this as having the minimum energy in a perfect crystal lattice; the pattern is distorted by lattice imperfections and grain boundaries. By counting the number of vortices and measuring the flux it was found that each vortex carried one quantum of magnetic flux, within the experimental error.

## 13.9. The Josephson effect

In a normal metal the electron wavefunctions are of the form $\exp j\mathbf{k}.\mathbf{r} = \exp j\phi$; we may regard $\phi$ as a 'phase factor', but only its gradient in space is important since the wave vector $\mathbf{k}$ is related to the rate at which $\phi$ changes with distance. For a superconductor, the Ginzberg–Landau theory introduced a complex order parameter $\psi$ such that the density of superconducting carriers is given by $\psi\psi^* = n_s$. Here $\psi(r)$ has the form of a wavefunction $\psi = |\psi| \exp j\phi$ in which phase coherence is maintained over macroscopic distances. The phase factor is indeterminate in a bulk sample, but in going round a ring it must increase by a multiple of $2\pi$, a condition that corresponds to the phenomenon of flux quantization discussed above.

A remarkable effect was discovered by Josephson (1962) in considering the problem of phase coherence in a superconducting ring interrupted by a thin insulating layer (Fig. 13.17). The thickness of the layer (of the order of 1 nm) is such that electrons can tunnel through it, and with a normal metal the current through the layer is proportional to the voltage across it (the effective resistance is only a few ohms). The wavefunction in the insulating layer is real, but it must be continuous at the two boundaries between the layer and the metal. When the metal becomes superconducting, the presence of phase coherence round the ring combined with the boundary conditions leads to the result that the phases $\phi_1$, $\phi_2$ at the two surfaces of the insulator in contact with the metal must have a constant difference $\theta = \phi_1 - \phi_2$, and that a current of Cooper pairs traverses the layer given by the relation

$$I_s = I_c \sin \theta. \tag{13.43}$$

This current requires no voltage across the layer so long as $I_s$ does not exceed the critical value $I_c$ which is determined by the properties of the barrier (in particular, its thickness). If this critical value is exceeded, current flow must occur through some other process involving a finite voltage across the insulating layer. The current–voltage characteristic is shown in Fig. 13.18. The insulating layer is known as a 'weak link' and the device is called a 'Josephson junction'. Currents of the order of 1–100 $\mu$A or more can flow; the non-linear characteristic and the low dissipation make it important in a range of applications, of which we

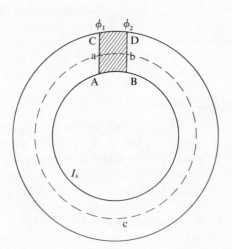

FIG. 13.17. A superconducting ring interrupted by a thin insulating layer ABCD. The current density inside the superconducting metal is zero, and the only contribution to the integral $\int \mathbf{J} \cdot d\mathbf{s}$ round the path abc comes from the part ab across the insulator.

FIG. 13.18. D.c. current–voltage characteristic for a Josephson junction. When the current exceeds the critical current $I_c$ for Cooper pairs, the passage of normal electrons occurs and a voltage must be present across the junction.

mention below only two. Niobium metal ($T_c = 9.5$ K) is a convenient superconducting material for use at 4·2 K, the boiling-point of liquid helium. Various forms of weak link have been designed; a niobium screw with a conical end gives an easily adjustable point contact with a low capacity (Fig. 13.19).

*Superconducting quantum interference device (SQUID)*

We now return to the question of a superconducting ring, with an insulating layer as a weak link, in the presence of an external magnetic

FIG. 13.19. A Josephson junction with a weak link at the point of contact between a niobium screw and niobium metal.

field. If the layer is so thick that no current can flow, the flux threading the ring is just that which would be expected from the applied field. In the opposite extreme, with zero thickness, a supercurrent flows in the ring which ensures that the flux threading it remains at zero if the applied field was zero when the ring became superconducting. In the presence of a weak link there is, however, a limit to the current of Cooper pairs which can flow through the link in the absence of any voltage. In general there will be a flux $\Phi$ through the ring, given by the relation

$$\Phi = \Phi_a + LI_s \qquad (13.44)$$

where $\Phi_a$ = flux due to applied field, $L$ is the self-inductance of the ring, and $I_s$ the supercurrent producing a flux $LI_s$ (see eqn (5.8)). The value of $I_s$ is given by the Josephson relation (eqn (13.43)), whose maximum value $I_c$ is the critical current of Cooper pairs which can cross the link at zero voltage.

The phase difference $\theta$ across the link can be related to the flux, using eqn (13.42), since $\Phi = \int \mathbf{A} \cdot d\mathbf{s}$. If we write $\Phi_0 = h/|2e| = h/q$ for the unit of flux, and divide eqn (13.42) by $h$, we have

$$N = \frac{m}{nqh} \int \mathbf{J} \cdot d\mathbf{s} + \frac{q}{h} \int \mathbf{A} \cdot d\mathbf{s}$$

$$= \frac{m}{nqh} \int \mathbf{J} \cdot d\mathbf{s} + \frac{\Phi}{\Phi_0}. \qquad (13.45)$$

Inside the superconductor $\mathbf{J} = 0$ and $\Phi$ must be an integral multiple of $\Phi_0$ for a complete superconducting ring. In a weak link $\mathbf{J}$ is finite, and a phase difference can occur across the link such that

$$\theta = 2\pi N - 2\pi(\Phi/\Phi_0). \qquad (13.46)$$

Since $N$ is an integer,

$$\sin \theta = -\sin 2\pi(\Phi/\Phi_0),$$

and on combining this with eqns (13.43), (13.44) we have

$$\Phi = \Phi_a + LI_s = \Phi_a + LI_c \sin \theta$$

$$= \Phi_a - LI_c \sin 2\pi(\Phi/\Phi_0). \qquad (13.47)$$

We can write this in a reduced form as

$$\frac{\Phi}{\Phi_0} = \frac{\Phi_a}{\Phi_0} - \frac{LI_c}{\Phi_0} \sin 2\pi\left(\frac{\Phi}{\Phi_0}\right), \qquad (13.48)$$

whose solution is shown in Fig. 13.20. As we increase $\Phi_a$ from zero, $\Phi$ also increases slowly, but at the point $P_1$, where $I_s$ reaches the critical value $I_c$, there is an instability and $\Phi/\Phi_0$ increases by one unit, jumping to the point $Q_1$. Further increases in $\Phi_a/\Phi_0$ give rise to corresponding jumps

FIG. 13.20. Graphical representation of eqn (13.48) for a particular ratio $(LI_c/\Phi_0) \sim 0\cdot 8$. Essentially the graph of $\Phi/\Phi_0$ against $\Phi_a/\Phi_0$ is a sine curve about the line AB corresponding to $\Phi = \Phi_a$, which it crosses whenever $\sin(2\pi\Phi/\Phi_0) = 0$. Energy is dissipated when a path such as $P_2Q_2$ RS is traversed in the presence of an oscillatory flux $\Phi_a$.

of one unit in $\Phi/\Phi_0$. If the process is reversed, and $\Phi_a/\Phi_0$ is decreased, there are again discontinuous drops of one unit in $\Phi/\Phi_0$, but these occur at the points where $I_s = -I_c$. The discontinuities occur when $I_s = I_c$ because the ring then momentarily ceases to be superconducting as this limit is exceeded, so that flux can enter; $I_s$ then drops below the critical value, and the ring is again superconducting until the supercurrent again reaches a critical value.

If a loop of wire is placed inside the ring, a voltage impulse is induced in the loop each time a flux jump occurs. The voltage impulse can be amplified, and by counting the impulses the total flux change through a ring of known geometry can be measured. If the ring has a diameter of about 1 mm, a change in flux density of order $10^{-9}$ T is sufficient to change the flux by one quantum. In practice, a better system is to couple the loop to a tuned circuit oscillating at about 30 MHz. The oscillations build up in amplitude, exciting an oscillating supercurrent across the weak link; when this exceeds the critical amplitude, the SQUID traverses a path such as $P_2Q_2RS$ in Fig. 13.20, involving a dissipation of energy which causes the level of oscillation to drop abruptly. The advantage of this system is that the radio-frequency voltage across the tuned circuit can be about 100 times that across the weak link.

In the presence of a steady magnetic flux $\Phi_a$ the sum of the direct current $I_s = (\Phi - \Phi_a)/L$ and the amplitude of the oscillatory component $I_s'$ cannot exceed $I_c$, so that the level of oscillation must vary periodically as $\Phi_a$ increases, the repetition rate being just equal to one quantum of flux.

Because of its high sensitivity, the device must be carefully shielded from extraneous changes in magnetic flux. In the two-hole system shown in Fig. 13.21, the metal forms a continuous superconducting circuit inside which the total flux must be constant. If the flux $\Phi_1$ in one hole increases, there must be an equal decrease in the flux $\Phi_2$ in the second hole, while the net current through the weak link depends on $\Phi_1 - \Phi_2$. The second hole is coupled inductively to the radio-frequency oscillatory circuit, while the first contains a loop of wire which is in series with a search coil placed in the field to be measured. The whole circuit formed by these two coils is also superconducting, so that the total flux through it is constant. A flux change $\Phi_S$ in the search coil induces a screening current $I = -\Phi_S/L'$, where $L'$ is the total self-inductance of the two-coil circuit. The corresponding flux change in the SQUID is $\Phi_a = +MI = -(M/L')\Phi_S$, where $M$ is the mutual inductance between the loop and the SQUID; the system acts as a 'flux transformer' which changes the sensitivity by varying the ratio $M/L'$. Application of the SQUID to magnetometry is discussed by Swithenby (1974).

FIG. 13.21. A squid formed from a cylinder of niobium metal with two holes and a niobium screw; this gives a system which is mechanically robust and can stand many thermal cycles. Typical diameters for the holes are about 3 mm. The r.f. coil is inductively coupled to the right-hand cavity.

*The alternating-current Josephson effect*

The flow of Cooper pairs at zero voltage across a weak link is known as the direct-current Josephson effect. When a voltage is maintained across the junction, Josephson predicted that there would also be a flow of alternating current. This arises in the following way. On one side of the junction the wavefunction has a phase factor which oscillates in time, given by

$$\psi_1 = |\psi_1| \exp(-j W_1 t/\hbar) = |\psi_1| \exp j\phi_1,$$

where $W_1$ is the energy. Similarly on the other side

$$\psi_2 = |\psi_2| \exp(-j W_2 t/\hbar) = |\psi_2| \exp j\phi_2.$$

The current through the junction is therefore

$$I_s = I_c \sin \theta = I_c \sin(\phi_1 - \phi_2) = I_c \sin\{(W_2 - W_1)t/\hbar\}.$$

If the energy difference is $W_2 - W_1 = |2e| V$, we can write

$$I_s = I_c \sin \omega t \tag{13.49}$$

where

$$\hbar\omega = h\nu = |2e| V. \tag{13.50}$$

This is known as the alternating-current Josephson effect. Essentially the passage of a Cooper pair over the voltage drop $V$ results, not in the dissipation of energy, but the transformation of the energy $|2e| V$ into an electromagnetic quantum $h\nu$. The frequency $\nu$ is linearly proportional to the voltage $V$, and has the value $483 \cdot 59$ MHz for a voltage of $1 \mu$V.

Power limitations stop the Josephson junction from becoming a useful radio-frequency generator, but it can be used as a detector. In the presence of microwave radiation, the d.c. characteristic (Fig. 13.22) shows a change in the current, with discontinuities at the points where the voltage is a multiple of $h\nu/2e$. The non-linear characteristic has been used for frequency multiplication up to about $10^6$ MHz, and for the determination of $h/e$ with great precision. High accuracy is readily available in the measurement of frequency (see § 22.5), and the small voltages across the junction are determined from comparison with a standard cell using Hamon's resistance ratio device (Problem 3.6). The accuracy of a few parts in $10^7$ (see Sanders and Wapstra 1972 (Part 9)) means that the method can be used to monitor changes in the voltage of standard cells, and to compare cells at different Standards laboratories.

## 13.10. Conclusion

In this chapter we have attempted to summarize the main experimental results which are basic to the understanding of the superconducting state.

FIG. 13.22. Current–voltage characteristic at 1·2 K of a Josephson junction in the presence of 10 GHz microwave radiation. The steps occur at voltages which are integral multiples of $h\nu/2e$. (Parker, Langenberg, Denenstein, and Taylor 1969.)

Much important work has been omitted; for example, there is no discussion of the intermediate state which arises when the field is greater than the critical field over some parts of the surface of the sample (for the sphere of Fig. 13.3, the value of **B** is larger at the equator than at the poles). Necessarily the BCS theory has been presented in qualitative terms. Despite the simplifying assumptions it contains, the theory works remarkably well, as can be seen from the data in Table 13.4 for the Type I superconducting metals aluminium, tin, and lead.

## References

ABRIKOSOV, A. A. (1957). *Sov. Phys.* **5**, 1174.
BARDEEN, J., COOPER, L. N., and SCHRIEFFER, J. R. (1957). *Phys. Rev.* **108**, 1175.
COOPER, L. N. (1956). *Phys. Rev.* **104**, 1189.
DEAVER, B. S. and FAIRBANKS, W. M. (1961). *Phys. Rev. Lett.* **7**, 43.
DOLL, R. and NÄBAUER, M. (1961). *Phys. Rev. Lett.* **7**, 51.

ESSMANN, U. and TRÄUBLE, H. (1971). *Scient. Am.* **224**, 75.
FINNEMORE, D. K. and MAPOTHER, D. E. (1965). *Phys. Rev.* A**140**, 507.
GLOVER, R. E. and TINKHAM, M. (1957). *Phys. Rev.* **108**, 243.
GORTER, C. J. and CASIMIR, H. B. G. (1934). *Physica* **1**, 305.
JOSEPHSON, B. D. (1962). *Phys. Lett.* **1**, 251.
LONDON, F. and LONDON, H. (1935). *Proc. R. Soc.* A**149**, 71.
MAXWELL, E. (1950). *Phys. Rev.* **78**, 477.
MEISSNER, W. and OCHSENFELD, R. (1933). *Naturwiss.* **21**, 787.
MENDELSSOHN, K. (1955). *Prog. low Temp. Phys.* **1**, 185.
MORSE, R. W. and BOHM, H. V. (1957). *Phys. Rev.* **108**, 1093.
PARKER, W. H., LANGENBERG, D. N., DENENSTEIN, A., and TAYLOR, B. N. (1969). *Phys. Rev.* **177**, 639.
PHILLIPS, N. E. (1964). *Phys. Rev.* A**134**, 385.
PIPPARD, A. B. (1953). *Proc. R. Soc.* A**216**, 547.
QUINN, D. J. and ITTNER, W. B. (1962). *J. appl. Phys.* **33**, 748.
REYNOLDS, C. A., SERIN, B., WRIGHT, W. H., and NESBITT, L. B. (1950). *Phys. Rev.* **78**, 487.
SANDERS, J. H. and WAPSTRA, A. H. (1972). *Atomic masses and fundamental constants 4.* Plenum Press, London.
SWITHENBY, S. J. (1974). *Contemp. Phys.* **15**, 249.

## Problems

13.1. If the critical flux density $B_c$ for the superconducting transition is given by eqn (13.1), show that the discontinuity in the specific heats $C_s - C_n$ should obey the relation

$$C_s - C_n = -\frac{B_0^2 T}{\mu_0 T_c^2}\left\{1 - 3\left(\frac{T}{T_c}\right)^2\right\}.$$

This suggests that $C_s - C_n$ should change sign at $T = T_c/\sqrt{3}$.

13.2. The reduced critical flux density $(B_c/B_0)$ can be expressed as a power series in $(T/T_c)^2$

$$B_c/B_0 = 1 - a_2(T/T_c)^2 + a_4(T/T_c)^4 + \dots .$$

By calculating the limiting value of $(C_s - C_n)/T$ as $T \to 0$ from eqn (13.9), show that

$$\gamma = 2a_2 \frac{V}{\mu_0} \frac{B_0^2}{T_c^2}$$

and that eqn (13.36) then leads to a value $a_2 = 1\cdot07$.

# 14. Paramagnetism

## 14.1. A general precession theorem

In Chapter 6 the origin of paramagnetism was discussed, and it was shown to exist in substances containing permanent magnetic dipoles. Such dipole moments are associated with moving charges, being due either to the motion of electrons in their orbits about the atomic nucleus or the spin of the electron about its own axis. From observations of hyperfine structure in atomic spectra it was inferred that the nuclei of many types of atom also possess 'spin' due to rotation about an internal axis, and that a magnetic dipole moment is associated with this spin. In all these cases the magnetic moment is associated with some units of angular momentum, and the direction of the magnetic moment $\mathbf{m}$ is parallel to that of the angular-momentum vector $\mathbf{G}$, and proportional to it. Thus we may write (cf. eqn (6.2))

$$\mathbf{m} = \gamma \mathbf{G}, \tag{14.1}$$

where $\gamma$ is a constant whose reciprocal $1/\gamma$ is known as the gyromagnetic ratio. For an electron of charge $-e$ and mass $m_e$ moving in an orbit, $\gamma$ is equal to the classical value $-e/2m_e$; the magnetic moment associated with the intrinsic spin of the electron is anomalously large, the ratio being in this case very nearly equal to $-e/m_e$. The minus sign in each of these cases arises from the fact that the charge carried by the electron is negative, and shows that the magnetic moment is oppositely directed to the angular-momentum vector. In the case of the nucleus, the angular momentum is of the same order as that of an electron, being either a small half-integral or integral multiple of $\hbar$, but the magnetic moment is about 1000 times smaller, corresponding to the greater mass of the particles (proton and neutron) in the nucleus. The value of $\gamma$ is then $g_N(e/2m_p)$, where $m_p$ is the mass of the proton, and $g_N$ is a number which is of the order of unity but is not in general an exact integer or a simple fraction.

When an atom or nucleus with a permanent magnetic dipole moment $\mathbf{m}$ is placed in a steady magnetic flux density $\mathbf{B}$, a couple is exerted on it which may be written in vector form as $\mathbf{m} \wedge \mathbf{B}$. The angular momentum must therefore change (either in magnitude or direction) at a rate equal to this couple, that is

$$\dot{\mathbf{G}} = \mathbf{m} \wedge \mathbf{B}.$$

Since **m** is proportional to and parallel to **G**, we have

$$\dot{\mathbf{G}} = \gamma \mathbf{G} \wedge \mathbf{B}. \tag{14.2}$$

This is a vector equation whose solution is easily found by writing down its components referred to Cartesian coordinates. If the magnetic field is assumed to act along the $z$-axis the components are

$$\left.\begin{array}{l} \dot{G}_x = \gamma B G_y, \\ \dot{G}_y = -\gamma B G_x, \\ \dot{G}_z = 0. \end{array}\right\} \tag{14.3}$$

Integration of the last equation shows that the component $G_z$ along the $z$-axis is a constant. It follows that the angle $\alpha$ which **G** makes with **B** is constant, and we may write $G_z = G \cos \alpha$. The equations for the $x$ and $y$ components may be solved by differentiating one of them and eliminating either $G_x$ or $G_y$. We find

$$\ddot{G}_x = \gamma B \dot{G}_y = -(\gamma B)^2 G_x,$$

with an identical equation for $G_y$. The solution is of the form

$$G_x = A \cos(-\gamma B t + \epsilon),$$

and from eqn (14.3) we find

$$G_y = A \sin(-\gamma B t + \epsilon).$$

Thus it will be seen that the projection of $G$ on the $xy$ plane is of constant magnitude $A = G \sin \alpha$, but rotates with the angular velocity $-\gamma B$, which we may write as $\omega_L$. Thus our solution for the components of **G** is

$$\left.\begin{array}{l} G_x = G \sin \alpha \cos(\omega_L t + \epsilon), \\ G_y = G \sin \alpha \sin(\omega_L t + \epsilon), \\ G_z = G \cos \alpha. \end{array}\right\} \tag{14.4}$$

The motion is such that the magnitudes of both **m** and **G** remain constant, but their directions 'precess' at a constant angle about the direction of **B** as in Fig. 14.1. The angular velocity of the precession depends only on the gyromagnetic ratio and the size of the magnetic field. The direction of precession is that of a right-handed screw progressing along **B** if $\gamma$ is negative, and vice versa. The angle $\alpha$ depends on the initial conditions prevailing when the magnetic field was switched on.

Although we have chosen to solve eqn (14.2) by the use of a Cartesian coordinate system, we could have derived a certain amount of information about the motion by inspection of the vector equation (14.2) itself. Since the vector product of two vectors is a vector perpendicular to both,

FIG. 14.1. Precession of **G** about **B**. The direction of precession shown is that for an electronic momentum ($\gamma$ negative).

it follows that $\dot{\mathbf{G}}$ is normal to **G** and to **B**. Thus, in Fig. 14.1, if instantaneously both **G** and **B** are in the plane of the paper, the motion of **G** must be normal to the paper. This means that if the momentum vector **G** is drawn from a fixed origin, then its tip must move out of the paper, and since $\dot{\mathbf{G}}$ always remains normal to **G**, the tip must move in a circle around **B**, that is, the angular momentum vector precesses around **B**.

This precessional motion, originally derived in a theorem due to Larmor, is a quite general result, depending only on the connection between angular momentum and magnetic dipole moment. In a quantum-mechanical system, such as the atom, the angular momentum plays an important role, as is well known from atomic theory. In the next section we turn to consideration of the magnetic moments of single atoms, making use of our general precession theorem. The chief difference we shall find from the classical case considered above is that the angle $\alpha$ is now fixed by the rules of quantization, only a small number of values being possible, instead of any value.

## 14.2. The vector model of the atom

A complete understanding of the behaviour of magnetic moments in atoms requires a thorough knowledge of quantum theory. Here we assume the reader is acquainted with atomic structure as presented in most elementary textbooks of atomic physics, and give only a résumé of the position, mainly in terms of orbits rather than wavefunctions. The results quoted will be those appropriate to the wave-mechanical theory.

The state of an electron in an atom is defined by four quantum numbers $n$, $l$, $m_l$, and $m_s$, whose significance is as follows. The principal quantum number $n$ has integral values from unity upwards, and the energy of the electron is mainly determined by the value of $n$. On the original Bohr theory the energy depended only on the value of $n$, its value being, for an atom with only one electron,

$$W_n = -(R_\infty Z^2/n^2),$$ (14.5)

where $R_\infty$ is a universal constant (Rydberg's number), and $Ze$ the charge on the nucleus. The same result is obtained by wave mechanics for a one-electron atom, but this result does not hold for atoms with several electrons owing to the electrostatic repulsion between the electrons. For most atoms it remains true that electrons with the lower values of $n$ have the lower energy, and the difference of energy for successive values of $n$ decreases as $n$ becomes larger (cf. Fig. 14.2).

The quantum number $l$ is defined by the value of the angular momentum which the electron possesses in its orbit around the nucleus; on the vector model the angular momentum $\mathbf{G} = l\hbar$, where $\hbar = h/2\pi$ and $h$ is Planck's constant. The allowed values of $l$ are integral, from 0 up to $(n-1)$ for an electron whose principal quantum number is $n$. Since the electron is a charged particle, its motion in an orbit is equivalent to a circulating current, and a magnetic dipole moment is associated with the orbit which has the same value as that expected on classical theory. This

FIG. 14.2. The energy levels of sodium compared with those of hydrogen. For sodium the levels are those of the single electron outside the closed shells $1s^2$, $2s^2$, $2p^6$. For the higher values of $n$ the levels approach closely those of the hydrogen atom. This is because at large distances from the nucleus the electric field is that of the nuclear charge $+Ze$ surrounded by $(Z-1)$ electrons, and hence (by Gauss's theorem) is that of unit positive charge. Orbits with lower values of $n$ penetrate the closed electron shell and so feel a greater positive charge, giving a lower energy. This is most marked for the 'penetrating' s orbits.

dipole moment is

$$\mathbf{m} = (-e/2m_e)\mathbf{l}\hbar = -(e\hbar/2m_e)\mathbf{l} = -\mu_B\mathbf{l}.$$

Here the minus sign shows that $\mathbf{m}$ and $\mathbf{l}$ have opposite directions, owing to the negative charge of the electron. The quantity $\mu_B = e\hbar/2m_e$ is known as the Bohr magneton and is the atomic unit of magnetic dipole moment; its magnitude is $0{\cdot}9274 \times 10^{-23}\,\mathrm{A\,m^2}$.

The precession theorem of § 14.1 shows that the dipole moment, and its associated angular momentum vector, precess at a constant angle about an applied magnetic field. The component $G_z$ of the angular momentum in this direction is therefore a constant. On quantum theory this component must be an integral multiple of $\hbar$; it is written $m_l\hbar$, where the 'orbital magnetic quantum number' $m_l$ can have all integral values, including 0, between $+l$ and $-l$. The components normal to this direction are not quantized, but the sum of their squares can be shown to be given by

$$G_x^2 + G_y^2 = \{l(l+1) - m_l^2\}\hbar^2.$$

Since $G_z^2 = m_l^2\hbar^2$, we see that the square of the angular momentum has the 'sharp' value $G^2 = l(l+1)\hbar^2$. It is a general feature of the quantum theory of angular momentum that one component has a value which is of the form $\hbar m_l$, corresponding to the stationary component in a field, while the square of the angular momentum is $l(l+1)\hbar^2$ and not $l^2\hbar^2$. These points are illustrated in Fig. 14.3.

Since the magnetic moment is proportional to the angular momentum, the moment associated with the orbit has a fixed component parallel to a magnetic field of magnitude $-m_l\mu_B$, and the square of the magnetic dipole moment has the value $l(l+1)\mu_B^2$.

FIG. 14.3. Quantization of orbital angular momentum. The figure is drawn for the case of $l\hbar = 2\hbar$ and the radius of the circle is $\hbar\sqrt{(2 \times 3)} = \hbar\sqrt{6}$.

In addition to the orbital motion, the electron also possesses a spin about its own axis, with angular momentum $\mathbf{G}_s = s\hbar$, where the electronic spin quantum number $s$ always has the value $s = \frac{1}{2}$. The square of the angular momentum is $s(s+1)\hbar^2$. With the spinning charge of the electron is associated a magnetic moment

$$-g_s(e\hbar/2m_e)\mathbf{s} = -g_s\mu_B\mathbf{s}.$$

Here the coefficient $g_s$ is inserted because the ratio of the magnetic moment to the angular momentum differs from the classical value (corresponding to $g_s = 1$). For a long time it was thought that the value of $g_s$ for the electron spin was exactly 2; it has now been shown both experimentally and theoretically that the value is $2(1\cdot00115966)$, but for our purpose it is sufficient to omit the correction and assume that $g_s$ is 2 for electron spin. In an atom there are relativistic and diamagnetic corrections to both the orbital and spin magnetic moments (of the order of $10^{-6}$–$10^{-4}$), which we shall neglect. The minus sign in the expression for the magnetic moment shows that it is oppositely directed to the angular momentum vector, owing to the negative charge of the electron, as in the orbital case. In a magnetic field both the spin angular momentum and its magnetic moment precess about the direction of the field, as in Fig. 14.4, the steady component of the angular momentum in this direction having one of the values $\pm m_s\hbar = \pm\frac{1}{2}\hbar$, and the corresponding steady component of the magnetic moment having the values $\mp\frac{1}{2}g_s\mu_B \approx \mp\mu_B$. Note that these components amount to 1 $\mu_B$, though the spin is half-integral.

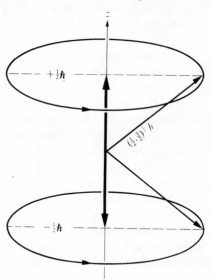

FIG. 14.4. Quantization of spin angular momentum $s = \frac{1}{2}$.

In an atom containing a number of electrons, the total angular momentum will be the vector sum of the individual momenta, both orbit and spin. In general this vector sum can be formed in a number of ways, with a number of different resultants. To know which of these is correct or, if several are allowed, which corresponds to the state of lowest energy (the ground or normal state of the atom), we need to know more about the mutual interactions between the various electrons. We shall see that these can be expressed in the form of a set of rules for coupling together the angular momenta in forming the vector resultant. These rules are subject to one overriding condition, expressed in the well-known Pauli principle: 'no two electrons in the same system can be in states with identical sets of quantum numbers'. When applied to an atom this means that no two electrons can have identical sets of values for $n$, $l$, $m_l$, and $m_s$. The Pauli principle sets a limit to the number of electrons in any one shell, and has the important result that the angular momentum of a filled shell containing $2(2l+1)$ electrons is zero: $L = S = 0$. Such a shell can have no resultant magnetic moment, so that the magnetic moment of an atom or ion must arise solely from its incompletely filled shells.

Our next problem is that of how to couple together the angular momenta in a partly filled subshell. This depends on the mutual interactions between the electrons, of which the two principal types are as follows:

1. *Mutual repulsion between the electrons, due to their electrical charge.* When this is treated by wave mechanics an unexpected result is found. The energy of the system contains two terms, one corresponding to the classical Coulomb interaction, the other known as an 'exchange energy', because it appears to be connected with an exchange of any pair of electrons between the states we assigned to them before including the effect of their mutual repulsion. These exchange forces have no analogue in classical theory, but play an important role in atomic theory. By means of the Pauli exclusion principle, their effect can be shown (see § 15.8) to correspond to a strong coupling between the electron spins, this coupling being such that within an atom the state with the spins parallel is more stable and has the lower energy. The energy of interaction of this coupling may be written in the form $W = -2\mathcal{J}_{ij}\mathbf{s}_i \cdot \mathbf{s}_j$, where $\mathbf{s}_i$, $\mathbf{s}_j$ are the spin vectors, and $\mathcal{J}_{ij}$ is called the 'exchange energy', being positive for any pair of electrons within a given atom but varying in magnitude, depending on their orbital quantum states. Thus with a number of electrons, the primary effect of the exchange forces is to couple together the various vectors $\mathbf{s}_i\hbar$, $\mathbf{s}_j\hbar$ to form a resultant $\mathbf{S}\hbar$, which in the state of lowest energy has the largest possible value consistent with the exclusion principle. The remaining orbital momenta are then coupled together by the electrostatic

forces to form a resultant $L\hbar$, which in the state of lowest energy again has the largest possible value consistent with the exclusion principle. (These two rules are known as Hund's rules.) Other values of $S\hbar$ and $L\hbar$ are possible, but correspond to states of higher energy. This method of coupling the angular momenta is known as $L$–$S$ or Russell–Saunders coupling. In wave mechanics $S$ and $L$ are quantum numbers and the absolute magnitudes of the squares of the total angular momenta associated with them are $\{S(S+1)\}\hbar^2$ and $\{L(L+1)\}\hbar^2$.

2. *Magnetic coupling between the magnetic moments of orbit and spin (the 'spin–orbit' interaction).* The orbital motion of the electron in an atom means that it sees the nucleus as a moving charge, producing a flux density **B** which acts on the magnetic moment associated with the electron spin. This is the origin of the 'spin–orbit interaction', which can be derived as follows. In the usual frame of reference where the nucleus is fixed, there is an electric field of strength **E** at the electron, and in this frame the electron is moving past the nucleus with velocity **v**. We need the value of **B** in a frame where the electron is fixed; in this frame the nucleus is moving with velocity $-\mathbf{v}$, and the electron experiences a magnetic flux density

$$\mathbf{B} = -(\mathbf{v} \wedge \mathbf{E})/c^2 = +(\mathbf{E} \wedge \mathbf{v})/c^2,$$

where $c$ is the velocity of light. This equation is derived from the special theory of relativity, and incidentally is a consequence of the experimental fact that the electric charge (unlike mass) is invariant of the velocity of the frame of reference.

In this field the electron precesses with an angular velocity $\boldsymbol{\omega}_s = -\gamma_s \mathbf{B}$, where $\gamma_s = -g_s(|e|/2m_e)$. This is, however, the precession velocity in the frame in which the electron is at rest, and from the theory of relativity Thomas showed that to transfer to the laboratory frame one must add a velocity of precession

$$\boldsymbol{\omega}_T = (\mathbf{f} \wedge \mathbf{v})/c^2$$

where **f** is the acceleration of the electron in the laboratory frame. Clearly $\mathbf{f} = -e\mathbf{E}/m_e$, so that the net precession frequency in the laboratory frame is

$$\boldsymbol{\omega}_s + \boldsymbol{\omega}_T = (g_s - 1)\frac{e}{2m_e} \cdot \frac{\mathbf{E} \wedge \mathbf{v}}{c^2}.$$

Over all, the effect is equivalent to a magnetic flux density

$$\mathbf{B} = \frac{(g_s - 1)}{g_s} \cdot \frac{\mathbf{E} \wedge \mathbf{v}}{c^2}.$$

In a hydrogen-like atom with nuclear charge $Ze$ and a single electron the field strength **E** at distance **r** from the nucleus is $\mathbf{E} = \mathbf{r}(Ze/4\pi\epsilon_0 r^3)$. In an atom with many electrons the electric field is still radial to a good approximation (the 'central-field' approximation), but the field is reduced through the screening effect of the other electrons. We may write $\mathbf{E} = \mathbf{r}(e/4\pi\epsilon_0)\langle Z'/r^3\rangle$, where $Z'e$ is the nuclear charge which gives the correct field at distance $r$ and $\langle Z'/r^3\rangle$ represents the mean value averaged over the electron orbit. Hence

$$\mathbf{B} = \frac{(g_s-1)}{g_s} \cdot \frac{e}{4\pi\epsilon_0 c^2}\left\langle\frac{Z'}{r^3}\right\rangle(\mathbf{r}\wedge\mathbf{v}).$$

But $m_e(\mathbf{r}\wedge\mathbf{v}) = \mathbf{l}\hbar$, the orbital angular momentum, and since $e\hbar/2m_e = \mu_B$, (the Bohr magneton), we have

$$\mathbf{B} = \frac{2(g_s-1)}{g_s} \cdot \frac{\mu_0}{4\pi}\left\langle\frac{Z'}{r^3}\right\rangle\mu_B\mathbf{l}. \tag{14.6}$$

The interaction energy with the spin magnetic moment is then

$$-\mathbf{B}.\mathbf{m}_s = g_s\mu_B(\mathbf{B}.\mathbf{s}) = \zeta(\mathbf{l}.\mathbf{s}). \tag{14.7}$$

The effect of this spin–orbit interaction tends to couple together the vectors $\mathbf{s}\hbar$ and $\mathbf{l}\hbar$ for each electron to form a resultant $\mathbf{j}\hbar$; the various values of $\mathbf{j}\hbar$ for the individual electrons would then be coupled together (vectorially) to form the total angular-momentum vector $\mathbf{J}\hbar$. However, the spin–orbit interaction is smaller in magnitude than the exchange interactions between the spins discussed in (1) above, except in the heaviest elements. We shall therefore confine ourselves to $L$–$S$ coupling, where the individual spins are coupled to form a resultant $\mathbf{S}\hbar$, and the individual orbital momenta to form a resultant $\mathbf{L}\hbar$. The spin–orbit interaction then couples $\mathbf{S}\hbar$ and $\mathbf{L}\hbar$ together with an energy

$$W = \lambda\mathbf{L}.\mathbf{S}, \tag{14.8}$$

which is similar in form to eqn (14.7). It can be shown that the relation between the two constants is $\lambda = \pm\zeta/2S$, where the positive sign is required for a shell that is less than half-filled, and the negative sign for one that is more than half-filled; the spin–orbit coupling parameter $\lambda$ vanishes for a half-filled shell. This coupling of $\mathbf{S}\hbar$ and $\mathbf{L}\hbar$ gives a resultant vector $\mathbf{J}\hbar$, the square of whose angular momentum is $J(J+1)\hbar^2$. The number of possible values of $J$ is either $(2S+1)$ or $(2L+1)$, whichever is the smaller. $L$ has only integral values, and the values of $J$ are therefore integral or half-integral according to whether the value of $S$ is integral or half-integral. The latter depends on whether the number of electrons involved is even or odd.

The nomenclature used to describe atomic energy states is mainly derived from pre-quantum attempts to analyse atomic spectra, and therefore does not possess the simple logical sequence which quantum theory could give it. Single-electron states for which the orbital quantum number $l$ has the values 0, 1, 2, 3, 4, 5,... are called s, p, d, f, g, h,... states, and similarly the levels of a many-electron atom for which $L = 0$, 1, 2, 3, 4, 5,... are denoted by the symbols S, P, D, F, G, H,.... The value of $n$ for a single-electron state is given by the number preceding the symbol, that is, 1s, 2s, 2p, 3s, 3p, 3d, etc. The number of electrons with given values of $n$ is denoted by a superfix; thus, three electrons with $n = 2$, $l = 1$ appear as $2p^3$. The spectroscopic state of the whole atom is defined by the values of $S$, $L$, and $J$; the value of the spin multiplicity $2S + 1$ is given by a superfix preceding the symbol for $L$, and the value of $J$ by a following suffix. Thus the symbol $^4F_{\frac{3}{2}}$ means that the state has $S = \frac{3}{2}$, $L = 3$, $J = \frac{3}{2}$; the other possible values of $J$ in this case are $\frac{5}{2}, \frac{7}{2}, \frac{9}{2}$, ranging in all from $L - S$ to $L + S$.

The coupling scheme for a many-electron atom or ion may be illustrated by reference to the energy-level diagram for the $Cr^{3+}$ ion, shown in Fig. 14.5. The triply charged chromium ion has the configuration $1s^2 2s^2 2p^6 3s^2 3p^6 3d^3$ with three electrons in the partly filled 3d shell. By

FIG. 14.5. Quartet energy levels of the free triply charged chromium ion, $Cr^{3+}$, $3d^3$. These are formed from the three electrons in the 3d shell: the splitting between the $^4F$ and $^4P$ terms is due to electrostatic repulsion between the electrons; the splittings between levels of different $J$ within each term are due to spin–orbit coupling. Doublet terms formed from $3d^3$ lie in the energy range $1 \cdot 4$–$3 \cdot 7 \times 10^6\,m^{-1}$. The next lowest levels are those belonging to the configuration $3d^2 4s$, and lie above $10^7\,m^{-1}$.

Hund's rules the energy is lowest when all three electrons have parallel spins, giving $S = \frac{3}{2}$. The electrons must then all have different values of $m_l$, by the Pauli principle, but since we have five possible values ($m_l = 2, 1, 0, -1, -2$) there are $(5!/3!2!) = 10$ possible arrangements. The largest possible value of $M_L = \sum m_l$ that we can have is $3 = 2+1+0$, and this belongs to an $L = 3$ state. This has $2L+1 = 7$ values of $M_L$, which therefore take up seven of the possible arrangements of electrons in the $m_l$ states; the other three belong to a state with $L = 1$. By Hund's rule, the $L = 3$ states will have lower energy than the $L = 1$ states. Both are shown in Fig. 14.5, the $^4P$ ($L = 1$) states being higher in energy than the $^4F$ ($L = 3$) states by about $1.4 \times 10^6 \, \text{m}^{-1}$. States of still higher energy (not shown) are formed by reversing one spin, giving $S = \frac{1}{2}$. Two electrons with opposite spin can now occupy the $m_l = 2$ state, so that the greatest possible value of $L_z = M_L$ is $5 = 2+2+1$, which belongs to a $^2H$ state. Altogether six doublet states, $^2H$, $^2G$, $^2F$, $^2D$ (twice), $^2P$ are allowed; they have energies ranging from about $1.4 \times 10^6 \, \text{m}^{-1}$ to $3.7 \times 10^6 \, \text{m}^{-1}$. All other states are much higher in energy, the $3d^2 4s$ configuration lying about $10^7 \, \text{m}^{-1}$ above the ground state $3d^3$, $^4F$.

The separation of the various quartet and doublet terms is determined by a combination of the exchange interaction and Coulomb interaction arising from the mutual repulsion of the electrons. The spin–orbit coupling splits the $^4F$ term into four levels with values of $J$ ranging from $|L-S|$ to $|L+S|$, and the $^4P$ term is similarly split, the only allowed values of $J$ in this case being $\frac{5}{2}$, $\frac{3}{2}$, and $\frac{1}{2}$. From the energy levels given in Fig. 14.5 it can be verified (see Problem 14.7) that the spin–orbit coupling constant $\lambda$ has the value $8.7 \times 10^3 \, \text{m}^{-1}$ for the ground states of $Cr^{3+}$. It can be seen that the splittings due to the 'magnetic' spin–orbit coupling are an order of magnitude smaller than those due to 'electrostatic' interactions.

## 14.3. Magnetic moments of free atoms

When we turn to consider the magnetic properties of atoms we find that the problem is simplified by the fact that these depend only on the partly filled electron shells, since completely filled shells (and of course empty shells) have $S$, $L$, and $J = 0$. The magnetic moment associated with each electron spin can be described by a vector parallel to and proportional to the angular-momentum vector $s\hbar$, and on forming the vector sum $S\hbar$ for a number of electrons the magnetic moments add in a similar way, so that the total magnetic moment of the spin is parallel to $S\hbar$ and has the same factor of proportionality to it. The same is true of the total orbital magnetic moment and the total orbital angular momentum $L\hbar$. When we come to make the vector addition of $S\hbar$ and $L\hbar$ the problem is not so simple because the factor of proportionality between the magnetic moment and the angular momentum is not the same for $S\hbar$ and $L\hbar$. The vector representing the total magnetic moment will not therefore be parallel to $\mathbf{J}\hbar$; this is illustrated by the vector diagram in Fig. 14.6. Here the magnetic-moment vector associated with $\mathbf{L}\hbar$ is drawn of the same length as $\mathbf{L}\hbar$, but on this scale the magnetic-moment vector associated with $\mathbf{S}\hbar$ must be drawn approximately twice as long as $\mathbf{S}\hbar$. The resultant magnetic-moment vector $\mathbf{m}$ is therefore at an angle to $\mathbf{J}\hbar$.

In considering this question further we must return to the discussion of the spin–orbit coupling between $\mathbf{L}\hbar$ and $\mathbf{S}\hbar$. This is primarily magnetic in origin, and arises from the magnetic moments of the orbit and spin. Each of these produces a magnetic field which interacts with the dipole moment of the other.

The interaction energy $\lambda\mathbf{L}.\mathbf{S}$ is equivalent to $-\mathbf{m}_S.\mathbf{B}_L$ or to $-\mathbf{m}_L.\mathbf{B}_S$, that is, to a flux density $\mathbf{B}_L = -\lambda\mathbf{L}/\gamma_S\hbar$ acting on the spin moment $\mathbf{m}_S = \gamma_S\hbar\mathbf{S}$, or to a flux density $\mathbf{B}_S = -\lambda\mathbf{S}/\gamma_L\hbar$ acting on the orbital moment $\mathbf{m}_L = \gamma_L\hbar\mathbf{L}$. Hence from eqn (14.2) the equations of motion will be

$$\left.\begin{aligned}\dot{\mathbf{L}} &= \gamma_L\mathbf{L}\wedge\mathbf{B}_S = \gamma_L\mathbf{L}\wedge(-\lambda\mathbf{S}/\gamma_L\hbar) = -(\lambda/\hbar)(\mathbf{L}\wedge\mathbf{S}),\\ \dot{\mathbf{S}} &= \gamma_S\mathbf{S}\wedge\mathbf{B}_L = \gamma_S\mathbf{S}\wedge(-\lambda\mathbf{L}/\gamma_S\hbar) = -(\lambda/\hbar)(\mathbf{S}\wedge\mathbf{L}).\end{aligned}\right\} \quad (14.9)$$

We note that the couples are equal and opposite, as they must be since no external couple acts on the system. Since $\mathbf{L}+\mathbf{S} = \mathbf{J}$, and

$$\mathbf{L}\wedge\mathbf{L} = \mathbf{S}\wedge\mathbf{S} = 0,$$

we have

$$\left.\begin{aligned}\dot{\mathbf{L}} &= -(\lambda/\hbar)(\mathbf{L}\wedge\mathbf{S}+\mathbf{L}\wedge\mathbf{L}) = -(\lambda/\hbar)(\mathbf{L}\wedge\mathbf{J}),\\ \dot{\mathbf{S}} &= -(\lambda/\hbar)(\mathbf{S}\wedge\mathbf{L}+\mathbf{S}\wedge\mathbf{S}) = -(\lambda/\hbar)(\mathbf{S}\wedge\mathbf{J}),\end{aligned}\right\} \quad (14.10)$$

showing that $\mathbf{L}$ and $\mathbf{S}$ each precess about $\mathbf{J}$ with angular velocity $\lambda\mathbf{J}/\hbar$.

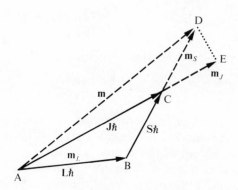

FIG. 14.6. Vector coupling of angular momentum vectors $\mathbf{L}\hbar$, $\mathbf{S}\hbar$ to form resultant $\mathbf{J}\hbar$ (represented by AB, BC, and AC), and of the associated magnetic moments $\mathbf{m}_L$, $\mathbf{m}_S$, and $\mathbf{m}$ (represented by AB, BD = $g_S$BC, AD). AE is the projection $\mathbf{m}_J$ of $\mathbf{m}$ on $\mathbf{J}\hbar$.

The vector addition of $L\hbar$ and $S\hbar$ to form $J\hbar$ is shown in Fig. 14.6. The magnetic moments $\mathbf{m}_L$, $\mathbf{m}_S$ are parallel and proportional to $L\hbar$ and $S\hbar$; in the figure they are represented by the lines AB (the same as $L\hbar$), and BD, which is $g_S \times BC = g_S S\hbar$. The resultant magnetic moment $\mathbf{m}$, given by AD, must precess around $J\hbar$ at the same rate as $L\hbar$ and $S\hbar$. It therefore has a component $\mathbf{m}_J$, given by AE, and a precessing component ED which averages to zero over a time long compared with the precession period. When a small external field of flux density $\mathbf{B}$ is applied, its main interaction is with the component $\mathbf{m}_J$, which is a fixed vector when $\mathbf{B} = 0$. We may find the value of $\mathbf{m}_J$ in quantum mechanics by the following method. The energy of interaction with the flux density $\mathbf{B}$ is given by

$$W = -\mathbf{m}_L \cdot \mathbf{B} - \mathbf{m}_S \cdot \mathbf{B} = \mu_B(\mathbf{L} + g_S\mathbf{S}) \cdot \mathbf{B}. \qquad (14.11)$$

If we write $\mathbf{m}_J = -g_J\mu_B\mathbf{J}$, where $g_J$ is a number yet to be found, its energy of interaction with $\mathbf{B}$ is

$$W = -\mathbf{m}_J \cdot \mathbf{B} = g_J\mu_B(\mathbf{J} \cdot \mathbf{B}), \qquad (14.12)$$

which is identical with (14.11) provided we write

$$\mathbf{L} + g_S\mathbf{S} = g_J\mathbf{J}. \qquad (14.13)$$

On multiplying this equation by $\mathbf{J}$ we obtain

$$\mathbf{L} \cdot \mathbf{J} + g_S(\mathbf{S} \cdot \mathbf{J}) = g_J(\mathbf{J} \cdot \mathbf{J}) = g_J J(J+1), \qquad (14.14)$$

and by squaring the quantities $\mathbf{S} = \mathbf{J} - \mathbf{L}$ and $\mathbf{L} = \mathbf{J} - \mathbf{S}$ we have

$$2(\mathbf{L} \cdot \mathbf{J}) = \mathbf{J} \cdot \mathbf{J} + \mathbf{L} \cdot \mathbf{L} - \mathbf{S} \cdot \mathbf{S} = J(J+1) + L(L+1) - S(S+1),$$
$$2(\mathbf{J} \cdot \mathbf{J}) = \mathbf{J} \cdot \mathbf{J} + \mathbf{S} \cdot \mathbf{S} - \mathbf{L} \cdot \mathbf{L} = J(J+1) + S(S+1 - L(L+1). \qquad (14.15)$$

Hence

$$g_J = \frac{(g_S+1)J(J+1) - (g_S-1)\{L(L+1) - S(S+1)\}}{2J(J+1)}, \qquad (14.16)$$

and if we make the approximation of writing $g_S = 2$,

$$g_J = \frac{3}{2} + \frac{S(S+1) - L(L+1)}{2J(J+1)}. \qquad (14.17)$$

It is easy to see that if $S$ or $L$ is zero, so that $J = L$ or $J = S$, then $g$ is 1 or 2 respectively, corresponding to the cases of 'orbit only' and 'spin only'. The value of $g_J$ is known as the 'Landé $g$-factor'.

When an atom such as we have been considering is placed in an external magnetic field, the behaviour of the angular-momentum vectors in general will be rather complicated. The reason is that each of the magnetic moments associated with orbit and spin is acted on by the magnetic field due to the magnetic moment of the other as well as the external magnetic field. No simple description of the motion is possible when these fields are of the same order of magnitude, but when one is much larger than the other an approximate treatment is possible. We shall consider only the case when the external field is very small compared with the field due to the spin–orbit coupling. The vectors **L**, **S** then precess round **J** as in the case of zero external field, but **J** is no longer stationary in space, its motion being a precession round the external field of flux density **B**. The precession of **L**, **S** about **J** is at a much higher frequency than that of **J** about **B**, since the external field is small compared with that due to the spin–orbit coupling, and we may therefore picture the components of the magnetic moment precessing around **J** as averaging to zero, leaving only the steady component along **J**. This is acted on by the external field to give the precession of **J** about **B**, at an angular velocity $\boldsymbol{\omega} = -g_J(-e/2m_e)\mathbf{B}$, where $g_J$ is the Landé g-factor. This is identical with the general result of § 14.1, if we take

$$\gamma = -g_J(e/2m_e) = -g_J\mu_B/\hbar.$$

The quantization rule for the projection of $J\hbar$ on the flux density **B** is similar to the previous rules for other angular momentum vectors. The projection has the values $M_J\hbar$, where $M_J$ takes the values $J, J-1, J-2,...,$ $-(J-1), -J$. The component of the magnetic dipole moment of the atom parallel to the field has the value $-M_Jg_J\mu_B$, and the energy is

$$W_{M_J} = -\mathbf{m}_J \cdot \mathbf{B} = M_Jg_J\mu_B B. \qquad (14.18)$$

Thus the $2J+1$ levels with different values of $M_J$ are split in energy by the application of a magnetic field, but have the same energy when $B = 0$. In the latter case they are said to be 'degenerate', and the application of a field 'lifts the degeneracy'. This 'Zeeman splitting' is illustrated in Fig. 14.7 for the $^4F$ states of the $Cr^{3+}$ ion in a flux density $B = 10$ T. The value of the spin–orbit coupling parameter $\lambda$ is about $8\cdot7\times10^3\,\mathrm{m}^{-1}$ for this ion (see Problem 14.7), and the frequency of precession of **L**, **S** about **J** is $\lambda J/h = 4\times10^6\,\mathrm{MHz}$ in the lowest state $J = \frac{3}{2}$. In contrast the frequency of precession of **J** about **B** in the $J = \frac{3}{2}$ state is only abut $6\times10^4\,\mathrm{MHz}$ in a field of $B = 10$ T. Thus our assumption of a very fast precession of **L**, **S** about **J**, and a much slower precession of **J** about **B**, will be valid for all fields of ordinary magnitude. In Fig. 14.7 this corresponds to the fact that the Zeeman splittings between the states of different $M_J$ (but the same $J$) are very small compared with the separation between states of different $J$. It is only when this inequality holds that the energy of a Zeeman sublevel is linearly proportional to the applied field (eqn (14.18)); when the Zeeman energy $(\sim\mu_B B)$ is comparable with $\lambda$ the behaviour of the energy levels is more complicated. In the opposite extreme when $\mu_B B \gg \lambda$, the coupling

FIG. 14.7. Zeeman splitting of the $^4F$ states of the $Cr^{3+}$ ion (see Fig. 14.5) in a magnetic flux density $B = 10$ T. Note that the Zeeman splittings are very much smaller relative to the spin–orbit splittings (separation of states of different $J$) than the figure suggests.

between **L** and **S** is broken down, and each tends to precess independently about the external field; this is known as the Paschen–Back effect, and can be observed only in very high fields for light atoms where the spin–orbit coupling is small.

## 14.4. The measurement of atomic magnetic moments—the Stern–Gerlach experiment

The spatial quantization of angular momentum (that is, the fact that $M_J$ can have only a discrete number of values, and not a continuous range, as in classical theory) was directly demonstrated in the celebrated atomic beam experiment of Stern and Gerlach in 1922, which also made possible the direct measurement of the magnetic moment of an atom. Though this experiment has been succeeded by more refined and accurate methods, it remains an historical landmark; developments of this method, originally due to Rabi and his colleagues, have made experiments with atomic and molecular beams the basis of the extremely accurate knowledge we now possess about atomic magnetic moments, and their interaction with the magnetic moment of the nucleus.

A molecular or atomic beam is a beam of molecules or atoms moving with thermal velocities in a given direction. It is formed by heating the substance in an oven until its vapour pressure is about 1 Pa, the oven being in a highly evacuated enclosure. Atoms or molecules effuse through a narrow orifice (see Fig. 14.8), and if the pressure is so low that the mean free path is large compared with its dimensions no collisions occur in the orifice and all molecules will be moving in substantially the same direction. The angular diameter of the beam is then further limited by the slits $S_1$, $S_2$. The total path traversed by the beam in the apparatus may be up to $0.5$ m, and the pressure must be so low ($<10^{-4}$ Pa) that very few collisions occur to scatter the molecules out of the beam.

In order to determine the magnetic moment of an atom, Stern and Gerlach deflected the beam in an inhomogeneous magnetic field. This was obtained from a magnet with one wedge-shaped pole piece, giving a field gradient $\partial B/\partial z$ in the $z$ direction, which is perpendicular to the path of the beam. The beam traversed the inhomogeneous field for a distance $l$ and then struck a detector, which in the early experiments was simply a target cooled in liquid air on which the molecules condensed. If $m_z$ is the component parallel to $\partial B/\partial z$ of the magnetic moment of the atom, then the force exerted on the atom in the $z$ direction is $m_z(\partial B/\partial z)$, and for

FIG. 14.8. Experimental arrangement for the Stern–Gerlach experiment.
M   end view of pole pieces producing inhomogeneous field,
D   density of trace on detector plate for atoms in doublet ground state, for example, Ag, $^2S_{\frac{1}{2}}$,
$S_1S_2$  narrow collimating slits.

a molecule of mass $M$ and velocity $v$ the deflection $s$ after traversing the field gradient is

$$s = \frac{\frac{1}{2}(l/v)^2 \mathfrak{m}_z(\partial B/\partial z)}{M} = \frac{l^2 \mathfrak{m}_z(\partial B/\partial z)}{2Mv^2}. \qquad (14.19)$$

To obtain appreciable deflections (of the order of a millimetre) fields of about 1 T with gradients of about 1000 T m$^{-1}$ are required. The deflection is inversely proportional to the thermal energy $\frac{1}{2}Mv^2$ of the molecules, and the traces are thus spread out owing to the distribution of velocities appropriate to the temperature of the oven. This limits the accuracy of this type of experiment, but with atoms such as sodium or silver, which are both in $^2S_{\frac{1}{2}}$ states, two distinct traces were obtained, with deflections appropriate to values of $\mathfrak{m}_z$ equal to $\pm 1 \mu_B$. Thus both the existence of spatial quantization corresponding to $M_J = \pm\frac{1}{2}$ and the magnetic moment of nearly 1 $\mu_B$ associated with electron spin of $\frac{1}{2}\hbar$ were confirmed. Later modifications of these experiments gave fairly precise values of atomic magnetic moments, but much higher accuracy has been obtained by the magnetic-resonance method, outlined in Chapter 24.

## 14.5. Curie's law and the approach to saturation

A theoretical derivation of Curie's law due to Langevin was given in Chapter 6. This was based on a classical approach in its use of Boltzmann statistics, but used the idea of the existence of permanent magnetic moments of fixed values. This latter assumption is not in accordance with classical theory, for we should then expect a continuous range of magnetic moments from $-\infty$ to $+\infty$. It was shown independently by Bohr and by Miss van Leeuwen that if such a continuous range is assumed, the paramagnetic and diamagnetic contributions to the susceptibility of any system should be exactly equal and opposite, and thus, in a strictly classical calculation, the susceptibility would be zero. The quantum-mechanical approach outlined in § 14.2 shows that finite permanent magnetic dipoles do exist in atoms, and we must now examine how the Langevin calculation must be modified to take account of the fact that only a finite number of projections of the moment on an external field are allowed.

We assume that all the atoms are in the same spectroscopic state $L$, $S$, $J$, this being the ground state of the atom. It was shown in § 14.3 that the potential energy $W$ of such an atom in a magnetic field is $M_J g_J \mu_B B$, where $M_J$ is the magnetic quantum number, and $g_J$ the Landé factor appropriate to the spectroscopic state of the atoms. As in classical theory, the probability of an atom being in a state with an energy $W$ is proportional to $\exp(-W/kT)$, and for a given value of $M_J$ this is therefore proportional to $\exp(-M_J g_J \mu_B B/kT)$. Thus the fraction of all atoms in this state is

$\exp(-M_J g_J \mu_B B/kT)/\sum \exp(-M_J g_J \mu_B B/kT)$, where the summation is over all values of $M_J$. The component of the atomic magnetic moment parallel to **B** is $-M_J g_J \mu_B$, and the total magnetic moment of a system of $n$ atoms will therefore be

$$n\bar{m} = n \frac{\sum (-M_J g_J \mu_B)\exp(-M_J g_J \mu_B B/kT)}{\sum \exp(-M_J g_J \mu_B B/kT)}, \qquad (14.20)$$

where the summation in each case is over all values of $M_J$ from $+J$ to $-J$. This expression is rather clumsy to handle but it may be shown by an algebraic reduction that it reduces to the form (see Problem 14.1)

$$n\bar{m} = n g_J \mu_B J \left\{ \frac{2J+1}{2J} \coth\left(\frac{2J+1}{2J}y\right) - \frac{1}{2J}\coth\left(\frac{y}{2J}\right) \right\}, \qquad (14.21)$$

where $y = g_J \mu_B JB/kT$. The expression in brackets in eqn (14.21) is called the Brillouin function. When $J$ becomes very large it approaches as a limit the Langevin function $\{\coth y - (1/y)\}$, as we should expect from the fact that a summation over a large number of terms can be replaced by an integration, as used in the derivation in § 6.3.

At normal field strengths and ordinary temperatures, the value of $y$ is very small; at $B = 1$ T and $T = 290$ K, with $g_J = 2$ and $J = \frac{1}{2}$, $y$ is about 0·002. It is then possible to make a series expansion of either eqn (14.20) or (14.21). To the first order the former equation becomes

$$n\bar{m} = -n g_J \mu_B \sum_{-J}^{+J} M_J(1-M_J g_J \mu_B B/kT) \bigg/ \sum_{-J}^{+J} (1-M_J g_J \mu_B B/kT)$$

$$= \frac{n g_J^2 \mu_B^2 B}{(2J+1)kT} \sum_{-J}^{+J} M_J^2.$$

The summation amounts to $\frac{1}{3}J(J+1)(2J+1)$, and the susceptibility is thus

$$\chi = \frac{n\bar{m}}{H} = \frac{\mu_0 n g_J^2 \mu_B^2 J(J+1)}{3kT} = \frac{C}{T}, \qquad (14.22)$$

which is the same as the classical expression (eqn (6.13)) if we write

$$m^2 = g_J^2 \mu_B^2 J(J+1).$$

This is just the value of the square of the atomic magnetic moment expected on the quantum-mechanical theory, but at higher field strengths there is a difference between the Langevin and Brillouin functions. In particular, the limiting saturation moment reached at high field strengths and low temperatures (large values of $y$) is $g_J \mu_B J$, and not $g_J \mu_B \{J(J+1)\}^{\frac{1}{2}}$. This is because the greatest component of each moment parallel to **B** is $g_J \mu_B J$, and the actual magnetic moment always precesses at a finite angle to the field. The correctness of the Brillouin function has been verified in a number of experiments, representative results being those of Henry

FIG. 14.9. Plot of average magnetic moment per ion m against $B/T$ for (a) potassium chromium alum $(KCr(SO_4)_2 \cdot 12H_2O)(J = S = \frac{3}{2})$, (b) iron ammonium alum $(NH_4Fe(SO_4)_2 \cdot 12H_2O)(J = S = \frac{5}{2})$, and (c) gadolinium sulphate octahydrate $(Gd_2(SO_4)_3 \cdot 8H_2O)(J = S = \frac{7}{2})$.

shown in Fig. 14.9. Note that the close approach to saturation is obtained by the combination of high field (5 T) and low temperature (4 K and lower).

## 14.6. Susceptibility of paramagnetic solids—the 4f group

The theory given above applies only to an assembly of free atoms, and the situation is rather different for matter in the aggregated state, because of the large forces exerted by the atoms on each other. These are mainly electrical in origin, and are generally far stronger than the interaction between the magnetic moment of an atom and an external magnetic field. We must therefore consider the effect of the interatomic forces first, and we shall find that, whereas most free atoms have permanent magnetic dipole moments, most bound atoms do not. This is due to the fact that the exchange forces between electrons in different atoms are nearly always of opposite sign to those between electrons in the same atom, and they therefore tend to make the electron spins line up antiparallel, giving no resultant spin whenever possible. Thus in the formation of a homopolar molecule such as $N_2$, the binding electrons from the nitrogen atoms are shared between the two atoms with their spins antiparallel; the orbits are also arranged so that the electrons have no resultant orbital angular momentum. The total angular momentum is therefore zero and the molecule has no permanent magnetic moment, though there will always be an induced negative moment when a magnetic field is applied, giving rise to diamagnetism. In heteropolar binding a molecule such as NaCl is formed of the two ions $Na^+$ and $Cl^-$, both of which have closed electron shells; thus again there is no resultant magnetic moment. Though this picture of molecule formation is oversimplified, and in general we have a mixture of homopolar and heteropolar binding, the general result of no permanent magnetic moment is still true. Thus the only common gases with permanent moments are NO and $NO_2$, which have an odd number of electrons, so that a resultant spin must remain (for NO there is also one unit of orbital angular momentum about the molecular axis); and $O_2$, where two electron spins are unpaired, giving oxygen gas a paramagnetism appropriate to $S = 1$, $g_J = 2$.

In the solid state most substances consist of ions with closed shells of electrons and are therefore diamagnetic. The main exceptions to this rule arise in compounds of the so-called 'transition elements', where an electron shell is in process of being filled. Such elements are marked by their possession of more than one chemical valency, and some (if not all) of their ions of different valency have unclosed shells, and hence a permanent magnetic moment. Since it is the electrons in the unclosed shell which determine the magnetic properties, the paramagnetism is typical of the ion, not the atom. Thus ions with the same electron configuration, even if formed from different atoms, have similar magnetic properties. These ions may be labelled by the spectroscopic description of the electron shell which is partly full; these are, 3d (iron group), 4d (palladium group), 4f (lanthanide group), 5d (platinum group), and 5f (actinide group). The titles in brackets are often used as being more descriptive, though less precise.

We consider first the 4f group, whose paramagnetism in the solid state is closest to that of an assembly of free ions. The spectroscopic states of the free ions of the 4f shell are shown in Table 14.1. It will be seen that they conform to Hund's rules, the values of first $S$ and then $L$ being the greatest possible consistent with the Pauli exclusion principle. The ground state has the smallest possible value of $J$ in the first half, and the largest value in the second half, as the spin–orbit coupling interaction changes sign when the shell is more than half full. The experimental values of $p^2$ have the following significance. If we assume that the susceptibility of a substance obeys Curie's law, we may write

$$\chi = \mu_0 n m^2 / 3kT = \mu_0 n p^2 \mu_B^2 / 3kT. \qquad (14.23)$$

Here $p$ is called the effective Bohr magneton number, and by comparison with eqn (14.22) we see that for an assembly of free ions

$$p^2 = g_J^2 J(J+1).$$

It is convenient to give the experimental results in terms of $p^2$, since this facilitates comparison with the theory, but it must be remembered that though we can always calculate a value of $p^2$ from the susceptibility at a given temperature, it has little significance if the susceptibility does not obey Curie's law. The latter can be established by measuring the susceptibility over a range of temperature. In this connection it must be emphasized that only measurements on 'magnetically dilute' salts are significant; by this phrase is meant salts where the paramagnetic ions are fairly far apart so that mutual interaction between them may be neglected (see Problem 14.2 and § 15.1). This condition is generally fulfilled for hydrated salts, and the values of $p^2$ in Tables 14.1, 14.2 are for salts where the effect of mutual interaction on the susceptibility is appreciable only at very low temperatures.

The average values of $p^2$ at room temperature measured experimentally and shown in the last column of Table 14.1 are in fair agreement with those calculated for an assembly of free ions with angular momentum $J$. The reason for this is that states of different $J$ are separated by energies corresponding to several thousand degrees Kelvin, so that (1)

TABLE 14.1

*Comparison of theoretical and measured values of $p^2$ for trivalent rare-earth ions*

| Number of electrons in 4f shell | Ion | Ground spectroscopic state | Theoretical values | | | | | Average experimental value of $p^2$ |
|---|---|---|---|---|---|---|---|---|
| | | | $S$ | $L$ | $J$ | $g_J$ | $p^2 = g_J^2 J(J+1)$ | |
| 0 | $La^{3+}$ | $^1S_0$ | 0 | 0 | 0 | — | 0 | 0 |
| 1 | $Ce^{3+}$ | $^2F_{\frac{5}{2}}$ | $\frac{1}{2}$ | 3 | $\frac{5}{2}$ | $\frac{6}{7}$ | 6·43 | 6 |
| 2 | $Pr^{3+}$ | $^3H_4$ | 1 | 5 | 4 | $\frac{4}{5}$ | 12·8 | 12 |
| 3 | $Nd^{3+}$ | $^4I_{\frac{9}{2}}$ | $\frac{3}{2}$ | 6 | $\frac{9}{2}$ | $\frac{8}{11}$ | 13·1 | 12 |
| 4 | $Pm^{3+}$ | $^5I_4$ | 2 | 6 | 4 | $\frac{3}{5}$ | 7·2 | — |
| 5 | $Sm^{3+}$ | $^6H_{\frac{5}{2}}$ | $\frac{5}{2}$ | 5 | $\frac{5}{2}$ | $\frac{2}{7}$ | 0·71 (2·5) | 2·4 |
| 6 | $Eu^{3+}$ | $^7F_0$ | 3 | 3 | 0 | — | 0 (12) | 12·6 |
| 7 | $Gd^{3+}$ | $^8S_{\frac{7}{2}}$ | $\frac{7}{2}$ | 0 | $\frac{7}{2}$ | 2 | 63 | 63 |
| 8 | $Tb^{3+}$ | $^7F_6$ | 3 | 3 | 6 | $\frac{3}{2}$ | 94·5 | 92 |
| 9 | $Dy^{3+}$ | $^6H_{\frac{15}{2}}$ | $\frac{5}{2}$ | 5 | $\frac{15}{2}$ | $\frac{4}{3}$ | 113 | 110 |
| 10 | $Ho^{3+}$ | $^5I_8$ | 2 | 6 | 8 | $\frac{5}{4}$ | 112 | 110 |
| 11 | $Er^{3+}$ | $^6I_{\frac{15}{2}}$ | $\frac{3}{2}$ | 6 | $\frac{15}{2}$ | $\frac{6}{5}$ | 92 | 90 |
| 12 | $Tm^{3+}$ | $^3H_6$ | 1 | 5 | 6 | $\frac{7}{6}$ | 57 | 52 |
| 13 | $Yb^{3+}$ | $^2F_{\frac{7}{2}}$ | $\frac{1}{2}$ | 3 | $\frac{7}{2}$ | $\frac{8}{7}$ | 20·6 | 19 |
| 14 | $Lu^{3+}$ | $^1S_0$ | 0 | 0 | 0 | — | 0 | 0 |

The values given in parentheses for $Sm^{3+}$ and $Eu^{3+}$ are those calculated by Van Vleck allowing for population of excited states with higher values of $J$, at $T = 293$ K.

only the ground state is occupied at ordinary temperatures, and (2) the precessing component of the magnetic moment represented by DE in Fig. 14.6 makes only a small contribution under these circumstances. The exceptions are the ions $4f^5$, $4f^6$ in which excited levels with higher values of $J$ are exceptionally low in energy (see Problem 14.7). Van Vleck has shown that much better agreement is obtained when allowance is made for the presence of the excited levels, and his values, calculated for $T = 273$ K, are shown in parentheses.

For the ions $Gd^{3+}$, $Eu^{2+}$ whose spectroscopic state is $4f^7$, $^8S_{\frac{7}{2}}$ the susceptibility follows Curie's law down to temperatures below 1 K, but for the other ions the susceptibility shows marked departures from this law at much higher temperatures. The reason for this is that we cannot neglect the influence of the charged ions which surround each paramagnetic ion in the solid state. In a magnetically dilute salt these immediate neighbours carry no permanent magnetic moment (they are diamagnetic ions such as $F^-$, $O^{2-}$, etc.), but they are electrically charged, and have an electrostatic interaction with the 4f electrons which are responsible for the paramagnetism. To a good approximation the 4f electrons can be regarded as

FIG. 14.10. Energy levels and optical transitions of the $Ce^{3+}$ ion ($4f^1$, $L = 3$, $S = \frac{1}{2}$) in $LaCl_3$. The splittings of both the $J = \frac{5}{2}$ and $J = \frac{7}{2}$ manifolds into series of doublet levels can be fitted within the experimental error by a crystal field of hexagonal symmetry and a spin–orbit coupling constant $\zeta = 6 \cdot 26 \times 10^4$ $m^{-1}$ in eqn (14.7). The over-all splitting of the $^2F_{5/2}$ levels is about 160 K.

moving in an electric field set up by the neighbouring ions, known as the 'crystalline electric field'. The energy of interaction with this field is smaller, for ions of the 4f group, than the Coulomb, exchange, and spin–orbit interactions within the paramagnetic ion itself; it gives rise to a 'Stark' splitting of the $2J+1$ levels of the ion. This is similar in nature to the effect of an external electric field on the spectrum of an atom, first investigated in detail by Stark, but is considerably more complex because the electrostatic potential set up by the neighbours varies in a complicated way over the space occupied by the 4f electrons. The over-all splittings of the $2J+1$ levels of a 4f ion are generally of the order of a few hundred degrees Kelvin, as shown in Fig. 14.10 for $Ce^{3+}$, $4f^1$ in $LaCl_3$. The susceptibility is rather insensitive to such splittings, and approaches that of the free ion at temperatures where most of the levels are appreciably populated. At low temperatures where only the very lowest levels are populated, the susceptibility can be very different from that of the free ion, and in a single crystal may be highly anisotropic.

At first sight it may appear surprising that the crystalline electric field can have such a marked effect on the magnetic properties. The basic reason is that the wavefunctions corresponding to different values of the orbital magnetic quantum number $M_L$ have different angular dependencies, that is, for each value of $M_L$ the distribution of electronic charge has a different shape, and hence acquires a different electrostatic energy in

FIG. 14.11. The values of $p^2$ (parallel and perpendicular) for a single crystal of erbium ethylsulphate ($Er(C_2H_5SO_4)_3 \cdot 9H_2O$). This forms hexagonal crystals, and the susceptibility is symmetrical about the hexagonal axis. The ground state of the $Er^{3+}$ ion is $^4I_{15/2}$, and this is split by the crystalline electric field into eight doublets lying at 0, 61, 108, 159, 249, 301, 375, and 438 K.

the crystalline electric field. Thus the primary interaction is associated with the electronic orbit, and the interaction is zero (except for a small residual effect due to a slight departure from pure $L$–$S$ coupling) for an ion such as $Eu^{2+}$ or $Gd^{3+}$ with a half-filled shell carrying no orbital angular momentum ($L = 0$). For the other ions the coupling together of $L$ and $S$ means that states of different $M_J$ have a different charge distribution, so that the crystalline electric field splits the $2J+1$ states which otherwise have the same energy in the absence of a magnetic field.

An important restriction on this splitting occurs for ions with an odd number of electrons, which have half-integral values of $S$ and hence also of $J$; in this case the states occur always in pairs which have the same charge distribution and differ only in the orientation of the magnetic moment. Each pair must thus retain the same energy in an electric field, though they can be split in a magnetic field. This result was proved in a theorem of Kramers and the double degeneracy of such states in an electric field is known as 'Kramers' degeneracy'. Typical examples are $Ce^{3+}$, $4f^1$ (Fig. 14.10), and $Er^{3+}$, $4f^{11}$. At low temperatures where just one such pair of states is occupied, the value of $p^2$ has a temperature variation of the form $A+BT$ (see Problem 14.3); in Fig. 14.11 the variation of $p^2$ for a compound of $Er^{3+}$ near $T = 0$ is of this form both for the perpendicular and for the parallel direction (in the latter case $g_{\parallel}^2$ is so small that $A$ appears to be zero in the figure). For a 'non-Kramers' ion, with an even number of electrons and an integral value of $J$, the ground state left by the crystal field may be a singlet state. This has no permanent

FIG. 14.12. Plot of $\chi^{-1}$ against $T$ for TmSb and for $Tm_{0.53}\,Y_{0.47}$ Sb, in which about half of the paramagnetic $Tm^{3+}$ ions are replaced by diamagnetic $Y^{3+}$ ions. The continuous line is a theoretical fit using a cubic crystal field. The vertical axis is in units where $\chi^{-1} = 1$ when a field of 1 T produces a magnetic moment of 1 $\mu_B$ per terbium ion.

magnetic moment, but when a field is applied it acquires an induced moment which is proportional to the field strength in small fields, but independent of temperature. The susceptibility then becomes independent of temperature, a phenomenon first explained by Van Vleck. A clear example of this effect is the cubic compound TmSb, where the ions occupy a regular crystal lattice with the NaCl structure (it is not an alloy). The Sb antimony ion is not magnetic, but the ion $Tm^{3+}$, $4f^{12}$ has $J = 6$. The inverse susceptibility $\chi^{-1}$ is plotted in Fig. 14.12; it can be fitted accurately with a cubic crystal field whose potential is of the form given in eqn (14.24).

## 14.7. Susceptibility of paramagnetic solids—the 3d group

The spectroscopic ground states of the free ions of the 3d shell, shown in Table 14.2, again follow Hund's rules. An assembly of free ions should give a susceptibility corresponding to $p^2 = g_J^2 J(J+1)$, but a comparison of the values of this quantity with the experimental values for solids shows a striking disagreement. In fact the experimental values lie much closer to $p^2 = 4S(S+1)$, the value which would be expected if there were no orbital angular momentum and the magnetism were due entirely to the electron spin. This is clearly brought out in Fig. 14.13, in which average experimental values of $p^2$ are plotted together with the quantities $g_J^2 J(J+1)$ and $4S(S+1)$.

TABLE 14.2

| Number of electrons in 3d shell | Ion | Ground state | $S$ | $L$ | $J$ | $g_J^2 J(J+1)$ | $p^2$ (experimental) | $4S(S+1)$ |
|---|---|---|---|---|---|---|---|---|
| 0 | $K^+, Ca^{2+}, Sc^{3+},$ $Ti^{4+}, V^{5+}$ | $^1S_0$ | 0 | 0 | 0 | 0 | 0 | 0 |
| 1 | $Ti^{3+}, V^{4+}$ | $^2D_{\frac{3}{2}}$ | $\frac{1}{2}$ | 2 | $\frac{3}{2}$ | 2·4 | 2·9 | 3 |
| 2 | $V^{3+}$ | $^3F_2$ | 1 | 3 | 2 | 2·67 | 6·8 | 8 |
| 3 | $V^{2+}, Cr^{3+}$ | $^4F_{\frac{3}{2}}$ | $\frac{3}{2}$ | 3 | $\frac{3}{2}$ | 0·6 | 14·8 | 15 |
| 4 | $Cr^{2+}, Mn^{3+}$ | $^5D_0$ | 2 | 2 | 0 | 0 | (23·3) | 24 |
| 5 | $Mn^{2+}, Fe^{3+}$ | $^6S_{\frac{5}{2}}$ | $\frac{5}{2}$ | 0 | $\frac{5}{2}$ | 35 | 34·0 | 35 |
| 6 | $Fe^{2+}$ | $^5D_4$ | 2 | 2 | 4 | 45 | 28·7 | 24 |
| 7 | $Co^{2+}$ | $^4F_{\frac{9}{2}}$ | $\frac{3}{2}$ | 3 | $\frac{9}{2}$ | 44 | 24·0 | 15 |
| 8 | $Ni^{2+}$ | $^3F_4$ | 1 | 3 | 4 | 31·3 | 9·7 | 8 |
| 9 | $Cu^{2+}$ | $^2D_{\frac{5}{2}}$ | $\frac{1}{2}$ | 2 | $\frac{5}{2}$ | 12·6 | 3·35 | 3 |
| 10 | $Cu^+, Zn^{2+}$ | $^1S_0$ | 0 | 0 | 0 | 0 | 0 | 0 |

The values of $p^2$ (at 300 K) are for double sulphates of the type $M''M_2'(SO_4)_2 \cdot 6H_2O$ or $M'''M'(SO_4)_2 \cdot 12H_2O$ (where $M''$ = divalent paramagnetic ion, $M'''$ = trivalent paramagnetic ion, $M'$ = monovalent diamagnetic ion). In these salts the distance between nearest paramagnetic ions is at least 0·6 nm, and interaction between them is negligible. The value in parentheses is for $CrSO_4 \cdot 6H_2O$: no double sulphate of $Cr^{2+}$ has been measured.

FIG. 14.13. Experimental (at 300 K) and calculated values of $p^2$ for the 3d group. In the second half of the group the orbital angular momentum is less effectively quenched than in the first half, so that the values of $p^2$ lie noticeably above the spin-only values.

This phenomenon, known as the 'quenching' of the orbital magnetism, is a result of the crystalline electric field. In the lanthanide group the 4f electrons, which are responsible for the paramagnetism, are fairly deep seated in the atom, but in the iron group the 3d electrons are in an outer shell which has a very much larger interaction with the crystalline electric field of neighbouring charged ions. On the other hand, the spin–orbit coupling in the 3d group is considerably smaller than in the 4f group. The result is that interaction between the orbit and the crystalline electric field is a good deal stronger than the spin–orbit coupling for 3d ions, so that the orbital momentum is primarily coupled to the crystal field, and it is no longer correct to regard $L$ and $S$ as coupled to form a resultant $J$. The quantitative expression of this situation is that the $2L+1$ orbital states are split in the crystal field, and have energies differing by about $10^6 \, \mathrm{m}^{-1}$, which is much larger than the spin–orbit splittings (of the order of $10^4$–$10^5 \, \mathrm{m}^{-1}$) between the states of different $J$ in the free ion. The simplest case to consider is that where the crystal field splitting of the orbital levels gives a singlet state as the lowest level. Such a state has no magnetic moment, and the orbital moment is completely 'quenched'. On the other hand, the electron spin has no direct interaction with the crystalline electric field, and remains free to orient itself in a magnetic field. Thus, in this case, the susceptibility would correspond exactly to the 'spin only' value of $p^2 = 4S(S+1)$ at all temperatures such that there is no appreciable population of an excited orbital state.

It can be seen from Fig. 14.13 that the values of $p^2$ do not follow exactly the 'spin only' values, particularly for ions with $d^6$ ($Fe^{2+}$) and $d^7$ ($Co^{2+}$) configurations. The basic reason for this is that the crystalline electric field does not always result in a singlet orbital state as the lowest state, but sometimes gives a group of low-lying orbital states which can make some contribution to the magnetic moment, though less than the full orbital contribution from a free ion. In principle we could calculate the splitting of the orbital levels, but in practice this is extremely difficult to do. However, the general features of the magnetic properties of salts of the 3d group are well understood, mainly through the work of Van Vleck, and we will now attempt to outline the main results of the crystal-field approach.

The size of an ion of the 3d group is such that six negatively charged ions can be packed round it; when these ions are identical, they are arranged in the form of an octahedron which is very nearly regular. In hydrated salts these six ions are commonly the oxygens of six water molecules, as shown in Fig. 14.14. In oxides such as MnO the octahedron is regular and formed by six $O^{2-}$ ions; in fluorides such as $KMnF_3$ there is a regular octahedron of $F^-$ ions. These ions are known as the 'ligand ions', and in general there is a small amount of sharing of electrons between them and the 3d ion. In the crystal field theory this is ignored, and the magnetic 3d electrons are assumed to be localized on the 3d ion, and to move in the electrostatic potential of the surrounding charged ligand ions. If the 3d ion is assumed to be at the point $(0, 0, 0)$, the ligand ions may be taken to lie at the points

$$(\pm a, 0, 0), \quad (0, \pm a, 0), \quad (0, 0, \pm a)$$

thus forming a regular octahedron. If we assign a charge $-2e$ to each oxygen ion, the electrostatic potential near the centre of the octahedron

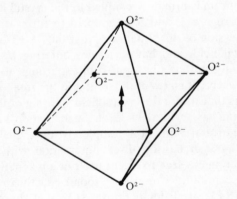

FIG. 14.14. Octahedron of oxygen ions round a paramagnetic 3d ion.

(see Problem 2.18) is

$$V = -\frac{12e}{4\pi\epsilon_0 a} - \frac{2e}{4\pi\epsilon_0}\left(\frac{35}{4a^5}\right)(x^4 + y^4 + z^4 - \tfrac{3}{5}r^4), \qquad (14.24)$$

and this will change the energy of an electron on the central ion by an amount $\int \psi^*(-eV)\psi \, d\tau$, where $\psi$ is the electronic wavefunction. This energy change is a quantitative expression of the fact that, since the electrons on the magnetic ion are negatively charged, they will have a lower energy in states where they avoid the negatively charged ligand ions as much as possible, and a higher energy when they do not, because of the electrostatic repulsion. Under the action of the potential (14.24), the correct $d$ orbitals are represented by real wavefunctions, which can be written as a radial function $f(r)$ times the following functions, which express the angular dependence in Cartesian coordinates instead of the spherical harmonics of § 2.2:

$$\left.\begin{aligned}
C_{2,0} &= \tfrac{1}{2}(2z^2 - x^2 - y^2)/r^2 \\[2mm]
\frac{1}{\sqrt{2}}(C_{2,2} + C_{2,-2}) &= \frac{\sqrt{3}}{2}(x^2 - y^2)/r^2
\end{aligned}\right\} \quad (d\gamma),$$

$$\left.\begin{aligned}
-\frac{j}{\sqrt{2}}(C_{2,2} - C_{2,-2}) &= \sqrt{3}\,xy/r^2 \\[2mm]
\frac{j}{\sqrt{2}}(C_{2,1} + C_{2,-1}) &= \sqrt{3}\,yz/r^2 \\[2mm]
-\frac{1}{\sqrt{2}}(C_{2,1} - C_{2,-1}) &= \sqrt{3}\,zx/r^2
\end{aligned}\right\} \quad (d\epsilon). \qquad (14.25)$$

The last three (known as $d\epsilon$ states) are each zero along two of the cubic axes (see Fig. 14.15), so that the charge density (which is proportional to the square of the wavefunction) is also zero along two of the axes. This gives a lower energy for these three states (by symmetry each must have the same energy in a cubic field, since $x$, $y$, $z$ are all equivalent if the octahedron is regular) than for the other two ($d\gamma$) states which have a finite density along all three cubic axes. Hence we get a splitting of the D(d) state as shown in Fig. 14.16 for $d^1$. The splitting is similar, but inverted, for $d^9$, which is one electron short of a filled d shell. A filled shell has a spherical charge distribution, and the charge distribution for $d^9$ is equivalent to a filled shell plus a 'positive hole', for which the electrostatic energy in the crystal field has the opposite sign. A half-filled shell also has spherical symmetry, with $L = 0$; for this reason the ligand field plays virtually no role in affecting the paramagnetism of $d^5$ ions, whose ground state is $^6S_{\frac{5}{2}}$. However, a $d^6$ ion, with one electron outside the half-filled shell, has a similar splitting to $d^1$, while $d^4$ corresponds to a positive hole in a half-filled shell and behaves like $d^9$ (see Fig. 14.16).

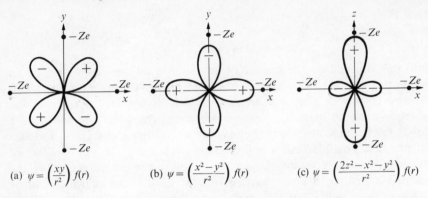

(a) $\psi = \left(\dfrac{xy}{r^2}\right) f(r)$      (b) $\psi = \left(\dfrac{x^2-y^2}{r^2}\right) f(r)$      (c) $\psi = \left(\dfrac{2z^2-x^2-y^2}{r^2}\right) f(r)$

FIG. 14.15. Angular variation of the d orbitals of eqn (14.25). (a) is a dε state (the other two dε states are similar but differently oriented); (b) and (c) are dγ states. dε and dγ functions have different symmetry properties: dγ functions do not change sign on reflection in any of the cubic planes (that is, $x \rightarrow -x$, or $y \rightarrow -y$, or $z \rightarrow -z$), while the dε functions change sign for two such reflections but not for the third.

Thus $d^1$, $d^4$, $d^6$, $d^9$ have a basically similar splitting pattern because each is equivalent to a single electron or single hole state, as far as the orbit is concerned (they do not all have the same spin).

Similarly, the remaining ions $d^2$, $d^3$, $d^7$, $d^8$ are orbitally equivalent to two-electron or two-hole states ($d^7 \equiv$ half-filled shell+2 electrons; $d^3 \equiv$ half-filled shell+2 holes; $d^8 \equiv$ filled shell+2 holes). They are all in F states, with $L = 3$, which are split by a cubic crystal field into a singlet and

FIG. 14.16. Splittings of D and F states in a crystal field of octahedral symmetry. The over-all splittings lie generally in the range $3 \times 10^6 \, \mathrm{m}^{-1}$. The strong colours of many paramagnetic compounds of the 3d group are due to absorption bands in or near the visible region of the spectrum which arise from transitions between the ground state and excited states shown above, combined with vibrational effects.

two triplet levels, as shown in Fig. 14.16. In $d^3$ and $d^8$ the lowest level is a singlet (it corresponds to a wavefunction $xyz$, which has zero density along all three cubic axes), but in $d^2$ and $d^7$ the splitting pattern is inverted.

Using these crystal field splittings, we can distinguish between two separate cases.

1. When the orbital ground state is a singlet, it has no component of angular momentum along any axis, so that the magnetism is due primarily to the spin. The susceptibility follows Curie's law very exactly (for example, the susceptibility of a chrome alum such as $CrKSO_4 \cdot 12H_2O$ does not deviate from Curie's law by more than 2 per cent between room temperature and 2 K). There are, however, two residual effects of the spin–orbit coupling: (*a*) the effective g-value differs from the free spin value by an amount of order $\lambda/\Delta$, where $\lambda$ is the spin–orbit coupling and $\Delta$ the splitting between the ground orbital level and the excited orbital level shown in Fig. 14.16. The spin–orbit constant $\lambda$ is positive if the d shell is less than half-filled, and negative if it is more than half-filled. It is also larger for the ions at the end of the group because of the increased nuclear charge. Thus the effective value of g is about 1 per cent smaller than the free spin value for $Cr^{3+}$, $d^3$, but 10 per cent higher for $Ni^{2+}$, $d^8$. (*b*) Where the spin is 1 or more, the $2S+1$ spin states may be split by amounts of order $(\lambda^2/\Delta)$, which is usually of the order of $10$–$10^3 \, m^{-1}$ (roughly $0 \cdot 1$–$1$ K). This gives a specific-heat anomaly, of which a typical example is shown in Fig. 14.17.

2. When the ground state is not a singlet, there is a first-order contribution from the orbit to the paramagnetism, but less than that for

FIG. 14.17. Magnetic specific heat anomaly of $\alpha$-nickel sulphate hexahyrate ($\alpha NiSO_4 \cdot 6H_2O$). The anomaly is associated with the spin triplet states, which lie at $0$ K, $6 \cdot 44$ K, and $7 \cdot 26$ K respectively. (After Stout and Hadley 1964.)

the free ion. When the ground state is a triplet in Fig. 14.16, it behaves like a P state with $L = 1$, and an effective $g_L$ which is $-1$ for $d^1$ and $d^6$, and $-\frac{3}{2}$ for $d^2$ and $d^7$. Thus it can interact with the spin through the spin–orbit interaction, giving states with an effective $J$ of $S-1$, $S$, and $S+1$ if $S \geqslant 1$, or $\frac{1}{2}$ and $\frac{3}{2}$ if $S = \frac{1}{2}$. The splittings between these levels are of order $10^4\,\mathrm{m}^{-1}$ ($\sim 100$ K), so that Curie's law is not obeyed because excited states become occupied as the temperature is raised. This is particularly noticeable for cobalt ($Co^{2+}$, $d^7$) salts, as shown in Fig. 14.18. The ions $d^4$, $d^9$ are exceptional because the doublet orbital states left as their ground states by the octahedral field have effectively $g_L = 0$. Thus they behave rather like case 1. Fig. 14.18 shows a typical cupric salt, $Cu^{2+}$, $d^9$, where Curie's law is obeyed closely, but the effective g-value is more than 10 per cent higher than the free spin value, giving $p^2 = 3\cdot76$ instead of the value $4S(S+1) = 3$ we would expect for a single hole ($S = \frac{1}{2}$).

A striking effect in many single crystals of paramagnetic substances of the 4f group is the high anisotropy of the susceptibility; this arises because the surroundings of the paramagnetic ion in such crystals have less than cubic symmetry. For the regular octahedron in Fig. 14.14 there would be no anisotropy, but in the 3d group this octahedron is normally somewhat distorted; as a result the orbital contributions to the magnetism depend on the direction in which the external field is applied relative to the crystal axes. The anisotropy is large when the splitting of the lowest orbital levels is small, that is, when there are also considerable departures from Curie's law, as in cobalt salts.

FIG. 14.18. Variation of $p^2$ with temperature for two iron-group salts. The change in $p^2$ with temperature for the cobalt salt is due to the presence of low-lying excited states (cf. Fig. 14.11).

## 14.8. Susceptibility of paramagnetic solids—strongly bonded compounds

Much less is known in detail about the magnetic properties of salts of the 4d and 5d groups, but in many cases it appears that the binding to the ligand ions is covalent rather than ionic in character. This is true also for a few salts of the iron group, notably the complex cyanides such as $K_3Fe(CN)_6$. Here, and in salts such as $K_2IrCl_6$, where the magnetic $Ir^{4+}$ ion has the configuration $5d^5$, the ligand ions (six CN groups in the former case, six $Cl^-$ ions in the latter) are again arranged in the form of a very nearly regular octahedron. We shall confine the discussion to this type of compound, as it affords an interesting comparison with the pure crystal field approach.

In a covalent bond the electrons are shared between the two ions concerned, in contrast with a purely ionic case where the electrons are localized on each ion. The latter is an oversimplification, and in practice there is always a small amount of covalent bonding, so that we distinguish only between weak bonding, and strong bonding. In a complex with octahedral symmetry, the $d\gamma$ states have maximum density along the cubic axes (towards the ligand ions), and can form $\sigma$-bonds with the ligand ions, while the $d\epsilon$ states can only form $\pi$-bonds. In the formation of a bond, a d-state on the magnetic ion is combined with the appropriate bonding state of the ligand ion, and the overlap of the electronic wavefunctions gives a splitting of the combined levels, in the same way as pointed out in § 12.2. The overlap is greater for the $d\gamma$ states (forming $\sigma$-bonds) than for the $d\epsilon$ states (forming $\pi$-bonds), giving the energy level diagram shown in Fig. 14.19. The lower bonding states are all filled with electrons, and behave as filled subshells. Thus the states available for magnetic electrons are the antibonding states, which are split in the same

FIG. 14.19. Splitting of the d states on the bonding model. The lowest (bonding) states are filled with electrons, and only the antibonding states are available for the magnetic electrons. In the crystal-electric-field approach the bonding states play no role, and an octahedral field splitting (see Fig. 14.16) of the d state is obtained similar to that for the antibonding states above.

way as by an octahedral crystal field. In weakly bonded compounds this splitting is about $10^6\,\mathrm{m}^{-1}$, as mentioned in the previous section, but in the strongly bonded compounds it is very much larger, so that the latter behave as though subjected to a very much stronger crystalline electric field. However, the approach from the bonding viewpoint is more correct, since it allows the magnetic electron wavefunctions to spread out from the central ion on to the ligand ions, for which there is direct experimental evidence from measurements of the hyperfine interaction between the magnetic electrons and the nuclear moments of the ligand ions.

In this more general approach, allowing for bonding, the splitting between the d$\epsilon$ states and the d$\gamma$ antibonding states is ascribed to the 'ligand field', and it is interesting to contrast the two cases of small and large ligand field. Here the comparison is with the electrostatic and exchange energy which is responsible for $L$–$S$ coupling, and when the ligand field splitting is large compared with this exchange energy we must regard the $L$–$S$ coupling as broken. The way in which the d$\epsilon$ and d$\gamma$ antibonding states are occupied by the magnetic electrons in the two cases is determined by the competition between the exchange energy (which favours parallel orientation of the electron spins) and the ligand field splitting (which favours electrons going into the d$\epsilon$ states because of their lower energy).

We may represent each orbital state by a pair of square boxes, as in Fig. 14.20, into each of which we can put one electron, with spin up or

FIG. 14.20. Single-electron model of filling of d states split by an octahedral crystal field. The exchange energy favours parallel spin arrangements (subject to the exclusion principle); the crystal field splitting favours electrons in the d$\epsilon$ states.

down. With one electron, the state of lowest energy will obviously be when this electron is in the $d\epsilon$ states. When further electrons are added they will also go into the $d\epsilon$ states, but with spins parallel in order to make the exchange energy a minimum; up to three electrons can be accommodated in this way. The $d\epsilon$ shell is then half full, and behaves like a state with zero orbital momentum (corresponding to the singlet orbital ground state for $d^3$ in Fig. 14.16) and $S = \frac{3}{2}$. When more electrons are added, they cannot go into the $d\epsilon$ shell with parallel spin, because this would violate the exclusion principle. There is therefore a competition between the exchange energy, which prefers parallel spin, and the ligand field energy, which prefers the $d\epsilon$ states. In the hydrated salts, the latter is less important, and the fourth and fifth electrons go into the $d\gamma$ states with parallel spin, making $d^5$ a state with $L = 0$, $S = \frac{5}{2}$ (a half-filled d shell). In the more strongly bonded salts the ligand field splitting is so large that no electrons go into the $d\gamma$ states, since they have a lower energy by occupying the $d\epsilon$ states with antiparallel spin, as shown in Fig. 14.20(c). Thus a strongly bonded $d^5$ salt behaves as if it had one hole in the $d\epsilon$ shell; for example, $K_3Fe(CN)_6$, where the $Fe^{3+}$ ion has a $d^5$ configuration, has magnetic properties quite different from hydrated ferric compounds, showing considerable departures from Curie's law, a susceptibility close to that of a single spin, and strong anisotropy in single crystals. With six electrons, $d^6$, strongly bonded, the $d\epsilon$ shell is completed, and the ion has no permanent magnetic dipole moment ($K_3Co(CN)_6$ has only a small temperature-independent paramagnetism). Any further electrons would have to go into the $d\gamma$ states, but these are so high in energy that such ions are usually chemically unstable. However, when the ligand field is less strong, the $d\gamma$ states are occupied, as shown in Fig. 14.20(d) for a hydrated $Ni^{2+}$, $d^8$, ion. Here the $d\epsilon$ states are completely filled and the $d\gamma$ half-filled, giving again a ground state with no orbital momentum (cf. Fig. 14.16(c)). On this single electron picture we can see that the orbital momentum is effectively quenched whenever the two sets of $d\epsilon$ and $d\gamma$ states are each either empty, half-filled, or completely filled with electrons; the reader can verify that a half-filled subshell, with all electron spins parallel, can be achieved by only one possible arrangement of the electrons in the various boxes, and therefore corresponds to a singlet orbital state. On the other hand, when the $d\epsilon$ states are occupied by one or two electrons, or four or five electrons, there are three equivalent ways of arranging them, giving a triply degenerate orbital state. This corresponds to the triplet state which is lowest for $d^1$, $d^2$, $d^6$, $d^7$ in Fig. 14.16.

## 14.9. Electronic paramagnetism—a summary

The magnetic properties of insulating paramagnetic compounds are the result of competition between (1) electrostatic and exchange interactions within the ion which favour $L-S$ coupling; (2) spin–orbit interaction within the ion; and (3) interaction with the ligand field of neighbouring (diamagnetic) ions. In this rather complex situation the largest interaction must be considered first, and the different behaviour of the various transition groups arises from the change in the size of the ligand interaction (3) relative to (1) and (2). In broad outline, the situation may be summarized as follows.

(*a*) Electrons of the 4f group (and, to a lesser extent, of the 5f group) are 'inner' electrons whose interaction energy with the ligand field is of order $10^4\,m^{-1}$. This is smaller than the spin–orbit coupling energy $(10^5\,m^{-1})$, and splits the $(2J+1)$ levels by amounts up to $2-5\times10^4\,m^{-1}$.

(*b*) Electrons of the 3d group are 'valence' electrons, and their interaction with the ligand field is of order $10^6\,m^{-1}$, comparable with the $L-S$ coupling energy. In a first approximation the ligand field splits the $(2L+1)$ orbital levels, and the spin states are then affected through the spin–orbit coupling.

(*c*) Electrons of the 4d, 5d groups (and of the 3d group in some compounds) are valence electrons for which the ligand interaction is stronger than $L-S$ coupling. The $(2l+1)$ orbital states of each electron are strongly split by the ligand field, generating new single-electron states on which the effects of electrostatic and spin–orbit couplings within the atom are evaluated subsequently.

The ligand field is electrostatic in origin and interacts directly only with the orbitals, which have different charge distributions, and not with the electron spin. The latter interacts with a magnetic field, either that applied from outside or that generated through the orbital motion (spin-orbit coupling). By Kramers' theorem, every energy level of an ion with an odd number of electrons must be associated with at least two electron states which have the same distribution of electric charge and hence the same energy in an electric field. Such pairs of states have equal and opposite magnetic moments, and are split in energy by a magnetic field, either externally applied or originating in the fields of neighbouring dipoles. The dipoles also interact with each other through exchange interaction, which may result in spontaneous magnetization, even at room temperature (see Chapters 15 and 16), and the onset of an ordered magnetic state.

The susceptibility is a primary magnetic quantity for the paramagnetic state, but it is in fact rather a poor guide to the ligand field splittings. At temperatures where all states have roughly equal populations, the susceptibility behaviour approaches that given by Curie's law. It departs markedly from this only at temperatures where the populations are changing rapidly, that is, where $kT$ is of the same order as the energy splitting (see

Problem 14.4). Optical and infrared spectroscopy give much more accurate information about the actual energy levels, and the distinctive colours of transition-ion compounds are due to absorption bands at or near optical frequencies. Electron spin resonance gives very detailed information on the properties of the lowest levels, such as the ground state and others within about 10 K in energy. Splittings of this order can also be determined calorimetrically.

## 14.10. Nuclear moments and hyperfine structure

For nuclei, as for atoms, magnetic moments are associated with the presence of a quantized angular momentum or 'spin', whose value is $I\hbar$. For both proton and neutron the value of $I$ is $\frac{1}{2}$, and if a nucleus contains an odd number of nucleons, $I$ is half-integral. For an even number of nucleons $I$ is an integer, but if the numbers of protons and neutrons are each individually even, the value of $I$ is zero for the state of lowest energy of the nucleus. Nuclear magnetic moments are measured in terms of a unit called the 'nuclear magneton', equal to $\mu_N = e\hbar/2m_p$; this is analogous to the Bohr magneton, but the mass in the denominator is that of the proton instead of that of the electron. If the proton obeyed a similar wave equation to that for the electron, we would expect it to possess a moment of one nuclear magneton associated with its spin $\frac{1}{2}$, just as the electron has a moment of 1 $\mu_B$ and spin $\frac{1}{2}$. In fact the moment of the proton is $+2\cdot793$ nuclear magnetons and the neutron, although uncharged, has a moment of $-1\cdot913$ $\mu_N$. Here the significance of the plus and minus signs is that the magnetic moments are respectively parallel and antiparallel to the spin. Since the magnetic moment of neither neutron nor proton is an integral number of nuclear magnetons we should not expect the moments of more complicated nuclei to be simple integers. They do, however, follow the trend which the nuclear shell model would indicate (see Problem 14.6); in general we write the nuclear magnetic moment as $\mathbf{m}_N = g_N\mu_N\mathbf{I}$, where $g_N$ is the nuclear magnetogyric ratio.

Interactions between a nuclear magnet and its surroundings are small. In a magnetic field each of the $2m_I+1$ states corresponding to different orientations of the nuclear moment takes up a different energy

$$W_{m_I} = -\mathbf{m}_N \cdot \mathbf{B} = -g_N\mu_N m_I B; \tag{14.26}$$

in a field of 1 T the separation between successive levels corresponds to a frequency of order 10 MHz. In an atom or ion which has a permanent electronic magnetic moment, the latter sets up a magnetic field at the nucleus which may be as much as $10^3$ T, being generally larger in the heavier atoms. This magnetic field is partly due to the electronic orbit and partly to the spin, but for most purposes we need consider only the steady component of the electronic field of flux density $\mathbf{B}_e$, which is parallel to and proportional to the resultant electronic angular momentum vector $\mathbf{J}$. Thus we have an additional 'hyperfine' energy

$$W = -\mathbf{m}_N . \mathbf{B}_e = A\mathbf{J}.\mathbf{I}, \tag{14.27}$$

where $A$ is a constant whose order of magnitude can be estimated as follows. The electronic flux density $\mathbf{B}_e$ is $\sim \mathbf{m}_e \langle R_e^{-3} \rangle = -g_J \mu_B \mathbf{J} \langle R_e^{-3} \rangle$, where $\langle R_e^{-3} \rangle$ is the mean inverse cube of the distance of the electron from the nucleus; since $\mathbf{m}_N = g_N \mu_N \mathbf{I}$, $A$ is of order $g_J g_N \mu_B \mu_N \langle R_e^{-3} \rangle$. In frequency units $A/h$ generally lies in the range $10^2 - 10^4$ MHz, so that the hyperfine energy may approach $100 \text{ m}^{-1} \equiv 1 \cdot 43$ K.

In addition to possessing a magnetic moment, a nucleus may have a non-spherical distribution of electric charge. Its electrostatic potential can then be expanded as in § 2.3, giving an energy of interaction with the electrons of the form (see eqns (2.30)–(2.34))

$$W = \frac{1}{4\pi\epsilon_0} \iint \frac{\rho_e \rho_N \, d\tau_e \, d\tau_N}{|R_e - r_N|}$$

$$= \frac{1}{4\pi\epsilon_0} \left\{ Ze \int \frac{\rho_e \, d\tau_e}{R_e} + \sum_{m=-2}^{+2} (-1)^{|m|} A_{2,m} B_{2,-m} + \ldots \right\}. \tag{14.28}$$

Here the subscripts e, N refer to the electrons and nuclei respectively; $Ze = \int \rho_N \, d\tau_N$ is the nuclear charge, so that the first term is the Coulomb interaction due to a point charge at the nucleus, and the quantities in the second term are

$$A_{2,m} = \int \rho_N r_N^2 C_{2,m}(\theta_N, \phi_N) \, d\tau_N,$$

$$B_{2,-m} = \int (-1)^{|m|} \rho_e R_e^{-3} C_{2,-m}(\theta_e, \phi_e) \, d\tau_e.$$

This term represents the interaction between the electric quadrupole moments of the nucleus and of the electrons, whose nature is that of a tensor. The charge distribution is spherical for nuclei with $I = 0$ or $\frac{1}{2}$, and for electronic shells with $J = 0$ or $\frac{1}{2}$, so that the quadrupole interaction vanishes in either case. Since the nuclear charge is symmetric about the axis of nuclear precession, the nuclear terms can be expressed in terms of a single quantity

$$A_{2,0} = \int \rho_N \frac{1}{2} r_N^2 (3 \cos^2 \theta_N - 1) \, d\tau_N = \frac{1}{2} e Q \left\{ \frac{3m_I^2 - I(I+1)}{I(2I-1)} \right\}, \tag{14.29}$$

where

$$Q = \frac{1}{e} \int \rho_N r_N^2 (3 \cos^2 \theta_N - 1) \, d\tau_N \tag{14.30}$$

is called the nuclear electric quadrupole moment and has the dimensions of an area of the same order as the (nuclear radius)$^2$. It is expressed in terms of the 'barn' $10^{-28}\,\text{m}^2$. The sign of $Q$ is positive for a prolate spheroid and negative for an oblate spheroid, as illustrated in Fig. 14.21. The expression in parentheses in eqn (14.29) gives the variation of $A_{2,0}$ with the nuclear magnetic quantum number $m_I$, and it is easily verified that in the states $m_I = \pm I$, $A_{2,0} = \frac{1}{2}eQ$.

In the absence of an external magnetic field, the electronic and nuclear angular-momentum vectors $J$, $I$ are coupled together by the magnetic hyperfine energy (eqn (14.27)) to form a resultant vector $F$. Different values of $F$ correspond to different energies, since the angle between $J$ and $I$ is changed, and from the vector model it can be shown that

$$W_F = \tfrac{1}{2}A\{F(F+1) - J(J+1) - I(I+1)\} \tag{14.31}$$

so that the energies of successive states form an arithmetical progression (cf. Problem 14.7, for the corresponding case of spin–orbit coupling). This simple rule no longer holds when an electric quadrupole interaction is present, but $F$ is still a good quantum number, and the size of the quadrupole interaction can be deduced from the departures from eqn (14.31). It can be shown that

$$W_F = \tfrac{1}{2}AC + B_Q \frac{\tfrac{3}{4}C(C+1) - I(I+1)J(J+1)}{2I(2I-1)J(2J-1)}, \tag{14.32}$$

where

$$C = F(F+1) - I(I+1) - J(J+1); \qquad B_Q = 2eQB_{2,0}/4\pi\epsilon_0 = eQ(\partial^2 V/\partial z^2),$$

where $\partial^2 V/\partial z^2$ is the electric field gradient set up by the electrons at the nucleus.

In the solid state an assembly of nuclear dipoles behaves as a simple

Positive quadrupole moment          Negative quadrupole moment

Fig. 14.21. Representation of non-spherical charge distribution in nucleus as combination of sphere and quadrupole.

paramagnetic substance, contributing an amount (cf. eqn (14.22))

$$\chi_N = \frac{\mu_0 n g_N^2 \mu_N^2 I(I+1)}{3kT},\tag{14.33}$$

which is only about $10^{-6}$ of that of any electronic paramagnetic substance, since the susceptibility depends on the square of the magnetic dipole moment. The nuclear contribution has been detected by static susceptibility measurements in solid hydrogen, where the nuclear paramagnetism just outweighs the electronic diamagnetism at about 1 K (see Problem 14.9).

The nuclear susceptibility follows Curie's law, eqn (14.33), only at temperatures such that $kT$ is large compared with any splittings of the nuclear levels. In substances without permanent electronic magnetic dipoles this means temperatures down to about $10^{-6}$ K, except where the nuclear levels are split through an electric quadrupole interaction with the electrostatic field gradient (the 'crystal field') set up by neighbouring ions. In such cases the gradient $(\partial^2 V/\partial z^2)$ is fixed, unlike in a free atom where it follows the precessing electronic angular-momentum vector for the orbit, which determines the orientation of the electronic charge cloud. Hence in a solid in which the local surroundings of a nucleus have symmetry about an axis (which we take to be the $z$-axis) the nuclear levels may be split according to the formula

$$W_{m_I} = \frac{1}{4\pi\epsilon_0} A_{2,0} B_{2,0} = eQ(\partial^2 V/\partial z^2)\frac{3m_I^2 - I(I+1)}{4I(2I-1)}.\tag{14.34}$$

The splittings range from a few kilohertz up to over 2000 MHz for $^{127}$I in I$_2$ (solid iodine).

In substances containing ions with both electronic and nuclear magnetic dipoles the two contributions to the susceptibility are additive at temperatures such that $kT$ is large compared with any hyperfine structure splittings (in practice this usually means down to about 1 K). Such splittings arise from both nuclear magnetic dipole and electric quadrupole interactions in the same way as for free atoms, but the effects are more complicated because of the complex interaction of the electrons with the crystal or ligand field discussed in §§ 14.6–14.9. The hyperfine splittings usually correspond to temperatures in the range 0·001–1 K, and affect both the electronic and nuclear contributions to the susceptibility in this temperature range and below.

## 14.11. Magnetism of conduction electrons

This chapter has been concerned mainly with paramagnetism of insulating compounds in which the electrons are localized; the compounds can be regarded as assemblies of atomic ions in which the paramagnetism arises from ions of the transition groups. Many metals in these groups are strongly magnetic; they are discussed in Chapters 15 and 16, and here we confine ourselves to the interaction of a magnetic field with the conduction electrons in 'normal' metals.

A magnetic field alters the translational motion of the conduction electrons, through the Lorentz force, and this gives rise to a diamagnetic susceptibility which is constant at ordinary field strengths. A formula for this was first derived by Landau, and on using an effective mass $m^*$ this can be written as

$$\chi_d = -\frac{\mu_0 e^2}{6\pi m^*}\left(\frac{3n}{8\pi}\right)^{\frac{1}{3}}, \qquad (14.35)$$

where $n$ is the number of conduction electrons per unit volume.

The second effect of a magnetic field arises from its interaction with the electron spin, the two orientations of the spin dipole having energies $+\frac{1}{2}g_s\mu_B B$ and $-\frac{1}{2}g_s\mu_B B$ respectively. In the latter state, with the lower energy, the magnetic dipoles are parallel to the field, and the probability of occupation is a little larger than for the antiparallel orientation, giving a net positive susceptibility. For conduction electrons this must be calculated using the Fermi–Dirac distribution function, and since this varies very little with temperature, the paramagnetic spin susceptibility is practically independent of temperature. At the absolute zero an expression for its value can be derived rather simply.

At the absolute zero, two electrons with oppositely directed spins occupy each translational energy level up to a certain energy $W_F$, the top of the Fermi distribution. When a magnetic field is applied, an electron can only reverse its spin magnetic dipole from an antiparallel to a parallel orientation if the decrease in its magnetic energy $g_s\mu_B B$ is sufficient to supply the extra kinetic energy required to raise it to an empty translational energy level. This follows from the Pauli principle, which shows that two electrons with parallel spin cannot occupy the same energy level. The effect on the distribution of electrons in the energy band is shown in Fig. 14.22. This differs from Fig. 11.2 in that the band is drawn in two halves, one containing the electrons whose spin dipoles are parallel to the flux density **B**, the other those with their spin dipoles antiparallel. The two half-bands are then separated in energy by $g_s\mu_B B$, the potential-energy difference in the magnetic field. For the total energy, magnetic plus kinetic, of the whole system to be a minimum, the electrons must fill the two displaced half-bands up to the same level. Any deviation from this would require a transfer of electrons from one half-band to higher vacant levels in the other half-band, and so increase the energy.

The total magnetic moment of the system is $2x(\frac{1}{2}g_s\mu_B)$, where $x$ is the number of electrons transferred from the antiparallel to the parallel

FIG. 14.22. Displacement of + and − bands of conduction electrons by an applied magnetic field. The displacement is equal to the difference of energy $g_s\mu_B B$ of a spin dipole parallel and antiparallel to the flux density $B$. The resultant magnetization is due to the excess of electrons in the + band.

orientation, since the excess in the latter is then $2x$ and the spin dipole moment is $\frac{1}{2}g_s\mu_B$. The value of $x$ can be found as follows. Suppose $w$ is the difference in kinetic energy between successive levels at the top of the Fermi distribution. To turn round the dipole on one electron requires that its kinetic energy increase by $w$, since we can take the electron from the highest filled level and transfer it to the next level, which is vacant. To turn round a second dipole requires an additional energy of $3w$, since the next two levels with parallel orientation are already filled. The third transfer requires $5w$ in energy, and the $x$th electron $(2x-1)w$, which may be taken as $2xw$ when $x$ is large. At equilibrium, $2xw$ must just equal $g_s\mu_B B$ in order that the two half-bands are filled to the same level.

Now the energy separation $w$ between successive kinetic energy levels is $[\frac{1}{2}\{g(W)\}_F]^{-1}$, and hence

$$2x = g_s\mu_B B/w = \tfrac{1}{2}g_s\mu_B B\{g(W)\}_F.$$

The susceptibility per unit volume is

$$\chi_P = 2x(\tfrac{1}{2}g_s\mu_B)/H = \tfrac{1}{4}\mu_0 g_s^2\mu_B^2\{g(W)\}_F, \qquad (14.36)$$

and for free electrons, using the value of $\{g(W)\}_F$ from eqn (11.11) this can be written as

$$\chi_P = \frac{\mu_0(3ng_s^2\mu_B^2)}{8\,W_F} \qquad (14.37)$$

This result was first obtained by Pauli, and the phenomenon is sometimes called 'Pauli paramagnetism'. For practical purposes $\chi_P$ is independent of temperature, the next term in a series expansion for $\chi_P$ being smaller by a factor of order $(kT/W_F)^2$, as in eqn (11.19). As in the case of the specific heat (§§ 11.6 and 12.6), the paramagnetic susceptibility depends on the density of states $\{g(W)\}_F$ at the Fermi surface, and is smaller by a factor $\sim(kT/W_F)$ than the corresponding quantity for particles obeying classical statistics. Similarly, instead of Curie's law (eqn (14.22)) we have an expression with $W_F$ in the denominator, instead of $kT$.

For free electrons, the value of $\chi_P$ given by eqn (14.37) with $g_s = 2$ is just 3 times the value of $\chi_d$ given by eqn (14.35) with $m^* = m$, as can be verified from eqn (11.7) for $W_F$ on substituting for $\mu_B$. Hence the net effect is a positive susceptibility. If $m^* > m$, the value of $\chi_d$ is diminished, while $\chi_P$ increases, since $W_F^{-1} \propto m^*$ by eqn (12.29). Transition metals show strong Pauli paramagnetism, corresponding to high values of $\{g(W)\}_F$. In addition, interaction effects between the electrons also cause an increase in $\chi_P$ (for a review see Van Vleck (1957)).

Static measurements of the susceptibility give the quantity $\chi = \chi_d + \chi_P + \chi_c$, where $\chi_c$ is the diamagnetic susceptibility of the electrons bound to the positive-ion cores. This can be estimated from values for neighbouring non-metallic elements, or from calculated values of $\sum (r^2)$ using eqn (6.7). Electron spin resonance can be used to determine $\chi_P$, and $\chi_d$ can then be found from the value of $(\chi - \chi_P - \chi_c)$. A comparison of theory and experiment is given in Table 14.3, based on Van Vleck (1957). Later measurements of $\chi_P$ give slightly different values, but a satisfactory comparison requires an experimental determination of $m^*/m$.

TABLE 14.3

*Experimental and theoretical values of the volume*
*susceptibility of lithium and sodium*

This table is in units of $10^{-6}$ e.m.u. cm$^{-3}$ (see Appendix D);
to convert to S.I. multiply by $4\pi$.

|  | Lithium | | Sodium | |
|---|---|---|---|---|
|  | Experiment | Theory | Experiment | Theory |
| $m^*/m$ |  | 1·46 |  | 0·985 |
| $\chi_P$ | 2·08±0·1 | 1·17† | 0·95±0·1 | 0·64† |
|  |  | 1·87‡ |  | 0·85‡ |
| $\chi$ | 1·89±0·05 |  | 0·70±0·03 |  |
| $\chi_c$ |  | −0·05 |  | −0·18 |
| $\chi_d$ | −0·14±0·15 | −0·19§ | −0·07±0·13 | −0·22§ |

(After Van Vleck 1957.)
† From eqn (14.37), using effective mass.
‡ Calculated by Pines, including interaction effects.
§ From eqn (14.35), using effective mass.

## References

COOPER, B. R. and VOGT, O. (1970). *Phys. Rev.* **B1,** 1211.
STOUT, J. W. and HADLEY, W. B. (1964). *J. chem. Phys.* **40,** 55.
VAN VLECK, J. H. (1957). *Nuovo Cim.* **6,** 857.

## Problems

14.1. In statistical mechanics the partition function $Z$ is defined as

$$Z = \sum_i \exp(-W_i/kT),$$

where $W_i$ is the energy of the $i$th state. Show that for an assembly of non-interacting paramagnetic ions, each of angular momentum $J$, in a field of flux density $B$

$$Z = \frac{\sinh\{(2J+1)y/2J\}}{\sinh(y/2J)},$$

where $y = Jg_J\mu_B B/kT$.

The magnetization $M$ for an assembly of $n$ such atoms is given by the formula

$$M = nkT\frac{d}{dB}(\ln Z).$$

Using this formula, derive the Brillouin function of eqn (14.21).

14.2. Show that the energy of interaction of two magnetic dipoles $m$ a distance $r$ apart is of the order $\mu_0 m^2/4\pi r^3$.

In potassium chrome alum, each $Cr^{3+}$ ion carries a magnetic moment of $3\mu_B$, and the mean distance apart is about $0.78$ nm. Assuming that serious departures from Curie's law will occur when the interaction energy between two neighbouring dipoles is $\approx kT$, show that this temperature is approximately $0.01$ K. (In fact the levels of each $Cr^{3+}$ ion are split by about $0.2$ K through a high-order effect of the crystalline field, and this is more important than the magnetic dipole interaction between neighbouring ions; it also gives a specific heat anomaly at about $0.1$ K of the type shown in Fig. 14.17.)

14.3. For a $Cu^{2+}$ ion, $S = \frac{1}{2}$ and the energy levels of the ground state in a magnetic flux density $B$ are of the form

$$W = \pm\tfrac{1}{2}g\mu_B B - \tfrac{1}{2}\alpha B^2.$$

Show that in small fields $(g\mu_B B/kT \ll 1)$ the partition function

$$Z = 2 + \alpha B^2/kT + g^2\mu_B^2 B^2/4(kT)^2 + \ldots,$$

and hence that

$$\chi/\mu_0 = n\{g^2\mu_B^2/4kT + \alpha\} = np^2\mu_B^2/3kT.$$

This shows that the term $B^2$ in $W$ gives rise to a temperature-independent contribution to the susceptibility. Note that $p^2$ is then of the form $A + BT$. ($B$ is very small for $Cu^{2+}$, but $Co^{2+}$ ions obey this relation below $100$ K—see Fig. 14.18.)

14.4. A pair of singlet levels have separation $\delta$ when $B = 0$, and in a flux density

$B$ their energies become $W_1 = +\frac{1}{2}\Delta$ and $W_2 = -\frac{1}{2}\Delta$ respectively, where $\Delta = \{\delta^2 + (g\mu_B B)^2\}^{\frac{1}{2}}$. Show from the relation $m = -dW/dB$ that the magnetic moments of the two states are $\mp\frac{1}{2}(g^2\mu_B^2 B/\Delta)$.

Verify that in a small field at temperature $T$ the susceptibility is given by

$$\chi/\mu_0 = \frac{1}{2}(g^2\mu_B^2/\delta)\tanh(\delta/2kT)$$

and that at high temperatures when $\delta/kT \ll 1$ this approaches $g^2\mu_B^2/4kT$, while at low temperatures when $\delta/kT \gg 1$ it approaches the constant value $g^2\mu_B^2/2\delta$.

14.5. Show that for a system where the magnetic moment associated with orbital angular momentum $\mathbf{l}\hbar$ is $g_l\mu_B\mathbf{l}$ and that associated with spin $\mathbf{s}\hbar$ is $g_s\mu_B\mathbf{s}$ and $\mathbf{l}$ and $\mathbf{s}$ are coupled together to form a resultant $\mathbf{j}$, the generalized Landé formula for the g-factor is

$$g = \frac{j(j+1)(g_l+g_s)+\{l(l+1)-s(s+1)\}(g_l-g_s)}{2j(j+1)}.$$

14.6. On the nuclear shell model the nuclear spin is due to the odd neutron or proton with spin $\frac{1}{2}$ moving in an orbit within the nucleus with angular momentum $l\hbar$. The observed nuclear spin $I$ is either $l+\frac{1}{2}$ or $l-\frac{1}{2}$. Apply the formula of the last question to calculate the magnetic moment, assuming that for a proton $g_l = 1$ and $g_s = 5\cdot586$, and for a neutron $g_l = 0$ and $g_s = -3\cdot826$. Show that the magnetic moment $m = g_N\mu_N I$ of a nucleus of spin $I$ is

(a) odd proton $\quad I = l+\frac{1}{2}, \quad m = \mu_N(I+2\cdot293),$

$\qquad\qquad\quad I = l-\frac{1}{2}, \quad m = \mu_N I(I-1\cdot293)/(I+1),$

(b) odd neutron $\quad I = l+\frac{1}{2}, \quad m = -1\cdot913\mu_N = m_N,$

$\qquad\qquad\qquad I = l-\frac{1}{2}, \quad m = 1\cdot913\mu_N I/(I+1) = -m_N I/(I+1).$

These formulae are known as the Schmidt limits. Observed nuclear moments follow the trend given by these formulae but generally lie between these limits.

14.7. Use the treatment of § 14.3 to show that the spin–orbit coupling leads to an energy for a state of total angular momentum $J\hbar$ of

$$W_J = \frac{1}{2}\lambda\{J(J+1)-L(L+1)-S(S+1)\}$$

and that $W_J - W_{J-1} = \lambda J$ (this is known as the Landé interval rule).

Show from the splittings of the $^4F$ multiplet in Fig. 14.5 that $\lambda$ is about $8\cdot7\times10^3\,\text{m}^{-1}$ for the $Cr^{3+}$ ion (slightly higher values are obtained from the $^4P$ states, but these are perturbed by doublet states which are not far away).

For the $Eu^{3+}$ ion, $4f^6$, $L = 3$, $S = 3$ the value of $\lambda$ is $2\cdot27\times10^4\,\text{m}^{-1}$. Verify that (a) the ground state is $J = 0$; (b) the first excited state $J = 1$ lies at $2\cdot27\times10^4\,\text{m}^{-1}$; (c) $g_J = \frac{3}{2}$ for all excited states.

14.8. Hydrogen molecules are of two types: (a) orthohydrogen, where the nuclear spin of $\frac{1}{2}\hbar$ of each proton is parallel to the other and the nuclear spin for the molecule is $I = 1$; (b) parahydrogen, where the two proton spins are antiparallel giving $I = 0$ for the molecule. Show that at high temperatures where the ratio of ortho- to parahydrogen molecules is $3:1$, the susceptibility due to the nuclear paramagnetism is identical with that of the same total number of hydrogen atoms with independent spin $I = \frac{1}{2}$. (Note that the equilibrium ratio of $3:1$ corresponds to the fact that there are three quantum states for $I = 1$, associated with three possible orientations of the spin, each with the same *a priori* probability as the single state for $I = 0$.)

14.9. Calculate the paramagnetic susceptibility of a mole of hydrogen at 1 K due to the nuclear moments, assuming that the ortho–para ratio is still 3 : 1. Show that it is of the same order as the diamagnetic susceptibility due to the electrons, assuming that each of the two electrons is in an orbit for which the mean radius is the Bohr radius.

(*Answers:* $2 \cdot 2 \times 10^{-11}$ and $-2 \cdot 0 \times 10^{-11}$ S.I. units.)

14.10. The ground state of sodium is $^2S_{\frac{1}{2}}$, and the yellow D lines are due to transitions to the ground state from the two lowest excited states $^2P_{\frac{1}{2}}$ and $^2P_{\frac{3}{2}}$. Show that the Landé g-factors of these two states are $\frac{2}{3}$ and $\frac{4}{3}$ respectively.

Hence show that one D line will be split in a magnetic field of flux density $B$ into four components, with frequencies $D_1 \pm \frac{2}{3}\delta$, $D_1 \pm \frac{4}{3}\delta$; and the other D line into six components with frequencies $D_2 \pm \frac{1}{3}\delta$, $D_2 \pm \delta$, $D_2 \pm \frac{5}{3}\delta$, where $\delta = \mu_B B/h$, when viewed normal to the field. (Only the $\Delta M = \pm 1$ components are seen when viewed parallel to the field.)

14.11. The arrangement of parallel cylindrical conductors carrying equal and opposite currents of Problem 4.3 is used to give a large field gradient and deflect atoms in an atomic beam. Each cylinder carries a current of 1000 A and the axes of the cylinders are 0·01 m apart. A beam of atoms in the $^2S_{\frac{1}{2}}$ state from an oven at 900 K travels parallel to the cylinders along the line where the inhomogeneous field is a maximum. Calculate the separation between the two components of the beam after travelling a distance of 0·2 m.

(*Answer:* $\approx 0 \cdot 1$ mm.)

# 15. Ferromagnetism

## 15.1. Exchange interaction between paramagnetic ions

IN the discussion in §§ 14.5–14.9 of paramagnetism in the solid state, it was assumed that interactions between neighbouring paramagnetic ions could be neglected. Such interactions are of two types: (1) magnetic dipole–dipole interaction, arising from the magnetic field due to one dipole acting on another; (2) exchange interactions between the electrons in different paramagnetic ions, of the same nature as those between electrons within the same atom (giving rise to $L–S$ coupling) or between the electrons of different atoms in chemical binding. Of these, exchange interaction greatly outweighs dipole–dipole interaction in ordinary substances. For example, the Curie point of nickel is 631 K (see Table 15.1). This is a rough indication of the temperature at which the interaction between neighbouring nickel ions (separation 0·25 nm) is of the order $kT$, whereas (Problem 14.2) the purely magnetic interaction of two atomic dipoles at this distance would be equivalent to $kT$ with $T$ less than 1 K. Exchange interaction decreases more rapidly than magnetic dipole interaction as the atomic separation is increased, though no simple law can be given for its rate of decrease. As an example, in paramagnetic salts of the 3d group the exchange interaction is more important than the magnetic dipole interaction until the separation between the paramagnetic ions is greater than about 0·6 nm, and then both are so small that they have an appreciable effect on the magnetic properties only well below 1 K.

The mechanism of exchange interaction, as originally proposed by Heisenberg in 1928, is one in which the forces involved are electrostatic in origin, but which, because of the constraints imposed by the Pauli exclusion principle, are formally equivalent to a very large coupling between the electron spins, of the type

$$W = -2\mathscr{J}\mathbf{s}_i \cdot \mathbf{s}_j. \qquad (15.1)$$

The quantity $\mathscr{J}$ is known as the exchange energy. Though several types of indirect exchange interaction have since been suggested (see § 15.8), they all lead to a basic coupling between the spins of this form, dependent on the cosine of the angle between the two spin vectors. For two separate atoms with total spin vectors $\mathbf{S}_i$, $\mathbf{S}_j$ we may use the vector summations to show that the total interaction energy is

$$W = -2\mathcal{J}\sum_i\sum_j \mathbf{s}_i\cdot\mathbf{s}_j = -2\mathcal{J}\sum_i \mathbf{s}_i\cdot\sum_j \mathbf{s}_j = -2\mathcal{J}\mathbf{S}_i\cdot\sum_j \mathbf{s}_j = -2\mathcal{J}\mathbf{S}_i\cdot\mathbf{S}_j, \quad (15.2)$$

which depends only on the relative orientation of the two total spin vectors $\mathbf{S}_i$, $\mathbf{S}_j$. An immediate result of eqn (15.2) is that the exchange interaction vanishes for any closed shell of electrons, since then $\mathbf{S} = 0$. Thus we need consider only the partly filled shells which are responsible for permanent magnetic dipole moments in atoms and ions.

For an ion in which $\mathbf{J}$ is a good quantum number (such as ions of the 4f group), we must project $\mathbf{S}$ onto $\mathbf{J}$; the reason for this is that $\mathbf{J}$ is a constant of the motion, and hence so also is the projection of $\mathbf{S}$ onto $\mathbf{J}$. The components of $\mathbf{S}$ normal to $\mathbf{J}$ are precessing rapidly, so that their contribution to the scalar product $\mathbf{S}_i\cdot\mathbf{S}_j$ is zero on a time average. From the equivalences $\mathbf{L}+2\mathbf{S}\equiv g\mathbf{J}$ (where $g$ is the Landé factor), $\mathbf{L}+\mathbf{S}\equiv\mathbf{J}$, we find at once that $\mathbf{S}\equiv(g-1)\mathbf{J}$. Thus, for a pair of such ions (assumed identical, with the same values of $\mathbf{J}$ and $g$) we have

$$W = -2\mathcal{J}\mathbf{S}_i\cdot\mathbf{S}_j = -2\mathcal{J}(g-1)^2\mathbf{J}_i\cdot\mathbf{J}_j = -2\mathcal{J}'\mathbf{J}_i\cdot\mathbf{J}_j. \quad (15.3)$$

This gives a coupling of the angular-momentum vectors of the same form as eqn (15.2), but with a modified value of the apparent exchange energy.

In a solid, any given magnetic ion is surrounded by other magnetic ions, with each of which it will have an exchange interaction. The total interaction for each ion will therefore be a sum of terms such as (15.3) taken over all pairs of ions; the energy for atom $i$ is thus

$$W_i = -2\mathbf{J}_i\cdot\sum_j \mathcal{J}'_{ij}\mathbf{J}_j.$$

The magnetic dipole moment of each ion is proportional to the angular momentum $\mathbf{J}$, since $\mathbf{m} = g\mu_B\mathbf{J}$, so that the exchange energy can be expressed in terms of the dipole moments, giving

$$W_i = -2\left(\frac{\mathbf{m}_i}{g\mu_B}\right)\cdot\sum \mathcal{J}'_{ij}\left(\frac{\mathbf{m}_j}{g\mu_B}\right),$$

assuming again that all ions have the same Landé $g$-factor. In a ferromagnetic substance, or a paramagnetic substance subjected to an external magnetic field, each ion will have an average dipole moment in the direction of magnetization, together with fluctuating components in other directions whose time average is zero. In summing over the interaction with neighbouring ions, that part associated with the fluctuating components will tend to average out, since at any instant the contributions from different neighbours will be as often positive as negative. To a fair approximation we can therefore replace the vector sum over the neighbouring dipole moments by a sum over the average moment per neighbour $\bar{\mathbf{m}}_j$, and if we assume further that the only important interaction is with $z$ equidistant neighbours, each having the same interaction energy $\mathscr{J}'$, we can write

$$W = -2\left(\frac{\mathbf{m}}{g\mu_B}\right) \cdot \sum \mathscr{J}'\left(\frac{\bar{\mathbf{m}}_j}{g\mu_B}\right) = -2\left(\frac{\mathbf{m}}{g\mu_B}\right) \cdot \left(\frac{z\bar{\mathbf{m}}}{g\mu_B}\right)\mathscr{J}'$$

$$= -\left(\frac{2z\mathscr{J}'}{ng^2\mu_B^2}\right)\mathbf{m} \cdot \mathbf{M} = -\mathbf{m} \cdot \mathbf{B}_{int}. \tag{15.4}$$

Here we have dropped the subscript $i$, since we assume all ions are identical, and the energy is the same for each; and we have replaced the mean moment per ion by the magnetization $\mathbf{M} = n\bar{\mathbf{m}}$, where $n$ is the number of ions per unit volume. The result is an equation formally identical with the potential energy of a dipole $\mathbf{m}$ in a field

$$\mathbf{B}_{int} = (2z\mathscr{J}'/ng^2\mu_B^2)\mathbf{M} = \lambda\mathbf{M}.$$

We may therefore represent the effect of the exchange forces, to a good approximation, by an effective 'internal field' of flux density $\mathbf{B}_{int}$ which is proportional to the intensity of magnetization. This concept was first introduced by Weiss to account for the occurrence of spontaneously magnetized substances (ferromagnetics).

As a preliminary, we shall discuss the effect of this internal field in a paramagnetic substance. The total field acting on an ion is then $\mathbf{B}_0 + \mathbf{B}_{int} = \mathbf{B}_0 + \lambda\mathbf{M}$, where $\mathbf{B}_0$ is the flux density of the external field. So long as the magnetization is small compared with the saturation value we may assume that Curie's law $\chi = C/T$ still holds if we replace $\mathbf{B}$ in our earlier theory by $\mathbf{B}_0 + \lambda\mathbf{M}$. Then we have

$$\mathbf{M} = \frac{(C/T)\mathbf{B}}{\mu_0} = \frac{C(\mathbf{B}_0 + \lambda\mathbf{M})}{\mu_0 T}$$

and hence

$$\chi = \frac{\mu_0 M}{B_0} = \frac{C}{(T - \lambda C/\mu_0)} = \frac{C}{(T - \theta)}. \tag{15.5}$$

This is known as the Curie–Weiss law, and represents the behaviour of paramagnetic substances at temperatures $T > \theta$ with fair accuracy; $\theta = \lambda C/\mu_0$ is often called the 'Weiss' constant.

The form of eqn (15.5) shows that some radical change in the magnetic properties is to be expected at the temperature $\theta$, and we may interpret the infinite susceptibility which is predicted by eqn (15.5) at this point in the following way. Since $\chi = \mu_0 M/B_0$, and the maximum value of $M$ is finite, being limited to the saturation moment obtainable when all the dipoles are aligned parallel to one another, we must assume $B_0 = 0$; in other words, the substance is magnetized even in the absence of an external field. This 'spontaneous magnetization', due to the internal field, is a characteristic of ferromagnetism, and the temperature $\theta$ is the boundary between paramagnetic behaviour at $T > \theta$ and ferromagnetic behaviour when $T < \theta$. The temperature below which spontaneous magnetization appears is known as the Curie point, and the experimental values for a number of substances are given in Table 15.1. The 'ferromagnetic Curie temperature' is defined as that below which spontaneous magnetization sets in, and it often differs by 10–20 K from the value of $\theta$ determined in the paramagnetic region by fitting the observed susceptibility to eqn (15.5). The latter value is sometimes called the 'paramagnetic Curie temperature'. On our simple theory there is no difference between the two Curie temperatures.

Since the Curie constant $C = \mu_0 n g^2 \mu_B^2 J(J+1)/3k$, the value of the Weiss constant in eqn (15.5) is

$$\theta = \frac{\lambda C}{\mu_0} = \frac{(2z\mathscr{J}'/ng^2\mu_B^2)\{\mu_0 n g^2 \mu_B^2 J(J+1)/3k\}}{\mu_0}$$

$$= \frac{2z\mathscr{J}'J(J+1)}{3k} \tag{15.6}$$

and on simple theory this is also the value of the Curie temperature $T_C$. More sophisticated methods of calculation produce a somewhat different value of the numerical constant, and Rushbrooke and Wood (1958) show that the results can be fitted remarkably well by the empirical formula

$$T_C = \frac{5\mathscr{J}'}{96k}(z-1)\{11J(J+1)-1\}. \tag{15.7}$$

This predicts somewhat lower values for the Curie point than eqn (15.6),

TABLE 15.1

*Saturation moment and Curie point of some ferromagnetic materials*

| Substance | Saturation moment at 0 K | Curie point (K) |
|---|---|---|
| | Bohr magnetons per atom | |
| Iron | 2·22 | 1043 |
| Cobalt (<670 K) | 1·715 | |
| (>670 K) | (1·76) | 1394 |
| Nickel | 0·605 | 631 |
| MnBi | 3·52 | 630 |
| MnAs | 3·40 | 318 |
| $Fe_2O_3$ | 1·20 (per atom of Fe) | 893 |

*Note:* Cobalt has a phase transition at about 670 K, being hexagonal in structure below that temperature, and face-centred cubic above. The values in brackets are obtained by extrapolation.

and conversely, gives higher estimates of the exchange interaction. For example, nickel has its Curie point at 631 K; its crystal structure is face-centred cubic (f.c.c.), for which the number of nearest neighbours is 12, which we take to be the value of z. If we make the further assumption that $J = S = \frac{1}{2}$, then we find that $\mathcal{J}'/k$ is 105 K from eqn (15.6), and 150 K from (15.7). Thus the exchange energy (there is no difference between $\mathcal{J}'$ and $\mathcal{J}$ when we are dealing with spin-only magnetism), is about $10^{-2}$ eV. The magnitude of this interaction can perhaps be appreciated best by expressing it in terms of the internal flux density $\mathbf{B}_{int}$ of eqn (15.4), which is found to be of order $10^3$ T. This is over 100 times larger than any field which can easily be produced in the laboratory, so that external fields would be expected to have little effect on the spontaneous magnetization below the Curie point.

Eqns (15.6) and (15.7) show that the sign of $\theta$ and $T_C$ is the same as that of $\mathcal{J}'$ (and hence also of $\mathcal{J}$, so long as we are dealing with identical ions). Thus a positive value of the exchange energy is required to give a vanishing denominator in the Curie–Weiss law (eqn (15.5)), and a co-operative state in which the electron spins are parallel to each other. This ferromagnetic state is a direct consequence of the fact that the exchange coupling (eqn (15.1)) gives a lower energy for any pair of electrons when their spins are parallel, provided the exchange energy $\mathcal{J}$ is positive. If it is negative, the state of lower energy is one with antiparallel spins; the Weiss constant is also negative, and the denominator of the Curie–Weiss law does not vanish at any real temperature. Nevertheless a co-operative state does then occur, but one in which the basic arrangement is of antiparallel spins. This phenomenon is called 'antiferromagnetism', and is discussed in Chapter 16.

## 15.2. The Weiss theory of spontaneous magnetization

Since the internal field in a ferromagnetic substance is so large, the magnetization will approach the saturation value even at ordinary temperatures. The assumption that the magnetization is small and proportional to the effective field, used in deriving eqn (15.5) for the susceptibility above the Curie point, thus cannot be used below the Curie point. If we retain the concept of an internal field, the magnetization may be calculated using the Brillouin function (see eqn (14.21)) which may be written in the form

$$M/M_s = \phi(y). \tag{15.8}$$

Here $M_s$ is the saturation magnetization per unit volume, and equals $ngJ\mu_B$, where $n$ is the number of atomic dipoles per unit volume. The argument of the Brillouin function may be written as

$$y = gJ\mu_B B/kT = M_s B/nkT$$

and $B$ must be taken as the sum of the external flux density $B_0$ and the internal field $\lambda M$. Hence we have

$$y = M_s(B_0 + \lambda M)/nkT, \tag{15.9}$$

which may be solved for $M$, giving

$$M/M_s = y(nkT/\lambda M_s^2) - (B_0/\lambda M_s). \tag{15.10}$$

The value of the magnetization under any given conditions of $B_0$ and $T$ may be found by eliminating the parameter $y$ between the two equations (15.8) and (15.10). This cannot be done analytically, but the general behaviour of the magnetization can be found from a graphical solution. We begin by equating $B_0$ to zero, and finding the value of the spontaneous magnetization $M_0$ in zero field. To obtain a graphical solution we then plot the two functions $M_0/M_s = \phi(y)$ (from eqn (15.8)) and $M_0/M_s = y(nkT/\lambda M_s^2)$ (from eqn (15.10)) against $y$, as in Fig. 15.1. The second function gives a straight line which passes through the origin and intersects the curve for $\phi(y)$ at this point. Thus one possible value of the magnetization is always zero. If the temperature $T$ is sufficiently high, the slope of the line $M_0/M_s = y(nkT/\lambda M_s^2)$ is so great that this is the only point of intersection, and the substance must therefore be unmagnetized in zero external field. This corresponds to the paramagnetic behaviour above the Curie point, discussed in the last section. As the temperature $T$ falls, the slope of the line given by eqn (15.10) decreases, until at a certain temperature $T_C$ it is tangential to the curve (a) at the origin. For small values of $y$,

$$\phi(y) = M/M_s = y(J+1)/3J,$$

FIG. 15.1. Graphical solution of eqns (15.8) and (15.10) for spontaneous magnetization. (a) is the Brillouin function $\phi(y)$ (eqn (15.8)); (b), (c), (d) are the straight lines $M/M_s = y(nkT/\lambda M_s^2)$ for temperatures $T > T_C$, $T = T_C$, and $T < T_C$ respectively, where $T_C$ is the Curie point (all with $B_0 = 0$); (e) is the function in eqn (15.10); an external flux density $B_0$ is applied, with the temperature the same as for (d).

and on equating this to the value of $M/M_s$ given by eqn (15.10) with $B_0 = 0$, the value of $T_C$ is found to be

$$T_C = \frac{\lambda M_s^2}{nk}\left(\frac{J+1}{3J}\right) = \frac{\lambda n g^2 \mu_B^2 J(J+1)}{3k} = \frac{\lambda C}{\mu_0} = \theta, \qquad (15.6a)$$

where $\theta$ is the Weiss constant defined by eqn (15.6). At still lower temperatures, the slope of the line is less than the initial slope of $\phi(y)$, and there will be two points of intersection, and two possible values of the magnetization, one zero and the other finite. It is easy to show that the former is unstable and the latter stable. For, if we imagine the magnetization at any instant to correspond to the point Q on $\phi(y)$, then the internal field produced by the magnetization corresponds to the point R, and this field will produce the greater magnetization corresponding to the point S on $\phi(y)$. Thus the magnetization will increase until the point P is reached where the two curves intersect. Above P, the two curves cross and any further increase in the magnetization would produce an internal field insufficient to sustain the increased magnetization. It thus appears that the state of spontaneous magnetization corresponding to the point P is stable, while the unmagnetized state is unstable.

Since the value of the spontaneous magnetization is determined by the intersection with $\phi(y)$ of the line corresponding to eqn (15.10) (with $B_0 = 0$), and the slope of this line depends on the temperature, it is obvious that the whole of the curve $\phi(y)$ will be traced out as we lower the temperature from the Curie point to the absolute zero. From eqn (15.6a), $\lambda M_s^2/nk = 3\theta J/(J+1)$, and hence we may express eqn (15.10) (with $B_0 = 0$) in the form

$$\frac{M_0}{M_s} = y\left(\frac{T}{\theta}\right)\left(\frac{J+1}{3J}\right).$$

Elimination of $y$ between this equation and eqn (15.8) shows that a 'reduced equation' may be found of the form

$$M_0/M_s = f(T/\theta), \qquad (15.11)$$

where the function $f(T/\theta)$ is the same for all substances with the same value of $J$. This function is plotted in Fig. 15.2 (broken line) for the special case of $J = \frac{1}{2}$; the curves for other values of $J$ lie slightly inside this curve at intermediate values of $(T/\theta)$. The experimental determination of $M_0/M_s$ and the verification of this 'law of corresponding states' will be discussed in § 15.6.

When a considerable external magnetic field of flux density $B_0'$ is applied the effect on the magnetization is found by a graphical solution of eqns (15.8) and (15.10), where the term in $B_0$ is retained in the latter equation. The straight line corresponding to a plot of $M/M_s$ against $y$ is

FIG. 15.2. Reduced equation of state for a ferromagnetic substance. (After Oliver and Sucksmith 1953.)

now displaced to the right compared to that for $B_0 = 0$ at the same temperature. The intersection with $M/M_s = \phi(y)$ occurs at the point P' in Fig. 15.1, and the magnetization is slightly increased over that corresponding to P, the value for zero external field. At temperatures well below the Curie point $M_0$ is already close to $M_s$ and $\phi(y)$ increases only very slowly, so that the effect of $B_0$ is small. At temperatures near the Curie point P is on the steeper part of the curve for $\phi(y)$ near the origin and the increase in $M$ produced by an external field is more noticeable.

The theory outlined above is similar to the original theory of Weiss, except that the Brillouin function has been substituted for the Langevin function. Its great success lies in the explanation of the presence of spontaneous magnetization in a ferromagnetic substance, but there are also difficulties. The fact that the unmagnetized state is unstable appears to be contrary to experience, since it is well known that a piece of iron can be demagnetized by dropping it. Moreover, in a single crystal the magnetization can be restored by applying an external field of less than $10^{-4}$ T, although the internal field is about $10^3$ T! We also require some explanation of the hysteresis curve. To overcome these difficulties Weiss introduced the concept of domains of magnetization within the specimen. Each domain contains some $10^{17}$–$10^{21}$ atoms, and a piece of unmagnetized iron contains many domains all spontaneously magnetized, but the directions of magnetization of different domains are oriented at random. The theory of spontaneous magnetization applies to a single domain, but the magnetization of the whole specimen depends on whether the domains themselves are aligned towards the field or whether they are randomly oriented. This theory, which was conceived before the nature of the exchange interaction which causes the spontaneous magnetization was known, is remarkably successful in explaining the main features of ferromagnetic substances. The existence of domains has been confirmed by the experiments of Bitter, briefly described in the next section, where we shall first consider what factors determine the size and shape of the domains.

### 15.3. Ferromagnetic domains

A considerable advance in the understanding of ferromagnetism occurred when it became possible to obtain single crystals of iron, cobalt, and nickel sufficiently large for their magnetization curves to be measured. In each case it was found that the crystals are anisotropic, that is, the magnetization depends on the direction the field makes with the crystal axes. Fig. 15.3 shows the curves for iron, which forms body-centred cubic (b.c.c.) crystals. The $M-B_0$ curve is found to rise more steeply when $B_0$ is

FIG. 15.3. Magnetization curves for a single crystal of iron. The directions of easy magnetization are the cube edges (for example, [100]). When the field is not along a cube edge, the initial process is of magnetization along the cube edges in directions nearest to that of the field; hence the curve for [110] breaks off roughly at $M_0/\sqrt{2}$, and that for [111] at $M_0/\sqrt{3}$, since further magnetization requires domain rotation against the anisotropy energy.

parallel to the edge of the unit cube [100] than any other direction, such as a face diagonal [110] or a body diagonal [111]. The energy of magnetization is $\int B_0\,dM$, and is represented by the area between the magnetization curve and the $M$-axis ($B_0 = 0$). This energy is least when the single crystal of iron is magnetized along the [100] direction (or its equivalents, [010] and [001]), and these are known as directions of easy magnetization. In the case of nickel, with a f.c.c. structure, the directions of easy magnetization are the body diagonals, while for cobalt, with a hexagonal structure at room temperature, there is only one direction of easy magnetization, the hexagonal crystal axis.

The excess energy required to magnetize the substance in a hard direction is known as the anisotropy energy. It is clear that the anisotropy energy cannot arise from the exchange interaction, for the latter depends only on the mutual orientation of the dipoles and not on the angle which they make with the crystal axes. Its origin is thought to be similar to that of paramagnetic anisotropy (see end of § 14.7), arising from the combined effect of spin–orbit coupling and the electric field of the neighbouring charged ions. The anisotropy energy has the same symmetry properties as the crystal, and is smallest for crystals of high symmetry. Thus it is less for iron or nickel, which are both cubic, than for cobalt, which has only axial symmetry. The anisotropy energy also causes a change of length on magnetization (magnetostriction).

In zero field the specimen, whether it is a single crystal or an aggregate of crystals, will be in equilibrium when its potential energy is a minimum. In an unstrained crystal the important contributions are the exchange energy, the anisotropy energy, and the magnetostatic energy (the energy stored in the magnetic field). If the crystal consisted of one single domain, as in Fig. 15.4(a), the 'free poles' at the ends would give rise to a large external magnetic field and to a large magnetostatic energy. This is reduced by having two domains oppositely magnetized as in Fig. 15.4(b), when the two poles partially cancel one another. If there are no free poles on any surface the magnetostatic energy is reduced still further. For this to be the case, the flux density **B** at the surface of the crystal must always be parallel to the surface, and the normal component of **B** must be continuous across the boundary between two domains. If the two domains are magnetized in perpendicular directions the wall between them must run at an angle of 45° to each direction of magnetization, and Fig. 15.4(c)

(a)          (b)          (c)          (d)

FIG. 15.4. Idealized domain structures in a single crystal, where the directions of easy magnetization are along the edges of a cube. (a) Single domain; external lines of field run from north to south pole and give large external field. (b) Double domain, where external lines of field run mostly between adjacent north and south poles, and the energy stored in external field is much reduced. (c) Arrangement with no free poles and no external field; the domains with perpendicular magnetization at top and bottom are called 'domains of flux closure'. (d) As in (c), but with further subdivision into smaller domains.

shows a possible arrangement. The little surface domains which produce a closed circuit of **B** are called domains of closure, and are generally much smaller than the inner domains. The size of the domains depends very much on the size and shape of the crystal, and this depends on the previous history of the substance. This fits in with the fact that the hysteresis curve is very sensitive to the composition and state of the specimen, since the domain structure must determine the shape of this curve.

The configuration in Fig. 15.4(d) is an alternative to that of Fig. 15.4(c), and we might expect the domains always to be very small in size and large in number; but energy is required to form the boundary between two domains, since the magnetization on either side is in opposite directions. The boundary between two domains is known as a 'Bloch wall'. It has a finite thickness, extending over a number of atoms whose spins change gradually in direction as we proceed through the wall (Fig. 15.5).

FIG. 15.5. Variation of spin orientation in a Bloch wall.

From eqn (15.3) the exchange energy between neighbouring identical spins is approximately $\Delta W_e = -2\mathscr{J}'J^2 \cos \psi$, where $\psi$ is the angle between the directions of the spin momentum vectors. Therefore the total exchange energy in going through the wall is

$$W_e = -\sum_{i>j} 2\mathscr{J}'J^2 \cos \psi_{ij}.$$

If the wall thickness extends over many atoms, and the angle between neighbouring spins is small, we may write $\cos \psi_{ij} \approx 1 - \psi_{ij}^2/2$, and the total increase in exchange energy because the spins are not exactly parallel is

$$W_e \approx \mathscr{J}'J^2 \sum \psi_{ij}^2.$$

For a wall which forms the boundary between two domains where the spins are antiparallel, the total change in angle in going through the wall is $\sum \psi_{ij} = \pi$. If there is a line of $n'$ atoms in the thickness of the wall, and $\psi_{ij}$ is the same for all adjacent pairs of atoms, $n'\psi_{ij} = \pi$ and

$$W_e \approx n'\mathscr{J}'J^2(\pi/n')^2 = \pi^2 \mathscr{J}'J^2/n'.$$

This equation shows that the exchange energy is reduced by making $n'$ large, and it would seem that the wall should be infinitely thick. This would increase the anisotropy energy $W_a$, however, since a number of spins in the wall are pointing at an angle to the direction of easy magnetization, and this number increases with the wall thickness. Thus $W_a$ is proportional to $n'$, and the total energy per unit area of wall in the substance is

$$W_e + W_a = \pi^2 \mathscr{J}' J^2 / n' a^2 + K n' a, \qquad (15.12)$$

where $K$ is a constant roughly equal to the anisotropy energy per unit volume. $a$ is the lattice constant of the substance, so that, for a simple cubic crystal, there are $1/a^2$ atoms per unit area of wall, and $n'a$ is the thickness of the wall.

The form of eqn (15.12) shows that there will be a minimum value of the total energy for some value of $n'$, which by differentiation is found to be $n' = (\pi^2 \mathscr{J}' J^2 / K a^3)^{\frac{1}{2}}$. For Ni, $J = \frac{1}{2}$, $\mathscr{J}$ is about $10^{-21}$ J, $K$ is about $10^4$ J m$^{-3}$, and $a^3$ is about $10^{-29}$ m$^3$. Hence $n'$ is of the order of 100 atoms, and the thickness of the wall is a few tens of nanometres. Substitution of the optimum value of $n'$ in eqn (15.12) gives the expression $2\pi(\mathscr{J}' K J^2 / a)^{\frac{1}{2}}$ for the wall energy per unit area, whose order of magnitude is found to be about $10^{-3}$ J m$^{-2}$.

As the domain width decreases in the flux closure arrangement shown in Fig. 15.4(d), the number of walls per unit area of the crystal surface increases, with a corresponding increase in the energy. The energy required to form a wall therefore tends to keep the domains small in number, and large in size. When $K$ is large, particles of about 1 $\mu$m diameter are found to consist of a single domain, because the energy required to form a wall is more than the reduction in the magnetostatic energy which would result from the subdivision into domains. In large crystals another factor which enters into the determination of domain size is that the domains of closure in Fig. 15.4(d) may require to be magnetized in a hard direction, thereby increasing the anisotropy energy. The volume occupied by the domains of closure decreases as the width of the domains decreases, and the anisotropy energy therefore tends to reduce the domain size, while the wall energy tends to increase it. The optimum domain size is determined by a compromise between these two effects.

The most striking evidence for the existence of domains is provided by the Bitter patterns which are obtained when finely powdered iron or cobalt, or colloidal magnetite, is spread on the surface of the crystal. The surface must be very carefully prepared and electrolytically polished to remove irregularities. The particles deposit themselves along the domain boundaries since here there are strong local inhomogeneous magnetic

fields which attract the particles. A typical Bitter pattern is shown in Fig.
15.6; the 'fir-tree' effect is obtained when the surface makes a small angle

FIG. 15.6. Domain patterns on a demagnetized single crystal of silicon–iron (the surface is
very nearly a (100) crystal plane). The magnetization is normal to the fine scratches visible
on the surface, and is directed as shown in the key diagram above. Domain walls labelled *a*
form the boundary between domains magnetized in directions differing by 90°, and those
labelled *b* are boundaries between domains differing by 180°. The 'fir-tree' closure domains
arise because the surface is not exactly a crystal plane. Two types of closure domain
(labelled 1 and 2) can be seen on the 90° wall. (Photograph by L. F. Bates and A. Hart.)

of 2° or 3° with the true (100) crystal plane. The branches of the tree are the domains of closure which close the flux circuit over the primary domains below. On looking through a microscope the patterns can be seen to change as a magnetic field is applied. The direction of magnetization in a domain is found by making a tiny scratch on the surface with a fine glass fibre. If the scratch is parallel to the magnetization the pattern is unchanged, but if it is normal to it the pattern is distorted. This is because a scratch parallel to the field behaves as a long narrow cavity, with no free poles at the ends; a scratch perpendicular to the field will have induced poles on its sides, and there will be a strong field in the cavity so that the pattern is distorted. Experiments of this type, and others, in which the scattering of beams of electrons or polarized neutrons have been used to investigate domain structure, show that the theory outlined above is correct in its main features.

The changes in the domain structure which occur when a magnetic field is applied, and the correspondence between these changes and the various parts of the magnetization curve, have already been outlined in § 6.4. The initial portions of the magnetization curve are associated with movements of the Bloch walls, which are reversible in small fields but irreversible after larger fields have been applied. Where there are strains or inclusions of impurities the energy depends on the position of the wall, as can be seen from considering the effect of a small particle embedded in the material. Such a particle will be a small domain magnetized in one of its own easy directions of magnetization, which do not in general coincide with those of the surrounding material, or it may be a particle of a non-ferromagnetic substance. In the latter case there will be free poles at its surface, as in Fig. 15.7(a), and the field of these poles gives extra

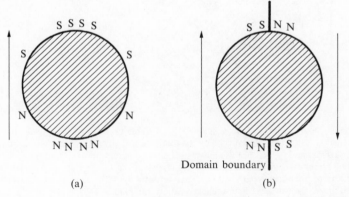

FIG. 15.7. Effect of a non-magnetic inclusion, (a) in the middle of a domain, and (b) when intersected by a domain boundary.

magnetostatic energy. If a Bloch wall intersects the particle, as in Fig. 15.7(b), this energy will be reduced, just as in the case of free poles on the surface of a ferromagnetic substance in Fig. 15.4(a) and (b). This gives a minimum of energy when a wall intersects as many inclusions as possible. In a small external field the wall is displaced slightly away from the minimum energy, but returns when the field is removed; this gives a reversible wall movement. In larger fields the wall may be shifted to a more distant position where the energy curve has passed through a maximum and then diminished; on removing the field the wall cannot cross the energy maximum and so is unable to return to its initial position. The displacement is then irreversible. The more free the material is from strains and inclusions, the greater the size of reversible wall movements, and the lower the field required to produce a movement, thus giving a large initial permeability, and a 'soft' magnetic material. With large strains and many inclusions the smaller is the possibility of boundary movement, and the higher the coercive force.

### 15.4. The gyromagnetic effect

It was pointed out in § 14.1 that the magnetic moment of an atom is proportional to the total electronic angular momentum of the atom. For a macroscopic system, the total magnetic moment $\mathbf{M}$ and the total electronic angular momentum $\mathbf{G_e}$ are formed by similar vector addition of the individual components, and they should therefore be related in the same way. Thus

$$\mathbf{M}/\mathbf{G_e} = \gamma = -g'e/2m_e,$$

where $g'$ is an effective Landé factor. It follows from this that if we could measure in some way the change in electronic angular momentum associated with the change in magnetization of a specimen, the value of $g'$ would be determined. Since $g'$ differs by a factor of 2 according to whether the magnetic moments are associated with orbital or spin angular momentum, this affords a method of verifying the assumption that ferromagnetism in the 3d group is associated with the electronic spins.

Since no external couple is exerted on a specimen by the act of changing its magnetization, the total angular momentum of the system must remain unaltered. The change $\Delta G_e$ in the electronic angular momentum must be accompanied by an equal and opposite change

$$\Delta G_{\text{lat}} = -\Delta G_e$$

in the angular momentum of the 'lattice', defined as the rest of the specimen, apart from the electrons responsible for the magnetization. It is this latter change in angular momentum which is observed, but it is very small. In a cubic centimetre of nickel there are some $10^{23}$ electrons whose individual momenta can be changed by $\hbar \approx 10^{-34}$ J s by reversal of the spin. The total angular momentum thus imparted to the lattice is only about $10^{-11}$ J s.

A variety of experimental methods have been used to determine $\gamma$, but only a short account will be given here (more details are given by the authors to whom references are made in this section). The methods fall into two classes. In one, an unmagnetized specimen is set into rotation and the resultant magnetization is measured. This is the Barnett effect, and typical experiments are those of Barnett (1944). Even with a large specimen, the magnetic moment induced is very small owing to the limited velocities of rotation which can be employed. In the second class, the magnetization is changed by a known amount and the change in angular momentum is determined; this is known as the Einstein–de Haas effect, though first suggested by Richardson. This method has the advantage that resonance can be used to enhance the effect. A ferromagnetic rod is suspended inside a long solenoid supplied with alternating current whose period is equal to the torsional oscillation period of the suspended rod. If $\mathscr{I}$ is the moment of inertia of the rod, $b$ the damping constant, $c$ the torsion constant of the suspension, and $M \sin \omega t$ the magnetic moment of the specimen at any instant, the equation of motion is

$$\mathscr{I}\frac{\mathrm{d}^2\theta}{\mathrm{d}t^2} + b\frac{\mathrm{d}\theta}{\mathrm{d}t} + c\theta = \frac{\mathrm{d}G}{\mathrm{d}t} = \frac{1}{\gamma}\omega M \cos \omega t.$$

At resonance, the amplitude of the angle of rotation is $(1/\gamma)(M/b)$; $b$ is found from the logarithmic decrement, and $M$ must be measured independently.

This method of measuring $\gamma$ was employed by Scott (1951) using a modification of the apparatus built for the determination of $e/m$ of the carriers of electric current (see § 3.1). The specimen, in the form of a rod, is suspended as a torsional pendulum. A coil is wound on the rod, and by reversing a current in this coil the magnetization of the rod can be reversed. The change in magnetization is measured by a null magnetometer placed half-way between the rod and a standard coil carrying a steady current which is simultaneously reversed with that in the specimen. This steady current is adjusted until a balance is obtained. The magnetometer is fitted with a mirror, and the light reflected from it falls on a pair of photocells each followed by an amplifier; by means of detection of the difference in the two outputs, a null magnetometer of great sensitivity is produced. A correction was made for the non-uniformity of magnetization of the rod, and the earth's field was neutralized by a system of Helmholtz coils. The period of oscillation of the rod was 26 s, and its rotation was observed by reflections of a beam of light from a mirror mounted immediately above the specimen. The procedure used was to reverse the magnetizing current at a moment when the specimen passed through the centre of its swing. The direction of reversal was chosen so that for 60 current reversals the amplitude was increased, and then for 60 reversals it was decreased. With small damping, the progressive change in amplitude was very nearly linear, and the amplitude change for one reversal was obtained from the two slopes of the plot for 120 reversals.

A number of experiments of high precision have been carried out using both the Einstein–de Haas and Barnett effects, and a mean of the results obtained between 1944 and 1960 is given in a survey by Meyer and Asch (1961), who show also that there is good agreement with results of ferromagnetic resonance experiments using microwave radiation (see § 24.7). The results are shown in Table 15.2, and are expressed in terms of two quantities g and g′, obtained from ferromagnetic resonance and gyromagnetic experiments respectively. When we have a mixture of orbit and spin, the magnetic moment and angular momentum may be written as $M = M_L + M_S = (e/2m_e)(G_L + 2G_S)$, $G = G_L + G_S$, and the ratio is

$$\gamma = \frac{M}{G} = \frac{e}{2m_e}\left(\frac{G_L + 2G_S}{G_L + G_S}\right) = g'\left(\frac{e}{2m_e}\right), \tag{15.13}$$

The 'spectroscopic splitting factor' g measured in a ferromagnetic resonance experiment has been shown by Kittel and Van Vleck to be defined by

$$\frac{M}{G_S} = \frac{e}{2m_e}\left(\frac{G_L + 2G_S}{G_S}\right) = g\left(\frac{e}{2m_e}\right), \tag{15.14}$$

from which it follows that

$$1/g + 1/g' = 1. \tag{15.15}$$

The results in Table 15.2 show that this relation is fulfilled within the experimental error for iron and nickel, and Meyer and Asch show that this is true also for a wide range of alloys of the 3d group. The fact that g, g′ are so close to 2 shows that the magnetism of the ferromagnetic

TABLE 15.2

*Some values of the quantities g′ and g*

| Substance | g′ | g | $1/g + 1/g'$ |
|---|---|---|---|
| Iron | 1·928±0·004 | 2·094±0·003 | 0·996±0·004 |
| Cobalt | 1·854±0·004 | — | — |
| Nickel | 1·840±0·008 | 2·185±0·010 | 1·001±0·009 |

The quantity g′ is derived from gyromagnetic (magnetomechanical) experiments, the quantity g from ferromagnetic resonance experiments; the values quoted are the means of a number of experimental results, given by Meyer and Asch (1961).

metals of this group is almost entirely due to spin. This result is similar to that found for the paramagnetism of salts of 3d group ions (Chapter 14), and there is little doubt that it is due essentially to the same cause, 'quenching' of the orbital moment by electrostatic interaction with the neighbouring (ligand) ions.

## 15.5. Thermal effects in ferromagnetism

When a substance is magnetized, with all the electron spins pointing in one direction, it is in a state of greater order than when it is unmagnetized, with the spins pointing in random directions. The magnetized state is therefore one of lower entropy than the unmagnetized state, and in passing from the former to the latter there will be an increase in the entropy of the spin system. If the transition is accomplished by heating a ferromagnetic substance through its Curie point, the entropy change appears as an anomaly in the specific heat. If it is accomplished by the sudden (adiabatic) removal of a magnetic field, the entropy change appears as a fall in the temperature of the substance; this is known as the magnetocaloric effect. Both this effect and the specific-heat anomaly have been used to obtain information about the ferromagnetic state.

The specific heat of a substance is $C = T(dS/dT)$, where the entropy change $dS$ is given by the relation

$$T \, dS = dU - \frac{1}{\rho} B \, dM; \tag{15.16}$$

$dU$ is the change in the internal energy, and $-B \, dM$ is the increase in the magnetic potential energy when the magnetization is increased by $dM$ at constant flux density $B$. The density $\rho$ appears because $M$ is the magnetization per unit volume, while the specific heat (and other thermal quantities) are per unit mass. In a ferromagnet, $B = B_0 + \lambda M$, and the specific heat is thus

$$C = T \frac{dS}{dT} = \frac{dU}{dT} - \left( \frac{B_0 + \lambda M}{\rho} \right) \left( \frac{dM}{dT} \right)$$

$$= C_M - \left( \frac{B_0 + \lambda M}{\rho} \right) \left( \frac{dM}{dT} \right). \tag{15.17}$$

Below the Curie point any external field $B_0$ is very small in comparison with the internal field $\lambda M$, so that we can write

$$C = C_M - \frac{1}{2} \left( \frac{\lambda}{\rho} \right) \frac{d(M^2)}{dT}. \tag{15.18}$$

Here $C_M$ is the specific heat of the substance at constant magnetization, while the second term arises from the change in magnetization with temperature. Since $M$ falls with increasing $T$, it gives a positive contribution to the specific heat (as the temperature rises, the degree of order in the magnetic system decreases, and the entropy associated with the magnetization increases). Fig. 15.2 shows that the rate of change of $M$ with temperature is greatest just below the Curie point, and the anomalous specific heat arising from the magnetic properties should be greatest at this point, followed by a sharp drop above the Curie point where $M$ becomes zero.

An experimental curve showing the variation of $C$ with $T$ for nickel is given in Fig. 15.8. The anomalous specific heat is appreciable only near the Curie point, where it appears as a sharp discontinuity. In order to obtain a value for $\lambda$ from the specific heat anomaly, $C_M$ must be estimated and subtracted from the measured specific heat, so that only the magnetic contribution remains. Measurements are made at constant pressure, so that we can write $C_M = C_V + (C_p - C_V) + C_e$. $C_V$ is obtained by extrapolation, using the Debye formula, from measurements at low temperatures; $(C_p - C_V)$ may be found from the expansion coefficient and compressibility using a standard thermodynamical formula; $C_e$ is the electronic specific heat. This is abnormally large in a ferromagnetic metal and difficult to estimate since it is associated with a high electron density in the 3d band (see § 12.6). $dM^2/dT$ must be found by plotting $M^2$ as a function of temperature, and then $\lambda$ is obtained. This is not a very accurate method of finding $\lambda$ and the value does not agree too well with

FIG. 15.8. The molar heats $C/R$ of copper and nickel, showing the excess in the ferromagnetic metal, and the sharp drop at the Curie point.

the value obtained from the magnetization curve, probably because of errors in $C_e$. However, the general form of the specific-heat curve is not incompatible with theory and this also applies to iron and cobalt, although the measurements on these metals are less certain (for experimental details, see Grew (1934)). The general form of the specific-heat anomaly ('lambda type') is typical of a 'co-operative' transition from an ordered to a disordered state.

If a field is applied to a magnetic substance, there is in general an increase in magnetization, and this results in a state of greater order than in zero field. In other words the entropy of the system has decreased, and the loss of (magnetic) potential energy of the dipoles in turning towards the field appears as heat of magnetization. If the field is switched off isothermally, heat is absorbed. If the field is switched off adiabatically the entropy of the system must remain constant; the increase in entropy due to increased disorder of the dipoles is then compensated by a decrease in the entropy associated with thermal agitation, and there is therefore a fall in temperature. This is the basis of the 'magnetic cooling' method for obtaining temperatures below 1 K using paramagnetic substances. This 'magnetocaloric' effect also has applications to ferromagnetics. Since $dS = 0$ in a reversible adiabatic process, we have from eqn (15.17)

$$dT = \left(\frac{B_0 + \lambda M}{\rho C_M}\right) dM. \tag{15.19}$$

Above the Curie point saturation effects are negligible and $M/B_0$ is a constant at a given temperature, so that in a finite change of the magnetization we have

$$\Delta T = \left(\frac{B_0/M + \lambda}{2\rho C_M}\right) \Delta(M^2). \tag{15.20}$$

Below the Curie point we can neglect $B_0$ in comparison with $\lambda M$, and we obtain

$$\Delta T = \left(\frac{\lambda}{2\rho C_M}\right) \Delta(M^2). \tag{15.21}$$

If the external field is initially zero, so that the magnetization of each domain has the spontaneous value $M_0$, the temperature rise on applying a field is

$$\Delta T = \left(\frac{\lambda}{2\rho C_M}\right)(M^2 - M_0^2). \tag{15.22}$$

If $\Delta T$ is plotted as a function of $M^2$, a curve of the form shown in Fig. 15.9 is obtained. It becomes a straight line in the region where the external field is large enough to change the magnetization of the domains, with a curved tail at lower fields where the magnetization of the substance

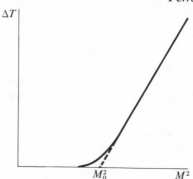

FIG. 15.9. Curve showing the variation of $\Delta T$ with $M^2$ in the magnetocaloric effect.
$\Delta T = (\lambda/2\rho C_M)(M^2 - M_0^2)$.

is mainly due to wall movements or the rotation of domains. Extrapolation of the straight portion to the axis $\Delta T = 0$ gives $M_0^2$ from the intercept.

The magnetocaloric effect may be used for a number of purposes, such as investigation of the hysteresis curve, one of the most important being the determination of the spontaneous magnetization $M_0$ near the Curie point. A good description of experimental technique is given by Oliver and Sucksmith (1953) in work on a copper–nickel alloy (24%Cu; 76%Ni).

## 15.6. Measurement of the spontaneous magnetization $M_0$ as a function of temperature

In § 15.2 it was shown that the spontaneous magnetization of a single domain should obey an equation of state which depends only slightly on $J$ (see Fig. 15.2). To test this relation it is necessary to determine the value of $M_0$ for a single domain at zero field over a wide range of temperature. Since in practice any specimen consists of a number of domains randomly oriented, so that (apart from remanence) the net magnetization will be zero, it follows that the spontaneous magnetization of a single domain cannot be directly measured. If we apply a sufficiently strong field, however, the various domains will rotate until they point in the direction of the external field, and the resultant magnetic moment will be close to the spontaneous magnetization of the individual domains. It will slightly exceed it, since the magnetization under these conditions corresponds, not to the point P in Fig. 15.1, but to the point P', the stable state in the presence of a magnetic field. In order to find the value corresponding to P, we must make measurements of $M$ for a range of values of the external field, and then extrapolate back to zero field. Since the fields which are applied are small compared with the internal field, the point P' is never

far from P, and the extrapolation required is not very great at temperatures well below the Curie point. Near the Curie point the magnetocaloric
effect is used as described in the previous section.

In a number of magnetic materials the nucleus of the magnetic ion
possesses a nuclear magnetic dipole moment $\mathbf{m}_N$ which interacts with the
magnetic flux density $\mathbf{B}_e$ of the electrons (see § 14.10). ($\mathbf{B}_e$ is the actual
magnetic flux density at the nucleus generated by the magnetic electrons,
and is nothing to do with the effective molecular field of flux density $\mathbf{B}_{int}$
introduced by Weiss to explain ferromagnetism.) The interaction energy

$$W = -\mathbf{m}_N \cdot \mathbf{B}_e$$

gives a hyperfine splitting of the nuclear levels, from observation of which
$B_e$ can be found if the nuclear moment is known. In a ferromagnetic
substance $B_e$ is parallel to the magnetization, and its time-average value is
proportional to the average magnetic moment on each ion (apart from
some small corrections). Thus $B_e$ is proportional to the magnetization,
and observation of the hyperfine structure separation as a function of
temperature gives a convenient and accurate method of determining the
saturation magnetization curve. This can be done in zero external field,
since it is not necessary to line up the domains, and many of the
difficulties of direct measurement of the bulk magnetization are avoided.

The magnitude of the electronic flux density $B_e$ lies generally between
$10$ T and $10^3$ T. The nuclear levels have energy $W_{m_I} = -g_N \mu_N m_I B_e$, where
$m_I$ is the nuclear magnetic quantum number, and are equally separated
by an amount corresponding to a frequency of $10^2 - 10^4$ MHz. Two
methods are available for measuring this separation over a range of
temperature. One of these is the Mössbauer effect, in which a low-energy
$\gamma$-ray is emitted from a nucleus in an excited state and absorbed by a
nucleus in the ground state. A $\gamma$-ray photon of energy $h\nu$ carries
momentum $h\nu/c$, so that the emitting or absorbing nucleus is given a
recoil momentum, and hence takes up energy which reduces the photon
energy. If the nuclei are in a solid the recoil momentum is generally taken
up by the solid as a whole, and the energy taken from the $\gamma$-ray is
negligible. Thus in a solid, unlike a gas (see Problem 15.3), there is no
spread in energy of the photon due to the varying amounts of energy
taken up by the recoil.

Only those $\gamma$-rays which are extremely narrow are of use, since the
width of the $\gamma$-ray must be smaller than the hyperfine splitting. For
magnetic purposes, the $14\cdot4$ keV transition between the excited state
($I = \frac{3}{2}$) and the ground state ($I = \frac{1}{2}$) of the isotope $^{57}$Fe has been especially
useful. This gives a linewidth of about 3 MHz, and the hyperfine levels
and structure of the Mössbauer $\gamma$-ray are shown in Fig. 15.10. The
structure is exactly analogous to the Zeeman effect in an atomic transition. The splittings are a very small fraction of the $\gamma$-ray frequency, and

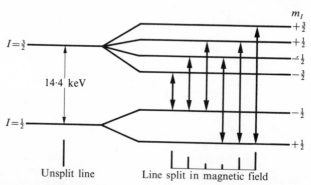

FIG. 15.10. Hyperfine splitting of the nuclear states of $^{57}$Fe in a magnetic field. The ground state $I = \frac{1}{2}$ has $g_N = +0\cdot18$, and the excited state $I = \frac{3}{2}$ has $g_N = -0\cdot010$; the allowed transitions are those for which $\Delta m_I = 0, \pm1$. Note the gross disparity in scale; the hyperfine splittings are about $10^{-11}$–$10^{-12}$ of the $\gamma$-ray frequency. In some substances there is also an electric quadrupole interaction in the $I = \frac{3}{2}$ state.

their analysis is made by means of the Doppler effect produced by a relative motion of the source and absorber. It is convenient to use a source with no hyperfine structure, such as $^{57}$Fe (derived from the nuclear decay of $^{57}$Co) in stainless steel, which is non-magnetic, since this gives a single emission line, as shown on the left of Fig. 15.10. For this to be absorbed by a $^{57}$Fe nucleus in a magnetic substance, where six transitions are allowed with slightly different frequencies as on the right of Fig. 15.10, a Doppler shift is needed of the correct velocity to bring one of the transitions to the same frequency as the single line on the left. Thus the entire hyperfine pattern can be scanned by systematically changing the relative velocity of source and absorber (the velocity required is of order a few millimetres per second). Of course source and absorber can be interchanged, and the choice is determined by experimental convenience.

The hyperfine field in metallic iron has been determined as a function of temperature by means of the Mössbauer effect; the results (see Fig. 15.11) of Nagle, Frauenfelder, Taylor, Cochran, and Matthias (1960) show close agreement with the saturation curve determined by conventional means. At sufficiently low temperatures for the magnetization to reach the saturation value, the hyperfine field is 33 T. On applying an external field parallel to the magnetization Hanna, Heberle, Perlow, Preston, and Vincent (1960) found that the net field at the $^{57}$Fe nucleus was reduced, showing that the hyperfine field was in the opposite sense to the external field, and hence also to the magnetization.

The hyperfine splitting of the ground nuclear levels in a ferromagnetic substance has also been measured by the method of nuclear magnetic resonance (see § 24.5). For $^{57}$Fe this gives a direct observation of transitions between the states $m_I = +\frac{1}{2}$ and $-\frac{1}{2}$ of the ground state $I = \frac{1}{2}$, at a

FIG. 15.11. The hyperfine magnetic field at a $^{57}$Fe nucleus in metallic iron, relative to that at room temperature, plotted against the reduced temperature $T/T_0$. The experimental points are measured by the Mössbauer effect, the solid line indicates the relative saturation magnetization as determined by a bulk measurement (see Nagle *et al.* 1960).

frequency such that

$$h\nu = W_{-\frac{1}{2}} - W_{+\frac{1}{2}} = g_N \mu_N B_e, \qquad (15.23)$$

where $g_N$ is the value for the ground state $I = \frac{1}{2}$. This gives a more precise measurement of $B_e$ than the Mössbauer method (in which the minimum linewidth is determined by the lifetime of the excited nuclear state). In iron at 0 K, the resonance frequency of 46·648 MHz has been determined to 1 kHz by Riedi (1973).

## 15.7. Spin waves

The molecular field theory introduced by Weiss is remarkably successful in giving a simple interpretation of the phenomenon of ferromagnetism. When combined with the Brillouin function it predicts a unique curve for the reduced magnetization as a function of reduced temperature for each value of **J**; the prediction is in good but not exact agreement with experiment (see Fig. 15.2). In this section we outline a more sophisticated approach, first formulated by Bloch (1932), in which the collective excitations of the assembly of magnetic dipole moments are analysed in terms of wave-like motions known as 'spin waves'. Formulae for the specific heat and magnetization are obtained in the form of expansions which are valid for the co-operative state at low temperatures.

At 0 K, where the magnetization has the saturation value, all the spins are rigorously parallel, but this is obviously not so at a non-zero temperature where the magnetization is smaller. However, it would be incorrect to regard the reduction in magnetization as due to the reversal of any given individual spin, because any such deviation would be passed on to neighbouring spins through the exchange interaction in a time of order $\hbar/\mathscr{J}'$, so that it would not remain localized on any given atom. In fact the average deviation of each spin from exact parallelism is small, and can be analysed in terms of sinusoidal spatial variations throughout the crystal, known as 'spin waves'. In a spin wave of wave vector $\mathbf{k}_s$, the angle between adjacent spins a distance $\mathbf{r}_0$ apart is $\mathbf{k}_s \cdot \mathbf{r}_0$, and since the exchange energy varies with the cosine of this angle, the extra energy required to excite a wave is proportional to $\mathscr{J}'(1-\cos \mathbf{k}_s \cdot \mathbf{r}_0) \sim \frac{1}{2}\mathscr{J}'(\mathbf{k}_s \cdot \mathbf{r}_0)^2 = \frac{1}{2}\mathscr{J}' k_s^2 r_0^2 \cos^2\theta_{k,r_0}$, where $\theta_{k,r_0}$ is the angle between $\mathbf{k}_s$ and $\mathbf{r}_0$. This must be summed over all neighbours, giving

$$W_k = \hbar\omega_k = Dk_s^2. \tag{15.24}$$

This is the dispersion relation connecting the frequency and wavelength of a spin wave. $D$ is proportional to the exchange energy $\mathscr{J}'$, and in a cubic lattice with $z$ equidistant neighbours with angular momentum $J$,

$$D = \frac{1}{3}\mathscr{J}' z J r_0^2. \tag{15.25}$$

The energy required to excite a spin wave is proportional to $k_s^2$, and at low temperatures only long waves will be excited; in the limit at $T = 0$ the only wave present will be $k_s = 0$, which corresponds to all the dipoles being parallel. As the temperature rises more spin waves of shorter wavelengths will be excited, and the energy required for this shows up as an additional term in the specific heat. This is quite distinct from the abnormal electronic specific heat discussed above, and would be present in a ferromagnetic insulator. It is purely magnetic in origin, and constitutes the low-temperature tail of the magnetic specific-heat anomaly discussed in § 15.5. Van Kranendonk and Van Vleck (1958) have shown that a spin wave behaves formally like a harmonic oscillator, its mean energy being

$$\bar{W}_k = \frac{\hbar\omega_k}{\exp(\hbar\omega_k/kT)-1} = \bar{n}\hbar\omega_k, \tag{15.26}$$

where $\bar{n} = \{\exp(\hbar\omega_k/kT)-1\}^{-1}$ is known as the 'occupation number'. The number of spin waves of wave vector $k_s$ (we have used $k_s$ to avoid confusion with Boltzmann's constant $k$ which occurs in eqn (15.26) is $g(k_s)\,dk_s = (V/2\pi^2)k_s^2\,dk_s$, (cf. eqn (11.5)). Hence the internal energy at temperature $T$ is

$$U = \int \bar{n}(\hbar\omega_k) g(k_s) \, dk_s$$

$$= \frac{V}{2\pi^2} \int \frac{(Dk_s^2) k_s^2 \, dk_s}{\exp(Dk_s^2/kT) - 1}$$

$$= \frac{VD}{4\pi^2} \left(\frac{kT}{D}\right)^{\frac{5}{2}} \int \frac{x^{\frac{3}{2}} \, dx}{\exp x - 1},$$

where we have made the substitution $x = (Dk_s^2/kT)$. The integral can be taken to infinity at low temperatures, and is therefore just a numerical constant. Differentiation with respect to temperature gives

$$C_V = dU/dT = c(kT/D)^{\frac{3}{2}}, \tag{15.27}$$

where $c$ is a numerical constant whose value depends somewhat on the crystal structure; for a simple cubic lattice ($z = 6$) we have (per mole)

$$C_V = 0 \cdot 113 kV(kT/D)^{\frac{3}{2}} = 0 \cdot 113 k(V/r_0^3)(kT/2\mathcal{J}'J)^{\frac{3}{2}}$$

$$= 0 \cdot 113 R(kT/2\mathcal{J}'J)^{\frac{3}{2}}. \tag{15.28}$$

For the ordinary ferromagnetic metals this magnetic specific heat is small and difficult to measure in the presence of the abnormally large electronic specific heat, and the lattice specific heat. It has been detected in some non-conducting ferrimagnetic compounds (see § 16.3).

The departure of the magnetization from the absolute saturation value $M_s$ can be computed similarly, using the fact that each spin wave reduces the magnetic moment by an amount $g\mu_B\bar{n}$. Thus

$$M_s - M_0 = g\mu_B \sum \bar{n} = \frac{g\mu_B V}{2\pi^2} \int \frac{k_s^2 \, dk_s}{\exp(Dk_s^2/kT) - 1}$$

$$= \frac{g\mu_B V}{4\pi^2} \left(\frac{kT}{D}\right)^{\frac{3}{2}} \int_0^\infty \frac{x^{\frac{1}{2}} \, dx}{\exp x - 1}.$$

On substituting for $D$ and using the fact that $g\mu_B(V/r_0^3) = M_s$, in a simple cubic lattice, we obtain the relation

$$M_0/M_s = 1 - a(kT/2\mathcal{J}'J)^{\frac{3}{2}}, \tag{15.29}$$

where $a = 0 \cdot 059/J$ in a simple cubic lattice. This result is the first term of a power series where the next terms contain $T^{\frac{5}{2}}$, $T^{\frac{7}{2}}$, and $T^4$. The terms in $T^{\frac{5}{2}}$ and $T^{\frac{7}{2}}$ have been verified in a special case where they are unusually large (see Gossard, Jaccarino, and Remeika 1961), using nuclear magnetic resonance of the $^{53}Cr$ nucleus in $CrBr_3$, where the nuclear resonance frequency is accurately proportional to the magnetization. In gadolinium metal, where the spontaneous magnetization has been measured by Elliott, Legvold, and Spedding (1953) by the bulk magnetization method, the $T^{\frac{3}{2}}$ law holds closely almost up to the Curie point, a result that has been explained as due to a near cancellation of the higher terms (Goodings 1962). In iron, Riedi (1973) has shown that nuclear magnetic resonance measurements are consistent with a variation in the magnetization which can be fitted accurately by a single $T^{\frac{3}{2}}$ term up to 32 K, and up to 400 K with one extra term in $T^{\frac{7}{2}}$. These experimental results confirm the validity of the spin-wave approach; the existence of spin waves is confirmed in magnetic resonance experiments using thin ferromagnetic films (§ 24.7).

## 15.8. Mechanisms of exchange interaction

In this chapter we have assumed that ferromagnetism is due to exchange interaction; certainly no other interaction of the correct form is known whose magnitude is large enough to produce a co-operative magnetic state at room temperatures. However, attempts to calculate the direct interaction from first principles have proved disappointing both as regards sign and size, and it appears that indirect mechanisms of exchange interaction play a predominant role.

The original treatment of Heisenberg and Dirac deals with the interaction between two electrons on the same atom. If the two electrons did not influence one another the solution of the wave equation would be a simple product of the two solutions for a single electron, of the form

$$\psi_{\mathrm{I}} = \phi_k^{(1)}\phi_m^{(2)}.$$

The physical interpretation of this is that electron (1) is in orbital $k$, and electron (2) is in orbital $m$. Since the two electrons are equivalent, the energy is unchanged if the two electrons are interchanged, giving another solution

$$\psi_{\mathrm{II}} = \phi_k^{(2)}\phi_m^{(1)}.$$

In general, any linear combination of these two solutions is allowed, the correct combination being determined when we include the electrostatic energy $e^2/r_{12}$ (where $r_{12}$ is the distance between the two electrons) of repulsion between the two negatively charged electrons. The correct solutions are then the symmetric and antisymmetric combinations

$$\psi_{\mathrm{sym}} = (2)^{-\frac{1}{2}}(\psi_{\mathrm{I}}+\psi_{\mathrm{II}}),$$
$$\psi_{\mathrm{ant}} = (2)^{-\frac{1}{2}}(\psi_{\mathrm{I}}-\psi_{\mathrm{II}}).$$

These two solutions no longer have the same energy, because the symmetrical solution allows the wavefunction to have a large amplitude if the two electrons are at the same point, while the antisymmetric wavefunction then vanishes because $\psi_{\text{I}} = \psi_{\text{II}}$. Thus the electrostatic repulsive energy between the two electrons is larger in the first case than in the second.

So far it has not been necessary to include the electron spin. With two electrons, the spin states (two for each electron, hence four in all) are divided between the triplet states ($S = 1$, $M_s = 1$, 0, $-1$) and the singlet state $S = 0$, according to whether the two individual spins are parallel or antiparallel. The triplet states are symmetrical with respect to interchange of the two electrons, the singlet state is antisymmetrical. Since only states whose over-all symmetry is antisymmetrical are allowed in nature, the spin triplet states can only be combined with the orbital $\psi_{\text{ant}}$, and the singlet spin state with the orbital $\psi_{\text{sym}}$. Thus the difference in electrostatic energy between the symmetrical and antisymmetrical orbital states carries with it a corresponding energy difference between the spin singlet and triplet. This is formally similar to the introduction of a coupling energy between the electron spins of the type assumed in § 15.1.

For two electrons within the same atom, the exchange energy is always positive, so that the state of lower energy is with the spins parallel. This coupling is 'ferromagnetic' in nature, and is the justification for Hund's rule (§ 14.2) which makes the ground state of the atom the one with maximum multiplicity in the spin. In the phenomenon of ferromagnetism proper, however, we are concerned with exchange interaction between electrons on different atoms, and the electrostatic energy involved contains terms arising both from the repulsive forces between the two nuclei and between the two electrons and from the attractive forces between an electron on one atom and the nucleus of the other atom. (The much larger energy of attraction between an electron and the nucleus of its own atom has already been included in the wave equation for each electron.) In simple molecules like $H_2$ the over-all exchange term is easily shown to be negative, in agreement with the ground state of the molecule being a singlet, but with more complex ions such as the 3d group, opinions have differed as to whether the net effect (which obviously varies with interatomic distance) would be positive or negative at the ionic separations typical of 3d-group metals. Since improved wavefunctions have become available from electronic computers, attempts have been made to carry out calculations which might be reasonably realistic. Stuart and Marshall (1960) obtained a positive energy, though two orders of magnitude too small, but Freeman, Nesbet, and Watson (1962) find a negative energy. Thus 'direct exchange', due to direct overlap of the electronic wavefunctions, appears incapable of accounting for ferromagnetism. A negative exchange energy would give rise to 'antiferromagnetism' (see Chapter 16), where neighbouring spins are arranged antiparallel rather than parallel.

A striking feature of ferromagnetism is that it is observed practically only in metals; very few ferromagnetic insulators have been discovered (one example is $CrBr_3$ whose Curie temperature is 33·84 K). This qualitative difference suggests that the conduction electrons play an important role, and measurements of electrical resistivity show there must be an interaction mechanism with the magnetic dipoles which scatters the conduction electrons. For example, the resistivity of gadolinium metal at room temperature is over 100 times that of copper, and in general appreciable changes in the resistivity are observed near the Curie point in ferromagnetic substances. We can go further and question whether the electrons responsible for the observed magnetic dipole moments are 'localized' on given ions, as in an insulator, or whether they are itinerant electrons which should be treated by a 'collective electron' model using Fermi–Dirac statistics, as in Chapters 11 and 12.

The consequences of the latter model have been investigated mainly by Bloch, Slater, Stoner, and Wohlfarth. The electrons occupy certain energy bands, and the exchange interaction is introduced as an internal field $\lambda M$, proportional to the magnetization, as in the Weiss treatment. This gives a difference in energy between spin dipoles pointing parallel and antiparallel to the internal field, which can be represented by dividing the band into two halves, as in Fig. 14.22. The important difference is that the effective field is now proportional to the magnetization. This depends on the number of electrons transferred from one half-band to the other, which thus determines the energy separation between the two half-bands which causes the transfer. This is a self-balancing situation in which a transfer occurs and an internal field is created only if this results in a lower energy for the magnetized state than for the unmagnetized state. The number of electrons transferred for a given energy difference is proportional to the density of states $g(W)$, and it turns out that a decrease of energy results only if the density of states is sufficiently large. This condition is satisfied for d-electron bands but not for s electrons. Thus sodium does not become ferromagnetic, and remains a Pauli paramagnet (§ 14.11), but transition metals with overlapping d and s bands may have co-operative magnetic states.

This collective model has the advantage that the saturation magnetization is determined by the net transfer of spins between the two half-bands, so that there is no reason why the net dipole moment per atom should correspond exactly to an integral multiple of a Bohr magneton. For example, the saturation moment of nickel is about 0·60 $\mu_B$ per atom. The 3d band is nearly full of electrons, and the saturation moment can be fitted by assuming that in the ferromagnetic state the half-band of lower energy is full as in Fig. 15.12, while that of higher energy has about 0·6 holes per atom. The collective model also gives a neat explanation of the change in magnetic properties which occurs on alloying with non-magnetic metals. For example, the copper atom has one more electron than nickel; when copper is alloyed into nickel, the extra electrons fill up

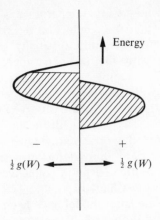

Fɪɢ. 15.12. Energy bands in nickel. The difference in energy of the two half-bands with opposite spin orientation is due to exchange interaction. The half-band of lower energy is full, while that with higher energy contains about 0·6 holes per atom.

the holes in the upper half-band, so that the average magnetic moment per nickel atom decreases steadily as the percentage of copper increases. Alloys with more than about 60 per cent of copper are not ferromagnetic, suggesting that at this point the 3d bands are full.

The collective model assumes that all the d electrons are 'itinerant' conduction electrons. The main evidence against this comes from neutron diffraction, where experiments indicate that the spatial distribution of the magnetic electrons in iron, cobalt, and nickel are similar to those in the free atoms. Furthermore, the excess magnetic specific heat in iron corresponds to an entropy of $R \ln 3$, that is, to a spin $S = 1$ on the iron atom. Obviously some d electrons must be itinerant to account for the electronic specific heat and various transport properties, but it has been suggested (Stearns 1973) that the fraction is only about 5 per cent. This corresponds to about half an electron per atom out of a d band which can accommodate ten electrons when full.

Whatever the exact degree of localization, there is general agreement that the itinerant electrons play a major role in the following way. As a result of direct exchange interaction with a local magnetic moment, the itinerant electrons in the vicinity become spin-polarized. The effect resembles the 'Pauli paramagnetism' discussed in § 14.11, but the polarization arises from a 'molecular field' (cf. § 15.1) and not from the true magnetic field of the local moment. The spin polarization of the itinerant electrons interacts with the spins on neighbouring local moments, producing an indirect exchange interaction between any pair of magnetic ions. This is a second-order interaction, but it proves to be more important than the direct exchange interaction between ions because it is a longer-range effect. For two ions separated by a distance $R$, a calculation involving a number of simplifying assumptions such as a spherical Fermi surface gives an indirect exchange interaction which falls off as

$$x^{-4}(x \cos x - \sin x),$$

where $x = 2k_F R$. The important features of this result are (1) the indirect exchange interaction is oscillatory and changes sign with distance, and (2) it decreases more slowly with distance than the direct exchange interaction, which drops exponentially with $R$. The order of magnitude is $\mathcal{J}_{mi}^2/W_F$, where $\mathcal{J}_{mi}$ is the exchange interaction between the spin of a localized moment $m$ and that of an itinerant electron $i$, and $W_F$ is the Fermi energy.

A mechanism of this type was first proposed for the interaction between nuclear spins in a metal, the hyperfine interaction between the nuclear spin and the itinerant electron spin appearing instead of the exchange interaction $\mathcal{J}_{mi}$; the oscillatory nature of the interaction, and its variation with distance were deduced by Rudermann and Kittel (1954). The corresponding effect for electron moments was proposed by Kasuya (1956) and de Gennes (1958) as the most important interaction in the lanthanide metals (see § 16.4), where the magnetic 4f electrons are strongly localized; it was extended to other systems by Yosida (1957).

*Exchange interaction in insulators*

In electrical insulators no conduction electrons exist to provide an exchange mechanism of the type just considered, but exchange interactions of considerable magnitude are found in compounds where the inter-ionic distance is so large that direct exchange between electrons localized on the magnetic ions must be negligible. As an example we consider a simple compound such as an oxide MO, where M is a dipositive ion of the 3d group (for example, $Mn^{2+}$, $Fe^{2+}$, $Co^{2+}$, $Ni^{2+}$). These have a f.c.c. structure similar to that of NaCl, a typical plane of atoms being shown in Fig. 15.13. Here the circles are drawn in proportion to the ionic radii, and it can be seen that the much larger anions separate almost completely even the nearest-neighbour cations. However, neutron diffraction results suggest that the stronger exchange interactions are between ions on next nearest neighbour sites such as A, B rather than between nearest neighbour sites such as A, C. Direct overlap between the wavefunctions of magnetic electrons on the cation sites is very small, but a purely ionic model where the magnetic electrons are localized on the cations is an oversimplification. Some degree of covalent binding is always present, whereby the wavefunctions of the magnetic electrons spread over onto the adjacent anions. Direct overlap of the magnetic electrons from

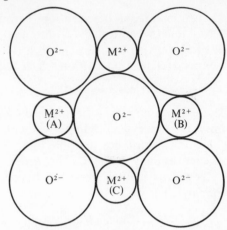

FIG. 15.13. A plane of atoms in an oxide MO, where $M^{2+}$ is a dipositive ion of the 3d group. The ions are drawn approximately to size, showing how the small cations ($M^{2+}$) are well separated by the large anions ($O^{2-}$).

neighbouring cations can therefore take place on the intervening anions. Of course the degree of overlap depends on the amount of covalent bonding, and so also does the size of this 'indirect' exchange interaction; it is very much smaller for ions of the 4f group, which take almost no part in covalent bonding, than for ions of the 3d group; for example, the transition temperature of the lanthanide oxides is below 10 K, while those of the 3d group oxides are over 100 K.

Although this solves the problem of interaction between ions at relatively large distances, there are two difficulties. The first is that the potential energy due to the electrostatic repulsion between the magnetic electrons in the overlap region leads to a ferromagnetic interaction (as for electrons within the same atom, where it results in Hund's rules—see § 14.2), whereas the vast majority of insulating magnetic compounds have an antiferromagnetic interaction. Second, the interacting magnetic electrons form a partly filled band, which should make the substance an electrical conductor. However, the energy band is narrow, and the fall in kinetic energy of an electron (cf. Fig. 12.2) on transferring from a localized state (which corresponds in energy to the centre of the band) to a conduction state at the bottom of the band is only ~1 eV. On the other hand, because of the electrostatic repulsion between the electrons, their potential energy is least when they are uniformly distributed, giving each ion the same number of electrons. This potential energy is part of the 'correlation energy' mentioned briefly in § 11.9. For s electrons, and to a smaller extent for p electrons, the correlation energy is small because their wavefunctions are extended and the charge density within the atom is small; however, the extended wavefunctions give a larger overlap and greater bandwidth in the solid (cf. Fig. 12.6). Thus such electrons have a lower energy over all when they are non-localized, and become conduction electrons. For d electrons the bandwidth is smaller and the correlation energy greater, making the latter correspondingly more important; in compounds the d-electron bandwidth is not more than about 1 eV, while about 10 eV is required to transfer an electron from one ion to another (that is, to create a pair of ions in $d^{n-1}$, $d^{n+1}$ states from a pair both in $d^n$ states). Thus it is energetically favourable for the electrons to remain localized and the substance is an electric insulator.

As a result of these conflicting energy considerations, the localization is not, however, absolutely complete. If $b$ is the reduction in kinetic energy which would result from moving from site to site, while $U$ is the potential energy required to overcome the electrostatic repulsion, the equilibrium state is one where the chance of such a movement is of order $b/U$, and the net reduction in kinetic energy is of order $b^2/U$. Through the exclusion principle this possibility of movement to adjacent sites is restricted nearly always to electrons with antiparallel spin, which can therefore acquire a lower energy than those with parallel spin. This is equivalent to an antiferromagnetic exchange energy of order $b^2/U$, which is of the order of $0.01-0.1$ eV (a few hundred degrees Kelvin).

The first explanation of how exchange interaction could arise between ions at the rather large interionic distances found in compounds was put forward by Kramers (1934), and a number of subsequent attempts were made to arrive at more explicit interpretations. The theory outlined above is due to Anderson (1963), and though difficult to explain in simple terms, appears to be the most satisfactory in its general approach.

## References

ANDERSON, P. W. (1963). *Adv. Solid St. Phys.* **14**, 99.
BARNETT, S. J. (1944). *Phys. Rev.* **66**, 224.
BLOCH, F. (1932). *Z. Phys.* **74**, 295.
DE GENNES, P. G. (1958). *C.r. hebd. Séanc. Acad. Sci., Paris* **247**, 1836.
ELLIOTT, J. F., LEGVOLD, S., and SPEDDING, F. H. (1953). *Phys. Rev.* **91**, 28.
FREEMAN, A. J., NESBET, R. K., and WATSON, R. E. (1962). *Phys. Rev.* **125**, 1978.
GOODINGS, D. A. (1962). *Phys. Rev.* **127**, 1532.
GOSSARD, A. C., JACCARINO, V., and REMEIKA, J. P. (1961). *Phys. Rev. Lett.* **7**, 122.
GREW, K. E. (1934). *Proc. R. Soc.* **A145**, 509.
HANNA, S. S., HEBERLE, J., PERLOW, G. J., PRESTON, R. S., and VINCENT, D. H. (1960). *Phys. Rev. Lett.* **4**, 513.
KASUYA, T. (1956). *Prog. theor. Phys.* **16**, 45.
KRAMERS, H. A. (1934). *Physica* **1**, 182.

MEYER, A. J. P. and ASCH, G. (1961). *J. appl. Phys.* **32**, 330.
NAGLE, D. E., FRAUENFELDER, R. D., TAYLOR, R. D., COCHRAN, D. R. F., and
   MATTHIAS, B. T. (1960). *Phys. Rev. Lett.* **5**, 364.
OLIVER, D. J. and SUCKSMITH, W. (1953). *Proc. R. Soc.* **A219**, 1.
RIEDI, P. C. (1973). *Phys. Rev.* **B8**, 5243.
RUDERMANN, M. A. and KITTEL, C. (1954). *Phys. Rev.* **96**, 99.
RUSHBROOKE, G. S. and WOOD, P. J. (1958). *Molec. Phys.* **1**, 257.
SCOTT, G. G. (1951). *Phys. Rev.* **82**, 542.
STEARNS, M. B. (1973). *Phys. Rev.* **B8**, 4283.
STUART, R. and MARSHALL, W. (1960). *Phys. Rev.* **120**, 353.
VAN KRANENDONK, J. and VAN VLECK, J. H. (1958). *Rev. mod. Phys.* **30**, 1.
YOSIDA, K. (1957). *Phys. Rev.* **106**, 893.

## Problems

15.1. Show that for a substance consisting of atoms or ions in the $^2S_{\frac{1}{2}}$ state, the Brillouin function becomes

$$M/M_s = \tanh y, \quad \text{where} \quad y = \mu_B B/kT.$$

Show that for such a ferromagnetic substance at low temperatures, where y is large, the Weiss internal field treatment of § 15.2 leads to the formula

$$M_0/M_s = 1 - 2 \exp(-2\lambda M_0 M_s/nkT)$$

for the spontaneous magnetization $M_0$ in zero field. Note that this does not give a simple power law such as in eqn (15.29).

15.2 Using the result of Problem 5.11, show that the magnetostatic energy of a small spherical particle of nickel, of radius b and magnetized to saturation ($M_s = 5 \cdot 1 \times 10^5$ A m$^{-1}$), is approximately $2 \times 10^5 b^3$ J (b in metres).

From the results of § 15.3, the energy required to form a Bloch wall increases with $b^2$ (for nickel the wall energy is about $10^{-3}$ J m$^{-2}$). Hence show that for particles whose radius is less than about $10^{-8}$ m, the reduction in magnetostatic energy obtained by division into two domains is less than the energy required to form a wall.

15.3. The nucleus of an atom of mass M moving with velocity v emits a γ-ray of energy hν in the forward direction. Show by considering the change in momentum and energy of the atom that the γ-ray energy is increased by a fraction $v/c$, provided that the γ-ray energy is small compared with the rest mass of the atom ($h\nu/Mc^2 \ll 1$). Note that this is the same as the classical Doppler shift.

# 16. Antiferromagnetism and ferrimagnetism

## 16.1. Antiferromagnetism

IN a paramagnetic substance the dipoles are free to orient themselves at random, and there is a correspondingly high entropy; if there are $2J+1$ levels having the same energy in the ground state, the entropy is $R \ln(2J+1)$. The substance would obey Curie's law down to 0 K if the ground state contains two or more levels with the same energy in the absence of an external field, but this would be a violation of the third law of thermodynamics, by which the entropy in a substance in thermodynamic equilibrium must be zero at 0 K. In practice there is always some mutual interaction between the dipoles (either through exchange or magnetic dipolar interaction), such that the internal energy $U$ of the system is lower when the dipoles are oriented in an orderly array than when they are randomly oriented. Thus, at the absolute zero, where the free energy $F = U - TS$ is equal to $U$, the equilibrium state of lowest *free energy* will be the ordered state with the lowest *internal energy*. At a sufficiently high temperature, on the other hand, the paramagnetic state, with its higher entropy corresponding to the random orientation of the dipoles, will have the lower free energy because of the second term in $F = U - TS$, and will thus be the equilibrium state. As the temperature falls, any substance where the dipoles still have some freedom of orientation (this excludes some paramagnetic substances which have a singlet ground state and a temperature-independent susceptibility) will make a transition from a disordered phase into an ordered phase. The ferromagnetic state discussed in the last chapter, in which all the dipoles are oriented parallel to one another at 0 K, is the state of lowest energy when the exchange energy $\mathcal{J}'$ has a positive sign. However, ferromagnetism is exhibited by relatively few substances, out of a large number which contain transition group ions.

It was suggested by Néel (1936) that in many substances the exchange interaction is large but negative, resulting in an ordered state where neighbouring dipoles are aligned in an antiparallel arrangement. Such an arrangement for a simple cubic lattice is shown in Fig. 16.1; the dipoles at adjacent corners of each cubic cell point in opposite directions. Another simple case is the body-centred cubic (b.c.c.) lattice, with an ion at the centre of each cube as well as at the corners; here all the ions at the

FIG. 16.1. Antiferromagnetic arrangement of dipoles in a simple cubic lattice.

corners have their dipoles parallel to each other, but antiparallel to the ions at the centres. In each case a given dipole is surrounded by a number of equidistant dipoles all pointing in the opposite direction, while the next nearest neighbours point in the same direction again. The system may be thought of as consisting of two interlocking sublattices, one of which is spontaneously magnetized in one direction, while the other is spontaneously magnetized in the opposite direction. As in ferromagnetism, this spontaneous magnetization of the sublattices sets in only below a certain transition temperature, in this case known as the 'Néel temperature'. Above this temperature the dipoles are randomly oriented, and the substance is paramagnetic, obeying a Curie–Weiss law with the Weiss constant of opposite sign to that found in ferromagnetism, as we should expect from the reversed sign of the exchange energy. The onset of spontaneous magnetization in the sublattices as the substance is cooled through the Néel temperature is accompanied by a specific heat anomaly of the co-operative type, as illustrated in Fig. 16.2. The substance as a whole exhibits no spontaneous magnetization in zero field, since the two sublattices are equally and oppositely polarized. When an external field is applied, a small magnetization occurs giving a positive susceptibility; the general behaviour of the susceptibility can be explained quite well on a molecular-field model, as outlined below.

### 16.2. The molecular field and two-sublattice model

Let the two sublattices be denoted by A and B. Then a dipole in lattice A is subject to an external field of flux density $\mathbf{B_0}$ and an internal field proportional to the magnetization of sublattice B; we write this internal

FIG. 16.2. Specific heat of $GdVO_4$. The lattice specific heat is almost negligible, unlike the case of nickel (see Fig. 15.8). (Cashion, Cooke, Hoel, Martin, and Wells 1970.)

field as $-\lambda \mathbf{M_B}$, where the minus sign appears because of the reversed sign of the exchange interaction. The effective fields acting on dipoles in lattices A and B are therefore

$$\mathbf{B_A} = \mathbf{B_0} - \lambda \mathbf{M_B},$$
$$\mathbf{B_B} = \mathbf{B_0} - \lambda \mathbf{M_A}. \tag{16.1}$$

At high temperatures where the dipoles are randomly oriented, we assume that the magnetization of each sublattice is proportional to the field acting on it. Then we have

$$M_A = \tfrac{1}{2}(\chi_0/\mu_0)B_A, \qquad M_B = \tfrac{1}{2}(\chi_0/\mu_0)B_B, \tag{16.2}$$

where the factor $\tfrac{1}{2}$ appears because only half the dipoles are in each sublattice, and $\chi_0$ is the value which the susceptibility would have in the absence of any exchange interaction. The total magnetization is then

$$M = M_A + M_B = (\chi_0/\mu_0)\{B_0 - \tfrac{1}{2}\lambda(M_A + M_B)\}$$
$$= (\chi_0/\mu_0)(B_0 - \tfrac{1}{2}\lambda M). \tag{16.3}$$

This equation has the simple solution

$$\frac{1}{\chi} = \frac{B_0}{\mu_0 M} = \frac{1}{\chi_0} + \frac{\lambda}{2\mu_0},$$    (16.4)

a result which shows that the inverse susceptibility $\chi^{-1}$ should be increased (that is, the susceptibility is smaller) in the presence of antiferromagnetic exchange interaction. This is the reverse of ferromagnetism, where a similar treatment shows that the susceptibility is increased by ferromagnetic exchange. Experimentally the result expressed in eqn (16.4) is illustrated in Fig. 16.3 for an alloy in which the value of $\lambda$ can be changed by altering the concentration of magnetic ions.

If we assume that $\chi_0$ obeys Curie's law so that $\chi_0 = C/T$, we find that

$$\chi = \frac{C}{T + (\lambda C / 2\mu_0)} = \frac{C}{T + \theta}.$$    (16.5)

The equation for the susceptibility above the Néel point is similar to eqn (15.5) for ferromagnetic exchange, except that the Weiss constant $\theta$ has the opposite sign.

FIG. 16.3. Inverse susceptibility $\chi^{-1}$ per $Tb^{3+}$ ion of the mixed compound $Tb_x Y_{1-x} Sb$. The values of $x$ are 0·05, 0·22, and 0·40, proceeding upwards. As the concentration of $Tb^{3+}$ ions rises, $\lambda$ increases, and the curve is displaced upwards. (After Cooper and Vogt 1970).

At low temperatures we cannot assume that the magnetization is linearly proportional to the effective field, since a large spontaneous magnetization will be created in each sublattice by the internal field. Instead of eqns (16.2) we must use modified forms of eqn (15.8):

$$\left.\begin{array}{l} M_A = \tfrac{1}{2}ng\mu_B J \ \phi(y_A), \\ M_B = \tfrac{1}{2}ng\mu_B J \ \phi(y_B). \end{array}\right\} \tag{16.6}$$

where $y = (g\mu_B/kT)\times$(effective flux density acting on each sublattice), and $\phi(y)$ is the Brillouin function given by eqn (14.21). The factor $\tfrac{1}{2}$ again appears because only half the dipoles are on each sublattice, and we have assumed that the dipole moment is $g\mu_B J$.

When $B_0 = 0$ the magnetization of each sublattice is given by the solution of eqn (16.6) with

$$B_A = -\lambda M_B = -\lambda(-M_0) = +\lambda M_0,$$
$$B_B = -\lambda M_A = -\lambda(+M_0) = -\lambda M_0,$$

since the two sublattices A and B will have equal and opposite magnetization, $+M_0$ and $-M_0$ respectively. The equation may be solved graphically, as in ferromagnetism. The spontaneous magnetization is given by the points A and B in Fig. 16.4, which correspond to the stable condition, while the other possible solution $M_A = M_B = 0$ is unstable. The value of $M_0$ at any temperature is given by the root of the transcendental equation

$$M_0 = \tfrac{1}{2}ng\mu_B J \ \phi\left(\frac{g\mu_B J}{kT}\lambda M_0\right).$$

FIG. 16.4. Graphical solution of equation for spontaneous magnetization of an antiferromagnetic substance. Curve CBOAD is the Brillouin function $\phi(y)$; BOA is the straight-line relation between $M$ and $y$ when the external field is zero, and B'OA' is a similar line for the case when an external field is applied parallel to the direction of the spontaneous magnetization.

As the temperature rises the line BOA becomes steeper, and the points A, B move back towards the origin; the spontaneous magnetization disappears at the Néel temperature $T_N$ where the line AB is tangential to the Brillouin function at the origin. Since $\phi(y) = y(J+1)/3J$ for small values of y, we have, when $T = T_N$,

$$M_0 = \tfrac{1}{2}ng\mu_B J\left\{\left(\frac{g\mu_B J}{kT_N}\right)\left(\frac{J+1}{3J}\right)\lambda M_0\right\},$$

or

$$T_N = \tfrac{1}{2}\lambda ng^2\mu_B^2 J(J+1)/3k = \lambda C/2\mu_0 = \theta. \tag{16.7}$$

Thus on this simple theory, due to Van Vleck (1941), the Néel point $T_N$ should have the same value as the Weiss constant $\theta$. The values of both $\theta$ and $T_N$ are given in Table 16.1 for a number of substances now established as being antiferromagnetic. It will be seen that in general $\theta$ and $T_N$ are different, and this can be accounted for by an extension of the theory given above where interactions with next nearest neighbours belonging to the same sublattice are included (Van Vleck 1951; see Problem 16.1). In addition other types of arrangement of the dipoles, where not all the nearest neighbours are antiparallel, are possible.

When an external field is applied at a temperature below the Néel point, a positive magnetization results whose magnitude can be estimated from the theory given above. In general, the effect of applying a field is to change the magnetization of each sublattice slightly, so that we may write

$$\left.\begin{aligned}\mathbf{M}_A &= \mathbf{M}_0 + \delta\mathbf{M}, \\ \mathbf{M}_B &= -\mathbf{M}_0 + \delta\mathbf{M},\end{aligned}\right\} \tag{16.8}$$

where these must be taken as vector equations if the external magnetic field is applied at an arbitrary angle to the direction of the spontaneous

TABLE 16.1

| Substance | Néel temperature $T_N$ (K) | $\theta/T_N$ | $\dfrac{\chi(T=0)}{\chi(T=T_N)}$ |
|---|---|---|---|
| Cr | 311 | — | — |
| $\alpha$-Mn | 100 | — | — |
| $MnF_2$ | 67 | 1·2 | 0·76 |
| $FeF_2$ | 78 | 1·5 | 0·72 |
| $CoF_2$ | 50 | — | — |
| $NiF_2$ | 73 | — | — |
| MnO | 116 | 4–5 | 0·69 |
| FeO | 198 | 3 | 0·75 |
| $CuCl_2 \cdot 2H_2O$ | 4·3 | — | — |
| $NiCl_2 \cdot 6H_2O$ | 5·3 | — | — |

magnetization $\mathbf{M}_0$. If $\mathbf{B}_0$ is parallel to $\mathbf{M}_0$, so also will be $\delta\mathbf{M}$, and, if we return to the graphical solution of our transcendental equation, we see that the magnetizations of the sublattices will be given by the intersection of the dotted line $B'A'$ in Fig. 16.4 with the Brillouin curve. For the effective fields become, using eqn (16.1),

$$\mathbf{B}_A = \mathbf{B}_0 - \lambda(-\mathbf{M}_0 + \delta\mathbf{M}) = \lambda\mathbf{M}_0 + (\mathbf{B}_0 - \lambda\delta\mathbf{M}),$$

$$\mathbf{B}_B = \mathbf{B}_0 - \lambda(+\mathbf{M}_0 + \delta\mathbf{M}) = -\lambda\mathbf{M}_0 + (\mathbf{B}_0 - \lambda\delta\mathbf{M}).$$

The resulting net magnetization $2\delta\mathbf{M}$ will depend on the slope of the Brillouin function (for small fields) at the point $\mathbf{M}_0$. As $\mathbf{M}_0$ increases, this slope decreases, reaching zero at saturation. It follows that the susceptibility $\chi_\parallel$ (in the direction parallel to $\mathbf{M}_0$) decreases to zero as the temperature falls to zero, as illustrated in Fig. 16.5. The exact shape of the curve depends only slightly on the value of $J$, as in the case of the saturation curve for a ferromagnetic substance.

If the external field is applied perpendicularly to $\mathbf{M}_0$, we can evaluate the susceptibility more easily. In this case the magnetizations $\mathbf{M}_A$, $\mathbf{M}_B$ of each of the sublattices turn through an angle $\alpha$ towards $\mathbf{B}_0$, as shown in Fig. 16.6. The effective field on a dipole now has the two components

$$B_x = B_0 - \lambda M_x, \qquad B_z = \lambda M_z,$$

and hence

$$\tan\alpha = M_x/M_z = B_x/B_z = (B_0 - \lambda M_x)/\lambda M_z.$$

This leads immediately to the result $2M_x = B_0/\lambda$, and hence the susceptibility in the perpendicular direction is

$$\chi_\perp = \mu_0(2M_x)/B_0 = \mu_0/\lambda = C/2T_N, \tag{16.9}$$

since $M_x$ is the component of each sublattice magnetization parallel to $B_0$. Thus $\chi_\perp$ should be constant below the Néel point and equal to the value

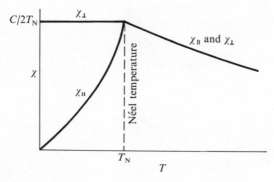

FIG. 16.5. Variation of $\chi_\parallel$ and $\chi_\perp$ on the simple theory of antiferromagnetism.

FIG. 16.6. Effect of applying a flux density $\mathbf{B}_0$ perpendicular to the spontaneous magnetization in an antiferromagnetic substance.

at the Néel point, as shown in Fig. 16.5. For a powder specimen consisting of microcrystals with random orientation, we have

$$\chi = \tfrac{1}{3}(\chi_\parallel + 2\chi_\perp), \tag{16.10}$$

and we should expect

$$\frac{\chi(T=0)}{\chi(T=T_N)} = \frac{2}{3}. \tag{16.11}$$

The values of this ratio for a number of powdered antiferromagnetics are also given in Table 16.1.

A more direct check on the theory is obtained from measurements on a single crystal, such as $GdVO_4$ (Fig. 16.7). Here the only ion with a dipole

FIG. 16.7.   Molar susceptibility of $GdVO_4$. (Cashion, Cooke, Hoel, Martin, and Wells 1970.)

moment is $Gd^{3+}$, $^8S_{\frac{7}{2}}$, and the susceptibility shows little anisotropy above the Néel temperature. Below this temperature $\chi_{\parallel}$ falls rapidly and approaches zero as $T \rightarrow 0$, while $\chi_{\perp}$ remains almost constant, as predicted by theory.

This model with two sublattices is valid for many antiferromagnetic substances, but in some cases there are more (in a f.c.c. lattice there are generally four). As in ferromagnetism, the exchange interaction itself gives no preference to any particular orientation of the spins relative to the crystal axes; this arises from the anisotropy energy. In a simple tetragonal crystal such as $MnF_2$ or $GdVO_4$, the spins are aligned along the tetragonal axis in a simple two sublattice antiparallel arrangement, but much more complicated arrangements are possible in which the vector sum of the dipole moments is zero—the distinctive feature of an antiferromagnetic.

## 16.3. Ferrimagnetism

The technical importance of magnetic materials in electrical industry has increased continuously, the ideal substance being one with a large magnetic moment at room temperature, which is also an electrical insulator. Ferromagnetic metals and alloys have been widely exploited, but their high electrical conductivity is a serious handicap in radio-frequency applications because of the eddy-current losses. For this reason a number of magnetic oxides ('ferrites', of which magnetite, $Fe_3O_4$, is the most famous as the original 'lodestone') have become of great technical interest because of their low electrical conductivity. They show spontaneous magnetization, remanence, and other properties similar to ordinary ferromagnetic materials, but the spontaneous moment does not correspond to the value expected for full parallel alignment of the dipoles.

In 1948 Néel put forward a theory for such materials; he suggested that they contain two sublattices in which the magnetizations are oppositely directed, but which give a net moment because the two sublattice moments are unequal. For this phenomenon he coined the word 'ferrimagnetism'. It can arise from a number of arrangements, of which the simplest are illustrated in Fig. 16.8. In (a) all the dipoles are equal in magnitude, but there are more on one sublattice than on the other; the

(a)                    (b)                    (c)

FIG. 16.8. Three possible arrangements of the dipole moments in a ferrimagnetic material. (a) unequal numbers of identical moments on the two sublattices, (b) unequal moments on the two sublattices, (c) two equal moments and one unequal.

most notable example is yttrium iron garnet (YIG). The simple arrangement (b) with ions of unequal moments occurs rather rarely. The arrangement (c), with two equal and opposite moments, and a third moment on one sublattice is typical of ferrites such as $MnFe_2O_4$.

*Ferrites*

These have the typical formula $M^{2+}Fe_2^{3+}O_4^{2-}$ (equivalent to $MO,Fe_2O_3$), where $M^{2+}$ is a dipositive ion, commonly $Mn^{2+}$, $Fe^{2+}$, $Co^{2+}$, $Ni^{2+}$, $Cu^{2+}$, $Zn^{2+}$, or $Mg^{2+}$; other tripositive ions such as $Cr^{3+}$ can replace $Fe^{3+}$. The crystallographic structure is cubic and similar to the mineral spinel ($MgAl_2O_4$), and the unit cell, with eight formula units, equivalent to $M_8Fe_{16}O_{32}$, contains eight cation sites with tetrahedral coordination (to four oxygen ions) and sixteen cation sites with octahedral coordination (to six oxygen ions), known as the A and B sites respectively. The division of cations between these sites is not unique, the limiting cases being:

|                    | A sites (8) | B sites (16)       |
| ------------------ | ----------- | ------------------ |
| 'Normal' structure | $8M^{2+}$   | $16Fe^{3+}$        |
| 'Inverse' structure| $8Fe^{3+}$  | $8M^{2+}+8Fe^{3+}$ |

Intermediate arrangements are also found, and we consider only the inverse structure. Each $Fe^{3+}$ ion is in a $^6S_{\frac{5}{2}}$ state with a moment of 5 $\mu_B$; however, the moments on the A and B sites are antiparallel. If $m_M$ is the moment of the $M^{2+}$ ion, the net saturation moment at 0 K for the unit $MFe_2O_4$ will be

$$m = m_M+(m_{Fe})_B-(m_{Fe})_A = m_M+5\mu_B-5\mu_B = m_M. \qquad (16.12)$$

The moments calculated thus (assuming that the M ion has a 'spin only' moment) are compared with the observed moments in Table 16.2. The agreement is satisfactory; some orbital moment would be expected in the $Fe^{2+}$, $Co^{2+}$, $Ni^{2+}$, $Cu^{2+}$ ions, and in magnesium ferrite the structure is not completely inverse.

TABLE 16.2

| M  | $m_M$ (spin only) | Observed moment | $T_N$ (K) |
| -- | ----------------- | --------------- | --------- |
| Mn | 5                 | 4·4–5·0         | 573       |
| Fe | 4                 | 4·0–4·2         | 858       |
| Co | 3                 | 3·3–3·9         | 793       |
| Ni | 2                 | 2·2–2·4         | 858       |
| Cu | 1                 | 1·3–1·4         | 728       |
| Mg | 0                 | 0·9–1·1         | —         |

The magnetic moments are in Bohr magnetons per unit $MFe_2O_4$.

Néel suggested that all the interactions in the ferrites are antiferromagnetic in sign, but that the A–B interaction is considerably stronger than the A–A or B–B interactions. Thus in the inverse structure the dominating A–B interaction makes the spins within each group parallel, despite their mutual antiferromagnetic interaction. This is supported by the fact that $ZnFe_2O_4$, which has the normal structure, has no net moment. Here the A sites are entirely occupied by zinc ions, with no moment, so the A–B interactions are zero. The ferric ions on the B sites are then aligned antiparallel through the antiferromagnetic B–B interaction, in equal numbers, so that the compound is antiferromagnetic. Its Néel temperature (9 K) is quite low, as would be expected if the B–B interactions are weak.

## Garnets

These have the typical formula $M_3Fe_5O_{12}$ (of which two units are equivalent to $5Fe_2O_3.3M_2O_3$), where both the M cation and the Fe are tripositive ions; the $M^{3+}$ ion is commonly yttrium or a member of the 4f group. The crystallographic structure is cubic and similar to the mineral garnet, though this has cations of other valencies. The unit cell is complex, containing eight units of $M_3Fe_5O_{12}$; for simplicity we shall discuss mainly YIG, where the $Y^{3+}$ ion has a closed shell and carries no magnetic moment. The ferric ions occupy two types of site; in each unit $Y_3Fe_5O_{12}$ two $Fe^{3+}$ ions occupy 'a' sites, coordinated to six oxygen ions, and three $Fe^{3+}$ ions occupy 'd' sites, coordinated to four oxygen ions. The magnetic moments of the two 'a' ions are antiparallel to those of the three 'd' ions, giving the arrangement shown in Fig. 16.8(a); the net moment per unit $Y_3Fe_5O_{12}$ is thus that of one $Fe^{3+}$ ion, or 5 $\mu_B$ (the best experimental value is 4·96 $\mu_B$). The Néel temperature is 545 K.

Amongst other ferrimagnetic materials we mention only $BaFe_{12}O_{19}$ (equivalent to $BaO.6Fe_2O_3$). This has a hexagonal structure, with a number of inequivalent sites for the ferric ions. Of the twelve ferric ions per formula unit, the moments of eight are antiparallel to the remaining four, giving a net moment of $4 \times 5 = 20$ $\mu_B$; the Néel temperature is about 820 K. Barium ferrite, as it is frequently called, has a high value of $(BH)_{max}$ and is used as a permanent magnet material (cf. Chapter 6). Being hexagonal, it has a high anisotropy energy; it is used in the form of pressed oriented fine particles.

## Discussion

Néel's theory of ferrimagnetism had considerable success in explaining the anomalous behaviour of the susceptibility above the Néel point. Using a molecular field approximation with three constants representing the

A–B, A–A, and B–B interactions he deduced the relation

$$\frac{1}{\chi} = \frac{T}{C} + \frac{1}{\chi_0} - \frac{\sigma}{T-\theta} \tag{16.13}$$

for a substance where all the magnetic ions have the same moment, such as YIG, or $MFe_2O_4$ when M carries no moment. Here $C$ is the usual Curie constant, but the other parameters are functions of the molecular field constants and the numbers of ions in each sublattice. The general behaviour of the inverse susceptibility $\chi^{-1}$ given by Néel's relation, fitted to experiments on YIG, is shown in Fig. 16.9. The theory also explains qualitatively the complex behaviour of the spontaneous magnetization curve below the Néel temperature. The magnetization does not always increase monotonically as the temperature falls, and in ferrimagnetic compounds containing more than one type of magnetic ion whose spontaneous magnetization varies in different ways with temperature a 'compensation point' may be observed, where the magnetization of the two sublattices is equal and opposite.

The magnetization curve of gadolinium iron garnet, $(Gd_3Fe_5O_{12})$ is shown in Fig. 16.10, together with that of $Y_3Fe_5O_{12}$. The latter is not unlike that of a ferromagnetic, but at low temperatures the former has a much higher magnetization, falling to zero at the compensation point at about 295 K. At 0 K we would expect each $Gd^{3+}$ ion to have a moment of $7 \mu_B$; if these are mutually parallel, but opposed to the net ferric moment, we would expect an over-all moment for the unit $Gd_3Fe_5O_{12}$ of $(3 \times 7 - \{3 \times 5 - 2 \times 5\}) \mu_B = (21 - 5) \mu_B = 16 \mu_B$; this is close to the observed moment. As the temperature rises the magnetization of the gadolinium

FIG. 16.9. Inverse magnetic susceptibility of the ferrimagnetic substance yttrium iron garnet (YIG), which has three and two $Fe^{3+}$ ions on the two sublattices (arrangement (a) in Fig. 16.8).

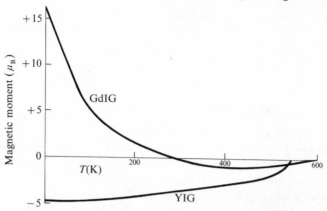

FIG. 16.10. Variation of the spontaneous moment with temperature for $Gd_3Fe_5O_{12}$ (GdIG) and $Y_3Fe_5O_{12}$(YIG) in Bohr magnetons per formula unit.

ions, which are subjected to a comparatively weak interaction with the ferric ions, falls much more rapidly than that of the iron lattice with its strong mutual interactions between the ferric ions. In fact the behaviour of the gadolinium ions is not far from that of paramagnetic ions with $S = \frac{7}{2}$, subjected to an internal field generated by the iron lattice. The Néel temperature of $Gd_3Fe_5O_{12}$ (564 K) is not appreciably different from that of $Y_3Fe_5O_{12}$ (545 K), as would be expected on this basis.

Apart from their technical importance, ferrimagnetic materials have played a major role in advancing our understanding of magnetic problems; for this purpose the garnets are more favoured than the ferrites, since the structure is unique and there are no uncertainties concerning the sites occupied by the magnetic ions. The absence of conduction electrons is a great asset, not only technically but also scientifically. On the one hand, we are dealing with localized magnetic moments, so the theory rests on a much firmer foundation; on the other many important experiments can be carried out to check the theory which would otherwise be impossible. An obvious example is magnetic resonance experiments (cf. Chapter 24) in the frequency range $10^4–10^7$ MHz, which have been a very fruitful field both for ferrimagnetics and for antiferromagnetics. The absence of conduction electrons plays a less direct but no less important role in measurements of the magnetic specific-heat contribution predicted by spin-wave theory. A term proportional to $T^{\frac{3}{2}}$ was first confirmed by Kouvel (1956) using $Fe_3O_4$ (it has also been measured in compounds such as YIG), whereas in the ordinary ferromagnetic metals it is obscured by the electronic specific heat. Another experimental achievement is the optical demonstration of the presence of domains, using the rotation of the plane of polarized light propagated parallel to the direction of magnetization (the Faraday effect); when a thin crystal of YIG is placed under a polarizing microscope the domains are visible as light and dark regions whose motion can be observed under the action of an applied field.

### 16.4. The lanthanide ('rare-earth') metals

Measurements of the susceptibilities of the lanthanide metals at high temperatures give generally a Curie–Weiss law where the size of the Curie constant agrees well with that expected for the tripositive ions. There are two notable exceptions to this: europium metal and ytterbium metal, which are cubic in structure with an ionic size indicating the presence of dipositive ions. In addition, cerium tends to show a phase transition at low temperatures to a cubic structure with $Ce^{4+}$ ions, dependent on the thermal history of the specimen. The $Ce^{4+}$ and $Yb^{2+}$ ions have closed shells and no magnetic moment, so they are not of interest here; the $Eu^{2+}$ ion has a half-filled shell, ground state $S = \frac{7}{2}$, but the magnetic behaviour of the metal shows unexpected complications. We shall therefore restrict ourselves to the metals containing tripositive ions, data for which are given in Table 16.3

No lanthanide metal shows a co-operative state above 300 K, so that exchange interactions are small compared with the spin–orbit coupling.

TABLE 16.3

*Magnetic data for the lanthanide metals containing $Ln^{3+}$ ions*

|      | gJ     | $(g-1)^2 J(J+1)$ | $T_N$ (K) | $T_C$ (K) |
| ---- | ------ | ---------------- | --------- | --------- |
| La   | 0      | 0                | —         | —         |
| Ce   | 2·14   | 0·18             | 12·5      | —         |
| Pr   | 3·2    | 0·8              | ?         | —         |
| Nd   | 3·17   | 1·84             | 7·5, 19   | —         |
| Sm   | 0·71   | 4·5              | 14        | —         |
| Gd   | 7      | 15·75            | —         | 293       |
| Tb   | 9      | 10·5             | 230       | 220       |
| Dy   | 10     | 7·1              | 176       | 88        |
| Ho   | 10     | 4·5              | 130       | 19        |
| Er   | 9      | 2·55             | 85        | 20        |
| Tm   | 7      | 1·17             | 57        | 32        |
| Lu   | 0      | 0                | —         | —         |

The value of gJ (from Table 14.1) gives the moment per ion at 0 K assuming no crystal field effects are present. Promethium has no radioactively stable isotopes; europium metal becomes antiferromagnetic below 91 K, the ion being $Eu^{2+}$, $^8S_{\frac{7}{2}}$; ytterbium metal contains $Yb^{2+}$ ions, with a filled 4f shell like $Lu^{3+}$.

We may therefore regard the spin and orbit as coupled to give a resultant angular momentum $J$, as in the paramagnetic salts (cf. § 14.6). On this basis the saturation moment per ion at 0 K should be $gJ\,\mu_B$, where $g$ is the Landé factor appropriate to the ground state $J$ of the free tripositive ion. Values of $gJ$ are given in column 2 of Table 16.3, and are generally substantiated by the magnetic evidence for gadolinium and the heavier metals.

In considering the exchange interaction, we have to project the spin vector $S$ onto the total angular momentum vector $J$, as pointed out in § 15.1. The exchange interaction $-2\mathscr{J}S_i.S_j$ between the spins thus becomes equivalent to a coupling $-2\mathscr{J}'J_i.J_j$ between the total angular momenta, with $\mathscr{J}'=(g-1)^2\mathscr{J}$, as given by eqn (15.3). If in fact $\mathscr{J}$ were the same for all the lanthanon metals, we should expect the Curie points to vary as $(g-1)^2J(J+1)$, from eqn (15.6). This quantity is largest for $Gd^{3+}$, with a half-filled shell, and this metal shows co-operative effects at a higher temperature (293 K) than any other lanthanide metal. Reference to Table 16.3 shows that the temperature at which co-operative effects appear varies qualitatively in accordance with this relation in the second half of the group. However, these metals show more than one ordered phase, being antiferromagnetic at higher temperatures and ferromagnetic at lower temperatures. This effect appears to require a reversal in sign of the exchange interaction as the temperature falls, but it turns out that the magnetic moments are arranged in very complex patterns which change with temperature.

In the lanthanons the 4f electrons belong to an inner shell, and their wavefunctions are less extended than those of d electrons. For this reason direct exchange, involving overlap of f-electron wavefunctions on adjacent ions, is unlikely to be important. Instead, the Rudermann–Kittel–Kasuya–Yosida (RKKY) indirect exchange mechanism, discussed in § 15.8, is regarded as dominant, giving a picture of the metals as consisting of ions with well-localized moments due to the 4f shells, imbedded in a 'sea' of conduction electrons formed from the valence electrons, $5d^1 6s^2$. The conduction electrons are polarized by exchange interaction with the 4f shells, and serve as the medium whereby the orientation of the moment on one ion can influence that on neighbouring ions. The conduction electrons contribute little to the magnetic moment, but the oscillatory nature of the RKKY interaction (see § 15.8) plays a major role in producing wave-like forms in the magnetic ordering observed in the ordered state.

Electrostatic interaction between the 4f electrons on a given ion and the charge density on neighbouring ions provides a 'crystal field' interaction in the metals of the same order as that in salts of the lanthanide group. Direct evidence for this comes from Schottky-type anomalies in the specific heats of the first members of the group. It is difficult to obtain reliable values of the crystal field splittings of the $2J+1$ levels from this excess 'magnetic' specific heat, but the overall splitting is about a few hundred degrees Kelvin for cerium, praseodymium, neodymium, and samarium. Since the values of $(g-1)^2 J(J+1)$ are rather low for these elements, the exchange interaction is less important than the crystal field and a co-operative phase sets in only below 20 K.

The normal crystal structure of the lanthanon metals is hexagonal, apart from those containing $Ce^{4+}$, $Eu^{2+}$, or $Yb^{2+}$ ions, which are cubic, but the structure changes slightly at gadolinium. In this and the heavier metals the crystal field is smaller than in the lighter metals, while the value of $(g-1)^2 J(J+1)$ tends to be larger; thus exchange interactions preponderate over crystal field splittings, and the co-operative magnetic phase sets in at temperatures where crystal field effects are relatively less important, though they give rise to a large anisotropy energy. In the ordered phase gadolinium appears to be a normal ferromagnet, though the direction of the magnetic moment in the crystal varies with temperature. The behaviour of the other lanthanon metals is much more complicated. In the heavy metals the moments of all ions in each layer normal to the hexagonal axis lie parallel to one another, but successive layers have different orientations so that the over-all magnetization is zero below $T_N$ and above $T_C$. The moments of the layers in some cases lie on the surface of a cone (Fig. 16.11); the turn angle between successive layers depends on the oscillatory exchange interaction, which thus determines the period at which the structure repeats along the $c$-axis. The angle $\theta$ which the moments make with the $c$-axis is determined by a number of competing

$c$–axis

$\theta$

$\frac{1}{2}qc$

FIG. 16.11. A conical spin arrangement. The moments of successive layers (distance $\frac{1}{2}c$ apart) perpendicular to the $c$-axis are indicated by arrows. The periodicity can ,be represented by a wave vector $\mathbf{q}$ parallel to the $c$-axis, and the turn angle between layers is $\frac{1}{2}qc$. When $q = 0$ the alignment is ferromagnetic.

interactions (exchange, crystal field, magnetostriction); the equilibrium state is that with the lowest free energy, and varies with temperature. At the lowest temperatures all the heavy metals enter a ferromagnetic phase; the saturation magnetization per ion is a little larger than $g\mu_B J$ for gadolinium, terbium, dysprosium, and holmium, the excess being attributed to polarization of the conduction electrons. In erbium, thulium, and the lighter metals the moments are all lower than the free-ion value, showing the importance of crystal field effects.

Detailed information about the complicated magnetic structures has become available only through the use of neutron diffraction, particularly on single crystals. In some cases neutron diffraction also suggests that the exchange interaction is itself anisotropic, the energy depending on the orientation of the moments as well as the distance between them.

## References

CASHION, J. D., COOKE, A. H., HOEL, L. A., MARTIN, D. M., and WELLS, M. R. (1970). *Colloques Internationaux du C.N.R.S.* No. 180, p. 417.
COOPER, B R., and VOGT, O. (1970). *Phys. Rev.* **B1**, 1218.
KOUVEL, J. S. (1956). *Phys. Rev.* **102**, 1489.
NÉEL, L. (1936). *Ann. Phys. (Fr.)* **5**, 256.
VAN VLECK, J. H. (1941). *J. chem. Phys.* **9**, 85.
—— (1951). *J. Phys. Radium* **12**, 262.

## Problems

16.1. The theory of antiferromagnetism can be extended by assuming that the molecular field acting on each sublattice contains a term due to exchange interaction with ions on the same sublattice as well as a term due to ions on the other sublattice. Show that in the paramagnetic state the equations

$$M_A = (C/2\mu_0 T)(B_0 - \lambda M_B - \lambda' M_A),$$
$$M_B = (C/2\mu_0 T)(B_0 - \lambda M_A - \lambda' M_B)$$

lead to a Curie–Weiss law for the susceptibility (eqn (16.5)) with

$$\theta = (C/2\mu_0)(\lambda + \lambda').$$

16.2. The Néel temperature $T_N$ can be found by putting $B_0 = 0$ in the preceding question, and finding the condition that the pair of equations for $M_A$, $M_B$ still have a solution. ($T_N$ is the temperature at which a vanishingly small magnetization can exist when $B_0 = 0$, and the Brillouin function is approximated by Curie's law.) Show that this gives $T_N = (C/2\mu_0)(\lambda - \lambda')$.

16.3. Show that eqns (16.1)–(16.3) lead to the result

$$(M_A - M_B)\left(1 - \frac{\chi_0 \lambda}{2\mu_0}\right) = 0.$$

Hence either $M_A = M_B$, as in the paramagnetic state, or the second bracket vanishes, determining the Néel point at which a finite sublattice magnetization can occur.

# 17. Semiconductors

## 17.1. Intrinsic and extrinsic conductivity

A SUBSTANCE in which the number of electrons is just sufficient to fill the lowest energy bands at 0 K is an insulator at very low temperatures. At a non-zero temperature a few electrons may have sufficient energy to be excited into the lowest unoccupied band (the 'conduction' band), leaving holes in the highest 'occupied' band (the 'valence' band). This gives a small electrical conductivity whose magnitude depends on the temperature and on the width of the energy gap $W_g$ between the full and empty bands. We shall find in § 17.3 that the number of electrons excited into the conduction band is proportional to $\exp(-W_g/2kT)$, and if the gap is not more than about 1 eV, which corresponds to a value of $kT$ with $T \approx 12\,000$ K, there will be a measurable conductivity at room temperature. This phenomenon is known as 'intrinsic conductivity' and is a characteristic of pure semiconductors; it has been observed in pure silicon, germanium, indium antimonide (InSb), and some other substances.

For each electron in the conduction band, there will be a corresponding 'hole' in the filled band. Both electrons and holes contribute to the conductivity $\sigma$, so that

$$\sigma = |e|(nu_e + pu_h), \qquad (17.1)$$

where $n$ and $p$ are the numbers per unit volume of electrons and holes respectively. The mobilities $u_e$, $u_h$ are taken as positive numbers and no sign is attached to $|e|$, though the holes and electrons drift in opposite directions under the influence of an electric field. For intrinsic conductivity $n = p$, since electrons and holes occur only in pairs. The equilibrium concentration rises rapidly as the temperature rises, but the mobilities vary much less rapidly with temperature; hence the increase in $n$ is the dominant factor and the conductivity rises as the temperature increases. This is the hall-mark of a semiconductor, and one of the features (together with its much smaller conductivity) which distinguishes it from a metal. Another difference is that electrons in the conduction band are in excited states, and have only a finite lifetime. An electron from the conduction band can drop down into the top of the valence band, recombining with a hole and releasing an energy $W_g$; conversely an electron–hole pair can be created by lifting an electron from the valence band to the conduction band. Both processes occur repeatedly, giving a dynamic equilibrium concentration which is a function of temperature.

The properties of a semiconductor are generally profoundly modified by the presence of an impurity, or some other cause of irregularity in the lattice. If these are present in not too great a concentration, they produce discrete energy levels. The reason for this is that the levels only spread out into bands when the impurity atoms are sufficiently close for their wavefunctions to overlap, and at low concentrations the impurity atoms are so far apart that any such overlap is negligible. These discrete energy levels are important when they lie in the forbidden band, and particularly so if they lie close to the conduction or the valence band, as illustrated in Fig. 17.1. In the former case electrons may occupy the impurity level D at low temperatures, and are then localized on the impurity atom and unable to partake in electrical conduction. As the temperature rises, these electrons are excited into the empty band. They behave then as conduction electrons, with negative charges; the material is known as an n-type semiconductor, and the impurity levels from which the electrons come are known as 'donor' levels. In the second case, the impurity levels A which lie just above the valence band will be unoccupied at 0 K, but as the temperature rises electrons are excited from the valence band into these levels, which are therefore known as 'acceptor' levels. This process leaves holes in the valence band, which behave as positively charged carriers, and the material is known as a p-type semiconductor.

If the impurity level lies close to a conduction or valence band, the temperature at which appreciable numbers of electrons or holes may be excited is relatively low, and the 'extrinsic conductivity' due to this cause may outweigh any intrinsic conductivity, even with small concentrations of impurities. In pure germanium at room temperature, for example, the number of intrinsic electrons in the conduction band is less than $10^{20}$ per cubic metre, whereas the number of germanium atoms per cubic metre is $4 \cdot 5 \times 10^{28}$. If an impurity atom which is easily ionized at room temperature is present to a concentration of one part in $10^8$ ($4 \cdot 5 \times 10^{20}$ impurity atoms

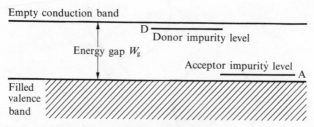

FIG. 17.1. Energy bands in a semiconductor, showing the gap between the valence band and the conduction band. At 0 K the valence band is full and the conduction band is empty, so that the substance behaves as an insulator. The discrete levels D, A are due to the presence of impurities in low concentration, where the impurity atoms are too far apart for their electronic wavefunctions to overlap.

per cubic metre), it can give rise to an extrinsic conductivity which exceeds the intrinsic conductivity.

The extrinsic conductivity increases as the temperature rises until all the donor impurity atoms are fully ionized, or all the acceptor levels fully occupied. The number of extrinsic conduction electrons, or holes, then becomes substantially constant; the conductivity becomes constant, or may fall with temperature because of a decrease in the mobility. This is known as the 'exhaustion range'.

Extrinsic and intrinsic conductivity may of course be present simultaneously, but the former will depend on the impurity content while the latter is a property of the pure material. In each case conduction depends on excitation into higher levels, and the charge carriers have a finite lifetime. All substances would be expected to show intrinsic conductivity at a sufficiently high temperature; 'insulators' are substances with such large energy gaps that appreciable conductivity sets in only at temperatures outside the normal laboratory range, and which may be above the melting-point of the substance.

## 17.2. Elementary and compound semiconductors

A number of elements are known to be semiconductors in their normal allotropic form; the principal ones are silicon, germanium, boron, selenium, and tellurium. By far the most important are silicon and germanium; they are used in many solid-state devices and their properties have been extensively investigated.

The elements carbon, silicon, germanium, tin, and lead belong to Group IV of the periodic table. Silicon, germanium, and the allotrope grey tin crystallize in the diamond structure (see Fig. 17.2) in which each

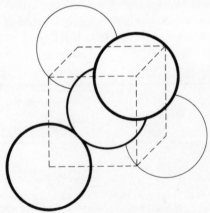

FIG. 17.2. The diamond structure, consisting of four atoms centred on alternate corners of a simple cubic lattice, bonded to one at the centre of the cube. The structure is repeated so that every atom is in identical surroundings.

TABLE 17.1

*Values of the energy gap (eV) at 0 K*

| Group IV elements | | III–V compounds | | II–VI compounds | |
| --- | --- | --- | --- | --- | --- |
| Diamond | ~5·3 | BN | ~10 | | |
| Silicon | 1·21 | AlP | 3 | | |
| Germanium | 0·78 | GaAs | 1·35 | ZnSe | ~5 |
| Grey tin | ~0·08 | InSb | 0·24 | CdTe | 1·6 |
| | | | | PbS | ~0·4 |

atom has four equidistant neighbours arranged in the form of a regular tetrahedron. Each atom forms four covalent bonds with these neighbours, donating one electron to each bond whose spin is paired off with that of the corresponding electron donated by the neighbour. These electrons can be regarded as being in a filled valence band, above which is an energy gap to the next band which is empty and forms a possible conduction band. This picture represents the position at 0 K, where the substances behave as insulators. The energy gaps are listed in Table 17.1. At a non-zero temperature some electrons may be excited from the valence band into the conduction band, making the substance an intrinsic semiconductor. On a localized electron model this corresponds to taking an electron out of a bond to behave like a 'free electron'. This leaves a 'hole' in one bond; an electron can migrate from an adjacent bond to fill this hole, thereby transferring the hole to another bond. In this way the hole can be pictured as moving in a random way through the crystal, and being mobile like the 'free' electron, though not necessarily with the same mobility.

Intrinsic conductivity is exhibited only by very pure specimens, and the properties of silicon and germanium are drastically modified by small amounts of impurity. If an impurity from Group V of the periodic table, such as phosphorus, is introduced, it enters 'substitutionally', occupying the place of a silicon or germanium atom. Like the atom it replaces, it forms four covalent bonds with its four immediate neighbours. This uses up four of its valence electrons, leaving one in excess, which experiences an electrostatic attraction to the phosphorus because of the extra positive charge of its parent ion. This system of a singly charged ion and excess electron resembles a hydrogen atom, but one in which the potential due to the positively charged nucleus is modified by the presence of the surrounding silicon or germanium ions. These are electrically polarized by the excess positive charge of the phosphorus nucleus, and their effect on the electrostatic potential can be crudely approximated by the introduction of a relative permittivity, making the potential $V = e/4\pi\epsilon_r\epsilon_0 r$. Obviously this device of using a relative permittivity is only realistic at distances large compared with the interatomic distance, where the electron orbit is so large that it embraces many atoms, and whose effect then resembles that of a continuous medium. For an electron of effective mass $m^*$ in an orbit of principal quantum number $n$ the energy is then found to be (see Problem 17.1)

$$W = -\frac{m^*}{m}\frac{R}{\epsilon_r^2 n^2} = -\frac{m^*}{m\epsilon_r^2 n^2}\times 13\cdot5\text{ eV},\qquad(17.2)$$

where $13\cdot5$ eV is the ionization potential of an electron in the $n=1$ state of a free hydrogen atom. The value of $\epsilon_r$ for germanium is 16, and using this and a value of $m^*/m = 0\cdot2$ (an average of values obtained from other evidence), we find $W \sim 0\cdot01$ eV for the lowest state $n=1$. This is fairly close to that actually observed for Group V donors in germanium (for phosphorus the observed value is $0\cdot012$ eV). The corresponding orbit radius is over 4 nm, which is quite large compared with the interatomic distance of $0\cdot245$ nm. Since $0\cdot01$ eV is equivalent to a temperature of only 120 K, it is evident that such donor impurities will release nearly all their electrons into the conduction band at room temperature, a process similar to that of ionization of free hydrogen atoms at the very high temperatures in the interior of stars. For Group V donors in silicon the corresponding binding energy is about $0\cdot04$ eV (see Problem 17.1).

If a Group III element is added as an impurity we have a different situation. The impurity atom now has one electron too few to fill the four bonds which it should make on replacing a silicon or germanium atom, and we are therefore left with a hole in one bond. If this hole moves away from its parent impurity, all four bonds to the impurity atom become filled and it has one net negative charge. Since the hole is effectively a positive charge, it has an electrostatic attraction to the negatively charged impurity ion, and we have an 'inside-out' hydrogen atom consisting of a negatively charged 'nucleus' with the positive hole in orbit around it. This gives an 'acceptor' level just above the valence band, the height above the top of the valence band being again about $0\cdot01$ eV. At 0 K the valence band is full of electrons, and the hole occupies the impurity level; it is then localized on the impurity ion, forming a neutral atom in a bound state. At a finite temperature an electron may be excited from the valence band into the impurity level; this leaves a hole in the valence band which is free to move, and corresponds to ionization of the 'inside out' impurity atom.

This analysis shows that at 0 K the hole is in the highest level (the acceptor level) and as the temperature rises more and more holes are excited in the lower levels (the valence band). This behaviour is similar to that of electrons being excited from donor levels into the conduction band, except that for the holes energy must be measured downwards instead of upwards. In § 17.3 we shall find that holes obey similar equations to electrons provided we measure energy downwards from the top of the filled band, an example of which has already occurred in eqn (12.23b).

A wide range of compounds show semiconducting properties. Of these, only a few which illustrate general classes, and for which sufficient information exists to make their properties reasonably well understood, can be mentioned here. Following the Group IV compounds germanium and silicon, it is natural to discuss first the Group III–Group V compounds, taking as example indium antimonide, the most studied of such materials. The two elements, indium and antimony, come in the periodic table immediately before and after tin. They form a compound in which each atom is surrounded by four equidistant neighbours at the apices of a regular tetrahedron, as in the diamond structure, but with the difference that each of the four nearest neighbours is of the other type (this is known as the zinc-blende structure, after one form of the compound ZnS—see Fig. 17.3). These four bonds are mainly covalent in character, and link lattice sites which may be regarded as occupied by $In^-$ and $Sb^+$ ions in regular alternation. Each of these ions has the same electron configuration as tin, the Group IV element, and forms covalent bonds in a similar fashion. However, the fact that we now have ions of alternate negative and positive charge gives rise to some ionic binding. Indium antimonide can be prepared in a sufficiently pure state to behave as an intrinsic semiconductor, with an energy gap of about 0·24 eV at 0 K, decreasing to 0·18 eV at room temperature.

The next binary compounds in sequence are the II–VI and I–VII compounds; these grow progressively more ionic in character, with larger energy gaps. This is illustrated in the sequence formed from atoms in the

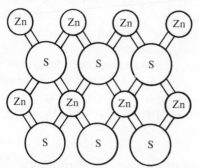

FIG. 17.3. The zinc-blende structure (two-dimensional representation). It is similar to the diamond structure, except that zinc and sulphur ions alternate.

seventh row of the periodic table:

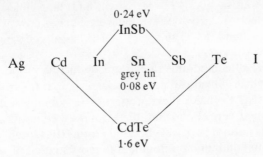

Silver iodide (AgI) is a good insulator, and clearly has a large energy gap. Though not as pure a polar compound as NaCl, we may regard it as consisting of $Ag^+$ and $I^-$ ions, with closed shells of electrons. With a II–VI compound such as CdTe we have the dilemma of whether to regard it as a polar compound, consisting of $Cd^{2+}$ and $Te^{2-}$ ions, or a covalent compound formed of $Cd^{2-}$ and $Te^{2+}$ ions. However, the crystal structure resembles that of zinc-blende, suggesting a covalent compound; on the other hand, the group of salts PbS, PbSe, PbTe have the NaCl structure, suggesting a polar compound. These difficulties illustrate the reserve with which such extreme classification should be regarded. In fact the group of lead salts have smaller energy gaps than CdTe, and their electrical properties are more typical of semiconductors.

The energy gaps of a number of substances are given in Table 17.1. General (but not invariable) rules are that the energy gap diminishes the heavier the atoms involved (that is, reading downwards in the table), but increases on moving from covalent to polar compounds (that is, from left to right). Much less information is available about the polar semiconductors, owing to the difficulty of preparing them in the pure state. Apart from foreign atoms, which act as acceptors or donors according to their group in the periodic table, such crystals may be non-stoichiometric. For example, PbS may have an excess of lead, producing donor levels, or of sulphur, producing acceptor levels.

The energy gap varies considerably with temperature, and most of the above are rounded values. In the case of silicon, germanium, and indium antimonide the values given in the table are those for 0 K; at room temperature they are approximately 1·12 eV, 0·66 eV, and 0·18 eV respectively.

### 17.3. Electron distribution and the Fermi level

Under conditions of thermal equilibrium the number of electrons with energy between $W$ and $W+dW$ can be calculated by means of statistical mechanics, using of course the Fermi–Dirac statistics appropriate to particles of half-integral spin. This number is

$$dn = f(W)g(W)\, dW, \qquad (17.3)$$

where $f(W)$ is the Fermi–Dirac function (written as $f$ in eqn (11.15))

$$f(W) = \frac{1}{\exp\{(W-W_F)/kT\}+1} \qquad (17.4)$$

and $g(W)$ is the density of states. $W_F$ is the Fermi level defined as the energy at which the function $f(W) = \frac{1}{2}$. In a metal, $W_F \gg kT$ at ordinary temperatures, and the only electrons which are thermally excited or can take part in conduction processes are those very close to the Fermi level.

In a semiconductor, it is not so obvious where the Fermi level lies with respect to the conduction and valence bands. We shall consider first an intrinsic semiconductor, where at low temperatures only a few electrons are excited into the conduction band. In the limit of extremely few electrons the chance of an electron occupying a given state is very low, and the restrictions imposed by the exclusion principle play little role. We are thus in a situation where the classical Maxwell–Boltzmann statistics are a good approximation, so that we can write

$$f_c(W) = \exp\{-(W-W_F)/kT\} \quad \text{very nearly.} \qquad (17.5)$$

This is the approximation of eqn (17.4) in the limit where $W-W_F \gg kT$, so that we can neglect the second term in the denominator; this approximation is appropriate for electrons in the conduction band, which is empty of electrons at 0 K. At this temperature the valence band is full, and thus corresponds to energies well below the Fermi level. In the region where $W_F-W \gg kT$ the first term in the denominator of eqn (17.4) is now very small, and we can write

$$1-f_v(W) = \exp\{-(W_F-W)/kT\} \quad \text{very nearly.} \qquad (17.6)$$

The quantity $1-f_v(W)$ is the chance of finding a hole in the valence band with energy $W$ (cf. § 12.4).

We must now consider the density of states $g(W)$. This is zero at the edge of a band and we assume that it varies as the square-root of the distance from the edge of the band. That is, we can modify eqn (11.8) and write (using $C' = C_m^{-\frac{3}{2}}$)

$$g_c(W) = C'(m_c^*)^{\frac{3}{2}}(W-W_c)^{\frac{1}{2}} \qquad (17.7)$$

and

$$g_v(W) = C'(m_h^*)^{\frac{3}{2}}(W_v-W)^{\frac{1}{2}}, \qquad (17.8)$$

where $W_c$, $W_v$ are the energies at the bottom of the conduction band and the top of the valence band respectively.

The total number of electrons in the conduction band is thus

$$n = \int_{W_e}^{\infty} f_c(W) g_c(W) \, dW$$

$$= C'(m_e^*)^{\frac{3}{2}} \int_{W_e}^{\infty} (W-W_c)^{\frac{1}{2}} \exp\{-(W-W_F)/kT\} \, dW.$$

On writing $y = (W-W_c)/kT$, this integral becomes

$$n = C'(m_e^* kT)^{\frac{3}{2}} \exp\{-(W_c-W_F)/kT\} \int_0^{\infty} y^{\frac{1}{2}} \exp(-y) \, dy$$

$$= (\pi^{\frac{1}{2}}/2) C'(m_e^* kT)^{\frac{3}{2}} \exp\{-(W_c-W_F)/kT\}$$

$$= N_c \exp\{-(W_c-W_F)/kT\}. \tag{17.9}$$

This result is the same as if we had a number $N_c$ of states at energy $W_c$; thus $N_c$ is the effective density of states at the bottom of the conduction band, and on substituting for $C' = C_m^{-\frac{3}{2}}$ as in eqn (11.9) we find

$$N_c = 2(2\pi m_e^* kT/h^2)^{\frac{3}{2}}. \tag{17.10}$$

Similarly for the number of holes in the valence band

$$p = \int_{-\infty}^{W_v} \{1 - f_v(W)\} g_v(W) \, dW$$

$$= N_v \exp\{-(W_F-W_v)/kT\}, \tag{17.11}$$

where

$$N_v = 2(2\pi m_h^* kT/h^2)^{\frac{3}{2}}. \tag{17.12}$$

For the product $np$ we find

$$np = N_c N_v \exp\{-(W_c-W_v)/kT\} = N_c N_v \exp(-W_g/kT), \tag{17.13}$$

and since in an intrinsic conductor $n = p = n_i$ we obtain

$$n_i = n = p = (N_c N_v)^{\frac{1}{2}} \exp(-W_g/2kT), \tag{17.14}$$

where $W_g = W_c - W_v$ is the width of the energy gap between the valence and conduction bands.

To find the position of the Fermi level $W_F$ we must equate the formulae for $n$ and $p$, which yields

$$\frac{N_v}{N_c} = \exp\left(\frac{2W_F - W_c - W_v}{kT}\right),$$

whence, since $N_v/N_c = (m_h^*/m_e^*)^{\frac{3}{2}}$,

$$W_F = \tfrac{1}{2}(W_c + W_v) + \tfrac{3}{4}kT \ln(m_h^*/m_e^*). \tag{17.15}$$

This result shows that for most intrinsic semiconductors, where $m_h^*$, $m_e^*$ are nearly equal, the Fermi level lies in the middle of the energy gap, as shown in Fig. 17.4. In some cases, such as indium antimonide, where $m_h^*/m_e^* \approx 20$, the level varies markedly in position with temperature, and at room temperature is shifted well towards the bottom of the conduction band.

When impurities are present, and conduction is partly intrinsic and partly extrinsic, the position is a good deal more complicated. There is, however, one important general result which holds provided the numbers of electrons in the conduction band and holes in the valence band are small compared with the density of states. In that case the relations (17.9)–(17.13) are still valid, since they do not depend on any supposition

FIG. 17.4. The Fermi–Dirac distribution of electrons and holes in an intrinsic semiconductor; the figure is drawn for a case where $m_h^* = m_e^*$, so that the Fermi level $W_F$ is in the centre of the forbidden band. The shaded areas indicate the numbers of electrons in the conduction band, and holes in the valence band.

about the position of the Fermi level. Thus we have in eqn (17.13) an important relation between the numbers of electrons and holes, and in the light of eqn (17.14) we have also $np = n_i^2$, where $n_i$ is the number of intrinsic electrons which would exist at the same temperature. The ratio of numbers of electrons and holes depends on the position of the Fermi level, for which we shall quote some results only for extreme cases.

When donors or acceptors (but not both together) are present, which produce discrete levels lying close to the conduction or valence band respectively, the conductivity at low temperatures is dominated by the ionization of the impurity levels. The Fermi level at 0 K then lies between the donor level and the conduction band, or between the acceptor level and the top of the valence band (see Fig. 17.5). As the temperature rises the Fermi level shifts because of a term similar to the second term in eqn (17.15), but involving $\ln(N_d/N_c)$ or $\ln(N_a/N_v)$, where $N_d$ and $N_a$ are the number of donor and acceptor levels per unit volume respectively. Thus for donors we have

$$W_F = \tfrac{1}{2}(W_d + W_c) + \tfrac{1}{2}kT \ln(N_d/N_c) \quad (n \ll N_d), \tag{17.16}$$

showing that as the temperature rises the Fermi level will rise if $N_d > N_c$, or fall if $N_d < N_c$.

In the exhaustion range the donor levels are fully ionized and $n = N_d$; in this case the Fermi level is given approximately by

$$W_F = W_c + kT \ln(N_d/N_c) \quad (n = N_d). \tag{17.17}$$

If $N_d < N_c$ the Fermi level lies below the conduction band, the number of electrons excited into the conduction band is small compared with the number of available states and they obey the classical statistics. This condition is called 'non-degenerate'. On the other hand, if $N_d > N_c$ the

FIG. 17.5. Variation of Fermi level when *either* donors *or* acceptors are present. At very low temperatures the Fermi level lies midway between the impurity level and the conduction or valence band. For small impurity concentrations ($N_a < N_c$ or $N_a < N_v$) the level moves towards the centre of the forbidden band with rising temperature. At higher temperatures the electron distribution is dominated by the intrinsic contribution, and $W_F$ is at the centre of the forbidden gap if $m_h^* = m_e^*$.

Fermi level lies in the conduction band, and since $n = N_d$, the number of conduction electrons is greater than the number of available states; this condition is called 'degenerate'. This situation is similar to that in a metal, where the exclusion principle limits the number of electrons in a given energy range, and the electron distribution must be treated by Fermi–Dirac statistics instead of classical statistics.

Similar results are obtained for acceptor impurities, provided we count energy as increasing downwards from the top of the valence band rather than upwards from the bottom of the conduction band (cf. Fig. 17.4).

The importance of the Fermi level lies in the fact that its value is equal to that of the thermodynamic potential $G = U - TS + |e| V$ of the electrons. If the electron distributions in two substances are in thermal equilibrium with each other, then the values of the thermodynamic potential in the two substances are equal, and hence so also are the Fermi levels. Thus the position of the Fermi level plays an important role in discussing the properties of junctions (cf. § 11.3).

We end this section by giving some numerical values, assuming $m^* = m$. For intrinsic material at room temperature, the values of $n_i$ are roughly $10^{20} \, \mathrm{m}^{-3}$ for germanium and $10^{16} \, \mathrm{m}^{-3}$ for silicon. The latter contains $5 \times 10^{28}$ atoms per cubic metre, and if one atom in $5 \times 10^6$ is replaced by a donor atom which is fully ionized at room temperature, the density of electrons in the conduction band becomes $n = N_d \sim 10^{22} \, \mathrm{m}^{-3}$. The density of holes is $p = n_i^2/n \sim 10^{10} \, \mathrm{m}^{-3}$. In this material electrons are the 'majority carriers' and holes the 'minority carriers'. In each case the density is small compared with $N_c = N_v = 2 \cdot 5 \times 10^{25} \, \mathrm{m}^{-3}$, so that classical statistics are a valid approximation.

## 17.4. Optical properties

Semiconductors such as germanium and silicon look very much like metals; they are opaque to visible light and have a high reflectivity. This is because the quantum carried by a visible photon, which corresponds to an energy roughly between 1·5 eV and 4 eV, is sufficient to excite an electron from the valence band right across the forbidden energy gap into the conduction band. If the absorption coefficient is measured at lower frequencies, a sharp drop in absorption would be expected when the photon energy becomes smaller than the energy gap $W_g$, that is, at frequencies such that $h\nu < W_g$. The change in the absorption coefficient can be quite dramatic, from $10^6 \, \mathrm{m}^{-1}$ or $10^7 \, \mathrm{m}^{-1}$ at frequencies higher than the absorption edge, down to $10 \, \mathrm{m}^{-1}$ at frequencies beyond the edge, as illustrated in Fig. 17.6. The absorption beyond the edge depends on the purity of the specimen, since impurities produce levels in the forbidden band from or to which electrons can still be excited by photons of lower energy.

FIG. 17.6. Absorption coefficient for electromagnetic radiation in pure germanium, showing the steep rise at the band edge. The curves for two different temperatures show that the energy gap is slightly higher at the lower temperature.

An optical determination of the position of the absorption edge gives in principle a direct measurement of the energy gap, whose accuracy is limited by the fact that the drop in absorption is spread out over a small but finite range of frequency. Careful analysis of the experimental results in the light of a detailed theory of how the absorption coefficient should vary with frequency in the vicinity of the absorption edge has given quite accurate measurements of the energy gap, and shows directly how it varies with temperature. In the interpretation allowance must be made for the presence of selection rules analogous to those discussed in § 10.9 for optical absorption and lattice waves. The simple absorption of a photon can only excite an electron from the valence band into a state in the conduction band with the same crystal momentum $\hbar\mathbf{k}$, so that we have the selection rule $\Delta\mathbf{k} = 0$. If the maximum of the valence band and the minimum of the conduction band both occur at the same value of $\mathbf{k}$, then the lowest absorption frequency gives the energy gap. This is the case in indium antimonide, where the valence-band maximum and conduction band minimum both occur at $\mathbf{k} = 0$. In germanium and silicon, however, the conduction band has only a subsidiary minimum at $\mathbf{k} = 0$, and a deeper minimum occurs at a finite value of $\mathbf{k}$, as shown in Fig. 17.7. Thus

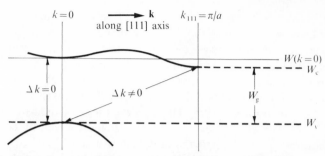

FIG. 17.7. Shape of the band edges against crystal momentum **k** for germanium. The momentum of a photon is negligible, so that there can be no net transfer of momentum on absorption. Either $\Delta \mathbf{k} = 0$ for the electron, or the difference in momentum when $\Delta \mathbf{k} \neq 0$ must be taken up by the creation or destruction of a phonon. For germanium $W_g = 0.75$ eV at 0 K but $W(\mathbf{k} = 0) - W_c \approx 0.14$ eV, so that the $\Delta \mathbf{k} \neq 0$ transitions give structure on the absorption edge.

for these substances transitions in the vicinity of $\mathbf{k} = 0$ do not determine the minimum value of $W_g$, defined as the difference in energy between the highest point in the valence band and the lowest point in the conduction band. (This is the energy required for thermal excitation of electrons across the gap, when no selection rules are involved.) In fact optical transitions such as that marked $\Delta \mathbf{k} \neq 0$ in Fig. 17.7 are allowed through the simultaneous creation or destruction of a phonon. Such transitions are called indirect or phonon-assisted transitions, while transitions in which $\Delta \mathbf{k} = 0$ are known as direct transitions.

*Excitons*

When an electron is excited from the valence band into the conduction band by a direct transition, a hole is created in the valence band whose momentum must be equal and opposite to that of the electron in the conduction band in order to make $\Delta \mathbf{k} = 0$. The electron and hole therefore move apart in opposite directions. In the vicinity of $\mathbf{k} = 0$, they move apart rather slowly, and their mutual Coulomb attraction begins to play a role; finally at $\mathbf{k} = 0$ itself the electron and hole stay together. Under these conditions their behaviour resembles that of an electron and proton in a hydrogen atom; a better comparison is with an electron and donor impurity atom, as discussed in § 17.2. Electron and hole may move in discrete orbits about the mutual centre of mass, giving rise to a series of energy levels

$$W = -\frac{m_r}{m} \frac{R}{\epsilon_r^2 n^2}, \qquad (17.18)$$

where $m_r$ is the reduced mass given by the relation

$$\frac{1}{m_r} = \frac{1}{m_e^*} + \frac{1}{m_h^*}. \qquad (17.19)$$

In eqn (17.18) $W = 0$ corresponds to separation of the electron and hole by such a large distance that their mutual attraction is negligible (that is, to 'ionization' of the electron–hole 'atom') and thus corresponds to the bottom of the conduction band at $\mathbf{k} = 0$. A lower energy is obtained when the electron and hole are together, so that the levels of eqn (17.18) lie just below the conduction band (like those of a donor impurity and bound electron), as shown in Fig. 17.8. The energy is also of the same order; if $m_e^* = m_h^* = 2m_r$, the energy levels are just half those given by eqn (17.2), and lie therefore very close to the conduction band.

The electron–hole bound pair is known as an 'exciton' and has been identified through its hydrogen-like spectrum in some semiconductors with large energy gaps, together with cuprous oxide and germanium. In germanium excitons are associated both with the direct and indirect transitions. In the former case a sharp line spectrum would be expected at frequencies just short of the energy gap at $\mathbf{k} = 0$; in the indirect transitions the electron–hole pair can be formed with finite momentum and possess kinetic energy, so that the exciton levels are not sharp but broadened into a band.

### The spectra of impurity atoms

When an impurity atom is substituted for a 'host' atom, it produces discrete levels in the energy gap, as discussed in § 17.2. For shallow donors (donors with a low ionization energy) the levels obey a relation

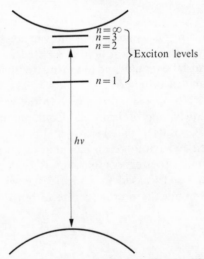

FIG. 17.8. Exciton levels lying just below the conduction band in the vicinity of $\mathbf{k} = 0$. The quantum of energy shown is that required to excite an electron from the valence band to the $n = 2$ level, and is slightly smaller than that corresponding to the absorption edge, which requires excitation to the bottom of the conduction band ($n = \infty$).

FIG. 17.9. Absorption spectrum at T = 20 K of phosphorus as a donor impurity in silicon. The sharp lines are due to transitions from the ground state to excited states of the donor electron bound to the impurity atom, ending at about 0·045 eV in the photo-ionization continuum. Note that in the highest quantum states the radii of the electron orbits are over 100 nm. (Courtesy Dr. R. A. Stradling).

close to that given by eqn (17.2), and transitions can be induced between them. Fig. 17.9 shows the spectrum of phosphorus in silicon, with a number of discrete lines, ending in the continuum corresponding to excitation of an electron from the donor atom into the conduction band.

*Photoconductivity*

When electromagnetic radiation excites electrons into the conduction band (or creates holes in the valence band), the electrical conductivity of a semiconductor changes appreciably; this phenomenon is known as photoconductivity. For small intensities of illumination the increase in conductivity is roughly proportional to the intensity, and forms an important method of detecting radiation of wavelengths shorter than the absorption edge. Cadmium sulphide gives a flat response between 0·5 $\mu$m and 0·65 $\mu$m, while lead sulphide has its peak response at about 2 $\mu$m and indium antimonide at 6 $\mu$m. Photoconductive detectors have a high sensitivity, and a short response time (from 1 $\mu$s to 100 $\mu$s) because the excess carriers quickly disappear through recombination, etc. This makes it possible to modulate the incoming radiation (using, for example, a mechanical chopping device) to produce an a.c. signal which can be amplified and detected.

The use of a semiconductor doped with suitable impurities makes it possible to construct detectors which are sensitive to longer wavelengths, the photoconductive effect then being due to electrons excited into the conduction band from donor impurity levels (or holes created by electrons being lifted into acceptor levels from the valence band). If the smallest separation of such discrete levels from the adjacent bands is about 0·01 eV, suitably doped samples are sensitive to wavelengths up to about 100 μm. Such detectors must of course be cooled to a temperature where thermal ionization of the impurity levels is unimportant.

*Emission of radiation*

Electrons excited into the conduction band may lose their energy in a collision, or they may make optical transitions in the reverse direction to those considered above in absorption processes. This results in the emission of electromagnetic radiation which lies in the visible region when a semiconductor with a suitable gap is used. Gallium arsenide (GaAs), and alloys of GaAs and gallium phosphide (GaP) are used for this purpose. A p–n junction (§ 17.7) is biased to inject a large density of electrons into the conduction band of p-type material, where they recombine with the holes to emit radiation with high efficiency. This process is the basis of the light-emitting diode (LED), and laser action can also be obtained under suitable circumstances.

More detailed discussions of the optical properties of semiconductors, and their applications, are given by Houghton and Smith (1966).

## 17.5. Transport properties

In addition to the energy gap $W_g$, the most important quantities we need for a semiconductor are the numbers of charge carriers of either sign, their effective masses, and their mobilities. The best method of measuring the effective mass is by means of cyclotron resonance, which is discussed in Chapter 24. Determination of $W_g$ by optical methods is possible only on very pure specimens, and in many cases it can only be deduced indirectly, using eqn (17.13). To bring out the temperature-dependence explicitly we rewrite this in the form

$$np = (2 \cdot 33 \times 10^{43})(m_e^* m_h^*/m^2)^{\frac{3}{2}} T^3 \exp(-W_g/kT), \qquad (17.20)$$

where the units are (metre)$^{-6}$. At first sight the simplest method of finding the number of charged particles would be measurement of the Hall coefficient $R_H$, which from eqn (12.51) is inversely proportional to the number of carriers. Since this number is so much smaller in a semiconductor than in a metal, the Hall effect is much larger ($\approx 0 \cdot 1$–1 m$^3$ C$^{-1}$ for pure silicon and germanium at room temperature) and correspondingly easier to measure. However, we have the difficulty that the Hall coefficient changes sign according to whether the current is carried by electrons or positive holes; in particular, in an intrinsic semiconductor, with equal numbers of each, we would expect the Hall coefficient to vanish. In practice this does not happen. because the electrons and holes have different mobilities, and so carry different fractions of the current. With two sets of carriers the expression for the Hall Coefficient is

$$R_\text{H} = -\frac{B}{|e|}\frac{(nb^2-p)}{(nb+p)^2},\qquad(17.21)$$

where $b = \mu_\text{e}/\mu_\text{h}$ is the ratio of the mobilities of electrons and holes, and $B$ is a coefficient not far from unity. The presence of $B$ arises from the fact that different averages over the distributions of velocity and relaxation time are involved in calculating the conductivity and the Hall effect. For a metal or a degenerate semiconductor, $B = 1$; for a non-degenerate semiconductor the value of B depends on how the scattering cross-section varies with carrier velocity.

Measurement of the Hall coefficient $R_\text{H}$ and the conductivity $\sigma$ makes it possible to determine both $n$ and $p$, using eqns (17.1) and (17.21), provided that the value of $b$ is known. The direct measurement of mobility is discussed below, but this is possible only with certain substances such as silicon and germanium. In other cases we can proceed only by making some assumptions about $b$. In an intrinsic semiconductor, $n = p = n_\text{i}$, and we have the relations ($n_\text{i}$ per cubic metre)

$$n_\text{i} = (4\!\cdot\!8\times10^{21})(m_\text{e}^* m_\text{h}^*/m^2)^{\frac{3}{4}} T^{\frac{3}{2}} \exp(-W_\text{g}/2kT)\qquad(17.22)$$

and

$$R_\text{H} = -\frac{B}{n_\text{i}|e|}\frac{(b-1)}{(b+1)}.\qquad(17.23)$$

Although the mobilities $u_\text{e}$ and $u_\text{h}$ both vary with temperature, their ratio $b$ does not vary rapidly in comparison with $n_\text{i}$, which is dominated by the exponential factor in eqn (17.22). Thus a plot of $\ln(R_\text{H}T^{\frac{3}{2}})$ against $T^{-1}$ should be a straight line, and this is found to hold for silicon and germanium. The slope of the straight line yields a value of $W_\text{g}$, but since $W_\text{g}$ is itself temperature dependent, care must be exercised in the interpretation (see Problem 17.2).

Another approximate method of finding $W_\text{g}$ involves only measurement of the conductivity. For an intrinsic semiconductor (or an impurity semiconductor at high temperatures where the conductivity is dominated by the intrinsic electrons), we have

$$\sigma = |e|n_\text{i}(u_\text{e}+u_\text{h}),\qquad(17.24)$$

and from measurement of the conductivity over a range of temperature we can find the variation of $n_\text{i}$ and hence determine the energy gap provided we known how the mobilities vary with temperature. When the mobility is determined by scattering processes due to lattice vibrations (phonons) the mobility (see below) should vary as $T^{-\frac{3}{2}}$. Since from eqn (17.22) $n_\text{i}$ varies as $T^{\frac{3}{2}}\exp(-W_\text{g}/2kT)$, we should expect that the conductivity would follow the law

$$\sigma = \sigma_0 \exp(-W_\text{g}/2kT),\qquad(17.25)$$

so that the energy gap can be found from a plot of ln $\sigma$ against $T^{-1}$. This method was used in early work, but such plots show a slight curvature, indicating that the mobility does not follow a $T^{-\frac{3}{2}}$ law exactly.

Determination of the energy gap from the optical absorption edge needs careful experimentation and interpretation, but it is the most satisfactory method and the only one which gives directly the energy gap at a given temperature. Use of the Hall effect or the conductivity depends on assumptions about the mobility, whose experimental determination will now be discussed.

### Measurement of mobility

The mobility of electrons and holes in semiconductors can be measured directly by a method due originally to Shockley and Haynes. The specimen is in the form of a narrow rectangular bar, about 0·5 mm square and a few centimetres long. A steady voltage is applied between the ends to give a field of order $10^3\,\mathrm{V\,m^{-1}}$ along the bar; two electrodes A, B are applied to the specimen, as shown in Fig. 17.10. A short voltage pulse of duration about 1 $\mu$s is applied to electrode A; if the semiconductor is n-type, and A is made positive in the pulse, electrons are withdrawn from the semiconductor by the electrode; some of these may come from the valence band, creating an excess of holes in the semiconductor. To preserve electrical neutrality of the specimen, electrons enter at any terminals which are negative with respect to A. One of these is B, which is connected (through an amplifier) to the Y-plates of an oscilloscope; this registers a voltage because of the flow of such electrons through the resistance R. This pulse does not quite coincide in time with the pulse at A, but the time delay is that required by an electromagnetic wave set up by the disturbance at A to travel to B, which is of the order of $10^{-4}$ $\mu$s and quite negligible. The holes injected at A are swept by the field towards B,

FIG. 17.10. Apparatus for direct measurement of mobility. The steady voltage applied at extreme left produces a uniform field in the semiconductor bar under whose influence a microsecond pulse of minority carriers drift from A to B. The mean drift velocity is $L/t$, where $t$ is measured on the oscilloscope.

and arrive at time $t = L/v$ later, where $L$ is the distance between electrodes A, B and $v$ is the drift velocity in the steady field. On arrival at B they appear as a second voltage pulse on the oscilloscope, and the time interval $t$ between the two pulses can be determined from the oscilloscope trace by calibrating the time-base. The voltage $V$ between the electrodes A, B due to the steady field is measured independently. Since the drift velocity $v = uE = uV/L$, the time $t = L/v = L^2/uV$ and the mobility is found from the relation

$$u = L^2/Vt. \qquad (17.26)$$

In a typical experiment the value of $t$ is about $30\ \mu$s. The second pulse at B due to the arrival of holes is broader and smaller than the first for two reasons: (1) diffusion of the holes in random directions; for accurate results the holes must be swept from A to B in a time $t$ short enough to make diffusion effects small, and the voltage pulse applied at A must be short compared with $t$; (2) holes are lost by recombination with electrons within the specimen.

To measure the mobility of electrons, all that would seem necessary at first sight would be to use n-type material and apply a negative pulse at A. This would inject electrons, creating a local excess, which is, however, dissipated in an extremely short time (see Problem 17.3), restoring the equilibrium concentration of electrons everywhere in the semiconductor. In this case only the first pulse is observed at B, and no second pulse. When a positive pulse is applied at A, with an n-type semiconductor, holes are created, and although electrical neutrality is restored by an inflow of electrons, we have now a non-equilibrium distribution with an excess of holes and a corresponding excess of electrons. Equilibrium is restored only when the holes flow out at B, or are annihilated within the semiconductor by recombination with electrons. The net result is that we can measure directly the mobility only of 'minority carriers'; that is, of holes in an n-type semiconductor, or electrons in a p-type semiconductor. By observing the spread in the second pulse at B the diffusion constant can be determined, and by observing its size as a function of $L$ or of electric field strength $E$ the rate of recombination can be found.

The quantity measured directly in such experiments is called the 'drift mobility', but in many semiconductors it cannot be so determined because of rapid diffusion of the carriers. In that case the mobility, when only one type of carrier is present, can be found from the conductivity and Hall effect, since then $|R_H| = B/(n\,|e|)$ and $|R_H\sigma| = Bu$. The quantity $|R_H\sigma|$ is often written $u_H$ and called the 'Hall mobility' to distinguish it from the drift mobility.

## Variation of mobility with temperature

The mobility of electrons or holes in semiconductors is limited by scattering processes which are basically similar to those in metals, but the temperature-dependence of the mobility is very different for two reasons. In a metal only electrons at the Fermi surface contribute to the conduction current, and their velocity $v$ is substantially independent of temperature; hence we do not need to take into account any velocity-dependence in a scattering cross-section. Similarly the fact that the scattering cross-section determines the free path $l$, while the mobility depends on the relaxation time $\tau = l/v$, does not of itself introduce any temperature-dependence. In a semiconductor, however, the average kinetic energy of the charge carriers is $\approx kT$, and the velocity variation as $T^{\frac{1}{2}}$ plays an important role.

Measurements of mobility at various temperatures and different impurity concentrations show that it is a function of both. In a pure material the charge carriers are scattered by the lattice vibrations (phonons); at all but the lowest temperatures (when impurity scattering dominates in any case) the scattering cross-section is proportional to the mean-square amplitude of the thermal fluctuations, and hence is proportional to $kT$. This gives a mean free path $l$ proportional to $T^{-1}$ which is the same for all carrier velocities, and the relaxation time $\tau = l/v$ and hence also the mobility should vary as $T^{-\frac{3}{2}}$. Table 17.2 shows that this law is not very well obeyed, except for electrons in germanium. The discrepancies may be due either to scattering by short-wavelength lattice vibrations where adjacent atoms vibrate in antiphase ('optical modes'), or to the complicated band structure (see Fig. 17.7), both of which allow scattering processes with a large change in electron wave vector. The observed mobilities in silicon and germanium at room temperature are found from Table 17.2 to lie between about $0\cdot05\ \mathrm{m^2\,V^{-1}\,s^{-1}}$ and $0\cdot4\ \mathrm{m^2\,V^{-1}\,s^{-1}}$, and are thus considerably higher than those in metals (for copper the value is about $0.004\ \mathrm{m^2\,V^{-1}\,s^{-1}}$). The high mobility is partly due to low values of the effective mass of electrons and holes in semiconductors. In InSb and GaAs, where the electron masses (see Table 24.2) are particularly small, the mobilities are exceptionally high ($\sim 8\ \mathrm{m^2\,V^{-1}\,s^{-1}}$ for electrons in InSb).

The scattering cross-section due to neutral impurities is inversely proportional to carrier velocity, giving a relaxation time independent of

TABLE 17.2

*Mobilities in elemental semiconductors*

|     | Electrons | Holes |
| --- | --- | --- |
| Si | $(40 \times 10^4)T^{-2 \cdot 6}$ | $(2 \cdot 5 \times 10^4)T^{-2 \cdot 3}$ |
| Ge | $(0 \cdot 35 \times 10^4)T^{-1 \cdot 6}$ | $(9 \cdot 1 \times 10^4)T^{-2 \cdot 3}$ |

The units are $\mathrm{m^2\,V^{-1}\,s^{-1}}$, and the values are taken from Ziman (1960).

velocity and hence also of temperature. Charged (that is, ionized) impurities will scatter carriers by a process analogous to Rutherford scattering of alpha particles; the cross-section is inversely proportional to the square of the carrier energy and hence varies as $T^{-2}$, so that the mean free path varies as $T^2$. Since the mean velocity varies as $T^{\frac{1}{2}}$, the relaxation time varies as $T^{\frac{3}{2}}$. A special case of scattering by charged particles is the mutual scattering of electrons and holes.

TABLE 17.3

*Summary of dependence of electron (hole) scattering on velocity $v$ and temperature $T$*

| Scattering mechanism | Cross-section $\sigma$ | Free path $l \propto \sigma^{-1}$ | Relaxation time $\tau = l/v$ |
|---|---|---|---|
| Phonons (at ordinary temperatures) | $T$ | $T^{-1}$ | $T^{-\frac{3}{2}}$ |
| Neutral impurities | $v^{-1}$ | $v$ | constant |
| Ionized impurities | $v^{-4}$ | $v^4$ | $v^3 \equiv T^{\frac{3}{2}}$ |

The velocity- and temperature-dependence of these scattering mechanisms are summarized in Table 17.3. In a first approximation the rates of scattering by different processes are additive, that is, we can write (cf. eqn (12.31).

$$\frac{1}{\tau} = \sum_i \frac{1}{\tau_i}. \tag{17.27}$$

At low temperatures the lattice scattering decreases as the lattice vibrations die away, and the mobility is eventually dominated by the impurity scattering.

### Recombination and diffusion

At any given temperature there is an equilibrium concentration of electrons and holes in a semiconductor, the two concentrations being equal in intrinsic material and generally unequal in extrinsic material. In the experiment of Shockley and Haynes we have seen that an excess of minority carriers can be injected at a contact, and to preserve electrical neutrality there will be a corresponding injection of majority carriers, possibly at another electrode. Any abnormal charge distribution caused thereby vanishes in about $10^{-5}\,\mu\text{s}$ (see Problem 17.3), so that we can write $\Delta p = \Delta n$, where $\Delta p$, $\Delta n$ are the local excesses in the number of holes and electrons respectively per unit volume. If we did not have this equality, a space charge $\rho = |e|(\Delta p - \Delta n)$ would be set up, which by Poisson's equation

$$\text{div } \mathbf{E} = \rho/\epsilon = (e/\epsilon)(\Delta p - \Delta n)$$

would give rise to strong electric fields. These would cause current to flow which would neutralize the space charge in the time given above. Obviously this current flow consists mainly of the more numerous majority carriers, and the controlling factor is the departure from the equilibrium value of the number of minority carriers. The fact that $\Delta p = \Delta n$ (in practice the equality is not exact, but departures from it are very small and can be neglected for present purposes) means that the *fractional* change in the number of minority carriers may be appreciably greater than the *fractional* change in majority carriers. A number of important devices described in later chapters depend on changes in the minority carrier concentration; such changes represent departures from equilibrium, and the mechanisms by which they decay play an important role in the design of such devices.

As mentioned in § 17.1, the charge carriers in a semiconductor have a finite lifetime, but this varies widely with the purity of the crystal. Simple recombination of an electron and a hole can only take place if certain restrictions on momentum and energy are satisfied, and measurements on very pure germanium show that this process would give a lifetime greater than $10^{-2}$ s. The observed lifetimes are generally much shorter, owing to the presence of chemical impurities which provide extra levels, known as 'traps'. In an n-type material, for example, electrons may drop from the conduction band into such traps; holes may then collide with these electrons to give recombination. The energy and momentum considerations involved in this 'indirect' process are much less restrictive than for the direct process of recombination, and the lifetime of the carriers is correspondingly shorter. These processes take place in the body of the semiconductor, but discrete levels at the surface (see § 17.6) may also act as traps which promote recombination through 'indirect' processes. A further loss of minority carriers also occurs at the electrodes.

The chance of a hole and an electron recombining is proportional to the concentration of each species; hence the rate of annihilation is $-apn$, where $a$ is a constant which depends on the mechanism involved. In thermal equilibrium the loss by recombination is balanced by the creation of new carriers through thermal excitation into the conduction band; if we denote the rate of creation by $c$, then clearly $c = ap_0n_0$ where $p_0$, $n_0$ are the equilibrium concentrations. When a departure from equilibrium takes place we have

$$dp/dt = dn/dt = c - apn = a(p_0n_0 - pn).$$

As mentioned above, the *fractional* change in the majority carrier concentration is much smaller than the *fractional* change in the minority carrier concentration, and to a first approximation the former may be neglected. Hence taking, for example, the holes in an n-type material, we may write

$$dp/dt \sim an_0(p_0-p) = -\Delta p/\tau_h, \tag{17.28}$$

showing that the rate of decay of the minority carrier concentration is simply proportional to the excess of minority carriers. The quantity $\tau_h$ is the 'recombination lifetime' of the minority carriers.

Recombination at the surface or extraction at an electrode causes a local diminution in the excess of minority carriers; this is counteracted by the movement of carriers from regions where they are more numerous. This is a process of diffusion, and for small field strengths diffusion currents are much larger than conduction currents. For simplicity, we consider a case where the minority carrier density $p$ (taking again holes in n-type material) varies only in one dimension. Then the number crossing unit area per second is $-D_h(\partial p/\partial x)$, and the net rate of increase in a thickness $dx$ is $d\{-D_h(\partial p/\partial x)\} = -D_h(\partial^2 p/\partial x^2)\,dx$, where $D_h$ is the diffusion coefficient for the minority carriers. In the steady state this equals the rate of loss by recombination, which in thickness $dx$ is $\{(p_0-p)/\tau_h\}\,dx$; hence we obtain the relation

$$\frac{\partial^2 p}{\partial x^2} = \frac{p-p_0}{D_h\tau_h}. \tag{17.29}$$

This has a solution (writing $\Delta p$ for $(p-p_0)$)

$$\Delta p = (\Delta p)_0 \exp(-x/L_h), \tag{17.30}$$

where $(\Delta p)_0$ is the excess concentration at $x = 0$, and $L_h = (D_h\tau_h)^{\frac{1}{2}}$ is a measure of the mean distance a minority carrier will move under the action of diffusion before it is lost by recombination. It is known as the 'diffusion length' and is an important quantity in transistor design. A typical value for germanium is 1 mm for both holes in n-type material and electrons in p-type material; typical values in silicon are smaller by a factor of about 3.

### Hot carriers and the Gunn effect

As the electric field in a bulk semiconductor rises, the carrier drift velocity at first increases linearly with the field, as in a metal. At higher field strengths the additional energy acquired by the carriers through acceleration by the field becomes comparable with or greater than their thermal energy. The relaxation rate is a function of carrier energy, so that the mobility changes. In most cases the mobility increases, but in gallium arsenide (GaAs) the drift velocity decreases as shown in Fig. 17.11. The explanation for this effect, due to Gunn, is based on the band structure. The main minimum in the conduction band is at $\mathbf{k} = 0$ (Fig. 17.12), but at certain values of $\mathbf{k}$ along the [001] axes subsidiary minima occur with an

Electric field strength

FIG. 17.11. Electron drift velocity against field strength for gallium arsenide. The velocities are about $10^5$ m s$^{-1}$ and the minimum occurs at a field strength of about $10^6$ V m$^{-1}$.

energy about $0 \cdot 36$ eV greater than the central minimum. In the presence of an electric field sufficiently large to raise the carrier energy to $0 \cdot 36$ eV, electrons are scattered by optical phonons into the subsidiary minima. In these minima the effective mass $m^*$ is much higher than it is in the central minimum, so that carriers transferred into the subsidiary minima have a lower mobility. As the electric field increases, a greater proportion of the electrons are scattered into the subsidiary minima, and the mean drift velocity decreases. The velocity–field plot of Fig. 17.11 leads to a negative resistance effect, but the current–voltage characteristic is complicated by instabilities which lead to rapid current oscillations.

## 17.6. Metal–semiconductor junctions

To investigate the electrical properties of semiconductors it is necessary to make electrical connections to them. The behaviour of a junction between a metal electrode and a semiconductor depends on the nature and geometry of the connection, as well as on the properties of the metal and semiconductor, making a full treatment very complex. The most important property of such a junction is that the current flow for a given

Wave vector **k** (along [001] direction)

FIG. 17.12. Plot of electron energy against **k**-vector for the [001] axes in gallium arsenide, showing the subsidiary minimum at about $0 \cdot 36$ eV.

voltage is quite different in opposite directions, so that it acts as a rectifier.

When contact is made between a metal and a semiconductor, a potential difference is set up between the two in a similar manner to that between two metals (the contact potential). For an n-type semiconductor whose Fermi level is above that of the metal, electrons pass from the semiconductor to the metal until the two Fermi levels are equal. This process is illustrated in Fig. 17.13. The excess negative charge on the metal repels electrons near the surface of the semiconductor, creating a layer which is depleted of conduction electrons and so has a higher resistance than the bulk of the semiconductor. This layer is known as the 'barrier layer' and is a region of positive space charge because it contains the ionized donor impurities without the compensating charge of the negative conduction electrons. By Poisson's equation (eqn (2.1)) the potential will vary through the space-charge layer, so that there will be a potential difference $V_0$ between the position of the bottom of the conduction band at the surface and in the bulk semiconductor; the energy bands are therefore distorted near the surface, as shown in Fig. 17.13.

In equilibrium there will be no net current flow across the barrier, but this is a dynamic equilibrium between a current $-I_d$ of electrons flowing out of the metal into the semiconductor, and an equal current of electrons (and, to a much lesser extent, holes) leaving the semiconductor for the metal. The latter may be written in the form $I_0 \exp(-|e| V_0/kT)$, since the fraction of the electrons which have sufficient energy to surmount the barrier $V_0$ is proportional to $\exp(-|e| V_0/kT)$. If a voltage $V$ is now

Before contact                                                After equilibrium is established

FIG. 17.13. Energy bands and Fermi levels at a metal to semiconductor contact, before and after equilibrium is established.

applied which makes the metal positive with respect to the semiconductor, this extra voltage appears almost entirely across the barrier since this has a much larger resistance than either the metal or the bulk semiconductor. The current leaving the semiconductor then becomes $I_0 \exp\{-|e|\,(V_0-V)/kT\}$, because the barrier height is reduced from $V_0$ to $V_0-V$; on the other hand the current leaving the metal is still $-I_d$, since the barrier which the electrons in the metal have to surmount is unaltered (see Fig. 17.14). Since

$$I_d = I_0 \exp(-|e|\,V_0/kT),$$

the net current flow is

$$I = I_d\{\exp(|e|\,V/kT)-1\}. \tag{17.31}$$

If $V$ is positive and greater than $(kT/|e|)$, large forward currents can flow, while if $V$ is negative $I$ approaches $-I_d$; the current–voltage characteristic therefore has the form shown in Fig. 17.15(a), and the junction is an efficient rectifier.

(a) Metal positive
(forward direction)

(b) No applied voltage

(c) Metal negative
(backward direction)

FIG. 17.14. Effect of applied voltage $V$ at metal–semiconductor junction. In (a) a forward voltage $V$ is applied, reducing the barrier height to $V_0-V$, and giving a large current flow of electrons from the semiconductor into the metal; in (c) a reverse voltage is applied, increasing the barrier and reducing the current flow. Note that there is a change with voltage in the effective thickness of the barrier

$$\left(\delta = \left\{\frac{2\epsilon(V_0-V)}{eN_d}\right\}^{\frac{1}{2}}\right),$$

and that the Fermi level in the metal is depressed when the metal is made positive because of the negative sign of the electronic charge.

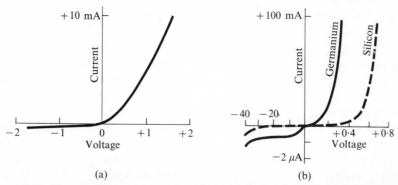

FIG. 17.15. Current–voltage characteristics of (a) silicon–tungsten diode for centimetre wavelengths, and (b) p–n junction. Note differences in scales.

This treatment gives a satisfactory qualitative treatment of the rectifying properties, though agreement with experiment is by no means exact. A satisfactory feature is that it gives the right sign for the forward direction, that is, that the direction of easy flow of electrons (for an n-type semiconductor) is from semiconductor to metal. In an early theory the flow of electrons through the barrier was ascribed to the tunnel effect, so that the easy flow was from metal (where the electron density is high) to semiconductor. However, the tunnel effect is only appreciable when the barrier thickness is comparable with the electron wavelength in the metal, that is, less than a nanometre, whereas the theory of a barrier due to a depletion layer gives a thickness of order 10 nm. A simple version of a theory for the barrier thickness put forward by Schottky is as follows.

We assume that all conduction electrons are removed from the barrier layer, so that the charge density $\rho = eN_d$, where $N_d$ is the number of donors per unit volume (for simplicity we take these all to carry unit charge, and all to be positively ionized, that is, we are in the exhaustion region). Let the surface of the semiconductor be the plane $x = 0$, and the barrier extend from the surface to the plane $x = \delta$, so that the space charge becomes zero for $x \geqslant \delta$. Then the potential is constant for $x \geqslant \delta$, and the electric field vanishes at $x = \delta$. In the barrier, Poisson's equation reduces to

$$\nabla^2 V = d^2 V/dx^2 = -eN_d/\epsilon \qquad (17.32)$$

Integration, using the boundary conditions $dV/dx = 0$ (no electric field) at $x = \delta$, and $V = 0$ at $x = 0$, gives

$$V = -(eN_d/2\epsilon)\{(x-\delta)^2 - \delta^2\}. \qquad (17.33)$$

Hence the difference of potential at the surface $x = 0$ from that in the interior of the semiconductor $(x \geqslant \delta)$ is

$$V_0 = eN_d\delta^2/2\epsilon. \qquad (17.34)$$

If we take $\epsilon \approx 12$ (as in silicon) and $N_d = 10^{24}\,\text{m}^{-3}$, $\delta$ is found to be about 30 nm if $V_0$ is about 1 V.

A difficulty in the Schottky theory is that $V_0$ should be equal to the difference in the work functions of metal and semiconductor, and hence $I_d$, which is proportional to $\exp(-eV_0/kT)$, should depend on the metal used, whereas experimentally it does not. To overcome this difficulty Bardeen put forward the idea of 'surface states'. At the surface of the semiconductor the atomic arrangement is different because there are no atoms on the one side with which to form bonds. Thus the surface atoms have different energy levels, and these levels are discrete because bands are only formed from the levels of atoms which are identical. There may also be impurity atoms absorbed on the surface. Those surface levels which lie below the Fermi level of the semiconductor will be filled by electrons from the conduction band, giving a negative charge on the surface which repels electrons near the surface. This gives a depletion layer just inside the surface, which acts as the barrier. If $N_s$ is the number of surface states per unit area, and $N_d$ the number of donors per unit volume, then the thickness $\delta$ of the barrier will be determined by the relation $N_s = N_d\,\delta$ if we assume that all the conduction electrons for a distance $\delta$ are drawn into the surface states. Application of Poisson's equation to the barrier layer yields eqn (17.34) as before, but on writing $\delta = N_s/N_d$ we have

$$V_0 = 2N_s^2/2\epsilon N_d. \tag{17.35}$$

This equation shows that $V_0$ depends on $N_s$ and $N_d$ for the semiconductor and is independent of what metal is used to make the contact; in fact the barrier layer exists in the absence of any contact.

The discussion above has been on the basis of an n-type semiconductor, but similar arguments apply to a p-type semiconductor if holes are substituted for electrons (thereby reversing the direction of easy current flow). Thus any metal–semiconductor junction will act as a rectifier. However, in order to use it two junctions must be made to complete a circuit, and since the forward direction will be in the opposite sense at the two junctions, no rectifying action will result unless the two junctions are different in nature. At microwave frequencies one junction must be very small in cross-section, since otherwise the capacitance between the metal and bulk semiconductor across the barrier layer acts as a short-circuit, the current flowing as displacement current through this capacitance instead of real current through the barrier. This small contact is made by a thin metal whisker (usually tungsten) pressed against the semiconductor (silicon); the other contact is soldered and of large area, so that it offers little resistance (and large capacitance) to the flow of current in what would otherwise be its 'backward' direction (see Fig. 18.16).

Such fine contacts have a relatively high resistance, since the current has to spread out from a fine point through the interior of the semiconductor, and they cannot carry more than a few milliamperes of current, as in Fig. 17.15(a). At low frequencies metal–semiconductor junctions with a larger area can be used, the metal and semiconductor being separated by an insulating oxide layer sufficiently thin that electrons can readily tunnel through it. Such 'Schottky' diodes' can carry large currents, comparable with those in the p–n junction described below. They have a low forward resistance because electrons crossing the barrier are majority carriers with relatively high velocities, arriving in the metal with the extra energy equivalent to the difference in work functions.

## 17.7. Theory of the p–n junction

Single crystals of a semiconductor (silicon or germanium) can be prepared in which one end is doped to make it p-type, and the other end n-type, the change from p-type to n-type taking place in a region whose thickness is $\sim 10$ nm. Such a unit is called a p–n junction. The p-type material is made by doping with a concentration $N_a$ of acceptors, the n-type by doping with a concentration $N_d$ of donors; obviously there will be a narrow region at the junction where the doping concentration varies from one extreme to the other, but provided this region is small in width compared with the thickness of the barrier region estimated below, we can regard the change from p-type to n-type as discontinuous. In germanium doped with Group III and Group V impurities the ionization potential is so small that at room temperature we can regard the donors and acceptors as fully ionized (that is, we are in the exhaustion region).

The energy level situation shown in Fig. 17.16 is not stable and could only exist if the n-type and p-type material were in separate crystals. There is an excess hole concentration in the p-type, and excess electron concentration in the n-type, so that when they are in the same crystal holes will diffuse to the right and electrons to the left, each producing a positive current to the right. This gives a positive potential to the n-type material, so that the energy levels of its electrons are lowered because of their negative charge. This process continues until the Fermi levels of the two halves are equalized, as in Fig. 17.17. The difference of potential between the two halves means that strong electric fields exist near the junction, and these sweep out the mobile carriers in the vicinity of the junction, giving a barrier layer of high resistance. The whole of the potential drop occurs across this barrier layer; removal of the holes in the barrier layer on the p-type side leaves a negative space charge of $-eN_a$ per unit volume, and removal of electrons on the n-type side leaves a positive space charge of $+eN_d$ per unit volume, as shown in Fig. 17.18. We now apply Poisson's equation to the barrier layer, making the same

FIG. 17.16. Energy bands and Fermi level at p–n junction, before equilibrium is established.

kind of simplifying assumptions as in the treatment of the metal–semiconductor junction. Let the change from p- to n-type occur discontinuously at the plane $x = 0$, and let the barrier thickness be $\delta_p$ and $\delta_n$ on each side respectively. Then, taking $V = 0$ at $x = 0$ we have, using the boundary conditions $\mathrm{d}V/\mathrm{d}x = 0$ at $x = -\delta_p$ and at $x = +\delta_n$:

$$x < 0 \qquad\qquad\qquad x > 0$$

$$\frac{\mathrm{d}^2 V}{\mathrm{d}x^2} = +\frac{|e|\,N_a}{\epsilon} \qquad\qquad \frac{\mathrm{d}^2 V}{\mathrm{d}x^2} = -\frac{|e|\,N_d}{\epsilon}$$

$$V = \frac{|e|\,N_d}{2\epsilon}\{(x+\delta_p)^2 - \delta_p^2\} \qquad V = -\frac{|e|\,N_d}{2\epsilon}\{(x-\delta_n)^2 - \delta_n^2\}$$

Thus the over-all potential difference $V_0$ between the n-type material and the p-type material is the difference between $V_n$, the value of $V$ at $x = \delta_n$, and $V_p$ at $x = -\delta_p$. This is

$$V_0 = V_n - V_p = \frac{|e|}{2\epsilon}(N_d \delta_n^2 + N_a \delta_p^2). \tag{17.36}$$

FIG. 17.17. Electron energy bands and the Fermi level at a p–n junction, after equilibrium is established. The n-type becomes positively charged, so that its electrons have lower potential energy.

FIG. 17.18. Space-charge density at a p–n junction, on the simplified model used in the text.

From the fact that $dV/dx$ must be continuous at $x = 0$ we obtain the additional condition

$$N_a \delta_p = N_d \delta_n. \tag{17.37}$$

This result makes the total charge on either side equal but opposite in sign, as must be the case if no lines of electric field emerge from the barrier layer. Finally, comparison of Figs 17.16 and 17.17 shows that

$$|e|(V_n - V_p) = W_{Fn} - W_{Fp} \approx W_g, \tag{17.38}$$

where $W_{Fn}$, $W_{Fp}$ are the Fermi levels in n- and p-type as shown in Fig. 17.16; in the exhaustion range these lie close to the bottom of the conduction band and top of the valence band respectively, so that the energy difference is nearly equal to the energy gap. If we take

$$N_a = N_d = 10^{24} \, \text{m}^{-3},$$

then for germanium the width of the barrier $(\delta_n + \delta_p)$ is found to be about 50 nm.

In the p-type material, holes form the 'majority carriers' but there are a small number of electrons, known as the 'minority carriers'. Similarly in the n-type material, electrons are the majority carriers and holes the minority carriers. If $p$ is the density of the majority carriers (holes) in the p-type material, then the density of holes (minority carriers) in the n-type material is $p_0 = p \exp(-|e| V_0 / kT)$. There is a similar ratio for the electrons as minority and majority carriers, and the product of the hole and electron densities on each side is the same (see § 17.3).

The fact that the barrier layer has a very much higher resistivity than the bulk material means that any external voltage applied appears almost wholly across the barrier. In the bulk material the potential variation is practically zero, and currents near the barrier are the result of diffusion, being determined by the gradients in the numbers of holes and electrons. If an external voltage $V$ is applied which makes the n-type material less positive, the voltage across the barrier is reduced to $(V_0-V)$ and there is a larger number of holes with sufficient energy to cross from left to right. Since the minority carrier density is normally only about $10^{-5}$ of the majority carrier density, the latter does not change appreciably as a result of any transfer, and can be assumed to be constant. On the other hand, the minority carrier density is increased, so that the hole density at the right-hand edge of the barrier becomes

$$p_V = p \exp\{-|e|\,(V_0-V)/kT\} = p_0 \exp(+|e|\,V/kT).$$

A similar equation holds for electrons at the left-hand edge of the barrier, as shown in Table 17.4.

TABLE 17.4

*Densities of holes and electrons at barrier edges (Fig. 17.18), without and with an applied voltage*

|  | p-type edge | n-type edge | Applied voltage |
|---|---|---|---|
| Density of holes | $p$ | $p_0 = p \exp(-|e|\,V_0/kT)$ | 0 |
| Density of electrons | $n_0 = n \exp\{-(|e|\,V_0/kT)\}$ | $n$ | 0 |
| Density of holes | $p$ | $p_V = p \exp\{-|e|(V_0-V)/kT\}$ | $V$ |
| Density of electrons | $n_V = n \exp\{-|e|\,(V_0-V)/kT\}$ | $n$ | $V$ |

The densities $p_V$ and $n_V$ at the edges of the barrier are greater than the equilibrium densities $p_0$ and $n_0$ in the bulk material, and it follows from eqn. (17.30) that there is a density gradient

$$\frac{dp}{dx} = \frac{-(p_V-p_0)}{L_h} = -\left(\frac{p_0}{L_h}\right)\left\{\exp\left(\frac{|e|\,V}{kT}\right)-1\right\}. \tag{17.39}$$

The resulting current density resulting from diffusion of the holes is

$$J_h = -|e|\,D_h\left(\frac{dp}{dx}\right) = \frac{|e|\,D_h}{L_h}p_0\left\{\exp\left(\frac{|e|\,V}{kT}\right)-1\right\},$$

and there is a similar equation for electrons diffusing across the barrier to the left. Hence the total current density across the junction is

$$J = J_e + J_h = |e| \left( \frac{n_0 D_e}{L_e} + \frac{p_0 D_h}{L_h} \right) \left\{ \exp\left( \frac{|e| V}{kT} \right) - 1 \right\}, \qquad (17.40)$$

where $D_h$, $L_h$ are the diffusion coefficient and diffusion length for holes in the n-type material and $D_e$, $L_e$ the corresponding quantities for electrons in the p-type material. The entire current density across the junction arises from the flow of minority carriers, thus emphasizing the importance of their role.

This equation, due originally to Shockley, is similar to that for the metal–semiconductor junction, and the current–voltage characteristic (Fig. 17.15(b)) differs from Fig. 17.15(a) only in that the reverse current is much lower and the forward current much higher. There is a limit to the back voltage which can be applied because of breakdown (see § 18.3). The materials normally used for p–n junctions are silicon and germanium. Germanium has a smaller energy gap, giving a larger carrier density; the mobility is also rather higher, so that the forward resistance is lower than for silicon, but the reverse current is greater. Silicon has the advantage that it can be operated at higher temperatures before thermal ionization across the energy gap increases the back current and drops the rectification efficiency.

*Capacitance of a p–n junction*

The depletion layer with its high resistance acts as a capacitance (see Problem 17.4). The charge distribution shown in Fig. 17.18 assumes an abrupt change in the impurity concentration, and eqn (17.36) gives a voltage drop that varies with the square of the barrier thickness. In fact the process used in making p–n junctions produces an impurity density which changes almost linearly with distance, and the voltage across the barrier varies more nearly with the cube of the barrier thickness. Conversely the barrier thickness depends on the voltage, so that the barrier behaves as a capacitance which is proportional to $V^{-\frac{1}{3}}$. A p–n junction, with reverse bias to keep the flow of current small, acts as a non-linear capacitance, and is used in applications such as automatic frequency-control of radio receivers.

At high frequencies the barrier capacitance reduces the rectification efficiency of a p–n junction, since it takes a finite time to alter the charge density within the depletion layer when the applied voltage is reversed.

## References

HOUGHTON, J. T. and SMITH, S. D. (1966). *Infra-red physics*. Clarendon Press, Oxford.
ZIMAN, J. M. (1960). *Electrons and phonons*. Clarendon Press, Oxford.

## Problems

17.1. Show that the energy levels of an 'atom' consisting of a positive charge equal to that of the proton, with an effective mass $m_h^*$, and a negative charge equal to that of the electron, with an effective mass $m_e^*$, moving in a medium of relative permittivity, $\epsilon_r$ are those given by eqns (17.18) and (17.19).

Show that the binding energy of a donor impurity level in silicon ($m_h^* = \infty$, $m_e^*/m = 0\cdot4$, $\epsilon_r = 11\cdot5$) is about $0\cdot041$ eV, and that the Bohr radius is about $6\cdot5$ times the interatomic distance ($0\cdot235$ nm).

17.2. In many semiconductors the energy gap varies linearly with $T$ over a fair temperature range, though it tends to a constant value at low temperatures. Show that if $W_g = W_g^0 - aT$, then a plot of $\ln(R_H T^{\frac{3}{2}})$ against $T^{-1}$ gives a straight line whose slope gives the value of $W_g^0$. Show also that the effect of the temperature variation of $W_g$ on eqn (17.22) is to replace $W_g$ by $W_g^0$ and to increase the apparent value of the products $(m_e^* m_h^*)$ by $\exp(2a/3k)$.

17.3. Using eqn (3.16) show that any abnormal charge distribution in germanium where $\epsilon_r = 16$, $\sigma = 10\ \Omega^{-1}\,\mathrm{m}^{-1}$ vanishes in a time of order $10^{-11}$ s.

A sample of germanium contains $10^{18}$ holes per $\mathrm{m}^3$. Show that the presence of a net space charge equivalent to 1 per cent of the hole concentration would give rise to an electric field gradient of about $10^7\ \mathrm{V\,m}^{-2}$.

17.4. When an external bias voltage $V$ is applied to a p–n junction with a high resistance barrier, the voltage across the barrier in eqn (17.36) becomes $V_n - V_p = V + V_g$, where $|e|\,V_g = W_g$. Show that the total charge $\pm Q$ in each of the space-charge regions of Fig. 17.18, when $N_a = N_d = N_I$, is given by

$$Q = (\epsilon e N_I)^{\frac{1}{2}}(V + V_g)^{\frac{1}{2}} \text{ per unit area.}$$

This is a function of the applied voltage, and the barrier acts as a capacitance for a.c. voltages of magnitude

$$C = (\mathrm{d}Q/\mathrm{d}V) = \tfrac{1}{2}(\epsilon e N_I)^{\frac{1}{2}}(V + V_g)^{-\frac{1}{2}} \text{ per unit area.}$$

Verify that the capacitance is the same as that of a parallel-plate capacitor with plate separation equal to the barrier thicknesss and filled with dielectric of permittivity $\epsilon$.

17.5. In a crystal the velocity of a particle is given by the relation $\hbar \mathbf{v} = \mathrm{grad}_k\,W$, where $\mathrm{grad}_k = \mathbf{i}_x(\partial/\partial k_x) + \mathbf{i}_y(\partial/\partial k_y) + \mathbf{i}_z(\partial/\partial k_z)$, in which $\mathbf{i}_x$, etc., are unit vectors along the $x$, $y$, $z$ directions. Show that in general $\mathbf{v}$ is not parallel to $\mathbf{k}$ unless it is along one of the axes, for a particle whose energy $W$ is given by eqn (12.25).

Show that with respect to axes $(x', y', z')$ which are derived from the $(x, y, z)$ axes by a rotation through an angle $\theta$ about the $y$-axis, eqn (12.25) becomes

$$W = W_0 \pm \tfrac{1}{2}\hbar^2 \left\{ k_x'^2 \left( \frac{\cos^2\theta}{m_x^*} + \frac{\sin^2\theta}{m_z^*} \right) + \frac{k_y'^2}{m_y^*} + k_z'^2 \left( \frac{\sin^2\theta}{m_x^*} + \frac{\cos^2\theta}{m_z^*} \right) + \right.$$
$$\left. + 2k_x' k_z' \sin\theta \cos\theta \left( \frac{1}{m_x^*} - \frac{1}{m_z^*} \right) \right\}.$$

Hence show that the effective mass for a particle for which $\mathbf{k}$ is along the $z'$-axis is

$$\frac{1}{\hbar^2} \frac{\partial^2 W}{\partial k_z'^2} = \frac{\sin^2\theta}{m_x^*} + \frac{\cos^2\theta}{m_z^*}$$

($m^{*-1}$ is a tensor quantity, and cross-product terms such as $k'_x k'_z$, etc., are absent only when a suitable choice of axes (usually dictated by the crystal symmetry) is made).

17.6. For a semiconductor in the infrared, where $\omega\tau \gg 1$, the effective conductivity (see eqn (10.51)) becomes $\sigma = \sigma_0/(1 + \omega^2\tau^2)$. Show that an electromagnetic wave will fall in intensity inside the semiconductor as $W = W_0 \exp(-\alpha x)$, where

$$\alpha \sim \frac{\sigma}{n\epsilon_0 c(1+\omega^2\tau^2)}$$

provided that $k \ll n$ in the complex refractive index $n - jk$. Calculate the value of $\alpha$ for a sample of Ge in which $n = 4$, $\sigma_0 = 10\,\Omega^{-1}\,\mathrm{m}^{-1}$, at a frequency where $\omega\tau = 100$.

(*Answer*: $\alpha = 10^{-1}\,\mathrm{m}^{-1}$, approximately.)

# 18. Solid-state diodes

## 18.1. Introduction

ALMOST the first reasonably reliable detectors used in the reception of radio waves were 'crystal rectifiers', where a rectifying contact was made by a flexible wire 'cat's whisker' held lightly in contact with a semiconducting crystal. Good performance was obtained only from 'sensitive' spots on the crystal surface, requiring frequent readjustment of the contact. The invention of the thermionic diode by Fleming in 1904, and the development of a vast family of thermionic vacuum tubes, led to the crystal rectifier being entirely superseded, apart from laboratory work with very short wavelengths where thermionic devices failed, mainly because of the transit time of the electrons. Shortly before 1940, the invention of the cavity magnetron and klystron led to the widespread use of centimetre waves, where the crystal rectifier (§ 17.6) once more came into its own. The solid-state diode, with a junction between two different semiconducting substances, was developed just before 1950, and in turn has superseded the thermionic diode, being more versatile, smaller, cheaper, and more rugged.

Solid-state electronic devices are constructed from semiconducting materials, and their operation is dependent on the unique properties of semiconductors, discussed in Chapter 17. In this and the following chapters the operation of such devices is outlined in terms which do not require a detailed knowledge of the theory of semiconductors, apart from the introductory sections §§ 17.1–17.2. On this basis we start with a summary of the nomenclature commonly used and an elementary explanation of the action of the p–n junction. This is essential to the understanding of transistor action, and is a simplified version of the treatment in § 17.7.

A 'hole' is created when one electron is absent from a covalent bond of the pure crystal. A hole behaves as a positively charged carrier of current.

The 'intrinsic' energy gap is the energy required to remove an electron from such a chemical bond in a 'pure' semiconductor, creating one conduction electron and one hole. At room temperature, for silicon this energy is about $1\cdot1$ eV; for germanium it is about $0\cdot7$ eV. In a 'doped' semiconductor, a fraction of the atoms are replaced by atoms of a different valency. In silicon and germanium (valence 4), the addition of 'donor' atoms of valence 5 (such as phosphorus) produces an n-type semiconductor, while 'acceptor' atoms of valence 3 (such as aluminium) produce a p-type semiconductor. The fraction of impurity atoms varies from 1 in $10^9$ to 1 in $10^6$. The 'extrinsic' energy gap is the energy required to remove the excess electron from a donor atom (creating a conduction electron) or to move an electron to an acceptor atom (creating a hole) in a 'doped' semiconductor. For silicon this energy is of order of 0·04 eV, depending on the doping impurity.

In all semiconductors some carriers of opposite sign ('minority carriers') are present as the result of thermal excitation. In n-type semiconductors, electrons are the majority current carriers and holes the minority carriers. In p-type semiconductors, holes are the majority carriers and electrons the minority carriers. Both n- and p-type semiconductors ·are uncharged, each atom being on average electrically neutral.

## 18.2. The p–n junction

The p–n junction has been discussed in detail in § 17.7 and is shown diagrammatically in Fig. 18.1, where the region of the change in chemical composition from p-type to n-type is assumed to be infinitely narrow. The electron concentration on one side of the boundary and the hole concentration on the other side are each assumed to be constant. The electrons on the right and the holes on the left will each diffuse across the boundary until a potential difference is built up between the two sides, such that its effect on the charge carriers exactly balances the effect of the concentration gradient across the boundary. Then the net flow of charge (that is, the current) across the boundary is zero.

The electric potential $V$ as a function of distance from the boundary is shown in Fig. 18.2. The region in which $V$ is changing is only about 50 nm in width, so that in this region a strong electric field exists which sweeps away all mobile charges; this region is known as the 'depletion layer'. On either side of the boundary, but within the depletion layer, the lines of electric field terminate on layers of charge due to the fixed ions, positive in the n-type and negative in the p-type material. (The charge density $\rho$ is related to $V$ by Poisson's equation $d^2V/dx^2 = -\rho/\epsilon$, so that the curvature of $V$ has opposite sign in the two charge layers.) Outside the depletion layer the conductivity is sufficiently high that the charge density

| Majority carriers: | Holes | Electrons |
|---|---|---|
| | p-type | n-type |
| Minority carriers: | Electrons | Holes |

FIG. 18.1. Schematic diagram of a junction between a p-type semiconductor and an n-type semiconductor. The boundary is here assumed to be infinitely sharp; in practice it is about 10 nm thick.

Fig. 18.2. Variation of the electrostatic potential near the boundary between p-type and n-type material in a p–n junction. Outside the depletion layer the potential is constant when no current is flowing. The potential energy is $+|e|\ V$ for holes and $-|e|\ V$ for electrons. A positive voltage applied to the p-type material reduces the barrier height (forward bias).

is zero and V is constant except for a small ohmic drop when current is flowing.

In equilibrium there is no net current, but both electrons and holes are crossing the boundary in each direction, giving a dynamic equilibrium. We consider first the holes. In spite of the potential barrier, some holes in the p-type material (coming predominantly from acceptor impurities) have sufficient energy to cross into the n-type material, where they disappear eventually by recombination with electrons. This constitutes the hole 'annihilation current'. In the n-type material, electron–hole pairs are being created, at a constant rate on average, through thermal excitation. In this region, the holes are minority carriers, and on arriving at the barrier they fall over the energy drop into the p-type material, giving a 'generation current'. Under equilibrium conditions the annihilation current and the generation current of holes are equal and opposite. Similar generation and annihilation currents exist for the electrons, which also balance in equilibrium, so that there is no *net* flow across the barrier either of electrons or of holes.

If the energy jump at the barrier is $W_B$, the chance that an electron has sufficient energy to cross it is proportional to $\exp(-W_B/kT)$, and the annihilation current $I_A$ is therefore proportional to this quantity. If in thermal equilibrium the annihilation current is $I_A^0$ and the energy jump is $W_B^0$, then we have

$$\frac{I_A}{I_A^0} = \exp\left(-\frac{W_B}{kt}\right)\bigg/\exp\left(-\frac{W_B^0}{kT}\right) = \exp\left\{-\frac{W_B-W_B^0}{kT}\right\}. \qquad (18.1)$$

On connecting an external battery, the extra potential appears almost wholly across the high-resistance depletion layer, so that $W_B = W_B^0 - |e| V$ (see Fig. 18.3) for electrons. The barrier is smaller when $V$ is positive, and the annihilation current becomes

$$I_A = I_A^0 \exp(|e| V/kT). \qquad (18.2)$$

The presence of the exponential means that the annihilation current can change enormously under the influence of an external voltage. On the other hand, the generation current is determined solely by the rate at which electron–hole pairs are generated (which is determined by the temperature) and not by the energy change near the boundary, since the carriers are falling over into a region of lower energy. Hence $I_G$ is independent of $V$, and since $I_G = I_A^0$ when $V = 0$, the net electron current for finite $V$ is

$$I_e = I_A - I_G = I_G\{\exp(|e| V/kT) - 1\}.$$

A similar argument is valid for holes, giving a net current

$$I_h = I_G'\{\exp(|e| V/kT) - 1\}$$

and a total current

$$I = I_e + I_h = I_0\{\exp(|e| V/kT) - 1\}. \qquad (18.3)$$

The way in which $I$ varies with $V$ is shown in Fig. 17.15(b) for both positive and negative values of $V$. The exponential relation gives a very steeply rising curve for positive $V$, but the reverse current for negative $V$ is very small and substantially independent of $V$. The reason for this is that when $V$ is negative the exponential becomes vanishingly small and the current is simply the generation current arising simply from the random production of electron–hole pairs, giving charge carriers which fall over the energy drop at the barrier. On the other hand, the

Reverse bias              No bias                    Forward bias
($V$ negative)                                       ($V$ positive)

FIG. 18.3. Potential-energy rise at the boundary in a p–n junction, with and without biasing voltages. For an electron, with negative charge, the potential-energy rise occurs when moving to the left, as shown. For a hole, with positive charge, the potential-energy rises when moving to the right, as in Fig. 18.2.

forward current for positive $V$ depends on the fraction of current carriers which have sufficient energy to surmount the barrier; this depends exponentially on $W_B$ and is very sensitive to the value of $V$. Thus the junction acts as an efficient rectifier. Unlike the thermionic diode, the reverse current is finite, but normally is only a fraction of a microampere.

When the forward voltage is sufficient to make the exponential the dominant term in eqn (18.3), we can write to a good approximation that

$$r_0 = \frac{dV}{dI} = \left(\frac{kT}{e}\right)\frac{1}{I} = \frac{25 \cdot 6}{I_{mA}} \, \Omega \text{ at } T = 300 \text{ K}. \tag{18.4}$$

This expression was derived by Shockley for the input resistance of a junction diode in the forward direction, and it is found to hold for any p–n junction provided the current does not become so large that it is limited by ohmic resistance. Clearly, the forward resistance is very low, being just a few ohms for a current of about 10 mA. This forward resistance is only realized provided that the external connections to the semiconductor needed to complete the circuit are good 'ohmic' contacts, that is, they must have no rectifying action and negligible resistance. This is achieved by heavy doping of the area around the contact, using impurity concentrations of $\sim 10^{26}$ atoms per cubic metre (over one part in $10^3$), so that electrons can readily tunnel between semiconductor and metal.

The symbol used to denote a solid-state diode is shown in Fig. 18.4, together with the batteries needed for forward and reverse bias. Fitted with large metal cooling fins, with forced air or water cooling to keep the junction temperature within safe limits, solid-state diodes can be used for currents up to 1000 A, and they have largely replaced thermionic diodes. An external resistance is sometimes used in series with the diode to limit the forward current to a safe value, since at 1000 A the value of $r_0$ in eqn (18.4) is $25 \times 10^{-6} \, \Omega$, and the power dissipation is about 25 W.

(a)    (b)    (c)

FIG. 18.4. The solid-state diode rectifier: (a) the symbol; (b) forward-bias connection, with resistance $R$ to limit the forward current; (c) reverse-bias connection. The direction of 'easy' current flow is that indicated by the direction of the apex of the triangle regarded as an arrow head.

## 18.3. The Zener diode

At low reverse voltages the current in a p–n junction is substantially constant; for silicon it is of order $0{\cdot}001\ \mu\text{A}$ and for germanium $0{\cdot}1\ \mu\text{A}$. However, if the reverse voltage exceeds a certain critical value, the current suddenly increases enormously and the resistance of the diode falls to a very low value. The current–voltage characteristic in this region is shown in Fig. 18.5. Two separate processes are involved in this rapid increase in the current and it is not always clear which is predominant.

The first effect which causes a rapid increase in current is an 'avalanche process'. In the depletion layer at a p–n junction, if the current carriers are accelerated by a high electric field, they gain sufficient energy to cause 'ionization' and produce further electron–hole pairs. The critical energy for this process is somewhat greater than the minimum energy required to create an electron–hole pair, since momentum must be conserved as well as energy in the collision. When a few electrons have produced new pairs of electrons and holes, these are accelerated to produce further ionization, giving an exponential increase of current by an 'avalanche' effect similar to that in the Townsend discharge in a gas (§ 3.8). The energy acquired by an electron between collisions is determined by the product of the electric field and the mean free path, which at room temperature is substantially independent of the degree of doping, and the avalanche occurs at a certain critical field which is of the order of $10^7\ \text{V m}^{-1}$. By adjusting the doping to control the barrier thickness, the breakdown voltage can be as much as 100 V, but it is temperature-dependent because the mean free path varies with temperature. To allow sufficient ionizing collisions to occur, the barrier thickness must be large, and it is increased by using only light doping (see eqns (17.36)–(17.37)).

FIG. 18.5. Current–voltage characteristic of a Zener diode.

FIG. 18.6. With sufficient reverse bias the valence energy band in the p-type material is raised above the bottom of the conduction band in the n-type material, and electrons can tunnel from the latter into the former.

The second process is a 'tunnel effect' which was suggested by Zener in 1934. Electrons can tunnel through a thin barrier provided that an empty state of the same energy exists on the other side. This occurs if the applied potential is large enough to raise the top of the valence band on one side of a junction above the bottom of the conduction band on the other side, as in Fig. 18.6. The probability of tunnelling is greatly increased if the barrier is very thin, and it then predominates over the avalanche effect because electrons do not gain sufficient energy to cause ionizing collisions within the depletion layer. The breakdown voltage is rather low (a few volts), and a thin barrier is produced by using heavy doping of either side.

Above the critical voltage the current increases very rapidly but the voltage drop across the diode remains essentially constant. 'Zener diodes' (a generic title which includes those utilizing the avalanche effect) are used as control elements, either as switches to control a circuit where the voltage must not exceed a certain value, or to produce a constant reference voltage. The symbol used to distinguish a Zener diode from an ordinary diode is shown in Fig. 18.7, and this diagram illustrates how a

FIG. 18.7. A Zener diode, as used to keep a constant voltage across the resistance R. The resistance $R_1$ protects the diode from excessive current flow.

Zener diode is used to maintain a constant voltage across the load resistance R.

## 18.4. Rectification

If an alternating voltage is applied to a diode, a high current flows in the forward direction, but there is very little current in the reverse direction provided that the voltage amplitude does not exceed the break-down value. A diode thus produces a current flow almost wholly in one direction, and acts as a rectifier. A single diode gives only 'half-wave rectification' (as in Fig. 18.8(a)), and the output contains large a.c. components at the applied frequency, so that considerable smoothing is needed to produce a steady output. A much better circuit, shown in Fig. 18.9, consists of four diodes arranged in the form of a bridge. Both halves of the alternating waveform are rectified (see Fig. 18.8(b)) in this 'full-wave bridge rectifier'. An alternating voltage $V_0 \cos \omega t$, whose amplitude is a little larger than the required steady output voltage, is applied across the terminals PQ in Fig. 18.9 from a suitable transformer. When P is positive and Q negative, diodes $D_1$ and $D_3$ conduct and current charges the large capacitance C. When P is negative and Q positive, diodes $D_2$ and $D_4$ conduct, so that current again charges the capacitance in the same direction. Charge continues to accumulate on the capacitance until the voltage across it becomes equal to $V_0$. In alternate half-cycles the peak reverse voltage across each diode, when it is not conducting, reaches $V_0$, and the diodes must be able to withstand this reverse voltage without breakdown.

If the capacitance C has no leakage and the reverse current in the diodes is zero, C would remain charged to the greatest peak voltage applied to the system. Measurement of this steady voltage determines the amplitude of the alternating voltage, but such a 'peak voltmeter' requires

FIG. 18.8. Half-wave and full-wave rectification. V = voltage across load resistance R.

FIG. 18.9. Full-wave bridge rectifier circuit. Normally the d.c. output terminals are connected to a smoothing circuit to eliminate the ripple voltage.

a thermionic diode with zero reverse current. If the voltmeter used has a
finite resistance $R$, which is in parallel with the capacitance $C$, the circuit
is able to follow changes in the peak voltage so long as they do not occur
within a time of order $RC$, the time constant of the $R–C$ combination.

In general the circuit of Fig. 18.9 is used to deliver direct current into a
load $R$. Under these conditions the capacitor will discharge slightly
through the resistance during that part of the cycle when the diode is not
conducting, being recharged to the peak voltage when the diode conducts.
The voltage across the load is therefore not constant, but contains a
component fluctuating at the frequency of the applied alternating voltage,
as in Fig. 18.8. This component is not sinusoidal, owing to the asymmetrical nature of the capacitor charge–discharge system. The 'ripple voltage',
as the fluctuating component is generally called, becomes larger if $R$ is
reduced, since the time constant of the $R–C$ combination is then smaller,
and the capacitor discharges to a lower voltage before being recharged.
The ripple voltage is therefore more serious when the diodes are on load.
The ripple voltage must be eliminated or very considerably reduced if the
system is used to supply d.c. power for an amplifier or other electronic
device, since any alternating voltage applied to the early stages will be
greatly magnified at the output.

The advantage of full-wave rectification over half-wave rectification
can be seen from a comparison of the two when the capacitance $C$ is in
parallel with a load resistance $R$. With the full-wave rectification the
capacitor is charged to the peak voltage by the passage of current through
one pair of diodes during the first half of the cycle, and is then again·
charged during the second half-cycle by current flowing through ·the
second pair. If the time constant of the $R–C$ combination is long compared with the period of the supply voltage, so that the voltage drop
during the discharge period is only a small fraction of the initial peak
voltage, then the voltage on $C$ will fall nearly linearly with time during
the discharge interval. With full-wave rectification, this interval is only

half a cycle, and the size of the ripple is thus only half as great as with half-wave rectification (see Fig. 18.8). In addition its fundamental frequency is now twice the supply frequency, making the filtering problem easier.

A simple type of filter consists of a low-pass section as shown in Fig. 9.1. For supply frequencies of 50–60 Hz this requires $L \sim 25$ H and $C \sim 10 \, \mu$F. Such an inductance is heavy, bulky, and expensive, and the filter circuit is often replaced by a voltage regulator such as a Zener diode whose breakdown voltage is 1 or 2 V lower than $V_0$, using the circuit of Fig. 18.7.

### 18.5. Amplitude modulation and detection

The transmission of intelligence by a radio wave requires some form of modulation. The radio wave is normally at a high frequency known as the carrier frequency; the modulation is at a much lower frequency, such as an audio-frequency, when the information to be conveyed is speech or music. One system used is called amplitude modulation, since the amplitude of the carrier signal is made to vary with the period of the audio-frequency, and by an amount which is proportional to the strength of the audio-frequency information. For simplicity we shall consider only a single audio-frequency of constant strength. The amplitude-modulated radio signal may then be written in the form

$$V = A(1+m \cos pt)\cos \omega t. \qquad (18.5)$$

Here $A$ is the amplitude of the carrier signal in the absence of modulation, and $\omega/2\pi$ its frequency. The constant $m$ is known as the depth of modulation, and cannot be greater than unity, and $p/2\pi$ is the audio-frequency.

The nature of an amplitude-modulated signal can be seen from Fig. 18.10(a), which shows the variation of the voltage $V$ with time. Its amplitude fluctuates slowly between a maximum value of $A(1+m)$ and a minimum of $A(1-m)$, the period of a complete cycle of this fluctuation being $2\pi/p$. Manipulation of eqn (18.5) shows that it may be rewritten as

$$V = A \cos \omega t + \tfrac{1}{2}mA \cos(\omega+p)t + \tfrac{1}{2}mA \cos(\omega-p)t. \qquad (18.6)$$

This indicates that the modulated signal may also be regarded as composed of the carrier signal $A \cos \omega t$, together with two other frequencies, higher and lower by $p/2\pi$, which are known as side-bands and whose amplitude is proportional to the product of the carrier strength and the depth of modulation. The presence of these side-bands shows that any receiver with r.f. circuits must be designed to have a pass band which will accept the frequencies $(\omega \pm p)/2\pi$ as well as the carrier frequency $\omega/2\pi$; otherwise the audio-frequency modulation will be cut out. The presence

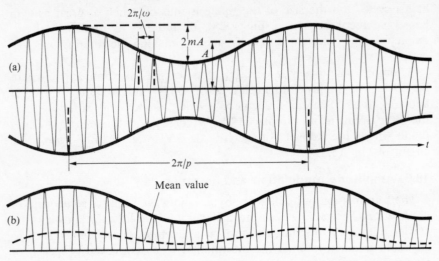

FIG. 18.10. (a) Amplitude-modulated signal, before detection; (b) amplitude-modulated signal, after detection. Normally $\omega$ is much greater than $p$.

of the side-bands may be demonstrated by applying the modulated signal to a sharply tuned frequency-meter, which will show responses at the three frequencies $(\omega-p)/2\pi$, $\omega/2\pi$, and $(\omega+p)/2\pi$.

In general the modulation will not consist of a single audio-frequency, but of a whole range of frequencies. For speech or music these cover the range from about 50 Hz to several kilohertz, while for television over a million pieces of information are transmitted per second, and a band of several megahertz is required. By Fourier analysis any modulation can always be resolved into a set of sinusoidal oscillations, and our analysis can therefore proceed in terms of one such frequency, bearing in mind that the various parts of a receiver must have the bandwidth required to accommodate all modulation frequencies up to the highest.

The mean value of the signal voltage $V$ is zero over any period long compared with that of the carrier frequency, and it will therefore produce no effect in a receiver designed to accept only audio-frequencies. If the signal is passed through a rectifier stage so that the portions where $V$ is negative are wiped out, as in Fig. 18.10(b), the mean value of the resultant is not zero and fluctuates at the audio-frequency rate corresponding to the modulation. This process is known as detection, since the information which is conveyed by the modulation can now be detected by the ear if the signal from the rectifier, after suitable amplification, is applied to headphones or a loudspeaker. An obvious requirement of a detector is that its output signal shall be nearly as possible a true reproduction of the original modulation, that is, the output voltage should be linearly proportional to the depth of modulation $m$, and the constant of proportionality should be the same for all modulation frequencies.

A circuit using a diode for the detection of amplitude-modulated waves is shown in Fig. 18.11. It is a half-wave rectifier circuit, and certain limitations must be placed on the values of $R$ and $C$ to obtain efficient and distortionless detection. These may be summarized as follows:

1. The load resistance $R$ should be large compared with the effective output resistance $r$ of the diode. The latter is approximately equal to the reciprocal of the slope of the diode characteristic, and forms a voltage divider with $R$. Since $r$ varies with the size of the applied signal, the condition $R \gg r$ not only makes the fraction of the possible output voltage appearing across $R$ nearly unity (high efficiency) but also makes this fraction nearly independent of $r$ and hence of the magnitude of the applied signal (low distortion).

2. The time constant of the $RC$ circuit should be long compared with the period of the carrier voltage, to avoid voltages of this frequency appearing in the output (that is, $1/\omega C \ll R$).

3. An upper limit to $RC$ is set by the requirement that the voltage across the capacitance $C$ shall change sufficiently rapidly to follow the modulation. This requires $1/pC < R$, but this is only approximate; a more accurate treatment shows that the inequality must depend on the size of $m$, because the rate of change of the carrier amplitude depends on the depth of modulation.

4. $C$ should be several times as large as the electrode capacitance $C_{ca}$ of the diode, since $C$ and $C_{ca}$ form a voltage divider for the r.f. voltage applied to the diode.

The condition $R \gg r$ (see point 1, above) makes it necessary for the size of the input voltage to be of the order of a volt or so, in order to work on

FIG. 18.11. Diode circuit for detection of amplitude-modulated signal. For audio-frequency modulation, typical values of $R$, $C$ are $R = 10\,\text{k}\Omega$, $C = 10\,\text{nF}$, giving a time constant of $10^{-4}\,\text{s}$. Thus $C$ behaves as a low-impedance by-pass to $R$ for the radio-frequencies but not for the audio-frequencies. The single diode behaves as a linear 'half-wave' rectifier provided that the amplitude of the input signal is sufficient (see § 18.5).

a portion of the diode characteristic where the slope is fairly high. In the reception of broadcast signals ranging from millivolts down to microvolts, it is therefore necessary to amplify the signal before detection.

### Linear and square-law detection

The use of the diode described above, where the applied signal is large enough to operate the diode on the straight part of its characteristic, is known as 'linear detection', since the output voltage is linearly proportional to the amplitude of the input voltage. If the input voltage is very small, as would be the case if a broadcast signal were applied to the diode directly without previous amplification, the diode is operated only over a very tiny portion of its characteristic, and detection or rectification results from the curvature of the characteristic of this region. The output current or voltage is proportional to the square of the input voltage, and the process is known as 'square-law detection'. An approximate analysis may be made by assuming the load resistance is small compared with the mean output resistance of the diode; the latter is very high when the applied signal is small. Then the current through the diode when a small signal voltage $v$ is applied may be written as

$$I = I_0 + \left(\frac{\partial I}{\partial V}\right)v + \frac{1}{2}\left(\frac{\partial^2 I}{\partial V^2}\right)v^2 + \dots = I_0 + av + bv^2 + \dots, \tag{18.7}$$

where $I_0$ is the current flow (if any) when $v = 0$, and $a$ and $b$ are determined by the slope and curvature of the characteristic near the point $I = I_0$. If $v = v_1 \cos \omega t$, then

$$i = I - I_0 = av_1 \cos \omega t + \tfrac{1}{2}bv_1^2(1 + \cos 2\omega t) + \dots, \tag{18.8}$$

showing that there is a change $\tfrac{1}{2}bv_1^2$ in the mean current, which is proportional to the square of the applied signal. If the latter is modulated, so that $v_1 = B(1 + m \cos pt)$, then the low-frequency current change is $\tfrac{1}{2}bB^2(1 + 2m \cos pt + \tfrac{1}{2}m^2 + \tfrac{1}{2}m^2 \cos 2pt)$, showing that the detected signal will have harmonic distortion owing to the presence of the term in $\cos 2pt$. For this reason, and because of the very low efficiency, square-law detection is not used in radio reception.

## 18.6. Frequency changing

Since square-law detection is very inefficient compared with linear detection (see Problem 18.1), it is always desirable that a signal be amplified sufficiently, before being applied to the detector, to work the latter in its linear region. Often it is undesirable, and sometimes impossible, to provide sufficient amplification for this purpose at the carrier frequency. A device known as frequency-changing is then used, in which, as the name suggests, the carrier frequency is altered to another more convenient frequency, the modulation being preserved intact. In eqns (18.7) and (18.8) for a modulated signal, this means that $\omega$ is changed to another value, but that the terms in $m$ remain the same.

This change of frequency is accomplished by adding to the original signal an alternating voltage of another frequency ($\omega_1/2\pi$) generated locally, and passing the two into a rectifying stage known as the mixer. The output from the mixer then contains voltage components which fluctuate at, apart from the modulation frequencies, ($\omega_1 \pm \omega)/2\pi$ (see below). If the difference frequency ($\omega_1 - \omega)/2\pi$ lies in the audible range it may be amplified and made to work headphones or a loud-speaker. This system is known as heterodyne reception. In a superheterodyne system, the sum and difference frequencies are outside the audible range, and one of them is selected and amplified. This frequency is known as the intermediate frequency (i.f.) and the i.f. amplifier magnifies the signal, with its original modulation, to a level at which it can be detected by a diode operating in the linear region. Since the mixing stage must incorporate a non-linear device, it is often called the 'first detector', while that following the i.f. amplifier is called the 'second detector'.

The operation of frequency-changing (or 'frequency-conversion') can be readily understood as follows. Suppose an alternating voltage $v_1 \cos \omega_1 t$ is supplied by a 'local oscillator', and to this is added a small signal voltage $v \cos \omega t$, where $v \ll v_1$ and $\omega_1$ is close to $\omega$. Then the total amplitude of the alternating voltage will fluctuate between ($v_1 + v$), when the two components are in phase, and ($v_1 - v$) when they are out of phase. The time interval between instants at which the two are in phase is $2\pi/(\omega_1 - \omega)$; the amplitude therefore fluctuates sinusoidally at a frequency equal to the difference of the two original components, and the size of the fluctuation is the same as that of the signal voltage $v$. This constitutes an amplitude-modulated voltage which can be detected as described in the last section, the difference being that the 'modulation' frequency is determined by the difference between the signal and local oscillator frequencies. The amplitude of the local oscillator voltage $v_1$ may be adjusted so that the detector is worked on the linear portion of its characteristic, and the 'modulation' of the local oscillator voltage pro-duced by the signal appears in the output as a component at the i.f. frequency whose amplitude is proportional to that of the original signal. Any slow fluctuation of the latter, such as that due to an audio-frequency amplitude modulation, is preserved, and the signal at the i.f. amplifier differs from the original only in the frequency of the carrier voltage.

For mathematical analysis, the action of the local oscillator voltage on the detector may be assumed to produce a periodic variation of its slope conductance $dI/dV$ which may be analysed as a Fourier series of the form

$$g = g_0 + g_1 \cos \omega_1 t + g_2 \cos 2\omega_1 t + \dots . \tag{18.9}$$

The effect of adding a small voltage $v \cos \omega t$ is to change the detector current by an amount

$$gv \cos \omega t = g_0 v \cos \omega t + g_1 v \cos \omega_1 t \cos \omega t + \dots$$

$$= g_0 \cos \omega t + \tfrac{1}{2} g_1 v \{\cos(\omega + \omega_1)t + \cos(\omega - \omega_1)t\} + \dots, \tag{18.10}$$

showing that there are Fourier components at both the sum and difference of the signal and local oscillator frequencies. The i.f. amplifier may be tuned to accept either of these; components at $(n\omega_1 \pm \omega)$ also exist, but they are usually small because the coefficients $g_n$ decrease in magnitude as $n$ increases. This analysis shows that either the sum or the difference frequency may be used, although in the previous discussion only the difference term was considered. In work at very high frequencies the difference is generally used, since this is more convenient in building an i.f. amplifier. In addition, where selectivity is required, it is easier to get a narrow pass band from circuits at the lower frequency. For example, a $Q_F$ of 100 would give a pass band of about 10 kHz in the circuits of an i.f. amplifier at 1 MHz, whereas to obtain the same limited band at a radio-frequency of say 100 MHz, would require a $Q_F$ of $10^4$. In a superheterodyne receiver using such frequencies, the local oscillator might be at 99 MHz or 101 MHz, and the r.f. circuits would have to be sufficiently sharply tuned to reject any signal at 98 MHz or 102 MHz respectively, which would produce the same beat frequency. This requirement is known as 'second channel suppression'. In general it means that the tuned r.f. circuits for the signal frequency $\omega/2\pi$ must be sufficiently selective to reject any unwanted signal at the 'image frequency' $(2\omega_1 - \omega)/2\pi$, which is separated from the local oscillator frequency $\omega_1/2\pi$ by the same amount, and hence would also be accepted by the i.f. amplifier. A circuit using a diode rectifier for frequency-changing is shown in Fig. 18.12.

FIG. 18.12. Diode frequency-changer circuit.

## 18.7. Frequency modulation

'Frequency modulation' is an alternative method of conveying intelligence by means of a radio wave, in which the signal wave has a constant amplitude, but its frequency is varied periodically in accordance with the modulating signal. The amount of frequency variation is proportional to the amplitude of the modulating voltage, and the rate of variation is proportional to the modulating frequency. The unmodulated carrier wave, for which $\omega$ is constant, may be written as

$$V = A \cos \phi(t) = A \cos \omega t,$$

where the function $\phi(t) = \int \omega \, dt$. If the frequency of this carrier wave is modulated by a single audio-frequency $(p/2\pi)$ of constant amplitude, then the instantaneous angular frequency becomes

$$\omega_i = \omega + \Delta\omega \cos pt,$$

where $\Delta\omega$ is the maximum deviation of $\omega_i$ from $\omega$, the frequency of the unmodulated carrier. Then for the frequency-modulated wave

$$V = A \cos\left(\int \omega_i \, dt\right) = A \cos\left(\omega t + \frac{\Delta\omega}{p} \sin pt\right) = A \cos(\omega t + m_f \sin pt).$$

$$(18.11)$$

The quantity $m_f = \Delta\omega/p$ is called the modulation index. For example, if the unmodulated carrier wave has a frequency $\omega/2\pi = 10^8$ Hz, and the modulation is at a frequency $p/2\pi = 500$ Hz, and the modulation index is $m_f = 0.04$, then the modulated carrier wave will vary in frequency from $(10^8 + 20)$ Hz to $(10^8 - 20)$ Hz and back again 500 times a second. On the other hand, if the modulation index is 20, the frequency of the carrier varies from $(10^8 + 10^4)$ Hz to $(10^8 - 10^4)$ Hz and back again 500 times a second.

Since a frequency-modulated wave is not a simple sine wave, it contains side-bands, which are more complicated than those for an amplitude-modulated wave. By Fourier analysis it may be shown that the voltage waveform of eqn (18.11) can be written as

$$V = AJ_0(m_f)\cos \omega t + AJ_1(m_f)\{\cos(\omega+p)t - \cos(\omega-p)t\} +$$

$$+ AJ_2(m_f)\{\cos(\omega+2p)t + \cos(\omega-2p)t\} + \ldots$$

$$= A\left[J_0(m_f)\cos \omega t + \sum_{n=1}^{\infty} J_n(m_f)\{\cos(\omega+np)t + (-1)^n \cos(\omega-np)t\}\right].$$

$$(18.12)$$

Here the numerical coefficients $J_n(m_f)$ can be found from tables of Bessel functions, for $J_n$ is a Bessel function of order $n$. Although the side-band frequencies stretch to infinity, the more distant side-bands have small intensity. If the modulation index $m_f = 0.5$, the first-order side-bands $(\omega \pm p)$ have amplitude $0.24$, and the second-order side-bands $(\omega \pm 2p)$ have amplitude $0.03$ relative to the unmodulated carrier; higher-order side-bands are negligible. If $m_f = 5$, the amplitudes of the side-bands are larger, as shown in Fig. 18.13; the amplitude of the carrier is markedly reduced, and most of the energy is in the side-bands (the total energy is independent of $m_f$). This represents an economy in transmitter power over amplitude modulation, where the carrier wave is fixed in amplitude and carries half the energy even with 100 per cent depth of modulation. As a rough rule the width of the frequency band over which the side-bands have appreciable amplitude is approximately

$$2\left\{\left(\frac{\Delta\omega}{2\pi}\right)+\left(\frac{p}{2\pi}\right)\right\}.$$

In a typical system for transmitting speech and music, the maximum frequency deviation $(\Delta\omega/2\pi)$ is $\pm 75$ kHz, and the maximum audio-modulation frequency 15 kHz, so the bandwidth required is

$$2(75+15) = 180 \text{ kHz}.$$

Though the bandwidth required is thus considerably greater than for transmission of an amplitude-modulated wave with the same maximum audio-frequency, a frequency-modulation system has the great advantage in that it cuts out all amplitude-modulated disturbances caused by interference and noise, and so gives much improved reception. The carrier frequencies used for frequency-modulation transmission are high ($\approx 100$ MHz), partly because the fractional frequency deviation $(\Delta\omega/\omega)$ is then small and easier to realize in transmission, and partly because only the direct ray from the transmitter is then received. Any ray received

FIG. 18.13. Side-bands in a frequency-modulated wave with modulation index $0.5$ and $5$ respectively. OA = amplitude of unmodulated carrier.

indirectly (for example, by reflection from the ionosphere) would be more seriously distorted by selective fading (unequal transmission of different frequencies) than in an amplitude-modulated system, because of the greater bandwidth required.

In the reception of frequency-modulated transmission it is necessary to convert the frequency modulation into an amplitude modulation, and this is accomplished by a 'discriminator'. Several types of discriminator are in use, the essential ingredient being a circuit whose impedance depends on frequency. The system shown in Fig. 18.14 depends on the variation in phase of the voltage across a circuit, tuned to a centre frequency $\omega$, as the frequency departs from this value. The secondary winding (inductance $L$) of a transformer is tuned to frequency $\omega$ by the capacitance $C$, across which the voltage is $V_S$. If $I_P$ is the primary current, and $M$ the mutual inductance between primary and secondary, then in the secondary circuit

$$0 = j\omega M I_P + \mathbf{Z} I_S = j\omega M I_P + j\omega C \mathbf{Z} V_S,$$

so that

$$V_S = -(M/C\mathbf{Z})I_P. \tag{18.13}$$

If $M$ is kept below the value for critical coupling (§ 7.4), the secondary resonance curve has just a single peak (see Fig. 7.10). For simplicity we neglect the impedance reflected from the primary into the secondary circuit; then near resonance, where $(\delta\omega/\omega)Q_F$ is small, we have, from eqn (7.20),

$$\mathbf{Z}^{-1} = R^{-1}\{1 - 2j(\delta\omega/\omega)Q_F\},$$

Fig. 18.14. Full-wave diode detection circuit for frequency-modulated signal. The secondary winding of the transformer is tuned by capacitance $C$ to the centre frequency $\omega$, and at this frequency the voltage across the secondary $V_S$ is in quadrature with the voltage $V_T$ from the tertiary coil (see text and Fig. 18.15). Unwanted amplitude-modulation signals are suppressed by $R_1$, $C_3$ for which typical values are $R_1 = 20\,\text{k}\Omega$, $C_3 = 10\,\mu\text{F}$, giving a time constant $R_1 C_3 = 0 \cdot 2\,\text{s}$.

giving

$$V_S = -(M/CR)\{1-2j(\delta\omega/\omega)Q_F\}I_P.\qquad(18.14)$$

This result shows that the secondary voltage has an imaginary component which varies linearly with $\delta\omega$ and vanishes when $\delta\omega = 0$.

In Fig. 18.14 the secondary winding is centre-tapped, so that equal and opposite voltages $\pm\frac{1}{2}V_s$ are applied to the diodes $D_1$, $D_2$. A third circuit, coupled to the primary by the mutual inductance $M'$, develops a voltage $V_T = j\omega M'I_P$ which is applied equally to the two diodes, being connected to the centre-tap. At frequency $\omega$, the net voltages $V_1$, $V_2$ across the diodes are equal, as in Fig. 18.15(a). When $\delta\omega \neq 0$, the voltages $V_1$, $V_2$ vary in opposite directions as in Fig. 18.15(b), oscillating with $\delta\omega$. Since $\delta\omega = \Delta\omega \cos pt$, the variation is at the modulation frequency $p$. From the relation

$$V_1^2 = (V_T+\tfrac{1}{2}\delta V_S)^2+(\tfrac{1}{2}V_S)^2$$

we have for small changes

$$2\,\delta V_1 = (V_T/V_1)\,\delta V_S = \frac{V_T}{V_1}\Big(\frac{M}{CR}\Big)2j\Big(\frac{\delta\omega}{\omega}\Big)Q_F I_P,\qquad(18.15)$$

showing that the voltage variation is linear with $\delta\omega$.

The voltages $V_1$, $V_2$ are rectified by the diodes, and the modulation-frequency signal is taken from across $C_1$ or $C_2$, as in Fig. 18.14. If an unwanted amplitude modulated signal is present, its effect is periodically to increase and decrease both $V_T$ and $V_S$ in phase; the circuit $R_1$, $C_3$ is arranged to have a long time constant to suppress such modulation in the output signal.

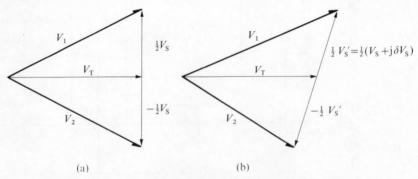

(a)                                (b)

FIG. 18.15. Vector diagram showing the voltage $V_T$ from the tertiary coil and the voltages $\pm\frac{1}{2}V_S$ from the centre-tapped secondary coil of Fig. 18.14. In (a) $\frac{1}{2}V_S$ is exactly in quadrature with $V_T$ when the secondary circuit is tuned exactly to resonance with the centre frequency $\omega$. In (b) the secondary voltage has a component $\frac{1}{2}\delta V_S$, proportional to the frequency deviation $\delta\omega$; $\frac{1}{2}\delta V_S$ is in quadrature with $V_S$ and hence in phase with $V_T$. $V_1$, $V_2$ are the voltages applied to the two diodes.

(a)                                      (b)

FIG. 18.16. Crystal diodes. (a) Capsule type: (b) waveguide mounting for millimetre
wavelengths.

## 18.8. The point-contact diode

As mentioned in § 17.7, the barrier capacitance of a p–n junction
reduces its efficiency at high frequencies. For a barrier thickness of 50 nm
and an area of 1 mm² the capacitance is ~1000 pF, which has an
impedance of about 1 Ω at a frequency of 100 MHz, so that little voltage
appears across the barrier. By using a fine tungsten wire in light contact
with the semiconductor surface, the capacitance can be reduced to 0·1 pF
or less, and efficient rectification can be obtained up to about 10⁵ MHz.
With such a small contact area there is an appreciable 'spreading resistance' (~20 Ω) to current from the contact flowing into the bulk semiconductor. This limits the slope conductance in the forward direction to
values much smaller than in the p–n diode (see Fig. 17.15); the back
current is also considerably higher. For good rectification efficiency, about
a milliwatt of power is needed, and for small signals the diode is normally
used in a frequency-changing circuit. Fig. 18.16 shows two typical mount-
ings; a capsule type for wavelengths of 100 mm and above (an alternative
coaxial construction is preferred for wavelengths ~10 mm), and a
waveguide mounting for millimeter waves. The theory of the metal–
semiconductor junction is discussed in § 17.6.

## Problem

18.1. An amplitude-modulated voltage signal $v = B(1 + m \cos pt)\cos \omega t$ is applied
to two different receivers: (1) a diode detector with the characteristic given by eqn
(18.7), working in the square-law region, followed by an audio-frequency amplifier with overall amplification $A$; (2) a signal frequency amplifier with overall
amplification $A$, followed by the same diode working in the linear region. Show
that the output voltages (assuming that the diode works into a resistance $R$ in
each case) from the two systems are in the ratio $bB : a$, and show that if
$B = 10^{-5}\,\mathrm{V}$, $a = 10^{-3}\,\mathrm{A\,V^{-1}}$, $b = 10^{-5}\,\mathrm{A\,V^{-2}}$ the ratio is $10^{-7}$. This illustrates the
inefficiency of square-law detection.

# 19. Solid-state triodes

## 19.1. The junction transistor

THE junction transistor is a single crystal of semiconductor material in which different regions have been doped by a diffusion process in such a way that a thin layer of p-type material is produced between two regions of n-type material, or vice versa. For convenience, we shall restrict our discussion to the former, but it may readily be applied to the latter by interchanging the roles of holes and electrons. It is clear from Fig. 19.1 that a transistor may be regarded as two p–n junctions back-to-back, the material (p-type in our example) in the thin central section being common to both junctions. It is important that this central section (known as the 'base') be so thin that nearly all the carriers which cross one junction also cross the other; in practice this involves a thickness which may be as small as $0 \cdot 1 – 0 \cdot 2 \, \mu \text{m}$. The transistor is not normally symmetrical, that is, the two n-regions (known as the 'emitter' and the 'collector') have different impurity concentrations; also, the collector is usually more highly doped than the base.

The symbols used to denote npn junction transistors are shown in Fig. 19.2, where the letters are used to label the electrodes and the arrows

FIG. 19.1. The npn junction transistor, showing the electron energy levels under normal bias. The emitter–base diode is an n–p junction with forward bias, while the base–collector diode is a p–n junction with reverse bias.

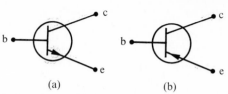

show the direction of current flow at the emitter. In operation the device is connected to batteries, as shown in Fig. 19.3, the larger voltage being applied to the collector. The emitter–base junction is a forward-biased diode, and a reduction in the voltage applied to the base produces a large increase in the emitter current, electrons flowing into the base from the emitter. The current is given by eqn (18.3). Because the base is very thin, practically all the electrons reach the collector provided that it is at a positive voltage with respect to the emitter. The current is primarily determined by the base voltage, and depends very little on the collector voltage. If a load $z_L$ is placed in the collector–emitter circuit, as in Fig. 19.3, a voltage $I_c z_L$ is developed across it whose value is primarily determined by the voltage $V_b$ applied to the base.

In the base region, which is p-type material, the electrons are minority carriers. They are therefore in 'enemy territory', and in danger of being lost by recombination with the majority carriers which are holes. The electrons move across the base by diffusion to the collector junction, where they fall down the energy drop which is in their favour, and eventually arrive at the collector electrode. The currents at each electrode are $I_e$, $I_b$, and $I_c$, as shown in Fig. 19.3, and from conservation of charge

$$I_e = I_b + I_c. \tag{19.1a}$$

To keep $I_b$ small, the thickness of the base must be small compared with the diffusion length $L_e$ for electrons (cf. eqn (17.30)). Normally $I_b$ is less than 2 per cent of the emitter current; this is the fraction lost by recombination in the base, into which, to maintain charge neutrality, a current of holes flows from the base electrode to balance the charge lost by recombination.

If, as in § 7.6, small changes in the three currents are denoted by $i_e$, $i_b$, and $i_c$, then clearly

$$i_e = i_b + i_c. \tag{19.1b}$$

For the circuit shown in Fig. 19.3, the current gain is defined as the ratio of the current change through the collector to that through the base, that is,

$$\beta = i_c/i_b. \tag{19.2}$$

Fig. 19.3. The common emitter (CE) connection for a transistor with a load $z_L$ in the collector–emitter circuit. The arrows indicate the direction of positive current flow for an npn transistor, with the biasing batteries, whose values are about $V_b - V_e = 0.6$ V (for silicon), $V_c - V_e = 10$ V. A is the common terminal.

A typical value of $\beta$ is 50 or more, so that the current gain is high. The arrangement shown in Fig. 19.3 is primarily a 'current amplifier'. Since the emitter electrode is common to the input and output circuits, it is called the 'common emitter' connection, often denoted by the letters CE.

An alternative circuit is shown in Fig. 19.4; the common electrode is the base, and this 'common base' connection is denoted by CB. In this case the current gain is defined as the ratio of the current change through the collector to that through the emitter. That is,

$$\alpha = i_c/i_e, \tag{19.3}$$

and since $i_b$ is very small compared with the other currents it follows that the current gain $\alpha$ in this case is close to unity. Clearly there is a third alternative, the 'common collector' connection (CC), often called the emitter follower circuit because of its similarity to the cathode follower circuit used with triode vacuum tubes; this connection is discussed in § 19.3. The batteries which maintain the required steady voltages on emitter, collector, and base are shown in Figs 19.3 and 19.4; for a pnp transistor the batteries must be reversed.

By combining the equations above we obtain the relations

$$\beta + 1 = 1/(1-\alpha); \qquad \alpha = \beta/(\beta+1); \qquad \beta = \alpha/(1-\alpha). \tag{19.4}$$

The quantity $\beta+1$ occurs frequently in transistor calculations and is often denoted by $\beta'$; clearly $\beta'$ is very nearly equal to $\beta$. $\alpha$ is substantially constant and can be taken as equal to unity, but $\beta$ can vary considerably between transistors bearing the same type number because of small differences in $(1-\alpha)$. An essential feature of transistor circuit design is to achieve an over-all performance which is not appreciably affected by this variation. The quantities $\alpha$, $\beta$ listed in transistor handbooks are the ratios $i_c/i_e$, $i_c/i_b$ measured with zero impedance in the collector lead (sometimes known as 'short-circuiting load'). Since $(1-\alpha)$ is the fraction of mobile carriers lost in the base through recombination, it is practically independent of the size of the current flowing, and so are $\alpha$ and $\beta$.

The battery connections in Fig. 19.4 emphasize that the junction transistor consists of two diodes back to back; the emitter–base diode, with forward bias; and the base–collector diode, with reverse bias. It is the combination of these two diodes with forward and reverse bias that is the essence of transistor action: electrons swept into the collector by the forward-biased emitter–base diode have a marked effect on the collector current, but the latter is almost independent of the collector voltage because the base–collector diode has reverse bias. This means that the output resistance of the transistor is extremely high (several megohms), whereas the input resistance is that of a forward-biased diode.

From eqn (18.3) we have

$$I_c = \alpha I_e = I_0\left\{\exp\left(\frac{|e|\,V_{eb}}{kT}\right)-1\right\},$$

where the second term in the bracket is normally very small compared to the exponential term. Neglecting it, we have for the input conductance

$$\partial I_e/\partial V_{eb} = I_e(|e|/kT) \sim 40 I_e,$$

so that the input resistance is $(40\,I_e)^{-1}$, or about 25 $\Omega$ when $I_e = 1$ mA. In the same approximation we have

$$g_m = \partial I_c/\partial V_{eb} = I_c(|e|/kT) \sim 40 I_c.$$

The quantity $g_m$ will be used in the circuit analysis of § 19.3, but qualitatively the action of the transistor can be described in a simple way.

FIG. 19.4. The common base (CB) connection for a transistor with a load $z_L$ in the collector–base circuit. the arrows indicate the direction of positive current flow for an npn transistor, with the biasing batteries, whose values are about $V_b - V_e = 0\cdot6$ V; $V_c - V_e = 10$ V.

It translates a small change in the emitter current flowing through the small input resistance into a nearly equal change in the collector current; when this flows through a high resistance load, both voltage and power amplification are obtained. The ratio $i_c/i_e$ or $i_c/i_b$ gives the current gain when the input is applied respectively to the emitter electrode and the base electrode, and the load is zero. The load in practical circuits is of the order of 1–10 kΩ, but because of the high output impedance, the actual current gain can still be taken as simply $\alpha$ or $\beta$ to a very good approximation.

## 19.2. Characteristics of the junction transistor

The behaviour of a transistor can be assessed from its static characteristics. Since we have three electrodes, there are a number of these, but the most useful is the relation between the output current and the output voltage for various values of the base current in the CE connection. A circuit for obtaining such a characteristic is shown in Fig. 19.5. The resistances $R_1, R_2$ form a potentiometer by means of which $V_{eb} = V_e - V_b$ is adjusted, thus altering the value of the base current $I_b$; by varying the value of $V_0$, a series of curves such as those given in Fig. 19.6 is obtained. Here $V_c = V_0 - I_c R_L$; the right-hand battery must have sufficient voltage to allow for the drop in $R_L$, and $V_c$ can be calculated from the other quantities.

In operation the values of $V_0$, $V_{eb}$, and $R_L$ are chosen so that in the absence of any alternating voltage the operating point (or 'quiescent point') Q is near the centre of a characteristic where $I_c$ varies almost linearly with both $V_c$ and $V_b$. This permits relatively large swings in $V_c$ without distortion, that is to say, the change in output voltage or output current is linearly related to the change in the input. All changes must be

FIG. 19.5. Circuit for measuring the static characteristics of an npn transistor in the CE connection.

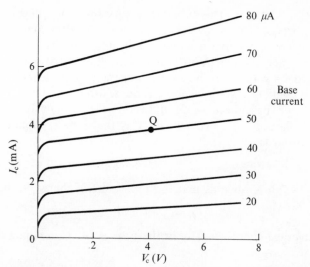

FIG. 19.6. Characteristics of a junction transistor; collector current versus collector voltage for various values of the base current. The straight portions are well represented by $I_c = h_{21}I_b + h_{22}V_o$ (cf. eqn (7.50b), with $h_{21} \sim 100$ and $h_{22} \sim 5 \times 10^5 \ \Omega^{-1}$.

confined to the 'straight' portion of the characteristic, since distortion will occur if they extend into the curved region.

In practice, transistor handbooks give suitable values of the currents which enable an approximate Q point to be selected. The steady bias voltages are then adjusted until a satisfactory alternating output waveform is obtained. For example, in the circuit of Fig. 19.5 using the transistor whose characteristic is shown in Fig. 19.6, to obtain the operating point Q, $I_c$ must be 3·8 mA when $I_b$ is 50 $\mu$A. If $R_L$ is 500 $\Omega$, the voltage drop across it is 1·9 V, so that the battery voltage $V_0$ must be 5·9 V to make $V_c = 4$ V.

In a silicon npn transistor the voltage drop across the emitter–base diode is usually about 0·6–0·7 V; in germanium the corresponding figure is about 0·2 V (see Fig. 17.15(b)). In a practical circuit the left-hand battery is eliminated by joining the points A and B in Fig. 19.5, and adjusting the resistances $R_1$, $R_2$ to give the correct voltage at $V_b$, allowing for the fact that a stabilizing resistance is also needed in the emitter lead (see § 20.2).

### Frequency-dependence of $\alpha$ and $\beta$

In a transistor, a sudden increase in emitter current is followed by a corresponding increase in the collector current $I_c$, but $I_c$ does not attain its final value until (for an npn transistor) electrons from the emitter have crossed the base and holes have entered the base to replace those lost by recombination within it. This time-lag is very short, but becomes important at high frequencies, where it results in $\alpha$ and $\beta$ becoming functions of the frequency. A simple treatment indicates how they are affected.

We assume that the electron density varies linearly over the base as in Fig. 19.7, from a value $n_f$ at the emitter edge to zero at the collector edge. The density gradient is $n_f/d$, where $d$ is the width of the base. The rate of flow across the base because of diffusion is $D(n_f/d)$ where $D$ is the diffusion coefficient for electrons. If $A$ is the cross-sectional area, the collector current is

$$I_c = A\,|e|D(n_f/d). \tag{19.5}$$

From Fig. 19.7 it is obvious that the total electric charge in the base is

$$Q_b = A\,|e|(\tfrac{1}{2}n_f)d, \tag{19.6}$$

since the average charge density in the base is $\tfrac{1}{2}n_f$. Combining these two equations we have

$$Q_b = I_c(d^2/2D) = I_c\tau_0, \tag{19.7}$$

where $\tau_0 = (d^2/2D)$ has the dimensions of time. It is called the mean carrier transit time and is determined by the values of $d$ and $D$.

From the conservation of charge and eqn (19.1a) we have

$$dQ_b/dt = I_e - I_b - I_c, \tag{19.8}$$

and hence from eqn (19.7)

$$\frac{dQ_b}{dt} + \frac{Q_b}{\tau_0} = I_e - I_b = \alpha_0 I_e, \tag{19.9}$$

where $\alpha_0$ is the ratio of $I_c$ to $I_e$ in the steady state. The solution of this equation is

$$Q_b/\tau_0 = \alpha_0 I_e\{1 - \exp(-t/\tau_0)\}$$

and

$$I_c = \alpha_0 I_e - (dQ_b/dt) = \alpha_0 I_e\{1 - \exp(-t/\tau_0)\}, \tag{19.10}$$

FIG. 19.7. Variation of electron density over the base, thickness $d$, from $n_f$ at the emitter edge to 0 at the collector edge.

showing that there is a time-delay in the build up of the current at the collector to its steady-state value $I_c = \alpha_0 I_e$.

If there is an alternating variation $i_e \exp(j\omega t)$ in the value of $I_e$, the corresponding change $q_b$ in $Q_b$ is given by the differential equation

$$\frac{dq_b}{dt} + \frac{q_b}{\tau_0} = \alpha_0 i_e \exp(j\omega t), \tag{19.11}$$

whose steady-state solution is

$$\frac{q_b}{\tau_0} = \frac{\alpha_0 i_e \exp(j\omega t)}{1+j\omega\tau_0}. \tag{19.12}$$

This is similar to the steady-state solution $Q_b/\tau_0 = \alpha_0 I_e$ of eqn (19.9), except that we have

$$\alpha(\omega) = \frac{\alpha_0}{1+j\omega\tau_0} = \frac{\alpha_0}{1+j(\omega/\omega_\alpha)} \tag{19.13}$$

instead of $\alpha_0$. Here $\omega_\alpha$ is the angular frequency at which $\alpha$ falls to $\alpha_0/\sqrt{2}$; in practice it is not exactly equal to $1/\tau_0$ because we have neglected the change in the distribution of electron density across the base when the currents are fluctuating. It is easy to show that

$$\beta(\omega) = \frac{\alpha(\omega)}{1-\alpha(\omega)} = \frac{\beta_0}{1+j(\omega/\omega_\beta)}, \tag{19.14}$$

where $\beta_0 = \alpha_0/(1-\alpha_0)$ and

$$\omega_\beta = (1-\alpha_0)\omega_\alpha = \left(\frac{\alpha_0}{\beta_0}\right)\omega_\alpha. \tag{19.15}$$

Here $\alpha_0$, $\beta_0$ are the values of $\alpha$, $\beta$ in the steady state, that is, at angular frequencies low compared with $\omega_\alpha$ and $\omega_\beta$. The corresponding frequencies are $f_\alpha = \omega_\alpha/2\pi$, $f_\beta = \omega_\beta/2\pi$, and $f_\beta$ is lower than $f_\alpha$ by the factor $\alpha_0/\beta_0$. A typical value for $f_\beta$ is 100 MHz, above which $\beta$ decreases and is frequency-dependent, but CB transistors can be used in the microwave region, although the gain is lower and frequency-dependent. The transit time across the base is $10^{-9}$–$10^{-10}$ s, and must be small compared to $1/f_\beta$ for constant gain. This low value is achieved by making the base as thin as possible and by increasing the mobility of the minority carriers in the base by varying the doping concentration, in order to produce an electric field in the base. This electric field accelerates the carriers, thus increasing their average speed and reducing the transit time across the base.

Since $f_\alpha$ is considerably higher than $f_\beta$, the current gain of the CB transistor is constant up to higher frequencies than that of the CE transistor, but in the microwave region effects due to small capacities and other frequency-dependent circuit elements can cause instability and loss of gain.

*Temperature-dependence*

In a CE transistor the current to the collector is actually

$$I_c = \beta I_b + I_{ce0}.$$

The last term, which has previously been omitted because it is very small compared with $\beta I_b$, represents the current flowing from base to collector when the emitter is on open-circuit. It arises from minority carriers produced in the base by thermal ionization, and increases as the temperature rises. So also does $I_b$; this is again controlled by the minority carrier density, which varies as $T^{\frac{3}{2}} \exp(-W_g/2kT)$ (see § 17.5). For germanium $W_g$ is about 0·66 eV at 300 K, and an increase in temperature can result in instability and 'thermal runaway': a temperature rise caused by ohmic heating produces an increase in current, giving a further temperature rise; this may continue to a point where the transistor is destroyed. For silicon $W_g = 1·12$ eV; the rate of current rise is much smaller and silicon transistors can operate up to 420 K. If stabilization is required, a suitable resistance is included in the emitter lead to produce negative feedback (see § 20.2).

## 19.3. Equivalent circuits for the junction transistor

A general analysis of a four-terminal active network was given in § 7.6. The transistor is a three-terminal device, with a common terminal between input and output, marked A in Figs 19.3–19.4 and in Fig. 19.8, which shows the equivalent circuit for a transistor. Here the box is the active network, and the common input and output terminals are each marked A. This circuit can be drawn in terms of eqns (7.50) which use the **h**-parameters; both these and the **y**-parameters of eqns (7.51) can be related to the physics of the transistors. The quantities

FIG. 19.8. Equivalent circuit for the junction transistor in terms of eqns (19.16) and (19.17).

which are most useful in circuit design are those which can most conveniently be measured. One such quantity is the mutual conductance $g_m$ defined in eqn (7.72), which is the same as the parameter $y_{21}$. In Fig. 19.8 the equivalent circuit is drawn to represent the eqns (7.50a) and (7.51b), which we repeat as

$$v_1 = h_{11}i_1 + h_{12}v_2, \tag{19.16}$$

and

$$i_2 = y_{21}v_1 + y_{22}v_2 = g_m v_1 + v_2/r_0. \tag{19.17}$$

Here the parameters $h$, $y$ are real and need not be written as vectors.

The resistance $h_{11}$ includes two contributions. For the CE and CB connections the first is the resistance of the forward-biased input diode, given by eqn (18.4), while the second is the 'spreading' resistance within the base (normally 10–100 $\Omega$), encountered by current flowing to or from the base electrode. The second term $h_{12}v_2$ represents the influence exerted by the output circuit on the input circuit. $g_m v_1$ is the current generator in the output circuit, proportional to the input voltage $v_1$, in parallel with an internal conductance $y_{22}$ which is the reciprocal of the resistance $r_0$ of the back-biased output diode. This is normally of the order of megohms, and large compared with the load, for which a typical value is a few kilohms.

## The CE connection

If $r_0$ is very large compared with $z_L$, it follows from eqn (19.2) and Fig. 19.3 that the current gain $A_i$ is simply

$$A_i = i_2/i_1 = \beta = +i_c/i_b.$$

On combining this with eqns (19.16) and (19.17) we obtain

$$v_1(y_{22} + h_{12}g_m) = i_1(h_{11}y_{22} + \beta h_{12}).$$

Since $r_0$ is very large it is a good approximation to put $y_{22} = 0$, in which case the input resistance is just

$$z_i = v_1/i_1 = \beta/g_m. \tag{19.18}$$

The output voltage is $v_2 = -i_2 z_L$, measured from the common terminal as zero, and the voltage amplification $A_v$ is

$$A_v = v_2/v_1 = -i_2 z_L/i_1 z_i = -\beta(z_L/z_i) = -g_m z_L. \tag{19.19}$$

This agrees with the result expressed in eqn (7.66b); the presence of the negative sign shows that the voltages at the input and output terminals vary in opposite directions with respect to the common terminal. If the output resistance is required, it can be found from eqn (7.56); the assumptions made above are equivalent to treating it as infinite.

*The CB connection*

With the same approximation $r_0 = \infty$, and using eqn (19.3) applied to Fig. 19.4 the current gain is

$$A_i = i_2/i_1 = -\alpha = -i_c/i_e,$$

where the negative sign appears because the input current $I_e$ in Fig. 19.4 flows in the opposite direction to $I_b$ in Fig. 19.3. Following similar algebra as for the CE connection, it is found that the input impedance is

$$\mathbf{z}_i = v_1/i_1 = \alpha/g_m, \tag{19.20}$$

and the voltage amplification is

$$A_v = +\mathbf{z}_L i_2/\mathbf{z}_i i_1 = +\alpha \mathbf{z}_L/\alpha g_m^{-1} = +g_m \mathbf{z}_L. \tag{19.21}$$

Since $\alpha$ is nearly unity while $\beta$ lies between 50 and 300 we see that the CB connection has a very low input resistance compared with the CE connection, and this is one reason why it is used much less frequently. The change in sign of $A_v$ compared with the CE connection is a simple consequence of the fact that the input currents $I_b$ and $I_e$ flow in opposite directions in Figs 19.3 and 19.4.

*The CC connection*

The third possible arrangement is with the collector as the common electrode. This is known as the common collector connection; it is shown in Fig. 19.9. An important difference from the other two connections is that the large differential resistance of the back biased diode between base and collector is now in series with the input, so that we anticipate that the input impedance will be high.

FIG. 19.9. Circuit diagram for the CC (emitter follower) connection. The correct voltage on the base is normally attained by making $R_1$ a resistance of the correct size to give the required voltage drop between collector and base by the flow of the base current through it.

The voltage gain can be calculated in a simple way as follows. The voltage across the load is

$$V_L = \mathbf{z}_L I_e = \mathbf{z}_L I_c (1 + \beta^{-1}), \tag{19.22}$$

while the voltage applied between B and A is $V_{bc} = V_L + V_{eb}$. For small changes the voltage amplification is $\partial V_L / \partial V_{bc}$, and hence

$$A_v^{-1} = \frac{\partial V_{bc}}{\partial V_L} = 1 + \frac{\partial V_{eb}}{\partial V_L}$$

$$= 1 + \frac{\partial V_{eb}}{\partial I_c} \frac{1}{\mathbf{z}_L(1 + \beta^{-1})}$$

$$= 1 + \frac{1}{g_m \mathbf{z}_L(1 + \beta^{-1})}.$$

Since $\beta^{-1}$ is very small, we have to a good approximation,

$$A_v = g_m \mathbf{z}_L / (1 + g_m \mathbf{z}_L). \tag{19.23}$$

This is always less than unity, but can be made to approach it by using a load $\mathbf{z}_L$ of a few kilohms. The current gain is large, since

$$A_i = i_e / i_b = 1 + \beta. \tag{19.24}$$

The input resistance is given by

$$\partial V_{bc} / \partial I_b = (1 + \beta)(\partial V_{bc} / \partial I_e) = (1 + \beta)\mathbf{z}_L, \tag{19.25}$$

and is therefore very high. We can find the output impedance by noting that eqn (7.66a) can be written as

$$|A_v| = \frac{g_m \mathbf{z}_L}{1 + \mathbf{z}_L / r_0}$$

and from comparison with eqn (19.23) it is clear that $r_0 = g_m^{-1}$ and is comparatively low. The CC connection can therefore be used for matching a high-impedance source to a low-impedance load; it is much cheaper for this purpose than a transformer, and has the additional advantage of giving a power gain $\sim(1 + \beta)$.

## Comparison of CB, CE, and CC connections

The three methods of connection give considerable flexibility in the design of transistor circuits, as can be seen from the summary of their properties in Table 19.1. Provided that the impedance of the source is small, both CB and CE connections have high voltage gain, but the current gains are $\alpha$ and $\beta$ respectively, so that the *power gain* of the CE connection is much higher than that of the CB connection. Since it also has a much higher input resistance, the CE circuit is much more widely used than the CB circuit. The performance depends on the transistor parameters, particularly $\beta$; this is a drawback because $\beta$ may vary widely from one transistor to another. This is overcome by the addition of negative feedback, as described in § 20.2.

At high frequencies the performance of the CE circuit becomes less good than that of the CB circuit because of feedback through the interelectrode capacitance between input and output. This capacitance is $C_{cb}$ for the CE circuit, and is larger than the corresponding capacitance $C_{ce}$ for the CB circuit, so that the Miller effect (§ 20.2) becomes relatively more important at any given frequency.

The CC circuit is used to match a high-impedance source to a lower-impedance load. Its voltage gain is less than unity but it can have a considerable power gain.

TABLE 19.1

*Comparison of current gain $A_i$, voltage gain $A_v$, input resistance $z_i$, and output resistance $r_0$ for the CE, CB, and CC connections (the approximations assume $|z_L|$ is a few kilohm).*

|  | CE | CB | CC |
|---|---|---|---|
| $\|A_i\|$ | $\beta$ | $\alpha\,(\sim 1)$ | $1+\beta$ |
| $A_v = (z_L/z_i)A_i$ | High (negative) | High | $\sim 1$ |
| $z_i$ | $\beta/g_m$ | $(\alpha/g_m) \sim (1/g_m)$ | $(1+\beta)z_L$ |
| $r_0$ | High | High | Low |
| Phase change with resistive load | $\pi$ | 0 | 0 |

## 19.4. The field-effect transistor, or FET

Field-effect transistors were discovered at about the same time as junction transistors, but manufacturing problems for the junction transistor were solved much more rapidly than for the former. It was only in 1970 that the field-effect transistor (FET) and the insulated gate device (IG), also known as the metal oxide–semiconductor device (MOS), became sufficiently consistent in their characteristics from mass production for them to be widely used commercially. They differ from the junction transmitter described previously in that the current is carried only by one type of carrier—the majority carrier; they are therefore known as unipolar devices, whereas the junction transistor is a bipolar device in which both types of carriers are involved.

The FET consists of two regions of p-type semiconductor (normally silicon) on either side of a slice of n-type semiconductor, the two regions being connected together as in Fig. 19.10, and forming the gate G.

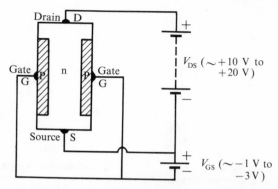

FɪɢG. 19.10. Schematic diagram of a field effect transistor (FET), showing the biasing batteries for a device with an n-type channel (the batteries must be reversed when the channel is p-type and the gate n-type).

Connections known as the source S and the drain D are made to each end of the n-type slice. The voltage applied to the drain is positive with respect to the source, as in the figure, while that applied to the gate is negative and normally considerably smaller in magnitude than that on the drain. The n-type slice provides a channel in which electrons can flow from the source to the drain, but the current flow is restricted to a central section whose width depends on the voltage applied to the gate. Thus the resistance of the conducting channel, and the current flowing through it, can be controlled by the gate voltage. In this respect the device resembles a thermionic triode, the gate being analogous to the grid, the source to the cathode, and the drain to the anode.

Field-effect transistors can equally well be made with p-type channel and n-type gate, giving additional flexibility in circuit design; the bias voltages must then be reversed. The symbol for a junction FET is shown in Fig. 19.11, where the arrow indicates whether the channel is n-type or p-type. For simplicity we shall describe only the action of an FET with an n-type channel.

The surface between gate and channel forms a p–n junction, with a 'depletion layer' at the boundary. There are no free carriers in this layer apart from a few electron–hole pairs formed by dissociation under the influence of the thermal energy, so that there is a 'space charge' arising from the donor ions in the n-type and the acceptor ions in the p-type material. The n-type channel is lightly doped, while the p-type regions are heavily doped. It therefore follows from eqn (17.37) for the relative widths of the depletion layers that $\delta_n \gg \delta_p$, that is, the depletion region lies almost entirely in the channel and very little in the p-type region. In the depletion region in the n-type material the space charge density is $\rho = |e|N_d$, where $N_d$ is the number of positively charged donor

(a)                                    (b)

Fig. 19.11. Symbols for a junction FET. The direction of the arrow distinguishes between (a) n-type channel, (b) p-type channel; it shows the direction of conventional current flow when the gate-channel diode is forward-biased.

atoms per unit volume, and the variation of the electric potential $V$ is given by Poisson's equation

$$\nabla^2 V = -\rho/\epsilon = -|e|N_d/\epsilon. \tag{19.26}$$

We choose a set of Cartesian coordinates in which the $z$-axis is parallel to the line joining the source to the drain, while the $x$-axis is normal to the two gate electrodes, assumed to be parallel planes $x = \pm\frac{1}{2}H$. The channel is thus a rectangular parallelepiped, extending indefinitely in the $y$ direction. Hence $\partial^2 V/\partial y^2 = 0$ and, when there is no voltage difference between source and drain, $\partial^2 V/\partial z^2 = 0$. The value of $V$ is then given by the solution of

$$\partial^2 V/\partial x^2 = -|e|N_d/\epsilon, \tag{19.27}$$

and if $N_d$ is uniform throughout the channel, the solution for positive values of $x$ is

$$V - V_G = \left(\frac{|e|N_d}{8\epsilon}\right)(H - 2x)(2x + H - 2h). \tag{19.28}$$

Here the constants of integration have been chosen to make $V = V_G$ at $x = \frac{1}{2}H$ (that is, we have ignored the depletion layer inside the p-type region), and $\partial V/\partial x = 0$ at $x = \frac{1}{2}h$. This equation for $V$ holds for $\frac{1}{2}H > x > \frac{1}{2}h$; the region $x = 0$ to $\frac{1}{2}h$ is the conduction channel, in which $V$ is constant. The variation of $V$ is shown in Fig. 19.12, which includes the mirror image region for negative values of $x = 0$ to $-\frac{1}{2}H$. In the conducting channel $V$ has the same value as at $x = \frac{1}{2}h$, that is,

$$V - V_G = (|e|N_d/8\epsilon)(H - h)^2 \tag{19.29}$$

from which we see that the value of $h$ depends on the potential difference $V - V_G$. Thus the width of the conducting channel depends on this potential difference, as can be seen from Fig. 19.12(a) and (b).

FIG. 19.12. Variation of the electric potential $V$ in the n-type channel of an FET. $V$ rises roughly quadratically within the channel near the surface of the gate, becoming constant within the conducting channel. This channel is narrower in (a), where $V - V_G$ is higher than in (b).

When a small voltage difference $V_{DS}$ is set up between D and S, initially the current flow rises linearly with $V_{DS}$. However, as $V_D$ rises, the width of the conducting channel decreases markedly nearer the drain, because the potential in the conducting channel is large there, and, from eqn (19.29), an increase in $V$ causes a reduction in $h$. Thus in Fig. 19.12, (b) corresponds to a point near the source, while (a) corresponds to a point near the drain. In the conducting channel the current is independent of $z$, but the current density and the field strength $E_z = -\partial V/\partial z$ driving it must be larger where the channel is narrower. The potential $V$ must therefore vary non-linearly with $z$, and no simple solution for $V$ as a function of $x$ and $z$ can be obtained. The way in which the widths of depletion layer and conducting channel vary with $z$ is shown in Fig. 19.13. As the conducting channel becomes narrower, the resistance between source and drain increases, and the current rises less rapidly than the voltage $V_{DS}$, as shown in the characteristic curves of Fig. 19.14. Essentially the size of the current is determined by the electric field strength in the depletion layer, giving the name 'field-effect transistor'.

*Pinch-off*

In eqn (19.29), if $(V - V_G)$ becomes equal to $(|e|N_d/8\epsilon)H^2$, then $h$ must fall to zero; the conducting channel vanishes, and so must the current flow. In practice this does not happen; as the conducting channel narrows, the current density increases and so does the electric field driving it. The value of $I_D$ becomes almost constant, nearly independent of $V_D$ but not of $V_G$, as shown in the characteristic curves of Fig. 19.14. In this region the value of $I_D$, for a given value of $V_{DS}$, is given approximately by the

FIG. 19.13. Cross-section of an n-type channel junction FET. The upper gate and lower substrate are connected internally, both in front of the paper (not shown) and behind. The depletion regions extend further into the n-channel nearer the drain.

expression

$$I_D = (I_D)_0 \left(1 - \frac{V_{GS}}{V_p}\right)^n,$$
(19.30)

where the exponent $n$ is close to 2. Thus a plot of $I_D^{\frac{1}{2}}$ against $V_{GS}$ (both $V_{GS}$ and $V_p$ are negative for an n-type channel) gives a straight line cutting the axis $I_D = 0$ at $V_{GS} = V_p$, which is known as the 'pinch-off' voltage.

The mutual conductance (or transconductance) $g_m = (\partial I_D / \partial V_{GS})$, $V_{DS}$ being kept constant. In the pinch-off region, from eqn (19.30) with $n = 2$,

$$g_m = -\frac{2(I_D)_0}{V_p}\left(1 - \frac{V_{GS}}{V_p}\right)$$

$$= (g_m)_0\left(1 - \frac{V_{GS}}{V_p}\right),$$
(19.31)

where $(g_m)_0$ is the value of $g_m$ when $V_{GS} = 0$ (since $V_p$ is negative for an n-channel device, $(g_m)_0 = -2(I_D)_0/V_p$ is positive). To make $g_m$ large, we need to make $(I_D)_0$ as large as possible; since it is proportional to the width $W$ of the channel in the y direction and varies inversely with the distance $L$ between source and drain, this means increasing the ratio $W/L$. However, the capacitance between the gate and the conducting channel increases with the area $WL$, so that in practice $L$ is made as small as possible and $W$ adjusted accordingly.

A typical value of $(g_m)_0$ is $0.4$–$1.2 \, \text{mA} \, \text{V}^{-1}$. This is high, but it depends on the characteristics of the device, which in manufacture vary from sample to sample. As with junction transistors, the practice is to introduce sufficient feedback to make the amplifier gain independent of the actual

(a)                                                                (b)

FIG. 19.14. Characteristics of an n-type channel junction FET. (a) $I_D$ against $V_{DS}$ for five values of $V_{GS}$; (b) $I_D$ against $V_{GS}$ for $V_{DS} = +2$ and $+15$ V. The value of the pinch-off voltage $V_p$ is $-6$ V. the curves of $I_D$ turn up sharply at the higher values of $V_{DS}$ because of avalanche breakdown of the p–n junction between gate and channel; in operation this region must be avoided, to prevent damage. Q is a typical quiescent point.

value of $g_m$. Clearly, in the pinch-off region where $I_D$ is practically independent of $V_{DS}$, the output resistance is very high (typically, the output conductance is about $10^{-2}g_m$). The input resistance is also high, because the gate is reverse-biased with respect to the source. This contrasts strongly with the junction transistor, which has a low input impedance because the input diode is forward-biased.

The structure of a planar junction FET is shown in Fig. 19.15. Starting with a single crystal, which is highly doped p-type material to form the substrate, the n-type channel region is grown on epitaxially (that is, as a single crystal whose axes have the same orientation as in the substrate). The other zones are formed by diffusing in from the surface the required impurity. These include the upper p$^+$ (highly doped) gate (as in Fig. 19.13), and the highly conducting n$^+$ regions which form source and drain. Metal is deposited by evaporation onto these to give good ohmic contacts, surface insulation being provided by the oxide layer of silica (SiO$_2$). The channel length $L$ in Fig. 19.13 is only a few $\mu$m in length, while the width $W$ can be made as large as required by extending the structure in the plane out of the paper, and by the use of repeated structures such as SGDGSGD..., where the source fingers are all connected together at one end and the drain fingers at the other. The channels between them are then in parallel with one another.

The advantage of the planar construction is that it gives a high component density; many transistors can be constructed on a single slice of semiconductor crystal. The slice may be sawn up to produce individual

FIG. 19.15. Structure of a planar junction FET with n-type channel. The layer of silica ($SiO_2$) is used only to insulate the electrodes.

transistors, or made into an 'integrated circuit' where both active and passive elements are manufactured by successive diffusions. The latter are widely used in digital circuitry where a large number of elements are required, since the packing density can be high and the circuit connections are very reliable. The individual components can be made with fewer processing steps and smaller dimensions for unipolar transistors than for bipolar transistors (see Hittinger 1973).

## 19.5. The MOS field-effect transistor, or MOSFET

The insulated-gate FET is an important variant of the junction FET described in the last section. The difference is that the metal gate electrode is insulated from the semiconductor crystal by a thin film ($\sim 0.1 \mu$m thick) of Silica ($SiO_2$), as illustrated in Fig. 19.16.

FIG. 19.16. Structure of a metal–oxide–semiconductor field effect transistor (MOSFET). The silica layer, about $0.1 \mu$m thick, insulates the metal gate electrode G from the semiconductor. When G is made positive, electrons are attracted to the interface between oxide and semiconductor, providing a conducting channel in which electrons can flow from source to drain when a voltage $V_{DS}$ is applied. When $V_{GS} = 0$, there is no conducting channel in the 'normally-off' mode.

The name of MOSFET is derived from the sequence metal–oxide–semiconductor. In construction, two strongly doped ($n^+$) zones are diffused into the p-type substrate to form the source and the drain, connected to metal electrodes. The metal gate electrode is on top of the oxide film; it forms, with the p-type substrate, a parallel-plate capacitor, in which the silica is the dielectric. The presence of this oxide layer makes the leakage current to the gate extremely low (of the order of $10^{-9}\,\mu A$), but the gate voltage still controls the current flowing underneath between source and drain. When the voltage on the gate is positive with respect to the substrate, electrons in the latter are attracted towards, and holes repelled from, the interface between the substrate and the oxide layer. This produces an n-type layer (the 'inversion layer' or 'n-channel') through which the current $I_D$ flows between source and drain when a voltage $V_{DS}$ is applied. Variation of the voltage on the gate changes the electron density in the n-channel, and controls the size of the current $I_D$. Basically the action is the same as in the junction FET, but the gate voltage can be of either sign, whereas in the junction FET a gate voltage of the wrong sign will cause a large gate current to flow when the gate–channel p–n junction becomes forward-biased.

The additional flexibility in the gate voltage shows itself in the two normal methods of operation for the MOSFET. These are as follows.

### 'Normally-off' or 'enhancement' mode

At zero gate voltage there is no conducting channel. If a positive voltage is applied to the drain, the p–n junction between drain and substrate becomes reverse-biased so that no current (apart from leakage) flows between source and drain. It is necessary to apply a positive gate voltage in order to create a conducting channel with mobile charge carriers, as described above. The positive voltage between gate and source at which appreciable current starts to flow is called the pinch-off voltage $V_p$ (or the threshold voltage). A typical set of characteristics is shown in Fig. 19.17 (a), (b), and the symbol is shown in (c). The electrode B ('bulk') is the substrate terminal, which is generally connected to the source (often internally). The arrow shows the direction of current flow in the p–n junction.

### 'Normally-on' or 'depletion' mode

When positively charged impurity ions are present in the silica insulating layer, an electric field is set up, even when $V_{GS} = 0$, in the same direction as if there were a positive voltage on the gate. An n-type inversion layer is therefore present, giving a conduction channel except when the gate voltage is made sufficiently negative. The pinch-off voltage at which the current is cut off is thus negative, and the characteristics are

(a)                    (b)                              (c)

FIG. 19.17. (a), (b) Characteristics, and (c) symbol for a MOSFET in the 'normally-off' or enhancement mode. The value of $V_p$ is positive, and no current flows when $V_{GS} = 0$. The arrow indicates the direction of current flow in the p–n junction, the direction shown being for an n-channel device.

as shown in Fig. 19.18(a), (b), and the symbol as in (c), where the vertical line is now continuous instead of being broken.

These diagrams apply only to the MOSFET with n-type channel. With a p-type channel the arrow in the symbol is reversed, and so must be the biasing voltages applied to the electrodes. An n-channel device is faster in action than a p-type channel device, corresponding to the greater mobility of electrons compared with holes. In practice, the MOSFET is normally used in the enhancement mode, which allows greater flexibility in circuit design.

Because of the high input impedance, the equivalent circuit of an FET has the simple form shown in Fig. 19.19. At high frequencies the various interelectrode capacities must be included; also, a resistance $R_G$ is normally connected between gate and source to stop charge accumulating on the gate. This allows the gate current to drain away and maintains $V_{GS}$ close to zero. The drain current varies with temperature for two reasons:

(a)                    (b)                              (c)

FIG. 19.18. (a), (b) Characteristics, and (c) symbol for a MOSFET in the 'normally-on' or depletion mode. The value of $V_p$ is negative, and the device can be used with $V_{GS}$ either positive or negative. The arrow indicates the direction of current flow in the p–n junction, and the direction shown corresponds to an n-channel device.

FIG. 19.19. Equivalent circuit of an FET.

(1) the mobility of the charge carriers in the channel decreases as the temperature rises; (2) the pinch-off voltage $V_p$ decreases with increasing temperature. The first effect causes $I_D$ to fall with increasing temperature, while the second (see eqn (19.30)) causes it to rise. At a certain (negative) value of $V_{GS}$ the two effects balance, but in practical circuits it is difficult to stabilize the operating point so as to make use of this property. At higher values of $I_D$, effect (1) predominates, so that there is no danger of thermal run-away with an FET.

## 19.6. Summary of transistor properties

1. The junction transistor behaves as a current amplifier. Its input impedance is low, being that of a forward-biased p–n junction.

2. The FET has a high input impedance, being that of a reverse-biased diode for the junction FET, and that of the oxide layer for a MOSFET. The high input impedance simplifies the design of circuits with a large number of stages.

3. Distortion is smaller for the FET and MOSFET, because of the square-law relation of $I_D$ to $V_{GS}$, than for the junction transistor with its exponential relation.

4. Variations in manufacture and with temperature are smaller for the FET than for the junction transistor; in either case they are overcome in amplifier circuits by the use of negative feedback (see § 20.2).

5. Power consumption for the FET is smaller than for a junction transistor, because for the latter a biasing potentiometer (see Fig. 20.4) is not required.

## Reference

HITTINGER, W. C. (1973). *Scient. Am.* **229**, 48.

# 20. Amplifiers and oscillators

## 20.1. Negative feedback

NEGATIVE feedback is an essential constituent in the stabilization of any system employing automatic control. The system may be biological (for example, the concentration of a particular cation in the blood must be maintained within certain limits), mechanical, or electrical. Here we shall be concerned with electronic circuits. The essential requirement is that an adverse change in the output must trigger a correction at the input whose effect is to offset the original deviation. The more sensitive the trigger mechanism to small changes, and the more rapid its response, the closer will be the limits within which the output is maintained. For such stability it is of course essential that the correction be in the direction which reduces any unwanted change; hence the name 'negative feedback'. A correction in the wrong direction will produce a further departure from equilibrium; this is 'positive feedback', which leads to instability and oscillation.

A circuit using feedback must contain an amplifier, together with components that transfer a fraction of the output power back to the input. Such components are linear circuit elements such as resistors, or inductances and capacitances (frequently an interelectrode capacitance), whose impedances are frequency-dependent.

There are two types of feedback; (1) series-voltage feedback, in which a feedback *voltage* proportional to the output is injected in *series* with the input voltage; and (2) parallel-current feedback, in which a feedback *current* proportional to the output is injected in *parallel* with the original current. The two classes are illustrated in Figs 20.1 and 20.2, and are discussed in more detail below. Positive feedback occurs if the sign (or

FIG. 20.1. Series voltage feedback. A fraction $Fv_2$ of the output voltage $v_2$ is fed back in series with the input voltage.

FIG. 20.2. Parallel current feedback. A current $i$ is fed back to the input through an admittance $y_F$ connected between output and input terminals.

phase) of the feedback is such as to augment the signal; the over-all amplification is thereby increased, and with sufficient feedback the circuit will oscillate spontaneously without any external input signal being applied. Positive feedback is essential in a self-oscillating system. Systems in which a stable level of amplification is required make use of negative feedback, where the feedback voltage or current is of such phase as to reduce the injected signal.

*Series-voltage feedback*

An example, in schematic form, of series-voltage feedback, is shown in Fig. 20.1. The four terminal network denotes an amplifier, whose (voltage) gain in the absence of feedback is $A$; a fraction $F$ of the output voltage $v_2$ is fed back in series with the input circuit. The output voltage is now $v_2 = A(v_1 + Fv_2)$, giving for the gain

$$G = v_2/v_1 = A/(1-AF). \tag{20.1}$$

Differentiation gives

$$\frac{\delta G}{G} = \frac{\delta A}{A} \cdot \frac{1}{(1-AF)}, \tag{20.2}$$

a relation which shows that a fractional change in the amplification factor $\delta A/A$ produces a fractional change in the gain $\delta G/G$ which is smaller by the factor $1/(1-AF)$. $F$ is called the feedback parameter, and $AF$ the loop gain; both $A$ and $F$ may be functions of frequency. When $AF$ is negative, the gain is reduced, but so also is the distortion. A distortion voltage appearing at the output is fed back to the input, and is reduced in the same ratio as the amplification. If now the input voltage is increased by means of a preceding amplifier until the output is restored to its former level, the distortion voltage will remain at the reduced level provided that the preceding amplifier is free from distortion. This condition is usually fulfilled, because the signal in the preceding amplifier is at a lower level, and distortion is most likely to occur where the signal level is high, in the final stages of amplification.

If the quantity $(-AF)$ is made large compared with unity, the gain becomes simply $1/F$; it depends only on the feedback ratio and not on $A$, so that the gain is independent of the characteristics of the active components. This is an important consideration in use of the transistor, which is inherently a non-linear device.

The series feedback voltage clearly alters the effective input impedance of the network. If the feedback voltage has the same sign as $v_1$ (positive feedback) then a smaller value of $v_1$ is needed to drive the same current $i_1$ and the input impedance is smaller. If the feedback voltage is of opposite sign to $v_1$, a larger value of $v_1$ is required to maintain the same $i_1$, so that negative feedback increases the effective input impedance. The current gain $A_i = i_2/i_1$ is unchanged, but the voltage gain $A_v = v_2/v_1$ is decreased by negative feedback since a larger value of $v_1$ is required to maintain the same value of $v_2$.

If the quantity $AF$ in eqn (20.1) becomes equal to $+1$, the gain becomes infinite. This means that a finite output exists when no external input voltage is present, since the series voltage feedback in Fig. 20.1 is just sufficient and in the correct phase to produce the output voltage. The system becomes unstable and, if $AF > +1$, the currents increase to a point at which they are limited by non-linearities in the circuit; that is, the quantity $AF$ is a function of current and, assuming it falls as the current increases, the latter attains a steady value such that the mean value of $AF$, averaged over a cycle, is just equal to $+1$. In general both $A$ and $F$ may be complex, so that the quantity $AF$ is complex, and also a function of frequency. The criterion for stability, which has been derived by Nyquist, is as follows. If the real and imaginary parts of $AF$ are plotted against one another, a curve is obtained from the points for different frequencies: the condition for stability is that this curve must not enclose the point $(+1, 0)$ on the real axis. In practice this may be ensured by making the gain, at any frequency where the feedback is real and positive, too small to make the real part of $AF$ greater than $+1$.

*Parallel-current feedback and the Miller effect*

An example of parallel-current feedback is presented by the circuit shown in Fig. 20.2, where an admittance $\mathbf{y}_F$ is connected between the output and input terminals. The current flowing through this admittance is

$$i = \mathbf{y}_F(v_1-v_2) = \mathbf{y}_F(1-A)v_1,$$

where $A$ is the voltage amplification factor of the four terminal network. The total input current is now $i_1+i$, and the additional current $i$ is equivalent to an admittance in parallel with the input admittance of the network, of magnitude

$$i/v_1 = (1-A)\mathbf{y}_F. \qquad (20.3)$$

If this is positive, the extra admittance results in a decrease in the over-all input impedance. If $A$ is large and negative, the extra admittance can greatly exceed $\mathbf{y}_F$; the reason for this is clearly that the voltage across $\mathbf{y}_F$ is $(v_1-v_2) = (1-A)v_1$ instead of just $v_1$.

In general, the input admittance in the absence of feedback will contain a capacitive component, so that it will be of the form $\mathbf{y}_i^0 = g_i+j\omega C_i^0$. If the feedback admittance is capacitive, $\mathbf{y}_F = j\omega C_F$, then the total input admittance becomes

$$\mathbf{y}_i = g_i+j\omega\{C_i^0+(1-A)C_F\} = g_i+j\omega C_i,$$

where we have assumed that $A$ is real. The effective capacitance at the input has been increased to

$$C_i = C_i^0+(1-A)C_F,$$

which can be much greater than $C_i^0$ if $A$ is negative. An effect of this kind occurs both in vacuum tubes and in solid-state devices through the interelectrode capacitance, which appears as a larger capacitance across the input terminals. At high frequencies this represents such a small impedance that only a fraction of the voltage from a source with a finite source resistance may actually be applied to the following input of an amplifier. The resultant drop in over-all gain is known as the 'Miller effect'.

If the input is fed from a voltage source $v_S$ with an internal resistance $r_S$, as in Fig. 20.2, the voltage $v_1$ applied to the input terminals can be written as

$$\frac{v_1}{v_S} = \frac{r_i}{r_i+r_S}\left\{\frac{1}{1+j(\omega/\omega_0)}\right\}. \qquad (20.4)$$

Here $\omega_0 = (C_i r_i')^{-1}$, where $r_i' = r_i r_S/(r_i+r_S)$ and $r_i = g_i^{-1}$, so that $\omega_0$ is determined by $C_i$ and the resistance formed by $r_i$ and $r_S$ in parallel. $\omega_0$ is the angular frequency at which $|v_i/v_S|$ falls by a factor $1/\sqrt{2}$ below its value at $\omega = 0$, and is sometimes known as the 'mid-band' frequency.

*Feedback and bandwidth*

If the input is to an amplifier whose gain at $\omega = 0$ is $A_0$, then the effective gain at a finite frequency is

$$A = \frac{A_0}{1+j(\omega/\omega_0)}. \qquad (20.5)$$

If feedback is introduced, with $F$ independent of frequency, the gain becomes

$$A_F = A/(1-AF)$$

$$= \frac{A_0}{(1-A_0F)}\left\{\frac{1}{1+j(\omega/\omega_1)}\right\}, \qquad (20.6)$$

where $\omega_1 = \omega_0(1-A_0F)$. Thus the bandwidth is increased by a factor $(1-A_0F)$, while the gain is reduced by the same factor; the product (gain × bandwidth) is unchanged by the introduction of feedback.

The circuits discussed in §§ 20.2–20.3 use resistive components, and the feedback is also provided by resistance. The circuits are therefore suitable for wide-band amplifiers.

### 20.2. Feedback and stabilization of a transistor amplifier

Fig. 20.3 shows a circuit in which a resistance $R_e$ is included in the emitter lead of a CE circuit. The voltage across $R_e$ is common to the input and output circuits, and provides an example of feedback. If the value of $\beta$ for the transistor is high, the base current is a very small fraction ($\leqslant 1$ per cent) of the collector or emitter currents, and we can take $i_c = i_e$. The voltage fed back to the input is $-i_eR_e = -(v_2/R_L)R_e = -(R_e/R_L)v_2$, and the feedback ratio $F = -(R_e/R_L)$ is negative.

As indicated in Fig. 20.3, the resistance $R_e$ may be shunted by a capacitance $C$, so that the combination is a low impedance for alternating current. This has no effect at zero frequency, so that the d.c. current and voltage are still stabilized by the presence of $R_e$, and the Q point remains

FIG. 20.3. A transistor circuit with a feedback resistance $R_e$ in the emitter lead.

the same, but the a.c. gain is not reduced if $C$ is suitably chosen. If $\omega_1$ is the lowest frequency which it is desired to amplify, the condition which $C$ must satisfy is not simply $\omega_1 C R_e = 1$. Analysis of the circuit shows that other parameters are involved, and for thermionic vacuum tubes or an FET we have

$$\omega_1 C = (g_m + r_s^{-1}),$$

while for junction transistors

$$\omega_1 C = h_{21}/(h_{11} + r_s),$$

where $r_s$ is the source resistance.

The effect of $R_e$ on the input impedance can readily be found by comparison with the equivalent circuit of Fig. 19.8. It provides a negative-feedback voltage $R_e i_e$ in the input circuit, so that

$$h_{12} v_2 = R_e i_e = R_e (i_b + i_c) = R_e (1 + \beta) i_b = R_e (1 + \beta) i_1.$$

This shows that the presence of $R_e$ results in an extra resistance $(1 + \beta) R_e$ in the input circuit, and the input impedance is increased by this amount. Taking as typical values $\beta = 100$, $R_e = 3 \text{ k}\Omega$, we find an input resistance of order $300 \text{ k}\Omega$, a value which is much larger than for the simple CE connection in which $R_e = 0$.

The effect of $R_e$ in stabilizing the performance can be seen from an analysis of the circuit of Fig. 20.4, which includes the provision of bias for the base by means of the potentiometer formed by $R_1$, $R_2$. For a silicon transistor, the voltage $V_{be}$ across the base–emitter diode is $0.6$–$0.7$ V; to avoid the effect of this spread, it is desirable that the voltage across $R_e$ should be large compared with $V_{be}$, a value of about 3 V being used in

FIG. 20.4. A transistor circuit with feedback resistance $R_e$, and potentiometer formed by $R_1$, $R_2$ to provide the correct voltage on the base. With the correct choice of resistances, the current gain, voltage gain, and input impedance can be made virtually independent of the transistor characteristics.

practice. If the emitter current at the quiescent point is 1 mA, then $R_e$ must be 3 kΩ. The voltage required on the base is then about 3·6 V, and if the supply voltage $V_S = 7$ V, roughly equal values of $R_1$, $R_2$ will provide the required base voltage.

We now examine how the value of the emitter current $I_e = (1+\beta)I_b$ depends on the circuit parameters. It is readily shown that

$$I_e\left\{R_e + \frac{1}{(1+\beta)}\left(\frac{R_1R_2}{R_1+R_2}\right)\right\} = \left(\frac{R_2}{R_1+R_2}\right)V_S - V_{be}. \qquad (20.7)$$

If the quantities on the right-hand side are constant, the variation of $I_e$ with $\beta$ is given by

$$\frac{dI_e}{I_e} = -\frac{d(1+\beta)^{-1}}{(R_e/R_B)+(1+\beta)^{-1}} \qquad (20.8)$$

showing that, if $(1+\beta)^{-1} = 0\cdot01$, we need $R_e/R_B$ to be large compared with $0\cdot01$ to lessen the dependence of $i_e$ on $\beta$. Here $R_B = R_1R_2/(R_1+R_2)$, and if $R_1 = R_2 = 15$ kΩ, so that $R_B = 7\cdot5$ kΩ, and $R_e = 3$ kΩ, the fractional variation in $I_e$ is reduced by a factor 41 compared with its variation when $R_e = 0$. A similar factor applies to its dependence on $V_{be}$. Thus the circuit fulfils two of the requirements: it supplies stabilization against fluctuations in $\beta$ and in $V_{be}$, and it helps to reduce the dependence of $I_e$ on $V_S$, though $I_e$ still drops from 1 mA to 0·37 mA if $V_S$ drops to half its normal value.

For an alternating current the battery maintaining the supply voltage $V_S$ is essentially a short-circuit, and to a signal source $v_1$ the resistances $R_1$, $R_2$ are in parallel. The input impedance is therefore $R_B$ in parallel with $r_i'$, where $R_B = 7\cdot5$ kΩ while $r_i' = (1+\beta)R_e = 300$ kΩ. Since $r_i'$ is proportional to $\beta$, it shows the same variation as $\beta$, but for the circuit of Fig. 20.4 the input impedance is virtually just $R_B$, which is independent of $\beta$.

Of the current $i_1$ drawn from the source, only a fraction $R_B/(R_B+r_i')$ flows through the base, and the current gain is therefore reduced from $i_c/i_b = \beta$ to

$$A_i = \frac{i_c}{i_1} = \beta\frac{R_B}{R_B+R_e} \sim \frac{R_B}{R_e}, \qquad (20.9)$$

where we have used the approximations of neglecting $\beta^{-1}(R_B+R_e)$ in comparison with $R_e$, and taking $r_i' = (1+\beta)R_e$. In the same approximation the voltage gain $A_v$ is

$$A_v = \frac{i_cR_L}{i_br_i'} = \frac{\beta R_L}{(1+\beta)R_e} = \frac{R_L}{R_e}. \qquad (20.10)$$

Thus we see that to a good approximation the current gain, the voltage gain, and the input impedance are all independent of the characteristics of the transistor.

For alternating voltages, the emitter voltage is $v_e = i_e R_e = i_b(1+\beta)R_e$ $= i_b r_i'$, which is the voltage applied to the base. Hence this circuit behaves in the same way as the 'emitter follower' circuit, and the approximation that $v_b = v_e$ is often useful in circuit analysis.

## 20.3. The small-signal low-frequency amplifier

A small-signal amplifier is one where the input voltage is sufficiently small that the resulting change in output current is only a small fraction of the mean output current; usually, a parameter such as $g_m$ can be assumed to be constant within the operating range. To provide more amplification than is available from a single transistor, several stages are coupled together 'in cascade'. The coupling must allow the alternating voltages to be transferred with minimum loss, while preserving the correct steady potential on the transistor electrodes. For audio-frequencies, the most common method employs resistance–capacitance $(RC)$ coupling. In Fig. 20.4, the output voltage is fed through the capacitance $C_2$ to the input of the next stage (that is $C_2$ of one stage is identical with $C_1$ of the next stage). To ensure that uniform amplification is maintained over a range of frequencies, the susceptance $1/\omega C$ of the coupling capacitor must be small compared with the input impedance of the next stage down to the lowest frequency at which the amplifier is required to operate. The high-frequency limit is usually set by the Miller effect (§ 20.1), which reduces the input impedance of each stage. As a result of the coupling, the load for each stage is now $R_L$ in parallel with the input impedance of the following stage, which for Fig. 20.4 is $R_1$, $R_2$ in parallel ($R_B$ in the previous section). By making $R_B$ large compared with $R_L$, nearly all the output voltage is transferred to the input of the next stage.

*Direct coupling*

Solid-state devices do not require large differences between the steady potentials on the electrodes, and it is possible to couple two stages together directly, provided that each transistor can be maintained at a suitable Q point. No capacitor is required, reducing the number of components with no loss of gain, and making d.c. amplification possible. Two direct-coupled amplifiers are shown in Figs 20.5 and 20.6, the first using two npn transistors in the CE connection, the second an npn followed by a pnp transistor. It is advantageous to arrange that the major part of the current output of the first transistor flows into the second transistor; this requires the input impedance of the second stage to be small compared with the load resistance $R_L$ in the collector lead of the first transistor. Then, for the pair, the over-all current gain $A_i$ is equal to the product of the individual current gains. The voltage gain $A_v$ $= A_i \times$(load resistance of second circuit/input resistance of first circuit).

FIG. 20.5. A direct-coupled amplifier using two npn transitors in the CE connection. The capacitance $C'$ reduces the negative feedback due to the resistors in the emitter lead of the second transistor.

The base voltage for the first stage is provided by the potentiometer in the emitter circuit of the second stage. The resistance of the potentiometer also gives d.c. stabilization, but the capacitor $C'$ shunted across it means that there is no negative feedback for the alternating current.

The ability to use either npn or pnp transistors gives flexibility in the design. With two identical stages, as in Fig. 20.5, whether the phase change in each be 0 or $\pi$, the input and output voltages must be in phase. It is readily verified that the phase change is also zero for Fig. 20.6. In

FIG. 20.6. A direct-coupled amplifier using an npn and a pnp transistor.

this circuit, the resistance connected between the collector of stage 2 and the emitter of stage 1 provides negative feedback. As we have seen, the 'emitter follower' nature of stage 1 means that its base voltage follows its emitter voltage, so that $R_c$ is effectively connected between the collector of stage 2 and the base of stage 1, and the feedback is carried across two stages.

### The Darlington compound

A simple and much used combination of transistors is the 'Darlington compound' circuit of Fig. 20.7, consisting of two transistors in which the second forms the load for the first. The over-all current gain is approximately $\beta^2$, and can be as high as several thousand. The base current of the second transistor is the emitter current of the first, and it follows that the first must be operated at comparatively low emitter current, while the second has a much higher emitter current. The circuit is therefore suitable for applications requiring a fairly high output current.

The Darlington compound is used in a circuit very much like that for a single transistor, but with a higher current gain. In the circuit of Fig. 20.7, negative feedback is provided by the load $R$, which plays a role similar to $R_e$ in Fig. 20.4. Essentially the transistors are in two 'emitter follower' circuits, so that the voltage gain is near unity, provided that $R$ is large compared with the input resistance of the emitter–base diode given by eqn (18.4). The current gain $\sim R_B/R$, where $R_B = R_1 R_2/(R_1 + R_2)$.

### Direct coupling with the MOSFET

The very low gate current drawn by a MOSFET makes it suitable for direct coupling. A simple amplifier is shown in Fig. 20.8, in which negative feedback is carried back over three stages by means of the resistance $R_F$. This resistance carries no direct current, so that the gate

FIG. 20.7. The 'Darlington compound' in the CE connection.

Fig. 20.8. A direct-coupled amplifier using three MOSFETs with negative feedback.

voltage of the first FET is equal to the drain voltage, which is the same for all three FETs provided that $R_L$ is identical in each stage. The MOSFETs are used in the normally-off (enhancement) mode.

Essentially the circuit is the same as that of Fig. 20.2; the feedback admittance is a pure conductance, and amplification is provided by the three FETs. From eqn (20.3) the input admittance is $(1-A)/R_F$ in parallel with the input impedance of the first stage. The latter is so high that the input resistance is effectively $R_F/(1-A)$, or approximately $R_F/(-A)$, even though the value of $R_F$ may be as high as 50 MΩ. There is a phase change of $\pi$ at each section, since an increase in the input voltage causes the current flowing through $R_L$ to rise, so that the drain voltage falls. To give a positive input resistance, an odd number of stages is required; in practice three must be used to give sufficient amplification, which may be several thousand.

### 20.4. Amplification at radio-frequencies

At high frequencies the Miller effect, giving a low input impedance, becomes very important. Because of the capacitance between the electrodes, the input impedance is usually capacitative; inductance is added in parallel to produce a parallel tuned circuit, which has a fairly high impedance (see § 7.3). Other electrode capacitances, and stray capacity of lead wires are treated similarly; a typical circuit is the tuned-gate, tuned-drain amplifier shown in Fig. 20.9. The use of tuned circuits means that only a narrow range of frequencies near the resonance frequency is amplified, since the admittance of each tuned circuit is approximately (see Problem 7.1)

$$\mathbf{Y} = G + 2\mathrm{j}\,\Delta\omega C, \tag{20.11}$$

where $G$ is the conductance at resonance, $\Delta\omega$ is the departure of $\omega$ from the resonance value $\omega_0 = (LC)^{-\frac{1}{2}}$. If the FET has mutual conductance $g_m$ and output resistance $r_0$, the amplified voltage across the output tuned circuit is

$$\frac{-g_m v_G}{r_0^{-1} + G + 2j\,\Delta\omega C_2} = A v_G. \tag{20.12}$$

The amplification therefore falls by a factor $1/\sqrt{2}$ at frequencies deviating from the resonant frequency such that

$$\pm 2\,\Delta\omega C_2 = r_0^{-1} + G.$$

If the selectivity is defined as $f/(2\,\Delta f) = \omega/(2\,\Delta\omega)$, so that it is analogous to the $Q_F$ of a resonant circuit, we see that the selectivity is that of the tuned circuit shunted by the output resistance of the FET. The presence of more than one tuned circuit increases the over-all selectivity, since each reduces the response at frequencies away from resonance, and the over-all selectivity must be adequate to accommodate the range of frequencies comprised in a modulated signal (see §§ 18.5 and 18.7).

If $C_{GD}$ is the electrode capacitance between gate and drain, it follows from eqn (20.3) that the input admittance of the FET is

$$(1 - A)j\omega C_{GD}.$$

From eqn (20.12) above, the amplification is of the form $(-A_0 + jA'\,\Delta\omega)$, and the input admittance therefore contains a real part $+\omega C_{GD}(A'\,\Delta\omega)$ which vanishes at resonance ($\Delta\omega = 0$) but is negative when $\Delta\omega$ is negative. At frequencies below the resonant frequency there is a negative conductance in parallel with the input-tuned circuit, and if this is sufficiently negative to outweigh the positive conductance corresponding to the losses, the circuit will break into oscillation. This is avoided by balancing the feedback through the interelectrode capacitance by providing another

FIG. 20.9. A tuned-gate, tuned-drain amplifier for 100 MHz, with neutralizing feedback circuit.

FIG. 20.10. A high-frequency amplifier using an FET in the 'common gate connection'. $z_L$ and $z_i$ are parallel tuned circuits, and may be quarter-wave tuned lines.

feedback path, a process known as 'neutralization'. In the circuit of Fig. 20.9, the inductance $L'$ provides this extra path, $C'$ being simply the capacitance of a 'blocking capacitance' of low impedance to maintain the required steady bias voltages on drain and gate. The inductance $L'$ forms a parallel tuned circuit with the electrode capacitance $C_{GD}(\sim 1 \text{ pF})$, making the feedback admittance small and primarily inductive rather than capacitative. Any negative component in the input admittance is thus very much reduced, and oscillation is avoided.

Amplification at still higher frequencies can be obtained by using an FET in the 'common gate' connection illustrated in Fig. 20.10. Here the gate–drain capacitance $C_{GD}$ becomes part of the tuned circuit forming the load $z_L$. As in the CC connection for the junction transistor (§ 19.3), negative feedback is provided because the output current flows through $z_i$ in the source–drain circuit, and the voltage amplification is simply $z_L/z_i$, since the gate current is negligible. Both $z_L$ and $z_i$ are parallel tuned circuits, and may be quarter-wave tuned lines (see § 9.5).

## 20.5. Power amplifiers

Because of the non-linear characteristics of active devices, the output will always contain some distortion. If in eqn (18.8) we take $i$ to be the current output and $v$ the signal input, it is clear that distortion produces harmonics of the signal frequency in the output current. For an r.f. amplifier whose load is a tuned circuit, the harmonic content of the *voltage* output is small because the load has a low impedance for the harmonics and a high impedance for the fundamental frequency. In amplifiers for a range of frequencies the even harmonics can be eliminated by the use of a 'push–pull' circuit, consisting of two matched devices to which the same input voltage is applied but with opposite

FIG. 20.11. Push–pull amplifier circuit using a pair of matched transistors, one npn and the other pnp.

phase. The output currents then have the form

$$i' = g_0 + g_1 v + g_2 v^2 + g_3 v^3 + g_4 v^4 + \ldots,$$
$$i'' = g_0 - g_1 v + g_2 v^2 - g_3 v^3 + g_4 v^4 + \ldots,$$

(20.13)

and by arranging the circuit so that the output current is the difference between these two currents, we obtain an output

$$i' - i'' = 2(g_1 v + g_3 v^3 + \ldots),$$         (20.14)

from which the even harmonics are missing. A simple circuit, using a matched pair of transistors, one npn and the other pnp, is shown in Fig. 20.11. The same signal voltage $v_1$ is applied to the base of each transistor, but because of the reversed polarity, the difference $(i''_e - i'_e)$ between the two emitter currents gives the current $i_2$ through the load, and all even harmonics are missing. The most important remaining harmonic is the third; this is much smaller for FETs, with their square-law characteristics, than for junction transistors with exponential characteristics.

### Efficiency of power amplifiers

In an amplifier the ultimate source of the a.c. output power delivered into the load is the supply; this is not obvious, because the current drawn from the supply does not change when a signal is being amplified if there is no distortion. If the load is a resistance $R_L$ in the collector lead carrying a current $I_c$, the power drawn from the supply (voltage $V_s$) is $I_c V_s$, of which an amount $i_c^2 R_L = I_c(V_s - V_c)$ is dissipated in the load. The remainder $I_c V_c$, where $V_c$ is the collector voltage, is dissipated in the transistor and appears as heat. When a signal voltage is applied to the base, the collector current becomes $I_c + i_c \sin \omega t$, and the collector voltage becomes $V_S - R_L(I_C + i_c \sin \omega t)$. The mean power dissipated in the transistor is given by the product of these two, and contains terms in $\sin \omega t$ which average to zero over a cycle, the remainder being

$$V_S I_c - R_L I_c^2 - R_L i_c^2 \sin^2 \omega t.$$

The first part of this expression, independent of $i_c$, is the power dissipated in the transistor in the absence of a signal, while the last, which averages to $\frac{1}{2} R_L i_c^2$, represents the reduction in transistor heating when a signal is present. It is just equal to the signal power in the load. In physical terms, the reduction in heating of the transistor arises from the fact that the current through it increases when the collector voltage decreases, and vice versa, so that on average $V_c I_c$ is diminished.

The efficiency of an amplifier is equal to the ratio of the a.c. power output to the power drawn from the supply. In a typical example with 5 W output, $I_c = 1$ A, $i_c = 0.9$ A, $V_S = 12$ V, and $R_L = 15 \, \Omega$. The efficiency $\frac{1}{2} R_L i_c^2 / I_c V_S$ is close to 50 per cent.

*Class A, B, and C operation*

An amplifier in which the transistor is operated on the linear portion of the characteristic is known as a Class A amplifier. Higher efficiency can be obtained by biasing the transistor to a point such that the main current flow occurs when the output electrode is at the lowest voltage point during the alternating voltage swing. In a Class B amplifier, the transistor is biased close to the point where the steady current from the output electrode is zero. Current then flows only during one half of the r.f. cycle, giving higher efficiency but considerable distortion. This is reduced by using a pair of transistors in push–pull, one transistor providing current during one half of the cycle and the other in the other half.

In Class C operation the transistor is biased even further back, so that current flows only for a small fraction of a cycle. The amplitude of the applied alternating voltage must be slightly greater than the amount by which the input electrode is biased beyond cut-off. The efficiency is high, but the output is very distorted, consisting only of short pulses of current repeated at the signal frequency. Class C amplifiers are used only for power output stages; a large drive voltage is needed, and distortion in the output voltage is reduced by use of a tuned circuit for the output load. If harmonics of the input signal are required, the load can be tuned to a multiple of the input frequency; the distorted output current has a high harmonic content, each harmonic being an exact multiple of the fundamental. Care must be taken to avoid unwanted oscillations at such higher frequencies because of positive feedback through interelectrode and stray capacity.

## 20.6. Oscillators

If the feedback in an amplifier is such that the factor $AF$ in eqn (20.1) is positive and greater than unity, the circuit becomes unstable. The power fed back to the input is enough to compensate for all losses in the input circuit, so that the net resistance becomes negative. Reference to § 5.3 shows that any transient signal in the circuit then increases exponentially instead of decaying, power from the d.c. supply being converted into a.c. power. No external input signal is needed; the impulse from switching on the supply will start oscillations if sufficient feedback is present.

The design of the oscillator depends on the desired waveform, frequency, and output; some typical examples are described below. Most solid-state oscillators are adapted from circuits devised for thermionic vacuum tubes, and some still bear the names of their originators. At radio-frequencies a tuned circuit is used, either in the input or the output. At high frequencies sufficient feedback may be present through the Miller effect, but more generally it is provided by mutual inductance coupling. An equivalent circuit for such an oscillator, with the tuned circuit at the output, is shown in Fig. 20.12. For the input circuit, if $r_i$ is large, the flow of current in the circuit can be neglected, with considerable simplification in the analysis. We have then

$$v_1 = M(\mathrm{d}i/\mathrm{d}t)$$

and

$$i_2 = g_m v_1 + v_2/r_0,  \tag{20.15}$$

so that

$$g_m M(\mathrm{d}i/\mathrm{d}t) = i_2 - v_2/r_0.  \tag{20.16}$$

For the output circuit

$$i_2 - i = -C(\mathrm{d}v_2/\mathrm{d}t),$$

which can be combined with (20.16) to eliminate $i_2$, giving

$$g_m M(\mathrm{d}i/\mathrm{d}t) = i - v_2/r_0 - C(\mathrm{d}v_2/\mathrm{d}t).  \tag{20.17}$$

FIG. 20.12. Equivalent circuit for an oscillator with mutual inductance coupling between input and output.

Finally we have also

$$v_2 + L(di/dt) + ri = 0, \tag{20.18}$$

so that $v_2$ can be eliminated to give

$$LC\left(\frac{d^2i}{dt^2}\right) + \frac{di}{dt}\left(rC + \frac{L}{r_o} - g_m M\right) + i\left(1 + \frac{r}{r_o}\right) = 0. \tag{20.19}$$

This differential equation represents oscillations which will be sustained provided that

$$g_m M = rC + \frac{L}{r_0} = L\left(\frac{1}{R} + \frac{1}{r_0}\right), \tag{20.20}$$

where $R = L/Cr$ is the parallel impedance of the tuned circuit, with which the output impedance $r_o$ is in parallel. The frequency of oscillation is given by

$$\omega^2 LC = 1 + r/r_o, \tag{20.21}$$

and hence depends slightly on the value of $r_o$. In practice $g_m M$ must exceed the value given by (20.20), and the level of oscillation builds up until curvature in the characteristic of the active device reduces the average value of $g_m$ to the level required by (20.20). The resonance frequency is also influenced by the coupling (not included here) needed to extract power from the output circuit, and it may drift if the electrical constants of the components change with temperature. For high-frequency stability, a quartz-crystal oscillator (see § 22.5) is used. This is a low-power oscillator (for example, a few watts), which is followed by r.f. amplifiers (Class C for efficiency) to increase the level to the required output.

FIG. 20.13. Basic circuit for a Hartley oscillator.

*The Hartley oscillator*

When a large power output is required, but good frequency stability is not essential, an oscillator run under Class C conditions is used. Like the Class C amplifier it gives high efficiency, the input electrode being biased back well beyond cut-off. The excitation voltage must have sufficient amplitude to carry the device into the conducting region at the positive peaks, a convenient circuit being the Hartley oscillator shown in Fig. 20.13. A tapped inductance, used as an autotransformer, supplies the feedback, the emitter of a junction transistor being connected to a point near the centre of the inductance through a blocking capacitor which provides a low-impedance path for the radio-frequency. The other two electrodes are connected to opposite ends of the inductance, where the alternating potentials are in antiphase and provide feedback of the right sign. The Q point is fixed by the emitter and base resistors, which may have to be adjustable so that oscillations can start under Class A conditions, since sufficient input swing to make current flow may not be available for starting under Class C conditions.

*Resistance–capacitance oscillators*

At low frequencies the inductance needed for a tuned circuit becomes tiresomely large, and it can be avoided by the use of a resistance–capacitance oscillator, using a circuit for which feedback in the correct phase is obtained only at one frequency. In the network shown in Fig. 20.14, the voltage ratio $v_2/v_1$ is

$$\frac{v_2}{v_1} = 1 + \frac{R}{R'} + \frac{C'}{C} + j\left(\omega C'R - \frac{1}{\omega CR'}\right) \qquad (20.22)$$

and the phase difference is zero at the frequency which satisfies the relation $\omega^2 CC'RR' = 1$. If $v_2$ is the output and $v_1$ the input of an amplifier with zero phase shift between input and output, oscillations will take place at the frequency just given provided the gain is greater than $(1 + R/R' + C'/C)$, a condition which is readily satisfied for, say, $R = R'$, $C = C'$. By switching in a number of capacitors, increasing in size by a factor 10, and varying $R$, $R'$, oscillations can be obtained over a range of

FIG. 20.14. A resistance–capacitance network for providing feedback in a low-frequency sine-wave oscillator.

frequencies from about 10 Hz to over 100 kHz, while the feedback ratio can be kept constant by using ganged resistors to keep the ratio $R/R'$ constant. This circuit, known as a 'Wien bridge' oscillator from its resemblance to the Wien bridge (§ 22.5), is often used in audio-frequency signal generators.

## 20.7. The multivibrator

The oscillators considered so far produce sinusoidal oscillations whose frequency is controlled almost entirely by the circuit parameters; even in a Class C oscillator the voltage across the tuned circuit is nearly sinusoidal in form, though the current is not.

The question arises, what happens if feedback is introduced into an untuned amplifier, of sufficient size and the correct sign to produce instability. The device shown in Fig. 20.15 consists of a two-stage aperiodic amplifier with feedback from the output of the second stage to the input of the first stage. If the voltage amplification of each stage is $A$, the over-all amplification is $A^2$. The feedback factor $F$ is practically unity, so that if $A^2 > 1$, the device should be unstable.

In analysing this system we must take account of the electrode capacitance and other stray capacitances in the amplifier; they are represented by the small capacitance $C$ shunting the load resistance $R$ in each stage. The output current of the first stage is

$$i_1 = g_m v_1 + v_2/r_0,$$

where lower-case symbols indicate we are dealing only with fluctuating components. Here $v_1$ is the voltage change applied at the input of stage 1, and its output voltage change $v_2$ gives the voltage change at the second

FIG. 20.15. Two-stage aperiodic amplifier with feedback.

input. Since the current $i_1$ flows through $R$ and $C$ in parallel, we have

$$-i_1 = C(dv_2/dt) + v_2/R.$$

Elimination of $i_1$ gives a relation between $v_1$ and $v_2$, and on applying an identical analysis to the second stage a similar relation is obtained with $v_1$ and $v_2$ interchanged. Hence, writing $R'$ for the resistance formed by $R$ and $r_0$ in parallel,

$$-g_m v_1 = C(dv_2/dt) + v_2/R'$$
$$-g_m v_2 = C(dv_1/dt) + v_1/R' \tag{20.23}$$

The voltage amplification in each stage is $A = -g_m R'$, and the equations can be written as

$$Av_1 = \tau(dv_2/dt) + v_2,$$
$$Av_2 = \tau(dv_1/dt) + v_1. \tag{20.24}$$

Here $\tau = CR'$ is the time constant of the capacitance $C$ in parallel with $R$ and $r_0$. The solution of these equations is

$$v_1 = -v_2 = v_0 \exp\{-(A+1)t/\tau\}. \tag{20.25}$$

If the value of $-A$ for each stage is greater than unity, this solution shows that any disturbance of the input potential will increase exponentially. The input to one stage rises while the other decreases at the same rate; current increases in the former until it 'saturates', and decreases in the other until it cuts off. The situation remains until a short pulse applied in the input of the stage which is cut off brings it into the conducting region; then the exponential rise in its input voltage carries it to saturation while the other stage is changing back from saturation to cut-off. The time required for the reversal is very short. If $-A \gg 1$, the time constant in the exponential of eqn (20.25) is approximately $\tau/(-A) = C/g_m$, which is of order $10^{-8}$ s for $C = 15$ pF, $g_m = 1 \cdot 5$ mA V$^{-1}$.

Essentially the circuit acts as a very fast current switch; the two stages are respectively 'on' and 'off', but a rapid change to 'off' and 'on' can be triggered by either a short positive pulse to the input of the 'off' stage or a negative pulse to the 'on' stage. In the multivibrator, direct coupling between the stages is replaced by $RC$ coupling, as in Fig. 20.16, and a periodic reversal now occurs automatically. Suppose that at a given instant stage 1 is 'on,' and 2 is 'off'. The voltage at input 2 is negative, but it is falling in size as the current through $R_1$ required to charge the capacitor $C_1$ declines towards zero. When conduction in stage 2 becomes possible, the exponential voltage 'landslide' described above takes place. The voltage at point Y in Fig. 20.16 drops suddenly, and this drop is transferred through $C_2 R_2$ to the input of stage 1, which therefore switches from 'on' to 'off'. The voltage across $R_2$ now decays as $C_2$ recharges to its

FIG. 20.16. The multivibrator. With a MOSFET in each stage, and $R_1 = R_2 = 68\ M\Omega$, $C_1 = C_2 = 1\ \mu F$, square waves with a period of 2 min can be obtained.

equilibrium value, and the reverse landslide occurs when it has fallen sufficiently to permit conduction in stage 1. The cycle is now complete. The point Y is alternately at the supply voltage when the current through the second FET is zero, and at a low voltage determined by the resistance R and the current through it; point X undergoes a similar cycle in antiphase. The output waveform is very nearly rectangular in shape, and the period of a complete cycle is mainly determined by the sum of the times required for $C_1$ to charge through $R_1$, and $C_2$ through $R_2$.

The multivibrator is readily synchronized with a sinusoidal signal applied to the input of one stage, provided its natural period is close to that of the sine wave. The effect of the latter is to delay the return of the input voltage to the conducting point if it is arriving early, and to speed it up if its natural period would cause it to return too late. This property of synchronization is of use in frequency measurement; the square waves have a high harmonic content, and the harmonics occur at exact multiplexes of a standard frequency. The multivibrator may also be used for frequency division, for it will synchronize with a signal whose period is close to an exact fraction of its own; that is, a frequency of 5–10 times its own.

By changing the methods of coupling between the two stages a useful variety of switching units can be produced. The 'monostable' circuit has a stable state in which stage 1 is normally 'on' and stage 2 is 'off', but a negative pulse to the input of stage 1 will trigger a short-lived state of pre-determined duration in which 1 is 'off' and 2 is 'on'. In the 'bistable' circuit or 'flip–flop' both the 'on–off' and the 'off–on' states are stable; a transition from one to the other is triggered by a negative pulse to the 'on' stage. A complete cycle requires two pulses, one steered to each stage, and a succession of such units can form a 'scaler' in which the number of pulses is divided by 2 in each unit.

# 21. Thermionic vacuum tubes

THE initial development of modern electronics followed the invention of the thermionic vacuum tube. In the diode electrons are emitted from a heated cathode; they are attracted to an anode maintained at a positive voltage with respect to the cathode, under high-vacuum conditions inside a glass, ceramic, or metal envelope. In the triode the flow of electrons is controlled not only by the potential on the anode but, more importantly, by that on a metal grid inserted between cathode and anode. A heater supply is required for the cathode, consuming appreciable power. Thermionic vacuum tubes are more expensive to make, larger in size, and less rugged than solid-state devices, which have therefore replaced them except in certain special applications.

## 21.1. The thermionic diode

The emission of electrons from hot solids was discussed in § 11.5. The emission current per unit area of a cathode surface at temperature $T$ (in degrees Kelvin) is given by eqn (11.34),

$$J = AT^2 \exp(-\phi/kT), \tag{21.1}$$

where the constants $A$ and $\phi$ depend on the material, some representative values being listed in Table 11.1 (p. 335). The temperature at which adequate emission is obtained is determined primarily by the value of the work function $\phi$. Practical considerations, such as long life, limit the materials used in general to tungsten (2500 K), thoriated tungsten (1900 K), and a barium–strontium oxide mixture (1100 K), where the temperatures are those required for current densities of order $10^4$ A m$^{-2}$.

The oxide-coated cathode, with its low working temperature, has the additional advantage that it can be heated indirectly, giving an equipotential cathode surface. However, it is easily 'poisoned' by the presence of gas, especially oxygen, and it is essential to maintain a good vacuum ($<10^{-4}$ Pa) for the whole life of the tube. Most surfaces contain occluded gas, which is gradually evolved in a vacuum at room temperature, but is rapidly evolved at high temperatures. Nickel is commonly used for the anode and larger electrodes, and tungsten wire for grids; they are 'out-gassed' by heating to ~1300 K *in vacuo* before the tube is assembled.

At high values of the anode voltage, all electrons emitted from the cathode reach the anode; the current density is determined solely by the

FIG. 21.1. Current–voltage curve for a diode with plane-parallel electrodes, 10 mm apart, for two different cathode temperatures $T_1$ and $T_2$ ($T_2 > T_1$).

temperature of the cathode (see Fig. 21.1) and is given by eqn (21.1). At lower anode voltages the current is smaller because of the presence of 'space charge' near the cathode (see § 3.7), and instead of being 'temperature limited' it is said to be 'space-charge limited'. Under these conditions, which correspond to the part of Fig. 21.1 nearer the origin, the current density is given by Child's law,

$$J = b V_a^{\frac{3}{2}}/d^2. \tag{21.2}$$

For a diode with plane parallel electrodes the value of $b$ is given in eqn (3.45) and $d$ is the separation between cathode and anode. From dimensional considerations a similar relation is expected for other geometries, with $d$ an appropriate distance.

## 21.2. The thermionic triode

In the thermionic triode a third electrode, known as the grid, is interposed between the cathode and the anode. If the electrodes are planar, the grid takes the form of a coarse wire mesh; if the electrodes are cylindrical the grid is wound in the form of a helix. The total current drawn from the cathode will now depend on the potentials of both grid and anode, since both control the field at the cathode. In the absence of space charge, the charge on the cathode, assumed to be at zero potential, is equal to

$$-(C_{cg}V_g + C_{ca}V_a) = -C_{cg}\left(V_g + \frac{V_a}{C_{cg}/C_{ca}}\right) = -C_{cg}\left(V_g + \frac{V_a}{\mu}\right),$$

where $C_{cg}$, $C_{ca}$ are the coefficients of capacitance between cathode and grid (at potential $V_g$) and anode (at potential $V_a$) respectively. These capacitances are in the same ratio as the division of lines of force from the cathode between the grid and the anode when $V_a = V_g$. The ratio $\mu = C_{cg}/C_{ca}$ is known as the 'amplification factor' of the triode; (the values of $C_{cg}$, $C_{ca}$ given in tube manuals do not normally fit this relation because they include the capacitances of the leads). If the current from the cathode is limited by space charge, so that the field at the cathode (as in the assumption made in deriving Child's law) is zero, then no lines of electric field reach the cathode, but all terminate on the space charge. Since the latter is mainly located very close to the cathode, it is affected equally by lines of force from the grid and from the anode, and the current leaving the cathode depends on the equivalent voltage $V'_a = (V_g + V_a/\mu)$. Experimentally it is found that the current increases very nearly with the three-halves power of this equivalent voltage in the region of complete space charge limitation, corresponding to the fact that near the cathode the potential distribution is the same as in an 'equivalent diode'. We may therefore write

$$J = b(V_g + V_a/\mu)^{\frac{3}{2}}/d^2, \qquad (21.3)$$

where $d$ is known as the 'equivalent diode spacing'; for triodes with fairly high values of $\mu$ (that is, when nearly all the lines of force terminating on the space charge come from the grid) $d$ is nearly equal to the cathode–grid spacing in a plane triode. The constant $b$ has the same numerical value as for the diode. In general the triode is used with the grid voltage negative, so that no current flows to it, and the total current leaving the cathode (to which eqn (21.3) applies) is also the anode current. Current will leave the cathode so long as the equivalent voltage $(V_g + V_a/\mu)$ is positive, and a small negative grid voltage must be combined with a large positive anode voltage. This has the advantage that a source of voltage applied to the grid will influence the anode current without any current being drawn from the source by the grid.

For half a century the thermionic triode was the work-horse of electronics, but for small-power applications it has been superseded by solid-state triodes, its main use now being for high-power applications, such as radio transmitters and industrial purposes such as eddy-current heating. One set of characteristic curves, anode current against grid voltage, is shown in Fig. 21.2 for a power triode; since the anode current depends on the quantity $V_g + V_a/\mu$, we expect all the curves to be similar, but displaced to more negative values of $V_g$ as $V_a$ is increased, the shift in $V_g$ being $-(1/\mu)$ of that in $V_a$. A second set of characteristics is formed by plotting $I_a$ as a function of $V_a$ for various values of $V_g$, and a third set by plotting $V_g$ against $V_a$ for various values of the anode current. These characteristic curves are given in vacuum-tube manuals, and must be used in a numerical analysis when working at full power. In regions where they

Fig. 21.2. Anode current against voltage characteristics for a power triode with $\mu = 32$. $g_m = 17$ mA V$^{-1}$ at $V_a = 4$ kV, $I_a = 1$ A. The cathode, a thoriated tungsten filament, requires 400 W heating power. The output power is 3–5 kW, depending on conditions of operation.

are approximately straight lines, the quantities

$$g_m = (\partial I_a / \partial V_g)_{V_a}; \qquad r_0 = (\partial V_a / \partial I_a)_{V_g}; \qquad -\mu = (\partial V_a / \partial V_g)_{I_a},$$

introduced in § 7.6, are meaningful parameters so long as the voltage swings are not excessive; they are also quoted in the manuals. In general the value of $\mu$ ranges from 20 to 35. Since $\mu = C_{cg}/C_{ca}$, it is substantially independent of the working conditions, being determined by the closeness of the winding of the grid and the relative distances of grid and anode from the cathode. If the grid is closely wound, the number of lines of force reaching the cathode from the anode is small compared with the number from the grid, and the magnification factor $\mu$ is high. If the grid is rather openly wound, $\mu$ is low.

Values of $g_m$ range roughly from 3 mA V$^{-1}$ to 40 mA V$^{-1}$, and depend on the working conditions. For example, at $V_a = 4$ kV and $I_a = 0 \cdot 12$ A the value of $g_m$ may be $3 \cdot 3$ mA V$^{-1}$, while at $V_a = 1$ kV, $I_a = 2 \cdot 3$ A, $g_m = 10$ mA V$^{-1}$ for the same triode. This variation with anode current is in line with the three-halves power law; from eqn (21.3) we expect that $(\partial I_a / \partial V_g)_{V_a}$, and hence also $g_m$, should vary as $I_a^{\frac{1}{3}}$.

The most important use of the triode is as an amplifier or oscillator, with the load in the anode lead as in Fig. 21.3. Small changes in the anode current obey an equation such as (7.71).

$$i_a = g_m v_g + v_a/r_0 \tag{21.4}$$

and the voltage amplification (see eqn (7.66a)) is given by

$$A_v = \frac{v_a}{v_g} = \frac{-i_a \mathbf{z}_L}{v_g} = -g_m \frac{r_0 \mathbf{z}_L}{r_0 + \mathbf{z}_L}. \tag{21.5}$$

Maximum power output is obtained by making $\mathbf{z}_L = r_0$. With a parallel tuned circuit as load, the maximum voltage swing can be equal to the supply voltage $V_S$ without driving the anode negative at any point. The power into the load, with $\mathbf{z}_L = r_0$, is then $\frac{1}{2}V_S^2/r_0$, showing that a triode with a low value of $r_0$ is required to obtain the highest power.

For low values of the grid swing, no electron current flows to the grid, but this may not be the case in high-power applications. For the triode whose characteristics are shown in Fig. 21.2, a typical working condition is with $V_g$(mean) $= -300$ V, but with a grid swing of amplitude 600 V. The average grid current is about $0 \cdot 3$ A, with a grid dissipation of about 200 W, and an output power of $3 \cdot 5$ kW. The Hartley circuit (Fig. 20.13), with minor modifications, can be used for a power oscillator.

At frequencies above about 100 kHz the effects of the interelectrode capacitances denoted by broken lines in Fig. 21.3 become important. The cathode–anode capacitance $C_{ca}$ (which is usually less than 1 pF) may be regarded as part of the load $\mathbf{z}_L$, but the effect of the grid–anode capacitance is magnified because it has the amplified signal voltage across it (see

FIG. 21.3. The triode with its interelectrode capacitances. Since the high-tension (H.T.) supply must have zero impedance at the radio-frequency, the cathode–anode capacitance $C_{ca}$ is effectively in parallel with $\mathbf{z}_L$ and can be considered part of it. The small cathode lead inductance $L_c$ is important only at very high frequencies (see Problem 21.1).

Fig. 21.4). This is an example of the Miller effect, discussed in §§ 20.1 and 20.4 but originally derived for the thermionic triode; it produces an input admittance

$$y_{gc} = j\omega C_{gc} + j\omega C_{ga}(1 - A_v).$$  (21.6)

For a triode with $\mu = 30$ and $z_L = r_0$, $A_v = -15$, typical values are $C_{gc} = 5$ pF, $C_{ga} = 10$ pF, so that

$$C_g = 5 + 10(1 + 15) = 165 \text{ pF}.$$

At a frequency of 10 MHz, this corresponds to a capacitive input impedance of only 100 Ω.

The undesirable results of the Miller effect can be avoided by neutralization (cf. § 20.4), or by using the triode in the 'grounded grid connection' which is equivalent to the 'common gate' connection of Fig. 20.10. An alternative method is to reduce the value of $C_{ga}$ by introducing another electrode between grid and anode, maintained at a fixed potential, which intercepts the lines of force between grid and anode. By this means $C_{ga}$ can be reduced to less than 1 pF, and in many cases to less than 0·1 pF. In the 'beam power tetrode' the additional electrode is a 'screen grid' whose wires are exactly in the 'shadow' of the wires of the control grid, so that electrons flow in beams between the wires, and a minimum of current is intercepted. A special geometry is used to ensure that any secondary electrons emitted from the anode return to the anode and do not reach the screen grid, which is typically at half the mean anode potential, and may be at a higher potential than the anode for part of the cycle. Such tetrodes can be used in amplifiers up to frequencies of about $10^3$ MHz in specially designed circuits. Dielectric losses in the tube holder are reduced by the use of insulating materials with a low loss tangent; radiation losses are avoided by the use of coaxial lines, all leads being kept as short as possible. It is also necessary to keep the cathode lead as short as possible, since otherwise its inductance, combined with the cathode–grid capacitance, reduces the input admittance (see Problem 21.1). Another problem is the finite transit time of the electrons, which is discussed in the next section.

FIG. 21.4. Equivalent circuit of the triode with its interelectrode capacitances. The cathode–anode capacitance is not shown, but regarded as part of $z_L$.

## 21.3. Effect of transit time on input conductance

While an electron is leaving one electrode of a tube and approaching another it induces a charge on each of these electrodes. As it moves, the induced charge on the electrode which it has left diminishes, while that on the electrode which it is approaching increases. This can be seen quite simply by considering two plane parallel electrodes which are maintained at voltages 0 and $V_a$ respectively by means of a battery. Suppose a charge $-q$ is emitted from the plane of zero voltage. It will be accelerated towards the other plane, and when it has moved through a potential $V$ the work done on the electron will be $qV$. This work must be supplied by the battery, whence it follows that a charge $q(V/V_a)$ must have flowed through the battery. The direction of flow is such that the plane at potential $V_a$ will have acquired a charge $+q(V/V_a)$, while the charge on the other plane, which was $+q$ at the moment after the electron was emitted, is reduced to $+q(1-V/V_a)$. Thus the movement of the charge is accompanied by changes in the induced charges on the two planes, corresponding to the change in the number of lines of field from the electron which terminate on either plane (see Fig. 21.5). The duration of these changes is equal to the transit time of the charge between the two planes, and a current pulse flows for this length of time.

Similar arguments hold if one plane is replaced by a grid, and the passage of a charge through a grid therefore causes a momentary flow of charge to the grid which reverses in sign as the charge passes through. It is not necessary for the charge to hit the grid to create an induced charge, and the current flow accompanying the passage of the charge is shown in Fig. 21.6. The area under the curve up to any point represents the charge induced at that moment. The total area (that is, the net accumulated charge) is zero, since the positive and negative sections annul one another provided that all the charge flows through the grid and none is intercepted. The low-frequency components of the induced current are negligible provided the period is much longer than the transit time. This is about $10^{-9}$ s for electrons in a normal tube, and at frequencies up to 10 MHz the induced current is virtually zero. At higher frequencies,

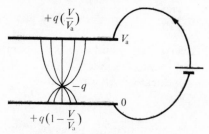

FIG. 21.5. Induced charges on electrodes.

FIG. 21.6. Current flow to grid during transit of an electron. $t$ is the instant at which the electron passes through grid. Transit time is $\tau_1 + \tau_2$.

where the transit time is an appreciable fraction of an r.f. cycle, the effect of the passage of the electrons in inducing an r.f. current to flow to the grid is appreciable. Full analysis of the effect is complicated, but an estimate of its order of magnitude can be obtained by the following method.

At a time $t$, let the voltage applied to the grid be $V = V_0 \sin \omega t$. Let the transit time from cathode to grid be $\tau_1$, and that from grid to anode be $\tau_2$. Then the current induced in the grid by the electrons approaching it will be approximately

$$I_1 = g_m V_0 \sin \omega(t - \tau_1)$$

since the size of the current is determined by the value of the grid voltage at the time $(t - \tau_1)$ when the electrons left the cathode (or, more strictly, the space-charge region). Similarly the current induced in the grid by the electrons leaving for the anode may be written

$$I_2 = -g_m V_0 \sin \omega(t - \tau_1 - \tau_2),$$

the minus sign arising from the reversal of the current for departing electrons. The net current is therefore

$$I_1 + I_2 = 2 g_m V_0 \cos \omega \{t - (\tau_1 + \tfrac{1}{2}\tau_2)\} \sin \tfrac{1}{2}\omega\tau_2$$

$$= 2 g_m V_0 \sin \tfrac{1}{2}\omega\tau_2 \{\cos \omega t \cos \omega(\tau_1 + \tfrac{1}{2}\tau_2) + \sin \omega t \sin \omega(\tau_1 + \tfrac{1}{2}\tau_2)\}$$

$$= 2 g_m \sin \tfrac{1}{2}\omega\tau_2 \left\{ \frac{dV}{dt} \frac{\cos \omega(\tau_1 + \tfrac{1}{2}\tau_2)}{\omega} + V \sin \omega(\tau_1 + \tfrac{1}{2}\tau_2) \right\}.$$

This contains both a capacitative and a resistive component. The latter is more important since it causes a loading of the input circuit. If both $\omega\tau_1$ and $\omega\tau_2 \ll 1$ the input conductance may be written

$$G = g_m \omega^2 (\tau_1 \tau_2 + \tfrac{1}{2}\tau_2^2). \tag{21.7}$$

A full analysis by North shows that for tubes of common size the input conductance $G$ is approximately equal to $g_m \omega^2 \tau_1^2 / 10$.

If $g_m = 5 \, \text{mA} \, \text{V}^{-1}$ and $\tau_1 = 0\cdot5\times10^{-9}\,\text{s}$, the value of $G$ is found to be roughly $5\times10^{-21} \, f^2 \, \Omega^{-1}$ at a frequency $f$, and it is therefore comparable with the input conductance due to cathode-lead inductance (see Problem 21.1).

## 21.4. The klystron

We have seen already that the finite time which an electron takes to pass from cathode to anode causes difficulty in the operation of conventional tubes below a metre wavelength. At centimetre wavelengths the problem of reducing the cathode-grid clearance so as to keep the transit time down to a small fraction of a cycle becomes practically insuperable. One obvious method of reducing the transit time is to shoot the electrons at a large velocity through a grid which then acts as an effective cathode. For example, if electrons are accelerated by a potential of 2500 V, their velocity is about $3\times10^7 \, \text{m} \, \text{s}^{-1}$, and they will traverse a distance of 1 mm in $3\times10^{-11}\,\text{s}$, which is only one-tenth of a period at a wavelength of $0\cdot1$ m. Thus, if the cathode–grid system is replaced by two grids, between which the high-frequency voltage is imposed, and the electrons are shot through these grids at a high velocity, the transit time can be kept short.

Such an arrangement must depend on some different principle for its working from that of a conventional vacuum tube, where the grid voltage influences the space charge in the potential minimum just in front of the cathode, and thus causes a change in the number of electrons flowing to the anode. When an alternating voltage is applied between grid and cathode a periodic 'density modulation' is set up in the electron stream flowing to the anode. If, now, the cathode, which emits electrons with an average energy corresponding to about $0\cdot1$ eV, is replaced by a grid through which electrons are injected at high voltage, there will be practically no space charge between this pseudo-cathode and second grid. Application of a small r.f. voltage between the two grids will not cause any change in the density of electrons leaving this space. Instead, it produces a 'velocity modulation'; some of the electrons will be accelerated by the r.f. field, while others which go through this field half a period later, when it is reversed in sign, will be retarded.

The principle of 'velocity modulation' rather than 'space-charge modulation' is fundamental in the working of the klystron and other centimetre-wave tubes. Velocity modulation is not of itself sufficient to produce amplification or oscillation, since for this we require a density modulation of the electron beam. However, if a velocity-modulated beam is allowed to 'drift' along in a field-free space, a density modulation will be set up in the following way. The electrons which were accelerated by the r.f. field will gradually overtake the slower electrons in front of them, which were retarded by the field. In this way 'bunching' of the electrons

FIG. 21.7. Bunching of electrons after velocity modulation.

will occur, as illustrated in Fig. 21.7. Here the distances covered by a number of electrons, initially uniformly spaced in the beam, are plotted against time. Lines corresponding to fast electrons overtaking slow electrons converge, while at points appropriate to half a period earlier or later in the r.f. field, the lines diverge. The former gives a 'bunch' since the convergence of the lines means that more electrons occupy a given volume, while the latter gives a 'rarefaction'. If now the beam traverses a second pair of grids, between which an r.f. field of the same frequency is applied in such a phase that a bunch is retarded by the field, while a rarefaction is accelerated, energy will be transferred from the beam to the field because more electrons are slowed down than are speeded up. This constitutes a conversion of energy from the high-tension supply used for the initial acceleration of the beam into energy in the alternating electromagnetic field, in a similar manner to that in a conventional radio tube.

There, in an ordinary amplifier or oscillator, the denser current flow to the anode coincides with the moments at which the anode potential is low, so that these electrons are slowed up by the alternating component of the electric field in front of the anode, thus doing work against this field (cf. § 20.5).

A schematic diagram of a klystron is shown in Fig. 21.8. Electrons accelerated and formed into a beam by a suitable gun pass through a resonator B where a velocity modulation is imposed on them by the r.f. field. They then travel through the field-free 'drift-space' in which bunching occurs, and enter a second resonator C called the 'catcher', tuned to the same frequency as B. Finally, the electrons are collected on an electrode at high-tension potential, which plays no essential role in the action. If some of the r.f. signal in the catcher is fed back through a coupling loop to the buncher, oscillations will be set up if the phase is correctly adjusted, and if more energy is extracted from the beam by the catcher than is dissipated in the combined resistances of the catcher and buncher. In this connection it should be noted that the velocity modulation of the beam by the r.f. field in the buncher requires no net power if the transit time is short, for as many electrons are slowed down as are speeded up. For high efficiency, it is necessary to develop the greatest r.f. electric field in the resonator at the point where the beam traverses it, with the smallest dissipation of power in the resistive walls. The cavity resonator gives the best type of circuit in this respect, and its shape is determined primarily by the requirement of a short transit time for the electrons. If the latter travel with one-tenth of the velocity of light, and their transit time is to be not more than one-tenth of the period of oscillation, the gap which they traverse must be about one-hundredth of a wavelength. This requires an indented cavity of the shape shown in Fig. 21.8. It may be regarded either as a short section of coaxial line, whose inductance resonates with the capacitance across the gap, or as an indented waveguide resonant cavity.

As with most oscillators, the full mathematical theory is rather complex, but it is possible to derive the starting condition for oscillations from elementary considerations as follows. We assume that the electrons leave the gun with potential $V_0$, and the path they traverse in the resonant cavity B has a potential difference $V \cos \omega t_0$ across it at the instant $t_0$. Then if $V \ll V_0$, as will be the case for small amplitudes of oscillation, we may use the differential relation $\delta v/v = \frac{1}{2}(\delta V/V)$ to find the fractional change in their velocity after traversing the cavity; their final velocity may then be written as

$$v = v_0(1 + V \cos \omega t_0/2V_0), \qquad (21.8)$$

where $v_0 = (2eV_0/m)^{\frac{1}{2}}$ is the velocity with which an electron leaves the gun. An electron which leaves B at time $t_0$ reaches the second resonator

FIG. 21.8. Two-resonator klystron oscillator. Collector, buncher B, and catcher C at high-tension positive (earth); electron gun at high-tension negative potential.

C, at a distance $x$ away, at a time

$$t = t_0 + x/v = t_0 + (x/v_0)(1 + V \cos \omega t_0/2V_0)^{-1}$$
$$\approx t_0 + (x/v_0)(1 - V \cos \omega t_0/2V_0), \qquad (21.9)$$

where we have again used the approximation $V/V_0 \ll 1$.

Now the current at any point is equal to $dq/dt$, the rate at which charge passes that point. To find the current, we shall consider a small section of the beam containing charge $dq$, and follow it along the beam. Owing to the velocity modulation, the front and rear portions of this section travel at different speeds, and the time $dt$ which the section takes to pass a given point therefore changes with the distance it has travelled. The beam current at the time $t$ is therefore

$$I = dq/dt = (dq/dt_0)(dt_0/dt) = I_0(1 + \omega x V \sin \omega t_0/2v_0 V_0)^{-1}$$
$$\approx I_0\{1 - (\omega x V/2v_0 V_0)\sin \omega(t - x/v_0)\}, \qquad (21.10)$$

where we have used the relations $(dq/dt_0) = I_0$, the initial beam current, and

$$(dt_0/dt) = (dt/dt_0)^{-1} = (1 + \omega x V \sin \omega t_0/2v_0 V_0)^{-1}$$

obtained by differentiation of eqn (21.9).

Eqn (21.10) shows that we have now a density-modulated current, whose amplitude of modulation increases linearly with $x$, the distance travelled, so long as we restrict ourselves to small velocity modulation. If this current now passes through a second resonator, with a potential difference $V_2 \cos \omega t$, the mean power extracted from the beam will be

$$\overline{-IV_2 \cos \omega t} = -\frac{\omega}{2\pi} \int_0^{2\pi/\omega} I_0 V_2 \cos \omega t \left\{ 1 - (\omega x V/2 v_0 V_0) \sin \omega \left( t - \frac{x}{v_0} \right) \right\} dt$$

$$= \frac{I_0 V V_2 \omega x}{2 v_0 V_0} \frac{\omega}{2\pi} \int_0^{2\pi/\omega} \cos \omega t \sin \omega \left( t - \frac{x}{v_0} \right) dt = -\frac{I_0 V V_2 \omega x}{4 v_0 V_0} \sin \frac{\omega x}{v_0},$$

since only terms in $\cos^2 \omega t$ contribute to the mean power. The power extracted from the beam will be greatest when $-\sin \omega x/v_0 = +1$, that is, $\omega x/v_0 = 2\pi(n + \frac{3}{4})$, where $n$ is an integer. In other words, if oscillations in B and C are in phase, the beam must take $(n + \frac{3}{4})$ periods to travel from B to C. The time of travel can be adjusted by altering the initial accelerating voltage $V_0$.

So far we have considered the power extracted from the beam when r.f. signals from an external source are fed into resonators B and C. It is clear, however, that if a little of the power that is fed from the beam into C is returned to B to act there as the source of signal, oscillations can be sustained so long as the power extracted from the beam is greater than that dissipated in resistive heating of resonators B and C. We may represent this dissipation by a resistance $R$ for each resonator. The oscillations will be sustained if

$$\frac{I_0 V V_2 \omega x}{4 v_0 V_0} > \frac{V^2}{2R} + \frac{V_2^2}{2R}.$$

If $V = \alpha V_2$, where $\alpha$ is small, the power dissipated in the buncher may be neglected, and this relation may be expressed in the form

$$\text{minimum starting current } I_0 = \frac{2 V_0 (v_0/\omega x)}{\alpha R}$$

$$= \frac{V_0}{\alpha R \pi (n + \frac{3}{4})}. \tag{21.11}$$

This shows that the smallest beam current required for sustained oscillations is of the order of the beam voltage divided by the parallel impedance of the resonator. The current required is diminished if $\alpha$ is increased, or if the time of drift $x/v_0$ between the resonators is increased, since the density of the bunches reaching the second resonator is enhanced in either case.

With large electrode voltages and currents, the klystron is an efficient and powerful oscillator, and can be used as a transmitter, but its principal

FIG. 21.9. Reflex klystron. Electrons from the gun pass through the gap in the resonator, and are returned back through the gap by the reflector which is at a potential negative with respect to the cathode.

use is as a low-power local oscillator, for which an output of a few milliwatts is sufficient, in a superheterodyne receiver. For this purpose the klystron must be tunable, and this is not feasible when two separate resonators of high $Q_F$ (several thousand) must be adjusted simultaneously. A single resonator or reflex klystron is therefore employed, as in Fig. 21.9, where the electron stream after passage through the resonator is confronted by an electrode whose potential is negative with respect to the cathode potential. The electrons are thereby halted and reflected back through the same resonator. Bunching occurs because the electrons which are speeded up in the first passage through the resonator travel further towards the reflector electrode and so return later (as in the case of a ball thrown into the air), together with the electrons which passed through the resonator half a period later and were retarded by the r.f. field. Thus the bunches occur at points in the returning beam which the faster electrons reach later rather than earlier, but as they are retarded on their return passage through the resonator when the r.f. field is directed away from the cathode instead of towards it (as in the two-resonator klystron), the required transit time for oscillation is still $(n+\frac{3}{4})$ periods. The minimum starting current is now given by eqn (21.11) with $\alpha = 1$. In an early type of klystron for $0\cdot1$ m wavelength, $V_0 = 1200$ V, $R = 70$ k$\Omega$, $(n+\frac{3}{4}) = 1\frac{3}{4}$, giving a starting current of about 3 mA. The running current is about 8 mA, and the power output of an average tube $\approx$300 mW, corresponding to an efficiency of 3 per cent. Later types run at a beam voltage of 300 V, and a current of 20 mA, with a rather lower power output.

Reflex klystrons of this type have been made to oscillate at wavelengths down to about 5 mm, which seems to be about the limit. The main difficulties in making such oscillators for shorter wavelengths arise from the smaller size of the resonant cavity, with its correspondingly lower parallel impedance $R$, which means that a higher beam current is required to start oscillations. This higher current needs to be sent through a smaller hole in the cavity, but the current density which can be obtained in the beam is limited by the mutual repulsion of the electrons.

## 21.5. The magnetron

Oscillations of high power at centimetre wavelengths are produced by the magnetron; an outline diagram of a typical tube is shown in Fig. 21.10. Electrons are emitted from a central cylindrical cathode, and are accelerated towards a coaxial cylindrical anode consisting of a solid copper block with a number of resonant cavities. These may have the shape shown in the figure, but other shapes are also possible. Essentially they form a set of quarter-wave resonant lines or cavities, the open end of the line being at the inner surface of the anode block. Thus, when oscillations take place, a strong r.f. electric field is set up at the inner

Output side arm

Output coupling loop

Resonator system

Oxide-coated cathode

Insulated heater

Cooling fins

Oxide-coated cathode

Tungsten heater and cathode leads (cathode connected to one side of heater)

(a)                              (b)

FIG. 21.10. A typical magnetron: (a) from side, (b) along axis (parallel to external magnetic field). (After Willshaw *et al.* 1946.)

surface, the field lines running mainly in the circumferential direction across the open end of the cavity, as shown in Fig. 21.11.

The magnetron operates with a strong axial magnetic flux density of a few tenths of a tesla, which is normally provided by a permanent magnet, and a potential of 10–50 kV on the anode. An electron on its way from the cathode to the anode experiences a magnetic force perpendicular to its direction of motion and a radial electric force. Its trajectory under the action of these forces can best be pictured by reference to a case with simple plane geometry (see Problem 4.10). The trajectory is then a cycloid, the path traced out by a point on the circumference of a cylinder rolling along a plane with angular velocity $\omega = eB/m$; the radius of the cylinder is $E/\omega B = mE/eB^2$, and the linear velocity of its axis is $v = E/B$.

In the case of cylindrical geometry, the electron orbit is approximately an epicycloid generated by rolling a cylinder on the cylindrical cathode, and so is rather similar to the case of the parallel planes if we imagine the latter to be given a curvature. The approximation arises because we are neglecting the radial decrease in the electric field which occurs with cylindrical geometry as we go from cathode to anode. In addition we are neglecting the mutual repulsions of the various electrons ('space charge') in either case. If the difference between the radius $a$ of the cathode and the radius $b$ of the anode is small compared with either, at a point midway between cathode and anode we may write the angular velocity of the electron cloud as approximately

$$\frac{v}{\frac{1}{2}(a+b)} = \frac{2E}{B(a+b)} = \frac{V}{b-a} \cdot \frac{2}{B(a+b)} = \frac{2V}{B(b^2-a^2)}, \tag{21.12}$$

FIG. 21.11. Electron trajectories and lines of r.f. electric field (arrows show direction of force on electrons) in the '$\pi$-mode' magnetron.

where $V$ is the steady voltage applied between anode and cathode. This expression shows that to give the electron cloud a certain angular velocity of rotation, we must maintain a certain linear relation between the anode voltage and the magnetic field. The importance of this angular rotation of the electron cloud arises from the necessity of synchronizing the movements of the electrons with the alternation of the r.f. electric fields in the resonators, in order to preserve the right phase relationships. This is essential for the efficient transfer of energy from the electron cloud to the r.f. field, the basis of any oscillator.

Before considering the mechanism of this transfer, it is necessary to discuss the resonator system. Each resonant cavity behaves like a tuned circuit but the system is more complicated than that of a simple oscillator because of the presence of $N$ such circuits, all coupled together. This coupling is partly electric, lines of r.f. electric field originating from one cavity terminating in another, but it is predominantly magnetic; the lines of r.f. magnetic field going down through one cavity are completed by returning up through another cavity. With a system of $N$ circuits all tuned to the same frequency, and strongly coupled together, the natural frequencies of oscillation are split apart in the same way as with two circuits (see § 7.4), but analysis of the system is much more complex. The different resonant frequencies correspond to oscillations with varying change of phase between successive resonators; they can be analysed into systems of standing waves, or of travelling waves, or a mixture of the two. The simplest system is a standing wave where the phase difference between successive cavities is $\pi$ (the so-called '$\pi$-mode'), and this is also one of the most efficient modes of operation of the magnetron. At any instant the direction of the lines of force in successive cavities is exactly reversed, as would arise from a simple potential distribution where alternate segments are just plus and minus in the r.f. voltage (see Fig. 21.11).

The interaction between the electrons and the r.f. field must now be considered. Under normal conditions of operation, but in the absence of oscillation, the electrons would travel (approximately) in circles such that their farthest point from the cathode is about half-way across the cathode–anode space. At the moment when they return to the cathode, their velocity would be zero, since the magnetic field does no work on them, and that done by the electrostatic field as they move initially away from the cathode is all regained on the return path. Suppose now that an electron is just moving tangentially at the farthest point in its trajectory from the cathode. Then the force exerted on it by the magnetic field is towards the cathode, while that exerted by the electrostatic field is towards the anode. If the tangential velocity of the electron is increased at this moment through being accelerated by the fringing field of one of the cavities, the magnetic force on it (which is proportional to its velocity) will be increased, while the electrostatic force is unchanged, so that the effect is to return it towards the cathode. If, on the other hand, the electron is retarded by the r.f. field, the magnetic force is decreased and the electron will move in a path which brings it closer to the anode than it would have got in the absence of the r.f. field. If now it arrives opposite another cavity at the moment when it is again retarded, it again gives up energy to the r.f. field, and moves still closer to the anode. Note that as it does so, it moves into positions where the r.f. field is stronger and so a greater proportion of the kinetic energy of the electron is transferred to the r.f. field. On the other hand, an electron which is speeded up returns towards the cathode where its interaction with the r.f. field is smaller. Thus, if the right phase relationship can be maintained, some of the electrons will give up energy to several cavities in succession, and eventually reach the anode with kinetic energy much less than that corresponding to $eV$, while others will be returned to the cathode. On the whole, the latter take much less energy from the r.f. field than the former give to it, and the net transfer of energy will maintain oscillation.

To get the right phase relationship, the angular velocity of the electron cloud must coincide with the angular velocity of rotation of one of the Fourier components of the r.f. field system. For the $\pi$-mode, this simply means that the electron cloud must rotate through the angle $2\pi/N$, between successive cavities, in $(n+\frac{1}{2})$ cycles (where $n$ is an integer). Its angular velocity must therefore be

$$\frac{(2\pi/N)}{(n+\frac{1}{2})\tau} = \frac{2\pi f}{N(n+\frac{1}{2})},$$

where $f = 1/\tau$ is the frequency of the oscillations. Equating this to (21.12) gives

$$V = \frac{\pi f}{k} b^2 B\left(1 - \frac{a^2}{b^2}\right), \tag{21.13}$$

where $k = N(n+\frac{1}{2})$. Assuming the ratio $a:b$ is roughly constant, it will be seen that to maintain operation at a given frequency $f$ in a given mode $k$, at a fixed flux density $B$, the anode voltage must be increased with the square of the anode diameter. If it is desired to keep the voltage fixed and to construct a magnetron of higher frequency (shorter wavelength) but with equivalent operating conditions, then the resonator system and anode diameter must be scaled in proportion to the wavelength ($b \propto 1/f$), and $B$ must be increased in proportion to $f$.

High power output from the magnetron can be achieved only if high anode voltages and high anode currents are used. By running the tube in short pulses roughly of $1~\mu s$ duration, with a repetition rate of about

1000 Hz, the power in the pulse can be made over 1000 times as great as can be obtained under continuous operation. These high powers are mainly due to three factors:

(1) the electronic conditions are such that high efficiency is attained at high level;

(2) oxide-coated cathodes can give very high currents per unit area, 100 times greater under pulsed conditions than under continuous running;

(3) the mean power dissipated on the anode is reduced, and is easily removed by conduction through the solid copper anode.

An important factor under (1) is focusing action by the r.f. field, which helps to concentrate the space charge into a number of narrow spokes (see Fig. 21.11). Each spoke then passes through the r.f. field at the moment when it is a maximum, giving the equivalent of Class C operation in ordinary triodes. The main technical difficulties have been the construction of rugged cathode surfaces, which can withstand the heavy bombardment by the returning electrons accelerated by the r.f. field, and avoiding 'mode jumping', where the frequency changes as the tube jumps from one value of $k$ to another. Power is extracted by means of a loop coupling in one of the resonators, or through a waveguide slit in one resonator.

Typical operating conditions for a medium high power magnetron operating at $0 \cdot 1$ m wavelength are: magnetic flux density $B = 0 \cdot 28$ T, anode voltage 31 kV, anode current during pulse 35 A, output power in pulse 750 kW. In this tube the cathode diameter is $6 \cdot 0$ mm, and the inside diameter of the anode is $16 \cdot 1$ mm; the length of the anode block is 20 mm, and the over-all length of the tube is 32 mm. The dimensions of the tube are thus comparatively small, and the high power obtainable in the pulse is due to the high efficiency (70 per cent), which also reduces the dissipation on the anode block to only 30 per cent of the input power. At 100 mm wavelength, output pulse powers of a few megawatts can be achieved, but the power decreases rapidly as the wavelength is reduced, owing to a number of factors. Experimental tubes have been made to operate at wavelengths of a few millimetres, and the short wavelength limit is about the same as or a little lower than that of the klystron. Most cavity magnetrons are fixed-frequency tubes, but some magnetrons tunable over a range of 10–20 per cent in frequency have been constructed, the variation being obtained by plungers moving into the resonators from one end.

## 21.6. Travelling-wave tubes

An important class of electronic tubes for centimetre wavelengths, which we shall not discuss in detail, is that of the 'travelling-wave' tube. This is a velocity modulation device in which the electron beam interacts continuously with an electromagnetic wave, instead of only locally, as in the klystron. To make this possible the wave velocity must be reduced to coincide with the beam velocity; this is accomplished by means of a 'slow wave structure', such as a wire helix surrounding the beam, which behaves as an artificial transmission line with wave velocity $1/(LC)^{\frac{1}{2}}$ (see § 9.3). Energy is transferred to the electromagnetic wave from a slow space-charge or plasma wave on the electron beam (see Pierce 1974).

The device is untuned and amplification is possible over a wide range of frequencies. The (gain×bandwidth) product is $(100 \times 5000) = 5 \times 10^5$ MHz; this is 10 times better than can be achieved with a transistor amplifier, and over 100 times better than with vacuum-tube triodes. The main use of the travelling-wave tube is as a wide-band amplifier, capable of giving 100 W or more power output at a frequency of 5 GHz.

## References

WILLSHAW, E. E., RUSHFORTH, L., STAINSBY, A. G., LATHAM, R., BALLS, A. W. and KING, A. H. (1946). *J. Inst. Elect. Engrs.* **93**, 985.

PIERCE, J. R. (1974). *Almost all about waves.* M.I.T. Press, Cambridge, Massachusetts.

## Problems

21.1. Show that the effect of the inductance in the cathode lead of Fig. 21.12(a) can be represented by the equivalent circuit of Fig. 21.12(b), giving an input admittance

$$\mathbf{y} \sim j\omega C_{gc} + g_m \omega^2 L C_{gc},$$

provided that $\omega L g_m \ll 1$ and the anode current can be taken as $g_m v_g$.

If $g_m = 5$ mA $V^{-1}$, $C_{gc} = 5$ pF, $L = 5 \times 10^{-8}$ H, show that the input conductance $R_g^{-1}$ is about $5 \times 10^{-20} f^2 \ \Omega^{-1}$

(a)                                          (b)

FIG. 21.12. Effect of cathode lead inductance at high frequencies; $R_g = (g_m \omega^2 L C_{gc})^{-1}$.

21.2. Calculate the anode voltage $V'_a$ of the equivalent diode for a triode with. plane parallel electrodes in which the grid–cathode spacing is $0\cdot3$ mm and the current density is $200$ A m$^{-2}$, assuming that the equivalent diode spacing is the same as the actual grid–cathode spacing.

If $V_g = -3$ V, $V_a = +150$ V on the triode, what must be the value of $\mu$?

(*Answer:* $V'_a = 3\cdot9$ V; $\mu = 22$.)

21.3. Prove that in a plane parallel diode the transit time of an electron is $\frac{3}{2}$ times as long under space-charge limited conditions as it would be in the absence of space charge.

What will be the transit time between cathode and grid in the triode of Problem 21.2?

(*Answer:* $7\cdot7 \times 10^{-10}$ s.)

# 22. Alternating current measurements

## 22.1. Measurement of voltage, current, and power

IF an alternating voltage is applied to the terminals of a d.c. instrument such as a moving-coil galvanometer, the reading observed is usually zero. The movement of the galvanometer is too sluggish to follow the alternations of the applied voltage if these occur at more than a few cycles per second. The instrument therefore records only the mean value of the current over many cycles, which is zero for a symmetrical waveform. Thus the measurement of alternating currents and voltages requires the use of special instruments, which may be divided into three classes according to the principle involved in their construction. In the first class are instruments with very rapid responses so that they can follow the alternating

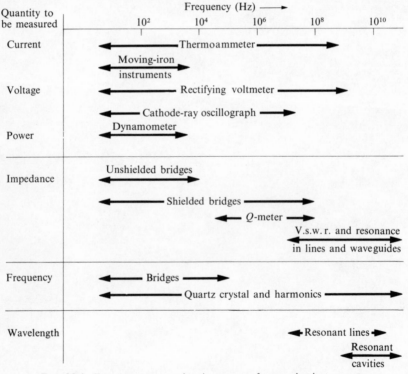

FIG. 22.1. Frequency ranges of various types of measuring instruments.

waveform; second, 'square-law' instruments, so called because they respond to the square of the current or voltage applied; and third, rectifier instruments, where the alternating voltage is converted to a steady voltage which can be measured on a d.c. instrument.

The instruments with the greatest frequency range (see Fig. 22.1) are the thermoammeter, a square-law instrument, for current; the voltmeter, incorporating a rectifier; and the cathode-ray oscillograph, a short-time-constant instrument for the display of waveform and measurement of voltage. The dynamometer is a moving-coil instrument which measures power directly, but its use is confined to supply frequencies. At radio-frequencies power is normally determined from the voltage across a known resistance, or from the current through it. In general, current and voltage are the primary quantities measured. Many commercial instruments contain electronic circuits for protection against overloading or wrong connections. Digital instruments contain solid-state electronic circuits to convert the output to digital read-out.

*The cathode-ray oscillograph*

This is an instrument whereby the waveform of an alternating voltage can be displayed on a screen. It consists of a number of electrodes (Fig. 22.2) inside an evacuated glass envelope. An electron gun, with cathode, grid, and various accelerating electrodes is used to focus a beam of electrons onto a small spot on a screen. This is coated with a mixture of compounds which become fluorescent (or phosphorescent, in 'after-glow' tubes) and emit light when excited by the electron beam. The brightness depends on the beam current (of the order of 10 $\mu$A), which is controlled by the voltages on the grid and the final anode (the screen). The latter ranges from 5 kV to 10 kV on large tubes (0·1 m diameter or more) down to 500 V on small tubes. The electrodes $XX$ and $YY$ are two pairs of plates oriented at right-angles to one another, and various voltages may be applied across the plates of either pair to deflect the electron beam. These deflections in the $x$ and $y$ directions are normal to the axis of the tube and proportional to the voltages applied to the $X$ and $Y$ plates. The magnitude of the deflection, and hence the sensitivity (defined as the deflection per unit voltage applied to the $X$ or $Y$ plates), is inversely proportional to the accelerating voltage (see Problem 22.1). The sensitivity is increased by reducing the separation between the two members of a pair of deflecting plates, and they are therefore splayed in order not to intercept the beam at large deflections. Electrostatic deflection, as this system is called, causes a certain amount of distortion, and magnetic deflection, using fields generated by small coils placed outside the tube, is usual for television tubes, where very large deflection angles are employed. Electrostatic deflection is used for most laboratory work, and the

Fig. 22.2. Cathode-ray tube. A beam of electrons is deflected by the XX and YY plates to give a luminous spot on the fluorescent screen.

pattern observed on the screen is then determined by the voltages applied to the two sets of deflecting plates. It is usual to apply a known voltage waveform to the X plates (the 'time-base'), while the unknown voltage is applied to the Y plates. The most useful type of time-base is one where the spot moves to the right across the screen in the x direction at constant velocity, followed by a rapid 'fly-back' to the left-hand side. This is called a linear time-base, and requires a sawtooth waveform which can be generated by special electronic circuits. Basically, a steadily increasing voltage is obtained by charging a capacitor through a constant-current device such as an FET biased to run at its saturation current.

To obtain a stationary picture, the repetition frequency of the time-base must be an exact multiple of the basic frequency of the waveform applied to the Y plates. Thus the time-base must be adjustable, and 'synchronization' is usually obtained by injecting a fraction of the voltage from the Y plates into the time-base circuit, so that the latter is 'locked in'. Other types of time-base include the single sweep for observing transient phenomena, the sweep being triggered by the onset of the transient; and circular or elliptical time-bases, obtained by applying sinusoidal voltages differing in phase by $\frac{1}{2}\pi$ to the X and Y plates. These are useful in the measurement of frequency. If an unknown frequency is applied to the anode of the cathode-ray tube, modulating the sensitivity, and if its frequency is $n$ times the time-base frequency, a stationary picture with $n$ loops is obtained (Fig. 22.3(a)). Alternatively, the unknown frequency may be applied to the grid of the tube, thus modulating the intensity. The pattern on the screen is broken up (as in Fig. 22.3(b)) into dots whose number gives the frequency ratio. Grid modulation is more sensitive than anode modulation, an amplitude of a few volts being sufficient for the unknown frequency.

The oscillograph may be used to determine the amplitude of an alternating voltage by measurement of the deflection on the screen from peak to peak. For this purpose it must be calibrated using a known d.c. or low-frequency a.c. voltage. The sensitivity of a $0 \cdot 1$ m diameter tube is usually of the order of a few tenths of a millimeter deflection per volt.

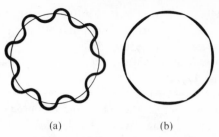

(a)                    (b)

FIG. 22.3. Comparison of frequency with circular time-base. Time-base frequency = $\frac{1}{8}$ (unknown frequency). (a) Anode modulation; (b) grid modulation. If the ratio of the frequencies is not exactly an integer, the pattern is not stationary but rotates.

The range may be extended by the use of an amplifier of known gain, and signals of the order of microvolts can be made to give an observable deflection. This technique may also be used for current waveform, by passing the current through a low resistance and amplifying the voltage developed across it.

The great advantage of the cathode-ray oscillograph is its ability to portray the waveform of an alternating voltage up to frequencies of a few hundred megahertz. At this point limitations arise from the difficulty of making suitable time-bases and amplifiers, as well as from inherent drawbacks in the tube itself (see Problem 22.2).

*Square-law instruments*

Any d.c. instrument whose deflection depends on the square of the current or voltage can be used for a.c. measurements, and the reading obtained by calibration with d.c. will give the r.m.s. value. Thus the electrostatic voltmeter can be used for alternating voltages and currents; so can the dynamometer (§ 5.6), though it is confined to audio-frequencies. The thermoammeter has a greater frequency range; the current passes through a resistive coil which heats a copper disc; this is soldered to a thermojunction, the current from which is read on a moving-coil galvanometer. The instrument can be calibrated by d.c. or low-frequency a.c., and is used for currents of the order of milliamperes. The readings are independent of frequency up to about 1 MHz, but above that there is coupling between the coil and the thermocouple due to stray capacitance. The sensitivity can be increased by mounting the thermo-junction in an evacuated glass envelope to improve the thermal insulation; the copper disc is sometimes joined to the junction by a small glass bead, which provides thermal contact but insulates the r.f. circuit from the galvanometer. The heater is very thin, to avoid any change of resistance with frequency due to skin effect, and the deflection is very nearly proportional to the square of the current over a large range of frequencies. The frequency range may be greatly extended by using a separate thermo-junction which may be inserted in the circuit quite apart from the instrument used to measure its d.c. output voltage. Leads to the latter instrument must be carefully decoupled.

The dynamometer wattmeter can be used for measuring power, by connecting it as in Fig. 5.17. The scale reading is then proportional to the average value of $V_0 I_0 \sin \omega t \sin(\omega t + \alpha)$ over a cycle, where $V_0 \sin \omega t$ is the voltage across the load, and $I_0 \sin(\omega t + \alpha)$ is the current through it. The scale reading gives $\frac{1}{2}(V_0 I_0)\cos \alpha$, and this is the power consumed by the load, so that the instrument can be calibrated to read power directly, and no determination of the phase angle or power factor is required.

The dynamometer is suitable only for audio-frequencies up to about 1 kHz. At radio-frequencies power can readily be measured from the voltage across or the current through a known resistance. This method can be used at all frequencies where the calibration of the voltmeter or ammeter is reliable, but the direct measurement of power can be extended, into the gigahertz region. For low powers ($\sim$ a watt down to microwatts) the 'thermistor' is a convenient element for this purpose. It consists of a tiny bead of various semiconducting oxides (usually of transition elements), whose resistance falls steeply with increasing temperature and hence also with power input. The resistance can be adjusted in manufacture to be $\sim 100 \; \Omega$, which can readily be matched to a coaxial line with characteristic impedance of this order. The change in resistance is found with a Wheatstone's bridge; if this is balanced with the r.f. power on, and the direct current increased to restore the balance when the radio-frequency is switched off, then the r.f. power can be calculated from the change in the d.c. power dissipated in the thermistor.

Powers of more than a few watts are measured by dissipating the power in a suitable load and observing the temperature rise; at centimetre wavelengths water, with its high absorption coefficient (see § 10.8), is a convenient load. A continuous-flow method is generally employed.

*Rectifier instruments*

Rectification is the process of turning an a.c. voltage into a d.c. voltage, and clearly this may be used as the basis of methods of measuring alternating voltages. A very simple system makes use of a solid-state diode, as in Fig. 22.4, with a fairly high resistance and microammeter in the output circuit. Current flows through the meter during alternate half-cycles, and it records the mean current flow. For a sine wave this should be $I_0/\pi$, where $I_0$ is the peak current, but this is not quite true because the diode is not exactly linear when forward-biased, and it passes a small leakage current when backward-biased. If a capacitor is placed

FIG. 22.4. A simple type of diode circuit for measuring alternating current.

across the terminals, X, Y, it becomes charged up to the peak voltage $V_0$ of the alternating waveform (see the discussion in § 18.4). Many commercial instruments are basically peak voltmeters, but calibrated to read $V_0/\sqrt{2}$, which is the r.m.s. value for a pure sine wave. Caution must therefore be used in interpreting such readings when the waveform is more complicated, as observed on a oscillograph.

The primary requirement of a voltmeter is an input impedance so high that it does not appreciably alter conditions in the circuit to which it is connected. If, to improve the sensitivity, the voltmeter circuit is used in combination with an amplifier, this must be chosen to have a high input impedance. A variety of commercial instruments are available which use solid-state circuitry to give a digital output instead of a pointer and scale reading.

We often require to measure the voltage of a particular frequency separate from other frequencies which may simultaneously be present. This can be done using a 'phase-sensitive detector', whose principle resembles that used in frequency changing (§ 18.6). A reference signal of frequency $\omega_1/2\pi$ is applied to a detector circuit, producing a fluctuating conductance of the form (eqn (18.9))

$$g = g_0 + g_1 \cos \omega_1 t + \ldots.$$

Suppose a small voltage $v \cos(\omega_2 t + \delta)$ is also applied to the circuit, where $\omega_2$ is close to $\omega_1$. The change in the detector current contains a low-frequency component

$$\tfrac{1}{2} g_1 v \cos\{(\omega_2 - \omega_1)t + \delta\}.$$

If $\omega_2 = \omega_1$, this represents a steady change $\tfrac{1}{2} g_1 v \cos \delta$, whose magnitude and sign depend on the phase difference $\delta$. Normally two identical detector circuits in push–pull are used; this doubles the output, and helps to eliminate spurious fluctuations. If $\omega_2$ is not exactly equal to $\omega_1$, the rectified output fluctuates at frequency $(\omega_2 - \omega_1)/2\pi$, but this is recorded only if it falls within the bandwidth of the meter or following circuits.

Thus only signals within a narrow range around $\omega_1/2\pi$ are recorded. If the over-all input due to noise or interference is large, care must be taken to ensure that any amplifier preceding the detector is not overloaded; for example, a filter circuit at the input can be used to reduce signals at unwanted frequencies. Overloading the detector can also produce spurious signals.

## 22.2. Alternating-current bridges

The measurement of resistance using direct current is usually accomplished most precisely by means of a bridge, either Wheatstone's bridge or one of its modifications. The measurement of a complex impedance is also readily achieved with high precision by means of an a.c. bridge. To determine a complex impedance fully, two quantities must be measured—its real and imaginary parts. At first sight this might seem to require two separate experiments, but in fact the balancing of an a.c. bridge requires that two separate conditions be satisfied simultaneously. These two conditions involve the real and imaginary components of the unknown impedance, and thus both are determined when both conditions are fulfilled. The reason for this extra complexity in the balancing of an a.c. bridge can readily be seen from consideration of a simple network such as the generalized Wheatstone's bridge shown in Fig. 22.5, with complex impedances in each of the arms. For a balance the voltage applied to the detector must be zero. This voltage is equal to the difference of voltage between the points A and B, which is truly zero only if the voltage at these points has not only the same amplitude but also the same phase. In other words, the voltage at each of these points must be represented by a vector with two components, and for the vectors at the two points to be identical, their components must be individually the same.

FIG. 22.5. Generalized Wheatstone's bridge.

The presence of two balance conditions which must be fulfilled simultaneously has an important bearing on the design of a.c. bridges. To avoid disturbing one balance condition when adjusting the other, it is essential that the two balance conditions shall be independent of one another. This can be achieved by choosing a bridge where each balance condition can be met by adjusting a variable impedance which does not appear in the other balance condition. The final balance can be obtained relatively quickly by first adjusting one variable until a minimum detector reading is obtained, and then the other. On returning to the first a finer balance is obtained, and so on. A second highly desirable quality is that the balance conditions shall be independent of frequency. The reason for this is that the source of power employed for the bridge never produces a pure sine wave, but contains some distortion which can be represented in a Fourier analysis of the waveform by harmonics of the fundamental frequency. The presence of quite a small harmonic content will be important if the balance conditions depend on frequency because the sensitivity of the bridge depends on the ability to detect a small fraction of the applied voltage, and this will be obscured by the harmonic content unless this is balanced out simultaneously. In practice it is often found that the harmonics do not vanish even in a bridge where the balance conditions are independent of frequency, because the impedances used may vary with frequency, usually because of stray reactances (see Problems 7.10 and 7.11). In this case it is advantageous to use either a tuned detector, such as a phase-sensitive detector with phase shifts of 0 and $\frac{1}{2}\pi$ so that signals can be observed both in phase and quadrature, or to insert a filter at the input to the detector to eliminate the harmonics.

The driving voltage for the bridge is usually provided by a small oscillator, a few volts being sufficient for most purposes. The detector consists of an amplifier followed by a voltmeter or by a cathode-ray oscilloscope. The latter has the advantage that it shows the waveform reaching the detector, and the presence of harmonics near the balance point is readily observed. To some extent it is possible to separate visually the fundamental and harmonics and to reduce the former to one-fifth or so of the harmonic. The amplifier gain must be variable (a potentiometer before the input is sufficient) in order to avoid saturation of the later stages when the bridge is far from balance. The gain is increased as the balance is approached and the amplifier has the advantage that it is not readily damaged by an overload.

When an electronic generator and amplifier are both used it may happen that one terminal of each is earthed, or has a large capacitance to the mains supply which is common to both. This would throw either a short-circuit or a large capacitance across one arm of the bridge, and an isolating transformer (preferably one with an electrostatic screen between primary and secondary) should be used between the bridge and either the generator or the detector-amplifier.

Variable impedances are required to balance the bridge, and a good general rule is that the impedances in all the arms should be of the same order of magnitude for optimum sensitivity. A great variety of bridges have been designed, of which we shall discuss only three.

### Schering's bridge

The generalized Wheatstone's bridge shown in Fig. 22.5 is the basis of most bridge circuits, and the balance condition is similar to that for the d.c. bridge:

$$\mathbf{Z}_1/\mathbf{Z}_2 = \mathbf{Z}_3/\mathbf{Z}_4. \tag{22.1}$$

The two balance conditions are contained in this complex equation, since the real and imaginary parts must be satisfied simultaneously. This will be seen in the following application to Schering's bridge, which is commonly used for the determination of capacitance.

The circuit diagram of Schering's bridge is shown in Fig. 22.6. The unknown (lossy) capacitor is represented by the series combination of $C$ and $R$. $C_1$ is a good standard capacitance, whose magnitude should be of the same order as that of the capacitor under test. $R_1$ is a fixed resistance, and $R_2$ is a variable resistance shunted by a variable capacitance $C_2$. The balance condition is

$$j\omega C_1\left(R + \frac{1}{j\omega C}\right) = R_1\left(\frac{1}{R_2} + j\omega C_2\right).$$

The real and imaginary parts of this give

$$C = C_1(R_2/R_1)$$

and

$$R = R_1(C_2/C_1). \tag{22.2}$$

FIG. 22.6. Schering's bridge for the measurement of capacitance.

These conditions fulfil the requirements outlined earlier. They are independent of each other, provided that $R_2$ and $C_2$ only are varied, and independent of frequency. In addition the unknown capacitance is obtained in terms of a known standard capacitance and the ratio of two resistances; the variable capacitance $C_2$ enters only into the equation for the apparent series resistance of the unknown capacitor. With a good capacitor this will be small, and high accuracy in the determination of $R$ is seldom required. A small variable air capacitor usually suffices for $C_2$, and its leakage resistance under normal conditions will be so high that it does not affect $R_2$, with which it is in parallel.

### Anderson's bridge

A self-inductance has an appreciable resistance; at audio-frequencies the quality factor $Q_F$ is generally not better than about 30, while the loss tangent $(= 1/Q_F)$ of a good mica capacitor is about $10^{-4}$. It is often preferable to determine an unknown inductance in terms of standard capacitances and resistances. This can be done by Maxwell's $L/C$ bridge, which uses a network of the Wheatstone type (see Problem 22.3), but to make the two balance conditions independent of each other a standard variable capacitance is required. A modification of this bridge, due to Anderson, has the advantage that only variable resistances are required, together with a standard fixed capacitance. The circuit is shown in Fig. 22.7. The unknown self-inductance $L$, with resistance $r$, forms one arm of the bridge when placed in series with a variable resistance $S$. The fixed capacitance $C$ is in series with a variable resistance $T$, the combination

FIG. 22.7. Anderson's bridge for the measurement of self-inductance $L$ in terms of a standard capacitance $C$.

being shunted by a resistance $P$. The detector is connected from B to the junction E of $C$ and $T$, instead of to the point A. The balance condition is most readily found by calculating the voltages across FE and FB as fractions of the driving voltage $V$. The voltage across FE is a fraction $1/(1+j\omega CT)$ of that across FA, while that across FA is $V\mathbf{Z}/(\mathbf{Z}+Q)= V/(1+Q/\mathbf{Z})$, where $\mathbf{Z}$ is the total impedance between $F$ and $A$. Since

$$\frac{1}{\mathbf{Z}}=\frac{1}{P}+\frac{j\omega C}{1+j\omega CT},$$

the voltage across FE is

$$\frac{V}{\{(1+j\omega CT)(1+Q/\mathbf{Z})\}}=\frac{V}{\{(1+j\omega CT)(1+Q/P)+j\omega CQ\}},$$

while the voltage across FB is $VR/(R+j\omega L+r+S)$. Equating these voltages, we have

$$1+(j\omega L+r+S)/R = (1+j\omega CT)(1+Q/P)+j\omega CQ.$$

The real and imaginary parts of this equation give separately

$$(r+S)/R = Q/P$$

and

$$L/R = CT(1+Q/P)+CQ.$$

It is generally convenient to make $Q = P$, in which case the balance conditions reduce to

$$\begin{aligned} r &= R-S, \\ L &= CR(2T+P). \end{aligned} \tag{22.3}$$

It is obvious from this that no balance is possible unless $CRP < L$; if this condition is being violated it will be indicated by the fact that the nearest approach to balance is obtained when $T$ is zero. Inspection of the balance conditions shows also that they are independent of frequency, and of each other if $S$ and $T$ are made the variables.

### The transformer-ratio bridge

Standard capacitors and resistors can be constructed with a reproducible accuracy of about one part in $10^4$. This is sufficient for many purposes, but not for monitoring small changes, for automation procedures, or the maintenance of standards. For such purposes toroidal transformers are used; they can be wound with such precision that a given number of turns can be tapped off to give a voltage divider accurate to one part in $10^7$. By using such transformers, with accurately calibrated ratios, in two arms of a bridge network, the ratio of the impedances of the other two arms can

FIG. 22.8. The transformer ratio bridge.

be determined to a few parts in $10^7$. The use of a toroidal shape reduces the leakage of magnetic flux to extremely small values. The coils are wound on a special core, such as a tape of supermumetal, 25 $\mu$m thick (to reduce eddy currents), which has an initial relative permeability of about $7 \cdot 5 \times 10^4$.

A bridge of the simplest type is shown in Fig. 22.8. If $\mathbf{Z}_1$, $\mathbf{Z}_2$ are the impedances of the two sections of the secondary winding of the transformer, the balance condition is simply given by eqn (22.1). The transformer coils will always have some resistive loss, although this is kept as small as possible. We assume that the two sections have self-inductance $L_1$, $L_2$ and resistance $R_1$, $R_2$, and the mutual inductance between them is $M$. When the bridge is balanced, the current through each section is the same, so that

$$\frac{\mathbf{Z}_1}{\mathbf{Z}_2} = \frac{R_1 + j\omega(L_1 + M)}{R_2 + j\omega(L_2 + M)}. \tag{22.4}$$

If we can assume that the flux leakage is zero, then $M^2 = L_1 L_2$. The turns ratio is $N_1/N_2 = n$, and the coils are wound so that $L_1 = nM$ and $M = nL_2$; if also $R_1$ is made equal to $nR_2$, then $\mathbf{Z}_1/\mathbf{Z}_2 = n$. The bridge can be balanced either by adjusting one of the impedances $\mathbf{Z}_3$ or $\mathbf{Z}_4$, or by using a transformer with a number of tappings to give an adjustable turns ratio. For example, if $\mathbf{Z}_3$ and $\mathbf{Z}_4$ are capacities, one a standard, the other can be compared with it to a few parts in $10^7$; small changes in capacity can also be measured. Since $\mathbf{Z}_1/\mathbf{Z}_2$ is real and positive, only like impedances can be compared. However, no capacitor is completely free from loss, and the circuit of Fig. 22.9 can then be used to obtain a balance. The T-network between points A and D provides a small extra current in the right phase. By converting the T-network to a $\pi$-network, as in Problem 7.15, it can be shown that if $R_a C_a = R_b C_b$ the resistance between A and D is $(R_a + R_b)$, and the current flowing from A to D will be in quadrature

FIG. 22.9. Transformer ratio bridge with additional network to balance out-of-phase current if capacitance $C_3$ has some loss.

with that through the capacitance $C_3$. In practice the resistances $R_a$, $R_b$ are fixed, so that for a given voltage between A and D the current through the resistances is constant and independent of frequency. Balance is obtained by adjusting the turns ratio at the point A.

There are many varieties of transformer-ratio bridges, specially designed to meet different needs. At radio-frequencies great care is required to shield all components from electric and magnetic pick-up, particularly between the points of highest potential and earth. With correct design they can be used up to 100 MHz.

## 22.3. The Q-meter

The careful design and high precision needed in the manufacture of a bridge for use at radio-frequencies are unnecessary when an accuracy of the order of 1 per cent is sufficient. A cheaper and simpler alternative is the Q-meter, a general-purpose instrument whose basic circuit is shown in Fig. 22.10. A small current $I$ from an oscillator is read on a thermoammeter and then flows to a known low resistance $r$. A series tuned circuit is connected in parallel with $r$, and the voltage $V$ developed across the capacitance $C$ is read on a high-resistance voltmeter when the capacitor is adjusted for resonance. The latter is indicated by a maximum reading of the voltmeter. The $Q_F$ of the circuit under test is then equal to the ratio $V/Ir$, since $Ir$ is the voltage introduced in series with the tuned circuit. It is necessary that the resistance $r$ shall be small compared with the series resistance $R$ of the tuned circuit, in order that substantially all the current registered by the milliammeter shall flow through $r$. This requirement may be stated in another way: $r$ must be small compared with $R$ in order not

FIG. 22.10. *Q-meter circuit.*

to load the circuit under test. (It is easy to show that the measured $Q_F$ is that for a circuit whose total series resistance is the sum of $r$ and $R$.) In a commercial instrument an internal oscillator of calibrated variable frequency is provided; the current from it may be adjusted to bring the milliammeter reading always to a fixed mark, and an internal voltmeter is calibrated directly to read the $Q_F$ of an unknown coil. The variable capacitance $C$ is included in the instrument, and is calibrated so that the inductance of the unknown coil may be calculated from the known oscillator frequency and the tuning capacitance. An unknown capacitance may also be measured by the substitution method: a suitable coil $L$ is inserted and the reading of $C$ required to tune it to resonance with and without the unknown capacitor in parallel with $C$ is found. An unknown resistance can be measured by finding the effect on the $Q_F$ of a circuit when it is placed in series (if a low resistance) or in parallel (for a high resistance) with the tuned circuit.

## 22.4. Measurements on lines

At frequencies above about 100 MHz (wavelengths of 3 m and less) the leads to the impedance under test are not negligible in length compared with the wavelength and errors may be introduced because the current and voltage at the measuring instrument are not the same as those at the unknown impedance. These errors may be eliminated by making the leads part of a transmission line of known and constant impedance, the unknown impedance being placed at the end of this line and acting as its termination. The impedance $Z_0$ of the transmission line may be calculated from its dimensions (see § 9.3) and the unknown impedance is determined as a ratio to $Z_0$. This may be carried out either by determining the voltage standing-wave ratio (v.s.w.r.) on the line, or by a resonance method.

The first of these methods has the advantage that the results do not depend on the generator impedance, and the generator may therefore be connected directly to the line. From the theory of transmission lines (Chapter 9) it follows that the voltage at any point on the line may be regarded as due to an incident wave of amplitude $A$ and a wave reflected from the terminating impedance of amplitude $A_1$. The resultant voltage amplitude is a maximum $(A+A_1)$ at an anti-node where the incident and reflected waves are in phase, and a minimum $(A-A_1)$ where they are 180° out of phase, these points being a quarter of a wavelength apart. From a measurement of the v.s.w.r.

$$(A+A_1)/(A-A_1),$$

and the position of the nodes or anti-nodes, the ratio of the terminating impedance $\mathbf{Z}$ to the characteristic impedance $Z_0$ may be found using eqns (9.19):

$$\left.\begin{aligned}
\frac{A_1}{A} &= \left(\frac{Z_1^2+Z_0^2-2Z_1Z_0\cos\phi}{Z_1^2+Z_0^2+2Z_1Z_0\cos\phi}\right)^{\frac{1}{2}}, \\
\tan\delta &= \frac{2Z_1Z_0\sin\phi}{Z_1^2-Z_0^2},
\end{aligned}\right\}$$

where $\mathbf{Z}=Z_1\exp j\phi$, and $\delta$ is the difference in phase between the reflected and incident waves at the point of reflection (the termination of the line). This phase constant can be found from the position of a voltage node, this being generally more accurate than the location of an anti-node (especially if the v.s.w.r. is high) because the sensitivity of the detector can be increased as the node is approached. If the end of the line is at $x=0$, and the voltage node at a point $x=-l$, then the phase of the incident wave at this point is $-\omega(-l/v)=2\pi l/\lambda$, and that of the reflected wave is $\delta+\omega(-l/v)=\delta-2\pi l/\lambda$. For a node these must differ by $\pi$, whence $\delta=(4\pi l/\lambda)+\pi$. To determine $l$ accurately, it is best to replace the unknown impedance $\mathbf{Z}$ by a short-circuit and find the distance between the previous node and the new one; the latter is (electrically) exactly an integral number of half-wavelengths from the end of the line.

The v.s.w.r. can be measured by moving any loosely coupled voltage detector along the line. Since only a ratio of the maximum and minimum readings is required, the absolute calibration of the indicator is unnecessary, and a knowledge of the rectifying characteristic (d.c. current or voltage output against r.f. voltage input) is sufficient. With a coaxial line, a section of air-spaced line is made up with known dimensions, and a narrow slot is cut lengthwise along the outer conductor. Since this slot is parallel to the direction of current flow in the line, it does not disturb conditions on the line materially. In this slot (see Fig. 22.11) is inserted a small radial probe, which is parallel to the lines of electric field inside the coaxial line; it picks up a small fraction of the voltage on the line and feeds it to a detector. The intrusion of the probe is made as small as possible to minimize disturbance on the line, and for this purpose a sensitive detector is required to obtain adequate sensitivity. The probe is mounted on a movable carriage, carefully machined so that the intrusion

FIG. 22.11. Standing-wave detector on coaxial line.

of the probe does not change as it moves along. This may be checked by observing the constancy of the detector reading when the line is terminated by its characteristic impedance, when the v.s.w.r. should be unity. With a parallel-wire line a similar arrangement may be used with a probe near the wires, but the indicator and its leads must be shielded and kept well away from the line, since the electric and magnetic fields around the line are not now rigorously confined as they are in the coaxial line.

From the measurement of the v.s.w.r. and of $\delta$, the ratio of the real and imaginary parts of $\mathbf{Z}$ to $Z_0$ can be found from eqns (9.19) (see above), but these are algebraically so clumsy to handle that graphical methods are normally employed. 'Impedance diagrams' can be obtained from which the real and imaginary parts of $\mathbf{Z}/Z_0$ can be read off at once.

When the impedance to be measured has only a small dissipative component the standing-wave ratio becomes very large and is difficult to measure accurately, principally because the detector law must be known over a wide range. It is often then more convenient to use a resonance method. The unknown impedance is connected across the end of a line as in Fig. 22.12, and an oscillator and detector are loosely coupled to it. A movable bridge, which should make such good contact as to be essentially a short-circuit, is adjusted until resonance is indicated by maximum deflection of the detector. If the load $\mathbf{Z}$ is represented by a resistance $R$ in parallel with a reactance $jX$, then resonance occurs when this reactance is equal and opposite to the line reactance. If the length of the line at this point is $l$, then the line reactance is

$$jX' = -jX = jZ_0 \tan 2\pi l/\lambda,$$

which may be positive or negative according to the value of $l/\lambda$. Thus the value of $X$ is determined from the resonant length. The value of $R$ may be found by measuring the sharpness of resonance; for example, by

Loose coupling from oscillator

Movable short-circuit—

$R$    $jX$

$l$

Loose coupling to calibrated detector

FIG. 22.12. Measurement of impedance using resonance method on transmission line.

varying the length of the line until the detector reading shows that the voltage (or current) on the line has fallen to $1/\sqrt{2}$ of the maximum. At this point the susceptance formed by $-X^{-1}$ in parallel with $Z_0^{-1} \cot 2\pi(l\pm\delta l)/\lambda$ has risen from zero to be just equal to $1/R$ (cf. the theory of the parallel tuned circuit in § 7.3). If the change in length is $\delta l$, then the value of the susceptance is

$$\left| -X^{-1} + Z_0^{-1} \cot \frac{2\pi(l\pm\delta l)}{\lambda} \right| = Z_0^{-1}\left(\frac{2\pi\delta l}{\lambda}\right)\operatorname{cosec}^2 \frac{2\pi l}{\lambda} = \frac{1}{R},$$

where $R$ can be determined. If loss in the line cannot be neglected, as has been assumed here, then it can be found by a separate measurement with the line short-circuited at both ends (or open-circuited at one end) and a correction applied. The calculation is rather complicated, but from eqn (9.28) it can be seen that the length of line can be represented by a complex admittance $\mathbf{Y}'$ which is in parallel with $1/\mathbf{Z}$. If $\mathbf{Y}'$ is separated into its real and imaginary parts $G'$ and $S'$, then the calculation proceeds as before. As the resonant lengths of line unloaded and terminated by $\mathbf{Z}$ will be different, the loss on the line must be expressed in terms of the attenuation coefficient $\alpha$ (eqn (9.24)).

The advantage of the resonance method over the v.s.w.r. method is that the detector law need only be known over a small range, the other measurements being those of lengths. Both types of measurements may also be used with waveguides at centimetre wavelengths, though here the concept of a lumped impedance loses most of its meaning.

## 22.5. Measurement of frequency and wavelength

The measurement of an audio-frequency can be made in terms of known impedances by the use of a bridge whose balance is dependent on frequency. It is obvious that the oscillation to be measured must be constant in frequency and free from harmonics, since the latter would be out of balance in the bridge (two uses of a frequency bridge are the suppression of a given frequency such as a troublesome harmonic and the analysis of harmonic content). A large number of bridges have been devised which satisfy the desiderata that the balance conditions should be mutually independent and only one of them should depend on the frequency. A simple bridge using only resistances and capacitances is due to Wien and is shown in Fig. 22.13. The balance conditions are

$$\omega^2 = \frac{1}{RSC_1C_2},$$
$$\frac{C_1}{C_2} = \frac{Q}{P} - \frac{S}{R}, \tag{22.5}$$

which are mutually independent if the variables $R$, $S$ and also $C_1$, $C_2$ are ganged so that their ratio is kept constant. Then the second condition will remain satisfied once it has been set up and measurement of a wide range of frequencies is obtained by a single adjustment.

At wavelengths less than 1 or 2 m the measurement of wavelength directly becomes quite convenient, and high accuracy may be attained because of the high $Q_F$ of resonant transmission lines and waveguide cavities. Parallel-wire lines may be used for the longer wavelengths, but coaxial lines are better at wavelengths of 0·1 m to 1 m, and cavity resonators at centimetre wavelengths. A simple type of coaxial-line

FIG. 22.13. Wien's bridge for the measurement of frequency.

FIG. 22.14. Coaxial-line wavemeter and equivalent circuit. Note that $L$, $C$, $r$ all vary with the length of the coaxial line.

wavemeter is shown in Fig. 22.14. The centre conductor is variable in length, and moves through a spring contact which forms the closed end of the line. Power is introduced by means of a small loop which intersects some of the magnetic lines of force at the closed end, and a second loop takes power to a detector (usually a diode rectifier). These loops must be kept small to give loose coupling, and to avoid pulling the oscillator whose wavelength is to be measured through coupled-circuit effects. The equivalent circuit of the wavemeter is shown in Fig. 22.14, from which it can be seen that the detector reading is a maximum when the line is resonant. Since the loops introduce small impedances which alter the electrical length of the line, it is preferable to measure successive points of resonance, which are exactly half-wavelengths apart. The wavemeter then needs no calibration, the wavelength being found directly from a scale and vernier attached to the moving part. The accuracy is usually about one part in $10^3$, the main difficulty being in making a good contact between the moving conductor and the stationary end.

A basic type of cavity wavemeter is shown in Fig. 22.15. A section of circular waveguide is closed at one end, the other end being formed by a plunger driven by a micrometer head. Power is fed into the cavity from a waveguide through a small hole, and resonance is detected by coupling a little power out through a second hole to a detector. The holes should be kept as small as possible, subject to getting a finite detector reading, in order to avoid lowering the natural $Q_F$ of the resonator, which may be of the order of $10^4$. The wavelength in the cavity is found from the distance $\lambda_g/2$ between successive resonance points. The wavelength in free space may be calculated from the diameter of the cavity, and the mode of resonance. To avoid the difficulty of making a good contact between the moving plunger and the walls, a particular waveguide mode (TE$_{01}$ or $H_{01}$)

FIG. 22.15. Resonant-cavity wavemeter.

is often used, where there is no current flow across this contact (see § 9.7). Other modes of resonance may then also be present, since the $TE_{01}$ mode needs rather a large cavity diameter (the cut-off wavelength is equal to $0.82$ times the diameter). These may be avoided by using special arrangements of the coupling holes (see Bleaney, Loubser, and Penrose 1947). An accuracy of one or two parts in $10^4$ may be attained, but a correction is then needed for the refractive index $(1.0003)$ of the air in the cavity.

*The quartz-crystal oscillator*

Where high accuracy of frequency control or measurement is required, use is made of the properties of a piezoelectric crystal, quartz being the most satisfactory for this purpose. A quartz crystal grows in the form of a hexagonal prism with pointed ends, the cross-section of the prism being as shown in Fig. 22.16(a). If an electric field is applied to the crystal in the $X$ direction, the crystal contracts or elongates in the $Y$ direction according to the sign of the electric field. Similarly, if a mechanical stress is applied in the $Y$ direction, an electric polarization is set up in the $X$ direction and charges appear on the faces of the crystal. These effects are reversible and are very nearly linearly related; their importance lies in the fact that they couple together an electrical and a mechanical system. If an alternating voltage is applied in the $X$ direction, an alternating stress appears in the $Y$ direction and the amplitude of the resulting mechanical

(a)                                    (b)

FIG. 22.16. (a) Quartz crystal ($x$, $y$ are one of the three pairs of $X,Y$-axes). (b) Equivalent circuit of crystal and its electrodes.

vibrations is large if the frequency of alternation coincides with a natural mechanical vibration of the crystal. A number of different modes of oscillation exist, but those most commonly used are longitudinal and shear vibrations. The damping of the mechanical vibrations is very low, and the sharpness of the resonance makes them very suitable for use as frequency standards. The desirable properties for this purpose are:

(1) zero temperature coefficient of frequency of oscillation;

(2) high piezoelectric effect;

(3) a single mode of mechanical resonance well separated in frequency from other modes, so that there is no tendency to jump from one mode to another.

The greatest piezoelectric effect is obtained when the electrical and mechanical stresses are applied along the electrical or $X$-axis and mechanical or $Y$-axis respectively, but oscillations can be excited by any stress which has a component parallel to these axes. (Note that, owing to the high symmetry of the crystal, there are three sets of $X$- and $Y$-axes, related to one another by rotations of $120°$ and $240°$ about the $Z$-axis, the optic axis of the crystal.) An $X$-cut crystal consists of a thin slab with faces parallel to the $YZ$ plane, and the temperature coefficient of the frequency of oscillation is negative, about $-22 \times 10^{-6} \, K^{-1}$. A $Y$-cut crystal is a thin slab with its faces parallel to the $XZ$ plane, and the temperature coefficient is positive with a number of discontinuities due to couplings between different modes of oscillation. In general it is more important to obtain zero temperature coefficient than high piezoelectric activity, and intermediate cuts are used such as the AT-cut, a thin plate whose faces contain the $X$-axis and a line in the $YZ$ plane making an angle of about $35 \cdot 5°$ with the $Z$-axis. When a voltage is applied between the large faces, a shear vibration is set up whose frequency in megahertz is $1 \cdot 675/$(thickness in millimetres). This is suitable for frequencies from roughly $0 \cdot 5 \, MHz$ to $10 \, MHz$. Lower frequencies may be obtained from modes where the frequency is determined by one of the long dimensions of the slab, the full range of quartz crystals being roughly from $25 \, kHz$ to $15 \, MHz$. To apply the alternating voltage the slab is mounted between the plates of a capacitor, generally formed by sputtering a metallic film on to the large faces. This minimizes the loading on the mechanical vibrations, which is further reduced by mounting the crystal *in vacuo* between light supports touching the crystal at a mechanical node. For the highest frequency stability the crystal is kept in an oven thermostatically controlled to $0 \cdot 1 \, K$ or better, because the temperature coefficient is zero only over a narrow range of temperature.

The mechanical system of a quartz crystal may be represented by the equivalent electrical circuit shown in Fig. 22.16(b). The mechanical resonance is equivalent to a series tuned circuit and this is shunted by the capacitance $C_1$ of the electrodes. Typical values are:

$X$-cut quartz (lengthwise vibration)
  *Dimensions*: rectangular bar,

$$X = 1{\cdot}4\,\text{mm}, \quad Y = 30{\cdot}7\,\text{mm}, \quad Z = 4{\cdot}1\,\text{mm}$$

$$R = 15\,\text{k}\Omega \qquad\qquad C_1 = 3{\cdot}54\,\text{pF}$$

$$L = 137\,\text{H} \qquad\qquad Q_F = 5150$$

$$C = 0{\cdot}0228\,\text{pF} \qquad\quad f_0 = 89{\cdot}87\,\text{kHz}$$

AT-cut quartz

  *Dimensions*: disc, 25 mm diameter, thickness $1{\cdot}10$ mm

$$R = 24.2\,\Omega \qquad\qquad C_1 = 17.9\,\text{pF}$$

$$L = 0.119\,\text{H} \qquad\quad Q_F = 46500$$

$$C = 0{\cdot}0945\,\text{pF} \qquad\quad f_0 = 1500\,\text{kHz}$$

  (From **Cady** 1946)

The presence of $C_1$ causes the circuit to behave as a parallel tuned circuit at a frequency just above that of the series resonance (see Problem 22.6). The difference between these two frequencies is very small so that the phase angle of the circuit varies very rapidly. A simple circuit (due to Pierce) for maintaining the crystal in oscillation is shown in Fig. 22.17. Energy is fed back to the gate of an FET from the drain circuit through the quartz crystal; the feedback is largest when the crystal has a low impedance near the series resonance frequency. Here the equivalent

FIG. 22.17. Pierce circuit for quartz-crystal oscillator, using an FET. Feedback from drain to gate circuit takes place through the crystal at the frequency where it is a low impedance.

electrical circuit of the crystal is inductive; the condition for oscillation (gate and drain in antiphase) then requires that both the tuned circuit in the drain lead and the gain circuit should be capacitive. In such a circuit an FET with its high input impedance is better than a junction transistor because it produces less loading on the quartz crystal, improving the $Q_F$ and the frequency stability.

*Frequency standards*

The primary standard of frequency is a hyperfine transition of the caesium atom (see § 24.6), which since 1964 has been defined to be exactly

$$9\ 192\ 631\ 770\ \text{Hz.}$$

This is a more convenient and a more precise standard than previous ones based on the mean length of the solar day or year, because the motion of the earth is known to be subject to fluctuations. The excellent frequency stability of the quartz-crystal oscillator makes it extremely useful as a secondary standard in the convenient frequency region of about 100 kHz to 10 MHz, which can be extended upwards by multiplication and downwards by subdivision to control a clock. In one system of intercomparison with the caesium standard, the output from a quartz-crystal oscillator close to 5 MHz in frequency is multiplied by electronic circuits to give a harmonic within 10 Hz of the caesium frequency, and the difference frequency is fed to a counting circuit. In this way the frequency of radio transmitters can be linked to the caesium standard, and already by 1960 standard frequency transmissions could be compared internationally and shown to be consistent to one part in $10^{11}$.

For most laboratory measurements of frequency a digital instrument is both precise and convenient. An internal crystal oscillator is used to control a 'gate' which can be opened for precise lengths of time varying from say 1 $\mu$s to 10 s. The number of cycles which occur while the gate is open, converted into pulses, is counted electronically.

## 22.6. Measurement of electric permittivity

The relative permittivity $\epsilon_r$ of a substance affords some valuable information as to the structure of its constituent molecules (see Chapter 10), and its accurate measurement is therefore of some importance. Since $\epsilon_r$ is defined by the ratio of the capacitance of a capacitor filled with the substance under test to that of the empty capacitor, it is obvious that in general two measurements of capacitance will suffice. For solids and liquids the relative permittivity varies from about 2 to 100, and any of the bridges designed to measure capacitance may be used to give accurate results. For high values of $\epsilon_r$ care must be taken to avoid stray capacitance which may seriously affect the reading obtained with the empty capacitor, if this has a rather small capacitance.

For gases the value of $\epsilon_r$ differs from unity only by about $10^{-3}$ at atmospheric pressure, and an accurate bridge such as the transformer-ratio bridge is needed to measure the difference with reasonable precision. At centimetre wavelengths Essen and Froome (1951) used the high $Q_F$ of a resonant cavity to obtain accurate measurements. The cavity was cylindrical, with a diameter of about 0·05 m, and resonated in the $TE_{01}$ mode at about 24 GHz. The frequency of resonance was determined first with the cavity evacuated, and then filled with gas, by plotting out the resonance curve using a klystron oscillator whose frequency could be determined to one part in $10^8$ by comparison with the N.P.L. frequency standard. The frequency of resonance is given by eqn (9.34)

$$\frac{f^2}{v^2} = \frac{\mu_r \epsilon_r f^2}{c^2} = \frac{1}{\lambda_c^2} + \frac{1}{\lambda_g^2},$$

where $\mu_r$ and $\epsilon_r$ are the relative permeability and relative permittivity of the gas filling the resonator and $\lambda_c$, $\lambda_g$ are fixed by the diameter and length of the cavity respectively. Thus if $f'$ is the resonant frequency of the empty cavity, and $f''$ that of the gas-filled cavity, $(f'/f'')^2 = \mu_r \epsilon_r$. Hence the ratio of the two frequencies determines $n = (\mu_r \epsilon_r)^{\frac{1}{2}}$, the refractive index of the gas. A correction must be applied for the permeability, which differs slightly from unity for air and oxygen, since the latter is paramagnetic. A comparison of the measurements of $\epsilon_r$ at different frequencies, together with the square of the optical refractive index, is given in Table 10.4 (p. 306).

The cavity resonance method may also be used for measurement of the permittivity of liquids and solids, provided that their loss tangent is fairly small (see Faraday Society Conference on Dielectrics 1946). A partly filled cavity must be used for solids or liquids of high loss tangent, but for non-polar liquids a filled cavity was employed by Bleaney, Loubser, and Penrose (1947). A tunable cavity resonant in the $TE_{01}$ mode of the same type as described earlier (§ 22.5) was adjusted to resonance with a klystron oscillator of fixed frequency, first with the cavity empty, and then filled with liquid. By measuring a number of successive resonant points, the wavelength in the guide was found in each case, and the relative permittivity calculated from the equations

$$\frac{\epsilon_a}{\lambda^2} = \frac{1}{\lambda_a^2} + \frac{1}{\lambda_c^2}, \qquad \frac{\epsilon_r}{\lambda^2} = \frac{1}{\lambda_d^2} + \frac{1}{\lambda_c^2}, \tag{22.6}$$

where $\epsilon_a$ is the relative permittivity of air and $\epsilon_r$ that of the liquid, $\lambda$ the wavelength in free space, and $\lambda_a$, $\lambda_d$ the wavelengths in the air- and liquid-filled cavity respectively. The loss tangent of the liquid was found from the width of the resonance curve determined by detuning the cavity. Thus only measurements of length, depending on a micrometer thread, were involved. Typical results at a temperature of 293 K are given in Table 22.1. When two measurements are given at 13·5 mm wavelength, they were made with cavities of different diameter.

TABLE 22.1

| | Relative permittivity $\epsilon_r$ | | Loss tangent $(\tan \delta)$ | |
|---|---|---|---|---|
| | $\lambda = 32$ mm | $\lambda = 13\cdot5$ mm | $\lambda = 32$ mm | $\lambda = 13.5$ mm |
| Cyclohexane | 2·0244 | 2·0246, 2·0251 | 0·00005 | 0·00019 |
| n-Heptane | 1·9220 | 1·9223 | 0·00037 | 0·00076 |
| n-Hexane | 1·9016 | 1·9016 | 0·00034 | 0·00076 |
| $CS_2$ | 2·6476 | 2·6477 | 0·00024 | 0·00072 |
| $CCl_4$ | 2·2386 | 2·2390 | 0·00031 | 0·00078 |

All the samples except those of n-hexane and $CCl_4$ were specially purified. The loss tangent is considerably affected by small traces of polar impurities, but it is not certain that such impurities would account for all the dielectric loss.

## 22.7. Measurement of the velocity of radio waves

The velocity of electromagnetic radiation has long been regarded as one of the fundamental constants of physics, and much effort has been devoted to its accurate determination. Apart from one measurement of the velocity of radio waves on a transmission line by Mercier (1924), most of the early work used light waves. The results showed a good deal of scatter; since 1945 a number of new determinations have been made, of greater accuracy, which show that the true value is nearly 299 793 km s$^{-1}$. These methods have made use of radio techniques to improve the accuracy, and in some cases the wavelength of radiation used has been a few centimetres.

In § 22.5 it was pointed out that both frequency and wavelength can be measured at centimetre wavelengths. The product of these two quantities gives the phase velocity, and this has been the basis of a number of modern measurements. Frequency can be determined with an accuracy of one part in $10^{12}$ (see Bradley, Edwards, Knight, Rowley, and Woods 1972), and the main problem is the determination of wavelength. Resonant cavities have been used at centimetre wavelengths (Essen and Gordon-Smith 1948; Essen 1950), but more accurate results have been obtained by interferometer experiments, which approximate closely to a free-space method. Froome (1952) has used a microwave analogue of the

FIG. 22.18. Froome's microwave Michelson interferometer.

Michelson interferometer, as shown in Fig. 22.18. Power from a stabilized klystron oscillator flowing along a waveguide was divided into two portions at a hybrid junction B (the analogue of a half-silvered plate). One half traversed a short length of waveguide and was reflected from a shorting piston. The other was fed to a horn and launched as a wave in space. Part of this radiation was reflected back to the horn by a 0·15 m square metal plate M which could be placed at points from 6·5 m to 21·5 m away. This reflected wave on returning to the hybrid junction interferes with that reflected from the shorting piston in the second arm, and the vector sum of the two amplitudes is passed along the fourth arm to a detector (a superheterodyne receiver). The latter is used to detect when the two reflected waves are exactly in antiphase and so give a null at the detector. The metal plate M is then moved through successive null points, which occur every half-wavelength. The total distance moved was 1·62 m, and this could be measured with an accuracy of ±0·003 mm. At the same time the frequency of the klystron oscillator was measured against the quartz-crystal standard with an accuracy of one part in $10^8$. Thus the wavelength in air and the frequency were determined simultaneously. In the former case two important corrections must be applied to find the wavelength *in vacuo*:

(1) a correction for the refractive index of the air, based on the measurements of Essen and Froome (see § 22.6);

(2) a correction for the fact that the wave front reaching the mirror is not a plane, but has a small curvature, and similarly for the

reflected wave; this correction was calculated from diffraction theory, using data from different mirror distances.

The final value obtained for the velocity *in vacuo* was $(299\ 792 \cdot 6 \pm 0 \cdot 7)\ \text{km s}^{-1}$. In later experiments Froome (1954, 1958) has used a four-horn interferometer of symmetrical design, first at a wavelength of $12 \cdot 5$ mm, then at 4 mm. The final results are

$$(299\ 792 \cdot 75 \pm 0 \cdot 3)\ \text{km s}^{-1} \quad \text{(frequency 24 GHz)},$$

$$(299\ 792 \cdot 5 \pm 0 \cdot 1)\ \text{km s}^{-1} \quad \text{(frequency 72 GHz)}.$$

The latter was still the accepted value in 1972, but the accuracy has since improved by a factor of about 100. Starting from a klystron oscillator, harmonics are generated by means of point-contact diodes or Josephson junctions, and the frequencies of a sequence of laser lines such as

$$HCN \quad f = \quad\ \ 890\ \text{GHz} \qquad \lambda = 333 \quad \mu\text{m}$$

$$H_2O \quad f = 10\ 700\ \text{GHz} \qquad \lambda = \ \ 28 \quad \mu\text{m}$$

$$CO_2 \quad f = 32\ 176\ \text{GHz} \qquad \lambda = \ \ \ 9 \cdot 3\ \mu\text{m}$$

are determined in terms of the caesium standard. A helium–neon laser line is then used as an intermediate step in an interferometric measurement of wavelength in terms of the krypton line at $\lambda = 0 \cdot 606\ \mu\text{m}$, adopted in 1960 as the primary standard of length. Essentially both the frequency and the wavelength of one stabilized laser line are measured, and the values found for $c$ are

$$(299\ 792 \cdot 4562 \pm 0 \cdot 0011)\ \text{km s}^{-1} \quad \text{(Evenson } et\ al.\ 1972)$$

$$(299\ 792 \cdot 4590 \pm 0 \cdot 0008)\ \text{km s}^{-1} \quad \text{(Blaney } et\ al.\ 1974).$$

## References

BLANEY, T. G., BRADLEY, C. C., EDWARDS, G. J., JOLLIFFE, B. W., KNIGHT, D. J. E., ROWLEY, W. R. C., SHOTTON, K. C., and WOODS, P. T. (1974). *Nature, Lond.* **251,** 46.

BLEANEY, B., LOUBSER, J. H. N., and PENROSE, R. P. (1947). *Proc. phys. Soc.* **59,** 185.

BRADLEY, C. C., EDWARDS, G. J., KNIGHT, D. J. E., ROWLEY, W. R. C., and WOODS, P. T. (1972). *Phys. Bull.* **23,** 15.

CADY, W. G. (1946). *Piezoelectricity,* McGraw-Hill, New York.

ESSEN, L. (1950). *Proc. R. Soc.* A**204,** 260.

—— FROOME, K. D. (1951). *Proc. phys. Soc.* B**64,** 862.

—— GORDON-SMITH, A. C. (1948). *Proc. R. Soc.* A**194,** 348.

EVENSON, K. M., WELLS, J. S., PETERSEN, F. R., DANIELSON, B. L., DAY, G. W., BARGER, R. L., and HALL, J. L. (1972). *Phys. Rev. Lett.* **29,** 1346.

FARADAY SOCIETY CONFERENCE ON DIELECTRICS (1946). *Trans. Faraday Soc.* **42A.**

FROOME, K. D. (1952). *Proc. R. Soc.* A**213**, 123.
—— (1954). *Proc. R. Soc.* A**223**, 195.
—— (1958). *Proc. R. Soc.* A**247**, 109.
MERCIER, J. (1924), *J. Phys. Radium* **5**, 168.

## Problems

22.1. A cathode-ray tube has plane parallel deflecting plates of separation $a$ and length $b$ parallel to the axis of the tube; the distance from the centre of the plates to the screen is $L$. If the electrons are initially accelerated by a voltage $V_0$, show that their deflection on the screen due to a voltage $V_1$ on the deflector plates is

$$s = \tfrac{1}{2}(LbV_1/aV_0),$$

assuming that $L \gg b$, that the field is uniform between the plates, and that edge effects can be neglected.

If $a = 5$ mm, $b = 40$ mm, $L = 0.3$ m, and $V_0 = 1.3$ kV, show that the deflection sensitivity is $0.92$ mm V$^{-1}$.

22.2. If the frequency limit of the cathode-ray tube of Problem 22.1 were set by the finite transit time of the electrons through the deflector plates, show that the deflection would fall to zero at about $540$ MHz.

22.3. Maxwell's bridge for comparing an inductance and a capacitor has the circuit of Fig. 22.5, with the following impedances:

$\mathbf{Z}_1$ an inductance $L$ in series with a resistance $R_1$,

$\mathbf{Z}_2$ a resistance $R_2$,

$\mathbf{Z}_3$ a resistance $R_3$,

$\mathbf{Z}_4$ a capacitance $C$ is parallel with a resistance $R_4$.

Show that the balance conditions are

$$R_1/R_3 = R_2/R_4, \qquad L = R_2R_3C.$$

To make the two balance conditions independent, $R_4$ and $C$ must be varied.

22.4. At higher audio-frequencies, resistances may possess a small inductive component; in Anderson's bridge this may be allowed for by writing the components as $\mathbf{P} = P + jP'$, $\mathbf{Q} = Q + jQ'$, $\mathbf{R} = R + jR'$, $\mathbf{T} = T + jT'$ (we neglect any inductive component in $S$ as this will be added to $L$ at all frequencies). If $\mathbf{P}$ and $\mathbf{Q}$ are identical impedances, show that the balance conditions are

$$r + S = R - \omega C(2RT' + 2R'T + RQ' + R'Q),$$

$$L = C(2RT + QR - 2R'T' - Q'R') + R'/\omega.$$

These equations show that it is important to make $R'$ as small as possible. If $R' = 0$, the error in the determination of $L$ is zero, while that in the resistance $r$ of the inductance is $\omega CR(2T' + Q')$.

22.5. In the equivalent circuit (Fig. 22.14) of the coaxial-line wavemeter, the source is taken to be a generator of voltage $V_1$ with internal resistance $R_1$, and the detector has a resistance $R_2$. If the series impedance of the tuned circuit by itself is $\mathbf{Z}$, show that the ratio of the voltage $V_2$ across the detector to the input voltage is

$$\frac{V_2}{V_1} = -\frac{\omega^2 M_1 M_2}{R_1(\mathbf{Z} + \omega^2 M_1^2/R_1 + \omega^2 M_2^2/R_2)}.$$

This equation shows that $V_2$ is a maximum when $\mathbf{Z}$ is a minimum, that is, when the wavemeter is on tune and $\mathbf{Z}$ is just the resistance $r$. It shows also that the coupled impedances $\omega^2 M_1^2/R_1$ and $\omega^2 M_2^2/R_2$ lower the effective $Q_F$; by writing $\mathbf{Z} = r + 2\mathrm{j}\,\delta\omega L$ near resonance, show that the

$$\text{'loaded } Q_F' = \frac{L^{\frac{1}{2}}}{C^{\frac{1}{2}}(r + \omega^2 M_1^2/R_1 + \omega^2 M_2^2/R_2)},$$

and that it may be measured by finding the fractional change in the frequency required to reduce $V_2$ to $1/\sqrt{2}$ of its maximum value (neglect changes in the coupled impedance when varying $\omega$).

22.6. In the equivalent circuit (Fig. 22.16(b)) of a quartz crystal, the components for a particular crystal are $L = 3\cdot3$ H, $C = 0\cdot042$ pF, $R = 4500\ \Omega$, $C_1 = 5\cdot8$ pF. Show that it behaves as a parallel resonant circuit at a frequency approximately 8 Hz above the series resonance frequency (the natural mechanical resonance frequency).

# 23. Fluctuations and noise

## 23.1. Brownian motion and fluctuations

THE irregular motion of small particles suspended in a fluid was first observed by Brown in 1828. This 'Brownian motion' never ceases and is a result of the random motion of the molecules both of the particles themselves and of the fluid. If the motion is observed over a long time, it is found that the average component of the velocity in any direction is zero, since positive and negative values occur with equal probability. The mean-square value of the velocity is not zero, and from classical statistical mechanics it may be shown that the average value of each of the terms $\frac{1}{2}m\dot{x}^2$, $\frac{1}{2}m\dot{y}^2$, $\frac{1}{2}m\dot{z}^2$ of the translational kinetic energy is $\frac{1}{2}kT$, where $k$ is Boltzmann's constant (approximately $1\cdot38\times10^{-23}$ J K$^{-1}$) and $T$ is the temperature in degrees Kelvin. This is a special case of the theorem of equipartition of energy: if the energy of a system can be written as the sum of a number of terms each containing only the square of a variable, then the average energy of each of these terms is $\frac{1}{2}kT$. This theorem applies just as much to macroscopic objects as to microscopic ones or molecules, but the magnitude of the fluctuations in the dynamical variable becomes smaller as the inertia of the object increases, since the average energy is independent of size. Given sufficient magnification, the motion can always be observed, and it sets a limit to the sensitivity of any measuring instrument; the fluctuations give a random signal which masks any applied signal of smaller magnitude.

If this theorem is applied to a suspension galvanometer, the following result is obtained. The suspension has one degree of freedom, a rotation measured by the angle $\theta$. The total energy may be written as the sum of two terms, the potential energy of the suspension due to work done against the restoring torque, and the kinetic energy, so that

$$W = \tfrac{1}{2}c\theta^2 + \tfrac{1}{2}\Im\dot{\theta}^2, \tag{23.1}$$

where $c$ is the restoring torque per unit angle of twist and $\Im$ is the moment of inertia of the system. To each of these terms we must assign an average energy $\frac{1}{2}kT$, so that fluctuations in the angle $\theta$ and the angular velocity $\dot{\theta}$ will occur whose mean-square values are given by

$$\tfrac{1}{2}c\overline{\theta^2} = \tfrac{1}{2}\Im\overline{\dot{\theta}^2} = \tfrac{1}{2}kT. \tag{23.2}$$

A system which is mathematically similar is the electrical tuned circuit, consisting of an inductance, capacitance, and resistance connected together. The total electrical energy of such a system, where $I$ is the instantaneous current and $q$ the instantaneous charge on the capacitor, is

$$W = \tfrac{1}{2}q^2/C + \tfrac{1}{2}LI^2. \tag{23.3}$$

If the theorem of equipartition of energy applies also to electrical systems, as we would expect in view of its general nature, then the mean-square values of the fluctuating charge and current will be given by

$$\tfrac{1}{2}\overline{q^2}/C = \tfrac{1}{2}L\overline{I^2} = \tfrac{1}{2}kT. \tag{23.4}$$

These relations give only the mean-square values of the total fluctuations, and tell us nothing about the frequency distribution of the fluctuations. If we imagine that we perform a Fourier analysis of the fluctuations, and postulate that they are due to some random force acting on the system, then for the electrical tuned circuit we write

$$L(d^2q/dt^2) + R(dq/dt) + q/C = V_f \exp j\omega t, \tag{23.5}$$

where $V_f$ is the amplitude of the component of the random e.m.f. causing the fluctuations at the frequency $f = \omega/2\pi$. We now make the following assumptions about $V_f$: its mean-square value $\overline{V_f^2}$ is independent of frequency, but voltages of different frequency are entirely uncorrelated, so that the average value of the product $\overline{V_f V_{f'}}$ is zero. The justification for these assumptions will not be discussed here, but it is obvious that they are plausible in view of the random nature of the fluctuations. On solving eqn (23.5) to find the mean-square amplitude $\overline{q_f^2}$ of the fluctuating charge at the frequency $f$, we have

$$d(\overline{q_f^2}) = \frac{d(\overline{V_f^2})}{(L\omega^2 - 1/C)^2 + R^2\omega^2}. \tag{23.6}$$

The frequencies are continuously distributed, and the differentials are used since this expression gives the mean-square amplitude of the fluctuations in the frequency range between $f$ and $f+df$. The total mean-square fluctuation must be given by eqn (23.4), and hence, integrating over all frequencies, we must have

$$\tfrac{1}{2}kT = \frac{\tfrac{1}{2}\overline{q^2}}{C} = \frac{1}{2C}\int d(\overline{q_f^2}) = \frac{1}{4\pi C}\frac{d(\overline{V_f^2})}{df}\int_0^\infty \frac{d\omega}{(L\omega^2 - 1/C)^2 + R^2\omega^2}.$$

This integral may be evaluated as follows. On making the substitution $\omega = x(LC)^{-\frac{1}{2}}$, it becomes

$$\left(\frac{C^3}{L}\right)^{\frac{1}{2}}\int_0^\infty \frac{-d(1/x)}{(x-1/x)^2 + R^2C/L} = \left(\frac{C^3}{L}\right)^{\frac{1}{2}}\int_0^\infty \frac{dx}{(x-1/x)^2 + R^2C/L},$$

where the second form is obtained by replacing $x$ by $1/x$. Hence the integral may be written as

$$\frac{1}{2}\left(\frac{C^3}{L}\right)^{\frac{1}{2}} \int_0^\infty \frac{d(x-1/x)}{(x-1/x)^2+R^2C/L} = \frac{1}{2}\left(\frac{C^3}{L}\right)^{\frac{1}{2}} \int_{-\infty}^\infty \frac{dz}{z^2+R^2C/L} = \frac{\pi C}{2R}.$$

Hence

$$\tfrac{1}{2}kT = \frac{\tfrac{1}{2}\overline{q^2}}{C} = \frac{1}{8R}\frac{d(\overline{V_f^2})}{df},$$

or

$$d(\overline{V_f^2}) = 4kTR\,df, \tag{23.7}$$

and

$$d(\overline{q_f^2}) = \frac{4kTR\,df}{(L\omega^2-1/C)^2+R^2\omega^2}. \tag{23.8}$$

It is easily verified (see Problem 23.1) from these results that $\tfrac{1}{2}L\overline{I^2}=\tfrac{1}{2}kT$ in accordance with eqn (23.4). The equations lead to the interesting result that, whereas the total mean-square values of the fluctuations depend only on $L$ and $C$, the expression for the distribution of the voltage fluctuations with frequency involves only $R$. The result given by eqn (23.7) may be expressed by saying that the mean-square voltage $d(\overline{V_f^2})$ of the fluctuations in the frequency range $df$ is $4kTR\,df$, and is thus proportional to the bandwidth $df$. The existence of such fluctuations was first verified by Johnson, and they are known as resistance or 'Johnson' noise. They will be considered in more detail in § 23.3.

## 23.2. Fluctuations in galvanometers

We return now to the case of the galvanometer, and consider first a moving-coil suspension galvanometer when the coil is on open-circuit. Then the equation of motion is

$$\Im(d^2\theta/dt^2)+b(d\theta/dt)+c\theta = F_f \exp j\omega t, \tag{23.9}$$

where $\Im$ is the moment of inertia, $b$ the mechanical damping constant, and $c$ the restoring torque per unit angle of twist. We assume that the fluctuations are caused by a random torque, whose Fourier component at the frequency $f = \omega/2\pi$ has the amplitude $F_f$. Our further postulates about the nature of $F$ are similar to those made about $V$ in the last section. Then the analysis is exactly similar to the previous case of the electrical tuned circuit, so that by comparison we obtain at once

$$d(\overline{F_f^2}) = 4kTb\,df \tag{23.10}$$

and

$$d(\overline{\theta_f^2}) = \frac{4kTb\,df}{(\Im\omega^2-c)^2+b^2\omega^2}. \tag{23.11}$$

By integration it may be shown that these expressions satisfy eqn (23.2).

In general the galvanometer will be used for observing a current and will therefore be connected to a circuit whose total resistance (including the galvanometer coil) is $R$. Then we have two equations

$$\left. \begin{array}{c} \Im(d^2\theta/dt^2)+b(d\theta/dt)+c\theta = \Phi I+F_f \exp j\omega t, \\ RI = -\Phi(d\theta/dt)+V_f\exp(j\omega t+j\delta), \end{array} \right\} \quad (23.12)$$

where $\Phi = nAB$ and $I$ is the instantaneous current through the circuit. Two sources of fluctuations have been included; a random torque due to Brownian motion of the suspended coil, and a random voltage associated with the electrical circuit. In eqns (23.12) the Fourier components of these two sources of fluctuations at the frequency $f = \omega/2\pi$ have been used, with a phase difference $\delta$ between them. Since the two sources are independent, we do not expect any correlation in phase, and for different frequencies the phase difference $\delta$ will have random values. Elimination of the current $I$ between the two equations gives

$$\Im(d^2\theta/dt^2)+(b+\Phi^2/R)(d\theta/dt)+c\theta = (\Phi/R)V_f\exp(j\omega t+j\delta)+F_f \exp j\omega t,$$

and the square of the amplitude of the fluctuations at the frequency $f$ is found to be

$$\theta_f^2 = \frac{(\Phi/R)^2 V_f^2+F_f^2+2(\Phi/R)V_fF_f\cos\delta}{(\Im\omega^2-c)^2+(b+\Phi^2/R)^2\omega^2}.$$

On summing over a range of frequencies, $\delta$ takes all values between 0 and $2\pi$ and the mean value of $\cos\delta$ is therefore zero. Hence the mean-square angular amplitude in the frequency range $f$ to $f+df$ is

$$d(\overline{\theta_f^2}) = \frac{\{(\Phi/R)^2 d(V_f^2)\}+d(F_f^2)}{(\Im\omega^2-c)^2+(b+\Phi^2/R)^2\omega^2}. \quad (23.13)$$

On substituting the expressions for $d(\overline{V_f^2})$ and $d(\overline{F_f^2})$ given by eqns (23.7) and (23.10), we find

$$d(\overline{\theta_f^2}) = \frac{4kT(b+\Phi^2/R)\,df}{(\Im\omega^2-c)^2+(b+\Phi^2/R)^2\omega^2}. \quad (23.14)$$

This equation is similar to that obtained for the galvanometer on open circuit except that the total damping constant $(b+\Phi^2/R)$ appears instead of just the mechanical damping $b$. Integration of eqn (23.14) over all frequencies will obviously give the same result, $\frac{1}{2}c\overline{\theta^2} = \frac{1}{2}kT$, as for the galvanometer on open-circuit, since the result is independent of the magnitude of the damping. Thus, although there are now two independent sources of random fluctuations, and these add in the squares as shown by the numerator of eqn (23.13), each is associated with a damping term so that the total mean energy $\frac{1}{2}c\overline{\theta^2}$ stored in the suspension remains unaltered, provided that each source is at the same temperature. This argument could be extended by separating the mechanical damping $b$ into two parts, one due to imperfect elasticity of the suspension and the other to damping by the viscosity of the air. Then it follows that the total mean-square angular fluctuations have the same value whether the galvanometer is evacuated or not; the admission of air provides an extra source of fluctuations owing to the molecular bombardment, whose tendency to increase the mean-square deflection is just counterbalanced by the viscous air damping which accompanies it. The frequency distribution of the fluctuations is of course changed because of the increase in the damping, but it is important to realize that the Brownian motion is inherent in the suspended coil and is not caused by the bombardment by the gas molecules. If it were, and the suspension had an imperfect elasticity, then the molecular bombardment would result in the suspension being heated, through the dissipation of energy in it, and the gas would be cooled, even though both were originally at the same temperature. This is contrary to the second law of thermodynamics.

The processes which we regard as 'damping' in the galvanometer represent a degradation of mechanical energy into heat energy; in viscous damping, into kinetic energy of the gas molecules; in electromagnetic damping, ultimately into the vibrational energy of the lattice of the resistance in the external circuit (the coil moving in the magnetic field acts as a transducer, converting mechanical motion into electrical voltage). At the level of the molecular fluctuations, the damping processes are just the mechanisms by which thermal equilibrium is established; without them, an individual component of the system (galvanometer suspension, gas molecules, lattice of the resistor) would have no means of knowing what the temperatures are of the other components. In the electrical case, resistance arises from the conversion of electrical energy into heat energy, and at the fluctuation level is the mechanism by which the electrical fluctuations reach thermal equilibrium with the lattice fluctuations. The nature of the carriers of the electric current is no more important in this process than that of the molecules of the gas causing viscous damping.

It is convenient to define the minimum observable current for a galvanometer as that current which would produce a deflection equal to the r.m.s. value of the total Brownian angular motion. For a steady current $I$ the deflection $\theta = I(nAB)/c = I\Phi/c$, and hence the minimum observable current $I_m$ would be

$$I_m = (ckT)^{\frac{1}{2}}/\Phi. \tag{23.15}$$

In general the electromagnetic damping term $\Phi^2/R$ is much larger than the mechanical damping term $b$, and the critical damping resistance $R_c$ is given by eqn (5.50),

$$R_c = \tfrac{1}{2}\Phi^2/(\Im c)^{\frac{1}{2}},$$

while the period $\tau = 2\pi(\Im/c)^{\frac{1}{2}}$. Using these two relations the minimum observable current and voltage are conveniently expressed in the form

$$I_m = (\pi k T / R_c \tau)^{\frac{1}{2}}, \qquad V_m = (\pi k T R_c / \tau)^{\frac{1}{2}}, \qquad (23.16)$$

since $V_m = R_c I_m$ if the galvanometer is critically damped. Taking room temperature as 290 K, so that $kT = 4 \times 10^{-21}$ J, for a galvanometer of period 2 s and critical damping resistance 100 $\Omega$, we find that the minimum observable current and voltage are approximately $8 \times 10^{-12}$ A and $8 \times 10^{-10}$ V.

The correctness of the expressions derived above has been verified experimentally by Jones and McCombie (1952). The deflections of an ordinary galvanometer of about 2 s period (sensitivity 1 mm deflection at 1 m distance for 0·01 $\mu$A) were magnified by an optical lever. The beam of light reflected from the galvanometer mirror fell on a split photocell, so that rotation of the mirror transferred light from one cell to the other. The difference in the currents from the two photocells was observed on a second galvanometer; a deflection of 15 mm on this instrument corresponded to a voltage of about $10^{-3}$ $\mu$V (or a current of $10^{-5}$ $\mu$A) applied to the first galvanometer. To make use of this amplification, all external sources of disturbance such as vibration had to be eliminated. Typical traces obtained were similar to those shown in Fig. 23.1. With the first galvanometer on open-circuit the damping is small, and the frequency distribution of the angular deflections is large only in the region around the normal frequency of the suspension. Consequently the fluctuations resemble bursts of oscillation at the natural frequency, the number of oscillations in each being in inverse ratio to the damping (and roughly equal to the '$Q_F$' of the suspension). When the galvanometer is just critically damped, $(b + \Phi^2/R)^2 = 4\Im c$ and the denominator of eqn (23.14) can be written as $(\Im \omega^2 + c)^2$, showing that the frequency distribution of the

(a)

(b)

FIG. 23.1. Fluctuations of a galvanometer: (a) On open-circuit; (b) nearly critically damped. (After Jones and McCombie 1952.)

fluctuations now has its maximum value at zero frequency. The appearance of the fluctuations is now that of a random disturbance without any sinusoidal character (Fig. 23.1($b$)). The voltage sensitivity of the system was found by applying a voltage of about $10^{-2}\,\mu$V, obtained by attenuating a known voltage $\approx 1$ V through a resistance chain, and a thorough statistical analysis of the results showed that the magnitude of the fluctuations agreed with the theoretical value within 1 per cent.

### 23.3. The relation between resistance noise and thermal radiation

In an evacuated enclosure containing thermal radiation at an absolute temperature $T$ the energy density in the frequency range $f$ to $f+df$ is given by Planck's law,

$$dU = \frac{8\pi hf^3\,df}{c^3\{\exp(hf/kT)-1\}},\qquad(23.17)$$

where $h$ is Planck's constant and $k$ is Boltzmann's constant. For all radio-frequencies $hf \ll kT$ at room temperature, since $290k$ corresponds to a quantum of energy for a wavelength of approximately $50\,\mu$m. We may therefore expand the exponential, obtaining

$$dU = 8\pi f^2 kT\,df/c^3,\qquad(23.18)$$

which is simply the Rayleigh–Jeans law of classical theory. Since the polarization of the radiation is random, on the average only one-third of this energy corresponds to radiation whose electric vector is parallel to a given direction (say the y-axis), and only such radiation will induce a voltage in a short dipole aerial inserted in the enclosure parallel to the y-axis. From § 8.3 the mean-square electric field component is then given by $E_y^2 = cZ_0(\tfrac{1}{3}U)$, where $Z_0 = (\mu_0/\epsilon_0)^{\frac{1}{2}}$ is the intrinsic impedance of free space. Hence the mean-square voltage induced in an aerial of length $s$ will be

$$d(V_r^2) = s^2\,d(E_y^2) = 8\pi f^2 s^2 kTZ_0\,df/3c^2.\qquad(23.19)$$

Even if the aerial consists of a perfectly conducting wire, the resulting current which flows will be finite, since energy will be re-radiated by this oscillatory current, and this energy must just be equal to that picked up by the aerial. The radiation must therefore behave as a generator of open-circuit voltage $V_r$ with an internal impedance $R_r$, as in the equivalent circuit of Fig. 23.2(a). This drives a current $I_r$ when short-circuited, and the power dissipated is $V_r^2/R_r = I_r^2 R_r$; this power is lost by re-radiation, and from § 8.9 it follows that $R_r$ is just the radiation resistance given by eqn (8.71) as

$$R_r = 2\pi Z_0 f^2 s^2/3c^2.\qquad(23.20)$$

FIG. 23.2. Equivalent circuit of an aerial. $V_r$ = voltage induced by thermal radiation; $R_r$ = radiation resistance of aerial. (a) Aerial short-circuited at centre; (b) aerial with resistance $R$ at centre; (c) as (b) but showing noise voltage due to $R$.

If the aerial is not a perfectly conducting wire, and has a real ohmic resistance $R$, the equivalent circuit will be as shown in Fig. 23.2(b), and the energy dissipated in the load $R$ will be $V_r^2 R/(R_r+R)^2$. This will heat the resistance $R$, while less energy is re-radiated to the enclosure. If $R$ is initially at the same temperature $T$ as the radiation in the enclosure, the apparent result will be that $R$ is heated and the enclosure cooled, which is contrary to the second law of thermodynamics. In order that the net exchange of energy between $R$ and the enclosure be zero, we must postulate that there is a fluctuation voltage associated with $R$, as in Fig. 23.2(c), of mean-square voltage $V^2$ and internal resistance $R$. This will send a power $V^2 R_r/(R_r+R)^2$ back into the aerial which must just equal that drawn from the enclosure and dissipated in $R$. Thus $V_r^2 R = V^2 R_r$, and in the frequency range from $f$ to $f+df$

$$\frac{d(\overline{V^2})}{R} = \frac{d(\overline{V_r^2})}{R_r} = \frac{8\pi f^2 s^2 kTZ_0\, df}{3c^2} \times \frac{3c^2}{2\pi Z_0 f^2 s^2} = 4kT\, df. \qquad (23.21)$$

This result is identical with that obtained earlier (eqn (23.7)) by considering a simple tuned circuit. The voltage fluctuations have a constant distribution with frequency so long as the energy quantum $hf \ll kT$; this limitation corresponds to our use of the classical expression (Rayleigh–Jeans law) for the energy density in the enclosure. The fluctuations associated with a resistance $R$ can be represented by inserting a voltage generator $V$, whose mean-square open-circuit voltage is given by eqn (23.21), for which $R$ acts as the internal impedance as in Fig. 23.2(c). The equivalent current generator will have a mean-square current

$$d(\overline{I^2}) = 4kT\, df/R, \qquad (23.22)$$

and it will be shunted by the resistance $R$.

Let us suppose that we are able to connect to our aerial a load of resistance $R$ which itself produces no noise (for example, a resistance kept at a temperature very close to 0 K). Then the maximum power which can be drawn from the enclosure and dissipated in $R$, obtained by making $R$ equal to $R_r$, is $d(\overline{V_r^2})/(4R_r) = kT\, df$; this is the 'available noise power'.

If $R$ is in fact a radio receiver, this power drawn from the thermal radiation incident on the aerial forms a source of 'noise', and can be heard as a hiss from a loudspeaker, or viewed on a cathode-ray oscillograph. It will obscure any signal which it is desired to receive unless the signal power in the aerial is larger than that picked up from the radiation background. This difficulty cannot be overcome by increasing the over-all amplification of the receiver, since both noise and signal will be amplified together. Thus the radiation noise sets a limit to the useful sensitivity of a receiver. If a theoretically perfect receiver is defined as one which itself introduces no noise, then the amplified noise output will be $AkT\,df$, where $A$ is the over-all amplification. The amplified signal output will be $AP$, where $P$ is the signal power incident on the aerial. Then the minimum detectable input signal may be defined as that which gives a signal output equal to the noise output, from which

$$\text{minimum detectable signal power } P_0 = kT\,df \qquad (23.23)$$

for a perfect receiver. It is clear that the only variable at our disposal here is the bandwidth $df$, and the reduction in noise obtained on narrowing the bandwidth can be seen in Fig. 23.3. This shows the noise output from an aperiodic receiver covering a wide band, before and after the insertion of a low-pass filter cutting out the higher frequencies. The change in character of the noise when the high-frequency components are absent can be seen as well as the reduction in amplitude. In general, however, the bandwidth cannot be reduced beyond a certain limit without impairing the quality of the reception, since the higher modulation frequencies will be cut out. If only audio-frequency modulation is involved, the bandwidth will be about $0\cdot01\,\text{MHz}$ and the minimum detectable signal power is $4\times10^{-17}\,\text{W}$. In a television receiver it is necessary to have a bandwidth of say $4\,\text{MHz}$ to include all the information necessary to form the picture, and the minimum signal power needed to equal noise in a perfect receiver is $1\cdot6\times10^{-14}\,\text{W}$.

In practice, all receivers generate a certain amount of internal noise, with the result that the noise output is greater than for a perfect receiver. The signal input $P_1$ required to give a signal output equal to the noise output is therefore greater than $P_0$. The quantity $P_1-P_0$ is a measure of the internal noise generated in the receiver, and by writing

$$P_1-P_0 = kT_e\,df$$

it may be expressed in terms of the 'excess noise temperature' $T_e$ of the receiver. In an ideal receiver $T_e=0$, but in practice little is gained by making it smaller than about $T/10$, where $T$ is the temperature of the thermal radiation being received in the application for which the receiver is designed. In laboratory applications the source to which the receiver is

(a)

(b)

Fig. 23.3. Electrical noise output from an aperiodic amplifier. The lower oscilloscope trace shows the reduction in root-mean-square amplitude and removal of higher-frequency components produced by insertion of a low-pass filter. (Photograph by C. A. Carpenter and J. Ward.)

connected is generally at room temperature, and it is then convenient to take a value of $T = 290$ K to define $P_0$, making it equal to $4 \times 10^{-21}$ d$f$ (watts). The ratio of $P_1$ to this value of $P_0$ is then defined as the 'noise figure' of the receiver, and denoted by $F$. Since $F$ is a ratio of two powers, it is often expressed in decibels.

## 23.4. Shot noise

For most purposes it is sufficient to consider the electron current in a thermionic diode as a uniform flow of charge to the anode. Since the current consists in fact of the arrival of a finite number of electrons per second, this cannot be true. The flow of electrons is a random process, and we may expect that there will be a fluctuation in the number arriving in a given time interval, if we measure over a number of such intervals. If the arrival of the electrons consists of a succession of completely random events, then the mean-square deviation from the average number $N$ per second is proportional to $N$. These fluctuations give rise to noise in the anode circuit, known as shot noise from the obvious analogy with the random patter of shot on a target. For simplicity, we discuss first shot noise in vacuum tubes, but a similar phenomenon occurs in solid state devices (see § 23.6).

In general we are interested not in the total deviation from the mean, but in the frequency distribution of the fluctuations. To find this it is necessary to carry out a Fourier analysis of the pulse of current due to the arrival of a single electron of charge $e$. We will assume that this pulse, occurring at time $t = 0$, has some irregular shape but is entirely confined within the time interval $-\tau/2$ to $+\tau/2$. Since the total charge arriving is $e$, we have

$$e = \int_{-\tau/2}^{+\tau/2} I \, dt.$$

We do not specify anything about the duration of the pulse $\tau$ except that it is very short ($\sim$ the transit time, see § 21.3). The Fourier series representing the frequency distribution of the current due to the arrival of $e$ is written

$$I = a_0 + \sum_{n=1}^{\infty} a_n \cos \frac{2\pi nt}{T} + \sum_{n=1}^{\infty} b_n \sin \frac{2\pi nt}{T}.$$

Here $a_0$, $a_n$, and $b_n$ are coefficients to be determined, and $T$ is an undefined large interval of time. In effect we regard all the frequencies we are interested in as multiples of the fundamental frequency $1/T$. To obtain a continuous frequency distribution we should make $T$ infinite, and replace the summations in the series by integrations. For the student who is more familiar with a Fourier series than a Fourier integral we use the former, and by making $T$ large we obtain a good approximation to a continuous frequency distribution.

From the ordinary formulae of Fourier analysis

$$a_0 = \frac{1}{T} \int_{-T/2}^{+T/2} I \, dt; \qquad a_n = \frac{2}{T} \int_{-T/2}^{+T/2} I \cos \frac{2\pi nt}{T} \, dt;$$

$$b_n = \frac{2}{T} \int_{-T/2}^{+T/2} I \sin \frac{2\pi nt}{T} \, dt.$$

To evaluate the coefficients we restrict ourselves to frequencies small compared with $1/\tau$. Then, since the current is finite only in the range $-\tau/2$ to $+\tau/2$, and zero outside this range, we can write $\cos 2\pi nt/T = 1$ and

$\sin 2\pi nt/T = 0$ over the range of integration for which the current is finite. Hence $b_n$ is zero, while

$$2a_0 = a_n = \frac{2}{T} \int\limits_{-T/2}^{+T/2} I \, dt = \frac{2e}{T}.$$

Thus we have

$$I = \frac{e}{T} + \sum_{n=1}^{\infty} \frac{2e}{T} \cos \frac{2\pi nt}{T},$$

and the mean-square value of the $n$th component is

$$\overline{I_n^2} = \tfrac{1}{2}(2e/T)^2 = 2e^2/T^2.$$

If $N$ electrons arrive in time $T$, then each will contribute an equal amount to the value of $I_n^2$. (The electrons arrive at random times, and their contributions to the Fourier series will all differ slightly in phase. Thus we must add intensities, and not amplitudes.) Then

$$\overline{I_n^2} = 2e^2N/T^2 = 2eI_0/T,$$

where $I_0 = Ne/T$ is the mean value of the current. Now the number of Fourier components whose frequencies lie within a range between $f$ and $f + df$ is $T \, df$, since the components are equally spaced in frequency by amounts $1/T$. Adding together the mean-square values of these components gives

$$d(\overline{I^2}) = 2eI_0 \, df \tag{23.24}$$

for the mean-square current fluctuation in the frequency range $f$ to $f + df$.

*Influence of space charge*

In this derivation of the formula for shot noise the arrival of an electron is considered as a random event, completely independent of the arrival of any other electron. We expect that the emission of electrons from the cathode has this property of complete randomness, but this is not necessarily true of their arrival at another electrode. In practice it is found that the value of the shot noise is materially lower than that given by the above equation unless the current to the anode is limited only by the rate of emission from the cathode. In general the anode current is only a fraction of the emission current because of the formation of 'space charge' outside the cathode which causes a large number of electrons emitted from the cathode to be turned back to the cathode. Since this is due to the mutual interaction of the electrons, we may expect that the flow of electrons to the anode is not now a succession of completely random events. The value of the fluctuations is greatest for random events, and falls as soon as they become not completely random. Physically, the action of the space charge may be envisaged as follows. Suppose that at some instant the number of electrons emitted from the cathode rises momentarily above the average. This will cause a temporary increase in the space charge, and a number of electrons greater than average will leave the space charge region for the anode. This number is smaller than the surge from the cathode because the space charge acts as a reservoir; the effect of the increased space charge is to turn some of the excess electrons back to the cathode. Similarly, at instants when the cathode emission falls momentarily below the average, the space charge also drops and less electrons are turned back. To allow for this 'space-charge smoothing', as it is called, a factor is inserted in the equation for the shot noise. Thus

$$d(\overline{I^2}) = 2\beta e I_0 \, df. \tag{23.25}$$

$\beta$ is called the space-charge smoothing factor, and may be as low as $0\cdot03$, showing that the smoothing effect is very considerable.

*Noise in multi-electrode tubes*

The presence of grids in a tube does not affect the validity of the equations given above for shot noise so long as they do not intercept any of the current on its way to the anode. Thus eqn (23.25) is still valid for a negative-grid triode, but this is not so for a screen-grid tetrode or a pentode, for then the positive screen grid intercepts a considerable portion of the anode current. Since the chance of an electron hitting the wire of the screen grid is purely random, the screen current will have the full shot noise appropriate to its magnitude. It is obvious that similar fluctuations, though of opposite sign, must be imposed on the current that goes through the screen grid to the anode. Assuming that less than half of the total current goes to the screen, we may write approximately for the anode current

$$d(\overline{I^2}) = 2\beta e I_a \, df + 2 e I_s \, df. \tag{23.26}$$

Since $\beta$ may be less than $0\cdot1$, while the screen current $I_s$ is $0\cdot2$ or $0\cdot3$ of $I_a$, the second term is often more important than the first. Hence screen-grid tetrodes and pentodes are generally more noisy than triodes. The additional noise is called 'partition noise'.

*Equivalent noise resistance*

It is often convenient to define the amount of noise by referring it to an equivalent resistance $R_n$ (at $290 \text{ K}$) in the grid circuit. The fluctuating voltage at the grid due to $R_n$ has the mean-square value

$$d(\overline{V^2}) = 4kTR_n \, df,$$

since the grid consumes no power and the equivalent noise resistance is therefore on open-circuit. This causes a fluctuating anode current whose mean-square value is

$$d(\overline{I^2}) = g_m^2 \, d(\overline{V^2}) = g_m^2 4kTR_n \, df, \qquad (23.27)$$

where $g_m$ is the mutual conductance of the tube. If $\overline{I^2}$ is due to the shot noise, the equivalent noise resistance may be calculated by means of this formula, $T$ being taken as room temperature. The advantage of this method of specifying the noise is that the value of $R_n$, unlike that of $d(\overline{I^2})$, is independent of the bandwidth, and it facilitates comparison of the shot noise with the resistance noise in the circuits attached to the grid. If partition noise is included by replacing eqn (23.26) by (23.25) with an effective value $\beta'$ instead of $\beta$, then

$$R_n = \beta' eI_0 / 2g_m^2 kT. \qquad (23.28)$$

An estimate of the relative importance of shot noise and resistance noise can be obtained from the formula for the equivalent noise resistance. For a typical triode, $g_m = 5$ mA V$^{-1}$, $\beta = 0 \cdot 03$, $I_0 = 10$ mA, $e = 1 \cdot 6 \times 10^{-19}$ C; this gives $R_n = 240 \, \Omega$. The value for a pentode would be somewhat higher, owing to partition noise. These figures apply at medium radio-frequencies (that is, of the order of megahertz); at higher frequencies (100 MHz upwards) $R_n$ rises owing to noise voltages induced in the grid which have period of oscillation comparable with the electron transit time (cf. § 21.3). At audio-frequencies the shot noise (particularly from tubes with oxide-coated cathodes) becomes abnormally large. This is known as the flicker effect, and is thought to be associated with changes in the state of the cathode surface which cause abnormal fluctuations in the anode current.

## 23.5. Design of receivers for optimum performance (minimum noise figure)

The correct design of a receiver is of great importance. If it is being used in an application where the signal strength is fixed, such as in r.f. astronomy or spectroscopy, then the limiting sensitivity attainable will depend entirely on the design of the receiver. In radio communications an improvement of a factor $n$ in signal–noise ratio can be achieved by increasing the transmitter power by a factor $n$, but a very much more economical method is to improve the receiver performance by the same factor instead. The following remarks illustrate only the basic principles, and do not go into any detail of receiver design.

In general all the stages of a receiver will contribute some noise, but if the amplification of each stage is high only the first stage or two is important. If stage $k$ gives noise power $N_k$, and the stage gain is $m$, then the signal–noise ratio after $n$ stages is

$$Sm^n/(N_1m^n+N_2m^{n-1}+...+N_n) = S/(N_1+N_2m^{-1}+N_3m^{-2}+...).$$

$$(23.29)$$

With a stage gain of 10–100 even the second stage will contribute little to the noise output. If not, the design of the second stage should follow the same principles as that of the first stage, and only the latter need be considered.

In the circuit of Fig. 23.4(a) $S$ represents a signal source of voltage $S$ with output resistance $R_1$. $R_1$ is assumed to be noisy, at temperature $T$, and its equivalent noise voltage is represented by $V_{n1}$, in series with $S$. The source $S$ may be a signal induced in an aerial, in which case $R_1$ is the radiation resistance of the aerial and $T$ is the ambient temperature which we take to be 290 K. The source is connected to the grid of the tube, and $R_2$ is the grid-bias resistance, or the first tuned circuit, in which case $R_2$ is its parallel impedance. In general, $R_2$ will also generate resistance noise, which is represented by the insertion of a voltage source $V_{n2}$ in series with $R_2$. In the first instance we shall assume that the tube contributes no shot noise ($R_n = 0$), and that noise from subsequent stages is negligible. Then the signal–noise ratio will be the same at the grid of the first tube as at any later point in the receiver, and we need only compute the ratio of the mean-square signal voltage on the grid to the mean-square noise voltage. For simplicity $R_1$ and $R_2$ are taken to have the same temperature, which in practice will not be far from true.

Since $V_{n1}^2/R_1 = V_{n2}^2/R_2 = 4kT\,df$, and $R_2$ acts as the load for the noise generator $V_{n1}$, and $R_1$ as the load for $V_{n2}$, the mean-square noise voltage on the grid is

$$4kT\,df\left\{R_1\frac{R_2^2}{(R_1+R_2)^2}+R_2\frac{R_1^2}{(R_1+R_2)^2}\right\} = 4kT\,df\times\frac{R_1R_2}{R_1+R_2}. \quad (23.30)$$

(a)            (b)

FIG. 23.4. (a) Equivalent input circuit of a receiver, showing noise voltages. (b) Actual input circuit, showing aerial tapped on to inductance of input tuned circuit.

Since $R_1$ and $R_2$ are random noise sources, the mean-square voltages
have been added; the result is the same as that for a resistance equal to
$R_1$ and $R_2$ in parallel, as we should expect.

The mean-square signal voltage on the grid is

$$S^2 R_2^2/(R_1+R_2)^2.$$

Hence the signal–noise ratio at the grid is

$$\frac{S^2}{4kT\,df}\frac{R_2}{R_1+R_2}\frac{1}{R_1}=\frac{P}{kT\,df}\frac{R_2}{R_1+R_2},\tag{23.31}$$

where $P = S^2/4R_1$ is the available signal power. If eqn (23.31) is put equal
to unity, we obtain the signal power $P_1$ required to give an output power
equal to the noise output power of the receiver. The noise figure $F$ is
defined as the ratio of this signal power to the value $kT\,df$ for a perfect
receiver, and hence the noise figure is

$$F = P_1/kT\,df = 1+R_1/R_2.\tag{23.32}$$

A simpler method of obtaining this result is as follows. In eqn (23.30) the
first term is the aerial noise and the second that from the resistance $R_2$. In
an ideal receiver the latter would be absent, and the only noise would be
that from the aerial. Hence we can determine $F$ by dividing eqn (23.30)
by its first term. This leads at once to eqn (23.32), and is the method used
below. If the aerial is matched to the first circuit, $R_1 = R_2$, and $F = 2$. But
if $R_2 > R_1$, $F$ is reduced and tends to its limiting value of unity as the ratio
of $R_2$ to $R_1$ is increased indefinitely. Hence to obtain the highest sensitivity it pays to mismatch the aerial to the receiver, since the reduction in
noise at the grid when $R_2$ is shunted by the lower resistance $R_1$ is greater
than the loss of signal at the grid due to the mismatch. We see also that,
in the absence of tube noise, it is possible to approach very closely to the
theoretical limit of sensitivity. In a practical case $R_1$ (for a half-wave
dipole aerial) would be 80 $\Omega$, while $R_2$ could be of the order of 100 k$\Omega$,
giving $F = 1 \cdot 001$.

We now allow for the presence of shot noise in the vacuum tube,
represented by the noise resistance $R_n$ in Fig. 23.4(a). On adding
$4kTR_n\,df$ to eqn (23.30), the mean-square noise voltage at the grid
becomes

$$4kT\,df\frac{R_1 R_2^2}{(R_1+R_2)^2}\left(1+\frac{R_1}{R_2}\right)+4kT\,df R_n.$$

On dividing this by the noise from the aerial we obtain

$$F = 1+\frac{R_1}{R_2}+\frac{R_n}{R_2}\left(\frac{R_1}{R_2}+2+\frac{R_2}{R_1}\right).\tag{23.33}$$

This result shows that there is now an optimum value for $R_1/R_2$, given by

$$\frac{R_2}{R_1} = \frac{R_1}{R_2} + \frac{R_1}{R_n}, \qquad (23.34)$$

at which $F$ has its minimum value. If $R_2 \gg R_1$, the optimum value of $R_1 = (R_n R_2)^{\frac{1}{2}}$; taking $R_n = 1000\ \Omega$, $R_2 = 100\ \text{k}\Omega$, this gives $R_1 = 10\ \text{k}\Omega$ and $F = 1 \cdot 22$. In general, $R_1$ is fixed (for example, it is equal to the radiation resistance of the aerial), but it can be transformed to the optimum value by the use of a circuit with variable coupling such as that shown in Fig. 23.4(b).

## 23.6. Noise in solid-state triodes

In solid-state devices noise arises from a number of causes, but these are basically the same as those already discussed. At very low frequencies flicker noise, whose power varies inversely with frequency, is of practical importance; it arises from imperfections and varies widely from sample to sample. In the frequency range between about 1 kHz and many megahertz (where transit-time effects set in), the effects which are dominant in a junction transistor are as follows.

1. Johnson noise in the ohmic resistance between the base terminal and the working region; because of the thinness of the base, this resistance may be as much as $200\ \Omega$.

2. Current fluctuations (shot noise) arising from the random nature of the flow of carriers across a barrier. The flow of carriers into the base from the emitter consists of those with sufficient thermal energy to surmount the barrier at the depletion layer. The absence of space charge means that there is no 'space-charge smoothing' to reduce the size of the fluctuations, and for the emitter current we may write

$$d(\overline{I_e^2}) = 2eI_e\, df. \qquad (23.35)$$

Not all of these carriers reach the collector junction, some being lost in the base through recombination. This is again a purely random process, analogous to the interception of electrons by a grid in a vacuum tube (partition noise). It is the origin of the base current, and the current fluctuations can be written as

$$d(\overline{I_b^2}) = 2eI_b\, df. \qquad (23.36)$$

For the collector current we have similarly

$$d(\overline{I_c^2}) = 2eI_c\, df. \qquad (23.37)$$

These two sets of current fluctuations are shown as current sources in the equivalent noise circuit of Fig. 23.5. It is unnecessary to include a third

FIG. 23.5. Sources of current fluctuations for a junction transistor in the CE connection.

independent source corresponding to eqn (23.35) for the following reason. The fact that fluctuations in $I_b$ and $I_c$ are uncorrelated means that $d(\overline{I_b I_c}) = 0$. Since $I_e = I_b + I_c$, we have

$$d(\overline{I_e^2}) = d(\overline{(I_b + I_c)^2}) = d(\overline{I_b^2}) + d(\overline{I_c^2}) = 2e(I_b + I_c)\,df$$
$$= 2eI_e\,df,$$

showing that eqn (23.35) can be derived from the two following equations and does not represent an independent source of fluctuations.

For many purposes it is convenient to represent the various sources of noise by a set of noise generators in the input circuit. For a transistor in the CE connection the mutual conductance is (see § 19.1)

$$g_m = eI_c/kT.$$

The current source $i_c$ in the collector circuit can be replaced by a voltage source $v$ in the base circuit such that $\overline{i_c^2} = g_m^2 \overline{v^2}$ (cf. eqn (23.27)). Hence

$$\overline{v^2} = 2eI_c\,df/g_m^2 = (2kT/g_m)\,df. \tag{23.38}$$

The equivalent input circuit is shown in Fig. 23.6. For simplicity we neglect here the base resistance, which is small, and the resulting noise voltage has a mean-square value $\overline{v_s^2} = 4kTR_s\,df$. Across the input resistance $R_i$ this develops a mean-square voltage

$$4kTR_s\,df R_i^2/(R_i + R_s)^2.$$

Similarly the current source $i_b$, driving a current that is divided between $R_s$ and $R_i$, develops a mean-square voltage across $R_i$ equal to

$$2eI_b\,df R_s^2 R_i^2/(R_i + R_s)^2.$$

F$_{IG}$. 23.6. Equivalent input noise circuit for a junction transistor in the CE connection.

In the CE connection the input resistance $R_i = \beta/g_m$, where $\beta = I_c/I_b$ is the current gain. $R_i$ is high, and we therefore assume that $R_s \ll R_i$, a simplifying assumption that will prove to be justified. Then, together with the mean-square voltage given by eqn (23.38), we have altogether a mean-square voltage across the base–emitter terminals equal to

$$4kTR_s\,df + 2eI_bR_s^2\,df + (2kT/g_m)\,df.$$

As in the previous section, we obtain the noise figure $F$ by dividing through by the first term, which represents the noise due to the source alone, obtaining

$$F = 1 + \frac{eI_bR_s}{2kT} + \frac{1}{2g_mR_s}.$$

Since $eI_c/kT = g_m$, this can be written as

$$F = 1 + \tfrac{1}{2}\left(\frac{I_b}{I_c}g_mR_s + \frac{1}{g_mR_s}\right),$$

from which it is clear that there is an optimum value of $R_s$, such that $I_b(g_mR_s)^2 = I_c$, or $R_s = \beta^{\frac{1}{2}}/g_m$. Since $\beta$ is large, this is clearly small compared with $R_i = \beta/g_m$, as we have assumed. Taking $I_c = 1$ mA and $\beta = 100$, $g_m^{-1} = 25\ \Omega$, so that $R_s = 250\ \Omega$ and $R_i = 2500\ \Omega$, we find that the optimum value of $F$ is 1·2.

The discussion given above follows a simple treatment by Robinson (1970); its validity is discussed in Appendix B of Robinson (1974).

## References

J$_{ONES}$, R. V. and M$_C$C$_{OMBIE}$, C. W. (1952). *Phil. Trans. R. Soc.* **244,** 205.
R$_{OBINSON}$, F. N. H. (1970). *Wireless World,* 339
R$_{OBINSON}$, F. N. H. (1974). *Noise and Fluctuations.* Clarendon Press, Oxford.

## Problems

23.1. Show, by differentiation of eqn (23.5) to obtain the differential equation for the current $I = dq/dt$, and following through an analysis similar to that of § 23.1, that $\frac{1}{2}L\overline{I^2} = \frac{1}{2}kT$.

23.2. A signal generator whose output impedance is 500 Ω is calibrated in terms of the power it will deliver into a matched load (that is, the available signal power). It is connected to a receiver whose bandwidth is 10 kHz, and whose first stage consists of a triode whose shot noise is negligible, with a 1000 Ω resistance connected between cathode and grid. What will the signal generator reading be when it is adjusted so that the signal output from the receiver is equal to the noise output?

(*Answer:* $6 \times 10^{-17}$ W.)

23.3. Referring to Problem 8.10, assume that the target is low over the sea and intercepts the power incident on an area $A_1$. This power is scattered with the same angular distribution as that of the radiation from a short horizontal dipole parallel to the transmitter dipole. Some of this scattered power falls on an aerial of effective area $A_2$ located at the transmitter, and is detected by a receiver of noise figure $F$ and bandwidth $df$. Show that the signal–noise ratio, for the signal returned from the target, is unity for a target distance

$$D = \left(\frac{36\pi^2 W A_1 A_2}{FkT\,df}\right)^{\frac{1}{8}} \left(\frac{Hh}{\lambda}\right)^{\frac{1}{2}}.$$

(This formula shows how difficult it is to increase the range by increasing the transmitter power $W$, and how much better it is to reduce the wavelength.)

23.4. By following the treatment of § 23.3 using Planck's law instead of the Rayleigh–Jeans law, show that the quantum-mechanical formula for resistance noise is

$$\frac{d(\overline{V^2})}{R} = \frac{4hf\,df}{\exp(hf/kT)-1}.$$

Verify from this formula that the transition from classical region to quantum-mechanical region occurs when the number of quanta per unit bandwidth in the noise power is of the order of unity.

23.5. In the electrical tuned circuit of § 23.1, show that, over a narrow range of frequencies $df$ at the resonant frequency $f$ of the circuit, the mean-square fluctuating current is

$$d(\overline{I_f^2}) = 4kTR^{-1}\,df,$$

and the mean-square fluctuating voltage on the capacitor is

$$d(\overline{V_f^2})_c = 4kT(L/CR)\,df.$$

Verify that the total mean-square fluctuating current $\overline{I^2}$ is equivalent to taking $d(\overline{I_f^2})$ over a range $df$ such that

$$df = R/4L \quad \text{or} \quad df/f = \pi/2Q_F,$$

and that an equivalent relation holds for the fluctuating voltages on the capacitor.

# 24. Magnetic resonance

## 24.1. The magnetic-resonance phenomenon

IT was shown in § 14.1 that when an atom or nucleus with a resultant angular momentum $\mathbf{G}$ and magnetic moment $\mathbf{m}$ is placed in a steady magnetic flux density $\mathbf{B}_0$ the equation of motion (obtained from eqn (14.2) by multiplying by $\gamma$) is

$$d\mathbf{m}/dt = \gamma \mathbf{m} \wedge \mathbf{B}_0, \qquad (24.1)$$

where $\gamma = \mathbf{m}/\mathbf{G}$ is the magnetogyric ratio. The motion represented by this equation consists of a precession of the angular-momentum vector $\mathbf{G}$ and hence also of $\mathbf{m}$ about the direction of $\mathbf{B}_0$ with a uniform angular velocity $-\gamma\mathbf{B}_0$, which we shall denote by $\boldsymbol{\omega}_L$. If the system is undisturbed it will continue indefinitely in this state of uniform precession with $\mathbf{m}$ at a fixed angle to $\mathbf{B}_0$, and it is convenient to make use of rotating coordinate systems in considering this motion. It is shown in § A.10 that the rate of change $d\mathbf{m}/dt$ of any vector quantity such as $\mathbf{m}$ in the laboratory coordinate system is related to the rate of change $D\mathbf{m}/Dt$ in a system rotating with angular velocity $\boldsymbol{\omega}$ relative to the laboratory system, by the equation

$$d\mathbf{m}/dt = D\mathbf{m}/Dt + \boldsymbol{\omega} \wedge \mathbf{m}.$$

Substitution of this in eqn (24.1) gives

$$\begin{aligned}
D\mathbf{m}/Dt &= \gamma \mathbf{m} \wedge \mathbf{B}_0 - \boldsymbol{\omega} \wedge \mathbf{m} \\
&= \gamma \mathbf{m} \wedge \mathbf{B}_0 + \mathbf{m} \wedge \boldsymbol{\omega} \\
&= \gamma \mathbf{m} \wedge (\mathbf{B}_0 + \boldsymbol{\omega}/\gamma). \qquad (24.2)
\end{aligned}$$

This result shows that in the rotating coordinate system the apparent magnetic field is $(\mathbf{B}_0 + \boldsymbol{\omega}/\gamma)$, and the apparent precession velocity is $-\gamma(\mathbf{B}_0 + \boldsymbol{\omega}/\gamma) = \boldsymbol{\omega}_L - \boldsymbol{\omega}$. Thus the apparent angular velocity is decreased by $\boldsymbol{\omega}$, as would be expected from simple considerations of relative angular velocity. If we write $\mathbf{B}' = -\boldsymbol{\omega}/\gamma$, the apparent field in the rotating system is $(\mathbf{B}_0 - \mathbf{B}')$, and it is reduced if $\mathbf{B}'$ is positive (that is, $\boldsymbol{\omega}$ has the same sign as $\boldsymbol{\omega}_L$) as shown in Fig. 24.1. Clearly, if $\mathbf{B}' = \mathbf{B}_0$, the apparent field $(\mathbf{B}_0 - \mathbf{B}')$ and precession velocity $-\gamma(\mathbf{B}_0 - \mathbf{B}')$ are both zero, and the vector $\mathbf{m}$ is stationary in the rotating coordinate system.

We shall now consider the effect of applying a small oscillating magnetic flux density $\mathbf{B}_1 \cos \omega t$ in the plane normal to the direction of the steady flux density $\mathbf{B}_0$. This oscillating field may be plane polarized or circularly

FIG. 24.1. Effective fields in a coordinate system rotating with angular velocity **ω**.

polarized; in the latter case **B**₁ is simply a vector, normal to **B**₀, which is constant in length but which rotates about **B**₀ with angular velocity **ω**. If the oscillating field is plane-polarized, with a component say in the $x$ direction (taking **B**₀ along the $z$-axis), it may be decomposed into two vectors rotating in opposite senses; thus there is no loss of generality in considering only the circularly polarized case.

If we now transfer from the laboratory system to a system rotating with the angular velocity **ω**, then the vector **B**₁ is stationary in this system, and can be represented by a constant vector **B**₁ as shown in Fig. 24.1. In this system, the atom or nucleus feels an apparent magnetic flux density (**B**₀−**B′**) parallel to the $z$-axis, together with the flux density **B**₁ normal to this axis; thus the resultant field in this rotating system is the vector sum of these two fields, which is denoted by **B**_eff in Fig. 24.1. To an observer in this system, the dipole moment **m** will appear to precess about **B**_eff with angular velocity −γ**B**_eff, and its projection on **B**₀ will change as time goes on. If **m** were initially parallel to **B**₀ (as we should expect in a macroscopic system), it would precess about **B**_eff and at some subsequent time would reach a maximum angle 2θ with **B**₀, where

$$\tan \theta = B_1/(B_0 - B').$$

If **B**₀−**B′** = 0, **B**_eff = **B**₁ and $\theta = \frac{1}{2}\pi$, so that **m** will turn right over to reach a position antiparallel to **B**₀ before commencing to return. This occurs only when

$$\omega = -\gamma \mathbf{B}' = -\gamma \mathbf{B}_0 = \omega_L,$$

so that the frequency of the applied field is then the same as that of the Larmor precession. Thus the precession about $\mathbf{B}_{\text{eff}}$ is a forced resonance phenomenon, whose amplitude is greatest when the applied frequency $\omega$ is equal to the natural frequency $\omega_L$.

In the case of an atomic or nuclear system, the angular momentum is quantized, and so are its projections on $\mathbf{B}_0$, so that the energy $W = -\mathbf{m} \cdot \mathbf{B}_0$ is also quantized. We shall consider first the nuclear case, assuming a nucleus of spin angular momentum $I\hbar$, and magnetic moment $\mathbf{m}$, where the magnetogyric ratio $\gamma = g_N(e/2m_p)$. Then the potential energy in a state whose magnetic quantum number is $m$ is

$$W_m = -\mathbf{m} \cdot \mathbf{B}_0 = -\gamma \hbar \mathbf{I} \cdot \mathbf{B}_0 = -\gamma \hbar m B_0. \tag{24.3}$$

Under the influence of an oscillating magnetic field polarized in the plane normal to $\mathbf{B}_0$, transitions between states with different values of $m$ may take place according to the selection rule $\Delta m = \pm 1$. The quantum of energy required is

$$\hbar \omega = W_m - W_{m-1} = -\gamma(h/2\pi)B_0. \tag{24.4}$$

This is the same for all transitions, as shown in Fig. 24.2. From this it follows that the frequency of the radiation must be

$$\nu = -\frac{\gamma}{2\pi} B_0 = +\frac{\omega_L}{2\pi}. \tag{24.5}$$

This is the resonance condition, which is the same as that given by the classical treatment above. The minus sign is significant only if circularly polarized radiation is used. If a system of nuclei with a positive value of $\gamma$ is to absorb energy from an applied oscillatory field, the selection rule for absorption is $\Delta m = -1$, and the vector $\mathbf{B}_1$ must rotate in the left-hand sense about $\mathbf{B}_0$, while if $\gamma$ is negative, the reverse holds. This gives a method of determining the sign of $\gamma$, but for many purposes this is immaterial and linearly polarized radiation may be used. Since this can be regarded as composed of two circularly polarized components rotating in opposite senses, transitions can be induced whatever the sign of $\gamma$. In the usual spectroscopic terminology these are 'magnetic dipole' transitions, corresponding to the fact that they are caused by the interaction of an oscillatory magnetic field with the magnetic dipole moments of the system. This phenomenon is generally known as 'magnetic resonance', and it offers a method of determining $\gamma$ directly from a measurement of a frequency and a magnetic field. The order of magnitude of the frequencies required may be found from the specific charge $(e/m_p)$ of the proton,

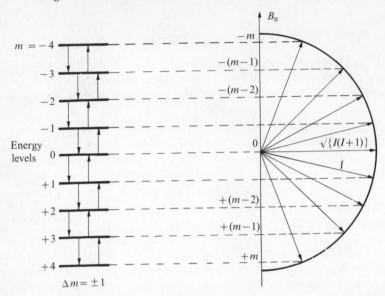

F$_{IG}$. 24.2. Diagram showing the nine energy levels and the allowed transitions between them for a nuclear spin $I = 4$ in a flux density $B_0$. Transitions can be induced by an oscillating field of frequency $\nu$ if $\nu = -(\gamma/2\pi)B_0$.

if we assume that $g_N$ is around unity. The value of $e/m_p$ for the proton is nearly equal to the Faraday, that is, it is about $10^8$ C kg$^{-1}$. Hence the frequency

$$\nu = -\frac{g_N}{2\pi}\left(\frac{e}{2m_p}\right)B_0$$

is $\approx 10^7$ Hz in a field of 1 T in the case of a nuclear magnetic moment. Electronic magnetic moments are some 2000 times larger, owing to the smaller mass of the electron, while the associated angular momentum is of the same order as in the nuclear case, so that the value of $\gamma$ and of the frequency required are higher in proportion. For an atom with a magnetic moment due only to electron spin, the frequency of the radiation required for resonance in a field of 1·07 T is 30 GHz (a wavelength of 10 mm).

The magnetic-resonance phenomenon has been used to investigate systems of both atomic and nuclear magnetic moments, and we shall discuss first the latter. It makes possible a direct determination of the value of $\gamma$, and hence, for a nucleus whose spin $I$ is known, of the nuclear magnetic moment. The chief experimental difficulty lies in the smallness of the effect, and we shall describe first an ingenious method due to Rabi, where the phenomenon is detected by its effect on the path of a molecule in a molecular beam.

## 24.2. Molecular beams and nuclear magnetic resonance

The use of atomic beams for the measurement of atomic magnetic moments has been mentioned in § 14.4. The method depends on deflecting the atoms by passing them through an inhomogeneous magnetic field; the deflection is proportional to the projection of the magnetic moment on the direction of the field gradient, and the initial beam is split into $(2J+1)$ beams if the total electronic angular momentum has quantum number $J$. The magnetic moment can be computed from the size of the deflections if the magnitude of the field gradient is known. The main difficulty in achieving high precision is the spread in velocity of the atoms in the beam, since the deflection is inversely proportional to the square of this velocity.

Application of this method to the determination of nuclear magnetic moments demands great refinements, since the size of the moment is some 2000 times smaller than that of an electron, and the deflection is correspondingly smaller. For a direct measurement molecules such as $H_2$ or NaCl, with no electronic magnetic moment, must be used. Owing to the great difficulty of working with purely nuclear moments, methods were devised using atoms with a hyperfine structure due to magnetic interaction between the electronic moment and the nuclear magnetic moment. These are rather complicated, and such deflection methods have now been superseded by others making use of magnetic resonance. The first of these was carried out by Rabi and his colleagues (1939).

FIG. 24.3. Schematic diagram of a molecular-beam apparatus and of some molecular paths. The two full curves (above) indicate the paths of two molecules having different moments and velocities and whose moments do not change during passage through the apparatus. The dotted curves indicate possible changes in path if the component of magnetic moment is changed. The motion in the $z$ direction is greatly exaggerated.

A schematic diagram of the apparatus is shown in Fig. 24.3. Molecules from an oven O emerge through slits $F_1$, $F_2$, moving at small angles with the axis of the apparatus. They enter a region of inhomogeneous magnetic field in the A-magnet, and are deflected by an amount proportional to the projection of their magnetic moments on the direction of the field; thus some of them will pass through the collimating slit S. Neglecting for the moment the C-magnet, we follow the molecules through the B-magnet, which produces an inhomogeneous field exactly like that of the A-magnet except that the gradient is reversed. The force on the nuclear magnets in a molecule is therefore also reversed (provided that the orientation of these magnets is the same as it was when passing through the A-magnet), and the molecules are deflected upwards and reach a detector D. The proviso about the orientation is important, because if it changes between A and B, then the deflection produced by B is different from that in A, and the molecules will not reach the detector. This gives a means of detecting a magnetic-resonance phenomenon, since it can be used to cause a change in the orientation after leaving A and before entering B. The C-magnet produces a uniform flux density $B_0$, and between its pole faces is a loop R carrying an r.f. current which produces an oscillating flux density $B_1 \cos \omega t$ in a direction perpendicular to $B_0$. When the resonance condition is fulfilled, transitions are induced within the Zeeman levels of the nuclear moment in the field $B_0$, so that the orientation of the nuclear moment is changed. The beam current at the detector then falls, a typical example being shown in Fig. 24.4. Here the radio-frequency is kept constant, and the magnetic flux density $B_0$ is varied through the resonance. The value of $\gamma$ is found from the observed values of $B_0$ (at the centre of the resonance) and the frequency. From the width of the resonance curve in Fig. 24.4 it can be seen that the accuracy is very much greater than could be obtained by a simple deflection method.

FIG. 24.4. Resonance curve for the $^{19}F$ nucleus in sodium fluoride (NaF) obtained by Rabi.

An estimate of the width can be obtained as follows. If the nuclear moment is initially parallel to $\mathbf{B}_0$, then at resonance it precesses about $\mathbf{B}_1$ in the rotating coordinate frame with angular velocity $\gamma \mathbf{B}_1$, and in a time $t$ it will exactly reverse its orientation provided that $\gamma B_1 t = \pi$. This corresponds to the maximum in the resonance curve. When we are off resonance, the moment precesses about $\mathbf{B}_{\text{eff}}$, and if $\mathbf{B}_{\text{eff}}$ is at an angle $\theta = 45°$ to $\mathbf{B}_0$ the moment will only reach a maximum angle of $2\theta = \frac{1}{2}\pi$ to $\mathbf{B}_0$. If we take this to define the half intensity points on the resonance curve, they correspond to

$$B_0 - B' = \pm B_1,$$

so that

$$\Delta \nu = \pm \gamma (B_0 - B')/2\pi = \pm 1/2t, \qquad (24.6)$$

showing that the linewidth is simply related to the time $t$ for which the dipole moment is subjected to the oscillatory field.

An experiment of this kind is by no means easy, as can be seen from the fact that the deflection of a molecule is only about $0 \cdot 05$ mm in a magnet with a flux gradient of the order of $10^3 \, \mathrm{T \, m^{-1}}$. This is using a magnet $0 \cdot 5$ m long with pole faces of the shape shown in Fig. 24.5; the curvature is adjusted to give a uniform value of $\mathrm{d}B/\mathrm{d}z$ over the width of the beam. Because of the small deflections the defining slits at S and the detector must be very narrow ($\approx 0 \cdot 01$ mm) and the beam intensity at the detector is very small, being determined by the solid angle which the detector slit makes with the oven, some $1 \cdot 5$ m away! Originally the difficulty of detecting a beam of uncharged molecules limited the method to hydrogen, deuterium, and the alkali metals, whose nuclear moments were measured with a precision of a few parts per thousand (see Table 24.1).

The spectra of molecules are complicated by the presence of: (1) a magnetic moment associated with rotation of the molecule as a whole; (2) interaction between the nuclear magnetic moment and the rotational magnetic moment; (3) interactions between the nuclear spins if more than

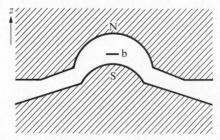

FIG. 24.5. Cross-section of the A and B magnets in Fig. 24.3, normal to the path of the beam (marked b in the figure). The curved parts of the pole pieces are cylindrical but the radius of curvature for the upper pole is larger than for the lower one.

TABLE 24.1

*Nuclear spins and magnetic moments of some common isotopes*

| Nucleus | Spin | Magnetic moment (nuclear magnetons) | |
| | | Molecular beam value | Nuclear resonance value |
|---|---|---|---|
| Neutron | $\frac{1}{2}$ | $-1\cdot913$ | — |
| $^1$H | $\frac{1}{2}$ | $+2\cdot789$ | $+2\cdot7927$ |
| $^2$H | 1 | $+0\cdot856$ | $+0\cdot8574$ |
| $^3$H | $\frac{1}{2}$ | — | $+2\cdot9788$ |
| Alkali metals: | | | |
| $^6$Li | 1 | $+0\cdot821$ | $+0\cdot8220$ |
| $^7$Li | $\frac{3}{2}$ | $+3\cdot253$ | $+3\cdot2563$ |
| $^{23}$Na | $\frac{3}{2}$ | $+2\cdot215$ | $+2\cdot2175$ |
| $^{39}$K | $\frac{3}{2}$ | $+0\cdot391$ | $+0\cdot3915$ |
| $^{41}$K | $\frac{3}{2}$ | $+0\cdot215$ | $+0\cdot2154$ |
| $^{85}$Rb | $\frac{5}{2}$ | $+1\cdot340$ | $+1\cdot3527$ |
| $^{87}$Rb | $\frac{3}{2}$ | $+2\cdot733$ | $+2\cdot7505$ |
| $^{133}$Cs | $\frac{7}{2}$ | $+2\cdot558$ | $+2\cdot5789$ |
| Halogens: | | | |
| $^{19}$F | $\frac{1}{2}$ | $+2\cdot62$ | $+2\cdot6285$ |
| $^{35}$Cl | $\frac{3}{2}$ | $+0\cdot819$ | $+0\cdot8218$ |
| $^{37}$Cl | $\frac{3}{2}$ | $+0\cdot681$ | $+0\cdot6841$ |
| $^{79}$Br | $\frac{3}{2}$ | $+2\cdot110$ | $+2\cdot1056$ |
| $^{81}$Br | $\frac{3}{2}$ | $+2\cdot271$ | $+2\cdot2696$ |
| $^{127}$I | $\frac{5}{2}$ | — | $+2\cdot8090$ |

one is present; (4) electric interaction between the electric quadrupole moment of the nucleus (if $I \geqslant 1$) and the electric field gradient in the molecule.

The detection of an atom or molecule is greatly simplified if it can be converted into a charged ion. There are two principal methods: (*a*) bombardment by a beam of electrons, and (*b*) surface ionization (see § 11.5), where the molecule falls on a hot surface whose work function is greater than its ionization potential, from which it 'evaporates' as a positive ion. Ions formed in these ways are passed through a mass spectrometer to separate them from the background ion current (and sometimes from other isotopes of the same element). They are then accelerated to strike a surface from which secondary electrons are ejected on impact; these form a current which is further amplified by an electron multiplier (see § 11.4).

To obtain high precision, a narrow resonance curve is required, and in a beam experiment (see eqn (24.6)) this requires a large value of *t*, that is, a long path through the C-field. The corresponding requirement of high uniformity in the C-field is made less rigorous by the use of two separate oscillatory fields, one at each end of the C-field. With this arrangement, due to Ramsey (1949), only the mean value of the C-field over the whole path is required to be the same as that at the positions of the oscillatory fields.

## 24.3. Nuclear magnetic resonance in bulk material

The molecular-beam method of detecting nuclear magnetic resonance is experimentally very difficult; it was first used at a time when there seemed no prospect of the direct detection of the emission or absorption of quanta of the oscillatory field. At radio-frequencies the position is quite different from optical spectroscopy, since the rate of spontaneous emission of quanta is negligibly small. Except at frequencies greater than $10^4$ MHz and temperatures less than 1 K, the pair of energy levels involved have a separation $h\nu$ small compared with $kT$. This means that transitions between the two levels are continuously being induced through the effect of thermal fluctuations. An electronic oscillator is effectively a very high-temperature source, and radiation from it will be absorbed provided it is close to the resonant frequency. The problem is to observe the transitions caused by the applied oscillatory field against the general background of electrical noise associated with the thermal fluctuations.

The magnitude of the absorption in nuclear magnetic resonance can be estimated in the following way. From the theory of anomalous dispersion (§ 10.6), the imaginary part of the susceptibility $\chi''$ at the centre of a narrow line is related to the static susceptibility $\chi_0$ by the formula (see Problem 24.1)

$$\chi''/\chi_0 = \omega/2\Delta\omega = \nu/2\Delta\nu, \tag{24.7}$$

where $\Delta\nu$ is the distance from the centre of the line to a point where the intensity has fallen to half its maximum value. Now

$$\chi_0 = \frac{\mu_0 n g_N^2 \mu_N^2 I(I+1)}{3kT},$$

and in a favourable case, such as the protons in water, the value of $\chi_0$ at room temperature is approximately $10^{-8}$ (S.I. units). To estimate $\chi''$ we need an approximate value of $\Delta\nu$. The main cause of line-broadening arises from the random magnetic fields of neighbouring nuclear magnetic moments. The effect of these is to change the actual field at a given nucleus by an amount depending on the orientation and number of neighbouring magnetic dipoles: this causes a spread in the magnetic field acting on different nuclei, and so gives a finite linewidth. The spread $\Delta B$ is of the order of $\mu_0 m/4\pi d^3$, where $m$ is the dipole moment of a neighbour, and $d$ its distance. The mean value of $d^3$ is just the reciprocal of the number of protons per unit volume, and for water this gives $\Delta B$ $\sim 10^{-4}$ T. Since $\nu/\Delta\nu = B_0/\Delta B$, at a field of $0\cdot2$ T $\chi''$ would be about $10^{-8} \times 10^3 \sim 10^{-5}$. For protons, whose resonant frequency would then be about 8·5 MHz, the power transmitted through 10 km of the substance would fall only by 1·8 per cent (see Problem 24.3). It is obviously out of the question to use a transmission method; instead, the substance is inserted in the r.f. magnetic field of a coil of a circuit tuned to 8·5 MHz. The magnetic-resonance phenomenon then causes a change in $1/Q_F$, where $Q_F$ is the quality factor of the circuit, of magnitude (see Problem 24.2)

$$\delta(1/Q_F) = \chi''/(1+\chi') \sim 10^{-5}. \tag{24.8}$$

The smallest value of $1/Q_F$, the reciprocal of the quality factor of the circuit in the absence of resonance, is $\sim 0·5 \times 10^{-2}$, so that to detect the resonance requires the measurement of a change in $Q_F$ of less than 1 per cent.

Basically, the method of detection is similar to the use of the $Q$-meter described in § 22.3. Power from an electronic oscillator is fed through a high resistance $R'$ to a tuned circuit (Fig. 24.6), whose parallel impedance at resonance is $Q_F/\omega C$. Provided that $R' \gg Q_F/\omega C$, a constant current is drawn by the circuit, irrespective of small changes in $Q_F$. A small fractional variation $\delta Q_F/Q_F$ therefore appears as a fractional change $\delta v/v$ in the voltage across the tuned circuit, which can be amplified and detected. From eqn (24.8)

$$\delta v/v = \delta Q_F/Q_F = -Q_F \, \delta(1/Q_F) \sim -Q_F\chi'',$$

and we can estimate the smallest observable value of $\chi''$ by setting the value of $\delta v$ equal to that which gives a signal–noise ratio of unity. If $R$ is the parallel resistance of the tuned circuit at resonance, and from eqn (23.7) we insert the value $(4kTR \, df)^{\frac{1}{2}}$ for $\delta v$, we have

$$\chi''_{min} = \frac{1}{Q_F v}(4kTR \, df)^{\frac{1}{2}}. \tag{24.9}$$

Taking as reasonable values $Q_F = 10^2$, $R = 10 \text{ k}\Omega$, $df = 0·1 \text{ MHz}$, $v = 1 \text{ V}$,

FIG. 24.6. Basic circuit for detection of nuclear magnetic resonance. $R'$ is a high resistance such that $R' \gg R = Q_F/\omega C$, the parallel impedance of the tuned circuit. This ensures that a constant current $i$ flows to the tuned circuit, and a change in $Q_F$ produces a corresponding change in $v = iR$, the voltage across the tuned circuit. The output from the amplifier and detector is taken to a display unit (oscilloscope or pen recorder).

we find that $\chi''_{min} = 4 \times 10^{-8}$, giving an ample margin of sensitivity compared with eqn (24.8).

The presence of $Q_F$ in the denominator of eqn (24.9) shows the advantage of using a tuned circuit. Only a small sample is needed to fill the coil; the oscillatory magnetic field is much larger than in an ordinary radiation field, and the magnetic absorption is correspondingly increased. (The tuned circuit is equivalent to a path length of $Q_F/2\pi$ wavelengths; see Problem 24.10.) A further advantage of a tuned circuit is that the noise voltages across it are limited to those in a band $df$ near the resonance frequency; this band has an effective width $df \sim f/Q_F$ (see Problem 23.5), and this gives the value used above. Higher sensitivity can be obtained by further limitation of the noise bandwidth; for example, a phase-sensitive detector at the modulation frequency can be combined with the system of field modulation described in the following paragraph.

In an experiment the sample is placed in a coil whose axis (the direction of the oscillating flux density $\mathbf{B}_1$) is perpendicular to the steady field $\mathbf{B}_0$. The latter is then swept through resonance, at which point the voltage $v$ across the tuned circuit momentarily diminishes. A low-frequency modulation is imposed on $\mathbf{B}_0$, so that the momentary drop in $v$ is repeated in synchronism with this modulation, giving a low-frequency modulation of $v$. After detection, the modulation can be amplified (if necessary) and applied to the $Y$ plates of an oscillograph. The $X$ sweep is derived from the low-frequency modulation, so that a given $X$ deflection corresponds to a given change in $B_0$ caused by the modulation. As the mean value of $B_0$ (the field corresponding to the midpoint of the modulation) is slowly varied, the resonance line appears at one end of the trace, moves across, and disappears at the other end.

The first experiment of this kind was carried out by Purcell, Torrey, and Pound (1946). Simultaneously, nuclear resonance was observed quite independently by Bloch, Hansen, and Packard (1946) using a slightly different principle known as nuclear induction. To understand this we shall return to the rotating coordinate system used in Fig. 24.1. If there is some damping mechanism by which energy can be transferred from the system of nuclear spins to the outside world, the equilibrium state will be one where the net magnetization $\mathbf{M}$ will be parallel to the steady flux density $\mathbf{B}_0$. When a rotating flux density $\mathbf{B}_1$ is applied, the magnetization vector precesses at an angle $\delta$ about $\mathbf{B}_0$ with the angular velocity $\omega$ of the applied rotating field. Thus it is a constant vector in the rotating coordinate system, though not coplanar with $\mathbf{B}_0$ and $\mathbf{B}_1$ (this is the steady-state solution of the forced precession of the damped system). Theory shows that, if $\Delta B$ is the linewidth, $\delta$ is given by

$$\tan \delta = \frac{B_1}{\{(B_0 - B')^2 + \Delta B^2\}^{\frac{1}{2}}}.$$

Hence $\delta$ is greatest at resonance $(\mathbf{B}_0 = \mathbf{B}')$, and the rotating component $M \sin \delta$ is then also a maximum, leading the rotating flux density $\mathbf{B}_1$ by an angle $\frac{1}{2}\pi$. This component will induce a voltage in a coil placed at right-angles to the main coil producing the driving field $B_1 \cos \omega t$. The induced voltage is an oscillatory one with the same frequency, and the detector coil must therefore be carefully oriented to reduce as far as possible any direct pick-up from the driving coil. Here again the steady flux density $\mathbf{B}_0$ is 'wobbled' at an audio-frequency, since this differentiates between such stray pick-up, which will not be modulated at the wobble frequency, and the resonance effect, which is.

## 24.4. Relaxation effects in nuclear magnetic resonance

The Purcell method is generally known as 'nuclear resonance', and the Bloch method as 'nuclear induction'; they are alternative methods of detecting the same phenomenon, and have the same ultimate sensitivity. Both have been pushed to the limits of sensitivity in applications such as measurement of the value of $\gamma$ for rare isotopes. In this connection linewidth is of great importance, since the intensity at the centre of an absorption line varies inversely as the linewidth (see eqn (24.7)), and the accuracy with which the position of the centre of the line can be determined is also higher for a narrower line. It turns out that the linewidth varies very considerably with the nature of the sample, and we shall discuss this briefly first.

In the estimate of the linewidth for $H_2O$ made in § 24.3 the field due to one neighbouring proton was found to be about $10^{-4}$ T. This was an underestimate of the width to be expected, since there is more than one neighbour, and we would expect the full width $(2\Delta B)$ to be over $10^{-3}$ T (it is found to be about $1{\cdot}6 \times 10^{-3}$ T in ice at low temperatures). In liquid water, on the other hand, the resonance is extremely narrow; so narrow, that its actual width is very difficult to determine, as variation of the field produced by the external magnet over the volume of the sample is usually the limiting factor in determining the breadth. By working at low field strength, however, Brown and Purcell (1949) were able to show that the over-all width was less than $7 \times 10^{-7}$ T. The explanation for this striking difference from the width in ice is as follows. In water the molecules are not stationary, but on the average change their positions once every $10^{-11}$ s or so, this figure being given by the relaxation time of the Debye absorption discussed in § 10.8. This means that at a given nucleus the field of a neighbouring nucleus will not be constant, but will change its value every $10^{-11}$ s. This is a very much shorter time than that of the precession period in a field of, say, $0{\cdot}2$ T, which is $\approx 10^{-7}$ s. At first sight the rapid fluctuations of the random fields of the neighbours might be expected to broaden the line, but in fact the nucleus cannot respond to changing fields whose duration is less than $T_2 = (\Delta\omega)^{-1} = (\gamma\Delta B)^{-1}$, the inverse linewidth; in the words of Purcell (1948) 'the nucleus rides out the storm like a well-balanced gyroscope on perfect gymbals.' It turns out that the more rapid the fluctuations the more closely does their effect average to zero. If the rate of fluctuation is lower, as it is in a liquid of high viscosity such as glycerine, the linewidth is not so effectively reduced, and increases rapidly if the viscosity is increased by lowering the temperature. When the Debye relaxation time becomes much longer than the characteristic time $T_2$, the full linewidth is attained. In many substances the linewidth does not have the expected width due to the random magnetic fields of other nuclear dipole moments even in the solid state. This is attributed to internal motion within the solid lattice, which 'averages out' the fields of the neighbours, and nuclear resonance has been used to investigate such internal motions in many cases.

The nuclear magnetic resonance spectrum of protons in ethyl alcohol (Fig. 24.7) shows that extremely narrow lines can be obtained in liquids under conditions of high resolution. This spectrum also illustrates an important chemical application (see § 24.5); the same nuclear species (in this case protons) at different sites in a chemical compound may give lines at slightly different field strengths at the same frequency, or different frequency at the same field.

A second question of considerable importance is the rate at which energy is transferred from the nuclear spin system to the lattice containing the nuclei. In the absence of an external magnetic field $\mathbf{B}_0$, the nuclear spins will point in random directions, and $\chi_0$ will be zero. If a field $\mathbf{B}_0$ is now applied, the energy levels corresponding to different nuclear spin orientations will be split, as shown in Fig. 24.8. For simplicity, a spin $I = \frac{1}{2}$ is assumed, giving just two levels. Originally, the populations of these two levels were equal, and after $\mathbf{B}_0$ is switched on they will remain so until a

FIG. 24.7. Nuclear magnetic resonance spectrum at 30 MHz of ethyl alcohol. Left, proton signal from OH; centre, proton signal from $CH_2$ group; right, proton signal from $CH_3$ group. The separations between these main groups arise from changes in the diamagnetic shielding ('chemical shift'), while splittings within the groups are due to spin–spin interactions between the protons. The high resolution achieved is illustrated by the separation of lines A, B; this is about 0·25 Hz.

FIG. 24.8. The energy levels and the relative populations of a nucleus of spin $I = \frac{1}{2}$ before and after a magnetic field is switched on. The populations approach the new equilibrium values exponentially, with time constant $T_1$, the 'relaxation time'. In thermal equilibrium $n_B = n_A \exp(-W/kT)$ and $n_A + n_B = n$.

number are transferred from the upper to the lower state to give the equilibrium Boltzmann distribution in which

$$n_B = n_A \exp(-W/kT).$$

This involves a transfer of energy from the system of spins to the lattice, and the magnetization approaches its equilibrium value $\mathbf{M}_0$ according to the exponential law (compare the corresponding equation for electric polarization in § 10.8)

$$d\mathbf{M}/dt = (\mathbf{M}_0 - \mathbf{M})/T_1,$$

which gives

$$\mathbf{M}_0 - \mathbf{M} = \mathbf{M}_0 \exp(-t/T_1) \tag{24.10}$$

if $\mathbf{M} = 0$ at $t = 0$. The parameter $T_1$ is known as the spin–lattice relaxation time, since it determines the rate at which energy is transferred from the magnetic dipoles of the system of nuclear spins to the crystal lattice in which they are embedded. Such a transfer requires that transitions be induced between the various nuclear levels corresponding to different orientations, and these can only be caused by the presence of an oscillating magnetic field whose frequency satisfies the condition for resonance. Furthermore, this oscillating field must originate in the thermal motion of the surroundings of the nucleus. In a liquid the Brownian motion of neighbouring molecules causes the local magnetic fields of their nuclei to fluctuate rapidly, and this random fluctuation contains components of the right frequency to cause transitions. For distilled water at 293 K Bloembergen, Purcell, and Pound (1948) found the value of $T_1$ to be about 2 s.

This can be shortened by dissolving a paramagnetic salt in the water, so that the protons interact with the much bigger electronic magnetic moments. Another feature is that one would expect the desired component of the randomly fluctuating magnetic fields to be greatest (and hence $T_1$ to be shortest) when the Debye relaxation time of the liquid $\tau$ is of the order $1/\omega_0$, where $\omega_0$ is the angular frequency of the magnetic resonance. This was verified by Bloembergen, Purcell, and Pound by measuring $T_1$ in glycerine over a range of temperatures, which gives a wide range of viscosity and hence of the Debye relaxation time $\tau$. In solids at low temperatures, where there is practically no thermal motion, we should expect $T_1$ to be very long, but in practice it turns out to be nothing like as long as the values predicted by theory. This is ascribed to the presence of paramagnetic impurities, where the electron spins turn over and so give a fluctuating field. Owing to their larger magnetic moments, very few such impurity ions are required, and the thermal contact between the electron spin (or orbit) and the lattice is much more intimate than that between a nuclear spin and the lattice (see § 24.7), so that a transfer of energy to the electron spins from the nuclei is effectively a transfer to the lattice itself.

The variation of $T_1$ with the Debye relaxation time $\tau$ for three substances is shown in Fig. 24.9. In ethyl alcohol, $\tau \ll 1/\omega_0$ and $T_1$ decreases as $\tau$ increases. In glycerin $T_1$ passes through a minimum and increases again, while in ice, where $\tau \gg 1/\omega_0$, $T_1$ is rising with increasing $\tau$. This variation of $T_1$ (with a corresponding increase in the linewidth towards the value calculated for an assembly of static dipoles as $\tau$ increases) confirms the internal molecular motion in ice indicated by the Debye relaxation. Internal motions exist in many other solids, and have

FIG. 24.9. The thermal relaxation time $T_1$ for protons in ethyl alcohol, glycerin, and ice, measured at 29 MHz, plotted against the Debye relaxation time $\tau$, obtained from dielectric dispersion data. Note the logarithmic scales.

been investigated by measurement of the relaxation time $T_1$ as well as the linewidth in nuclear magnetic resonance. Such experiments have added considerably to our knowledge of the solid state.

## 24.5. Applications of nuclear resonance

The most obvious application of nuclear resonance is the measurement of $\gamma$ for all possible isotopes. A direct measurement of $\gamma$ involves a precise measurement of the frequency, which is comparatively easy, and of the magnetic flux density $B_0$, which is difficult. For this reason it is usual to measure instead the ratio of $\gamma$ for the unknown isotope to that for a standard isotope such as $^1$H, the proton. This may be accomplished by measuring the two frequencies of magnetic resonance in the same magnetic field. To achieve high accuracy, a single sample is used, such as a solution of a substance containing the unknown isotope in water. From the discussion of linewidth in § 24.4 it will be appreciated that a solution gives the narrowest lines, and hence the greatest accuracy as well as the greatest intensity at the centre of the line. The sample is surrounded by two coils, one for each of the two resonance frequencies, which are usually arranged to be mutually perpendicular so as to minimize the mutual inductance between them. The accuracy which can be achieved in this way is illustrated by the ratio of the nuclear moments of the deuteron and the proton, found by Wimett (1953) to be

$$m_D/m_H = 0.307\ 012\ 192 \pm 0.000\ 000\ 015.$$

The measurement was made with compressed HD gas in order to avoid difficulties with 'diamagnetic shielding' (see below), and certain other small corrections.

The accuracy of absolute values of $\gamma$ is determined by results for the proton obtained in various Standards Laboratories, which lie close to $267.513 \times 10^6$ per tesla for a spherical sample of pure $H_2O$. This absolute determination for the proton is also important because nuclear resonance offers a simple and accurate laboratory method of measuring magnetic flux density (see § 6.6). Only a frequency measurement is required, which can readily be carried out to an accuracy of one part in $10^6$ or better (see § 22.5). The main requirement is that the flux density must be uniform over the proton sample in order to avoid broadening of the resonance line, but this is not often a serious limitation.

### *Diamagnetic shielding*

The reason why the nature of the sample is specified carefully in this measurement is because the resonance frequency is determined by the magnetic flux density at the proton, which may not be exactly the same as the external flux density. One correction arises from 'diamagnetic shielding', an effect closely allied to diamagnetism. It arises from the precession of the closed shells of electrons about $\mathbf{B}_0$, which sets up a small field at the nucleus with the opposite sense to $\mathbf{B}_0$, thus making the actual field acting at the nucleus slightly smaller than the external field. The apparent value of $\gamma$, if no correction is made, is therefore less than the true value by a fractional amount of about $2 \cdot 8 \times 10^{-5}$ for hydrogen, rising to $10^{-2}$ for the heaviest elements. This correction has been computed, but the probable error rises considerably in the heavy elements. This shielding effect, which makes the field in the interior of an atom different from that outside, must be distinguished from the 'demagnetizing field' (§ 4.4) and the 'local field', which can be evaluated sufficiently accurately for this purpose by the method due to Lorentz (cf. § 10.3). Each of these fields is proportional to the bulk diamagnetism of the sample, and in a spherical sample the demagnetizing field and the Lorentz field just cancel, so that the average local field is the same as the external field, apart from the diamagnetic correction.

*Chemical shielding*

The change in the local field due to diamagnetic shielding may be written as $\mathbf{B} = \mathbf{B}_0(1 - \sigma)$. The value of $\sigma$ depends on the local density of electrons near the nucleus, and this varies slightly from compound to compound. It results in a displacement of the resonant frequency for the same isotope; this is known as a 'chemical shift' and has proved to be a powerful analytical and research tool in chemistry. 'Paramagnetic shifts', opposite in sign, arise from the induced paramagnetic moment in compounds with a temperature-independent paramagnetism such as $K_3Co(CN)_6$. The diamagnetic shielding may also vary between different nuclear sites in the same compound; in Fig. 24.7 the spectrum of $CH_3CH_2OH$ shows resonances from the protons in the $CH_3$, $CH_2$, and OH groups which are separated in frequency by about one part in $10^6$. Since $\mathbf{B} - \mathbf{B}_0 = -\sigma \mathbf{B}_0$, greater resolution can be obtained at higher frequencies, requiring higher values of $\mathbf{B}_0$, provided that the requirement of higher homogeneity in the field strength can be satisfied. For protons, frequencies up to 220 MHz, using a flux density of over 5 T from a superconductive magnet, are utilized without increase in linewidth. Special methods have been developed to give a resolution in liquids of about one part in $10^8$; these make it possible to observe further splittings in the spectrum (typically of order 10 Hz) which arise from interactions between neighbouring nuclear spins.

*Solid-state applications*

In strongly magnetic compounds large shifts in the nuclear resonance frequency are observed because of the fields set up by the electronic magnetic dipoles. Such dipoles change their orientation very rapidly, either through relaxation effects due to the thermal fluctuations of the lattice or through interaction with neighbouring spins; so long as many reorientations occur in a time $T_2 = (\gamma_N \Delta B)^{-1}$ characteristic of the nuclear resonance linewidth $\Delta B$, relatively narrow nuclear magnetic resonance (n.m.r.) lines are obtained, shifted in frequency by a local field which is proportional to the time average of the electronic magnetic moment. Thus the shift is temperature-dependent and proportional to the average electronic magnetization; in a paramagnetic substance we can write for the nuclear precession frequency

$$\omega = -\gamma_N(\mathbf{B}_0 + a\mathbf{M}) = -\gamma_N \mathbf{B}_0 (1 + a\chi/\mu_0),$$

showing that the fractional change in frequency is proportional to the susceptibility. In an ordered magnetic substance, where the magnetization is finite in the absence of an external field, n.m.r. can be observed at a frequency which is very nearly proportional to the magnetization in a ferromagnetic substance, or to the sublattice magnetization in an antiferromagnetic or ferrimagnetic compound.

The nuclear magnetic moment of $^{19}$F is quite large, and signals of high intensity have been observed from nuclear magnetic resonance of this nucleus in a number of magnetic compounds. It was found by Shulman and Jaccarino (1956) that in $MnF_2$ in the paramagnetic state, the $^{19}$F

FIG. 24.10. Temperature-dependence of the $^{19}$F n.m.r. frequency in $MnF_2$ between 52 K and the Néel point 67·3 K. The lower curve shows the variation expected if the sublattice magnetization, relative to that at 0 K, followed the curve computed from molecular field theory for $S = \frac{5}{2}$. In this and other substances the resonance frequency varies near $T_N$ as $(T_N - T)^{\frac{1}{3}}$. (Heller and Benedek 1962.)

resonance was shifted by an amount proportional to the electronic paramagnetic susceptibility. However, the magnitude of the shift was greater than would be expected from the dipolar magnetic field at the $F^-$ site due to the electronic magnetic moments, assuming them to be localized on the $Mn^{2+}$ ions; the larger field is consistent with a spread of the wavefunctions of the magnetic electrons onto the $F^-$ ions, due to a small amount of covalent bonding. In the antiferromagnetic state the electronic moments are fixed in orientation, and the field which they produce at the $F^-$ nucleus is quite high even when $\mathbf{B}_0 = 0$. At 0 K in $MnF_2$ the precession frequency of $^{19}F$ in this electronic field is 159·99 MHz falling, as the temperature rises, to zero at the Néel point, 67·3 K. A comparison of the nuclear-resonance frequency, which should be proportional to the sublattice magnetization, with the magnetization calculated from the molecular field theory, is shown in Fig. 24.10.

## 24.6. Electron magnetic resonance in atomic beams

Magnetic-resonance experiments involving electronic magnetic moments can be carried out in ways analogous to experiments with nuclear magnetic moments. In general, experiments with electronic moments are easier, because the moments are so much larger and magnetic resonance is correspondingly easier to detect. In the analogue of Rabi's experiments (§ 24.2), beams of atoms with electronic magnetic dipole moments are used, for which appreciable deflections in the inhomogeneous magnetic fields A, B (see Fig. 24.3) can be achieved using relatively short magnets. The main interest of such experiments arises from the high precision which can be obtained by increasing the length of the C-field, since from eqn (24.6) the width of the resonance curve depends on the time $t$ during which the atom is subjected to the radio-frequency field. If the velocity of atoms in the beam is $10^3$ m s$^{-1}$ and the length of the C-field is 0·03 m, the value of $t$ is $3 \times 10^{-5}$ s, and the corresponding line width $2\Delta \nu$ is about 30 kHz. To achieve such a narrow line the C-field would have to be very homogeneous, since a variation of the field by as little as $10^{-6}$ T would change the electronic magnetic-resonance frequency of an atom with g = 2 by 30 kHz. The requirement of such high homogeneity is avoided by use of the two separated oscillatory fields method of Ramsey (1949), as mentioned in § 24.2.

For an atom with an electronic magnetic moment but no nuclear moment the magnetic-resonance transitions occur at a frequency (cf. eqn (24.5))

$$\nu = (\gamma/2\pi)B_0 = g_J(e/4\pi m_e)B_0 \tag{24.11}$$

and determination of $\nu$ in a known flux density $B_0$ gives a precise measurement of $g_J$. An important application of atomic-beam magnetic resonance is the precise measurement of hyperfine structure in atoms. As pointed out in § 14.10, in an atom with both electronic and nuclear moments the nuclear magnetic moment precesses in the electronic magnetic field of flux density $B_e$, and the electronic moment in the nuclear magnetic flux density $B_n$, the result being a precession of each moment about the resultant angular-momentum vector $F$. This gives rise to a set of hyperfine energy levels, as illustrated in Fig. 24.11. In magnetic resonance the transitions for which $\Delta F = \pm 1$ have an intensity associated with an electronic magnetic dipole moment, and can readily be observed; from measurements of the frequencies of such transitions the hyperfine constants $A$, $B_Q$ can be determined using eqn (14.32). These require no magnetic field in the C-magnet, but in practice a small field $B_0$ is generally used, whose effect (see Fig. 24.11) is to split each set of states with a given value of $F$ into $(2F+1)$ levels, provided that the Zeeman energy $g_J \mu_\beta (J \cdot B_0)$ is small compared with the hyperfine energies. This corresponds to a precession of the vector $F$ about $B_0$ at a rate slow compared with the precession of $J$, $I$ about $F$. The allowed transitions are then those for which $\Delta F = 0, \pm 1$; $\Delta m_F = 0, \pm 1$, but the $\Delta m = 0$ transitions occur only if the oscillatory field has a component parallel to $B_0$, while the $\Delta m = \pm 1$ transitions require a component $B_1$ perpendicular to $B_0$, as in Fig. 24.1.

The atomic-beam magnetic-resonance technique has been widely used to measure hyperfine constants of 'free' atoms with very high precision. A hyperfine frequency can be determined under conditions singularly free

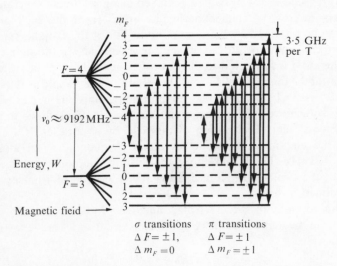

σ transitions       π transitions
$\Delta F = \pm 1$,     $\Delta F = \pm 1$
$\Delta m_F = 0$       $\Delta m_F = \pm 1$

FIG. 24.11. The hyperfine structure of $^{133}$Cs, $S = \frac{1}{2}$, $I = \frac{7}{2}$, showing the Zeeman effect in a small field (note the difference in scale between the Zeeman splittings and the overall splitting) and the allowed transitions.

from external influences, and we discuss here only the example of caesium, which is used to provide an atomic standard of frequency.

The ground state of the caesium atom is $^2S_{\frac{1}{2}}$, and the only stable isotope $^{133}Cs$ has nuclear spin $I = \frac{7}{2}$. Since the electronic ground state is $S = \frac{1}{2}$, there is no electric quadrupole interaction (see § 14.10), and in zero magnetic field there are two sets of hyperfine levels corresponding to $F = 3$ and $F = 4$, as in Fig. 24.11. Caesium is fairly volatile, so that an atomic beam of sufficient intensity can be emitted from an oven at 370 K, and this combined with the rather high atomic mass gives a low thermal velocity and increases the time required to traverse the C-field, where a distance of about 2·8 m is used between the separated oscillatory fields (see Essen and Parry 1959). A small field of $5 \times 10^{-6}$ T (this requires cancellation of the earth's field $\sim 5 \times 10^{-5}$ T) is maintained as the C-field, and the resonance observed is the transition $(F, m_F) = (4, 0) \leftrightarrow (3, 0)$, which has only a second-order Zeeman effect

$$\nu = \nu_0 + (426 \times 10^8) B_0^2 \text{ Hz} \quad (B_0 \text{ in tesla}).\qquad(24.12)$$

As a result of international collaboration between Standards Laboratories, the measurement of $\nu_0$ can be carried out with an agreed precision of one part in $10^{11}$, a considerable improvement on previous standards. These were based on astronomical observations (for a discussion, see Essen 1969); the difficulties are illustrated by the variation in the length of the day, as measured against a caesium 'atomic clock', shown in Fig. 24.12. Following an international recommendation in 1963, the

FIG. 24.12. Variation in the length of the day in terms of the atomic unit of time. (Essen 1969.)

caesium 'clock' has become the new standard of frequency, the second being defined as the time interval containing exactly

$$9\ 192\ 631\ 770\ \text{cycles}$$

of the caesium hyperfine frequency in zero magnetic field.

## 24.7. Electron magnetic resonance in solids

Magnetic-resonance experiments on substances containing permanent magnetic dipole moments due to electrons can be carried out in a manner analogous to those on nuclear dipoles, but there are a number of significant differences. If fields of a few tenths of a tesla are used, the resonance frequency for electrons is in the vicinity of 10 GHz, corresponding to wavelengths of a few centimetres. Frequencies of this order and higher are in fact used for a number of reasons.

1. The sensitivity is high; this is partly through eqn (24.7), and partly because a better filling factor can be obtained from a small sample by using it in a tuned circuit (a cavity resonator) which has similar dimensions.

2. The electronic levels may have splittings of many GHz due to crystal field effects (see § 14.7).

3. Linewidths in the solid due to magnetic fields of neighbouring ions are of the order of $0 \cdot 01 - 0 \cdot 1$ T, and to achieve reasonable accuracy in determining the centre of a line measurements must be made using external fields as large as possible. Linewidth and line-shape are also affected by exchange interaction between the ions; this can be avoided by making measurements on 'diluted' crystals—crystals in which most of the paramagnetic ions have been replaced by diamagnetic ions. For example, a crystal of $K_2Zn(SO_4)_2.6H_2O$ containing a few tenths of a per cent of $Cu^{2+}$ $3d^9$ ions replacing $Zn^{2+}$ $3d^{10}$ ions gives a linewidth of about $10^{-3}$ T; this residual width is due mainly to the nuclear magnetic moments of the protons in the water of crystallization, and can be further reduced by growing crystals with $D_2O$ instead of $H_2O$ because of the smaller nuclear moment of the deuteron.

Another important feature of electron spin resonance in paramagnetic substances is that spin–lattice relaxation times may be extremely short. Obviously the electron, with its larger magnetic moment, will be in more intimate contact with the lattice vibrations than a nuclear dipole, but a much more important effect is that the lattice vibrations distort the local surroundings of a paramagnetic ion, and so produce a fluctuating modulation of the crystal electric field or the ligand field. The extent to which this affects the magnetic dipole depends on the degree of 'quenching' of the orbital moment: for an ion in an S state, such as $Mn^{2+}$ ($3d^5$, $^6S_{\frac{5}{2}}$) or $Gd^{3+}$ ($4f^7$, $^8S_{\frac{7}{2}}$), there is no orbital moment except through small departures from $L–S$ coupling, and the spin–lattice relaxation time $T_1$ varies from $\approx 10^{-6}$ s at room temperature to $10^{-3}$ s at liquid–helium temperatures. For other ions the values of $T_1$ are very much smaller, and in many ions of the 4f group $T_1$ is so short and the lines so broad ($\Delta\omega = T_1^{-1}$) that magnetic resonance is unobservable, except at liquid-helium or liquid-hydrogen temperatures. The value of $T_1$ always increases as the temperature falls because the lattice vibrations die out, the variation being in general of the form

$$1/T_1 = aT + bT^n + c \exp(\hbar\omega/kT). \tag{24.13}$$

The first term $aT$ is due to 'direct transitions' in which magnetic quanta are exchanged with lattice vibrations of the same frequency as the magnetic-resonance frequency; the second term $bT^n$ (where $n = 5$, 7, or 9 according to the type of magnetic ion involved) is due to 'indirect' or 'Raman' processes in which any two lattice vibrations are involved whose frequency *difference* is equal to the magnetic-resonance frequency ($\omega_{lat} = \omega'_{lat} \pm \omega_{res}$); the exponential term is due to lattice vibrations whose quanta $\hbar\omega$ coincide with the difference in energy between the ground state and an excited state of the magnetic ion. The second two processes are weaker than the first, and at liquid-helium temperatures the first term almost always predominates.

A spin–lattice relaxation time $T_1$ will produce a linewidth of $2\Delta\nu$ between the half-intensity points of a line, where $2\pi\Delta\nu = \Delta\omega = T_1^{-1}$, and measurements of linewidth can be used to find $T_1$ when this is the dominant effect in the linewidth. For other purposes the need for high resolution in order to obtain accurate measurements makes it desirable, however, to work at temperatures where broadening due to spin–lattice relaxation is negligible, and a typical apparatus for low-temperature work is outlined in Fig. 24.13. Power from a microwave oscillator (usually a klystron) is carried by a waveguide or coaxial cable to a cavity resonator contained in a dewar vessel and placed between the poles of an electromagnet. When the latter is adjusted to resonance, power is absorbed in the paramagnetic sample, which is placed inside the cavity in a position of maximum oscillatory magnetic field. This additional power loss in the cavity produces a change in the signal reflected from the cavity, or in the signal transmitted through the cavity to another waveguide or coaxial line, which is detected by a silicon-crystal rectifier. Normally the field of the electromagnet is modulated at an audio-frequency, giving a corresponding modulation of the signal when the field for magnetic resonance is traversed; after detection, this modulation is amplified and displayed on

FIG. 24.13. Outline diagram of an electron spin resonance apparatus. Power from a klystron oscillator is fed through a waveguide to a loosely coupled resonant cavity containing the paramagnetic sample and immersed in a refrigerant between the poles of an electromagnet modulated at an audio-frequency. A small fraction of the cavity signal is fed through a second waveguide to a silicon diode detector. The modulation due to the absorption is amplified and displayed on an oscilloscope or fed through a phase-sensitive detector to a pen recorder.

FIG. 24.14. Energy levels and absorption curve at constant frequency for an ion with $S = \frac{3}{2}$ and a crystal field splitting:

$$W = g\mu_\text{B}S_zB_z + D\{S_z^2 - \tfrac{1}{3}S(S+1)\}$$

when the magnetic field is parallel to the principal axis ($z$-axis) of the splitting term (for other directions the levels do not diverge linearly with field). The intensity of the $S_z \leftrightarrow (S_z-1)$ transition is proportional to $\{S(S+1) - S_z(S_z-1)\}$ giving the $3:4:3$ intensity ratio shown in the figure.

FIG. 24.15. Hyperfine structure of the $Mn(I = \frac{5}{2})$ nucleus in the electron spin resonance spectrum of an $Mn^{2+}$ ion $(S = \frac{5}{2}$, transition $S_z = \frac{1}{2} \leftrightarrow -\frac{1}{2})$. The resonance condition is $h\nu = g\mu_B(B_0 + B_N)$, where $B_N$ is the magnetic field due to the nucleus. In first approximation $B_N$ is proportional to the nuclear magnetic quantum number $I_z$; this gives a pattern of $2I + 1 = 6$ lines, equally spaced and of equal intensity, since all nuclear orientations are equally probable at the temperature of the observation. The line-shape is the derivative of the absorption curve; it is obtained by a sinusoidal modulation of $B_0$ with amplitude small compared with the linewidth. This gives a corresponding modulation of the signal (measured by a phase-sensitive detector) whose amplitude is proportional to the slope of the absorption curve.

an oscilloscope or a recorder. The sensitivity achieved is quite high, and signals from as few as $10^{13}$ electronic dipoles can be seen if the lines are narrow.

The results of electron magnetic resonance in paramagnetic salts have greatly advanced the detailed understanding of the properties of paramagnetic ions subjected to ligand field interactions in solids. In general the resonance spectrum is very anisotropic, depending strongly on the angle between the external field and the crystal axes; for this reason single crystals must be used. In dilute salts the splittings of the levels due to the external field as a function of angle can be studied, together with any crystal field splittings of the same order as the microwave frequency (see Fig. 24.14). When the nucleus of the paramagnetic ion or of a ligand ion has a nuclear moment, a hyperfine structure may be observed, as in Fig. 24.15. A number of nuclear spins and moments have been determined from hyperfine structure in electron magnetic resonance, and the degree to which the wavefunctions of the magnetic electrons overlap onto the ligand ions because of covalent bonding effects (see § 14.8) can be estimated from the hyperfine structure due to interaction with the dipole moment of the ligand nucleus. In more concentrated salts the effect of magnetic dipole and exchange interaction between neighbouring dipoles can be studied, giving one of the few direct measurements of exchange interaction.

*Ferromagnetic resonance*

In substances where the exchange forces are strong magnetic resonance may be observed in the co-operative state below the transition temperature. Since all the dipoles are coupled together by the exchange forces it is convenient to work in terms of the magnetization **M**, which is the vector sum of the individual dipole moments **m**. By performing this vector sum over both sides of eqn (24.1) we obtain the equation of motion for the magnetization

$$\mathrm{d}\mathbf{M}/\mathrm{d}t = \gamma \mathbf{M} \wedge \mathbf{B}, \qquad (24.14)$$

where we have assumed that all dipoles have the same value of $\gamma$. Here we have written, not $\mathbf{B}_0$ the external flux density, but $\mathbf{B}$ the flux density within the sample, since in a ferromagnetic substance demagnetizing fields may be quite important. We shall assume that $\mathbf{B}_0$ is along the $z$-axis, and confine ourselves to certain sample shapes such that we can write for the components of $\mathbf{B}$:

$$B_x = -\mu_0 D_x M_x; \qquad B_y = -\mu_0 D_y M_y; \qquad B_z = B_0 - \mu_0 D_z M_z.$$
$$(24.15)$$

On substituting into eqn (24.14) we obtain

$$\mathrm{d}M_x/\mathrm{d}t = \gamma M_y \{B_0 + \mu_0 M_z (D_y - D_z)\},$$
$$\mathrm{d}M_y/\mathrm{d}t = -\gamma M_x \{B_0 + \mu_0 M_z (D_x - D_z)\}, \qquad (24.16)$$
$$\mathrm{d}M_z/\mathrm{d}t = \gamma \{\mu_0 M_x M_y (D_x - D_y)\},$$

which are no longer linear in $M$. However, the equations can be solved in the limit of small amplitudes of precession, when the product $M_x M_y$ becomes vanishingly small and can be neglected. Then $\mathrm{d}M_z/\mathrm{d}t = 0$, and $M_z$ is constant, its value being equal to the static magnetization; this is large in a ferromagnetic substance, and the corrections to $B_0$ are important in determining the resonance frequency. By solving the equations for $M_x$, $M_y$ it is easily shown that the precession frequency is

$$\omega_L = -\gamma [\{B_0 + \mu_0 M_z (D_y - D_z)\}\{B_0 + \mu_0 M_z (D_x - D_z)\}]^{\frac{1}{2}}. \qquad (24.17)$$

There are three simple cases of interest:

(1) a sphere, for which $D_x = D_y = D_z$; the precession angular velocity is $\omega_L = -\gamma B_0$, the same as if there were no demagnetizing fields;

(2) a thin plane film normal to $\mathbf{B}_0$, for which $D_x = D_y = -1$, $D_z = 0$, giving

$$\omega_L = -\gamma (B_0 - \mu_0 M_z);$$

(3) a thin plane film parallel to $\mathbf{B}_0$, for which $D_x = D_z = -1$, $D_y = 0$ (assuming the film to be normal to the $y$-axis), giving

$$\omega_L = -\gamma \{B_0 (B_0 + \mu_0 M_z)\}^{\frac{1}{2}}.$$

This treatment assumes that the magnetization (including the precessing components) is uniform throughout the sample; this requires that the dimensions be small compared with the wavelength in the sample, and in a conducting sample this means small compared with the skin depth.

Hence spherical samples of metal must be very small, and colloidal samples (where the particles are assumed to be spherical because of surface-tension effects in formation) have been used. Most work has been done on thin plane samples, which are attached to (but insulated from) one wall of the microwave cavity. The equations show that the magnetization must be known in order to determine $\gamma$; for simplicity it is usual to work at such high fields that the magnetization is equal to the saturation value. The phenomenon of ferromagnetic resonance was discovered experimentally by Griffiths (1946); the theory given above is due to Kittel (1948). Some values of g measured by ferromagnetic resonance are given in Table 15.2 (p. 493).

*Spin-wave resonance in ferromagnetic films*

The uniform precession mode assumed above corresponds to a spin wave with $k_s = 0$. It is possible to excite spin waves for which $k_s \neq 0$; since each spin wave corresponds to a unit change $\hbar$ in angular momentum, and hence to a change of $g\mu_B$ in magnetization, the energy required to excite a spin wave in a flux density $B$ (in the sample) is

$$\hbar\omega = g\mu_B B + D k_s^2, \qquad (24.18)$$

FIG. 24.16. Ferromagnetic spin-wave resonance in a thin film (600 nm thickness) of cobalt metal at room temperature and a frequency of 9370 MHz (32 mm wavelength). The line-shape is that corresponding to the differential of the absorption curve. The intensity of resonance decreases towards lower field strength (higher values of p) irregularly because of lack of uniformity in the film thickness. $B_0$ is normal to the surface of the film, so that rather high values of $B_0$ are needed to satisfy the resonance condition $\omega_1 = -\gamma(B_0 - \mu_0 M_s)$.
(Phillips and Rosenberg 1964.)

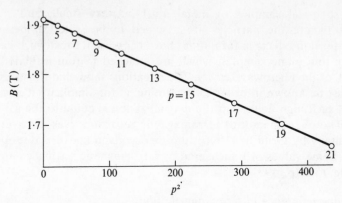

FIG. 24.17. Plot showing the linear relation between magnetic flux density and $p^2$ for ferromagnetic spin-wave resonance in a thin film of cobalt. (After Phillips and Rosenberg 1964.)

where the constant $D$ is the same as that in eqn (15.24). In a thin film of thickness $l$, the boundary conditions (assumed identical at the two faces of the film) limit the allowed values of $k_s$ to those for which the film thickness is an integral number of half-wavelengths; that is, $k_s = p\pi/l$, where $p$ is an integer. If magnetic resonance is observed at constant frequency, the value of the resonance flux density is found from eqn (24.18) above to be

$$B = B_{k_s=0} - \left(\frac{D}{g\mu_B}\frac{\pi^2}{l^2}\right)p^2, \qquad (24.19)$$

so that a series of resonances corresponding to different values of $p$ should be observed on the low field side of the ordinary ferromagnetic resonance flux density $B_{k_s=0}$. A spin wave resonance curve is shown in Fig. 24.16 for a thin film of cobalt metal, of thickness approximately equal to 600 nm. This is small compared with the skin depth, so that the oscillatory field is uniform within the sample, and the resonance intensity depends on the net magnetic moment in the direction of the oscillatory field. This is proportional to

$$\int_0^l \sin kx \, dx = \int_0^l \sin \frac{p\pi x}{l} \, dx = \left(\frac{l}{p\pi}\right)(1-\cos p\pi).$$

This vanishes for even values of $p$, and decreases as $1/p$ for odd values, giving the intensity change shown in Fig. 24.16. The resonance field decreases accurately as $p^2$ (see Fig. 24.17), and the value of $D$ can be found if the thickness $l$ is known.

*Ferrimagnetic and antiferromagnetic resonance*

The presence of two sublattices in these substances makes the magnetic-resonance phenomena in general much more complicated. One simple case occurs in ferrimagnetic substances with strongly coupled sublattices; the two sublattices can then precess together in such a way that the relative orientation of their two magnetic moments remains unaltered. This occurs at an angular velocity $\omega = -\gamma_{\text{eff}}B$; here $B$ is the flux density in the substance and $\gamma_{\text{eff}}$ is an average value obtained from the relation

$$\mathbf{M} = \sum_i \mathbf{M}_i = \sum \gamma_i \mathbf{G}_i = \gamma_{\text{eff}} \sum \mathbf{G}_i = \gamma_{\text{eff}} \mathbf{G},$$

where the summation is over all the individual ions. In both ferri- and antiferromagnetics more complicated modes of precession occur in which the relative orientation of the sublattice magnetic moments is not preserved; the frequency of precession then depends on a number of parameters, including the exchange and anisotropy energies.

## 24.8. Cyclotron resonance with free charged particles

When a charged particle of mass $M$ and charge $q$ is moving in a uniform magnetic flux density $\mathbf{B}$, its equation of motion is

$$\mathbf{F} = q\mathbf{v} \wedge \mathbf{B}.$$

Since this force is always normal to its instantaneous velocity $\mathbf{v}$, the particle will move in a circle of radius $r$ in the plane normal to $\mathbf{B}$ with angular velocity given by the equation

$$M\omega_c^2 r = q\omega_c r B,$$

that is,

$$\omega_c = (q/M)\mathbf{B}.$$

Thus if it is possible to determine the angular velocity $\omega$ in a known flux density $\mathbf{B}$, the ratio of charge to mass of the particle may be found. The frequency $\omega_c/2\pi$ is often called the 'cyclotron frequency' since it is the frequency of the r.f. electric field required to accelerate charged particles in the cyclotron. The success of this device, which depends on resonance between the frequency of the oscillating electric field and the frequency of rotation of the particles in the field $\mathbf{B}$, suggests that a similar principle may be used to determine the ratio of $q$ to $M$.

We begin by investigating the motion of a charged particle starting from rest under the action of a uniform flux density $\mathbf{B}$ (whose direction we take to be the $z$-axis of a system of Cartesian coordinates) and an oscillating electric field of frequency $\omega/2\pi$ polarized so that the lines of electric field are parallel to the $x$-axis. Then the equations of motion are

$$M\ddot{x} = qE \cos \omega t + q\dot{y}B,$$
$$M\ddot{y} = -q\dot{x}B, \qquad (24.20)$$
$$M\ddot{z} = 0.$$

The last of these equations shows that the $z$-component of the motion will be independent of $E$ and $B$, and does not appear in the other equations. The second equation can be integrated once giving

$$M\dot{y} = -qxB,$$

where the constant of integration has been equated to zero, corresponding to the assumption that the particle starts at rest from the origin. $\dot{y}$ may now be eliminated from the first equation giving

$$\ddot{x} + \omega_c^2 x = (q/M)E \cos \omega t,$$

where $\omega_c = (q/M)B$. The general solution of this equation is

$$x = \frac{(qE/M)\cos \omega t}{\omega_c^2 - \omega^2} + C \cos \omega_c t + D \sin \omega_c t.$$

If the initial conditions are $x = 0$, $\dot{x} = 0$ at $t = 0$, the unknown constants are determined and we have

$$x = \frac{(qE/M)(\cos \omega t - \cos \omega_c t)}{\omega_c^2 - \omega^2}$$

$$= \frac{2(qE/M)\sin \frac{1}{2}(\omega_c + \omega)t \sin \frac{1}{2}(\omega_c - \omega)t}{(\omega_c + \omega)(\omega_c - \omega)}$$

$$= \frac{qE}{M\omega'} \sin \omega' t \left( \frac{\sin \frac{1}{2}\Delta\omega t}{\Delta\omega} \right), \qquad (24.21)$$

where $\omega' = \frac{1}{2}(\omega_c + \omega)$, $\Delta\omega = \omega_c - \omega$. If $\Delta\omega \ll \omega_c$, the factor $\sin \frac{1}{2}\Delta\omega t$ varies very slowly with respect to time compared with $\sin \omega' t$, and $\omega'$ is very close to $\omega_c$, so that using the relation $\dot{y} = -\omega_c x$ we find approximately

$$y = \frac{qE}{M\omega'} \cos \omega' t \left( \frac{\sin \frac{1}{2}\Delta\omega t}{\Delta\omega} \right). \qquad (24.22)$$

Examination of the equations for $x$ and $y$ shows that the path of the particle is a spiral with angular velocity $\omega'$ and radius

$$\frac{qE}{M\omega'} \cdot \frac{\sin \frac{1}{2}\Delta\omega t}{\Delta\omega}.$$

If $\Delta\omega = 0$ (that is, $\omega = \omega_c$) then the value of the factor $(\sin \frac{1}{2}\Delta\omega t)/\Delta\omega$ is $\frac{1}{2}t$, showing that the radius increases linearly with $t$. On the other hand, if

$\Delta\omega \neq 0$, the radius has a maximum value $r_0$ (when the sine is unity) equal to $qE/M\omega'\,|\Delta\omega|$, which is very nearly equal to $qE/M\omega_c\,|\Delta\omega|$ when $\Delta\omega$ is small. Hence, if a collector is placed at a distance $R_0$ from the origin, only those ions will reach it for which $r_0 \geqslant R_0$, or

$$|\Delta\omega| \leqslant qE/M\omega_c R_0 = E/BR_0.$$

This is a measure of the precision with which $\omega_c$, and hence $q/M$, can be determined. The 'resolving power' will be

$$\frac{\omega_c}{|\Delta\omega|} = \frac{qB/M}{E/BR_0} = \frac{qB^2R_0}{ME}. \tag{24.23}$$

Hence, for a given ion and a given flux density $B$, the precision is increased by using a small amplitude of oscillating electric field strength $E$ and a large value of $R_0$. It can be shown that our expression for the resolving power is equal to $L/2R_0$ (see Problem 24.4), where $L$ is the total path traversed by the ion in its spiral journey from the origin to the collector. Thus the resolving power depends primarily on the number of revolutions which the ions make on their journey to the collector.

The apparatus used by Sommer, Thomas, and Hipple (1951) is shown in Fig. 24.18. An oscillatory voltage is applied between two parallel plates of size $3\,\mathrm{cm}\times5\,\mathrm{cm}$, and separation $2\,\mathrm{cm}$, with a number of parallel guard rings. These rings are equally spaced, and by means of a potentiometer system a fraction of the voltage proportional to the distance from one end-plate is applied to them so that a uniform r.f. field is obtained. A small steady positive voltage of about $0\cdot1\,\mathrm{V}$ is applied to the guard rings relative to the end-plates so as to retard the drift of positive ions in the direction parallel to **B**. The magnitude of **B** is determined by

FIG. 24.18. Apparatus for measuring the cyclotron resonance frequency of the proton. **B** is normal to the plane of the paper; $V$ is a steady voltage of about $0\cdot1\,\mathrm{V}$ for focusing the ion beam; **E** is an oscillatory electric field.

an n.m.r. experiment, using an r.f. coil containing a sample of oil. Ions are produced along the axis of the apparatus by firing in a narrow beam of electrons of about 70 V energy, which cause ionization by collision with the residual gas. The pressure must be kept low ($< 10^{-4}$ Pa) in order to prevent scattering of the ions by collision. The whole assembly is enclosed in a glass tube of 47 mm diameter, which fits between the poles of an electromagnet.

In a typical experiment $B = 0 \cdot 47$ T, and the oscillatory field strength $E$ is about $10 \, \text{V m}^{-1}$ at a frequency of about 7 MHz for the $H^+$ ion. With $R_0 = 10$ mm, the ions make about 7000 revolutions and attain an energy of about 1000 eV before reaching the ion collector, which is connected to an electrometer. The ion current at the peak of a resonance is about $3 \times 10^{-14}$ A, while the background fluctuations are about $4 \times 10^{-16}$ A. Owing to the small positive voltage on the guard rings, and to space charge, a small radial electric field exists which displaces the resonant frequency slightly. In a radial field strength $E'$ the equation of motion is

$$M\omega^2 r + qE' = q\omega rB,$$

whence approximately

$$\omega = \omega_c \left(1 - \frac{E'M}{rqB^2}\right). \tag{24.24}$$

In practice it turns out that $E'$ increases linearly with $r$, and hence the shift in the resonance is independent of $r$, but proportional to $M$. Thus by making measurements both with $H^+$ and $H_2^+$ ions ($H^+$ and $D_2^+$ ions were also compared) the size of the shift can be determined.

A number of measurements of the cyclotron resonance frequency of protons have been carried out, together with analogous experiments on electrons, where the resonance frequency is about $10^4$ MHz in a field approaching $0 \cdot 4$ T. In all such experiments the ratio of two frequencies in the same magnetic field is determined—for example, the cyclotron resonance frequency $\nu_c$ of free protons, and the n.m.r. frequency $\nu_p$ of protons in a standard liquid sample such as water. If a correction of 28 p.p.m. is applied for diamagnetic shielding of the protons in the water sample, the ratio

$$\frac{\omega_p}{\omega_c} = \frac{2\pi\nu_p}{2\pi\nu_c} = \frac{g_N(e/2m_p)B}{(e/m_p)B} = \tfrac{1}{2}g_N$$

gives the nuclear magnetic moment of the proton in nuclear magnetons, since the spin $I = \tfrac{1}{2}$. Over the period 1965–70 (see Cohen and Taylor 1972), four independent measurements (omitting the diamagnetic correction) lie within the limits $\nu_p/\nu_c = 2 \cdot 79277 \pm 0 \cdot 00002$, a precision of about one part in $10^5$.

## 24.9. Cyclotron resonance in semiconductors

It was pointed out in Chapter 12 that the equations of motion of electrons (and holes) in the periodic potential of a crystal lattice are similar to those of a free particle, provided that an effective mass $m^*$ is used instead of the true mass. This holds also for motion in a magnetic field, and the cyclotron resonance frequency becomes

$$\omega_c = (q/m^*)B,$$

if the effective mass is isotropic. Determination of this frequency is thus of great importance since it gives a direct measurement of $m^*$. In principle, the experiment is similar to those described in the previous section: an oscillatory electric field is applied normal to the steady magnetic field, and either its frequency or the strength of the magnetic field is varied while the power absorbed is measured. However, the charge carriers in a solid make collisions at a rate which is usually comparable with (and often much higher than) the cyclotron resonance frequency; this gives a very important damping term, and the equation of motion may be written as (cf. eqn (3.13))

$$m^*\left(\frac{d\mathbf{v}}{dt} + \frac{1}{\tau}\mathbf{v}\right) = q(\mathbf{E} + \mathbf{v} \wedge \mathbf{B}), \tag{24.25}$$

where $q = -e$ for electrons and $+e$ for holes.

To solve this equation we assume that $\mathbf{B}$ is along the $z$-axis of a Cartesian coordinate system and $\mathbf{E}$ is an oscillatory field along the $x$-axis. We therefore write $E_x = E_0 \exp j\omega t$, and look for the steady-state solution corresponding to the driven motion at angular frequency $\omega$; we can then replace $d/dt$ by $j\omega$, and the equations become

$$\left.\begin{array}{l}
\left(j\omega + \dfrac{1}{\tau}\right)v_x = \dfrac{q}{m^*}(E_x + v_y B), \\[3mm]
\left(j\omega + \dfrac{1}{\tau}\right)v_y = -\dfrac{q}{m^*}v_x B, \\[3mm]
\left(j\omega + \dfrac{1}{\tau}\right)v_z = 0.
\end{array}\right\} \tag{24.26}$$

The last equation shows that any momentary current in the $z$ direction dies away exponentially through collisions, and we may eliminate $v_y$ between the first two equations in order to find the oscillatory velocity $v_x$ in the direction of the applied electric field. This gives

$$v_x\left(\omega_c^2 - \omega^2 + \frac{1}{\tau^2} + \frac{2j\omega}{\tau}\right) = \frac{q}{m^*}E_x\left(\frac{1}{\tau} + j\omega\right),$$

where we have written $\omega_c$ for $(q/m^*)B$, the cyclotron resonance frequency. The conductivity of the solid at angular frequency $\omega$ in the $x$

direction is $\sigma_x = nqv_x/E_x$, where $n$ is the number of charge carriers per unit volume of mass $m^*$, and is given by the relation

$$\sigma_x = \frac{nq^2}{m^*}\left\{\frac{j\omega+1/\tau}{(\omega_c^2-\omega^2)+1/\tau^2+2j\omega/\tau}\right\}$$

$$= \sigma_0\left\{\frac{1+j\omega\tau}{1+2j\omega\tau+\tau^2(\omega_c^2-\omega^2)}\right\}. \qquad (24.27)$$

where $\sigma_0 = n(q^2/m^*)\tau$ is the ordinary conductivity of the substance at zero frequency in the absence of a magnetic field. This equation shows that the high-frequency conductivity is complex; on solving for the real part $\sigma_x'$ of the conductivity, we find

$$\frac{\sigma_x'}{\sigma_0} = \frac{1+\tau^2(\omega_c^2+\omega^2)}{\{1+\tau^2(\omega_c^2-\omega^2)\}^2+4\omega^2\tau^2}. \qquad (24.28)$$

The power absorption per unit volume of the sample is $\frac{1}{2}\sigma_x'E_x^2$; since it is usual to work at fixed frequency $\omega$ and measure the power absorption as $B$ (that is, $\omega_c$) is varied, it is useful to plot the quantity $\sigma_x'/\sigma_0$ as a function of $\omega_c/\omega$ for various values of the parameter $\omega\tau$. This is shown in Fig. 24.19. When $\omega\tau$ is appreciably less than unity, the mean time between collisions is a small fraction of an r.f. period, and little change occurs until $\omega\tau$ approaches unity. However, when $\omega\tau$ is rather greater than unity, a

FIG. 24.19. Plot of the ratio of the r.f. conductivity to the d.c. conductivity against $\omega_c/\omega$. Cyclotron resonance measurements are usually made at constant $\omega$ and variable field; since $\omega_c$ is proportional to $B$, the curves show the conductivity against $B$ (on a reduced scale). Well-resolved resonance curves are obtained when $\omega\tau$ is rather greater than unity.

FIG. 24.20. Waveguide cavity resonator used in cyclotron resonance experiments, showing the sample mounted at the centre of the cavity where the oscillatory electric field is a maximum. Carriers can be excited in the sample by light passed down the waveguide and through the coupling iris linking the cavity to the guide.

distinct resonance effect is observed, with maximum power absorption at a point close to the cyclotron resonance frequency.

In a semiconductor or metal at room temperature the value of $\tau$ is about $10^{-12}$–$10^{-14}$ s, so that even at a wavelength of 10 mm, where a magnetic field at resonance of about 1 T would be needed if $m^* = m$, the value of $\omega\tau$ is about $2\times10^{-1}$–$2\times10^{-3}$. However, the electron scattering is mainly due to phonons, and is reduced at low temperatures. In a semiconductor (see § 17.5) $\tau$ should vary as $T^{-\frac{3}{2}}$, and a factor of $10^3$ is gained in going from 300 K to 3 K, making $\omega\tau \sim 2$ to 200. As can be seen from Fig. 24.19, this is sufficient for a fairly accurate determination of the resonance frequency. However, the number of charge carriers $n$, which (see § 17.5) varies as $T^{\frac{3}{2}}\exp(-W_g/2kT)$ for a pure semiconductor, becomes vanishingly small at helium temperatures. Dexter, Zeiger, and Lax (1956) overcame this difficulty by irradiating the sample with light of sufficiently short wavelength to lift electrons across the energy gap from the valence to the conduction band, thus creating both holes and conduction electrons. The main features of their apparatus are shown in Fig. 24.20. The sample, in the form of a thin disc some 3 mm in diameter and 0·5 mm thick, is mounted at a point in a waveguide cavity where the oscillatory electric field is as large as possible without producing serious carrier heating effects through acceleration of the carriers. This cavity terminates a waveguide, and the change in the signal reflected by the cavity is a measure of the increased power absorption in the sample. The cavity is immersed in liquid helium in a dewar vessel placed between the poles of an electromagnet.

A simple, satisfactory method of detection is to modulate the light beam by passing it through a rotating disc pierced by a large number of holes. The lifetime of the carriers is short and they are present only for the duration of a light pulse; the reflected microwave signal is therefore modulated at the same frequency (usually 100–1000 Hz). Instead of using irradiation by light, carriers can also be created through ionization of impurity levels by application of an electric field across the sample, or by the oscillatory microwave electric field. The latter method gives distorted line-shapes, however, since the number of secondary carriers created depends on the carrier energy and this is a maximum at resonance. It has the advantage that only electrons are created in n-type material, and holes in p-type, since the microwave energy is only sufficient to cause ionization across the small gap (~0·01 eV in germanium) of impurity levels, and not across the main gap $W_g$.

In many substances the effective mass is anisotropic (see § 12.4), and the ratio of cyclotron resonance frequency to magnetic field is a function of the orientation of the field relative to the crystal axes. For this reason a single crystal must be used, with either a special device for rotating it in the cavity, or for rotating the external magnetic field, so that a whole plane of directions relative to the external magnetic field can be explored. An absorption curve for a given orientation of germanium is given in Fig. 24.21; it is due to Dresselhaus, Kip, and Kittel (1955), who made the first observations of cyclotron resonance in semiconductors in 1953. When anisotropy is present, the equations of motion are modified and must be solved to find the relation between the cyclotron resonance frequency and the effective mass parameters; the following method for this is due to Shockley (1953).

When the energy surfaces are not spherical in **k**-space, they can be

FIG. 24.21. Absorption curve for cyclotron resonance in a single crystal of germanium, at 24 GHz and 4 K. The static field is in a (110) plane at 60° from a [100] axis. Three electron lines are observed because $\mathbf{B}_0$ makes three different angles with the four [111] axes, which are the major axes of the four energy ellipses in germanium. (After Dresselhaus, Kip, and Kittel 1955.)

approximated near the band edges (see § 12.4) by equations such as

$$W = \tfrac{1}{2}\hbar^2\left(\frac{k_x^2}{m_x} + \frac{k_y^2}{m_y} + \frac{k_z^2}{m_z}\right) = \frac{1}{2}\left(\frac{p_x^2}{m_x} + \frac{p_y^2}{m_y} + \frac{p_z^2}{m_z}\right),$$

provided that the directions of the $x$-, $y$-, $z$-axes are chosen correctly. Along these axes the components of the equation of motion have their usual form, so that in a magnetic flux density with components $B_x$, $B_y$, $B_z$ we have

$$m_x(dv_x/dt) = q(v_y B_z - v_z B_y), \quad \text{etc.}$$

To find the cyclotron resonance frequency for a particular energy surface we assume a sinusoidal motion with frequency $\omega_c/2\pi$; on replacing $d/dt$ by $j\omega_c$ we obtain the set of linear equations

$$j\omega_c m_x v_x - q v_y B_z + q v_z B_y = 0,$$
$$j\omega_c m_y v_y - q v_z B_x + q v_x B_z = 0,$$
$$j\omega_c m_z v_z - q v_x B_y + q v_y B_x = 0,$$

which have an allowed solution only if the determinant

$$\begin{vmatrix} j\omega_c m_x & -qB_z & qB_y \\ qB_z & j\omega_c m_y & -qB_x \\ -qB_y & qB_x & j\omega_c m_z \end{vmatrix} = 0.$$

This gives either $\omega_c = 0$ (corresponding to **v** parallel to **B**), or

$$\omega_c^2 = \frac{q^2}{m_x m_y m_z}(m_x B_x^2 + m_y B_y^2 + m_z B_z^2). \tag{24.29}$$

This equation shows that the cyclotron resonance frequency depends on the orientation of the magnetic field with respect to the crystal axes; in any given plane a plot of $\omega_c^2$ against angle gives a $(\cos\text{ine})^2$ variation between the maximum and minimum values. When **B** is directed along one of the principal axes, such as the $z$-axis for an ellipsoidal energy surface, $(\omega_c)_z = qB/(m_x m_y)^{\frac{1}{2}}$; thus by measurement along each axis in turn, the principal values $m_x$, $m_y$, $m_z$ of the effective mass can be determined. The results for the elemental semiconductors silicon and germanium, together with those for InSb and GaAs are shown in Table 24.2. For silicon and germanium two of the principal values of the effective mass at the bottom of the conduction band are equal; this is known as the 'transverse mass' $m_T^*$, while the third (unequal) mass is called the longitudinal mass $m_L^*$. In InSb and GaAs the minimum of the conduction band occurs at **k** = 0 (see § 17.4), and the effective mass is isotropic.

The position at the top of the valence band is more complicated. Two energy surfaces coincide at **k** = 0, and at points near-by in **k**-space the

TABLE 24.2

*Effective masses in some semiconductors deter-mined by cyclotron resonance, relative to the free electron mass*

| Substance | Electrons | | Holes | |
|---|---|---|---|---|
| | $m_L^*$ | $m_T^*$ | 'light' | 'heavy' |
| Silicon | 0·98 | 0·19 | 0·16 | 0·5 |
| Germanium | 1·64 | 0·082 | 0·044 | 0·3 |
| InSb | 0·014 (isotropic) | | 0·02 | 0·4 |
| GaAs | 0·067 (isotropic) | | 0·084 | 0·45–0·6 |

energy surfaces for silicon and germanium are given by the relation

$$W = Ak^2 \pm \{B^2k^4 + C^2(k_x^2k_y^2 + k_y^2k_z^2 + k_z^2k_x^2)\}^{\frac{1}{2}}, \qquad (24.30)$$

which is also approximately correct for Group III–V semiconductors. If $C$ is small the two surfaces are nearly spherical, but with different curvature, corresponding to two different effective masses known as the 'light' and 'heavy' holes (see Table 24.2), giving resonance lines as in Fig. 24.21.

## 24.10. Azbel–Kaner resonance in metals

When a spectral line due to moving particles is observed, it is broadened through the Doppler effect, by an amount which is proportional to the random particle velocity. In a semiconductor at low temperatures, the electrons or holes have ordinary thermal velocities corresponding to energies of order $kT$, and broadening by the Doppler effect is not important. In a metal, on the other hand, the electron velocity is that at the Fermi surface; in copper, assuming $m^*/m = 1·5$ and $W_F = 4·7$ eV, this velocity is about $10^6$ m s$^{-1}$, while the phase velocity in the metal of an electromagnetic wave with a free-space wavelength of 10 mm is only about $4 \times 10^4$ m s$^{-1}$ (from eqn (8.30) it is equal to $\omega\delta$, where $\delta$ is the skin depth). Broadening through the Doppler effect thus makes it impossible to observe cyclotron resonance in metals by methods similar to those used for semiconductors. It can, however, be detected by a different method, originally due to Azbel and Kaner (1957, 1958).

As before, a reasonable degree of resolution is obtained only if $\omega\tau > 1$. This makes it essential to work at liquid-helium temperatures, using very pure samples in which the residual resistivity due to electron scattering by impurities and imperfections is as small as possible ($10^{-3}$–$10^{-5}$ of the room-temperature resistivity). In copper the radius of the electron orbit in the magnetic field required to make the cyclotron resonance frequency equal to $3 \times 10^{10}$ Hz is about $5 \times 10^{-6}$ m, and the mean path length of the electrons must be of this order in order to make $\omega\tau > 1$. This requires a conductivity greater than $10^{10}\ \Omega^{-1}\ m^{-1}$, and the 'classical' skin depth given by eqn (8.31) is about $3\times10^{-8}$ m, which is small compared with the mean path length. This is the region of the 'anomalous skin effect', where the conductivity is effectively reduced because only those electrons moving at a small angle to the surface such that their free paths lie wholly within the skin depth contribute fully to the oscillatory current (see § 12.10). However, even allowing for this, the 'anomalous' skin depth is about $2\times10^{-7}$ m, which is still small compared with the radius of the cyclotron orbit. If then a magnetic field of flux density **B** is applied parallel to the surface of the metal, a certain number of electrons moving in helical orbits about **B** will enter the skin-depth region once per cycle, and while in this region they can be accelerated by the oscillatory electric field component normal or parallel to **B**. The latter geometry is illustrated in Fig. 24.22; an important difference from the conventional cyclotron is that acceleration occurs only once per revolution instead of twice. Electrons will gain energy steadily if the frequency of the electromagnetic wave incident on the surface of the metal is synchronous with the cyclotron resonance frequency, or is an integral multiple of it. Hence the resonance condition is

$$\omega = p\omega_c = p(q/m^*)B, \qquad (24.31)$$

where $p = 1,\ 2,\ 3$, etc. It is usually convenient to work at a fixed frequency, making the metal sample one end of a cavity resonator as in ferromagnetic resonance (but with **B** normal or parallel to the oscillatory electric field instead of normal to the oscillatory magnetic field), and maxima in the absorption of energy then occur at values of $B$ given by the relation

$$B_p = \frac{1}{p}\left(\frac{m^*\omega}{q}\right). \qquad (24.32)$$

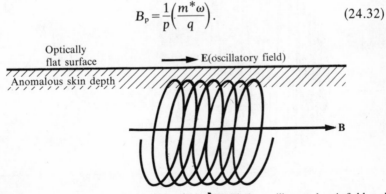

FIG. 24.22. Geometry of the steady magnetic flux density **B**, the oscillatory electric field, and the cyclotron orbits in a metal for Azbel–Kaner resonance. The electrons are accelerated by the electric field only when their orbits take them into the skin depth.

To obtain a resonance peak the electrons must complete a cyclotron orbit without experiencing a collision. This condition can be written as $\omega_c\tau \geqslant 1$, and since $\omega_c = \omega/p$ it is a more stringent condition than $\omega\tau \geqslant 1$. From eqn (24.32) it follows that this condition is more difficult to satisfy at low values of $B$, but in experimental charts similar to Fig. 24.23 peaks have been observed down to magnetic fields of about $0\cdot1$ T (values of $p$ up to 30).

Samples of high purity are needed to give good resolution; ideally they must be so flat that surface irregularities are small compared with the anomalous skin depth. If the effective mass is anisotropic, single crystals must be used, cut in special orientations so that cyclotron resonance can be observed in all the principal directions. The steady magnetic flux density **B** must be accurately parallel to the surface, or electrons will move away from the surface because of their velocity components parallel to **B**. As the electrons can drift distances $\sim 10\ \mu$m in one cyclotron period and the skin depth $\sim 0\cdot1\ \mu$m, the magnetic field must be aligned parallel to the surface to much better than $1°$; this is necessary to ensure that the electrons return to the same distance from the surface—otherwise they would experience a Doppler-shifted electric field. Grimes and Kip (1963) found that the effective mass is isotropic in sodium and potassium, with values of $m^*/m_e$ equal to $(1\cdot24\pm0\cdot02)$ and $(1\cdot21\pm0\cdot02)$ respectively. In copper (Koch, Stradling, and Kip 1964) the predominant absorption is due to electrons with $m^*/m_e$ about equal to $1\cdot4$, with only slight anisotropy, but other very anisotropic values ranging from $0\cdot4$ to $6$ are also observed, showing that the Fermi surface is rather complicated.

FIG. 24.23. Azbel–Kaner resonance at 4 K in a single crystal of copper. The magnetic flux is parallel to the surface and along a [100] direction; the frequency is 67 GHz (4·5 mm wavelength). (After Koch, Stradling, and Kip 1964.)

## References

AZBEL, M. YA., and KANER, E. A. (1957). *Sov. Phys. JETP* **5**, 730.
—— —— (1958). *J. Phys. Chem. Solids* **6**, 113.
BLOCH, F., HANSEN, W. W., and PACKARD, M. (1946). *Phys. Rev.* **69**, 127.
BLOEMBERGEN, N., PURCELL, E. M., and POUND, R. V. (1948). *Phys. Rev.* **73**, 679.
BROWN, R. M. and PURCELL, E. M. (1949). *Phys. Rev.* **75**, 1262.
COHEN, E. R. and TAYLOR, B. N. (1972). *Atomic masses and fundamental constants*, Vol. 4 (Eds. J. H. Sanders and A. H. Wapstra), p. 543. Plenum Press, New York.
DEXTER, R. N., ZEIGER, H. J., and LAX, B. (1956). *Phys. Rev.* **104**, 637.
DRESSELHAUS, G., KIP, A. F., and KITTEL, C. (1955). *Phys. Rev.* **98**, 368.
ESSEN, L. (1969). *Vistas Astron.* **2**, 45.
——, PARRY, J. V. L. (1959). *Nature, Lond.* **184**, 1791.
GRIFFITHS, J. H. E. (1946). *Nature, Lond.* **158**, 670.
GRIMES, C. C. and KIP, A. F. (1963). *Phys. Rev.* **132**, 1991.
HELLER, P. and BENEDEK, G. B. (1962). *Phys. Rev. Lett.* **8**, 428.
KITTEL, C. (1948). *Phys. Rev.* **73**, 155.
KOCH, J. F., STRADLING, R. A., and KIP, A. F. (1964). *Phys. Rev.* **133**, A240.
PHILLIPS, T. G. and ROSENBERG, H. M. (1964). *Phys. Lett.* **8**, 298.
PURCELL, E. M. (1948). *Science*, **107**, 433.
——, TORREY, H. C., and POUND, R. V. (1946). *Phys. Rev.* **69**, 37.
RABI, I. I., MILLMAN, S., KUSCH, P., and ZACHARIAS, J. R. (1939). *Phys. Rev.* **55**, 526.
RAMSEY, N. F. (1949). *Phys. Rev.* **76**, 996.
SHOCKLEY, W. (1953). *Phys. Rev.* **90**, 491.
SHULMAN, R. G. and JACCARINO, V. (1956). *Phys. Rev.* **103**, 1126.
SOMMER, H., THOMAS, H. A., and HIPPLE, J. A. (1951). *Phys. Rev.* **82**, 697.
WIMETT, T. F. (1953). *Phys. Rev.* **91**, 499.

## Problems

24.1. In a substance where the susceptibility is small and the Lorentz internal field can be neglected, show that eqn (10.19) can be written in the form

$$\chi' - j\chi'' = \frac{n_0 e^2}{m\epsilon_0} \frac{1}{(\omega_p^2 - \omega^2) + 2j\omega \, \Delta\omega}.$$

If $\chi_0$ is the static susceptibility, and $\chi_p''$ is the imaginary part of the susceptibility when $\omega = \omega_p$, prove that

$$\chi_p''/\chi_0 = \omega_p/(2\Delta\omega) = \nu_p/(2\Delta\nu),$$

where $\nu_p$ is the frequency at the centre of the absorption line, and $\Delta\nu = \Delta\omega/2\pi$. Although this formula was derived for electric susceptibility it is equally valid for the magnetic case.

24.2. The work done per unit volume to increase the magnetization of a substance by $dM$ in a field of flux density $B$ is $dW = B \, dM$. If $B$ is an alternating field

$$B_1 \cos \omega t = \text{Re}(B_1 \exp j\omega t),$$

the magnetization may be written as

$$M = \text{Re}\{(\chi' - j\chi'')(B_1/\mu_0)\exp j\omega t\},$$

where $(\chi'-j\chi'')$ is the complex susceptibility. Show that the rate of doing work per unit volume is

$$B(dM/dt) = dW/dt = -\omega\chi'(B_1^2/\mu_0)\cos\omega t\sin\omega t + \omega\chi''(B_1^2/\mu_0)\cos^2\omega t$$

and the mean power dissipated per unit volume is $\frac{1}{2}\omega\chi''B_1^2/\mu_0$.

Use the definition (6) for $Q_F$ given in § 7.3 to show that $1/Q_F = \chi''/(1+\chi')$ for a coil containing a magnetic substance in a tuned circuit with no other losses.

24.3. Adapt the results of Problem 8.6 to the case of the magnetic substance of the last problem (note that $\chi''/(1+\chi')$ is equivalent to $\epsilon''/\epsilon'$), and show that the power in an electromagnetic wave passing through such a medium would fall according to the law

$$P/P_0 = \exp(-2\pi\chi''x/\lambda)$$

if $\chi' \ll 1$.

Verify the figures given in § 24.3, that for $\lambda = 35$ m and $\chi'' = 10^{-5}$, the power will fall by about 1·8 per cent in a distance of 10 km.

24.4. Show, from eqn (24.21) and (24.22) that the instantaneous velocity of the charged particle in its spiral orbit is $(qE/M\,\Delta\omega)\sin(\frac{1}{2}\Delta\omega t)$. Hence show that the total length of path traversed by the particle in reaching its maximum radius $R_0$ when $\Delta\omega \neq 0$ is $L = 2qE/M(\Delta\omega)^2$, and verify that $L/2R_0$ is equal to the resolving power $\omega_c/\Delta\omega$.

24.5. Adapt the formula of Problem 14.5 to find the value of $g_F$ when $J$ and $I$ are coupled to form a resultant $F$, assuming that the nuclear magnetic moment can be neglected. Show that in the Zeeman splitting of Fig. 24.11,

$$g_{F=4} = -g_{F=3} = \tfrac{1}{8}g_J = \tfrac{1}{4}.$$

24.6. An atomic-beam determination of the ratio of the value of $g_J$ in the $^2P_{\frac{3}{2}}$ and $^2P_{\frac{1}{2}}$ states of the gallium atom gives it to be $2(1\cdot00172\pm0\cdot00006)$. Show, by writing $g_l = 1+\delta_l$ and $g_s = 2(1+\delta_s)$, that the ratio is equal to $2\{1+\frac{3}{2}(\delta_s-\delta_l)\}$, and hence that this result agrees within the experimental error with the accepted value $\delta_s = 0\cdot001160$ if $\delta_l$ is assumed to be zero.

24.7. A thin spherical shell of radius $r$, thickness $dr$ of electric charge density $\rho$ rotates with angular velocity $\boldsymbol{\omega}$ about a diameter. Show that the magnetic field $d\mathbf{B}$ at the centre is $\oplus\frac{2}{3}\mu_0\rho\boldsymbol{\omega}r\,dr$.

Use this result to show that the correction at the nucleus of a hydrogen atom in a magnetic field of flux density $\mathbf{B}$ due to diamagnetic shielding (see § 24.5) is

$$\delta B/B = -\mu_0 e^2/12\pi m a_0,$$

given that the charge density at distance $r$ is

$$\rho = (-e/\pi a_0^3)\exp(-2r/a_0).$$

24.8. Show from eqn (24.28) that the conductivity at zero frequency in the direction normal to a magnetic flux density $B$ should vary as

$$\sigma_x'/\sigma_0 = 1/(1+aB^2),$$

where $a = (e\tau/m^*)^2$. This is the magnetoresistance effect, which becomes appreciable only at low temperatures where $\tau$ increases. When only one type of carrier is present, the effect vanishes because the sideways force due to the magnetic field is exactly nullified by that due to the Hall voltage; when more than one type of charge carrier is present this cancellation does not occur (see § 12.11).

24.9. Show that in a cyclotron resonance experiment where $\omega\tau \gg 1$, the value of $\sigma'_x$ at resonance ($\omega_c/\omega = 1$) approaches $\frac{1}{2}\sigma_0$, and that the loss tangent of the specimen is then (writing $\Delta\omega_e$ for $\tau^{-1}$)

$$\tan\delta_e = \frac{ne^2}{2m^*\omega\epsilon_r\epsilon_0\,\Delta\omega_e}.$$

Use the result of Problem 24.1 to show that in an electron spin resonance experiment the magnetic loss tangent at resonance for a system of $n$ electrons with $S = \frac{1}{2}$, $g = 2$ is

$$\tan\delta_m = \frac{\chi''_P}{1+\chi_0} \approx \frac{\mu_0 n\mu_B^2\omega}{2kT\,\Delta\omega_m}$$

and hence prove that (taking $m^* = m$, $\Delta\omega_e = \Delta\omega_m$)

$$\frac{\tan\delta_e}{\tan\delta_m} = \frac{4mc^2kT}{\epsilon_r(\hbar\omega)^2}.$$

Hence verify that the inherent sensitivity of a cyclotron resonance experiment is very much higher than that of a spin resonance experiment, so that fewer electrons are needed.

Discuss whether the imaginary part of the conductivity (see eqn (24.27)) can justifiably be neglected in the formula for $\tan\delta_e$.

24.10. A tuned circuit of quality factor $Q_F$ is freely oscillating. Show that in the time taken by the stored energy to fall by a factor $e^{-1}$, a wave in free space would travel $Q_F/2\pi$ wavelengths.

# Appendix A: Vectors

## A.1. Definition of scalar and vector quantities

MANY physical quantities are completely defined by magnitude alone. Examples are temperature, time, or length. These are called scalar quantities. They obey the ordinary laws of algebra, and are represented in the text by a symbol printed in italic type. Other physical quantities are not completely defined unless the direction as well as the magnitude is given. Such quantities are called vectors, and are printed in bold-face type in the text.

Quantities such as velocity, force, or acceleration which involve a translation are known as 'polar vectors', to distinguish them from 'axial vectors' which are connected with rotations. They behave differently under reflection, as illustrated in Fig. 6.1.

## A.2. Vector addition and subtraction

A vector may be represented graphically by an arrow pointing in the direction of the vector and of length equal to its magnitude. The addition of two vectors $\mathbf{P}$ and $\mathbf{Q}$ is shown in Fig. A.1 (a), in which the vectors form two sides of a parallelogram. The vector $\mathbf{R}$ defined by the equation $\mathbf{P}+\mathbf{Q}=\mathbf{R}$ is the diagonal of this parallelogram, and its magnitude and direction can be found by trigonometry if $\mathbf{P}$ and $\mathbf{Q}$ are known. Similarly, the vector $\mathbf{D}=\mathbf{P}-\mathbf{Q}$ is obtained from Fig. A.1 (b). In the special case that the vectors $\mathbf{P}$, $\mathbf{Q}$ are parallel, $\mathbf{R}$ is equal to the scalar sum of $\mathbf{P}$ and $\mathbf{Q}$, and $\mathbf{D}$ is equal to the scalar difference, and $\mathbf{R}$ and $\mathbf{D}$ are parallel to $\mathbf{P}$ and $\mathbf{Q}$.

The converse process is often useful. That is, a vector $\mathbf{P}$ (Fig. A.2) can be resolved into two vectors $\mathbf{Q}$ and $\mathbf{R}$ such that $\mathbf{P}$ is the diagonal of a parallelogram, and $\mathbf{Q}$ and $\mathbf{R}$ are two adjacent sides. Generally, $\mathbf{Q}$ and $\mathbf{R}$ are chosen to be at right-angles, so that the parallelogram is then a rectangle. $\mathbf{Q}$ and $\mathbf{R}$ are called the components of $\mathbf{P}$. $\mathbf{P}$ may be resolved into three components parallel to the axes of Cartesian coordinates $x$, $y$, and $z$.

## A.3. Multiplication of vectors

1. Multiplication of a vector $\mathbf{P}$ by a scalar quantity $m$ changes the magnitude of the vector by the factor $m$, but the direction is unaltered. Multiplication by $-m$ gives a vector of magnitude $m\mathbf{P}$ in the opposite direction, that is, the vector $-m\mathbf{P}$. If $\mathbf{i}$, $\mathbf{j}$, and $\mathbf{k}$ are vectors of unit length parallel to the axes $x$, $y$, and $z$, we can write

$$\mathbf{P}=\mathbf{i}P_x+\mathbf{j}P_y+\mathbf{k}P_z,$$

(a)　　　　　　　　(b)

FIG. A.1.

FIG. A.2. If **Q**, **R** are at right-angles, then $Q = P \cos \theta$, $R = P \sin \theta$.

where $P_x$, $P_y$, and $P_z$ are scalar quantities giving the magnitude of the components of **P** parallel to the three axes (see Fig. A.3). Since $\mathbf{Q} = \mathbf{i}Q_x + \mathbf{j}Q_y + \mathbf{k}Q_z$ it follows that $\mathbf{P} + \mathbf{Q} = \mathbf{i}(P_x + Q_x) + \mathbf{j}(P_y + Q_y) + \mathbf{k}(P_z + Q_z)$.

2. *The scalar product.* The scalar product of two vectors **P** and **Q** is written **P.Q** and is a scalar quantity numerically equal to the magnitude of one vector multiplied by the component of the other parallel to the direction of the first one. If the angle between **P** and **Q** is $\theta$,

$$\mathbf{P} \cdot \mathbf{Q} = PQ \cos \theta = \mathbf{Q} \cdot \mathbf{P}$$

and

$$\mathbf{P} \cdot (\mathbf{Q} + \mathbf{R} + \mathbf{S} + \ldots) = \mathbf{P} \cdot \mathbf{Q} + \mathbf{P} \cdot \mathbf{R} + \mathbf{P} \cdot \mathbf{S} + \ldots$$

The scalar product of two perpendicular vectors is zero. Therefore, for the unit vectors **i**, **j**, and **k**, we have

$$\mathbf{i} \cdot \mathbf{j} = \mathbf{j} \cdot \mathbf{k} = \mathbf{k} \cdot \mathbf{i} = 0$$

and

$$\mathbf{i} \cdot \mathbf{i} = \mathbf{j} \cdot \mathbf{j} = \mathbf{k} \cdot \mathbf{k} = 1.$$

An example of a scalar product is the work $dW$ done on a charge $q$ in moving a distance $d\mathbf{s}$ in a region where the electric field strength is **E**, which is

$$dW = -q\mathbf{E} \cdot d\mathbf{s}. \tag{A.1}$$

3. *The vector product.* The vector product of two vectors **P** and **Q** is defined as a vector perpendicular to both **P** and **Q** of magnitude $PQ \sin \theta$, where $\theta$ is the angle between **P** and **Q**. If **P** is perpendicular to **Q**, the vector product is $PQ$ but if **P** and **Q** are parallel the vector product is zero.

The direction of the vector product $(\mathbf{P} \wedge \mathbf{Q})$ is that in which a right-handed screw would move if turned from the first vector **P** towards the second vector **Q**, as shown in Fig. A.4. Hence we have

$$(\mathbf{P} \wedge \mathbf{Q}) = -(\mathbf{Q} \wedge \mathbf{P}).$$

Also,

$$\mathbf{P} \wedge (\mathbf{Q} + \mathbf{R} + \mathbf{S} + \ldots) = (\mathbf{P} \wedge \mathbf{Q}) + (\mathbf{P} \wedge \mathbf{R}) + (\mathbf{P} \wedge \mathbf{S}) + \ldots$$

The formula for a vector product in terms of the vector components may be conveniently expressed as a determinant. For the unit vectors along a set of

FIG. A.3.

FIG. A.4. The vector $\mathbf{P} \wedge \mathbf{Q}$ is normal to the plane containing $\mathbf{P}$ and $\mathbf{Q}$.

right-handed Cartesian coordinates, we have

$$\mathbf{i} \wedge \mathbf{i} = \mathbf{j} \wedge \mathbf{j} = \mathbf{k} \wedge \mathbf{k} = 0,$$

$$\mathbf{i} \wedge \mathbf{j} = \mathbf{k} = -\mathbf{j} \wedge \mathbf{i}, \text{ etc.}$$

Hence

$$\mathbf{P} \wedge \mathbf{Q} = (\mathbf{i}P_x + \mathbf{j}P_y + \mathbf{k}P_z) \wedge (\mathbf{i}Q_x + \mathbf{j}Q_y + \mathbf{k}Q_z)$$

$$= \mathbf{i}(P_y Q_z - P_z Q_y) + \mathbf{j}(P_z Q_x - P_x Q_z) + \mathbf{k}(P_x Q_y - P_y Q_x),$$

which can be written as

$$\mathbf{P} \wedge \mathbf{Q} = \begin{vmatrix} \mathbf{i} & \mathbf{j} & \mathbf{k} \\ P_x & P_y & P_z \\ Q_x & Q_y & Q_z \end{vmatrix}.$$

If $\mathbf{P}$ and $\mathbf{Q}$ are polar vectors, then $(\mathbf{P} \wedge \mathbf{Q})$ is an axial vector.

An example of the use of a vector product is the equation for the force $d\mathbf{F}$ on an element $d\mathbf{s}$ of a wire carrying a current $I$ in a magnetic flux density $\mathbf{B}$. The force is normal to $d\mathbf{s}$ and to $\mathbf{B}$, and of magnitude $I \, ds \, B \sin \theta$. It is specified both in magnitude and direction by the vector equation

$$d\mathbf{F} = I(d\mathbf{s} \wedge \mathbf{B}).$$

Products of three vectors can be evaluated from the foregoing rules. The scalar triple product

$$\mathbf{P} . (\mathbf{Q} \wedge \mathbf{R}) = \text{scalar product of } \mathbf{P} \text{ and } (\mathbf{Q} \wedge \mathbf{R})$$

is a scalar quantity equal in magnitude to the volume of the parallelepiped whose sides are constructed from the three vectors $\mathbf{P}, \mathbf{Q}, \mathbf{R}$ (see Fig. A.5). Clearly,

$$\mathbf{P} . (\mathbf{Q} \wedge \mathbf{R}) = (\mathbf{P} \wedge \mathbf{Q}) . \mathbf{R},$$

and this is often written simply as $\mathbf{PQR}$. We have

$$\mathbf{PQR} = \mathbf{QRP} = \mathbf{RPQ} = -\mathbf{PRQ} = -\mathbf{QPR} = -\mathbf{RQP}.$$

The change of sign on inverting the order of any two of the vectors follows also from the determinantal form

$$\mathbf{PQR} = \begin{vmatrix} P_x & P_y & P_z \\ Q_x & Q_y & Q_z \\ R_x & R_y & R_z \end{vmatrix}.$$

The formula for the vector triple product may be expressed in the form

$$\mathbf{P} \wedge (\mathbf{Q} \wedge \mathbf{R}) = \mathbf{Q}(\mathbf{P} . \mathbf{R}) - \mathbf{R}(\mathbf{P} . \mathbf{Q}).$$

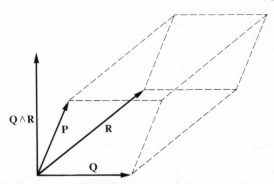

FIG. A.5. The scalar triple product $\mathbf{P}.(\mathbf{Q}\wedge\mathbf{R})$ is numerically equal to the volume of the parallelepiped whose sides are the vectors $\mathbf{P}$, $\mathbf{Q}$, and $\mathbf{R}$.

This may be verified by expressing the vectors in terms of their components along three Cartesian axes.

### A.4. Differentiation and integration of vectors

Vector quantities are often expressed as functions of scalar variables. For example, the electric field strength $\mathbf{E}$ can be expressed as a function of the position coordinates $x$, $y$, and $z$. The vector may be differentiated and integrated with respect to these variables. The differential of $\mathbf{P}$ with respect to a scalar variable $u$ is defined as

$$\frac{d\mathbf{P}}{du} = \lim_{\Delta u \to 0} \frac{\mathbf{P}(u+\Delta u)-\mathbf{P}(u)}{\Delta u}.$$

When a force $\mathbf{F}$ acts for a small distance $d\mathbf{s}$, the work done is $dW = \mathbf{F}.d\mathbf{s}$ and if the total work done over a finite distance is required, we can write

$$W = \int \mathbf{F}.d\mathbf{s} = \int F \cos\theta \, ds,$$

where $ds$ is the component of $d\mathbf{s}$ parallel to $\mathbf{F}$ at any point.

This integral occurs frequently and is called the line integral of $\mathbf{F}$ along the curve. The line integral along the curve AB is illustrated in Fig. A.6. If the integration is carried out round a closed path, returning to the original point A, it is written $\oint \mathbf{F}.d\mathbf{s}$.

The surface integral $\int \mathbf{F}.d\mathbf{S}$ is also important. $\mathbf{F}.d\mathbf{S}$ is the flux through the element of area $d\mathbf{S}$ due to $\mathbf{F}$, and the integral over a surface gives the total flux through that surface. If the vector $\mathbf{F} = \mathbf{v}$ represents the velocity of flow of a fluid, $\int \mathbf{v}.d\mathbf{S}$ gives the total volume of fluid passing through the area $S$ in unit time. If $\mathbf{F}$ is the electric displacement $\mathbf{D}$, the integral gives the number of lines of displacement crossing the surface $S$.

FIG. A.6.

In many problems in physics a scalar quantity is used which is a single-valued function of the position coordinates of the system. For example, in electrostatics the electric potential $V$ is a function of $x$, $y$, and $z$ in a Cartesian coordinate system. The change in potential corresponding to an infinitesimal displacement d**s** is given by Taylor's theorem, that is,

$$dV = (\partial V/\partial x)\,dx + (\partial V/\partial y)\,dy + (\partial V/\partial z)\,dz$$

and

$$d\mathbf{s} = \mathbf{i}\,dx + \mathbf{j}\,dy + \mathbf{k}\,dz.$$

The rate of change of $V$ with the displacement **s** is expressed in terms of a new quantity grad $V$ which is defined by the equation

$$dV = (\text{grad } V).d\mathbf{s}, \tag{A.2}$$

where

$$\text{grad } V = \mathbf{i}(\partial V/\partial x) + \mathbf{j}(\partial V/\partial y) + \mathbf{k}(\partial V/\partial z).$$

grad $V$ is a vector quantity and is an abbreviation for 'the gradient of $V$'. When grad $V$ is parallel to d**s**, $dV$ is a maximum, so that grad $V$ is in the direction of the greatest rate of change of $V$ with respect to the coordinates, and is normal to an equipotential surface. From Eqn (A.1) the work done on unit charge in moving a distance d**s** in a field of strength **E** is $-\mathbf{E}.d\mathbf{s}$ and this equal to $-dV$. Therefore we have $\mathbf{E} = -\text{grad } V$ and the electric field strength is equal to the gradient of the potential at any point, and is in the direction of the maximum rate of change of potential with respect to the space coordinates.

The operator $\mathbf{i}(\partial/\partial x) + \mathbf{j}(\partial/\partial y) + \mathbf{k}(\partial/\partial z)$ is often denoted by the symbol $\nabla$ (pronounced 'del'), so that

$$\text{grad } V \equiv \nabla V. \tag{A.3}$$

The operator $\nabla$ can be regarded as a vector operator, which operates on both scalar and vector quantities, and forms scalar and vector products. Thus eqn (A.2) can be written

$$dV = (\nabla V).d\mathbf{s}. \tag{A.4}$$

In general, any scalar potential function $\phi$, which is finite, single-valued, and free from discontinuities (these conditions must apply also to the first and second derivatives of $\phi$ with respect to the space coordinates), can be related to a field of force **F**, where

$$\mathbf{F} = -\text{grad } \phi,$$

so that once $\phi$ is everywhere determined, **F** is known at all points. Also, the line integral of **F** between any two points A and B is independent of the path taken between those points since

$$\int_A^B \mathbf{F}.d\mathbf{s} = -\int_A^B (\text{grad } \phi).d\mathbf{s} = -\int_A^B d\phi = \phi_A - \phi_B$$

by analogy with eqn (A.2). Similarly, the line integral round a closed path is zero.

### A.5. The divergence of a vector

The divergence of a vector **P** is written div **P**. It is an operator used to describe the excess flux leaving an element of volume in space. The flux may be flow of liquid in hydrodynamics, heat in a thermal field of varying temperature, or electric

FIG. A.7.

flux. In the latter case, the excess flux leaving the volume element is related to the total charge enclosed by Gauss's theorem. In Fig. A.7 there is a varying electric charge density $\rho_e$ throughout space. Gauss's theorem is applied to a volume element $dx\,dy\,dz$ at the point $(x, y, z)$ in a Cartesian coordinate system. The total charge enclosed is $\rho_e\,dx\,dy\,dz$. The total flux through the faces normal to the $x$-axis is

$$\left[\left\{D_x+\frac{\partial}{\partial x}(D_x)\,dx\right\}-D_x\right]dy\,dz=\frac{\partial D_x}{\partial x}dx\,dy\,dz,$$

where $D_x$ is the component of the electric displacement parallel to the $x$-axis at the point $(x, y, z)$. Writing similar expressions for the flux through the other two pairs of faces, Gauss's theorem becomes

$$\frac{\partial D_x}{\partial x}+\frac{\partial D_y}{\partial y}+\frac{\partial D_z}{\partial z}=\rho_e,$$

where $D_x$, $D_y$, and $D_z$ are the components of the electric displacement along the three axes at $(x, y, z)$.

The expression on the left-hand side of this equation is written div **D**, and is the divergence of the vector **D** at this point.

Now using the operator $\nabla$, we have

$$\nabla.\mathbf{D}=\left(\mathbf{i}\frac{\partial}{\partial x}+\mathbf{j}\frac{\partial}{\partial y}+\mathbf{k}\frac{\partial}{\partial z}\right).(\mathbf{i}D_x+\mathbf{j}D_y+\mathbf{k}D_z)$$

$$=\frac{\partial D_x}{\partial x}+\frac{\partial D_y}{\partial y}+\frac{\partial D_z}{\partial z}.$$

Therefore

$$\nabla.\mathbf{D}\equiv\mathrm{div}\,\mathbf{D}. \tag{A.5}$$

The divergence of a vector is a scalar quantity, since it represents the net amount of flux, or the number of lines of displacement, coming out of a volume element. If div **D** = 0, the total flux entering the element $dx\,dy\,dz$ is balanced by that leaving it. A vector satisfying this condition is said to be *solenoidal*.

### A.6. The curl of a vector

The curl (or rotation) of a vector **P** is written curl **P**. It arises in problems where a line integral of a vector round a closed path is related to the flux through the

FIG. A.8. Application of Ampère's law in Cartesian coordinates.

surface enclosed by the path of the line integral. For example, Ampère's law for the magnetic field due to a current is

$$\oint \mathbf{H} \cdot \mathbf{ds} = \int \mathbf{J} \cdot \mathbf{dS}.$$

Let us apply this equation to an element dy dz at the point $(x, y, z)$ in a Cartesian coordinate system (Fig. A.8). For the $x$ component $J_x$, of the current, the line integral $\mathbf{H}$ in the $yz$ plane is positive in an anticlockwise direction, and we have

$$J_x \, dy \, dz = \left( H_y - \frac{\partial H_y}{\partial z} \frac{dz}{2} \right) dy + \left( H_z + \frac{\partial H_z}{\partial y} \frac{dy}{2} \right) dz$$

$$- \left( H_y + \frac{\partial H_y}{\partial z} \frac{dz}{2} \right) dy - \left( H_z - \frac{\partial H_z}{\partial y} \frac{dy}{2} \right) dz$$

$$= \left( \frac{\partial H_z}{\partial y} - \frac{\partial H_y}{\partial z} \right) dy \, dz,$$

where $H_y$ and $H_z$ are the components of $\mathbf{H}$ parallel to the $y$- and $z$-axes respectively. Therefore

$$J_x = \frac{\partial H_z}{\partial y} - \frac{\partial H_y}{\partial z},$$

and similarly

$$J_y = \frac{\partial H_x}{\partial z} - \frac{\partial H_z}{\partial x} \quad \text{and} \quad J_z = \frac{\partial H_y}{\partial x} - \frac{\partial H_x}{\partial y}.$$

These equations are written

$$J_x = \text{curl}_x \, \mathbf{H}, \qquad J_y = \text{curl}_y \, \mathbf{H}, \qquad J_z = \text{curl}_z \, \mathbf{H}$$

or simply

$$\text{curl} \, \mathbf{H} = \mathbf{J},$$

where curl $\mathbf{H}$ is a vector quantity whose components are expressed by means of the determinant

$$\text{curl} \, \mathbf{H} = \begin{vmatrix} \mathbf{i} & \mathbf{j} & \mathbf{k} \\ \frac{\partial}{\partial x} & \frac{\partial}{\partial y} & \frac{\partial}{\partial z} \\ H_x & H_y & H_z \end{vmatrix}, \tag{A.6}$$

that is,

$$\operatorname{curl} \mathbf{H} = \mathbf{i}\left(\frac{\partial H_z}{\partial y} - \frac{\partial H_y}{\partial z}\right) + \mathbf{j}\left(\frac{\partial H_x}{\partial z} - \frac{\partial H_z}{\partial x}\right) + \mathbf{k}\left(\frac{\partial H_y}{\partial x} - \frac{\partial H_x}{\partial y}\right).$$

Also

$$\nabla \wedge \mathbf{H} = \left(\mathbf{i}\frac{\partial}{\partial x} + \mathbf{j}\frac{\partial}{\partial y} + \mathbf{k}\frac{\partial}{\partial z}\right) \wedge (\mathbf{i}H_x + \mathbf{j}H_y + \mathbf{k}H_z)$$

$$= \mathbf{i}\left(\frac{\partial H_z}{\partial y} - \frac{\partial H_y}{\partial z}\right) + \mathbf{j}\left(\frac{\partial H_x}{\partial z} - \frac{\partial H_z}{\partial x}\right) + \mathbf{k}\left(\frac{\partial H_y}{\partial x} - \frac{\partial H_x}{\partial y}\right)$$

$$= \operatorname{curl} \mathbf{H}. \tag{A.7}$$

## A.7. The divergence theorem

$S$ in Fig. A.9 is a closed surface in a region where there exists a vector field $\mathbf{F}$. The flux through an element of area $d\mathbf{S}$ is $\mathbf{F}.d\mathbf{S}$, and the total flux through the surface is $\int \mathbf{F}.d\mathbf{S}$. The total flux diverging from an element of volume $d\tau$ inside $S$ is, from § A.5 above, $\operatorname{div} \mathbf{F} \, d\tau$, where $\mathbf{F}$ is the value of the force field at this point. The integral $\int \operatorname{div} \mathbf{F} \, d\tau$ throughout the whole volume enclosed by $S$ must give the total flux through the surface, since for any two adjacent volume elements the flux through a common face gives equal positive and negative contributions. Hence

$$\int \mathbf{F}.d\mathbf{S} = \int \operatorname{div} \mathbf{F} \, d\tau, \tag{A.8}$$

which is the theorem of divergence. Again, the vector field $\mathbf{F}$ must be a well-behaved function. Conversely if the surface integral of a vector $\mathbf{F}$ is equal to the volume integral of a scalar function $m$ over the volume enclosed by the surface, whatever the surface, then we may conclude that

$$m = \operatorname{div} \mathbf{F}.$$

## A.8. Stokes's theorem

In Fig. A.10 the line integral of the vector $\mathbf{H}$ is taken round a closed path bounding an unclosed surface $S$. This integral is $\oint \mathbf{H}.d\mathbf{s}$. If the surface is divided up into small elements of area $d\mathbf{S}$ then, from § A.6,

$$\oint \mathbf{H}.d\mathbf{l} = \operatorname{curl} \mathbf{H}.d\mathbf{S},$$

where $\oint \mathbf{H}.d\mathbf{l}$ is the line integral of $\mathbf{H}$ round one small element of area $d\mathbf{S}$. If this equation is now summed over all the elementary areas, all the boundaries within

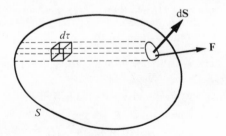

FIG. A.9. Illustrating the divergence theorem.

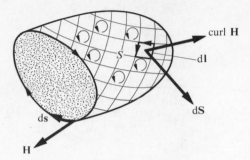

FIG. A.10. Illustrating Stokes's theorem.

the surface will cancel out on the left-hand side, and the result is the line integral round the circuit bounding the surface. Therefore

$$\oint \mathbf{H} \cdot \mathbf{ds} = \int \text{curl } \mathbf{H} \cdot \mathbf{dS}. \tag{A.9}$$

This is Stokes's theorem. It is necessary for $\mathbf{H}$ and its derivatives to be well-behaved continuous functions, but in the cases normally arising in electromagnetism, these conditions are satisfied.

Conversely, if the line integral of $\mathbf{H}$ round a closed curve is equal to the surface integral of $\mathbf{P}$ over a surface bounded by the curve, irrespective of what curve or surface are used, then $\mathbf{P} = \text{curl } \mathbf{H}$.

### A.9. Some useful vector relations

The vector operator $\nabla$ has been used above in expressions for grad $V = \nabla V$, div $\mathbf{D} = \nabla \cdot \mathbf{D}$ and curl $\mathbf{H} = \nabla \wedge \mathbf{H}$. Another useful operator has the form $(\mathbf{P} \cdot \text{grad}) = (\mathbf{P} \cdot \nabla)$; it operates on a vector

$$(\mathbf{P}.\nabla)\mathbf{Q} = \left( P_x \frac{\partial}{\partial x} + P_y \frac{\partial}{\partial y} + P_z \frac{\partial}{\partial z} \right)\mathbf{Q}$$

$$= \mathbf{i}\left( P_x \frac{\partial Q_x}{\partial x} + P_y \frac{\partial Q_x}{\partial y} + P_z \frac{\partial Q_x}{\partial z} \right) +$$

$$+ \mathbf{j}\left( P_x \frac{\partial Q_y}{\partial x} + P_y \frac{\partial Q_y}{\partial y} + P_z \frac{\partial Q_y}{\partial z} \right) +$$

$$+ \mathbf{k}\left( P_x \frac{\partial Q_z}{\partial x} + P_y \frac{\partial Q_z}{\partial y} + P_z \frac{\partial Q_z}{\partial z} \right). \tag{A.10}$$

If $\mathbf{Q}$ satisfies the relation curl $\mathbf{Q} = 0$, then $\partial Q_x/\partial y = \partial Q_y/\partial x$, etc., and (A.10) can be written with components $\mathbf{i}[P_x(\partial Q_x/\partial x) + P_y(\partial Q_y/\partial x) + P_z(\partial Q_z/\partial x)]$, etc. This type of vector can be used to represent the force on a dipole in a non-uniform field (cf. eqns (1.15) and (4.18b)).

The operator $\nabla$ may operate on a product of two quantities; a number of useful relations of this type are:

$$\text{grad}(mV) = \nabla(mV) = m \text{ grad } V + V \text{ grad } m = m\nabla V + V \nabla m. \tag{A.11}$$

$$\text{div}(m\mathbf{P}) = \nabla \cdot (m\mathbf{P}) = m \text{ div } \mathbf{P} + \mathbf{P}.\text{grad } m = m \nabla \cdot \mathbf{P} + \mathbf{P}.\nabla m. \tag{A.12}$$

$$\text{curl}(m\mathbf{P}) = \nabla \wedge (m\mathbf{P}) = m \text{ curl } \mathbf{P} - \mathbf{P} \wedge \text{grad } m = m \nabla \wedge \mathbf{P} - \mathbf{P} \wedge \nabla m. \tag{A.13}$$

grad $(\mathbf{P}.\mathbf{Q}) = \nabla(\mathbf{P}.\mathbf{Q}) = (\mathbf{P}.\text{grad})\mathbf{Q}+(\mathbf{Q}.\text{grad})\mathbf{P}+\mathbf{P}\wedge\text{curl }\mathbf{Q}+\mathbf{Q}\wedge\text{curl }\mathbf{P}$

$$= (\mathbf{P}.\nabla)\mathbf{Q}+(\mathbf{Q}.\nabla)\mathbf{P}+\mathbf{P}\wedge(\nabla\wedge\mathbf{Q})+\mathbf{Q}\wedge(\nabla\wedge\mathbf{P}). \quad (A.14)$$

$$\text{div}(\mathbf{P}\wedge\mathbf{Q}) = \nabla.(\mathbf{P}\wedge\mathbf{Q}) = \mathbf{Q}.\text{curl }\mathbf{P}-\mathbf{P}.\text{curl }\mathbf{Q} = \mathbf{Q}.\nabla\wedge\mathbf{P}-\mathbf{P}.\nabla\wedge\mathbf{Q}. \quad (A.15)$$

$$\text{curl}(\mathbf{P}\wedge\mathbf{Q}) = \nabla\wedge(\mathbf{P}\wedge\mathbf{Q}) = (\mathbf{Q}.\text{grad})\mathbf{P}-(\mathbf{P}.\text{grad})\mathbf{Q}+\mathbf{P}(\text{div }\mathbf{Q})-\mathbf{Q}(\text{div }\mathbf{P})$$

$$= (\mathbf{Q}.\nabla)\mathbf{P}-(\mathbf{P}.\nabla)\mathbf{Q}+\mathbf{P}(\nabla.\mathbf{Q})-\mathbf{Q}(\nabla.\mathbf{P}). \quad (A.16)$$

The operator $\nabla$ occurs twice in a number of relations. If $V$ is a scalar function, div grad $V = \nabla.(\nabla V)$, and from eqns (A.3) and (A.5)

$$\nabla.(\nabla V) = \left(\mathbf{i}\frac{\partial}{\partial x}+\mathbf{j}\frac{\partial}{\partial y}+\mathbf{k}\frac{\partial}{\partial z}\right).\left(\mathbf{i}\frac{\partial V}{\partial x}+\mathbf{j}\frac{\partial V}{\partial y}+\mathbf{k}\frac{\partial V}{\partial z}\right)$$

$$= \frac{\partial^2 V}{\partial x^2}+\frac{\partial^2 V}{\partial y^2}+\frac{\partial^2 V}{\partial z^2}. \quad (A.17)$$

It is convenient to write div grad as $\nabla^2$, where

$$\nabla^2 = \frac{\partial^2}{\partial x^2}+\frac{\partial^2}{\partial y^2}+\frac{\partial^2}{\partial z^2} \quad (A.18)$$

is known as the Laplacian operator (pronounced 'del squared'). Expressions for this operator in spherical polar and cylindrical polar coordinates are given in eqns (2.4) and (2.5).

Other vector identities in which $\nabla$ occurs twice are:

$$\text{curl grad } V = \nabla\wedge(\nabla V) = 0 \quad (A.19)$$

$$\text{div curl }\mathbf{P} = \nabla.(\nabla\wedge\mathbf{P}) = 0 \quad (A.20)$$

$$\text{grad div }\mathbf{P} = \nabla(\nabla.\mathbf{P}) \quad (A.21)$$

$$\text{curl curl }\mathbf{P} = \nabla\wedge(\nabla\wedge\mathbf{P}) = \text{grad div }\mathbf{P}-\nabla^2\mathbf{P}$$

$$= \nabla(\nabla.\mathbf{P})-\nabla^2\mathbf{P}. \quad (A.22)$$

The following formulae are useful in transformations from surface to volume integrals, and from line to surface integrals.

$$\int m \, d\mathbf{S} = \int \text{grad } m \, d\tau \quad (A.23)$$

$$\int \mathbf{P}\wedge d\mathbf{S} = -\int \text{curl }\mathbf{P} \, d\tau \quad (A.24)$$

$$\int \mathbf{P}.(\mathbf{Q}.d\mathbf{S}) = \int \{(\mathbf{Q}.\text{grad})\mathbf{P}+\mathbf{P}\text{ div }\mathbf{Q}\} \, d\tau \quad (A.25)$$

$$\int m \, d\mathbf{s} = -\int (\text{grad } m)\wedge d\mathbf{S} \quad (A.26)$$

$$\int \mathbf{P}\wedge d\mathbf{s} = \int (\text{div }\mathbf{P}) \, d\mathbf{S} + \int (\text{curl }\mathbf{P})\wedge d\mathbf{S} - \int (d\mathbf{S}.\text{grad})\mathbf{P} \quad (A.27)$$

## A.10. Transformation from a rotating coordinate system

When dealing with the effect of an applied magnetic field on an atomic system it is often convenient to transform to a rotating coordinate system. Vector methods make this transformation simple, as can be seen from the following treatment.

Suppose we are concerned with some vector quantity **A**, which to start with we will suppose to be fixed in the rotating coordinate system (a line on a spinning top is an example, but we do not have to restrict **A** to be simply a radius vector). The angular motion of the coordinate system is represented by a vector **ω**, whose magnitude is equal to the angular velocity and whose direction is parallel to the axis of rotation. Its sense is that in which a right-handed screw would advance if rotated in the same sense as the angular motion. If **A** is fixed in the rotating system, then in a time $\delta t$ the end point of the vector is displaced by an amount $\delta \mathbf{A}$ relative to a fixed coordinate system, as shown in Fig. A.11. It is clear that the motion of the end point is a simple rotation about the axis defined by **ω**. Hence

$$\delta \mathbf{A} = (\omega\, \delta t)A \sin \theta = (\boldsymbol{\omega} \wedge \mathbf{A})\, \delta t.$$

Hence the velocity of **A** relative to the fixed system is

$$\lim_{\delta t \to 0}(\delta \mathbf{A}/\delta t) = (d\mathbf{A}/dt) = (\boldsymbol{\omega} \wedge \mathbf{A}). \tag{A.28}$$

If we now suppose that **A** is not fixed in the rotating system, but has a velocity $(D\mathbf{A}/Dt)$ relative to that system, then we have, from the vector addition of the two velocities

$$d\mathbf{A}/dt = (D\mathbf{A}/Dt) + (\boldsymbol{\omega} \wedge \mathbf{A}). \tag{A.29}$$

This relation may be applied to any vector **A**, and hence it may be applied to the vector $d\mathbf{A}/dt$ to find the second differential of **A**. Retaining the notation that $d/dt$ refers to rate of change in the fixed coordinate system, and $D/Dt$ to rate of change relative to the rotating system, we have (since **ω** is a constant)

$$\frac{d^2\mathbf{A}}{dt^2} = \frac{d}{dt}\left(\frac{d\mathbf{A}}{dt}\right) = \left(\frac{D}{Dt} + \boldsymbol{\omega}\wedge\right)\left(\frac{d\mathbf{A}}{dt}\right) = \left(\frac{D}{Dt} + \boldsymbol{\omega}\wedge\right)\left(\frac{D\mathbf{A}}{Dt} + \omega \wedge \mathbf{A}\right)$$

$$= \frac{D^2\mathbf{A}}{Dt^2} + 2\left(\boldsymbol{\omega} \wedge \frac{D\mathbf{A}}{Dt}\right) + \boldsymbol{\omega} \wedge (\boldsymbol{\omega} \wedge \mathbf{A}). \tag{A.30}$$

### A.11. Larmor's theorem

Suppose that a charge $q$ is moving in a field of force (such as the attraction of a positively charged nucleus) whose value at any moment is described by the vector **F**. When a magnetic field is applied, the equation of motion is

$$m\frac{d^2\mathbf{r}}{dt^2} = \mathbf{F} + q\mathbf{v} \wedge \mathbf{B}, \tag{A.31}$$

FIG. A.11. Change $\delta A$ in time $\delta t$ of a vector **A** rotating with velocity **ω**.

where $m$ is the mass associated with the charge, and $\mathbf{v}$ is the instantaneous velocity ($= d\mathbf{r}/dt$). This is the equation of motion in vector form in a set of axes at rest with respect to the observer. Let us now change to a set of axes rotating with angular velocity $\boldsymbol{\omega}$ about the direction of $\mathbf{B}$. In transforming to rotating axes (see § A.10) we have the relation

$$m\frac{d^2\mathbf{r}}{dt^2} = m\frac{D^2\mathbf{r}}{Dt^2} + 2m\left(\boldsymbol{\omega} \wedge \frac{D\mathbf{r}}{Dt}\right) + m\{\boldsymbol{\omega} \wedge (\boldsymbol{\omega} \wedge \mathbf{r})\}, \tag{A.32}$$

where $D^2\mathbf{r}/Dt^2$, $D\mathbf{r}/Dt$ are the acceleration and velocity in the rotating coordinate frame, and $\boldsymbol{\omega}$ is the angular velocity expressed as a vector parallel to the axis of rotation, the direction of $\mathbf{B}$. The second term on the right-hand side is the 'Coriolis force' which appears if the particle is moving in the rotating system, and the last term is the centrifugal force normal to the axis of rotation. If $\boldsymbol{\omega} \wedge \mathbf{r}$ is small compared with $D\mathbf{r}/Dt$ (as we shall show below to be the case), the centrifugal force will be small compared with the Coriolis force, and in the first approximation eqns (A.31) and (A.32) give

$$m\frac{D^2\mathbf{r}}{Dt^2} = \mathbf{F} + q\mathbf{v} \wedge \mathbf{B} - 2m\boldsymbol{\omega} \wedge \frac{D\mathbf{r}}{Dt} = \mathbf{F} + q\mathbf{v} \wedge \mathbf{B} + 2m\mathbf{v} \wedge \boldsymbol{\omega}, \tag{A.33}$$

where we have neglected the small difference between $\mathbf{v}$ and $D\mathbf{r}/Dt$ (the velocity in the rotating frame) since they differ only by the quantity $\boldsymbol{\omega} \wedge \mathbf{r}$, which we have already assumed to be small in comparison. It is apparent that, if we choose the rate of rotation of the axes such that

$$\boldsymbol{\omega} = -(q/2m)\mathbf{B}, \tag{A.34}$$

the last two terms in (A.33) will vanish and the equation of motion is the same as if the magnetic field were absent. Thus to an observer rotating with the angular velocity given by (A.34) the motion of the charge appears to be the same as it would to a stationary observer in the absence of a magnetic field. Hence we may regard the motion of an electron of charge $-e$ in the field with flux density $\mathbf{B}$ as unchanged except for a precession with angular velocity $\boldsymbol{\omega} = +(e/2m)\mathbf{B}$ about the axis of $\mathbf{B}$. This is commonly known as the 'Larmor precession'.

The fact that it is justifiable to neglect the last term in eqn (A.32) can be seen as follows. When the electron is bound in the atom, it executes a periodic motion in its orbit whose frequency is of the same order as that of visible light. This corresponds to an angular frequency $\omega_0$ of the order of $10^{15}$ rad $s^{-1}$. The terms $D^2\mathbf{r}/Dt^2$ and $D\mathbf{r}/Dt$ are then of order of magnitude $\omega_0^2 r$ and $\omega_0 r$ respectively, so that successive terms in eqn (A.32) decrease in magnitude by the ratio $\omega/\omega_0$. Since $\omega$ is only about $10^{11}$ rad $s^{-1}$ even in a field of 1 T, the centrifugal force term is an order of magnitude smaller than the Coriolis force. In other words, the force on the electron due to the flux density $\mathbf{B}$ is small compared with the force exerted by the positively charged nucleus; if it were not, it would tear the atom apart.

The central force assumed above is that responsible for the orbital motion of an electron in an atom, and the angular velocity given by eqn (A.34) is identical with the angular velocity of precession (see § 14.1) of an electronic orbital magnetic moment in a magnetic flux density $\mathbf{B}$.

# Appendix B: Depolarizing and demagnetizing factors

SUPPOSE a uniform field of strength $\mathbf{E}_0$ exists in a medium of relative permittivity $\epsilon_2$. An ellipsoid with principal axes of length $a$, $b$, $c$ and relative permittivity $\epsilon_1$ is introduced with its principal $a$-axis parallel to $\mathbf{E}_0$ Then the field $\mathbf{E}_a$ inside the ellipsoid is also parallel to the $a$-axis; its value is

$$\mathbf{E}_a = \mathbf{E}_0 \frac{\epsilon_2}{\epsilon_2 + d_a(\epsilon_1 - \epsilon_2)}. \tag{B.1}$$

The quantity $d_a$ is a pure number, given by the integral

$$d_a = \tfrac{1}{2}abc \int_0^\infty \frac{ds}{(s+a^2)R_s}, \tag{B.2}$$

where

$$R_s^2 = (s+a^2)(s+b^2)(s+c^2).$$

Values of $d_a$, $d_b$, $d_c$ are tabulated by Osborn (1945) and Stoner (1945), and satisfy the identity

$$d_a + d_b + d_c = 1.$$

Two limiting sets of values are (for ellipsoids of circular cross-section)

|                       | $d_a$ | $d_b$ | $d_c$ |
|-----------------------|-------|-------|-------|
| short, fat ellipsoid  | 1     | 0     | 0     |
| long, thin ellipsoid  | 0     | $\tfrac{1}{2}$ | $\tfrac{1}{2}$ |

In the limit the short, fat ellipsoid is equivalent to a parallel-sided slab for which $\epsilon_1 \mathbf{E}_a = \epsilon_2 \mathbf{E}_0$; that is, the normal components of $\mathbf{D}$ are continuous, while the tangential components of $\mathbf{E}$ are continuous since $d_b = d_c = 0$. In the opposite limit the long, thin ellipsoid is equivalent to a circular cylinder; when $\mathbf{E}$ is parallel to the axis it is continuous, and when $\mathbf{E}$ is normal to the axis the formula agrees with the result of Problem 2.3. For a sphere $d_a = d_b = d_c = \tfrac{1}{3}$, and eqn (B.1) then reduces to eqn (2.43).

Eqn (B.1) cannot be used when a permanent moment is present, but it agrees with the definition of a 'depolarizing field' $\mathbf{E}_d = \mathbf{E}_a - \mathbf{E}_0$ given by eqn (2.48). The latter has the advantage that it is applicable when there also exists a permanent polarization parallel to the $a$-axis. This is an added complication, which is treated below for the magnetic case; the substitutions needed are $\mathbf{H}$ for $\mathbf{E}$; $\mathbf{M}$ for $\mathbf{P}/\epsilon_0$; $\mu$ for $\epsilon$.

Suppose the ellipsoid has a permanent magnetization $\mathbf{M}_0$ parallel to the $a$-axis, as well as an induced moment. Then

$$\mathbf{M}_1 = (\mu_1 - 1)\mathbf{H}_a + \mathbf{M}_0,$$

and

$$\mathbf{M}_2 = (\mu_2 - 1)\mathbf{H}_0. \tag{B.3}$$

The treatment given by Lowes (1974) is equivalent to the presence of a demagnetizing field $H_d = H_a - H_0$ given by

$$H_d = -\frac{M_1 - M_2}{1 + (d_a^{-1} - 1)\mu_2}. \qquad (B.4)$$

For a sphere $d_a = \frac{1}{3}$ and eqn (B.4) reduces to eqn (4.38c).

Lowes (1974) also derives the torque $T$ on a permanent magnet in the form of an ellipsoid with magnetization $M_0$ parallel to the $a$-axis in a flux density $B_0$,

$$T = V\frac{M_0 \wedge B_0}{\mu_2 + d_a(\mu_1 - \mu_2)} \qquad (B.5)$$

where $V = \frac{4}{3}\pi abc$ is the volume of the magnet. For a long, thin magnet ($d_a \rightarrow 0$), the torque is simply $T = \mu_0 V(M_0 \wedge H_0)$, in agreement with the experiment of Whitworth and Stopes-Roe (1971), but this is not true for other shapes. Eqn (B.5) gives only the torque associated with $M_0$; there is also a torque associated with induced magnetization which tends to align a paramagnetic ellipsoid with its longest axis parallel to $H_0$ (see Stratton 1941 (§ 3.29)).

## References

LOWES, F. J. (1974). *Proc. R. Soc. Lond.* **A337**, 555.

OSBORN, J. A. (1945). *Phys. Rev.* **67**, 351.

STONER, E. C. (1945). *Phil. Mag.* **36**, 803.

STRATTON, J. A. (1941). *Electromagnetic theory.* McGraw-Hill, New York.

WHITWORTH, R. W. and STOPES-ROE, H. V. (1971). *Nature, Lond.* **234**, 31.

# Appendix C: Numerical value of the fundamental constants

| | | |
|---|---|---|
| $c$ | velocity of light *in vacuo* | $2 \cdot 998 \times 10^8 \, \text{m s}^{-1}$ |
| $N_A$ | Avogadro's number | $6 \cdot 022 \times 10^{23} \, \text{mol}^{-1}$ |
| $e$ | electronic charge | $1 \cdot 602 \times 10^{-19} \, \text{C}$ |
| $m_e$ | electron rest mass | $9 \cdot 110 \times 10^{-31} \, \text{kg}$ |
| $m_p$ | proton rest mass | $1 \cdot 673 \times 10^{-27} \, \text{kg}$ |
| $m_p/m_e$ | ratio of proton to electron mass | $1 \cdot 836 \times 10^3$ |
| $h$ | Planck's constant | $6 \cdot 626 \times 10^{-34} \, \text{J s}$ |
| $\hbar$ | Planck's constant$/2\pi$ | $1 \cdot 055 \times 10^{-34} \, \text{J s}$ |
| $h/2e$ | | $2 \cdot 068 \times 10^{-15} \, \text{J s C}^{-1}$ |
| $F$ | Faraday's constant $(N_A e)$ | $9 \cdot 648 \times 10^4 \, \text{C mol}^{-1}$ |
| $e/m$ | charge/mass for electron | $1 \cdot 759 \times 10^{11} \, \text{C kg}^{-1}$ |
| $e^2/m$ | | $2 \cdot 818 \times 10^{-8} \, \text{C}^2 \, \text{kg}^{-1}$ |
| $a_0$ | Bohr radius $= 4\pi\epsilon_0 \hbar^2/m_e e^2$ | $5 \cdot 292 \times 10^{-11} \, \text{m}$ |
| $R_\infty c$ | Rydberg constant$\times c = m_e e^4/8\epsilon_0^2 h^3$ | $3 \cdot 290 \times 10^{15} \, \text{Hz}$ |
| $R_\infty$ | Rydberg constant | $1 \cdot 097 \times 10^7 \, \text{m}^{-1}$ |
| $k$ | Boltzmann's constant | $1 \cdot 381 \times 10^{-23} \, \text{J K}^{-1}$ |
| $R$ | gas constant $= N_A k$ | $8 \cdot 314 \, \text{J K}^{-1} \, \text{mol}^{-1}$ |
| $\mu_B$ | Bohr magneton $= e\hbar/2m_e$ | $9 \cdot 274 \times 10^{-24} \, \text{A m}^2$ |
| $\mu_N$ | nuclear magneton $= e\hbar/2m_p$ | $5 \cdot 051 \times 10^{-27} \, \text{A m}^2$ |
| $\alpha$ | fine structure constant $= e^2/4\pi\epsilon_0 \hbar c$ | $(137.0)^{-1}$ |
| $\epsilon_0$ | permittivity of a vacuum $= (\mu_0 c^2)^{-1}$ | $8 \cdot 854 \times 10^{-12} \, \text{F m}^{-1}$ |
| $4\pi\epsilon_0$ | | $10^7/c^2 = 10^{-9}/9$ approximately |
| $\mu_0$ | permeability of a vacuum (by definition) | $4\pi \times 10^{-7} \, \text{H m}^{-1}$ |
| $Z_0$ | intrinsic impedance of free space | $3 \cdot 767 \times 10^2 \, \Omega$ |
| $eV$ | electronvolt | $1 \cdot 602 \times 10^{-19} \, \text{J}$ |
| $kT$ | energy for $T = 290 \, \text{K}$ | $4 \cdot 004 \times 10^{-21} \, \text{J}$ |

1 electronvolt is equivalent to:
    wavelength $\lambda = 1 \cdot 240 \times 10^{-6} \, \text{m}$
    frequency $\nu = 2 \cdot 418 \ 10^{14} \, \text{Hz}$
    wave number $\bar{\nu} = 8 \cdot 065 \times 10^5 \, \text{m}^{-1}$
    temperature $T = 1 \cdot 160 \times 10^4 \, \text{K}$
    energy $W = 1 \cdot 602 \ 10^{-19} \, \text{J}$
1 $\text{m}^{-1}$ is equivalent to:
    wavelength $\lambda = 1 \, \text{m}$
    temperature $T = 1 \cdot 439 \times 10^{-2} \, \text{K} = hc/k$

# Appendix D: Some useful unit conversions

The system of units employed throughout this book is the International System (S.I.). However, other systems of units are still often used in the literature; these are usually m.k.s. (metre, kilogram, second) units or c.g.s. (centimetre, gram, second) units. In the theory of electricity and magnetism c.g.s. units may be either electrostatic (e.s.u.) or electromagnetic (e.m.u.).

The following is a brief list of equivalences which may be useful when reading the literature associated with topics in this book.

1 dyne = $10^{-5}$ newtons
1 erg = $10^{-7}$ joules
1 mm Hg = 1 torr ≈ 133·322 pascals
1 atmosphere = 101 325 pascals ≈ $10^5$ Pa = 1 bar
1 calorie = 4·184 joules
1 e.m.u. of charge = 10 coulombs
1 e.s.u. of charge ≈ 3·336×$10^{-10}$ coulombs
1 gauss = $10^{-4}$ teslas
1 oersted = $10^3/4\pi$ amperes per metre
1 maxwell = $10^{-8}$ webers
1 electronvolt ≈ 1·602×$10^{-19}$ joules
1 micron = 1 micrometre = $10^{-6}$ m
1 ångström = $10^{-1}$ nanometres = $10^{-10}$ m
1 hertz = 1 cycle per second
1 Bohr magneton = 0·9274×$10^{-20}$ e.m.u.
1 Faraday = 9·649×$10^3$ e.m.u. mol$^{-1}$.

In S.I. units the volume susceptibility (per cubic metre) is a number larger by a factor $4\pi$ than the corresponding number (per cubic centimetre) in c.g.s. units, the latter being expressed in e.s.u. for electric susceptibility and in e.m.u. for magnetic susceptibility.

# Index

This completely revised edition gives an up-to-date account at first-degree or introductory graduate level of the principles and experimental aspects of electricity and magnetism, together with an elementary account of the underlying atomic theory. S.I. units are used throughout and there are problems at the end of each chapter.

Reviewers of the previous edition have said:

'Electricity and magnetism have long had an important place in the training of a physicist. Here in one volume is practically all the material needed in those topics, up to a first degree and a bit beyond . . . This book should be very suitable as a text.'     *American Journal of Physics*

and describe it as 'a worthy investment for any enlightened under-graduate student'.                *Proceedings of the Physical Society*

The six main sections are (1) electrostatics, electric currents and magnetic fields, magnetostatics and magnetic effects, with emphasis on atomic applications; (2) alternating current theory and electromagnetic waves, including filters, transmission lines, and waveguides; (3) theory of dielectrics, free electron and band theory of metals, superconductivity; (4) paramagnetism, ferromagnetism, antiferromagnetism, and ferrimagnetism; (5) semiconductors, semiconductor diodes and triodes, vacuum tubes, electronics, and electrical noise; (6) cyclotron resonance and magnetic resonance.

**Professor B. Bleaney**, C.B.E., F.R.S., who is Dr. Lee's Professor of Experimental Philosophy at the University of Oxford, has taught physics at Oxford since 1945, apart from visits to MIT and the Universities of Harvard, Columbia, Pittsburgh, Manitoba, and Argentina. His wife, **Mrs. B. I. Bleaney**, has been tutoring in physics in Oxford since 1949.

SOME OTHER OXFORD BOOKS

Lectures on electromagnetic theory: A short course for engineers
L. Solymar   1976

Lectures on the electrical properties of materials
L. Solymar and D. Walsh   1970

DC and AC circuits   G. Lancaster   (*Oxford Physics Series*)   1974

Electromagnetism   F. N. H. Robinson   (*Oxford Physics Series*)   1973

Noise and fluctuations in electronic devices and circuits
F. N. H. Robinson   1974

OXFORD UNIVERSITY PRESS

**£9·95 net** in UK

*Also available in hardback*                ISBN 0 19 851141 8